Oceanic
Micropalaeontology

Oceanic
Micropalaeontology

Edited by

A. T. S. RAMSAY

Department of Geology and Oceanography
University College of Swansea, Wales

Volume 1

ACADEMIC PRESS · 1977
LONDON · NEW YORK · SAN FRANCISCO
A Subsidiary of Harcourt Brace Jovanovich, Publishers

ACADEMIC PRESS INC. (LONDON) LTD
24/28 Oval Road
London NW1

United States Edition published by
ACADEMIC PRESS INC.
111 Fifth Avenue
New York, New York 10003

Library of Congress Catalog Number: 76-1097
ISBN: 0–12–577301–3

PRINTED IN GREAT BRITAIN AT THE PITMAN PRESS BATH

Contributors to Volume 1

ALLAN W. H. BÉ: *Lamont–Doherty Geological Observatory of Columbia University, Palisades, New York 10964, U.S.A.*

W. A. BERGGREN: *Department of Geology and Geophysics, Woods Hole Oceanographic Institution, Woods Hole, Massachusetts 02543, U.S.A.*

RONALD W. GILMER: *Department of Zoology, University of California, Davis, California 95616, U.S.A.*

D. GRAHAM JENKINS: *Department of Geology, University of Canterbury, Christchurch 1, New Zealand*

BRUCE A. MASTERS: *Amoco Production Company, Research Center, P.O. Box 591, Tulsa, Oklahoma 74102, U.S.A.*

WILLIAM N. ORR: *Department of Geology, University of Oregon, Eugene 97403, U.S.A.*

W. F. RUDDIMAN: *Ocean Floor Analysis Division, United States Naval Oceanographic Office, Washington D.C. 20373, U.S.A.*

Preface

The intensive investigation of the ocean since the Second World War and the increasing availability of pelagic sediments from dredge hauls, gravity and piston cores, and more recently from the continuous or nearly continuous sequences recovered by the Deep Sea Drilling Project has opened up new areas for research in micropalaeontology. The micropalaeontological investigation of oceanic sediments has as a result of this stimulus developed into an expanding and exciting field of palaeontological research. Micropalaeontologists are now involved in investigating and testing both the accepted and latest concepts in marine geology. They are also utilising the information gleaned from the ocean to improve existing schemes or develop new schemes of biostratigraphical zonation, and to investigate the climatic and oceanographic history of the Earth since the Mesozoic.

It was not possible to cover all fields of oceanic micropalaeontology in this work. The papers that follow present a review of knowledge and recent developments in several aspects of oceanic micropalaeontology.

I am grateful to all the authors who contributed to this work for their cooperation and to Academic Press for their assistance and enthusiastic support.

March 1977   A. T. S. Ramsay

Contents

Contents of Volume 2

1: An Ecological, Zoogeographic and Taxonomic Review of Recent Planktonic Foraminifera

ALLAN W. H. BÉ

Lamont–Doherty Geological Observatory
of Columbia University, Palisades,
New York 10964, U. S. A.

A classification of 37 Recent species is presented and illustrated with scanning electron micrographs. One new subspecies, *Globorotalia menardii gibberula* is described.

Test growth in planktonic foraminifera occurs in two phases. The first is concerned with rapid formation of a relatively thin "adult" test. The second is directed to wall thickening, in the form of a "calcite crust", over the entire test in late ontogeny. Microstructural differences between several species are discussed.

Many factors influence the distribution patterns of planktonic foraminifera in the ocean waters as well as those of their death assemblages on the sea-floor. Biological factors (productivity, predation, food supply, interspecific competition, etc.), physical-chemical factors (nutrients, light, temperature, etc.), and artificial factors (field sampling and laboratory methods) influence our understanding of the spatial and time-series distributions of living planktonic foraminifera. The distribution patterns of their death assemblages are largely controlled by variations in long-term productivity (over thousands of years), the selective dissolution of their tests on the sea-floor, scouring and redeposition by bottom currents, and upward mixing of sediments and test destruction by burrowing organisms.

Recent planktonic foraminifera are grouped into five major faunal provinces. Many species occur in reciprocal latitudinal zones between the northern and southern hemispheres, and, hence, reflect bipolar and antitropical distribution patterns. Species diversity decreases from low to high latitudes. The distribution patterns of thirteen common species in surface sediments of the world's oceans are discussed.

Outline

1

Introduction

Planktonic foraminifera, as members of the marine plankton ecosystem, are small and rather inconspicuous protozoans. Yet, they are so ubiquitous and abundant in the open ocean, that a fine-meshed plankton net haul will invariably catch them. Their geologic significance lies in their immense productivity and ability to secrete calcite tests (shells). The constant "rain" of foraminiferal tests upon the sea-bed is a process which has occurred at varying rates over the past 130 million years. Today, their remains are preserved over extensive areas of the ocean basins. According to an estimate by Sverdrup *et al.* (1942), 47% of the total ocean-floor (equivalent to about 126 million square kilometres or the total area of all continents) is covered by so-called "globigerina ooze". A measure of their enormous productivity can be gathered from the fact that several thousand specimens of planktonic foraminifera ($>$ 200 μm diameter) are contained in a gram of globigerina ooze sediment (Correns, 1939).

Although planktonic foraminifera may appear insignificant in relation to larger and more commonly observed animal groups (such as the copepods, coelenterates, chaetognaths, or salps) in the plankton ecosystem, the selective enrichment of foraminiferal tests in bottom sediments have allowed micropalaeontologists, geochemists and biostratigraphers to utilise them in tackling problems of global significance.

During the current revolution in the earth sciences, planktonic foraminifera play a major role in such diverse investigations as palaeoclimatology, palaeo-oceanographic circulation patterns of surface and bottom currents, palaeo-geography, plate tectonics and sea-floor spreading, $CaCO_3$ cycle, and geochronology. We are witnessing today an unprecedented burst of activity in the study of planktonic foraminifera and their use in deciphering the late Mesozoic and Cainozoic history of the earth.

D'Orbigny (1826, 1839a, b, c) published the first significant descriptions o planktonic foraminifera which were found in beach sands and marine deposits of Cuba, the Canary Islands and other localities. He was, however, unaware of their pelagic life-style.

Later investigators, such as Ehrenberg (1861, 1873), Carpenter *et al.*, (1862) and Parker and Jones (1865) studied the foraminifera from deep-sea sediments collected in the northern Atlantic Ocean, but they assumed that all species of foraminifera dwelled on the ocean bottom.

In 1867, Owen drew attention to the fact that he found certain foraminifera in plankton tows and that therefore they had to live near the surface of the ocean. He described the planktonic habit of several species belonging to the genera *Globigerina*, *Orbulina* and *Pulvinulina* (=*Globorotalia*) and surmised that they had the ability to rise to and descend from the surface. Owen's discovery was either overlooked or ignored, for on the eve of the historical *Challenger* Expedition (1872–1876) the general concensus among leading naturalists was that all foraminifera were sea-floor dwellers (Wallich, 1862, 1876; Thomson, 1874; Carpenter, 1875).

One of the significant findings of the *Challenger* Expedition was the ample confirmation of Owen's observations. As planktonic foraminifera were routinely gathered in the surface waters during the expedition, their enormous abundance and widespread distribution soon become apparent. Brady is quoted in the cruise narratives by Tizard *et al.* (1885, p. 834) as noting:

> The relation of the pelagic Rhizopod-fauna of the ocean to that of the sea bottom, foreshadowed by the researches of Major Owen and others, has been placed on a broader and more intelligible footing by the discovery of numerous species in the surface water which were previously supposed to inhabit exclusively the bottom ooze. Furthermore, the whole subject of recent oceanic deposits and the organisms concerned in their production, of which the Foraminifera are amongst the most important, may almost be said to owe its initiation to data collected during the Challenger cruise.

Brady (1884, p. xiv) did not, however, rule out the possibility that these planktonic species could spend a part of their life cycle on the ocean floor, as he wrote:

> Taken by themselves, the facts that have been brought forward, as well as some others of less significance that might be adduced, tend to the inference that the Foraminifera which are found living in the open ocean have also the power of supporting life on the surface of the bottom-ooze; and further that, so far as our present knowledge goes, there is at least one variety of *Globigerina* which lives only at the sea-bottom.

Brady's (1884) monumental "Report on the Foraminifera dredged by H.M.S. *Challenger*" contains detailed information on the widespread occurrence of twenty planktonic species and the extraordinary abundance of the tests of these comparatively few species in deep-sea sediments.

Once the planktonic habit of these foraminifera was established, it was easy to comprehend how they could be widely dispersed by ocean currents. Murray and Renard (1891) and Murray (1897) observed that many species of planktonic foraminifera were distributed in global belts that were delineated from each other by the temperatures of the surface waters. Murray (1897) grouped the planktonic foraminifera into several faunal assemblages which clearly reflect latitudinal and climatic influences. He noted also that the majority of the species occurred in the tropical regions, that the number of species decreased in temperate regions and that only two or three species were found in the Arctic and Antarctic regions. Murray believed that water temperature was one of the prime factors causing the provincialism of most species.

The close relationship between faunal assemblages and environmental condition led to the next logical inference, i.e. that the stratified fossil remains of planktonic foraminifera in the ocean sediments might reveal changes in the climatic conditions during the geologic past. This was first suggested by Phillippi (1910) who interpreted the foraminifera-rich sediments overlying the red clays in the Indian Ocean as having been deposited after the last glacial period.

Another significant contribution was that of Schott (1935), who made the first quantitative study of planktonic foraminifera in the ocean waters as well as in the surface sediments of the equatorial Atlantic. On the basis of material collected by the *Meteor* Expedition (1925–1927), he was able to determine the regional variations in abundance of individual species of planktonic foraminifera, the ranges of their depth habitats, and the similarities and differences in the distribution patterns of life and death assemblages. He recognised the importance of temperature in influencing species distributions and attributed the presence of *Globorotalia menardii* in sediments that overlay deposits without *G. menardii* to the post-glacial change in climatic conditions. Schott's findings have been corroborated by many investigators of Quaternary deep-sea sediments during the nearly four decades following his publication (see References). The utilitarian value of planktonic foraminifera in palaeo-climatology and palaeontology is now widely accepted.

There is today a curious imbalance in our knowledge of the living *vis-à-vis* fossil planktonic foraminifera. A much smaller body of information is available on the distribution of living planktonic foraminifera than there is of their skeletons in the sediments (see References). While a vast quantity of information exists concerning the stratigraphic occurrence of planktonic foraminifera in nearly all Late Mesozoic and Cainozoic marine sediments on land as well as in the ocean, we still lack a great deal of knowledge of the spatial distributions of living species and especially of their biological processes.

Many of the underlying assumptions in palaeoecological analysis are based on the life processes of modern species. In order to close the gap between our biological and palaeontological knowledge of the planktonic foraminifera, we need to carry out more field studies of their population dynamics at fixed ocean stations. Planktonic foraminifera should also be cultured in the laboratory in order to obtain basic data on their life cycle, shell growth and environmental tolerances which cannot be gathered by sampling the ocean.

Classification

The planktonic foraminifera inhabiting the present-day oceans are the evolutionary products of a polyphyletic group which originated in the Late Mesozoic and Early Cainozoic and gave rise to many different lineages (Parker, 1967; Berggren, 1969). Radiational divergence patterns and iterative, parallel evolution of diverse morphotypic groups have occurred several times during the Cainozoic. According to Loeblich and Tappan (1964) the present-day species belong to three families of the superfamily Globigerinaceae, while five families have become extinct during the past 100 million years.

Contemporary students of planktonic foraminifera do not all agree on the criteria for defining a species. Mayr's (1942) concept that "species are reproductively separated groups of actually or potentially interbreeding populations" probably does not hold for protozoan groups like the planktonic foraminifera. Without the ability to delineate a species on a purely biological basis, we must content ourselves with the belief that similar phenotypes are a clue to close genetic relationship. Because of these difficulties we are not certain of the exact number of present-day specific, generic and suprageneric taxa.

Modern workers do not agree with each other on the relative taxonomic significance of the various test characters of phenotypes. Should greater taxonomic emphasis be placed on evolutionary lineages of fossil species or on the ontogenetic development of living species? Ideally both viewpoints should be considered, but this is not yet possible. The evolutionary record often lacks complete ontogenetic series of the fragile species, which are selectively destroyed or lose important test characters such as elongate spines. On the other hand, the ontogenetic development of present-day species is only beginning to be understood.

Our knowledge of the biology of planktonic foraminifera is still rudimentary and until they can be successfully cultured in the laboratory, our observations and interpretations are necessarily limited to preserved specimens from plankton tows. Investigations of living populations as well as the death assemblages in bottom sediments have now revealed the wide range of morphological variability of each species. The earlier, inherently simple belief that the designation of a type specimen and paratypes could provide a stable nomenclature by "defining" a species with all variations in growth and form of its population has proven to be grossly inadequate. In the past, excessive emphasis was placed on adult test characters, while similarities and dissimilarities of the juvenile test were overlooked.

Henceforth we must recognise and describe the total range of morphological variations, which are caused by environmental and hereditary changes on a local or regional scale. Although at present we must content ourselves with a classification system which is based on the morphology of hard parts of the test, future studies of living planktonic foraminifera will increasingly utilise characters of the soft parts and the biochemical composition of the test. The following aspects need to be considered in future classification schemes:

    I. Cytological composition of the foraminiferal protoplasm.
    II. Amino acid composition of the test.
    III. Evolutionary and phylogenetic relationships.
    IV. Test morphology and microstructure, mode of test secretion, and mineral composition.
    V. Ontogenetic development of test:
      (*a*) Presence or absence of elongate spines.
      (*b*) Mode of coiling.
      (*c*) Shape of chambers.
      (*d*) Shape and position of aperture(s), including presence or absence of a lip and tooth.
      (*e*) Pore concentration, pore diameter and test porosity.
      (*f*) Number of chambers per whorl; total number of chambers as an age index.
      (*g*) Surface texture.
      (*h*) Presence or absence of a keel.
      (*i*) Wall thickening and formation of "calcite crust" or "cortex".
      (*j*) Terminal chambers or structures (bulla, flap-like final chamber, etc.) and their significance as kummerform tests.

    Besides considering *Candeina* with its sutural apertures apart from all other planktonic species, the classification outlined in Table 1 places first-order importance in distinguishing between the planispiral tests of *Hastigerina*, *Hastigerinella* and the trochospiral tests of all other species. This taxonomic distinction is further supported by *Hastigerina pelagica*'s possession of completely triradiate spines (Plate 6, fig. 1) and a unique bubble capsule (Plate 1, fig. b) which surrounds the test (Bé, 1969). These two features are absent in all other planktonic foraminiferal species, with the possible exception of *Hastigerinella digitata*. A more typical spinose species is *Globigerinoides conglobatus* with its long spined, extended pseudopod network and symbiotic zooxanthellae (Plate 1, fig. a).

    Taxonomic importance is also attached to the presence or absence of elongate spines, mode of coiling, shape and position of chambers and number of chambers per whorl, secondary apertures, and surface texture. In addition, pore concentration, pore diameter and test porosity are microstructural criteria which can be used to differentiate genera (Bé, 1968).

Recent studies by King and Hare (1972a, b) have demonstrated that variations in the amino acid composition of the tests of 16 species of planktonic foraminifera from surface sediments parallel the variations in gross test morphology. A Q-mode factor analysis of the species in relation to two independent factor axes (glycine, alanine vs. aspartic acid) and three factor axes (alanine, proline, valine vs. aspartic, threonine vs. glycine, serine, glutamic) showed that the relative positions of the species reflect their degree of similarity in amino acid composition (see Figs 1 and 2). Non-spinose

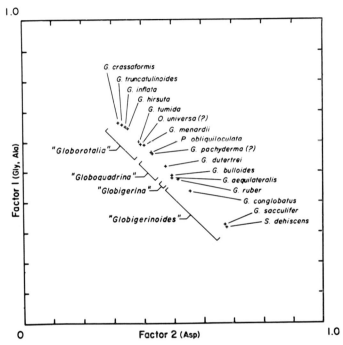

FIG. 1. Classification of 16 Recent planktonic foraminiferal species based on amino acid composition and using Q-mode factor analysis. This two-factor solution was generated by an analysis of 17 amino acids by King and Hare (1972b).

species belonging to the *Globorotalia* and *Globoquadrina* groups are more closely related to each other than to the spinose species of the *Globigerina* and *Globigerinoides* groups. In addition, species of the same genera have a high degree of similarity in amino acid composition and tend to cluster together. Further investigations of the amino acid composition of foraminiferal tests hold great promise that phylogenetic relationships and evolutionary lineages of fossil planktonic foraminifera can be resolved by an approach independent of the conventional method of comparing phenotype morphologies.

At present we must content ourselves with the classical approach. The generic classification outlined below is slightly modified from that of Bé (1967b), which agrees in many respects with that of Parker (1962). The reader is referred to the former paper for detailed taxonomic descriptions of the individual species. All Recent species are illustrated by scanning electron

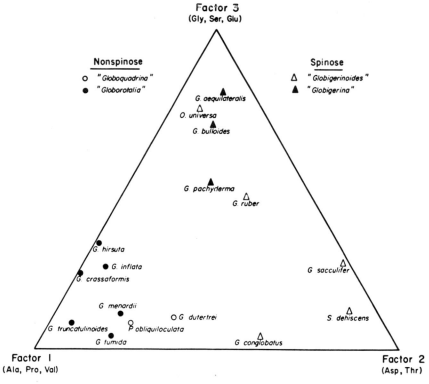

FIG. 2. Three-factor solution from analysis of transformed data for the eight most abundant amino acids. Each vertex of the triangle represents 100% of the indicated factor. The symbols represent groups of morphologically similar species listed in Table 1. (From King and Hare, 1972b.)

micrographs in Plates 6 to 12 from specimens collected mostly in the Indian Ocean. The morphological terms in the present key have been defined by Bolli *et al.* (1957) and Loeblich and Tappan (1964).

Order FORAMINIFERIDA

Superfamily GLOBIGERINACEA Carpenter *et al.*, 1862

Family HANTKENINIDAE Cushman, 1927

Description: Test planispiral or enrolled biserial; chambers spherical to elongate or clavate; primary aperture symmetrical and equatorial, single or multiple, and may have relict or areal secondary apertures; triradiate spines in living species.

Family GLOBIGERINIDAE Carpenter *et al.*, 1862

Description (after Parker, 1962): Test trochospiral in the adult or in ontogeny, streptospiral, or globular; chambers spherical, ovate or clavate; wall calcareous, perforate, radial in structure, hispid, spinose when living either in the adult or in ontogeny; primary aperture umbilical, umbilical-extra-umbilical, equatorial or spiroumbilical; may have secondary apertures; may have bullae with accessory infralaminal apertures.

Family GLOBOROTALIIDAE Cushman, 1927

Description (emended from that of Parker, 1962): Coiling of test trochospiral; chambers angular to ovate or spherical; may have a keel; wall calcareous, perforate, radial in structure, smooth, pitted; non-spinose when living both in the adult and in ontogeny; primary aperture extraumbilical-umbilical or umbilical; no secondary apertures.

Key to Genera

1a. Sutural apertures . . . . *Candeina*
1b. No sutural apertures 2
2a. Planispiral test . . . . Family Hantkeninidae: 3
2b. Trochospiral test 4
3a. Planispiral test throughout ontogeny . . . . . . *Hastigerina*
3b. Planispiral test becoming streptospiral in late ontogeny; bifurcate chambers . . . . *Hastigerinella*
4a. Trochospiral test without elongate spines Family Globorotaliidae: 5–7
4b. Trochospiral test with elongate spines . Family Globigerinidae: 8–11
5a. Test with angular to ovate chambers; spiral side flat or gently curved; keel usually present; aperture a narrow slit from umbilicus to periphery . . . . *Globorotalia*
5b. Test with spherical or hemispherical chambers; no keel; umbilical aperture; rounded periphery 6
6a. Streptospiral coiling in adult . . . . *Pulleniatina*
6b. Trochospiral coiling throughout ontogeny 7
7a. Spherical chambers; umbilical aperture frequently covered by bulla with infralaminal apertures; smooth surface texture . . *Globigerinita*
7b. Hemispherical chambers; umbilical aperture; reticulate surface texture; umbilical tooth . . . . *Globoquadrina*
8a. Primary aperture only 9
8b. Primary aperture and one or more secondary apertures 10

TABLE 1. Hierarchical subdivisions of the families, genera and species of Recent planktonic foraminifera. Figure numbers refer to those in Plates 1–9.

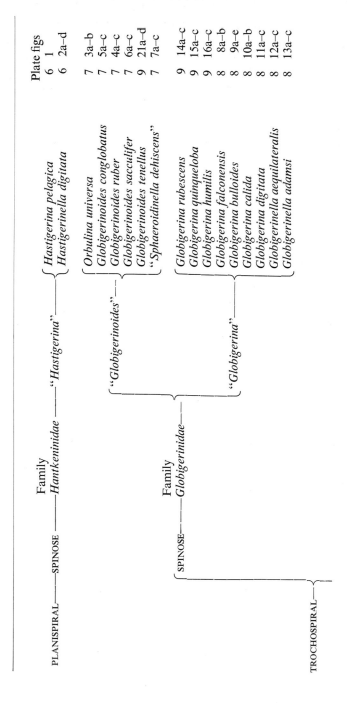

				Plate figs	
			Hastigerina pelagica	6	1
		"Hastigerina"	Hastigerinella digitata	6	2a–d
			Orbulina universa	7	3a–b
			Globigerinoides conglobatus	7	5a–c
			Globigerinoides ruber	7	4a–c
		"Globigerinoides"	Globigerinoides sacculifer	7	6a–c
			Globigerinoides tenellus	9	21a–d
			"Sphaeroidinella dehiscens"	7	7a–c
			Globigerina rubescens	9	14a–c
			Globigerina quinqueloba	9	15a–c
			Globigerina humilis	9	16a–c
			Globigerina falconensis	8	8a–b
		"Globigerina"	Globigerina bulloides	8	9a–e
			Globigerina calida	8	10a–b
			Globigerina digitata	8	11a–c
			Globigerinella aequilateralis	8	12a–c
			Globigerinella adamsi	8	13a–c

Family
Hantkeninidae

Family
Globigerinidae

PLANISPIRAL — SPINOSE

SPINOSE

TROCHOSPIRAL

TROCHOSPIRAL

Family Globorotaliidae — NON-SPINOSE

"Globigerinita"
- Globigerinita glutinata — 9 — 18a–d
- Globigerinita bradyi — 9 — 19a–b
- Globigerinita iota — 9 — 20a–c

"Globoquadrina"
- Globoquadrina pachyderma — 10 — 22a–i
- Globoquadrina dutertrei — 10 — 23a–f
- Globoquadrina hexagona — 10 — 24a–f
- Globoquadrina conglomerata — 10 — 25a–c
- Pulleniatina obliquiloculata — 10 — 26a–d

"Globorotalia"
- Globorotalia pumilio — 9 — 17a–c
- Globorotalia inflata — 11 — 27a–c
- Globorotalia scitula — 11 — 28a–c
- Globorotalia theyeri — 11 — 29a–c
- Globorotalia crassaformis — 11 — 30a–c
- Globorotalia crotonensis — 11 — 31a–c
- Globorotalia cavernula — 11 — 32a–c
- Globorotalia hirsuta — 11 — 33a–c
- Globorotalia truncatulinoides — 11 — 34a–c
- Globorotalia menardii menardii — 12 — 35a–c
- Globorotalia menardii gibberula — 12 — 36a–c
- Globorotalia menardii neoflexuosa — 12 — 37a–c
- Globorotalia tumida — 12 — 38a–c

TROCHOSPIRAL — NON-SPINOSE — Sutural apertures — "Candeina"
- Candeina nitida — 12 — 39a–b

9a. Aperture from umbilicus to periphery; trochospiral in early ontogeny
     becoming nearly planispiral in adult   .   .   .   .   .   *Globigerinella*
9b. Aperture umbilical; chambers spherical to ovate .   .   .   *Globigerina*
10a. One-chambered spherical test in adult stage; multi-chambered with
     secondary apertures in juvenile stages .   .   .   .   *Orbulina*
10b. Multi-chambered test   11
11a. Reticulate, honeycomb-like surface texture   .   .   .   *Globigerinoides*
11b. Adult test with translucent cortex and chamber flanges "*Sphaeroidinella*"

The generic classification of Recent planktonic foraminifera outlined above
and in Table 1 is a slightly modified version of an earlier scheme proposed by
Bé (1967b) and the reader is referred to this paper for detailed descriptions of
individual species. The modifications to the 1967 scheme are:

(1) The re-assignment of *Globigerina pachyderma* to the pitted, non-spinose
    genus *Globoquadrina*.
(2) The addition of the new species *Globorotalia cavernula* Bé (Bé, 1967a),
    Globorotalia crotonensis Conato and Follador (Parker, 1973b) and
    Globorotalia theyeri Fleisher (Fleisher, 1974), and the new sub-species
    Globorotalia menardii neoflexuosa proposed by Srinivasan *et al.* (1974)
    for *G. menardii flexuosa* described from the northern Indian Ocean by
    Bé and McIntyre (1970), and *Globorotalia menardii gibberula* Bé
    (this volume).

Test growth and microstructure

No direct observations of test secretion and growth of planktonic foraminifera
have been made to date. Therefore, our present knowledge is limited to deduc-
tions made from structural relationships of the test as observed from inner
and outer surfaces and cross-sections. In spite of the great detail which can be
observed by means of transmission and scanning electron microscopy, there
are still widely different interpretations on the structural units that make up
the test of planktonic foraminifera.

Towe and Cifelli (1967) have reviewed the problems in describing the
"granular" and "radial" structures of foraminifera and the difficulties which
arise when terms derived from either morphologic or optical crystallographic
criteria are equated to each other. The radial wall structure which Wood
(1949) described from polarised light observations as being built up of crystals
of calcite with their c-axes normal to the spherical surface is considered by
many workers to be synonymous with the morphological terms "fibrous",
"elongate" or "prismatic". Towe and Cifelli warn that:

   . . . similar optical orientations can be produced by different crystal morphologies and, conversely, similar crystal forms sometimes produce distinct

optical characteristics. The radial-granular concept must apply to either strictly polarized light optical phenomena or to crystal morphology. It cannot, except fortuitously, apply to both.

Another problem is that what appear to be uniform crystals under the lower resolving power of the polarising microscope are actually mosaics of much smaller units in the electron microscope. What is the size and shape of these basic building blocks of the test and how do they combine to produce larger structures?

Scanning electron microscopy shows that the inner calcite layers are made up of minute (0.5 μm) equidimensional microgranules (Plate 3, figs b, d; Plate 4, figs a, b). These microgranules may be the building blocks of the test. Although they are initially anhedral, they become subrhombic (subhedral) in the outer layers (Plate 4, figs c, d). With further wall thickening, the subrhombic units will, in turn, enlarge to become euhedral "crystals" (Plate 4, figs e, f). Whether the microgranules grow larger in size or whether they merely form parts of larger units is a question that still needs to be resolved. At present, it may suffice to theorise that, under the control of the organic matrix, a group of microgranules can form larger first-order units or "crystal mosaics", whose morphology varies considerably in successive layers.

It is the nature and configuration of the organic bonds between these first-order units that may be the underlying cause for the sometimes contradicting morphological appearances of the test walls. For example, the test wall of *Globigerinoides conglobatus* appears to consist of plate-like units that are vertically stacked because the *horizontal* separation of successive layers is emphasised (Plate 3, fig. c). In the test wall of *Globorotalia truncatulinoides*, the fibrous or columnar structure is evident, because the surfaces of separation run distinctly in a *vertical* direction (Plate 3, fig. a). In *Pulleniatina obliquilo-culata*, the entire test wall consists of equidimensional microgranules and neither a horizontal nor a vertical structural orientation is apparent (Plate 3, fig. b). Are these structural differences of a taxonomic nature or are they due to the preservational condition of the tests? In either case, it is conceivable that the microgranules have a preferred orientation and are regularly aligned to produce either a radial (fibrous, columnar, prismatic) structure if the bond separation between first-order units runs normal to the test surface or a plate-like structure if bond parting occurs parallel to the test surface.

Although a great deal is yet to be learned about the process of biomineral-isation in planktonic foraminifera, the author will nonetheless attempt to describe a working hypothesis of their test growth based on some recent observations. The formation of a new chamber begins with the emergence of a mass of protoplasm from the existing test. A proteinaceous primary organic membrane (POM) (Plate 3, fig. d) is produced just outside of the cytoplasm, followed by an outer and inner organic layer on the distal and proximal sides of the POM. These three organic layers are responsible for calcification and determine the form and shape of the test wall. They constitute the organic matrix "compartment" within which calcite secretion occurs. The initial chamber wall consists of two thin calcite layers, which are secreted between the outer and inner organic layers and the median POM (Fig. 3; Table 2). This two-layered wall or "bilamellar" test structure was first described by Reiss (1957, 1958, 1963) and observed also by Bé and Ericson (1963), Premoli-Silva (1966); McGowran (1968) and Hemleben (1969). Further wall thickening occurs by episodic calcification on the distal and proximal sides of the new chamber wall. A new calcite layer is formed by the secretory products of an organic layer which is always present on the inner and outer test surfaces. With each calcification episode, a new calcite layer is deposited upon an earlier organic layer, thus producing an alternating sequence of thin organic layers and thicker calcite layers.

Wall thickening is considerably less on the inner side than on the outer side of the POM. Thus as the test wall grows thicker, the position of the POM becomes increasingly proximal. In many species, this distinct organic layer is located in the inner third or fourth of the chamber wall and can be recognised by its continuity with the pore plate (Plate 3, fig. d).

An inner organic lining (IOL) is present between the cytoplasm and the inner test wall. The IOL of *Globorotalia inflata* shown in Plate 2 is probably homologous to the organic layer on the inside surface of the tests of benthic foraminiferal species, such as *Discorinopsis aguayoi* (Arnold, 1954). The IOL is, according to Moss, very likely a sulfated acid mucopolysaccharide (Bé and Ericson, 1953).

In summary, two main phases of calcification are believed to occur during test growth of most species of planktonic foraminifera. The two phases are illustrated by diagrams and photographs for spinose *Globigerinoides sacculifer* (Fig. 3) and non-spinose *Globorotalia truncatulinoides* (Plate 5, figs a, b) and are summarised in Table 2.

Phase 1. Formation of entire test to a certain adult size, during which the walls of previous chambers thicken slightly. Calcification is mainly concerned with the rapid construction of the test. The "bilamellar" wall of the final chamber shows two main structural units, separated by a primary organic membrane (POM) and is composed of anhedral microgranules. Additional calcite layers are secreted on the inner and outer test surfaces, but especially on the outer surface. With the formation of each new chamber, a new layer is also secreted over the outer surface of previous chambers of the test. This results in successive growth layers over earlier chambers of the test.

Phase 2. Formation of a thick "calcite crust" and, in some species, an

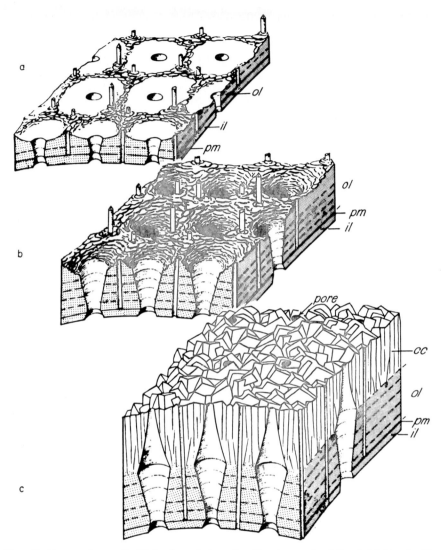

FIG. 3. Progressive wall thickening during the "'metamorphosis" of *Globigerinoides sacculifer* into "*Sphaeroidinella dehiscens*" according to Bé and Hemleben (1970). (a) and (b) represent Phase 1 and (c) represents Phase 2 of test growth.

(a), Bilamellar structure in early-stage *G. sacculifer* shows plate-like crystals forming over flat, smooth surface; (b), increase in wall thickness in mature specimens of *G. sacculifer* occurs mainly in the outer lamellar unit (*ol*); (c), euhedral "crystals" of the calcite crust (*cc*) cover the outer surface and constrict the pores in "*Sphaeroidinella dehiscens*" stage. An amorphous cortex (not shown) is eventually secreted over the calcite crust.

Broken lines in the outer lamellar unit (*ol*) represent approximate positions of the intermediate membranes. *pm*, primary organic membrane; *il*, inner lamellar unit.

additional "cortex" over the entire outer surface of the bilamellar test. In mature organisms, this wall thickening over the chambers of the last whorl

TABLE 2. Sequence of structural units of the tests of planktonic foraminifera. All species secrete a "bilamellar" test (a^1 and a^2), which consists of many smaller lamellae. *Sphaeroidinella dehiscens*, *Globorotalia inflata* and *Pulleniatina obliquiloculata* build in addition a calcite crust (b) and a "cortex" (c).

	Macrostructure	Microstructure
Test growth Phase 2	(c) Cortex (amorphous veneer)	Microgranules
	(b) Calcite crust (radial prisms)	Subhedral to euhedral "crystals"
Test growth Phase 1	a^2 Outer lamellar unit	Anhedral to subhedral microgranules
	------------- Primary Organic Membrane -------------	
	a^1 Inner lamellar unit	Anhedral microgranules

INNER ORGANIC LINING
(located against inner surface of test)

occurs with no or few additions of new chambers. The small, subhedral microstructural units form increasingly larger, euhedral crystals.

Cross-sections of two specimens of *Globorotalia truncatulinoides* of approximately equal size, one with and the other without a calcite crust are shown in Plate 5. The thinner specimen with the bilamellar test is in the first phase of calcification, while the thicker specimen shows a clear structural difference between the bilamellar test and the calcite crust of the second calcification phase (Bé and Lott, 1964; Pessagno and Miyano, 1968, Plates 3, 4 and 7). The calcite crust can add about 50% or more $CaCO_3$ by weight to the tests of *Globorotalia menardii*, *G. truncatulinoides*, "*Sphaeroidinella dehiscens*" and other species.

Factors influencing distribution and abundance of living planktonic foraminifera

The standing stock of planktonic foraminifera observed at a specific time and locale in the ocean is the product of a complex interaction between biological and physical-chemical factors. Compounding our problem of understanding this interaction are artificial, man-made factors. We can list these factors as follows:

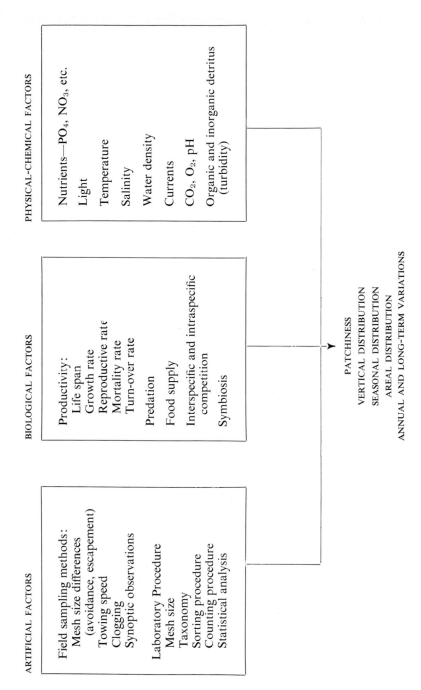

ARTIFICIAL FACTORS

Field sampling methods:
Mesh size differences
  (avoidance, escapement)
Towing speed
Clogging
Synoptic observations

Laboratory Procedure
Mesh size
Taxonomy
Sorting procedure
Counting procedure
Statistical analysis

BIOLOGICAL FACTORS

Productivity:
  Life span
  Growth rate
  Reproductive rate
  Mortality rate
  Turn-over rate

Predation
Food supply
Interspecific and intraspecific
  competition
Symbiosis

PHYSICAL-CHEMICAL FACTORS

Nutrients—PO_4, NO_3, etc.
Light
Temperature
Salinity
Water density
Currents
CO_2, O_2, pH
Organic and inorganic detritus
  (turbidity)

PATCHINESS
VERTICAL DISTRIBUTION
SEASONAL DISTRIBUTION
AREAL DISTRIBUTION
ANNUAL AND LONG-TERM VARIATIONS

We shall first review the artificial and biological factors which influence the distribution and abundance of the planktonic foraminifera and then discuss the interrelationships of some physical-chemical factors with the foraminiferal distribution patterns.

ARTIFICIAL FACTORS

The method of plankton collecting determines to a large extent the abundance and species composition of planktonic foraminifera in the samples. Plankton nets with mesh apertures ranging from 74 μm to over 500 μm in diameter have been used routinely to collect planktonic foraminifera and, naturally, the coarser the mesh aperture the greater the proportion of smaller foraminifera which escape through the net. Quantitative comparisons are rendered invalid when plankton samples have been obtained by nets of different mesh sizes. Berger (1969) has compared the average concentrations of planktonic foraminifera in different oceans as reported by various workers and plotted these values against the mesh openings of the nets used. He determined that there is an inverse relation between the cube of the mesh size and number of planktonic foraminifera caught, i.e. if the net mesh-opening is decreased by a factor of z, the catch usually increases by a factor of z^3 (Fig. 4).

Variations in towing duration and speed, clogging, ranges of sampling depths, flow meter reliability, and other idiosyncrasies in field sampling can

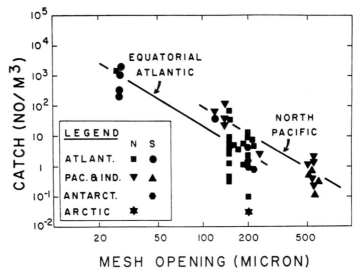

FIG. 4. Foraminiferal concentrations as a function of the mesh opening of the sampling gear of various authors in different oceanic regions. (From Berger, 1969.)

introduce additional artifacts in quantitative studies. There is also the problem of obtaining synoptic samples over extensive areas of the oceans as well as continuous year-round sampling at open-ocean localities.

Different methods in laboratory analysis of planktonic foraminifera also make quantitative comparisons difficult. Among these are differences in taxonomy, sorting and counting procedures, and statistical analysis among different workers.

BIOLOGICAL FACTORS

The productivity of planktonic foraminifera and their distribution in the ocean depend upon a number of biological factors, such as the life span and the rates of growth, reproduction, mortality and turnover of each species. In addition, the population of every species is subject to predation and requires a supply of food and organic detritus. Some species are interdependent with dinoflagellate symbionts.

Our knowledge of the biology of planktonic foraminifera is extremely meagre. Since Rhumbler's (1901, 1911) pioneering studies of their cytology, only a few studies on this aspect have appeared in recent years (Lee et al., 1965; Fèbvre-Chevalier, 1971; Zucker, 1973). Our ignorance is illustrated by the fact that we do not yet know the precise function of the unique fibrillar system of intertwining strands (Rhumbler's "gallertstränge"), inside the cytoplasm of the newest formed chambers.

Life span

No direct data from laboratory cultures are available concerning the physiology, life spans and the reproductive and mortality rates of planktonic foraminifera. How do planktonic foraminifera reproduce and how many offspring are produced by the parent(s)? Is the general assumption correct that reproduction terminates the life of the parent(s) and results in the production of empty tests?

Adshead (1967) has maintained planktonic foraminifera, mostly *Globigerina bulloides*, in the laboratory for periods up to three months, but reproduction did not occur. Bé et al. (in press) have obtained mean survival times in the laboratory for *Hastigerina pelagica, Globigerinoides conglobatus, Globigerinella aequilateralis, Orbulina universa, Globigerinoides sacculifer* and *G. ruber* of 21, 12, 11, 9, 7, and 6 days, respectively. Gametogenesis in planktonic foraminifera was first observed in *G. aequilateralis* and *G. sacculifer* by Bé and Anderson (1976).

Berger and Soutar (1967) have studied the rate of production of four species of planktonic foraminifera in the fertile California Current. By measuring their standing stocks in the upper 100 m of water and their output of empty tests (collected in a sediment trap) and assuming constant reproductive and mortality rates, they obtained turnover times for *Globigerinoides ruber*, *Globigerina bulloides*, *Globoquadrina dutertrei*, and *Globigerina quinqueloba* of 73, 58, 33 and 27 days, respectively. According to them the average life spans of these species should be shorter than the turnover times by a factor of 1 to 2 or in the order of 1 month.

Berger's (1970b) field observations of living populations off Baja California indicate that the minimum replacement time for 19 species ranges between 5 and 40 days. His data are based on output rates which are derived from a comparison of the standing stocks of living foraminifera and empty tests, assuming a constant settling velocity of 600 m/day. Berger (1970b) noted also that his output rates do not have a simple relationship to actual productivities or life spans of planktonic foraminifera, which depend to a large degree on predation and growth rates. Berger (1971a) believes that planktonic foraminifera in fertile regions have life spans of no more than about two weeks.

Predation

Again little information is available regarding the organisms which prey upon planktonic foraminifera. Who are their natural enemies? What is the mortality rate of planktonic foraminifera by predation? To what extent are their tests destroyed by mechanical crushing in predators' jaws or by enzyme dissolution in the guts of predators?

Bradbury *et al.*'s (1970) study of the gut contents of macroplankton and nekton in the Indian Ocean reported the presence of planktonic foraminifera in the guts of tunicates, pteropods, euphausids, and sergestids. Judkins and Fleminger (1972) also observed foraminifera in the foregut of prawns (*Sergestes similis*) in the eastern North Pacific. The tests were found to be intact and undissolved. On the other hand, Berger (1971b) found crushed foraminiferal tests in the guts of shrimps, salps and crabs (*Pleuroncodes*), but not of the copepods, collected in the oceanic waters off Baja California. The author has observed a large number of well-preserved, thin-walled tests of juvenile planktonic foraminifera in the fecal pellets of salps (*Pegea confoederata*) from the Gulf Stream off Bimini.

Food

Dinoflagellates and diatoms have been found in the food vacuoles of planktonic foraminifera (Zucker, 1973). We have also noted partly digested algae inside food vacuoles of *Globorotalia inflata* (Plate 2).

Rhumbler (1911) illustrated copepod remnants within the cytoplasm of *Hastigerina pelagica*, *Hastigerinella digitata*, and *Globigerinoides sacculifer* (= *G. triloba*). His observations are confirmed by the author who has seen living specimens of *H. pelagica*, collected in the Florida Current off Grand Bahama Island, which had incorporated copepods into their bubble capsule. When copepod specimens of the genera *Farranula* and *Oncaea* were placed onto the outer surface of the bubble capsule by pipette, it was noted that within 10 minutes the copepods were drawn into the capsule. After about 6–8 hours the copepods were digested and their undigested carapaces were later ejected. *Globigerinella aequilateralis* was also observed to capture copepods and amphipods. These observations indicate that some species of planktonic foraminifera have a carnivorous diet.

Symbionts

Symbiotic dinoflagellates occur both intracellularly and extracellularly in some species of planktonic foraminifera. It is believed that the foraminifera benefit from this relationship by obtaining oxygen from the algae, while the latter take advantage of the carbon-dioxide, protection and transport which are provided by their hosts. There is a possibility that the foraminifera may feed upon the algae in times of stress, but this has not been documented.

One of the earliest descriptions of small algae living on planktonic foraminifera is contained in the narratives of the *Challenger* Expedition (Tizard *et al.*, 1885):

> In *Orbulina* there are almost always a great number of yellow cells similar to those found in the Radiolaria; they are oval and about 0.01 mm in the longest diameter, and have a nucleus which colours quickly with carmine, before treatment with spirit. On several occasions they were seen to flow out from the interior of the shell with pseudopodia through the pores, and mount a considerable distance up the spines; when expanded in this way an *Orbulina* or *Globigerina* looks exceedingly like many of the Radiolaria. The yellow cells appeared to have an independent motion, as they were frequently observed turning about in the sarcode. Yellow cells are also present in *Globigerina bulloides*, var. *triloba*, *Globigerina sacculifera*, and *Globigerina conglobata*, but were not observed in *Hastigerina*, *Pullenia*, nor *Pulvinulina*.

Rhumbler (1911) illustrated zooxanthellae within the cytoplasm of *Orbulina universa* and *Globigerinoides sacculifer* (= *G. triloba*). Symbiotic zooxanthellae, identified as dinoflagellates similar to *Symbiodinium microadriaticum*, have been found to occupy as much as 80% of the cytoplasmic volume of *Globigerinoides ruber* (Lee *et al.*, 1965; Zucker, 1973).

Alldredge and Jones (1973) observed that *Hastigerina pelagica* (Plate 1, fig. b) harbours at least three species of dinoflagellate symbionts (*Pyrocystis fusiformis; P. noctiluca* and *Dissodinium elegans*) within its bubble capsule. They recorded an average number of 6 and maximum number of 79 dinoflagellates per foraminiferal host.

The author observed movement of the symbionts into and out of the tests of *Globigerinoides conglobatus* and within the bubble capsule of *Hastigerina pelagica*. In the former species, collected in the Sargasso Sea and observed on board ship, the symbionts vacated the test during the day time and retreated into the test after darkness. He observed that in *Hastigerina pelagica*, collected off Grand Bahama Island, the large dinoflagellates moved in and out of the bubble capsule, but not into the test. The passive movement of the symbionts is actuated by the pseudopodia and does not appear to be hindered by the compartments of bubbles.

The dinoflagellates in this facultative commensal or symbiotic association live in the euphotic zone. It is therefore not a coincidence that those species which harbour symbionts (*G. ruber, G. sacculifer, G. conglobatus, O. universa, H. pelagica*) usually occur in shallow depths, while those species without symbionts (*Globorotalia* species) occur over a greater depth range and frequently below the euphotic zone.

PHYSICAL-CHEMICAL FACTORS

The complex interrelationship between species distributions and environmental conditions can be appreciated when one attempts to correlate their quantitative abundance with physical–chemical parameters at each sample locality. An example of such a comparison for the planktonic foraminifera from the upper 300 m of the Indian Ocean is shown in Table 3. By calculating means and standard deviations for surface-water temperature, salinity, oxygen and phosphate values (from Wyrtki's 1971 Atlas) weighted according to each species abundance, it is possible to determine the optimum values and ranges of each environmental factor which each foraminiferal species favours.

Table 3 ranks each species according to their preferred levels of surface-water temperature, salinity, oxygen and phosphate. The results show that *Globoquadrina pachyderma* is characterised by low temperature and salinity and high oxygen and nutrient (phosphate) levels. *Globigerina bulloides* is generally associated with somewhat warmer, more saline, lower oxygen and nutrient levels than *G. pachyderma*. *Globoquadrina hexagona* prefers regions of high temperature, and low oxygen and salinity. *Globorotalia truncatulinoides*, *Globigerina rubescens* and *Orbulina universa* are indicators of the central subtropical regions of the Indian Ocean, where the waters are consistently low in phosphates, very high in salinity and intermediate in temperature and oxygen values.

Vertical distribution

Most species of planktonic foraminifera live in the euphotic zone where they can feed upon phytoplankton (Murray, 1897; Lohmann, 1920; Schott, 1935;

Phleger, 1951; Bradshaw, 1959; Bé, 1960a; Boltovskoy, 1964; Jones, 1967, 1968b; Berger, 1970b, 1971a, b; Bé and Tolderlund, 1971; Williams, 1971). The highest concentrations are generally found below the surface, approximately between 10 m and 50 m (Bradshaw, 1959; Bé, 1960a; Boltovskoy, 1964). Some species live in the sunlit waters for the maintenance of their symbiotic zooxanthellae and the olive-green or brownish colour of the protoplasm in epipelagic individuals is indicative of either freshly ingested phytoplankton or the presence of symbionts.

The planktonic foraminifera living in the surface layer are generally smaller and thinner-walled than those from deeper waters. Boltovskoy (1964) has shown that the larger tests of *Globoquadrina dutertrei*, *Globigerinita glutinata*, *Globigerinoides ruber*, *G. sacculifer* and *Globorotalia menardii* are less abundant in the upper 25 m of the equatorial Atlantic than they are in the 40–100 m depth range. Berger (1971b) also observed that small specimens ($< 200\ \mu$m) are more abundant than large ones in the upper 400 m of water off Baja California.

Planktonic foraminifera do occur in the mesopelagic (300–1000 m) and bathypelagic (1000–3000 m) zones (Schott, 1935; Bradshaw, 1959; Bé, 1960a; Jones, 1967, 1968b; and Berger, 1970b). Many tests encountered at these depths are empty, but those filled with protoplasm are presumed to be viable organisms. Most of them belong to species which are commonly found in the epipelagic zone. With the possible exception of *Hastigerinella digitata*, no other species are known to spend their entire life cycle below a depth of 1000 m.

Bé and Ericson (1963) and Bé (1965) believe that many species spend their earlier stages in the epipelagic zone and eventually seek deeper habitats. They noted that specimens collected from meso- and bathypelagic depths generally have thick-walled tests with a calcite crust and whitish protoplasm which is indicative of their isolation from the euphotic zone.

As the concentration of living planktonic foraminifera decreases exponentially with depth in the upper few hundred meters, the concentration of empty tests increases rapidly between 50–100 m, but increases only slightly between 100–500 m (Fig. 5). Empty tests outnumber living specimens below 500 m (Berger, 1971a).

DEPTH HABITATS OF INDIVIDUAL SPECIES

The vertical range of depth habitats for individual species is extremely wide and varies both regionally and seasonally. Some species probably adjust their depth habitat during their life cycle. Thus it is impossible to define the depth range of individual species within narrow limits, and any depth stratification of species is only valid in a statistical sense.

TABLE 3. Ranking of planktonic foraminiferal species according to four physical parameters at the surface of the Indian Ocean based on a comparison between the absolute species abundance in 154 plankton tows and physical-chemical environmental data from Wyrtki (1971).

	Temperature (°C)			Salinity (‰)	
Rank	Mean	S.D.	Rank	Mean	S.D.
1 Globoquadrina conglomerata	28.3	1.006	1 Globorotalia truncatulinoides	35.55	2.283
2 Globoquadrina hexagona	28.1	0.240	2 Globigerina rubescens	35.52	1.980
3 Globorotalia tumida	25.3	1.658	3 Globorotalia crassaformis	35.45	0.795
4 Globigerinoides sacculifer	25.2	3.256	4 Orbulina universa	35.40	1.632
5 Pulleniatina obliquiloculata	24.9	1.947	5 Globigerinita glutinata	35.38	4.403
6 Globigerinoides conglobatus	24.4	3.111	6 Globorotalia scitula	35.35	1.692
7 Globigerinoides ruber	24.2	3.544	7 Globigerinella aequilateralis	35.34	2.136
8 Globigerinella aequilateralis	23.5	4.446	8 Globoquadrina conglomerata	35.31	0.139
9 Globorotalia crotonensis—G. theyeri	23.4	4.671	9 Hastigerina pelagica	35.31	3.730
10 Hastigerina pelagica	23.3	4.098	10 Globigerinoides ruber	35.25	1.987
11 Globoquadrina dutertrei	23.2	2.991	11 Globorotalia inflata	35.23	3.797
12 Globorotalia menardii	23.1	2.758	12 Globoquadrina dutertrei	35.19	1.604
13 Globigerina rubescens	23.0	1.980	13 Globorotalia crotonensis—G. theyeri	35.11	2.018
14 Globorotalia scitula	22.7	3.713	14 Globorotalia menardii	34.99	1.547
15 Globigerinita glutinata	22.2	7.551	15 Globigerinoides conglobatus	34.99	1.325
16 Orbulina universa	21.7	2.393	16 Pulleniatina obliquiloculata	34.97	1.307
17 Globorotalia truncatulinoides	20.3	2.878	17 Globigerinoides sacculifer	34.94	1.476
18 Globorotalia crassaformis	18.1	3.455	18 Globigerina bulloides	34.76	5.116
19 Globorotalia inflata	16.5	3.819	19 Globorotalia tumida	34.70	1.053
20 Globigerina bulloides	13.4	7.114	20 Globigerina quinqueloba	34.51	3.579
21 Globigerina quinqueloba	9.8	5.411	21 Globoquadrina hexagona	34.49	0.064
22 Globigerinita bradyi	7.5	4.850	22 Globigerinita bradyi	34.09	1.882
23 Globoquadrina pachyderma	4.8	5.524	23 Globoquadrina pachyderma	34.05	2.901

Oxygen (ml/l)

Rank		Mean	S.D.
1	Globigerinita bradyi	7.520	0.069
2	Globoquadrina pachyderma	7.507	0.334
3	Globigerina quinqueloba	6.615	0.303
4	Globigerina bulloides	6.272	0.442
5	Globorotalia inflata	5.660	0.373
6	Globorotalia crassaformis	5.315	0.210
7	Globorotalia truncatulinoides	5.199	0.221
8	Globigerina rubescens	5.165	0.127
9	Globigerinita glutinata	5.134	1.451
10	Orbulina universa	5.085	0.197
11	Hastigerina pelagica	4.991	0.455
12	Globoquadrina dutertrei	4.953	0.186
13	Globorotalia scitula	4.952	0.170
14	Globorotalia crotonensis—G. theyeri	4.927	0.238
15	Globigerinella aequilateralis	4.880	1.000
16	Globorotalia menardii	4.862	0.279
17	Globigerinoides ruber	4.841	0.576
18	Globigerinoides conglobatus	4.808	0.219
19	Globigerinoides sacculifer	4.724	0.216
20	Pulleniatina obliquiloculata	4.660	0.213
21	Globorotalia tumida	4.652	0.293
22	Globoquadrina hexagona	4.530	0.014
23	Globoquadrina conglomerata	4.463	0.059

Phosphate-P (µg at/l)

Rank		Mean	S.D.
1	Globoquadrina pachyderma	1.517	0.451
2	Globigerinita bradyi	1.415	0.606
3	Globigerina quinqueloba	0.880	0.214
4	Globigerina bulloides	0.742	1.148
5	Globorotalia crassaformis	0.415	0.065
6	Globigerinita glutinata	0.369	1.931
7	Globorotalia inflata	0.365	0.420
8	Hastigerina pelagica	0.229	1.433
9	Globoquadrina conglomerata	0.223	0.718
10	Globorotalia scitula	0.222	0.878
11	Globigerinella aequilateralis	0.218	1.525
12	Pulleniatina obliquiloculata	0.216	1.136
13	Globoquadrina dutertrei	0.202	1.437
14	Globorotalia tumida	0.200	1.080
15	Globigerinoides sacculifer	0.198	1.386
16	Globorotalia crotonensis—G. theyeri	0.196	0.932
17	Globoquadrina hexagona	0.195	0.057
18	Globigerinoides ruber	0.193	1.577
19	Globorotalia menardii	0.191	1.158
20	Globorotalia truncatulinoides	0.191	0.366
21	Globigerinoides conglobatus	0.183	1.197
22	Orbulina universa	0.152	0.981
23	Globigerina rubescens	0.125	0.629

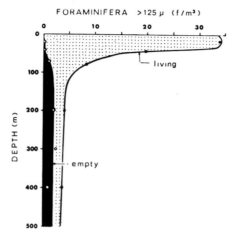

FIG. 5. Concentration of living planktonic foraminifera and empty tests at various depths in Santa Barbara Basin. Concentration expressed as number of foraminifera per m³ (f/m³). (From Berger, 1971a.)

Planktonic foraminifera may be considered in terms of three broad groups which characterise particular depth habitats (Table 4). The "shallow-water" species are mainly spinose forms and include all the species of *Globigerinoides* and some species of *Globigerina*. Adult tests of these species from subsurface depths have fewer spines or no spines. The "intermediate-water" species include both spinose (*G. bulloides, H. pelagica, O. universa, G. aequilateralis* and *G. calida*) and non-spinose species (*P. obliquiloculata, G. dutertrei, C. nitida* and *G. glutinata*). The "deep-water" group consists of 12 species which live in the euphotic zone as juveniles, and predominantly below 100 m as adults. All the species of *Globorotalia* and the non-spinose *Globoquadrina pachyderma* and *G. conglomerata* belong in this group. In subantarctic waters a large form of *Globorotalia scitula* is found at a depth range of between 500 and 1000 m, while *Globorotalia crassaformis* is a deep-water inhabitant of subtropical areas. Only three spinose species are considered to be deep-water forms, these are *Hastigerinella digitata, Globigerinella adamsi* and "*Sphaeroidinella dehiscens*". According to Bé (1965) and Bé and Hemleben (1970) "*S. dehiscens*" is an aberrant terminal stage of *Globigerinoides sacculifer*, which has secreted a thick calcite crust and cortex over its spinose test. The shell thickening process in many species is an adaptation to increasingly deeper habitats (Bé and Ericson, 1963).

Jones (1967) was able to distinguish the Equatorial Atlantic Undercurrent and the waters above and underneath it by means of their planktonic foraminiferal species compositions. *Globigerinoides sacculifer* and *G. ruber*

TABLE 4. Generalised depth habitats of planktonic foraminifera, based on plankton tows (Schott, 1935; Bradshaw, 1959; Bé, 1960a; Parker, 1960; Belyaeva, 1964b; Jones, 1964, 1967; Berger, 1969, 1971b; Bé and Tolderlund, 1971; Williams, 1971) and oxygen isotope composition (Emiliani, 1954; Lidz *et al.*, 1968; Oba, 1969; Hecht and Savin, 1973).

"Shallow-water" species living predominantly in the upper 50 m

> *Globigerinoides ruber*
> *Globigerinoides sacculifer*
> *Globigerinoides conglobatus*
> *Globigerina quinqueloba*
> *Globigerina rubescens*

"Intermediate-water" species living in the upper 100 m, but predominantly from 50–100 m

Globigerina bulloides	*Orbulina universa*
Hastigerina pelagica	*Candeina nitida*
Pulleniatina obliquiloculata	*Globigerinella aequilateralis*
Globoquadrina dutertrei	*Globigerina calida*
	Globigerinita glutinata

"Deep-water" species living in the upper few hundred meters and whose adult stages occur predominantly below 100 m

Globorotalia menardii	*Globoquadrina pachyderma*
Globorotalia tumida	*Globoquadrina conglomerata*
Globorotalia inflata	*Globoquadrina hexagona*
Globorotalia hirsuta	*Globigerinella adamsi*
Globorotalia truncatulinoides	*Sphaeroidinella dehiscens*
Globorotalia crassaformis	*Hastigerinella digitata*
Globorotalia scitula	

---

populate the upper 50 m of water above the Undercurrent, while *Globoquadrina dutertrei* and *Pulleniatina obliquiloculata* characterise the high salinity (36‰) Undercurrent between approximately 50 and 100 m. *Globorotalia crassaformis* and, to a lesser extent, *G. menardii* are predominantly found below the Undercurrent (>100 m depth).

Depth habitats of planktonic foraminifera can also be inferred from isotopic temperatures which are calculated from O^{18}/O^{16} ratios. Differences in the oxygen isotope composition of species from the same sediment assemblage have been interpreted in terms of differences in their depth habitats (Emiliani, 1954; Lidz *et al.*, 1968; Shackleton, 1968; Oba, 1969; Hecht and Savin, 1972; Savin and Douglas, 1973). Since the measured isotopic compositions are obtained from the total calcium carbonate of a large number of specimens, it is important to remember that the isotopically-determined depth habitat of a given species represents the average depth in which most of the $CaCO_3$ of tests assigned to that species was secreted. Depth stratification of species should thus be considered only as a statistical approximation. It does not mean that species of planktonic foraminifera are narrowly restricted to these depths and that they overlie each other in a rigidly stratified manner.

The statistical depth rankings of species based on isotopic data from different oceans by various investigators are generally in agreement with one another and with the data obtained from plankton tows. These depth rankings have been studied more thoroughly for tropical and subtropical species and the results shown on Table 4 are therefore more valid for warm-water species.

DAILY VERTICAL MIGRATION

Many nekton and plankton groups migrate upward at dusk, presumably to feed in the relatively rich surface waters, and usually descend to deeper levels before daybreak. This widely observed phenomenon of daily vertical migration has prompted a number of investigators to determine whether a similar relationship exists for the planktonic foraminifera. Do planktonic foraminifera adjust their depth habitats to changing light conditions and food availability? If so, what is the pattern of vertical migration and how does it vary according to species, geographic locale and season?

The answers to these questions are again shrouded in uncertainty, because of inadequate information. Comparisons of the data from different investigators is rendered difficult due to differences in the sampling methods (gear, mesh size, sampling depths), the region and season of sampling, the species involved and statistical methods of comparison.

Rhumbler (1911) suggested that planktonic foraminifera were more abundant in day tows than in night tows within the upper 200 m of the water column. This trend is completely opposed to the diurnal pattern of migration of most planktonic and nektonic organisms. The observations of Bradshaw (1959), Parker (1960), Bé (1960a) and Bé and Hamlin (1967) tend to support Rhumbler's suggestion.

Berger (1969) analysed Bradshaw's (1957) data statistically and came to the following conclusions:

(1) Species showing very significant diurnal variations were *Globigerina bulloides*, *Orbulina universa*, *Globigerinella calida*, *Globigerinita bradyi*, *Globorotalia menardii* and *Globoquadrina dutertrei*. Except for *G. menardii*, these species are abundant in the fertile cool-water regions.

(2) Species showing moderately significant diurnal variations were *Hastigerina pelagica*, *Globigerina quinqueloba*, *Globigerinella aequilateralis* and *Globorotalia hirsuta*.

(3) Species showing little variation in day–night concentrations were *Globigerinoides ruber*, *G. sacculifer*, *G. conglobatus*, *Globigerinita glutinata* and *Pulleniatina obliquiloculata*. These are mainly shallow-water species in the less fertile tropical regions.

(4) Shallow tows (0–100 m) taken with fine (70–140 μm mesh diameter) nets showed more pronounced day–night variations than those taken with coarse (500 μm) nets.

In another study of the diurnal variations of planktonic foraminiferal abundance, Tolderlund (1969) concluded that diurnal vertical migration is a species-specific and seasonally restricted phenomenon. He noted diurnal migration for only three species (*Globigerinoides ruber*, *Globigerinoides sacculifer*, *Globigerinella aequilateralis*) whose daytime density was significantly higher than their night-time abundance in the equatorial Atlantic during the summer, but not in the spring. In addition, *Orbulina universa* occurred in significantly greater abundance in day-time than in night surface tows in the Gulf Stream (Station Delta: 44°00′N, 41°00′W).

On the other hand, Boltovskoy (1973) did not observe any daily vertical migration in the planktonic foraminifera collected by pump in South Atlantic surface waters (0–5 m) between Argentina and South Africa. It is possible, however, that samples from a greater depth range are necessary for comparisons of diurnal fluctuations in abundance.

Seasonal distribution

Seasonal studies of planktonic foraminifera are rare, because of the difficulty of collecting plankton samples at frequent intervals at open-ocean localities. Except for the year-round observations off Bermuda and at four U.S. Coast Guard ocean stations in the western North Atlantic by Tolderlund and Bé (1971), the majority of seasonal investigations have been carried out at near-shore localities. Seasonal variations of the planktonic foraminifera were studied in the Sargasso Sea off Bermuda by Bé (1960a), in the Florida Straits by Jones (1964), and in the Gulf of Guinea by Eckert (1965). A novel approach to plankton sampling was undertaken in the Beaufort Sea by Vilks (1973), who used a helicopter to reach the sampling stations on the sea-ice from various bases along the Canadian Arctic Coast and drilled holes through the ice to lower the plankton nets.

Seasonal variations in the abundance of planktonic foraminifera are the general rule rather than the exception. Fluctuations in their standing stocks are closely linked to seasonal cycles in the nutrient chemistry and hydrography of the waters. Vertical mixing during winter leads to nutrient enrichment of the upper waters, and the resulting phytoplankton increase occurs at different times of the year depending upon latitude (Bogorov, 1960). The phytoplankton bloom coincides with the winter months in the tropical seas, whereas in subtropical and temperate latitudes spring and fall maxima of phytoplankton standing crops are commonly recognised. The spring bloom is progressively delayed towards the higher latitudes, until only a single phytoplankton peak occurs during August in the Arctic Ocean.

The seasonal fluctuations in abundance of planktonic foraminifera follow the phytoplankton productivity cycle. In mid-latitude regions, such as the northern Sargasso Sea, the maximum abundance of planktonic foraminifera takes place between December and March, whereas in the subarctic waters maximum concentrations are reached between May and October (Bé, 1960a; Tolderlund and Bé, 1971). The magnitude of the seasonal variations in abundance also increases from low to high latitudes. In the subtropical Sargasso Sea there may be a several-fold difference between the maximum and minimum standing stock, but this difference is an order of magnitude higher in the subarctic waters of the North Atlantic (Tolderlund and Bé, 1971). The seasonal contrast is most pronounced in the Arctic Ocean. Vilks (1973) reported that in the Beaufort Sea the summer (August to October, 1970) standing stock of planktonic foraminifera, consisting almost entirely of *Globoquadrina pachyderma*, was on the average 220 times greater than the winter (March–April, 1972) population.

Seasonal successions in planktonic foraminiferal populations are well-documented for the northern Sargasso Sea (Bé, 1960a; Tolderlund and Bé, 1971) and are illustrated in Fig. 6. Various species groups replace each other on a regular, seasonal schedule at Station Echo (35°00′N, 48°00′W) and off Bermuda. *Globorotalia truncatulinoides*, *G. hirsuta*, *G. inflata* and *Globigerina bulloides* are winter species which flourish between December and April. They are replaced by a group of summer species (*Globigerinoides ruber*, *G. sacculifer*, *Orbulina universa* and *Hastigerina pelagica*) which appear mostly between June and September. *Globigerinoides conglobatus* is found primarily in October, while *Pulleniatina obliquiloculata*'s peak abundance occurs in November.

In subpolar and polar regions, the seasonal succession of planktonic foraminiferal faunas is less evident or even absent. Tolderlund and Bé (1971) noted that at Station Charlie (52°45′N, 35°30′W), *Globigerina bulloides* and *Globorotalia inflata* were most abundant in September and October, while *Globigerina quinqueloba*'s peak abundance occurred in May. In subantarctic waters *Globorotalia truncatulinoides* is most common between May and

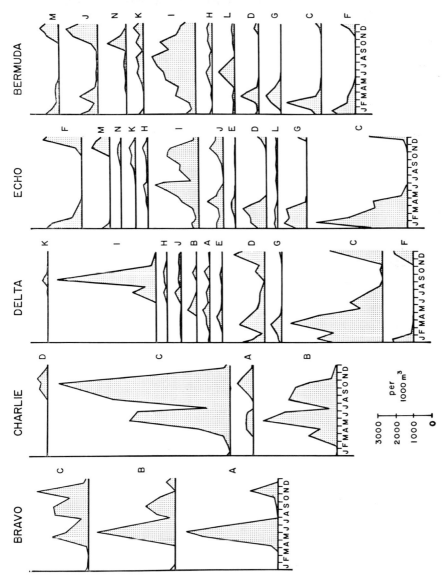

FIG. 6. Seasonal succession among the planktonic foraminifera in surface waters of the western North Atlantic, based on monthly averages of absolute abundance at Ocean Stations Bravo (56°30'N, 51°00'W), Charlie (52°45'N, 35°30'W), Delta (44°00'N, 41°00'W), Echo (35°00'N, 48°00'W), and off Bermuda. (From Tolderlund and Bé, 1971.) A, *G. pachyderma*; B, *G. quinqueloba*; C, *G. bulloides*; D, *G. inflata*; E, *G. glutinata*; F, *G. truncatulinoides*; G, *G. hirsuta*; H, *O. universa*; I, *G. ruber*; J, *G. aequilateralis*; K, *G. sacculifer*; L, *H. pelagica*; M, *P. obliquiloculata*; N, *G. conglobatus*.

October, while *G. inflata* proliferates between August and January (Bé, 1969).

The seasonal replacement of various species groups in the ocean has significant palaeoecological implications. It is important to remember that, although the empty tests of these species occur together in the same death assemblage on the sea-bed, they do not necessarily inhabit the overlying waters at the same time, and, therefore, each species may represent different ecologic and seasonal conditions.

Areal distribution and biogeographic divisions of life assemblages

Numerous plankton biogeographers, such as Giesbrecht (1892), Meisenheimer (1905), Brandt and Apstein (1901–1928), Steuer (1910, 1933), Russell (1939), Sverdrup *et al.* (1942), Bogorov (1967), Johnson and Brinton (1963), McGowan (1971), and Fleminger and Hülsemann (1973), have noted a close relationship between plankton communities and the major hydrographic regions of the world ocean. They recognised that circumglobal distributions are a common rule for many epiplanktonic groups and that their strongly latitudinal occurrences are indicative of significant interaction with surface-water temperatures.

The continental barriers of the Americas, Africa, Eurasia and Australia as well as islands and shallow seas interrupt the circumglobal continuity in the distribution of those epiplankton species living in the warm-water belt between 40°N and 40°S. Topographic barriers are deterrents to faunal mixing and the prevention of gene flow between adjacent populations for a sufficiently long duration ultimately results in the provincialism of the isolated populations.

In reviewing the biogeographic distribution of planktonic foraminifera in the epipelagic zone (upper 200 m of water) of the world ocean, we can recognise first-order patterns of circumglobal nature as well as second-order patterns which appear to be the result of genetic isolation and provincialism.

In broad terms three major biogeographical divisions can be distinguished for marine plankton communities: A Circumglobal Warm-water Region, and the Northern and Southern Cold-water Regions. The two latter which can be further subdivided into Polar (= Arctic and Antarctic) and Subpolar (= Subarctic and Subantarctic) Provinces. The Warm-water Region comprises two Subtropical Provinces with a single Tropical Province in between. Each

---

FIG. 7. Five major faunal provinces of living planktonic foraminifera. 1, Arctic and Antarctic Faunal Provinces; 2, Subarctic and Subantarctic Faunal Provinces; 3, Transition Zones; 4, Subtropical Faunal Provinces; 5, Tropical Faunal Province.

The Transition Zones (3) lie between the Northern and Southern Cold-water Regions (1 and 2) and the Circumglobal Warm-water Region (4 and 5). Compare with Table 5 and Fig. 8. (From Bé and Tolderlund, 1971).

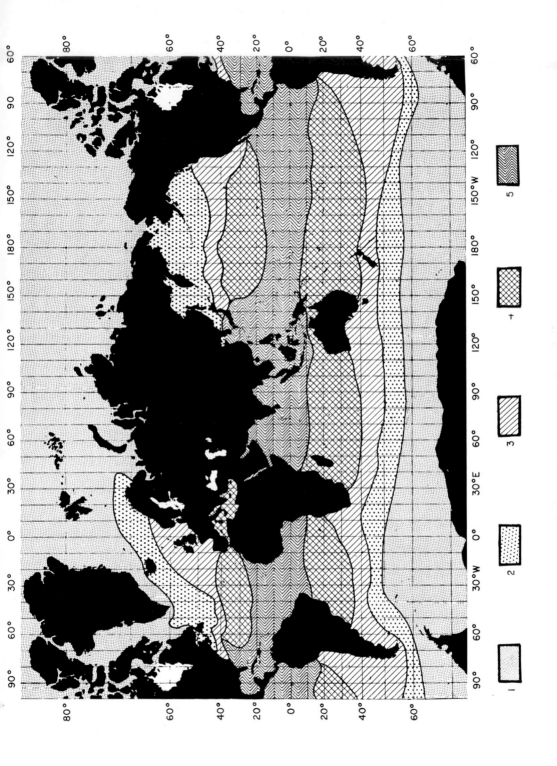

province has its reciprocal in the other hemisphere, so that although a total of nine geographic subdivisions can be distinguished there are faunistically only five discrete provinces. The faunal provinces shown in Fig. 7 correspond in general to the major hydrographic regions in the world ocean (Sverdrup *et al.*, 1942).

Species which belong to the warm-water region are transported out of their indigenous zone northwards by clockwise current gyres in the northern hemisphere and southwards by counterclockwise current gyres in the southern hemisphere. These currents are strongly developed along the western oceanic margins in the northern hemisphere, as evidenced by the extensive northerly displacement of warm-water species by the Gulf Stream and Kuro Shio. In the southern hemisphere, the southerly transport of warm-water species also occurs along the western oceanic margins via the Brazil Current, the Mozambique–Agulhas Current, and the East Australian Current.

The distributional continuity between Pacific and Atlantic warm-water species is completely interrupted by the North–South American landmass. The Indonesian archipelago and its shallow seas, however, allow relatively unrestricted communication between the Pacific and Indian Ocean populations. The African continent is a barrier of intermediate magnitude, allowing some faunal interchange between warm-water species at 35°S. Thus, at present the faunal contrast is greatest among warm-water species between the eastern Pacific and North Atlantic.

Species of the Northern Cold-water Region are carried to lower latitudes by southerly flowing currents, such as the Labrador and Portugal–Canaries Currents in the North Atlantic Ocean, and the Oya Shio and California Current in the North Pacific.

The species of the Southern Cold-water Region are distributed in continuous latitudinal belts around Antarctica, due to the uninterrupted flow of the West Wind Drift. Northerly transport of planktonic foraminifera from the Southern Cold-water Region is accomplished mainly by currents on the eastern margins of the oceans, such as the Benguela Current, the West Australian Current, and the Peru Current.

The integrity of the Warm-water and Cold-water Regions as major biogeographic divisions is best demonstrated by the distribution patterns of their species populations. Thus, cold-water species, such as *Globoquadrina pachyderma* and *Globigerina bulloides*, which have their highest concentrations in the polar and subpolar zones, decrease in abundance towards the lower latitudes. Vice versa, warm-water species, which are dominant in the tropical and subtropical zones, diminish rapidly towards higher latitudes.

The transport of warm-water species into the cold-water regions and, conversely, of cold-water species into the warm-water region results in the mixing of foraminiferal faunas with different ecological tolerances. These areas of mixing are called "Transition Zones", and they are frequently characterised by very high species diversities. The northern Transition Zones are narrow and well-defined on their western margins but widen and become more diffuse towards the east. The southern Transition Zones are more uniform in their latitudinal width than their northern counterparts. Some latitudinal broadening, however, occurs off the southwestern coasts of South America and Africa, due to the northern flow of the Peru and Benguela Currents, respectively.

FAUNAL PROVINCES AND SPECIES ASSEMBLAGES

By delineating the area where each species occurs most abundantly, we can group the planktonic foraminifera into several faunal provinces and transition zones. Table 5 and Fig. 8 list the species composition for each of the faunal provinces and transition zones. The reader is referred to the publications in the References which are followed by Roman and Arabic numbers for distributional data on individual species of planktonic foraminifera in various oceanic regions.

(a) Tropical province

This province has the highest species diversity and contains fourteen species and one subspecies. (Table 5 and Fig. 8.) The dominant species in the tropical waters are in approximate order of decreasing abundance: *Globigerinoides sacculifer*, *G. ruber*, *Pulleniatina obliquiloculata*, *Globoquadrina dutertrei* and *Globorotalia menardii*. Although *Globigerinoides ruber* is considered a subtropical species, it is prolific in tropical waters where it frequently outnumbers the tropical species.

Within this province *Globigerinella adamsi*, *Globoquadrina hexagona* and *Globigerinoides conglobatus* are absent in the Atlantic Ocean, though *G. hexagona* and *G. conglomerata* are recorded in the Pleistocene sediments of this ocean.

(b) Subtropical province

This province contains twelve species (Table 5; Fig. 8) four of which (*Globoquadrina dutertrei*, *Globigerinella aequilateralis*, *Hastigerina pelagica* and *Globorotalia menardii gibberula*) are associated with upwelling regions and boundary currents around the margins of the central watermasses. *G. menardii gibberula* is an indicator species of the South Equatorial Current in the Indian Ocean, and it also occurs in a broad belt across the South Atlantic between about 20°S and 35°S.

Tropical species which are also frequently encountered in the subtropical province are: *Pulleniatina obliquiloculata*, *Globigerinoides sacculifer* and

TABLE 5. Species composition of the planktonic foraminifera in five major faunal provinces shown in Figs 7 and 8.

NORTHERN AND SOUTHERN COLD-WATER REGIONS

1 *Arctic and Antarctic Provinces:*
Dominant, indigenous species (1)
Globoquadrina pachyderma (left-coiling variety)

Co-occurring species (4)
Globigerina quinqueloba
Globigerina bulloides
Globigerinita bradyi
Globigerinita glutinata

2 *Subarctic and Subantarctic Provinces:*
Dominant, indigenous species (4)
Globigerina bulloides
Globoquadrina pachyderma (right-coiling variety)
Globigerina quinqueloba
Globigerinita bradyi

Co-occurring species (4)
Globorotalia scitula
Globorotalia inflata
Globigerinita glutinata
Globorotalia truncatulinoides

TRANSITION ZONES

3 *Northern and Southern Transition Zones between Cold-water and Warm-water Regions:*
Dominant, indigenous species (1)
Globorotalia inflata

Co-occurring species (17)
6 subpolar species
11 subtropical species

WARM-WATER REGION

4 *Subtropical Provinces:*
Dominant, indigenous species (13)
Globigerinoides ruber
Globorotalia truncatulinoides
Globorotalia hirsuta
Globigerinella aequilateralis
Globigerina falconensis

Co-occurring species (7)
Globoquadrina dutertrei
Pulleniatina obliquiloculata
Globigerina bulloides
Globigerina rubescens
Globorotalia menardii

Globigerinoides sacculifer
Globorotalia inflata

Hastigerina pelagica
Globigerinoides conglobatus
Globigerinita glutinata
Globigerina humilis
Orbulina universa
Globorotalia crassaformis
Globorotalia menardii gibberula (Indian Ocean and South Atlantic)
Globorotalia crotonensis (Indo-Pacific only)

5 *Tropical Provinces:*
Dominant, indigenous species (14)
Globigerinoides sacculifer
Pulleniatina obliquiloculata
Globorotalia menardii menardii
Globorotalia tumida
Globigerina rubescens
Candeina nitida
Globoquadrina dutertrei
Hastigerinella digitata
"Sphaeroidinella dehiscens"
Globigerina digitata
Globoquadrina conglomerata
Globoquadrina hexagona
Globigerinella adamsi
Globorotalia theyeri
Globorotalia menardii neoflexuosa

} restricted to Indo-Pacific

Co-occurring species (10)
Globigerinoides ruber
Globigerinella aequilateralis
Globigerinoides conglobatus
Hastigerina pelagica
Globigerinoides conglobatus
Globigerinita glutinata
Globigerina cf. *bulloides* (upwelling region)
Orbulina universa
Globorotalia crassaformis
Globorotalia truncatulinoides

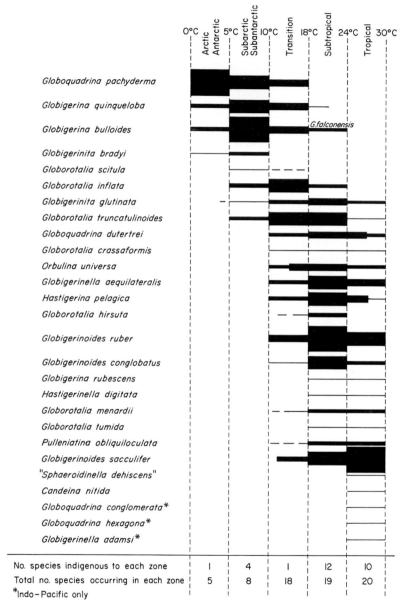

Fig. 8. Species assemblages and ranges in the five major faunal provinces. Varying thickness represents relative abundance within each zone. Compare with Table 5 and Fig. 7. (From Bé and Tolderlund, 1971.)

Globigerina rubescens. The fact that tropical and subtropical species do co-occur commonly in their respective provinces indicates that the faunistic as well as hydrographic differences between these regions are of a lesser order of magnitude than those between the subtropical and subpolar provinces.

(c) Transition Zones

The Transition Zones are regions where mixing of subtropical–tropical and subpolar species takes place and as a result completely different foraminiferal species assemblages may be encountered at stations within a few miles of each other, as for example at the boundaries of the Gulf Stream and Labrador Current (Bé, 1959; Bé and Hamlin, 1967; Bé *et al.*, 1971; Cifelli, 1962, 1967), the Kuro Shio and Oya Shio (Bradshaw, 1959) and the Falkland–Brazil Currents (Boltovskoy, 1966d). Although species diversity is therefore raised by the introduction of so-called "expatriate" species, there is only one species of planktonic foraminifera which is indigenous to the Transition Zones, namely *Globorotalia inflata*.

The sharp faunal boundaries, between subpolar and subtropical species of planktonic foraminifera which are encountered in the Transition Zones, are also evident among other zooplankton and phytoplankton groups (Hensen *et al.*, 1892–1926). In the Southern Transition Zone, our boundaries correspond very closely to Boltovskoy's (1966b) Subtropical–Subantarctic Convergence Zone, where *Globorotalia inflata* is the dominant species.

(d) Subpolar provinces

The subpolar fauna consists predominantly of *Globigerina bulloides*. Other subpolar species are *Globigerina quinqueloba*, right-coiling *Globoquadrina pachyderma*, *Globorotalia scitula* and *Globigerinita bradyi*. Some species appear to be more cold-tolerant in the South Pacific than in the northern oceans (Boltovskoy, 1966c; Bé, 1969). For example, *Globorotalia truncatulinoides*, *G. inflata* and *Globigerinita glutinata* are commonly encountered in subantarctic waters, whereas they inhabit subtropical or transitional waters in the northern hemisphere. *Globigerina quinqueloba* and *Globigerinita bradyi* are also frequent south of the Antarctic Polar Front, whereas they are predominantly found in subarctic waters in the North Pacific and North Atlantic. *Globorotalia cavernula* is a rare subantarctic species, which apparently occurs only in the southern hemisphere.

(e) Polar provinces

Left-coiling *Globoquadrina pachyderma* is the dominant species in the Arctic as well as Antarctic Oceans. Some subpolar species which are also found in the Antarctic Ocean are *Globigerina bulloides*, *G. quinqueloba* and *Globigerinita bradyi*. In the Arctic Ocean, *Globigerina quinqueloba* is the only other species which occurs together with *G. pachyderma* in the bottom sediments (Hunkins *et al.*, 1971). The larger number of species in the cold waters of the southern hemisphere can be attributed to the more continuous nature of the Antarctic Ocean circulation as compared with the landlocked nature and restricted circulation of the Arctic Ocean.

BIPOLARITY AND ANTITROPICALITY

When a species occurs in the middle and high latitudes of both hemispheres but not in the intervening low latitudes, its distribution is described as "bipolar" or "antitropical".

The present disjunct distributions of such species as *Globoquadrina pachyderma*, *Globigerina bulloides* and *Globorotalia inflata* have been variously explained. One of the most plausible explanations is that from a former continuous distribution, the recent isolation of bipolar species was caused by the warming of the ocean waters since the last glacial epoch (Würm). This has consequently resulted in the reduction of their geographical range.

The bipolarity and antitropicality of the foraminiferal species is also obvious from the reciprocal distribution patterns of the coiling "provinces" of *Globoquadrina pachyderma* and *Globorotalia truncatulinoides* (Bé and Tolderlund, 1971).

The absence of arctic and subarctic zones in the Indian Ocean may have had important repercussions in the dispersal patterns and evolution of warm-water planktonic foraminiferal faunas in that ocean. Thus, it may not be accidental that a number of "relict" species—*Globorotalia menardii neoflexuosa*, *Globoquadrina hexagona*, *Globigerinella adamsi*, and to a lesser extent, *Globoquadrina conglomerata*—are harboured in the northern Indian Ocean, and particularly in the Bay of Bengal.

GEOGRAPHIC VARIATIONS IN ABSOLUTE ABUNDANCE AND SPECIES DIVERSITY OF TOTAL PLANKTONIC FORAMINIFERA

The standing stock of total planktonic foraminifera, expressed in number of specimens per 1000 m³ of water, in the surface waters of the Atlantic and Indian Oceans is shown in Fig. 9. This composite map does not reflect the seasonal variations in productivity which are particularly strong in the middle and high latitudes.

The distribution pattern of absolute abundance of total planktonic foraminifera corresponds with the general pattern of oceanic fertility. Maximum foraminiferal concentrations occur in areas with high concentrations of nutrients, phytoplankton and microzooplankton. These areas occur in major current systems, boundary currents, divergences and upwelling zones. For example, high standing stocks ($> 10\ 000$ specimens/1000 m³) in the

Atlantic Ocean are observed in the Gulf Stream, North Atlantic Current, the North and South Equatorial Currents, Benguela Current, and along the Subtropical Convergence Zone of the South Atlantic.

In the Indian Ocean, the richest concentrations are found in antarctic–subantarctic waters and the Subtropical Convergence Zone during the austral summer months, the North Equatorial Current, the upwelling zones off Somalia and Java, and the South Equatorial Divergence.

In contrast, poor standing stocks (< 1000 specimens/1000 m^3) are found in the central waters of the North and South Atlantic and Indian Oceans (Fig. 9). Very sparse populations are also encountered over continental shelves, such as the North Sea and around the British Isles, along the eastern margin of North America, and the Argentinian–Brazilian continental shelf from Tierra del Fuego to Rio de Janeiro. It is conjectured that the lower salinities in these regions may inhibit incursions of holoplanktonic populations.

In general, species diversity decreases from low to high latitudes as well as from fertile to infertile areas within a water-mass (e.g. from the outer fringe to the central part of the Sargasso Sea). From the species compositions in each of the five major faunal provinces listed in Table 5 and Fig. 8, we note that the species diversity is highest in the tropical regions, where a total of 24 species are found. The species diversity of subtropical, transitional, subpolar and polar areas is 20, 18, 8 and 5 species, respectively. Within each province, the ratio of indigenous species to total species present is 1:5 in arctic–antarctic waters, 4:8 in subarctic–subantarctic waters, 1:8 in transitional waters, 13:20 in subtropical waters, and 14:24 in tropical waters. These observations suggest that evolutionary diversification has proceeded more rapidly in populations of the warm-water region than in those inhabiting the colder waters. Stehli *et al.* (1969) have postulated that the greater availability of solar energy in the tropics is responsible for more rapid rates of evolution among tropical invertebrates than those living in higher latitudes.

Factors influencing distribution and abundance of death assemblages of planktonic foraminifera

From the moment a planktonic foraminifer dies, its test is subject to a series of post-mortem alteration processes which determine whether it will be preserved and what its final resting place will be. The following variables play a role in the complex interaction between the empty tests and the biological, chemical and physical environment:

(1) Ingestion of dead planktonic foraminifera (with protoplasm-filled tests) by scavengers in the water column and on the sea-floor.

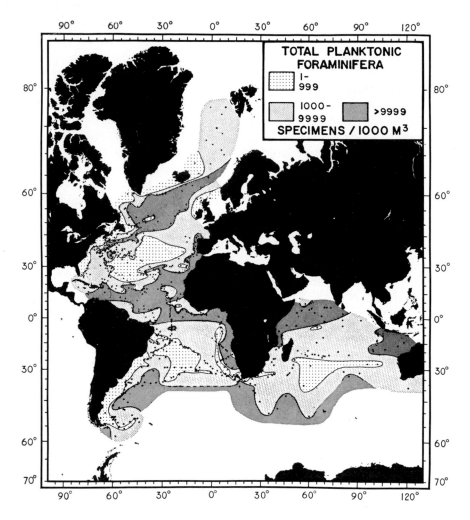

Fig. 9. Absolute abundance of total planktonic foraminifera in surface waters (0–10 m of water) of the Atlantic and Indian Oceans. (From Bé and Tolderlund, 1971.)

(2) Settling rate, current transport and selective dissolution of empty tests in the water column.

(3) Scouring and redeposition of sediments.

(4) Upward mixing of sediments by burrowing organisms.

(5) Long-term productivity of planktonic foraminifera.

(6) Selective dissolution of the tests on the sea-floor.

(1) Ingestion by scavengers

Dead individuals may have protoplasmic remains within their tests, which are a potential food source for heterotrophic organisms in the water column as well as on the sea-floor. The extent of test destruction through such predation is as yet unresolved.

(2) Settling rates and current transport

The average settling velocities of planktonic foraminifera, according to Berger and Piper (1972), range from 0.3 cm/s for tests of 62–125 μm size to 2.3 cm/s for tests > 250 μm. The descent time to an ocean depth of 5000 m is estimated to range from about 19 days (for tests of 62–125 μm size) to 2.5 days (for tests > 250 μm size). The settling rates vary for each species and non-spinose, thick-walled individuals may settle twice as fast as spinose, thin-walled forms of the same size.

In regions of strong surface and subsurface currents, the empty tests can be transported for considerable distances from their original habitat.

(3) Scouring and redeposition

Sediments containing death assemblages are subject to scouring and redeposition by strong bottom currents, as for example in the region between Australia and Antarctica (Watkins and Kennett, 1972). Turbidity currents are also important agents for transporting death assemblages from areas of high topography to deeper parts of the sea-floor and are particularly significant along the continental slopes (Heezen and Hollister, 1971).

(4) Long-term productivity

Because of our lack of data on the generation time and turnover rate of all species of planktonic foraminifera, any discussion of their long-term productivity in the oceans is fraught with speculation. The standing stock in a plankton sample at a specific time and place may be a measure of their productivity if conditions are considered constant (an assumption which is unlikely, but which is often accepted). Since each species probably differs in generation time, those with shorter life spans are likely to have greater turnover rates and, hence, higher rates of test production. According to Berger (1970b), spinose species (*Globigerinoides ruber*, *Globigerina bulloides* and *Orbulina universa*) produce empty tests twice as fast as non-spinose species (e.g. *Globoquadrina dutertrei*), but he cautions that this may be a sampling artifact, since spinose species probably settle more slowly than the non-spinose ones with thicker tests.

Long-term variations in productivity of planktonic foraminifera have been studied by Berger (1971a) from counts in the varved sediments in the Santa

Barbara Basin off California. He found that the productivity of *G. ruber*, *G. siphonifera* (= *G. aequilateralis*), *O. universa*, and *G. bulloides* varied greatly relative to each other; *G. bulloides* had the greatest variability and *O. universa* the least. Long-term variations were evident from the periods of low production of empty tests (1910–1920) and periods of high production (1950–1960) (Fig. 10).

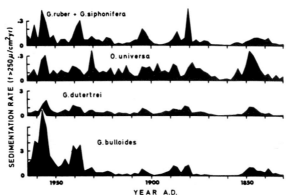

FIG. 10. Annual variations of foraminiferal sedimentation in Santa Barbara Basin based on quantitative varve counts (size fraction $> 250\,\mu$m) by Berger (1971a).

Although the standing stock of planktonic foraminifera in subpolar waters is roughly equivalent to that in tropical waters (Fig. 9), there are two reasons why foraminiferal carbonate probably accumulates more rapidly, above the calcite compensation depth, in tropical areas than in subpolar regions. Firstly, tropical planktonic foraminiferal production is a more continuous process than the seasonal nature of production in subpolar areas. Secondly, tropical species are generally larger than subpolar forms.

Some clues to the generation time and productivity of individual species may be gained from their relative abundance rankings in surface sediments, if due allowance is given to the influence of selective dissolution. Tables 6 and 7 rank the species in order of their relative abundance in surface sediments of two broad latitudinal regions of the Indian Ocean. The relative abundance of each species was calculated for 84 surface sediment samples located between 20°N and 20°S, and for 119 samples between 20°S and 45°S. It was obtained by dividing the total number of specimens of each species counted by the total number of specimens of all species in each region. The quantitative and geographic distributions of samples shown in Figs 11 and 16 indicate a much greater density of samples per 5° of latitude between 20°S and 45°S than between 20°N and 20°S. For this reason, as well as for faunistic comparability, we chose to rank the relative species abundances for two separate regions.

The results indicate that only a few species dominate the sediment assemblages. Between 20°N and 20°S, *Globigerinoides ruber*, *Globigerinita glutinata*, *Globorotalia menardii*, *Globigerinoides sacculifer* and *Globoquadrina dutertrei* make up about 68% of the foraminiferal assemblage (Table 6). Between 20°S and 45°S, *Globorotalia inflata*, *Globigerinoides ruber*, *Globoquadrina pachyderma* (right- and left-coiling), *Globigerina bulloides* and *Globigerinita*

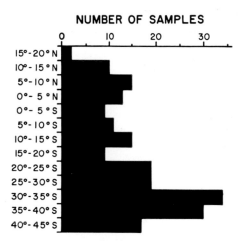

FIG. 11. Histogram of density distribution of core-top samples in the Indian Ocean. Locations of samples are shown in Fig. 16.

glutinata represent about 76% of the total planktonic foraminifera in the sediments (Table 7). Considering the fact that *Globigerinoides ruber*, *G. sacculifer* and *Globigerina bulloides* are highly susceptible to dissolution (Fig. 12), their original relative abundances in the sediments (before dissolution) were almost certainly higher. Thus, we conclude that for these fragile species the rankings of relative species abundances in Tables 6 and 7 correspond in a general way to their relative productivities in the present-day Indian Ocean.

(5) *Selective dissolution*

Dissolution of the calcareous tests occurs at depths below a few hundred metres in the Pacific Ocean and below about 1500 m in the Atlantic Ocean, because the waters are under-saturated with calcium carbonate. Berger (1968, 1970a, 1971a) has considered many aspects of the dissolution process of planktonic foraminifera. Among the factors that influence the dissolution of planktonic foraminiferal tests are:

TABLE 6. Ranking of planktonic foraminiferal species in order of decreasing relative
abundance in Indian Ocean surface sediments
between 20°N and 20°S, based on a total of 84
core-top samples shown in Fig. 16. Compare
also with Fig. 11.

Species	%
Globigerinoides ruber	22.3
Globigerinita glutinata	16.3
Globorotalia menardii	13.2
Globigerinoides sacculifer	8.7
Globoquadrina dutertrei	7.4
Globigerina bulloides	6.2
Globigerinella aequilateralis	5.3
Pulleniatina obliquiloculata	4.0
Globigerina rubescens	3.5
Globorotalia tumida	2.6
Globigerinoides conglobatus	1.4
Globigerinoides tenellus	1.3
Globoquadrina hexagona	1.3
Globoquadrina pachyderma (R)	1.1
Globoquadrina conglomerata	1.1
Globigerina humilis	0.8
Orbulina universa	0.6
Globigerina digitata	0.5
"*Sphaeroidinella dehiscens*"	0.5
Globorotalia scitula	0.3
Globorotalia theyeri	0.3
Globigerina falconensis	0.2
Globorotalia truncatulinoides	0.2
Globorotalia inflata	0.2
Globorotalia crassaformis	0.2
Globigerina calida	0.2
Globigerina quinqueloba	0.2
Candeina nitida	0.1
Globigerinita iota	0.1
Globorotalia menardii neoflexuosa	0.1
Globigerinella adamsi	trace
Globorotalia hirsuta	trace
Globorotalia pumilio	trace
	100.0

TABLE 7. Ranking of planktonic foraminiferal species in order of decreasing relative abundance in Indian Ocean surface sediments between 20°S and 45°S, based on a total of 119 core-top samples shown in Fig. 16. Compare also with Fig. 11.

Species	%
Globorotalia inflata	25.2
Globigerinoides ruber	14.4
Globoquadrina pachyderma (R)	12.0
Globigerina bulloides	11.6
Globigerinita glutinata	9.1
Globorotalia truncatulinoides	4.0
Globoquadrina pachyderma (L)	4.0
Globigerinella aequilateralis	3.0
Globigerinoides sacculifer	2.3
Globigerina quinqueloba	2.1
Globoquadrina dutertrei	1.9
Globigerina humilis	1.9
Globigerina rubescens	1.8
Globorotalia scitula	1.1
Globorotalia hirsuta	1.0
Globorotalia menardii	0.9
Orbulina universa	0.6
Globigerinoides tenellus	0.6
Globigerinoides conglobatus	0.4
Globigerina falconensis	0.4
Globigerina digitata	0.4
Pulleniatina obliquiloculata	0.3
Globorotalia tumida	0.2
Globigerinita iota	0.2
Globorotalia crotonensis	0.2
Globorotalia crassaformis	0.1
Globigerinita bradyi	0.1
"Sphaeroidinella dehiscens"	trace
Candeina nitida	trace
Globoquadrina hexagona	trace
Globigerina calida	trace
Hastigerina pelagica	trace
Globoquadrina conglomerata	trace
	100.0

(1) The degree of undersaturation of ocean water with respect to $CaCO_3$, which increases with greater depth and with latitude (Pytkowicz, 1965; Li, Takahashi and Broecker, 1969; Heath and Culberson, 1970; Berger, 1971a).

(2) The rate of bottom water flow, which determines the extent of replenishment of undersaturated water.

(3) The rate of oxidation of organic matter, which determines the production of carbon dioxide in the water column.

(4) The length of time that the tests are exposed to "corrosive" bottom water.

(5) Rate of supply of calcareous and non-calcareous particles.

(6) Size, wall thickness and microstructure of the planktonic foraminiferal tests.

Ruddiman and Heezen (1967), Berger (1967, 1968, 1970), and Parker and Berger (1971) have studied the dissolution effects on planktonic foraminiferal assemblages in the Atlantic and Pacific. These studies have shown that the dissolution process is selective with regard to species and their test size, microstructure, wall thickness and test porosity. The ranking of foraminiferal species according to their relative susceptibility to solution (Fig. 12) indicates that the more solution-susceptible species are relatively small in size and have large pores and thin walls (e.g. *Globigerinoides ruber*), whereas the less solution-susceptible species have a relatively large test with small pores and thick walls (e.g. *Globorotalia tumida*). In general, the spinose *Globigerina*, *Globigerinella* and *Globigerinoides* species are less resistant than the non-spinose *Globorotalia* and *Globoquadrina* species.

Selective dissolution has the effect of changing the species composition of the death assemblages with increasing depth. Thus, as a common rule, the shallower the sediment sample the greater the similarity in species composition between life and death assemblages. This rule does not hold for shallow regions near land masses where the oxidation of rapidly accumulated organic matter causes the dissolution of calcareous tests.

Berger (1968) has defined the "lysocline" as the depth range where the dissolution rate of planktonic foraminiferal tests increases rapidly. The "lysocline" is generally a few hundred metres above the calcium carbonate compensation depth (CCD), which is generally defined as the depth below which the amount of calcium carbonate is less than 10% (Lisitzin, 1972). The vertical distance between the lysocline and the CCD is less in highly fertile regions and is greater below less productive areas of the ocean. The bathymetric configuration of the CCD in the world ocean has been charted by Berger and Winterer (1974) and is shown in Fig. 13.

The CCD, in a crude analogy, can be likened to the snowline on land.

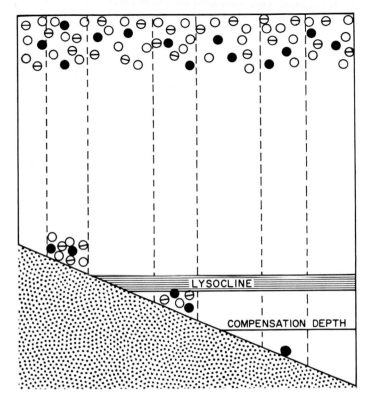

FIG. 12. Selective dissolution of planktonic foraminiferal species at depth. The species are ranked (by Berger, 1970a) in order of decreasing susceptibility to solution.

Low resistance	High resistance
1. *Globigerinoides ruber* ○	12. *Globorotalia hirsuta*
2. *Orbulina universa*	13. *Globorotalia truncatulinoides*
3. *Globigerinella aequilateralis*	14. *Globorotalia inflata*
(= *Globigerinella siphonifera*)	15. *Globorotalia menardii*
4. *Globigerina rubescens*	16. *Globoquadrina dutertrei*
5. *Globigerinoides sacculifer*	17. *Globoquadrina pachyderma* s.l.
6. *Globigerinoides tenellus*	18. *Pulleniatina obliquiloculata*
7. *Globigerinoides conglobatus*	19. *Globorotalia crassaformis*
8. *Globigerina bulloides* ⊖	20. *Sphaeroidinella dehiscens*
9. *Globigerina quinqueloba*	21. *Globorotalia tumida* ●
10. *Globigerinita glutinata*	22. *Globigerina humilis*
11. *Candeina nitida*	

Calcium carbonate particles blanket topographic highs under optimum conditions and gradually disappear downslope. The lysocline and CCD vary locally and regionally, becoming generally shallower towards the continents and poles.

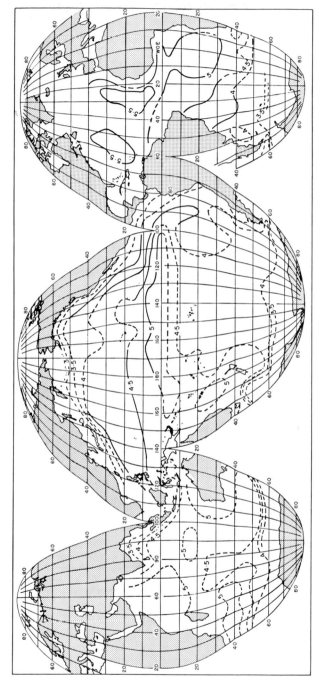

Fig. 13. The distribution pattern of calcium carbonate compensation depths in the world's ocean. (From Berger and Winterer, 1974.)

(6) *Planktonic foraminifera abundance and ocean bottom topography*

From the preceding discussion, it is obvious that a close relationship exists between planktonic foraminiferal abundance and ocean bottom topography. This is best shown by comparing Belyaeva's (1969c) map of the quantitative distribution of planktonic foraminifera in the world ocean (Fig. 14) and the map showing ocean regions below depths of 4500 m in the Indian and Atlantic Oceans and below 4000 m in the Pacific Ocean (Fig. 15). The North and South Atlantic have on the average larger areas with greater numbers of planktonic foraminifera per gram of surface sediment than the Indian or Pacific Oceans. Areas of high quantitative abundance ($>$ 1000 specimens/g sediment) are located over such relatively shallow regions as the mid-oceanic ridges, continental slopes, rises and plateaus of all ocean basins. By contrast, the deep ocean basins below 4500 m in the Atlantic and Indian Oceans and below 4000 m in the Pacific have greatly reduced numbers of planktonic foraminifera ($<$ 1000 specimens/g) or are devoid of them. Such conditions occupy approximately 64% of the Pacific Ocean, 46% of the Indian Ocean, and 32% of the Atlantic Ocean (Sverdrup *et al.*, 1942). Planktonic foraminifera are rare or absent in the Labrador–Baffin Basin, North America Basin, Canary–Cape Verde Basin, the Argentine Basin, Weddell Basin, Crozet Basin, Wilkes Basin, Mozambique Basin, Mascarene Basin, Arabian Basin, Bengal Basin, Wharton Basin, Tasman Basin, Philippine Basin and over the greater parts of the North and South Pacific Basins and Arctic Ocean Basin.

Preservation of significant amounts of planktonic foraminifera in the Pacific Ocean occurs only over the East Pacific Ridge, the equatorial belt, New Zealand Plateau, Lord Howe Rise, Fiji Plateau, Solomon Rise, Manihiki Plateau, Emperor Sea Mount Chain, Hawaiian Ridge and on other seamounts and archipelagoes (Belyaeva, 1969c).

Planktonic foraminifera decrease in abundance near land masses, because of dilution by terrigenous sediments or dissolution of their tests due to high concentrations of organic matter. Such conditions often prevail on continental shelves and such areas as the Ganges Cone (Bay of Bengal) and Indus Cone (Arabian Sea).

Areal distribution of individual species in surface sediments

The large number of quantitative studies of planktonic foraminifera in the surface layer of deep-sea sediments which have appeared during the last few decades have enabled us to piece together global views of the distribution patterns of individual species. Parker's (1971) generalised world distributions for five species represent one of the first attempts to synthesise the data from the literature.

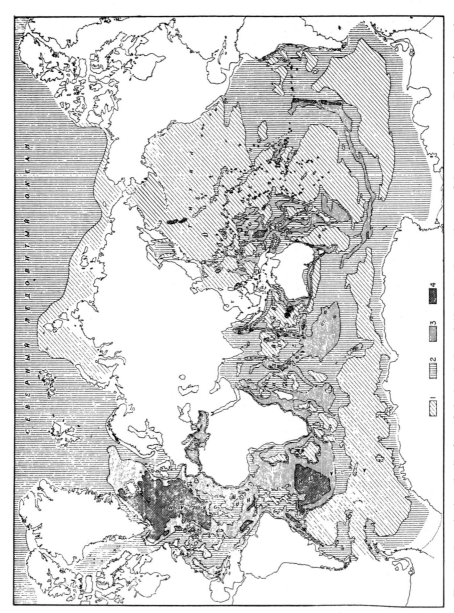

Fig. 14. Quantitative distribution of planktonic foraminifera in surface sediments of the world ocean. Density intervals: 1, absent; 2, 1 to 1000 specimens/g sediment; 3 = 1000 to 10 000 specimens/g sediment; 4, >10 000 specimens/g sediment. (From Belyaeva, 1969c.) Compare with Fig. 15.

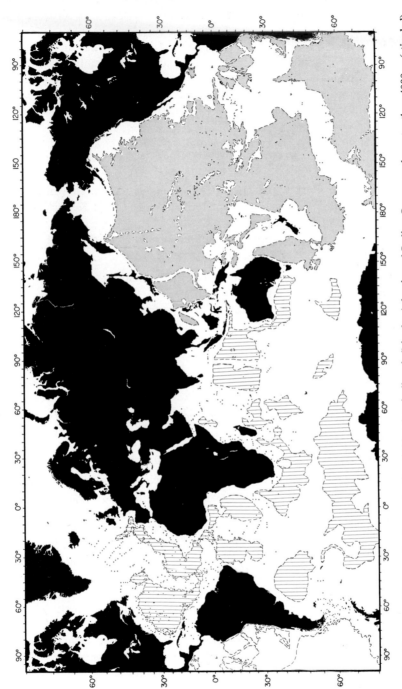

FIG. 15. Areas with ocean depths greater than 4500 m (vertically ruled) in the Atlantic and Indian Oceans and greater than 4000 m (stippled) in the Pacific Ocean. Compare with Fig. 14.

In this study the author presents global distributional maps for thirteen species which are common and ubiquitous in surface sediments. The maps in Figs 17 to 29 are based on the results of many investigators, whose publications are numbered and listed in the References. The author has relied especially upon the data from the following investigators some of whose study areas are shown in Fig. 16:

Arctic Ocean: Vilks (1973), *North Atlantic:* Phleger *et al.* (1953); Barash (1970); Ruddiman (1969b); Schott (1935, 1966); Imbrie and Kipp (1971), *Mediterranean Sea:* Parker (1955), *South Atlantic:* Loring (1966), *Equatorial and South Pacific:* Parker and Berger (1971), *Indian Ocean:* Bé (this study), Zobel (1973), *Antarctic Ocean:* Blair (1965); Herb (1968); Kennett (1969); Echols and Kennett (1973).

These above-mentioned publications are particularly significant, because they fulfill one or more of the following criteria:

1. they contain quantitative data on species composition;
2. they cover ocean-wide regions;
3. the quantitative analysis is based on the size fraction greater than 125 μm or 149 μm;
4. the taxonomy of Recent species is relatively stable and consistent among contemporary workers.

In addition, the author has attempted to fill some of the distributional gaps (e.g. in the North Pacific and the higher latitudes of the South Atlantic) by including unpublished data from Lamont-Doherty core-top samples.

POLAR SPECIES

Globoquadrina pachyderma (Ehrenberg) (Plate 10; figs 22a–j; Fig. 17)

Globoquadrina pachyderma is the dominant species in Arctic and Antarctic surface sediments. The species can be subdivided into two distinct populations of right- (Plate 10, figs 22f–j) and left-coiling (Plate 10, figs 22a–e) forms. Ericson (1959) noted that left-coiling forms make up 90% or more of the species assemblage in the Arctic Ocean, Labrador Sea, Greenland Sea and Norwegian Sea. Right-coiling *G. pachyderma* is an important constituent of sedimentary assemblages in the Subarctic and Transition Zones which extend from Cape Hatteras to the Bay of Biscay. In the North Atlantic the boundary between the right- and left-coiling provinces coincide with the 7.2°C April surface isotherm (Ericson, 1959).

A similar distributional relationship exists in the surface sediments of the Southern hemisphere, where 90% or more *G. pachyderma* tests are left-coiling south of the Antarctic Polar front (Echols and Kennett, 1973). In the New Zealand area the boundary between right- and left-coiling forms is

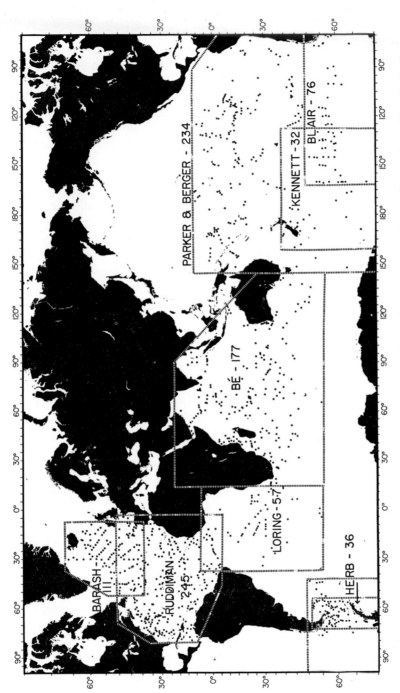

Fig. 16. Regions in which the planktonic foraminifera death assemblages have been studied by various authors. The distribution maps of Figs 17–29 are largely based on their data.

located a few degrees of latitude south of the Subtropical Convergence. A close relationship between coiling ratios and latitude is also observed in the living populations of the South Atlantic, South Pacific and Indian Oceans (Bé, 1969; Bé and Tolderlund, 1971). Since the distribution of sinistral and dextral populations of *G. pachyderma* are clearly influenced by latitudinal position and/or water temperature, coiling ratios of this species can be used to distinguish between polar conditions (sinistral tests dominant) and subpolar conditions (dextral tests dominant).

In Arctic and Antarctic regions the generally monospecific sediment assemblages consists almost entirely of *G. pachyderma*. In Subarctic and Subantarctic regions and the Transition Zones, this species makes up 20% to 50% of the total planktonic foraminifera in sediment assemblages. *G. pachyderma* rapidly decreases in relative abundance in the Subtropical areas, but it is still encountered in small percentages ($< 5\%$) in the eastern equatorial Atlantic and equatorial Indian Ocean. It is apparently absent in the Caribbean Sea, southwestern Sargasso Sea, western equatorial Atlantic, northern Arabian Sea, Bay of Bengal, and in the equatorial Pacific (Fig. 17).

SUBPOLAR SPECIES

Globigerina bulloides d'Orbigny sensu lato (Plate 8, figs 9a–e; Fig. 18)

Globigerina bulloides s.l. is one of the most successful and ubiquitous species of planktonic foraminifera. Although it occurs predominantly in subpolar regions, it is also commonly encountered in upwelling areas and boundary currents in low-latitude regions such as the Cariaco Trench area off Venezuela, off West Africa, the Canaries Current, the Somali Current, and off southwestern India, Java and Sumatra (Bé and Tolderlund, 1971).

In the distribution map (Fig. 18), we have included *Globigerina falconensis* which is rather similar both in morphology and in its ecological requirements to *G. bulloides*.

The highest densities of *G. bulloides* s.l. are found between 35°S and 50°S in the southern hemisphere. In the North Atlantic, the area of maximum abundance has an irregular pattern. In the Mediterranean, *G. bulloides* s.l. is also very common, especially in the western basin.

The poleward boundary for this species in the southern hemisphere lies south of the Antarctic Polar Front. Its northern boundary in the North Atlantic could not be delineated due to inadequate sample coverage. In the North Pacific, *G. bulloides* s.l. was only sporadically encountered in isolated localities of high topography.

The subpolar and subtropical–tropical forms of *G. bulloides* s.l. are virtually identical in test morphology. The subpolar variety (Plate 8, figs 9a–c) differs from the subtropical–tropical variety (Plate 8, figs 9d–e) in having a more deeply recessed aperture, a less pronounced apertural rim, and a test width which almost equals the length of the test.

TRANSITION ZONE SPECIES

Globorotalia inflata (d'Orbigny) (Plate 11, figs 27a–c; Fig. 19)

Globorotalia inflata is the only indigenous species of the Transition Zones which lie between the Subpolar and Subtropical Provinces. Its highest frequencies occur between 25°S and 50°S in the southern Atlantic, Indian and Pacific Oceans and between 35°N and 45°N in the North Atlantic. Its northern limit in the North Atlantic runs from the central Labrador Sea to the north of Iceland to the Norwegian Sea. The southern poleward limit is on the average several hundred miles south of the Antarctic Polar Front. It is absent in a belt between Antarctica and roughly 65°S.

Globorotalia inflata is relatively less abundant in the sedimentary assemblages of the North Atlantic than in all other ocean basins. This probably reflects the relatively favourable conditions for $CaCO_3$ preservation in the North Atlantic, where the selective preservation of more solution-susceptible species effectively masks and depresses the abundance of solution-resistant species such as *G. inflata*. This species is abundant at those depths in the Pacific and Indian Ocean basins where less resistant species have been removed by dissolution.

G. inflata's present-day northern and southern populations show a disjunct distribution in the Atlantic (Bé and Tolderlund, 1971) and probably also in the Pacific Ocean (Bradshaw, 1959; Parker, 1960). However, this is not the case in the surface sediments of the Atlantic Ocean, where *G. inflata* has an almost continuous north–south distribution except for parts of the Caribbean Sea and western Equatorial Atlantic.

In the Indian Ocean, *G. inflata* is only sporadically present in surface sediments north of 10°N. In the Pacific it appears to be absent in a broad equatorial belt.

It is not known whether the presence of *G. inflata* in the sedimentary assemblages and its absence from the life assemblages of the lower latitudes indicates a time-lag in the equilibration process of the sediment assemblages. The death assemblages may be different because they contain a fauna which is representative of past climatic conditions. Another possibility is that the difference reflects the upward mixing process of cold-water species assemblages from subsurface levels.

SUBTROPICAL SPECIES

Globigerinoides ruber (d'Orbigny) (Plate 7, figs 4a–c; Fig. 20)

Globigerinoides ruber is the most successful warm-water species both in terms of distribution and abundance in ocean waters as well as surface sediments. Although its peak abundance occurs in subtropical regions, it is also extremely common in tropical sediments. The success of this highly productive surface dweller may in part be due to its symbiotic relationship with zooxanthellae.

Maximum percentages of this species are encountered in three regions where the overlying surface waters are highly saline, i.e. the southern Sargasso Sea, the eastern Mediterranean, and the Brazil Current. Average surface salinities in the former region approach 37‰ and are greater than 38‰ in the eastern Mediterranean. Its great abundance and omnipresence in surface sediments between approximately 45°N and 45°S (Fig. 20) are even more impressive if one considers the fact that *G. ruber* is the least solution-resistant species among the planktonic foraminifera (Fig. 12). It is absent only in the very deep parts of the Brazil Basin, Angola Basin, Wharton Basin, etc., which lie below the calcite compensation depth. However, the author has observed well-preserved specimens of *G. ruber*, obtained from a depth of 5546 m from the Sohm Abyssal Plain. Their thickly encrusted tests were unusual for this species, but helped to explain its resistance to solution and its presence in deep basins.

Populations of *G. ruber* with red-pigmented tests are restricted to the North and South Atlantic and are absent in the Indian and Pacific Oceans (Bé and Tolderlund, 1971).

Globorotalia truncatulinoides (d'Orbigny) (Plate 11, figs 34a–c; Fig. 21)

Globorotalia truncatulinoides is a mid-latitude species, whose living populations show a distinct preference for winter conditions. In the Sargasso Sea it is predominantly found between December and April (Bé, 1960a; Bé, Vilks and Lott, 1971; Tolderlund and Bé, 1971). In the central South Atlantic it occurs abundantly in the austral winter (July and August) (Bé, unpublished data). *G. truncatulinoides* flourishes also between May and October in sub-antarctic waters of the Pacific and Atlantic (Bé, 1969). The strong seasonal nature of the productivity of this species accounts for its generally low frequencies in the surface sediments.

Globorotalia truncatulinoides is most abundant in the sediments beneath the subtropical central watermasses in the North and South Atlantic, Indian Ocean and southeastern Pacific (Fig. 21).

In the southern hemisphere, the predominantly left-coiling populations of this species are found in colder waters and higher latitudes than the right-coiling populations (Bé and Tolderlund, 1971), and the same relationship is also reflected in the sediment assemblages. In the North Atlantic, two distinct right-coiling provinces appear to be separated from each other by a central left-coiling province (Ericson, Wollin and Wollin, 1954). The left-coiling variety of *G. truncatulinoides* is apparently more successful than the right-coiling one, because in the regions where this species reaches frequencies between 10 and 20% the left-coiling forms greatly outnumber the right-coiling ones.

Globorotalia hirsuta (d'Orbigny) (Plate 11, figs 33a–c; Fig. 22)

Globorotalia hirsuta attains its maximum abundance (> 5%), but rarely more than 10% of the total fauna of planktonic foraminifera) in the surface sediments which accumulate beneath the central water masses of the North and South Atlantic (Fig. 22). It is almost entirely restricted to a belt between 25°S and 45°S in the Indian Ocean and southwestern Pacific, where it rarely exceeds 5% of the total planktonic foraminiferal assemblages. The species is almost completely absent between New Zealand and South America.

Globigerinella aequilateralis (Brady) (Plate 8, figs 12a–c; Fig. 23)

Globigerinella aequilateralis is a common subtropical–tropical species which prefers to live in boundary currents, areas of upwelling and near continental margins (Bé and Tolderlund, 1971). Its relative abundance in surface sediments is controlled to a large extent by selective dissolution. Since this species is highly solution-prone (Fig. 12), its distribution on the ocean-floor reflects the bottom topography, and higher frequencies of *G. aequilateralis* occur in relatively shallow regions such as the Mid-Atlantic Ridge, Caribbean Sea, the Atlantic continental slopes of North and South America, India, Melanesia, Micronesia, and close to other landmasses (Fig. 23).

Orbulina universa d'Orbigny (Plate 7, figs 3a–b; Fig. 24)

Orbulina universa is a subtropical species which is sparse but widely encountered in surface sediments between 65°N in the North Atlantic (55°N in the North Pacific) and 50°S in the southern oceans (Fig. 24).

A strongly inverse correlation exists between the test diameter of *O. universa* and latitude in the Indian Ocean waters and surface sediments (Bé *et al.*, 1973). Populations with large tests (600 μm to 800 μm in diameter) are found in tropical and subtropical areas. Intermediate mean test diameters (450 μm to 550 μm) occur between 25°S and 35°S in the surface sediments. Mean test diameters of less than 450 μm are encountered in the Transition Zone and Subantarctic sediments. Results of a similar study of *O. universa* from North Atlantic surface sediments show that its mean test diameter also varies inversely with latitude (or directly with surface-water temperature).

Orbulina universa occurs in higher frequencies in the North and South Atlantic than in the Indo–Pacific Ocean. Relatively high percentages (5–9.9%)

are encountered on the Walvis Ridge, the Mid-Atlantic Ridge, south of Hispaniola, off Senegal and the Bermuda Rise. This species occurs in extremely high percentages (25 to 78%) in the Biscay Abyssal Plain off Spain at 45°N (Ruddiman, 1969b).

Globigerinoides conglobatus (Brady) (Plate 7, figs 5a–c; Fig. 25)

Globigerinoides conglobatus is a subtropical species, whose living populations inhabit the central water masses of the North and South Atlantic and Indian Oceans (Bé and Tolderlund, 1971). Being moderately resistant to solution, this species tends to be preserved at relatively great depths in the ocean basins. Its highest abundance is encountered between about latitudes 15° and 30° in the oceans of both hemispheres, except in the North Pacific (Fig. 25).

TROPICAL SPECIES

Globigerinoides sacculifer (Brady) (Plate 7; figs 6a–c; Fig. 26)

Globigerinoides sacculifer is the most prolific tropical species, whose peak abundance occurs in a circumglobal belt between 20°N and 20°S (Fig. 26). Its distributional limits coincide approximately with the 50°N parallel in the North Atlantic, 40°N in the North Pacific and 40°S in the southern hemisphere.

Being a relatively solution-susceptible species (Fig. 12), *G. sacculifer* tends to be more abundantly preserved in the sediments of the Atlantic than in the other oceans. Its highest frequencies (> 20%) occur in the equatorial Atlantic belt, especially on the Mid-Atlantic Ridge, and in the Caribbean Sea.

In the Indian and Pacific Oceans, *G. sacculifer*'s highest frequencies (10–19.9%) are encountered in the Mozambique Channel, the Mid-Indian Ridge, Ninety-East Ridge, Indonesian Archipelago, Exmouth Plateau, Fiji Plateau, Solomon Rise, Manihiki Plateau and East Pacific Rise.

Globorotalia menardii menardii (d'Orbigny) (Plate 12, figs 35a–c; Fig. 27)

Globorotalia menardii sensu lato is predominantly a tropical species, whose highest densities occur between 20°N and 10°S in the Atlantic Ocean, north of the equator in the Indian Ocean, and in the isolated areas of the eastern tropical Pacific Ocean (Fig. 27). *G. menardii* appears to have its maximum abundance and greatest morphological variability in the Indian Ocean. Its quantitative superiority in the latter ocean may in part be due to fortuitous preservation conditions in relation to the Atlantic and Pacific Oceans.

Globorotalia menardii s.l. has considerable morphological variability and there are at least three subspecies known in the Indian Ocean. *Globorotalia menardii menardii* (Plate 12, figs 35a–c) is flat on both sides of its test and is the most typical and commonest form in the equatorial regions of all oceans. The two other subspecies are described below.

Globorotalia menardii (d'Orbigny) *gibberula* new subspecies (Plate 12, figs 36a–c; Fig. 27)

Globorotalia menardii gibberula is a new subspecies which can be easily identified by its humped, dome-shaped test. It is primarily found in the subtropical latitudes between about 15°S and 30°S in the Indian Ocean and South Atlantic. Its distribution appears to show little overlap with that of *Globorotalia menardii menardii* which occurs mainly in the equatorial regions. Thus, *G. menardii menardii* (and *G. menardii neoflexuosa* in the Indian Ocean) can be considered the tropical subspecies, while *G. menardii gibberula* is the subtropical subspecies. *G. m. gibberula* is found in large numbers in a belt from western Australia to Madagascar, and from the Walvis Ridge to the Rio Grande Rise. Its distribution in the Pacific is not yet delineated.

Description of the holotype: Test fairly large, consisting of about two and one-half whorls of rapidly enlarging chambers arranged in a high, dome-shaped trochospire. There are five chambers in the last whorl. The spiral (dorsal) side is characteristically humped or arched by the spiral arrangement of all chambers of the last whorl. The apertural (ventral) side is relatively flat with depressed sutures. Umbilicus small, open and deep. The aperture extends from the umbilicus to the periphery and is bordered by a broad lip. The wall is finely perforate except along the base of the aperture. The name "gibberula" stands for "small hump-back".

Dimensions of holotype: Maximum length is 1196 μm, maximum width is 938 μm.

Remarks: *Globorotalia menardii gibberula* new subspecies has a characteristically humped, dome-shaped appearance from side view or from the spiral side. The humped look involves all chambers of the last whorl. *G. menardii menardii* differs from the new subspecies in having flat spiral and apertural sides. *G. menardii neoflexuosa* has initially a flat test, whose last-formed chamber(s) is twisted towards the apertural side. Thus, *G. m. neoflexuosa* differs from *G. m. gibberula* in that the former's flexuose (twisted) condition is restricted to the last-formed chamber(s), whereas in the latter all the chambers of the last whorl contribute to the dome-shaped appearance.

Location of holotype: Collected alive in 0–500 m plankton sample, Vema 18/72 BPS, on June 30, 1962 at 24°52'S, 78°18'E in the Indian Ocean (Plate 12, figs 36a, b). The paratype on Plate 12, fig. 36c is from the same sample and location as the holotype.

Repository of types: U.S. National Museum, Washington, D.C. Holotype is USNM No. 208344 and illustrated in Plate 12, figs 36a, b. Paratype 1 is USNM No. 208345 and is illustrated in Plate 12, fig. 36c. Two other paratypes (USNM No. 208346 and 208347) are not illustrated here.

Globorotalia menardii neoflexuosa Srinivasan, Kennett and Bé (Plate 12, fig. 37; Fig. 27)

Globorotalia menardii neoflexuosa is distributed within the same but more restricted areas of the Indian Ocean as *G. menardii menardii*. It was originally described as *Globorotalia menardii flexuosa* by Bé and McIntyre (1970), but was renamed by Srinivasan *et al.* (1974) to differentiate it from the extinct Pleistocene *Globorotalia tumida flexuosa*.

Globorotalia menardii neoflexuosa is encountered sporadically in the death as well as life assemblages in the Bay of Bengal, Arabian Sea and the northern Indian Ocean.

Globoquadrina dutertrei (d'Orbigny) (Plate 10, figs 23a–f; Fig. 28)

Globoquadrina dutertrei is a tropical–subtropical species which occurs abundantly in active current systems, along continental margins and in upwelling regions. It is rare in the central oceanic regions that are remote from land masses (Fig. 28). In most of these areas, solution-resistant *G. dutertrei* is selectively preserved at depths approaching the lysocline, where differential solution has removed the more fragile species. *G. dutertrei*'s northern distributional limits coincide approximately with the 50°N parallel in the North Atlantic, and with a broad arc from northern Honshu to Vancouver Island in the North Pacific. Its southern limit runs roughly between 40°S and 45°S.

Globoquadrina dutertrei has a wide range of morphological variation (Zobel, 1968). In this study, we have chosen to lump these variants into a single species group. In the Indian Ocean, there are at least two distinct variants—one with $5\frac{1}{2}$ chambers per whorl (Plate 10, figs 23a–c) and the other with $6\frac{1}{2}$ to 7 chambers per whorl (Plate 10, figs 23d–f).

Pulleniatina obliquiloculata (Parker and Jones) (Plate 10, figs 26a–d; Fig. 29)

Pulleniatina obliquiloculata is a tropical species whose highest abundance in the surface sediments occurs in a relatively narrow belt between about 10°N and 10°S. This belt coincides generally with the equatorial current systems in the Atlantic, Indian and Pacific Oceans (Fig. 29). Outside the equatorial belt, *P. obliquiloculata* is sparse (0.1–4.9%) but ubiquitous. Somewhat higher frequencies (> 5%) occur in the northern Gulf of Mexico and the Gulf Stream region. It is absent in the Mediterranean Sea.

Pulleniatina obliquiloculata is present on the average in higher percentages in the Pacific than in the Atlantic or Indian Oceans. This can be due to either more favourable preservation conditions in the Pacific for this highly solution-resistant species or the masking effect of greater percentages of solution-susceptible species in the Atlantic and Indian Oceans.

The poleward distributional limits correspond approximately with the 45°N and 40°S parallels.

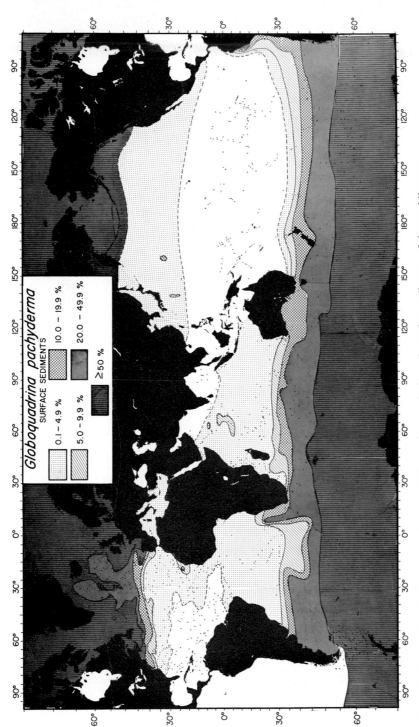

Fig. 17. Distribution of relative abundance of *Globoquadrina pachyderma* in surface sediments of the world's ocean.

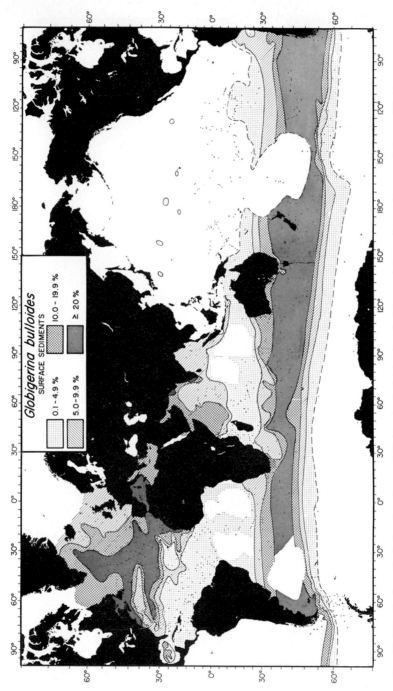

Fig. 18. Distribution of relative abundance of *Globigerina bulloides* plus *G. falconensis* in surface sediments of the world's ocean.

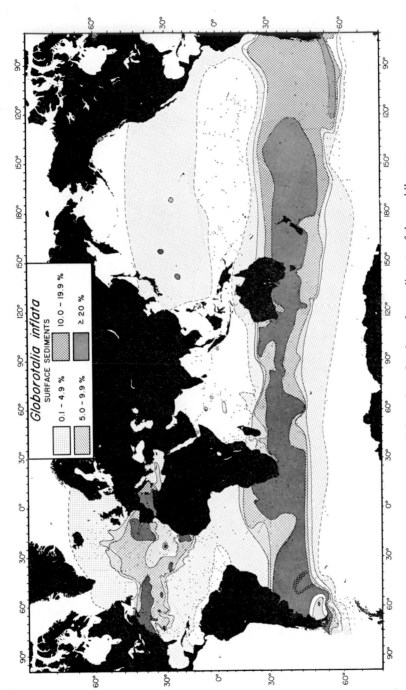

Fig. 19. Distribution of relative abundance of *Globorotalia inflata* in surface sediments of the world's ocean.

Fig. 20. Distribution of relative abundance of *Globigerinoides ruber* in surface sediments of the world's ocean.

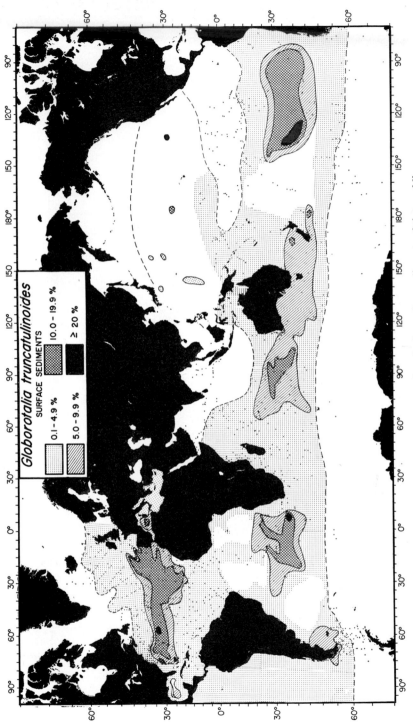

FIG. 21. Distribution of relative abundance of *Globorotalia truncatulinoides* in surface sediments of the world's ocean.

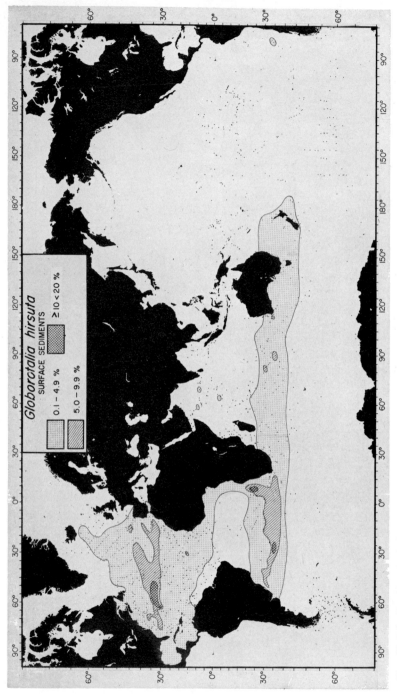

Fig. 22. Distribution of relative abundance of *Globorotalia hirsuta* in surface sediments of the world's ocean.

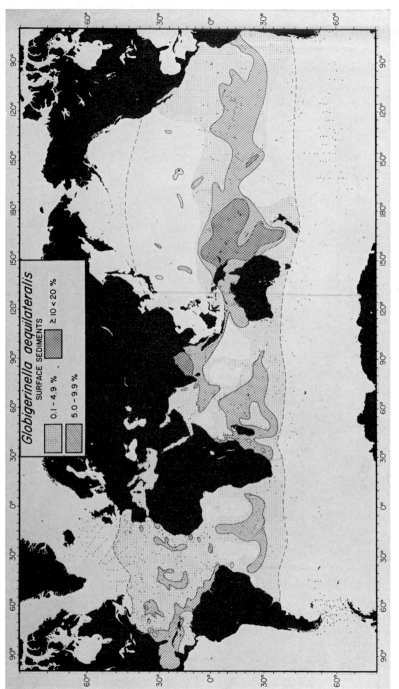

Fig. 23. Distribution of relative abundance of *Globigerinella aequilateralis* in surface sediments of the world's ocean.

Fig. 24. Distribution of relative abundance of *Orbulina universa* in surface sediments of the world's ocean.

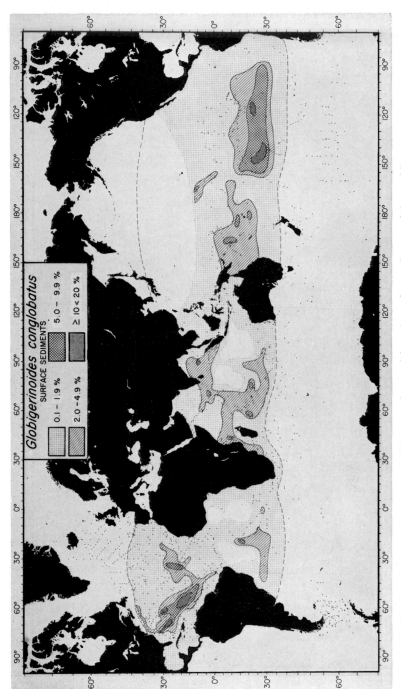

Fig. 25. Distribution of relative abundance of *Globigerinoides conglobatus* in surface sediments of the world's ocean.

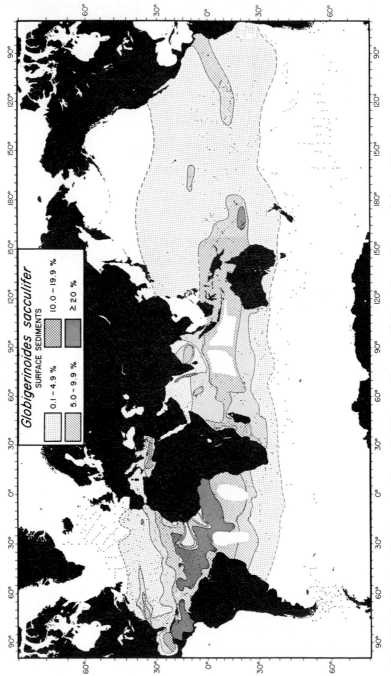

Fig. 26. Distribution of relative abundance of *Globigerinoides sacculifer* in surface sediments of the world's ocean.

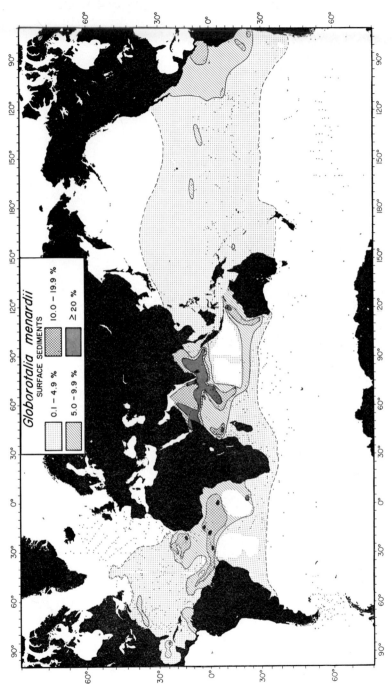

Fig. 27. Distribution of relative abundance of *Globorotalia menardii* in surface sediments of the world's ocean.

Fig. 28. Distribution of relative abundance of *Globoquadrina dutertrei* in surface sediments of the world's ocean.

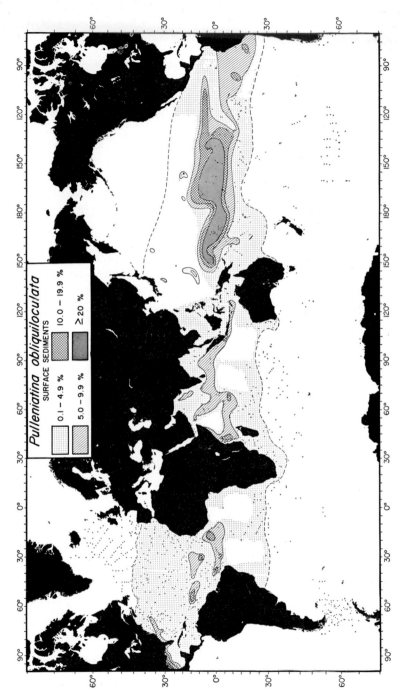

Fig. 29. Distribution of relative abundance of *Pulleniatina obliquiloculata* in surface sediments of the world's ocean.

Acknowledgements

This chapter is based largely on the publications of numerous investigators as well as some unpublished data from several Ph.D. dissertations. The author is grateful to all these investigators, particularly to those whose study areas are shown in Fig. 16 and whose data contributed to the distributional maps of Figs 17 to 29.

I am also grateful to my colleagues, too numerous to mention by name, at the Lamont-Doherty Geological Observatory, who helped directly and indirectly, in collecting the plankton samples and sediment cores on cruises of R/V *Vema* and R/V *Conrad*. Special thanks are due to Saijai Tuntivate and Stanley Harrison for taking the scanning electron micrographs; Rosemary Free and Mei Bé for micropalaeontological analyses; Leroy Lott and William Hutson for quantitative analysis of Indian Ocean data; Lawrence Madin for providing samples of salp faecal pellets; and William Hutson, Kenneth King, Boaz Luz, Tsunemasa Saito and Stephen Streeter for stimulating discussions.

This study received support from the U.S. National Science Foundation (Grants GA 38, 276X and GA 31,388X). The sediment cores used in the study are from the Lamont-Doherty Geological Observatory Core Library which is maintained by grants from the Office of Naval Research (NO 0014-75-C-0210) and the National Science Foundation (DES 72-01568 A03). This publication was completed in January 1974 and is contribution No. 2387 of the Lamont–Doherty Geological Observatory.

References

Publications that are followed by Roman and Arabic numbers in the right-hand column contain distributional data on recent species and are arranged as follows:

I, Planktonic foraminifera in surface sediments; II, Planktonic foraminifera in ocean waters.

1, Atlantic Ocean, including Gulf of Mexico, Carribbean Sea and Mediterranean Sea; **2,** Indian Ocean, including Red Sea and Persian Gulf; **3,** Pacific Ocean; **4,** Arctic Ocean and adjacent areas; **5,** Antarctic Ocean (regions roughly south of 45°S); **6,** World Oceans.

Adshead, P. C. (1967). Collection and laboratory maintenance of living planktonic Foraminifera. *Micropalaeont.* **13**(1), 32–40.

Alldredge, A. L. and Jones, B. M. (1973). *Hastigerina pelagica:* Foraminiferal habitat for planktonic dinoflagellates. *Mar. Biol.* **22**, 131–135.

Arnold, Z. M. (1954). *Discorinopsis aguayoi* (Bermudez) and *Discorinopsis vadescens* Cushman and Brönnimann—A study of variation in cultures of living Foraminifera. *Cushman Found. Foram. Res. Contr.* **5**(1), 4–13.

Bandy, O. L. (1961). Distribution of Foraminifera, Radiolaria and Diatoms in sediments of the Gulf of California. *Micropaleont.* **7**(1), 1–26.

Baracsh, M. S. (1964). Distribution of planktonic Foraminifera in sediments of the northern part of the Atlantic Ocean. *Okeanol. Issled.* **13**, 225–235. (In Russian.)
                                                                                                    (I-1)

Barash, M. S. (1970). Planktonic Foraminifera in North Atlantic sediments. *Akad. Nauk. SSSR, Inst. Okeanol.* 1–94. (In Russian.)   (I-1)

Barash, M. S. (1971). The vertical and horizontal distribution of planktonic Foraminifera in Quaternary sediments of the Atlantic Ocean. *In* (Funnell, B. M. and Riedel, W. R., eds), *The Micropalaeontology of Oceans*, 433–441. Cambridge University Press. (I-1)

Bé, A. W. H. (1959). Ecology of Recent planktonic Foraminifera. Part 1. Areal distribution in the western North Atlantic. *Micropaleont.* **5**(1), 77–100. (II-1)

Bé, A. W. H. (1960a). Ecology of Recent planktonic Foraminifera. Part 2. Bathymetric and seasonal distributions in the Sargasso Sea off Bermuda. *Micropaleont.* **6**(4), 373–392. (II-1)

Bé, A. W. H. (1960b). Some observations on Arctic planktonic Foraminifera. *Cushman Found. Foram. Res. Contr.* **11**(2), 64–68. (II-4)

Bé, A. W. H. (1965). The influence of depth on shell growth in *Globigerinoides sacculifer* (Brady). *Micropaleont.* **11**(1), 81–97. (II-1)

Bé, A. W. H. (1967a). *Globorotalia cavernula*, a new species of planktonic Foraminifera from the Subantarctic Pacific Ocean. *Cushman Found. Foram. Res. Contr.* **18**(3), 128–132. (II-5)

Bé, A. W. H. (1967b). Foraminifera, Families *Globigerinidae* and *Globorotaliidae*. Fiche No. 108. *In* (Fraser, J. H., ed.), *Fiches d'Identification du Zooplankton*. Conseil International pour l'exploration de la mer, Charlottenlund, Denmark.

Bé, A. W. H. (1968). Shell porosity of Recent planktonic Foraminifera as a climatic index. *Science* **161**(3844), 881–884.

Bé, A. W. H. (1969). Planktonic Foraminifera. *In* Folio 11: *Distribution of Selected Groups of Marine Invertebrates in Waters South of 35°S Latitude. Antarctic Map Folio Series, Am. Geogr. Soc.* New York, 9–12. (II-5)

Bé, A. W. H. and Anderson, O. R. (1976). Gametogenesis in planktonic Foraminifera. *Science* **192**(4242), 890–892.

Bé, A. W. H. and Ericson, D. B. (1963). Aspects of calcification in planktonic Foraminifera. *In Comparative Biology of Calcified Tissues. N.Y. Academy of Sciences, Ann.* **109**(1), 65–81.

Bé, A. W. H. and Lott, L. (1964). Shell growth and structure of planktonic Foraminifera. *Science* **145**(3634), 823–824.

Bé, A. W. H., McIntyre, A. and Breger, D. L. (1966). The shell microstructure of a planktonic Foraminifer, *Globorotalia menardii* (d'Orbigny), *Eclog. geol. Helv.* **59**, 885–896.

Bé, A. W. H. and Hamlin, W. H. (1967). Ecology of Recent planktonic Foraminifera: Part 3. Distribution in the North Atlantic during the summer of 1962. *Micropaleont.* **13**(1), 87–106. (II-1)

Bé, A. W. H. and Hemleben, C. (1970). Calcification in a living planktonic Foraminifer, *Globigerinoides sacculifer* (Brady). *N. Jb. Geol. Palaont. Abh.* **134**(3), 221–234.

Bé, A. W. H. and McIntyre, A. (1970). *Globorotalia menardii flexuosa* (Koch): an "extinct" foraminiferal subspecies living in the northern Indian Ocean. *Deep-Sea Res.* **17**, 595–601. (II-2)

Bé, A. W. H. and Tolderlund, D. S. (1971). Distribution and ecology of living planktonic Foraminifera in surface waters of the Atlantic and Indian Oceans. *In* (Funnell, B. M. and Riedel, W. R., eds), *The Micropalaeontology of Oceans*, 105–149. Cambridge University Press. (II-1, 2)

Bé, A. W. H., Vilks, G. and Lott, L. (1971). Winter distribution of planktonic Foraminifera between the Grand Banks and the Caribbean. *Micropaleont.* **17**(1), 31–42. (II-1)

Bé, A. W. H., Harrison, S. M. and Lott, L. (1973). *Orbulina universa* d'Orbigny in the Indian Ocean. *Micropaleont.* **19**(2), 150–192. (I, II-2)

Bé, A. W. H., Hemleben, C., Anderson, O. R., Spindler, M., Hacunda, J. and Tuntivate-Choy, S. (in press). Laboratory and field observations of living planktonic foraminifera. *Micropaleont.* **23**(2).

Belyaeva, N. V. (1963). The distribution of planktonic Foraminifera on the bottom of the Indian Ocean. *Voprosi Mikropaleontologii* **7**, 209–222. (In Russian.) (I-2)

Belyaeva, N. V. (1964a). Distribution of planktonic Foraminifera in the Indian Ocean. *Okeanol. Issled.* **13**, 205–210. (In Russian.) (I, II-2)

Belyaeva, N. V. (1964b). Distribution of planktonic Foraminifera in the water and on the floor of the Indian Ocean. *Trudy Inst. Okeanol. Akad. Nauk SSSR*, **68**, 12–83. (In Russian.) (I, II-2)

Belyaeva, N. V. (1966). Climatic and vertical zonality in the distribution of planktonic Foraminifera in the sediments of the Pacific Ocean. *Abstr. of Papers, Sec. Intern. Oceanogr. Congr. Publ. House Nauka, Moscow* 35–36. (I-3)

Belyaeva, N. V. (1967). Distribution of the shells of planktonic Foraminifera on the floor of the Bay of Bengal and some methodological aspects of the analysis of Foraminifera. *Okeanologiya* **7**(4), 645–654. (In Russian.) (I-2)

Belyaeva, N. V. (1968a). Planktonic Foraminifera in Atlantic Ocean sediments. *Dokl. Akad. Nauk SSSR* **183**(2), 445–448. (In Russian.) (I-1)

Belyaeva, N. V. (1968b). Quantitative distribution of planktonic foraminiferal tests in Recent Pacific sediments. *Okeanologiya* **8**(1), 111–115. (In Russian.) (I-3)

Belyaeva, N. V. (1969a). Thanatocoenoses of planktonic Foraminifera on the Pacific Ocean floor. *Okeanologiya* **9**(3), 500–504. (In Russian.) (I-3)

Belyaeva, N. V. (1969b). Planktonic Foraminifera in sediments of the surface layer. Chapter 3. *In* (Pergament, M. A., ed.), *The Pacific Ocean*, Vol. 8. *Microflora and Microfauna in the Recent Sediments of the Pacific Ocean.* Publ. House Nauka, Moscow. 203 p. (In Russian.) (I-3)

Belyaeva, N. V. (1969c). Planktonic Foraminifera in sediments of the world oceans. *Voprosi Mikropaleontologii* **12**, 168–172. (In Russian.) (I-6)

Belyaeva, N. V. (1970a). Distribution of planktonic foraminiferal tests in sediments from the rift zone of the Arabian-Indian Ridge. *Okeanologiya* **10**(4), 684–688. (In Russian.) (I-2)

Belyaeva, N. V. (1970b). Quantitative distribution of planktonic Foraminifera in Atlantic Ocean sediments. *Okeanologiya* **10**(6), 1016–1026. (In Russian.) (I-1)

Berger, W. H. (1967). Foraminiferal ooze: Solution at depths. *Science* **156**(3773), 383–385.

Berger, W. H. (1968). Planktonic Foraminifera: Selective solution and paleoclimatic interpretation. *Deep-Sea Res.* **15**, 31–43.

Berger, W. H. (1969). Ecologic patterns of living planktonic Foraminifera. *Deep-Sea Res.* **16**, 1–24. (II-6)

Berger, W. H. (1970a). Planktonic Foraminifera: Selective solution and the lysocline. *Mar. Geol.* **8**, 111–138.

Berger, W. H. (1970b). Planktonic Foraminifera: Differential production and expatriation off Baja California. *Limnol. Oceanogr.* **15**(2), 183–204. (II-3)

Berger, W. H. (1971a). Sedimentation of planktonic Foraminifera. *Mar. Geol.* **11**, 325–358. (I, II-6)

Berger, W. H. (1971b). Planktonic Foraminifera: Sediment production in an oceanic front. *J. Foram. Res.* **1**(3), 95–118. (II-3)

Berger, W. H. and Soutar, A. (1967). Planktonic Foraminifera: Field experiment on production rate. *Science* **156**(3781), 1495–1497. (II-3)
Berger, W. H. and Parker, F. L. (1970). Diversity of planktonic Foraminifera in deep-sea sediments. *Science* **168**, 1345–1347. (I-3)
Berger, W. H. and Soutar, A. (1970). Preservation of plankton shells in an anaerobic basin off California. *Geol. Soc. Amer., Bull.* **81**, 275–282. (II-3)
Berger, W. H. and Piper, D. J. W. (1972). Planktonic Foraminifera: Differential settling, dissolution and redeposition. *Limnol. Oceanogr.* **17**(2), 275–287.
Berger, W. H. and Winterer, E. L. (1974). Plate stratigraphy and the fluctuating carbonate line. *In* (Hsü, K. J. and Jenkins, H., eds), *Pelagic Sediments on Land and in the Ocean. Spec. Publs. Int. Assoc Sediment.* **1**, 11–48.
Berggren, W. A. (1969). Rates of evolution in some Cenozoic planktonic Foraminifera. *Micropaleont.* **15**(3), 351–365.
Blackman, A. (1966). Pleistocene stratigraphy of cores from the southeast Pacific Ocean. Ph.D. Thesis, University of California, San Diego, 200 p. (I-3)
Blair, D. G. (1965). The distribution of planktonic Foraminifera in deep-sea cores from the southern Ocean, Antarctica. *Dept. Geol., Florida State Univ.*, Contr. no **10**, Tallahassee, Florida, 141 p. (I-5)
Bogorov, B. G. (1960). Perspectives in the study of seasonal changes of plankton and of the number of generations at different latitudes. *In* (Buzzati-Traverso, A. A., ed.), *Perspectives in Marine Biology*, 145–158. University of California Press, Berkeley.
Bogorov, B. G. (1967). *The Pacific Ocean.* Vol. 7. *Biology of the Pacific Ocean.* Part 1. *Plankton.* Publ. House Nauka, Moscow, 266 p. (In Russian.)
Bolli, H., Loeblich, A. R. and Tappan, H. (1957). Planktonic foraminiferal families Hantkeninidae, Orbulinidae, Globorotaliide and Globotruncanidae. *U.S. Nat. Mus. Bull.* **215**, 3–50.
Boltovskoy, E. (1959a). Foraminíferos recientes del sur de Brasil, y sus relaciones con los de Argentina e India del Oeste. *Argent. Serv. Hidrogr. Naval* **H1005**, 1–124. (I-1)
Boltovskoy, E. (1959b). La corriente de Malvinas, un estudio en base a la investigación de Foraminíferos. *Argent. Serv. Hidrogr. Naval* **H1015**, 1–96. (I, II-1)
Boltovskoy, E. (1959c). Foraminifera as biological indicators in the study of ocean currents. *Micropaleont.* **5**(4), 473–481. (II-1)
Boltovskoy, E. (1961). Línea de la convergencia Subantárctica en el Atlántico sur y su determinacion usando los indicadores biológicos Foraminíferos. *Argent. Serv. Hidrogr. Naval* **H1018**, 1–35. (II-5)
Boltovskoy, E. (1962). Planktonic Foraminifera as indicators of different water masses in the South Atlantic. *Micropaleont.* **8**(3), 403–408. (II-1, 5)
Boltovskoy, E. (1964). Distribución de los Foraminíferos planctónicos vivos en el Atlántico ecuatorial, parte oeste (Expedición Equalant). *Argent. Serv. Hidrogr. Naval* **H639**, 1–54. (II-1)
Boltovskoy, E. (1966a). Resultados oceanograficos sobre la base del estudio del plancton recogido durante la campaña cosetri 2. *Argent. Serv. Hidrogr. Naval Bol.* **3**(2), 105–114. (II-1)
Boltovskoy, E. (1966b). Zonación en las latitudes altas del Pacífico sur según los Foraminíferos planctónicos vivos. *Mus. Argentino Cienc. Nat. "Bernardino Rivadavia", Rev.* **2**(1), 1–56. (II-5)
Boltovskoy, E. (1966c). Zonation of the high lattiudes of the South Pacific Ocean

according to the living planktonic Foraminifera. *SCAR, SCOR, IAPO, IUBS, Symp. on Antarctic Oceanography*, Santiago, Chile, 13–16, 1–6.   (II-5)

Boltovskoy, E. (1966d). La zona de convergencia subtropical/subantarctica en el Océano Atlántico, parte occidental, un estudio en base a la investigación de Foraminíferos indicadores. *Argent. Serv. Hidrogr. Naval* **H640**, 1–69.   (II-1, 5)

Boltovskoy, E. (1967). Campaña oceanográfica "Corrientes Drake VI". (Distribución de masas de aguas superficiales según el plancton). *Argent. Serv. Hidrogr. Naval Bol.* **4**(1), 5–16.   (II-1, 5)

Boltovskoy, E. (1968a). Living planktonic Foraminifera of the eastern part of the tropical Atlantic. *Rev. Micropaleont.* **11**(2), 85–98.   (II-1)

Boltovskoy, E. (1968b). Hidrologia de las aguas superficiales en la parte occidental del Atlantico sur. *Mus. Argentino Cienc. Nat. "Bernardino Rivadavia", Rev. Hydrobiol.* **2**(6), 119–224.   (II-1)

Boltovskoy, E. (1969a). Foraminifera as hydrological indicators. *In* (Brönnimann, P. and Renz, H. H., eds), *Proc. I. Plankt. Conf.* **2**, 1–14. E. J. Brill, Leiden.   (II-1)

Boltovskoy, E. (1969b). Tanatocenosis de Foraminíferos planctonicos en el estrecho de Mozambique. *Rev. Española Micropaleont.* **1**(2), 117–129.   (I-2)

Boltovskoy, E. (1969c). Living planktonic Foraminifera at the 90 degree E meridian from the equator to the Antartic. *Micropaleont.*, **15**(2), 237–255.   (II-2)

Boltovskoy, E. (1969d). Distribution of planktonic Foraminifera as indicators of water masses in the western part of the tropical Atlantic. *Proc. Symp. Oceanography and Fisheries Resources of the Tropical Atlantic*. UNESCO, Paris, 45–55.   (II-1)

Boltovskoy, E. (1970). Masas de agua (caracteristica, distribucion, movimientos) en la superfice del Atlantico sudoeste, segun indicadores biologicos-Foraminíferos. *Argent. Serv. Hidrogr. Naval* **H643**, 1–99.   (II-1)

Boltovskoy, E. (1971a). Ecology of the planktonic Foraminifera living in the surface layer of Drake Passage. *Micropaleont.* **17**(1), 53–68.   (II-5)

Boltovskoy, E. (1971b). Patchiness in the distribution of planktonic Foraminifera. *In* (Farinacci, A., ed.), *Proc. II Plankt. Conf.*, 107–115. Tecnoscienza, Roma.   (II-1)

Boltovskoy, E. (1973). Daily vertical migration and absolute abundance of living planktonic Foraminifera. *J. Foram. Res.* **3**(2), 89–94.   (II-1)

Boltovskoy, E. and Boltovskoy, D. (1970). Foraminíferos planctonicos vivos del mar de la flota (Antarctica). *Rev. Española Micropaleont.* **2**(1), 27–44.   (II-5)

Boltovskoy, E. and Lena, H. (1970). On the decomposition of the protoplasm and sinking velocity of the planktonic Foraminifers. *Int. Revue Ges. Hydrobiol.* **55**(5), 797–804.

Bradbury, M. G. *et al.* (1970). Studies in the fauna associated with the deep scattering layers in the equatorial Indian Ocean, conducted on R/V Te Vega during October and November 1964. *In* (Farquhar, G. B., ed.), *Proc. of an International Symposium on Biological Sound Scattering in the Ocean*, 409–452. U.S. Government Printing Office.

Bradshaw, J. S. (1957). Ecology of living planktonic Foraminifera in the North and Equatorial Pacific Ocean. Ph.D. Thesis, University of California, Los Angeles, 256 p.   (II-3)

Bradshaw, J. S. (1959). Ecology of living planktonic Foraminifera in the North and Equatorial Pacific Ocean. *Cushman Found. Foram. Res. Contr.* **10**(2), 25–64. (II-3)

Brady, H. B. (1884). Report on the Foraminifera dredged by H.M.S. *Challenger* during the years 1873–1876. *Rept. Voy. Challenger, Zool.* **9**, 814 p.   (I, II-6)

Brandt, K. and Apstein, C. (eds) 1901–(1928). *Nordisches Plankton* (many authors). Lipsius and Tischer, Kiel and Leipzig.

Carpenter, W. B. (1875). Remarks on Professor Wyville Thomson's preliminary notes on the nature of the sea-bottom procured by the soundings of H.M.S. *Challenger. Proc. Roy. Soc.* **23**, 234–245.

Carpenter, W. B., Parker, W. K. and Jones, T. R. (1862). Introduction to the study of the Foraminifera. The Royal Society, London.

Cifelli, R. (1961). *Globigerina incompta*, a new species of pelagic Foraminifera from the North Atlantic. *Cushman Found. Foram. Res. Contr.* **12**(3), 83–86.   (II-1)

Cifelli, R. (1962). Some dynamic aspects of the distribution of planktonic Foraminifera in the western North Atlantic. *J. Mar. Res.* **20**(5), 201–213.   (II-1)

Cifelli, R. (1965). Planktonic Foraminifera from the western North Atlantic. *Smithsonian Misc. Collections* **148**(4), 4599, 1–36.   (II-1)

Cifelli, R. (1967). Distributional analysis of North Atlantic Foraminifera collected in 1961 during cruises 17 and 21 of the R/V Chain. *Cushman Found. Foram. Res. Contr.* **18**(3), 118–127.   (II-1)

Cifelli, R. and Sachs, K. N. (1966). Abundance relationships of planktonic Foraminifera and Radiolaria. *Deep-Sea Res.* **13**, 751–753.   (II-1)

Cifelli, R. and Smith, R. K. (1969). Problems in the distribution of Recent planktonic Foraminifera and their relationships with water mass boundaries in the North Atlantic. *In* (Brönniman, P. and Renz, H. H., eds), *Proc. I Plankt. Conf.*, **2**, 68–81. E. J. Brill, Leiden.   (II-1)

Cifelli, R. and Smith, R. K. (1970). Distribution of planktonic Foraminifera in the vicinity of the North Atlantic current. *Smithsonian Contr. to Paleobiol.* **4**, 1–52.   (II-1)

Correns, C. W. (1939). Pelagic sediments of the North Atlantic Ocean. *In* (Trask, P. D., ed.), *Recent Marine Sediments—A Symposium.* 373–395. Amer. Assoc. Petr. Geol., Tulsa, Oklahoma.

Cushman, J. A. and Henbest, L. G. (1942). Geology and biology of North Atlantic deep-sea cores. Part 2. Foraminifera between Newfoundland and Ireland. *U.S. Geol. Surv. Prof. Paper* **196-A**, 35–54.   (I-1)

Echols, R. J. and Kennett, J. P. (1973). Distribution of Foraminifera in the surface sediments. *In* (Bushnell, V. ed.), *Marine Sediments of Southern Oceans*. Antarctic Map Folio Series 17, Am. Geogr. Soc., 13–17.   (I-5)

Eckert, H. R. (1965). Une station d'observation sur les Foraminifères planktoniques actuels dans le Golfe de Guinée. *Eclog. geol. Helv.* **58**(1), 1039–1058.   (II-1)

Ehrenberg, C. G. (1861). Über die Tiefgrund-Verhältnisse des Ozeans am Eingange der Davisstrasse und bei Island. *K. Preuss. Akad. Wiss. Berlin, Monatsber.* Jahr 1861 (1862), 275–315.

Ehrenberg, C. G. (1873). Mikrogeologische Studien über das kleinste Leben der Meeres-Tiefgründe aller Zonen und dessen geologischen Einfluss. *K. Preuss. Akad. Wiss. Berlin, Abh.* Jahr 1872, 131–397.

Emiliani, C. (1954). Depth habitats of some species of pelagic Foraminifera as indicated by oxygen isotope ratios. *Amer. J. Sci.* **252**, 149–158.

Enbysk, B. J. (1960). Distribution of Foraminifera in Northeast Pacific. Ph.D. Thesis, University of Washington, Seattle, 150 p.   (I-3)

Ericson, D. B. (1959). Coiling direction of *Globigerina pachyderma* as a climatic index. *Science* **130**(3369), 219–220.   (I-1)

Ericson, D. B., Wollin, G. and Wollin, J. (1954). Coiling direction of *Globorotalia truncatulinoides* in deep-sea cores. *Deep-Sea Res.* **2**, 152–158.   (I-1)

Ericson, D. B. and Wollin, G. (1956). Correlation of six cores from the equatorial Atlantic and the Caribbean. *Deep-Sea Res.* **3**(2), 104–125.   (I-1)

Ericson, D. B., Ewing, M., Wollin, G. and Heezen, B. C. (1961). Atlantic deep-sea sediment cores. *Geol. Soc. Amer. Bull.* **72**, 193–286.   (I-1)

Fèbvre-Chevalier, C. (1971). Constitution ultrastructurale de *Globigerina bulloides* d'Orbigny, 1826 (Rhizopoda-Foraminifera). *Protistologica* **7**(3), 311–324.

Fleisher, R. L. (1974). Cenozoic planktonic Foraminifera and biostratigraphy, Arabian Sea (Deep Sea Drilling Project, Leg 23A). *Initial Reports of the Deep Sea Drilling Project* **23**, 1001–1072.

Fleminger, A. and Hülsemann, K. (1973). Relationship of Indian Ocean epiplanktonic calanoids to the world oceans. In (Zeitzschel, B., ed.), *Ecological Studies. Analysis and Synthesis. The Biology of the Indian Ocean* **3**, 339–347. Springer–Verlag.

Frerichs, W. E. (1971). Planktonic Foraminifera in the sediments of the Andaman Sea. *J. Foram. Res.* **1**(1), 1–14.   (I-2)

Gardner, J. V. (1973). The eastern equatorial Atlantic: sedimentation, faunal and sea-surface temperature responses to global climatic changes during the past 200,000 years. Ph.D. Thesis, Columbia University, New York, 301 p.   (I-1)

Giesbrecht, W. (1892). Systematik und Faunistik der pelagischen Copepoden des Golfes von Neapel und der angrenzenden Meeresabschnitte. *Fauna und Golf. Neapel* **19**, 831 p.

Heath, G. R. and Culberson, C. (1970). Calcite: Degree of saturation, rate of dissolution and the compensation depth in the deep oceans. *Geol. Soc. Amer. Bull.* **81**, 3157–3160.

Hecht, A. D. and Savin, S. M. (1972). Phenotypic variation and oxygen isotope ratios in Recent planktonic Foraminifera. *J. Foram. Res.* **2**, 55–67.   (I-1)

Heezen, B. C. and Hollister, C. D. (1971). *The Face of the Deep*, 659 pp. Oxford University Press, New York.

Hemleben, C. (1969). Zur Morphogenese planktonischer Foraminiferen. *Zitteliana* **1**, 91–133.

Hensen, V. (ed.) (1892–1926). *Ergebnisse der Plankton-Expedition der Humboldtstiftung*. (Many volumes and parts by many authors). Lipsius and Tischer, Kiel and Leipzig.

Herb, R. (1968). Recent planktonic Foraminifera from sediments of the Drake Passage, Southern Ocean. *Eclog. geol. Helv.* **61**(2), 467–480.   (I-5)

Herman, Y. (1965). Etude des sediments Quaternaires de la Mer Rouge. *Ann. Inst. Oceanogr. Monaco* **42**(3), 339–415.   (I-2)

Heron-Allen, E. and Earland, A. (1932). Foraminifera, Pt. 1, The ice-free area of the Falkland Islands and adjacent seas. *Discovery Repts.* **4**, 291–460.   (I-5)

Huang, T. (1972). Distribution of planktonic Foraminifers in the surface sediments of the Taiwan Strait. *United Nations ECAFE, CCOP Tech. Bull.* **6**, 31–73.   (I-3)

Hunkins, K., Bé, A. W. H., Opdyke, N. D. and Mathieu, G. (1971). The Late Cenozoic history of the Arctic Ocean. In (Turekian, K. K., ed.), *The Late Cenozoic Glacial Ages*, 215–237. Yale University Press.

Imbrie, J. and Kipp, N. G. (1971). A new micropaleontological method for quantitative paleoclimatology, application to a Late Pleistocene Caribbean core. In

(Turekian, K. K., ed.), *The Late Cenozoic Glacial Ages*, 71–181. Yale University Press. (I-1)

Johnson, M. W. and Brinton, E. (1963). Biological species, water masses and currents. *In* (Hill, M. N., ed.), *The Sea* **2**, 381–414, John Wiley.

Jones, J. I. (1964). The ecology and distribution of living planktonic Foraminifera of the West Indies and adjacent waters. Ph.D. Thesis, University of Wisconsin, 193 p. (II-1)

Jones, J. I. (1967). Significance of distribution of planktonic Foraminifera in the Equatorial Atlantic Undercurrent. *Micropaleont.* **13**(4), 489–501. (II-1)

Jones, J. I. (1968a). Planktonic Foraminifera as indicator organisms in the eastern Atlantic equatorial current system. *Proc. Symp. Oceanography and Fisheries Resources of the Tropical Atlantic*. UNESCO, Paris, 213–230. (II-1)

Jones, J. I. (1968b). The relationship of planktonic foraminiferal populations to water masses in the western Caribbean and lower Gulf of Mexico. *Mar. Sci. Bull.* **18**(4), 946–982. (II-1)

Jones, J. I. (1971). The ecology and distribution of living planktonic Foraminifera in the Straits of Florida. *In* (Jones, J. I. and Bock, W. D., eds), *A Symposium of Recent South Florida Foraminifera, Memoir* **1**, Miami Geol. Soc. 175–190. (II-1)

Judkins, D. C. and Fleminger, A. (1972). Comparison of foregut contents of *Sergestes similis* obtained from net collections and Albacore stomachs. *Fish. Bull.* **70**(1), 217–223.

Keij, A. J. (1972). The relative abundance of Recent planktonic Foraminifera in seabed samples collected offshore Brunei and Sabah. *Borneo Reg. Malaysia Geol. Survey, Ann. Rept.* 146–155. (I-3)

Kennett, J. P. (1968a). Latitudinal variation in *Globigerina pachyderma* (Ehrenberg) in surface sediments of the south-west Pacific Ocean. *Micropaleont.* **14**(3), 305–318. (I-3, 5)

Kennett, J. P. (1968b). *Globorotalia truncatulinoides* as a paleo-oceanographic index. *Science* **159**, 1461–1463. (I-3, 5)

Kennett, J. P. (1969). Distribution of planktonic Foraminifera in surface sediments southeast of New Zealand. *In* (Brönnimann, P. and Renz, H. H., eds), *Proc. I Plankt. Conf.*, **2**, 307–322. E. J. Brill, Leiden. (I-3, 5)

King, K. Jr. and Hare, P. E. (1972a). Amino acid composition of planktonic Foraminifera: a paleobiochemical approach to evolution. *Science* **175**, 1461–1463.

King, K. Jr. and Hare, P. E. (1972b). Amino acid composition of the test as a taxonomic character for living and fossil planktonic Foraminifera. *Micropaleont.* **18**(3), 285–293.

Kustanowich, S. (1963). Distribution of planktonic Foraminifera in surface sediments of the south-west Pacific Ocean. *N.Z. J. Geol. Geophys.* **6**(4), 534–565. (I-3)

Le Calvez, Y. (1963). Contribution à l'étude des Foraminifères de la région d'Abidjan (Côte d'Ivoire). *Rev. Micropaleont.* **6**(1), 41–50. (I-1)

Lee, J. J., Freudenthal, H. D., Kossoy, V. and Bé, A. W. H. (1965). Cytological observations on two planktonic Foraminifera, *Globigerina bulloides* and *Globigerinoides ruber*. *J. Protozool.* **12**(4), 531–542.

Li, Y. H., Takahashi, T. and Broecker, W. (1969). Degree of saturation of $CaCO_3$ in the oceans. *J. Geophys. Res.* **74**(23), 5507–5525.

Lidz, B., Kehm, A. and Hendrick, M. (1968). Depth habitats of pelagic Foraminifera during the Pleistocene. *Nature* **217,** 245–247.

Lisitzin, A. P. (1972). Sedimentation in the world ocean. *Soc. Econ. Paleont. Mineral. Spec. Publ.* no. 17, 218 p.

Loeblich, A. R. Jr. and Tappan, H. (1964). *Treatise on invertebrate paleontology. Part C: Sarcodina, chiefly "Thecamoebians" and Foraminiferida.* R. C. Moore, (ed.). Geol. Soc. Amer. and University of Kansas Press.

Lohmann, H. (1920). Die Bevölkerung des Ozeans mit Plankton nach den Ergebnissen der Zentrifugenfänge während der Ausreise der Deutschland 1911, zugleich ein Beitrag zur Biologie des Atlantischen Ozeans. *Archiv. Biontol.* **4**(3), 617 p.

Loring, A. P. (1966). Distribution of planktonic Foraminifera in the South Atlantic Ocean. Ph.D. Thesis, New York University, New York, 99 p. (I-1)

Mayr, E. (1942). *Systematics and the origin of species from the viewpoint of a zoologist,* 334 p. Columbia University Press, New York.

McGowan, J. A. (1971). Oceanic biogeography of the Pacific. *In* (Funnell, B. M. and Riedel, W. R., eds), *The Micropaleontology of Oceans.* 3–74. Cambridge University Press, London.

McGowran, B. (1968). Reclassification of early Tertiary *Globorotalia. Micropaleont.* **14**(2), 179–198.

Meisenheimer, J. (1905). Die tiergeographischen Regionen des Pelagials, auf Grund der Verbreitung der Pteropoden. *Zool. Anz.* **29,** 155–163.

deMiro, M. D. (1965). Comparacion de la fauna de Foraminíferos de los sedimentos de la fosa de Cariaco con la del area oceanica adyacente. *Mem. Soc. Ciencias Nat. LaSalle* **25**(70–72), 223–260. (I-1)

deMiro, M. D. and Marval, J. A. (1967). Foraminíferos planctonicos vivos de la fosa de Cariaco y del talud continental de Venezuela. *Mem. Soc. Ciencias Nat. LaSalle* **27**(76), 11–34. (II-1)

Murray, J. (1897). On the distribution of the pelagic Foraminifera at the surface and on the floor of the ocean. *Nat. Science (ecology)* **11,** 17–27. (II-6)

Murray, J. and Renard, A. F. (1891). Deep-sea deposits based on the specimens collected during the voyage of H.M.S. *Challenger* in the years 1872–1876. *Rept. Voy. Challenger.* Longmans, London, 525 p.

Oba, T. (1969). Biostratigraphy and isotopic paleotemperature of some deep-sea cores from the Indian Ocean. *Sci. Rept. Tohoku Univ.* **41**(2), 129–195.

d'Orbigny, A. D. (1826). Tableau méthodique de la classe des Céphalopodes. *Ann. Sci. Nat. ser. 1,* **7,** 96–314.

d'Orbigny, A. D. (1839a). Foraminifères des Îles Canaries. *In* (Barker-Webb, P. and Berthelot, S., eds), *Histoire naturelle des Iles Canaries.* **2,** pt. 2, Zool., 119–146. Bethune, Paris.

d'Orbigny, A. D. (1839b). Foraminifères. *In* (de la Sagra, R., ed.), *Histoire physique, politique et naturelle de l'Ile de Cuba.* **8,** 1–224. Arthus Bertrand, Paris.

d'Orbigny, A. D. (1839c). *Voyage dans l'Amerique Méridionale.* **5,** pt. 5 (Foraminifères), 1–86. P. Bertrand, Strasbourg.

Owen, S. R. J. (1867). On the surface-fauna of mid-ocean. *J. Linn. Soc. Lond. (Zoology)* **9,** 147.

Parker, F. L. (1954). Distribution of the Foraminifera in the northeastern Gulf of Mexico. *Mus. Comp. Zool. Bull. Harvard* **3**(10), 454–588. (I-1)

Parker, F. L. (1955). Distribution of planktonic Foraminifera in some Mediterranean sediments. *Papers in Mar. Biol. and Oceanogr. Deep-Sea Res. Suppl. Vol.* **3**, 204–211. (I-1)

Parker, F. L. (1958). Eastern Mediterranean Foraminifera. *Rept. Swedish Deep-Sea Expedition* **8**(4), 217–283. (I-1)

Parker, F. L. (1960). Living planktonic Foraminifera from the equatorial and southeast Pacific. *Tohoku Univ., Sci. Repts. ser.* 2 (*Geol.*) *Spec. Vol.* **4**, 71–82. (II-3)

Parker, F. L. (1962). Planktonic foraminiferal species in Pacific sediments. *Micropaleont.* **8**(2), 219–254. (I-3)

Parker, F. L. (1965). Irregular distributions of planktonic Foraminifera and stratigraphic correlation. *Progress in Oceanogr.* **3**, 267–272. (I-6)

Parker, F. L. (1967). Tertiary biostratigraphy (planktonic Foraminifera) of tropical Indo-Pacific deep-sea cores. *Bull. Am. Paleont.* **52**(235), 115–208.

Parker, F. L. (1971). Distribution of planktonic Foraminifera in Recent deep-sea sediments. *In* (Funnell, B. M. and Riedel, W. R., eds), *The Micropalaeontology of Oceans*, 289–307. Cambridge University Press. (I-6)

Parker, F. L. (1973a). Living planktonic Foraminifera from the Gulf of California. *J. Foram. Res.* **3**(2), 70–77. (II-3)

Parker, F. L. (1973b). Late Cenozoic biostratigraphy (planktonic Foraminifera) of tropical Atlantic deep-sea sections. *Rev. Española Micropaleont.* **5**(2), 253–289.

Parker, F. L. and Berger, W. H. (1971). Faunal and solution patterns of planktonic Foraminifera in surface sediments of the South Pacific. *Deep-Sea Res.* **18**(1), 73–107. (I-3)

Parker, W. K. and Jones, T. R. (1865). On some Foraminifera from the North-Atlantic and Arctic oceans, including Davis Straits and Baffin's Bay. *Philos. Trans.* **155**, 325–441. (I-1)

Pessagno, E. A., Jr. and Miyano, K. (1968). Notes on the wall structure of the Globigerinacea. *Micropaleont.* **14**(1), 38–50.

Phillippi, E. (1910). Die Grundproben der deutschen Südpolar Expedition 1901–1903 *Deutsche Südpolar Expd.* **2**(6), 411–616.

Phleger, F. B. (1945). Vertical distribution of pelagic Foraminifera. *Amer. J. Sci.* **243**(7), 377–383. (II-1)

Phleger, F. B. (1951). Ecology of Foraminifera, northwest Gulf of Mexico: Part I Foraminifera distribution. *Geol. Soc. Amer. Mem.* **46**, 1–88. (I, II-1)

Phleger, F. B. (1954). Foraminifera and deep-sea research. *Deep-Sea Res.* **2**, 1–23. (I-1)

Phleger, F. B. and Parker, F. L. (1951). Ecology of Foraminifera, northwest Gulf of Mexico. Part 2, Foraminifera species. *Geol. Soc. Amer. Mem.* **46**, 1–59. (I-1)

Phleger, F. B., Parker, F. L. and Peirson, J. K. (1953). North Atlantic Foraminifera. *Swedish Deep-Sea Expedition, Rept.* **7**(1), 3–122. (I-1)

Premoli-Silva, I. (1966). La struttura della parete di alcuni foraminiferi planctonici. *Eclog. geol. Helv.* **59**(1), 219–233.

Pytkowicz, R. M. (1965). Calcium carbonate saturation in the ocean. *Limnol. Oceanogr.* **10**, 220–225.

Rao, K. Kameswara (1971). On some Foraminifera from the northeastern part of the Arabian Sea. *Proc. Indian Acad. Sci.* **73B**, 155–178. (I-2)

Rao, K. Kameswara (1972). Planktonic Foraminifera in sediment samples from the eastern Arabian Sea. *Indian J. Mar. Sci.* **1**, 1–7. (I-2)

Reiss, Z. (1957). The Bilameliidea, nov. superfam. and remarks on Cretaceous Globorotaliids. *Cushman Found. Foram. Res. Contr.* **8**(4), 127–145.

Reiss, Z. (1958). Classification of lamellar Foraminifera. *Micropaleont.* **4**(1), 51–70.

Reiss, Z. (1963). Comments on wall structure of Foraminifera. *Micropaleont.* **9**(1), 50–52.

Rhumbler, L. (1901). Nordische Plankton-Foraminiferen. *In* (Brandt, K., ed.), *Nordisches Plankton* **1**(14), 33 p. Lipsius and Tischer, Kiel and Leipzig.   (II-1)

Rhumbler, L. (1911). Die Foraminiferen (Thalamophoren) der Plankton-Expedition; Teil 1. Die allgemeinen Organisationsverhältnisse der Foraminiferen. *Plankton-Exped. Humboldt-Stiftung, Ergebn.* **3**, L, C, 1, 1–331.

Ruddiman, W. F. (1968). Historical stability of the Gulf Stream meander belt: Foraminiferal evidence. *Deep-Sea Res.* **15**, 137–148.   (I-1)

Ruddiman, W. F. (1969a). Recent planktonic Foraminifera: dominance and diversity in North Atlantic surface sediments. *Science* **164**, 1164–1167.   (I-1)

Ruddiman, W. F. (1969b). Foraminifera of the subtropical North Atlantic gyre. Ph.D. Thesis. Columbia University, New York, 282 p.   (I-1)

Ruddiman, W. F. and Heezen, B. C. (1967). Differential solution of planktonic Foraminifera. *Deep-Sea Res.* **14**, 801–808.   (I-1)

Russell, F. A. (1939). Hydrographical and biological conditions in the North Sea as indicated by plankton organisms. *J. Cons. Explor. Mer.* **14**, 171–192.

Savin, S. M. and Douglas, R. G. (1973). Stable isotope and magnesium geochemistry of Recent planktonic Foraminifera from the South Pacific. *Geol. Soc. Amer. Bull.* **84**(7), 2327–2342.

Schott, W. (1935). Die Foraminiferen in dem äquatorialen Teil des atlantischen Ozeans. *Deutsche Atlant. Exped.* "*Meteor*". 1925–1927, *Wiss. Ergebn.* **3**(3), 43–134.   (I–II-1)

Schott, W. (1966). Foraminiferenfauna und Stratigraphie der Tiefsee-Sedimente im nordatlantischen Ozean. *Swedish Deep-Sea Expedition, Rept.* **7**(8), 357–469   (I-1)

Shackleton, N. (1968). Depth of pelagic Foraminifera and isotopic changes in Pleistocene oceans. *Nature* **218** (5136), 79–80.

Smith, A. B. (1963). Distribution of living planktonic Foraminifera in the north-eastern Pacific. *Cushman Found. Foram. Res. Contr.* **14**(1), 1–15.   (II-3)

Smith, A. B. (1964). Living planktonic Foraminifera collected along an east-west traverse in the North Pacific. *Cushman Found. Foram. Res. Contr.* **15**(4), 131–134.   (II-3)

Srinivasan, M. S., Kennett, J. P. and Bé, A. W. H. (1974). *Globorotalia menardii neoflexuosa* new subspecies from the northern Indian Ocean. *Deep-Sea Res.* **21**(4), 321–324.

Stehli, F. G., Douglas, R. G. and Newell, N. D. (1969). Generation and maintenance of gradients in taxonomic diversity. *Science* **164**(3882), 947–949.

Stehman, C. F. (1972). Planktonic Foraminifera in Baffin Bay, Davis Strait and the Labrador Sea. *Maritime Sediments* **8**(1), 13–19.   (I-4)

Steuer, A. (1910). *Planktonkunde.* B. G. Teubner, Leipzig and Berlin.

Steuer, A. (1933). Zur planmässigen Erforschung der geographischen Verbreitung des Haliplanktons, besonders der Copepoden. *Zoogeographica* **1**, 269–302.

Steuerwald, B. A. and Clark, D. (1972). *Globigerina pachyderma* in Pleistocene and Recent Arctic Ocean sediments. *J. Paleont.* **46**(4), 573–580.   (I-4)

Stubbings, H. G. (1939). Stratification of biological remains in marine deposits.

John Murray Expedition 1933–1934, *Sci. Rept.* **3**(3), 159–192. (I-2)

Sverdrup, H. U., Johnson, M. W. and Fleming, R. H. (1942). *The Oceans.* Prentice-Hall, 1087 p.

Thiede, J. (1971). Planktonische Foraminiferen in Sedimenten vom ibero-marokkanischen Kontinentalrand. *"Meteor" Forsch.-Ergebn.* C(7), 15–102. (1-I)

Thiede, J. (1973). Planktonic Foraminifera in hemipelagic sediments: shell preservation off Portugal and Morocco. *Geol. Soc. Amer. Bull.* **84**, 2749–2754. (I-1)

Thomson, C. M. (1874). On dredgings and deep-sea soundings in the South Atlantic, in a letter to Admiral Richards, C.B., R.F.S. *Roy. Soc. Lond. Proc.* **22**, 423–428.

Tibbs, J. F. (1967). On some planktonic protozoa taken from the track of drift station Arlis, 1960–1961. *Arctic* **20**(4), 247–254. (II-4)

Tizard, T. H., Moseley, H. N., Buchanan, J. Y. and Murray, J. (1885). Narrative of the cruise of H.M.S. *Challenger*, with a general account of the scientific results of the expedition. *Sci. Res. H.M.S. Challenger* 1873–1876, *Rept.* **1**(2), 511–1110.

Todd, R. (1958). Foraminifera from western Mediterranean deep-sea cores. *Swedish Deep-Sea Expedition, Rept.* **8**(3), 169–215. (I-1)

Tolderlund, D. S. (1969). Seasonal distribution patterns of planktonic Foraminifera at five ocean stations in the western North Atlantic. Ph.D. Thesis, Columbia University, New York, 210 p. (II-1)

Tolderlund, D. S. and Bé, A. W. H. (1971). Seasonal distribution of planktonic Foraminifera in the western North Atlantic. *Micropaleont.* **17**(3), 297–329. (II-1)

Towe, K. M. and Cifelli, R. (1967). Wall ultrastructure in the calcareous Foraminifera: crystallographic aspects and a model for calcification. *J. Paleont.* **41**(3), 742–762.

Uchio, T. (1960). Planktonic Foraminifera of the Antarctic Ocean. *Spec. Publ. Seto Mar. Biol. Lab.* 3–10. (I, II-5)

Ujiié, H. (1968). Distribution of living planktonic Foraminifera in the southeast Indian Ocean. *Natl. Sci. Mus. Tokyo Bull.* **11**(1), 97–125. (II-2)

Ujiié, H. and Nagase, K. (1971). Cluster analysis of living planktonic Foraminifera from the south-eastern Indian Ocean. *In* (Farinacci, A., ed.), *Proc. II Plankt. Conf.*, 1251–1258. Tecnoscienza, Roma. (II-2)

Van Donk, J. (Vol. 2, Ch. 17). O^{18} as a tool for micropalaeontologists.

Vilks, G. (1969). Recent Foraminifera in the Canadian Arctic. *Micropaleont.* **15**(1), 35–60. (I-4)

Vilks, G. (1970). Circulation of surface waters in parts of the Canadian Arctic archipelago based on foraminiferal evidence. *Arctic* **23**(2), 100–111. (II-4)

Vilks, G. (1973). A study of *Globorotalia pachyderma* (Ehrenberg) *Globigerina pachyderma* (Ehrenberg) in the Canadian Arctic. Ph.D. Thesis, Dalhousie University, Halifax, Nova Scotia, 216 p. (I, II-4)

Vincent, E. S. (1972). Oceanography and Late Quaternary planktonic Foraminifera, southwestern Indian Ocean. Ph.D. Thesis, University of Southern California. Los Angeles, California, 372 p. (I-2)

Waller, H. O. and Polski, W. (1959). Planktonic Foraminifera of the Asiatic Shelf. *Cushman Found. Foram. Res. Contr.* **10**(4), 123–126. (I-3)

Wallich, G. C. (1862). The North-Atlantic sea-bed: comprising a diary of the voyage on board H.M.S. *Bulldog* in 1860, and observations on the presence of animal life and the formation and nature of organic deposits at great depths in the ocean. Pt. 1, 4–6, London.

Wallich, G. C. (1876). Deep-sea researches on the biology of *Globigerina*. 8 vols., London.

Watkins, N. D. and Kennett, J. P. (1972). Regional sedimentary disconformities and upper Cenozoic changes in bottom water velocities between Australasia and Antarctica. *Am. Geophys. Union, Antarctica Res. Ser.* **19**, 273–293.

Wilcoxon, J. A. (1964). Distribution of Foraminifera off the southern Atlantic coast of the United States. *Cushman Found. Foram. Res. Contr.* **15**(11), 1–24.   (I-1)

Williams, L. K. (1971). Selected planktonic Foraminifera as biological indicators of hydrological conditions in the eastern Gulf of Mexico. M.S. Thesis, Florida State University, 153 p.   (II-1)

Wiseman, J. D. and Ovey, C. D. (1950). Recent investigations on the deep-sea floor. *Geol. Assoc. Proc.* **61**(1), 28–84.   (I-1)

Wood, A. (1948). The structure of the wall of the test in Foraminifera: its value in classification. *Quart. J. Geol. Soc., London* **104**(2), 229–252.

Wyrtki, K. (1971). *Oceanographic Atlas of the International Indian Ocean Expedition.* National Science Foundation, Washington, D.C.

Zobel, B. (1968). Phänotypische Varianten von *Globigerina dutertrei* d'Orbigny (Foram.), ihre Bedeutung für die Stratigraphie in Quartären Tiefsee-Sedimenten. *Geol. Jb.* **85**, 97–122.

Zobel, B. (1971). Foraminifera from plankton tows, Arabian Sea: areal distribution as influenced by ocean water masses. *In* (Farinacci, A., ed.), *Proc. II Plankt. Conf.,* 1323–1334. Tecnoscienza, Roma.   (II-2)

Zobel, B. (1973). Biostratigraphische Untersuchungen an Sedimenten des indisch-pakistanischen Kontinental-randes (Arabisches Meer). *"Meteor" Forsch.-Ergebn.* C, **12**, 9–73.   (I, II-2)

Zucker, W. H. (1973). Fine structure of planktonic Foraminifera and their endo-symbiotic algae. Ph.D. Thesis, City University of New York, New York, 65 p.

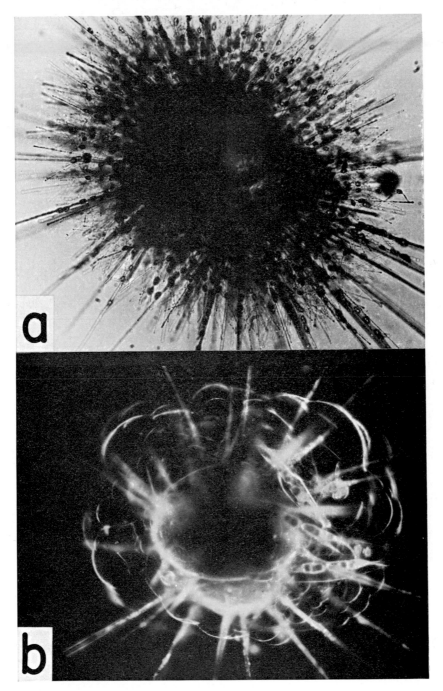

PLATE 1

fig. a. Living specimen of *Globigerinoides conglobatus* with many symbiotic zooxanthellae (small oval bodies) clinging to its pseudopods and spines (×83).

fig. b. Living *Hastigerina pelagica* with bubble capsule and several commensal dino-flagellates (*Dissodinium lunula*) (×60).

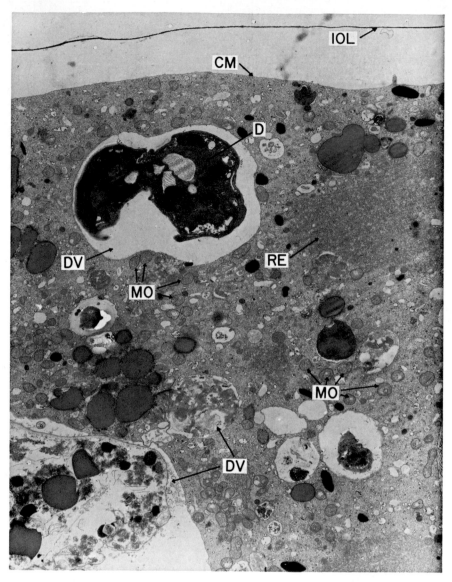

PLATE 2. The inner organic lining (IOL) is located between the cytoplasm and the inner test surface of *Globorotalia inflata*. The cytoplasm contains prominent digestive vacuoles (DV), tubular mitochondria (MO) and patches of endoplasmic reticulum (RE). A partly digested dinoflagellate (D) is inside one of the digestive vacuoles (×6000). Electron micrograph by O. R. Anderson.

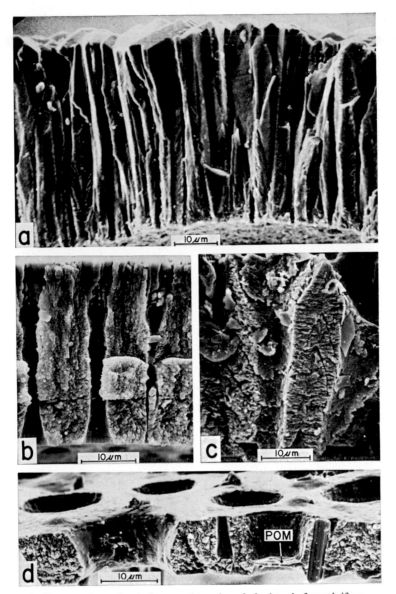

PLATE 3. Cross-sections of tests in several species of planktonic foraminifera.

fig. a. Vertical structural orientation of the calcite crust of *Globorotalia truncatulinoides*. Note that these elongate prisms appear as euhedral calcite "crystals" on the outer test surface.

fig. b. Random structural orientation of microgranules in *Pulleniatina obliquiloculata*.

fig. c. Horizontal structural orientation of brick-like units in *Globigerinoides conglobatus*. Note, however, that wall thickening does not occur in horizontal layers, but as successive growth increments of cone-like sheets draped on top of each other. Several "sheets" are visible along the relatively smooth inner surface of the centre pore.

fig. d. A primary organic membrane (POM) closes off a small pore of *Orbulina universa*, while the large pore on the left is open. Note the similarity of the random structural orientation of the microgranular wall in this species and *P. obliquiloculata*.

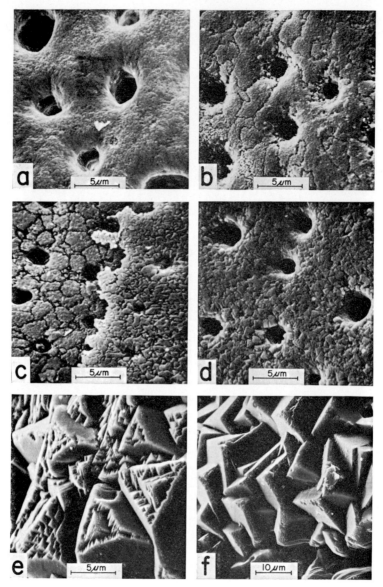

PLATE 4. Outer test surfaces of *Globorotalia truncatulinoides* at several stages of wall thickening.

fig. a. The microgranular texture is considered to be the earliest stage in the formation of a new chamber.

fig. b. Etching of microgranular surface reveals a "jigsaw puzzle" pattern of larger structural units.

fig. c. The "jigsaw puzzle" microgranular layer is overlain by a layer of small rhombic "crystals" to the right.

fig. d. The layer of small rhombic "crystals" is considered to be the intermediate stage in the wall thickening process.

fig. e. Incompletely formed euhedral "crystals" show that they consist of many smaller rhombic "crystals".

fig. f. Smooth faces of euhedral "crystals" represent ultimate stage of development. This coincides with the maximum thickening of the calcite crust.

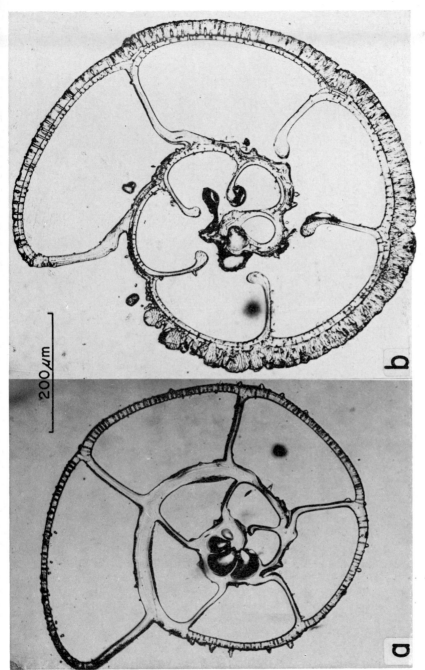

PLATE 5. Contrast in wall thickness in two horizontal sections of *Globorotalia truncatulinoides*. The left specimen, collected alive in 0–180 m plankton tow in the central
North Atlantic (35°04'N, 47°59'W), has a thin "bilamellar" test of uniform thickness. The
right specimen, from the surface sediment of the central North Atlantic (35°06'N, 45°56'W,
3190 m depth), has an additional calcite crust secreted over the "bilamellar" test. (From
Bé and Lott, 1964.)

PLATE 6. Most specimens in Plates 6–12 are from Indian Ocean surface sediment samples, except as otherwise indicated.

fig. 1. *Hastigerina pelagica* (d'Orbigny). Bermuda plankton tow (0–10 m depth) (×53).

figs 2a–d. *Hastigerinella digitata* (Rhumbler). (a) plankton tow (0–2850 m); 13°31′N, 18°03′W (×29); (b, c) plankton tow (0–360 m); 04°07′N, 80°44′E (×36); (d) plankton tow (0–830 m); 05°08′N, 42°24′W (×29).

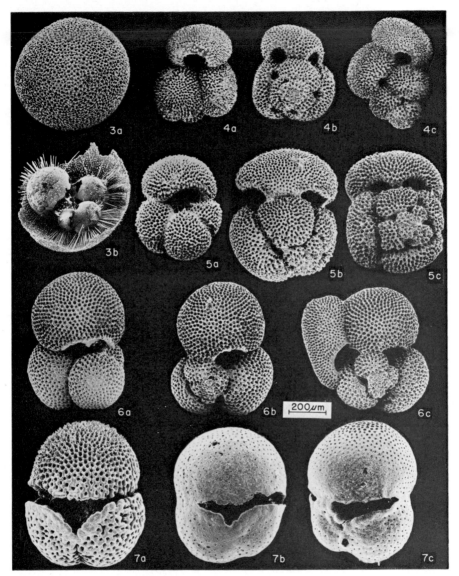

PLATE 7. All specimens on this plate are magnified ×48 (bar scale) and from surface sediment samples unless otherwise indicated.

figs 3a–b. *Orbulina universa* d'Orbigny. (a) spherical stage; 08°07′N, 73°15′E; (b) spiral stage. Plankton tow (0–10 m); 04°07′N, 80°44′E.

figs 4a–c. *Globigerinoides ruber* (d'Orbigny). (a, c) 25°38′S, 87°07′E; (b) 08°07′N, 73°15′E.

figs 5a–c. *Globigerinoides conglobatus* (Brady). (a, b) 08°07′N, 73°15′E; (c) 25°38′S, 87°07′E.

figs 6a–c. *Globigerinoides sacculifer* (Brady). (a, b, c) 08°07′N, 73°15′E; (c) with sac-like final chamber.

figs 7a–c. "*Sphaeroidinella dehiscens*" (Parker and Jones). (a) incipient development of cortex. Plankton tow (0–10 m); 4°07′N, 80°44′E; (b, c) fully developed cortex; 08°07′N, 73°15′E.

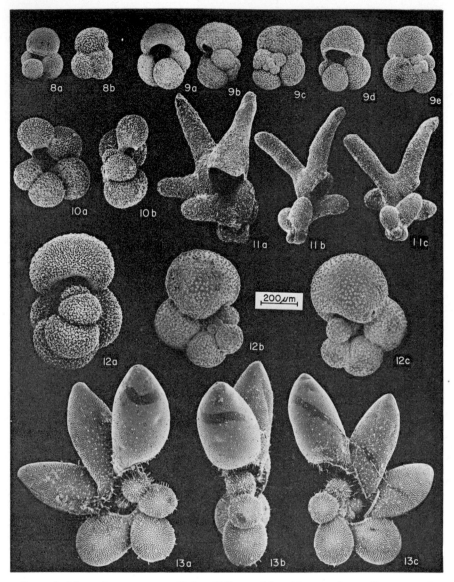

PLATE 8. All specimens are magnified ×48 (bar scale) and from surface sediment samples unless otherwise indicated.

figs 8a–b. *Globigerina falconensis* Blow. (a, b) 32°00′S, 98°52′E.

figs 9a–e. *Globigerina bulloides* d'Orbigny. (a, b, c) high-latitude form; 39°43′S, 82°15′E; (d, e) tropical-upwelling form; 08°07′N, 73°15′E.

figs 10a–b. *Globigerina calida* Parker. (a, b) 12°52′N, 48°24′E.

figs 11a–c. *Globigerina digitata* Brady. (a, b, c) 07°24′N, 72°48′E.

figs 12a–c. *Globigerinella aequilateralis* (Brady). (a) 08°07′N, 73°15′E; (b, c) 25°38′S, 87°07′E.

figs 13a–c. *Globigerinella adamsi* (Banner and Blow). (a, b, c) plankton tow (0–200 m); 05°58′S, 99°00′E.

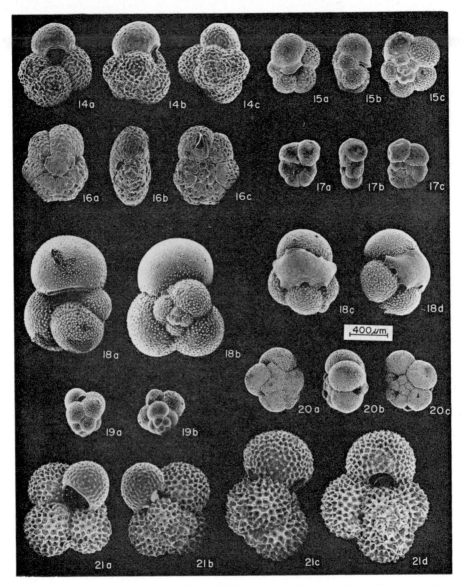

PLATE 9. All specimens are magnified ×96 (bar scale) and from surface sediment samples.
figs 14a–c. *Globigerina rubescens* Hofker. (a, b, c) 25°38′S, 87°07′E.
figs 15a–c. *Globigerina quinqueloba* Natland. (a, b, c) 55°20′S, 65°28′E.
figs 16a–c. *Globigerina humilis* (Brady). (a, b, c) 17°48′S, 62°40′E.
figs 17a–c. *Globorotalia pumilio* Parker. (a, b, c) 11°42′S, 109°43′W
figs 18a–d. *Globigerinita glutinata* (Egger). (a, b) without bulla; (a) 25°47′S, 93°43′E; (b) 08°07′N, 73°15′E; (c, d) with bulla; 08°07′N, 73°15′E.
figs 19a–b. *Globigerinita bradyi* Wiesner. (a, b) 55°20′S, 65°28′E.
figs 20a–c. *Globigerinita iota* Parker. (a, b, c) 30°12′S, 91°43′E.
figs 21a–d. *Globigerinoides tenellus* Parker. (a, b, c, d) 05°23′S, 72°41′E.

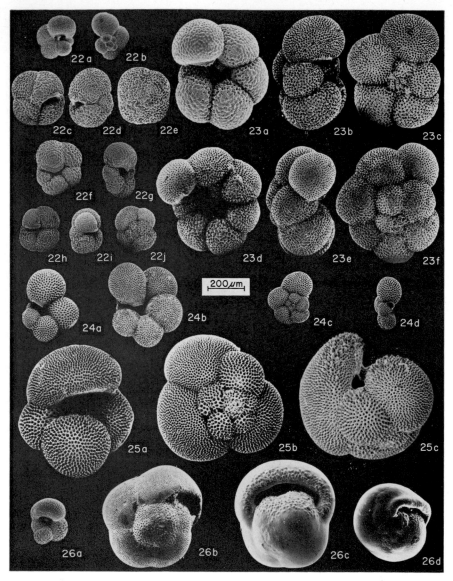

PLATE 10. All specimens are magnified ×48 (bar scale) and from surface sediment samples unless otherwise indicated.

figs 22a–j. *Globoquadrina pachyderma* (Ehrenberg). (a, b) left-coiling thin-walled form; 55°20′S, 65°28′E; (c, d, e) left-coiling thick-walled form; 55°20′S, 65°28′E; (f, g) right-coiling normalform; 28°01′S, 47°25′E; (h, i, j) right-coiling kummerform; 25°38′S, 87°07′E.

figs 23a–f. *Globoquadrina dutertrei* (d'Orbigny). (a, b, c) 5½ chambered form; (b) 28°26′S, 43°47′E; (b, c) 08°07′N, 73°15′E; (d, e, f) 6½ chambered form; 08°07′N, 73°15′E.

figs 24a–d. *Globoquadrina hexagona* (Natland). (a, c, d) 08°07′N, 73°15′E; (b) 02°59′N, 65°12′E.

figs 25a–c. *Globoquadrina conglomerata* (Schwager). (a, b) 08°07′N, 73°15′E; (c) plankton tow (0–250 m); 01°16′N, 60°07′E.

figs 26a–d. *Pulleniatina obliquiloculata* (Parker and Jones). (a) juvenile; 21°25′S, 112°48′E; (b) adult stage without cortex; 08°07′N, 73°15′E; (c, d) adult stage with cortex; (c) 06°42′N, 59°20′E; (d) 08°07′N, 73°15′E.

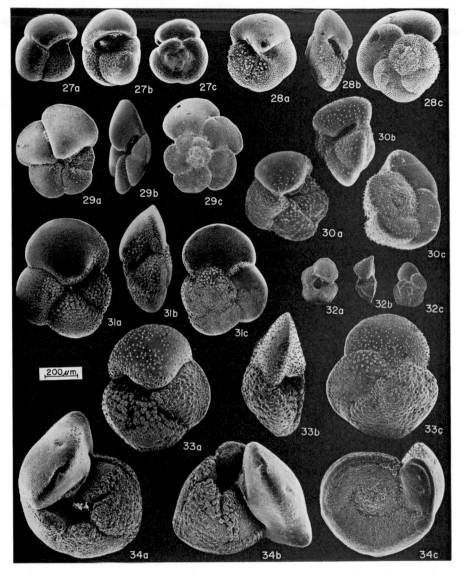

PLATE 11. All specimens are magnified ×43 (bar scale) and from surface sediment unless otherwise indicated.

figs 27a–c. *Globorotalia inflata* (d'Orbigny). (a, b, c) 25°38′S, 87°07′E.

figs 28a–c. *Globorotalia scitula* (Brady). (a, b, c) 08°07′N, 73°15′E.

figs 29a–c. *Globorotalia theyeri* Fleisher. (a, c) 08°07′N, 73°15′E; (b) plankton tow (0–200 m); 09°00′N, 83°00′E.

figs 30a–c. *Globorotalia crassaformis* (Galloway and Wissler). (a, c) plankton tow (0–400 m); 14°36′S, 03°40′E; (b) plankton tow (0–300 m); 14°49′N, 27°00′W.

figs 31a–c. *Globorotalia crotonensis* Conato and Follador. (a, b, c) 25°38′S, 87°07′E.

figs 32a–c. *Globorotalia cavernula* Bé. (a, b, c) 39°43′S, 82°15′E.

figs 33a–c. *Globorotalia hirsuta* (d'Orbigny). (a, b, c) 25°38′S, 87°07′E.

figs 34a–c. *Globorotalia truncatulinoides* (d'Orbigny). (a, b, c) 25°38′S, 87°07′E.

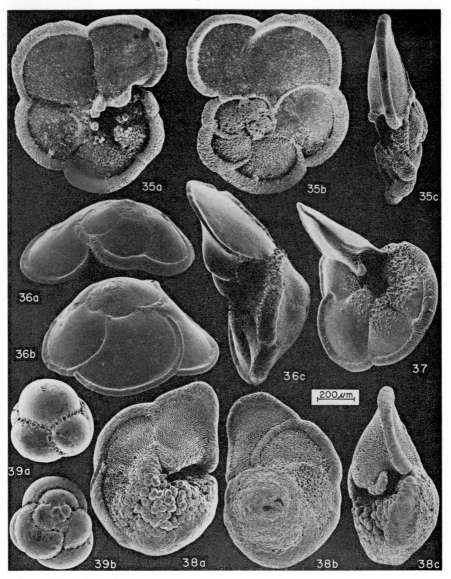

PLATE 12. All specimens are magnified ×48 (bar scale) and from surface sediment samples unless otherwise indicated.

figs 35a–c. *Globorotalia menardii* (d'Orbigny) *menardii* (d'Orbigny). (a, b, c) 08°07′N, 73°15′E.

figs 36a–c. *Globorotalia menardii* (d'Orbigny) *gibberula* n. subsp. Bé. (a, b, c) plankton tow (0–500 m); 24°52′S, 78°18′E.

figs 37. *Globorotalia menardii* (d'Orbigny) *neoflexuosa* Srinivasan, Kennett and Bé; plankton tow (0–1000 m); 17°54′N, 86°31′E.

figs 38a–c. *Globorotalia tumida* (Brady). (a, b, c) 06°42′N, 59°20′E.

figs 39a–b. *Candeina nitida* d'Orbigny. (a, b) 25°38′S, 87°07′E.

2: Investigations of Quaternary Climate based on Planktonic Foraminifera

W. F. RUDDIMAN

Ocean Floor Analysis Division,
United States Naval Oceanographic Office,
Washington D.C. 20373, U.S.A.

In palaeoclimatic analyses based on planktonic foraminifera, the present is the key to the past only within certain limits. With the assumption of a biota whose component species and environmental tolerances have been unchanged from the period of study to the present, ideally it is possible to transform the modern environmental preferences of the fauna or flora into quantitative estimates of past conditions recorded in the fossil record. Such palaeo-estimates are accessible only through the complex baffle of sedimentologic problems that alter the oceanic record: selective solution of fragile species, hiatuses in the record due to currents, and turbidite erosion or deposition. The achievable time resolution in any palaeoclimatic reconstruction is primarily dependent upon two factors: the sedimentation rate relative to the sample size, and the degree of sediment mixing by burrowing fauna.

Through large areas of the world's oceans, one or more of these deep-sea sedimentological problems interferes with the usually desired palaeoclimatic product: changes in surface water conditions at a time resolution of 10^3–10^4 years and over time intervals of 10^5–10^6 years. Turbidities disrupt much of the Indian Ocean record; solution alters or eliminates the history in large areas of the Pacific; sedimentation rates usually are prohibitively low in the Arctic and much of the Antarctic. Most detailed palaeoclimatology has been focused on the equatorial and subpolar North Atlantic Ocean and the surrounding seas, which record detailed and large-scale climatic oscillations in a relatively unaltered manner.

Recently, two approaches to foraminiferal palaeoclimatology have achieved considerable success: 1, objective quantitative techniques using factor analysis and linear regression to reconstruct unbiased palaeoestimates of sea-surface temperature and salinity; and 2, a regional approach to oceanic reconstruction, with cross-sections or areal maps revealing a full time-space sense of palaeooceanographic change such as migrations of frontal boundaries and ecological water masses, rather than just down-core variations analysed for a record of temporal change at one point in space. Applied to areas of minimal sedimentological disruption of the fossil record, the combined effect of these two approaches has made possible synoptic palaeooceanographic reconstruction of oceanic climates far different from the present; as a result, Quaternary foraminiferal studies are beginning to have an important impact on climatological theory.

Outline

Introduction

The approach of this study will be: 1, to discuss sedimentological problems which eliminate or prohibitively disrupt the orderly recording of a pelagic sequence of Quaternary climatic change; 2, to map the regions in which these processes are active; and 3, to examine in detail the results of foraminiferal studies in the remaining oceanic areas that are optimal for palaeoclimatic analysis.

Depositional and post-depositional alteration of palaeoclimatic records

SOLUTION EFFECTS

Variations in relative abundance of foraminiferal species in sediments deposited at shallow oceanic depths (< 3500 m) may give accurate indications of surface water temperature and salinity at the time of deposition. In deeper

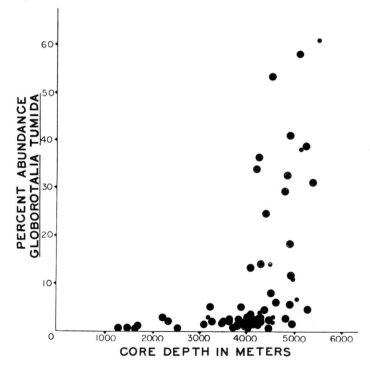

FIG. 1. Increasing percentage abundance of resistant species *Globorotalia tumida* in equatorial Atlantic core tops with increasing water depth. (After Ruddiman and Heezen, 1967.)

realms, such variations are more likely to be influenced by differential solution among the species. The possibility that corrosive bottom waters selectively
dissolve fragile species at the expense of robust resistant forms was recognised
in early deep-sea micropalaeontological studies (Schott, 1935; Arrhenius,
1952; Phleger *et al.*, 1953; and Hamilton, 1957). More recent studies have
quantitatively documented the selective effects of solution upon warm-water
species in low-latitude sediments (Kennett, 1966; Berger, 1967, 1968, 1970;
Ruddiman and Heezen, 1967; Parker and Berger, 1971; Valencia, 1973).

Differential solution of surface-sediment faunas

Because of prevailing undersaturation of calcium carbonate at most oceanic
depths, almost all calcareous deep-sea sediments are somewhat dissolved.
Only samples containing aragonitic pteropod shells or extremely fragile
foraminifera (such as *Hastigerina pelagica* d'Orbigny) cannot have been
appreciably dissolved.

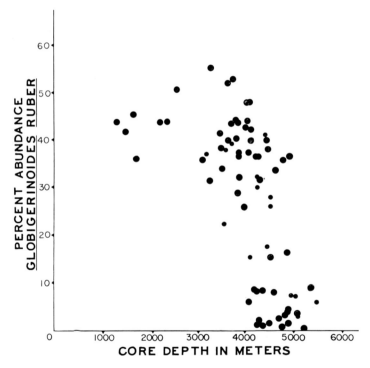

FIG. 2. Decreasing percentage abundance of fragile species *Globigerinoides ruber* in
equatorial Atlantic core tops with increasing water depth. (After Ruddiman and Heezen,
1967.)

Surface sediments are the most accessible time-stratigraphic datum for measuring the progressive effects of solution with increasing ocean depth. Ruddiman and Heezen (1967) documented the increasing percentages of resistant species like *Globorotalia tumida* Brady (Fig. 1) in deeper equatorial Atlantic surface sediments; this is caused by the selective removal of fragile species like *Globigerinoides ruber* d'Orbigny (Fig. 2). They noted that the depth of the inflection point within a 1000 meter zone of changing species, percentages corresponds to the contact of North Atlantic Deep Water and Antarctic Bottom Water (Fig. 3).

Berger (1968) further quantified the selection of the depth at which the maximum rate of change in species assemblage composition occurs he named it the lysocline and emphasised its abruptness. In equatorial Atlantic sediments, however, the lysocline is a broad zone of changing abundances rather than a specific depth boundary (Figs 1–3). Furthermore, a substantial loss of the original total population of forminifera (50–80%) is necessary for the

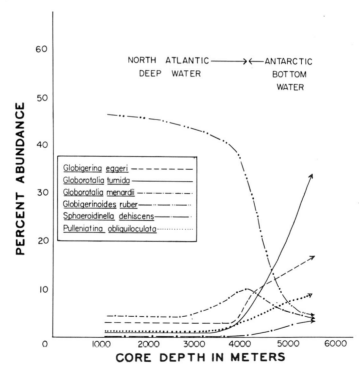

Fig. 3. Mean changes in abundance of several diagnostic planktonic foraminifera in equatorial Atlantic core tops with increasing water depth. (After Ruddiman and Heezen, 1967.)

species, percentages to show much change from the undissolved compositions (Berger, 1971). For these reasons, Berger (1971) later redefined a lysocline zone, with the lysocline level at its upper boundary. This lysocline thus marks the upper depth at which any solution-caused change can be detected, and not the level of maximum rate of assemblage change with depth (Fig. 4). Since this upper depth limit of potential alteration of the original palaeoclimatic record is critical to the palaeoclimatologist, the redefined lysocline is more useful. Parker and Berger (1971) have suggested that a bowl-shaped and sharply marked foraminiferal lysocline exists for South Pacific sediments, lying at greatest depths in the mid-basin, but rising and blurring towards both the equator and the eastern basin margin (Fig. 5).

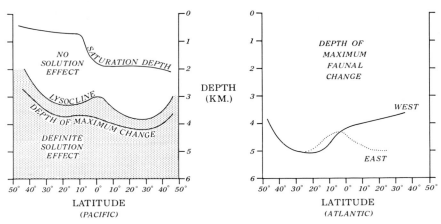

Fig. 4. Depth in km of saturation depth, lysocline, and level of maximum faunal change by latitudes in Atlantic and Pacific Oceans. (Modified from Berger, 1970, 1971.)

Both the south-eastern Pacific and equatorial Atlantic Oceans are favourable for studying differential solution because their undissolved faunas contain both a predominance of fragile species and a substantial minority of resistant forms, thus providing the contrast in solution susceptibility needed to register solution effects.

It has not yet been possible, however, to trace a sharply defined lysocline into either temperate or subpolar zones. In their investigation of samples from a wide range of ocean depths in the Southeast Pacific, Parker and Berger (1971) found a lower range of solution indices (quantitative measures of net selective solution) among low-temperature (high-latitude) samples than among high-temperature (low-latitude) sediments. They interpret these data as indicating moderately severe solution at almost all depths in high-latitude South Pacific sediments and hence a very shallow lysocline. Although solution of calcareous shells is likely to have occurred and may have been severe, it cannot be proved to have altered the fauna differentially because of the lack of any samples which could contain undissolved reference assemblages even at depths less than 2000 m. It is just as reasonable to infer that there is no lysocline or an ill-defined one at high southern (or northern) latitudes because solution has little or no differential effect upon the almost uniformly solution-resistant faunas. If this interpretation is correct, even a very large degree of dissolution of foraminiferal assemblages in high-latitude areas would not necessarily impair retrieval of palaeoclimatic information.

In view of this uncertainty, the alteration of foraminiferal assemblages by differential solution can only be considered as a limiting factor in palaeo-climatic studies for those low-latitude areas of the ocean where its effects have been recorded (see Fig. 6). In the author's opinion the 10% contour of calcium carbonate abundance (see Fig. 6), which was mapped and used by Lisitzin (1972) to define the level of the calcite compensation depth, is also a significant limiting factor in palaeoclimatic investigations. Sediments with less

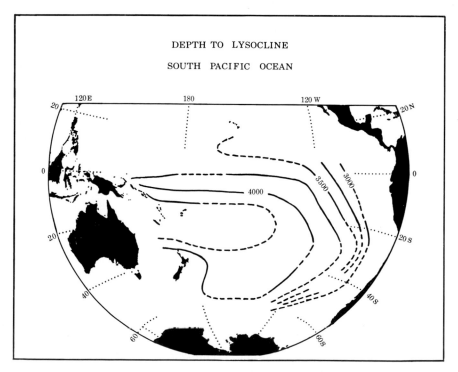

Fig. 5. Depth in metres of lysocline surface (level marking onset of detectable selective solution) in South Pacific sediments. (After Parker and Berger, 1971.)

than 10% calcium carbonate are produced either by a high rate of solution or significant levels of dilution by hemipelagic lutites. Consequently these sediments contain only a few percent of mature planktonic foraminiferal tests. In addition pelagic deposition is also disrupted by the introduction of terrigenous detritus via turbidities and bottom currents.

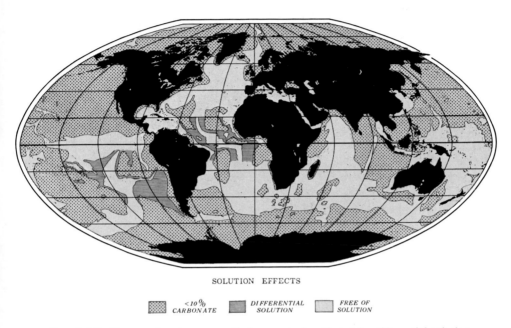

SOLUTION EFFECTS

| <10% CARBONATE | DIFFERENTIAL SOLUTION | FREE OF SOLUTION |

FIG. 6. World map of regions presently known to be affected by differential solution, and of areas containing less than 10% calcium carbonate in surface sediments. (After Lisitzin, 1972.)

Outside the carbonate-rich Atlantic basin, there are relatively few areas of open ocean with sediments unaffected by solution (Fig. 6). Many such areas are seamounts and insular or continental slopes, often subject to localised downslope sedimentation, current scour, and other effects associated with regions of abrupt relief which are usually detrimental to palaeoclimatic inference. The high-latitude North Atlantic and surrounding seas are the largest regions free of palaeoclimatically significant solution in the world's oceans.

Differential solution of pre-Holocene faunas

It is uncertain whether the present geographic pattern of surface-sediment solution effects can be projected back into the Quaternary record. Differential

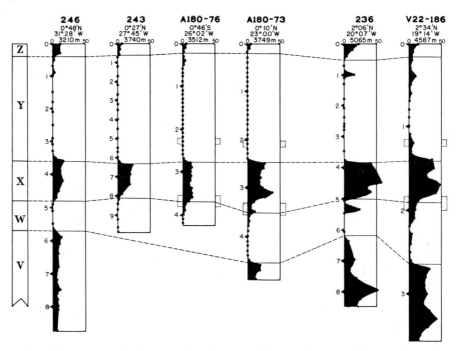

Fɪɢ. 7. Percentage abundance curves of resistant *Globorotalia menardii/G. tumida* species complex in late Quaternary sections (zones defined by Ericson and Wollin, 1968) of six equatorial Atlantic cores, showing generally increased percentages at correlative levels in cores from greater water depths. (After Ruddiman, 1971.)

solution is a difficult factor to measure precisely in the geologic record, since we often do not know what was originally deposited but most usually work with what is left after alteration by solution. For Quaternary sediments, moreover, the lysocline and compensation depth may be subject to vertical migration during glacial and interglacial periods, leaving secondary solution changes which are superimposed on primary palaeoclimatic oscillations only during particular intervals. In Quaternary sequences moreover, vertical migrations of the lysocline and calcite compensation depth during glacial intervals would, if they occur, lead to the superimposition of secondarily induced dissolution effects on the primary palaeoclimatic oscillations in foraminiferal assemblages. The influence of such effects, and of the different rates of solution at different depths can, however, be recognised by comparing faunas for the same interval of the Quaternary which has been repeatedly cored at a variety of ocean depths.

The solution of pre-Holocene Quaternary sediments has received the greatest attention in the equatorial Atlantic. Olausson (1960) commented on

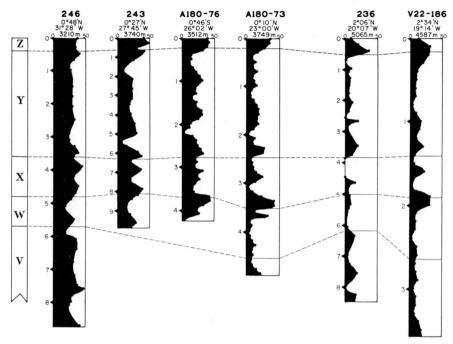

FIG. 8. Percentage abundance of fragile species *Globigerinoides ruber* in late Quaternary sections of six equatorial Atlantic cores, showing decreased percentages at correlative levels in deeper cores. (From data in Ruddiman, 1971.)

the increased percentage of solution-resistant species in late Quaternary deep-sea sediments described by Phleger *et al.* (1953). Berger (1968) studied sub-surface sample data from Schott (1935) for solution effects.

Ruddiman (1971), drawing in part on data from Phleger *et al.* (1953), reported on an east-west profile comprised of six correlative equatorial Atlantic cores taken from ocean depths ranging from 3210 to 5065 m. As in the surface sediments, resistant species like *Globorotalia tumida* were found to be residually concentrated in pre-Holocene sediments from cores at depths exceeding 4000 m (Fig. 7). Fragile species like *Globigerinoides ruber* decreased in percentage concentration in cores from increasing ocean depths (Fig. 8). This effect is also obvious for two resistant species at synchronous levels in cores spanning a far greater time interval (Fig. 9) than the cores illustrated in Figs 7 and 8. Ruddiman (1971) noted that the greatest net percentage increases in resistant species occurred during warm climates (the Z and X intervals defined by Ericson and Wollin, 1968, which are roughly equivalent to the Holocene and last interglacial, respectively). However, the net amount of visible percentage change in the fauna due to differential solution depends in part on the composition of the original undissolved assemblage. Variations in the mean resistance of the original population, and in the amount of contrast of its component species, will yield differing apparent solution effects. During warm periods the selective dissolution of the abundant but fragile warm species (e.g. *Globigerinoides sacculifer* Brady) will change the warm-water fauna more rapidly than an equivalent amount of absolute dissolution of a more homogeneous, moderately resistant, cold water assemblage of the cold X or Y intervals defined by Ericson and Wollin (1968).

Because of this initial contrast in resistance to solution, it would be wrong to infer that greater absolute solution occurred during interglacial than glacial intervals in the equatorial Atlantic. It does appear that a greater differential alteration of the foraminiferal assemblages occurred during interglacials, just as it does at present in low-latitude surface-sediment assemblages relative to high-latitude faunas.

To a first approximation, the ocean basin depth in the equatorial Atlantic which separates solution-altered and solution-unaffected foraminiferal populations is comparable during both glacial and interglacial intervals. It will thus be assumed in his paper that the rates of solution at the present (interglacial) depths (Fig. 6) are acceptable indicators of the solution rates at similar levels during the Quaternary.

VERTICAL MIXING

With the exception of small stagnant marginal basins, all oceanic faunal assemblages are substantially altered by mixing; as a result there is no unmixed sequence of oceanic sediments with which the effects of increased mixing can be compared. Nature has fortunately provided other means of assessing the effects of mixing; (*i*) the instantaneous deposition of mineralogical detritus which should be mixed in much the same way as foraminifera; and (*ii*) palaeoclimatically controlled faunal episodes of great brevity which set upper limits on the amount of mixing which could possibly have occurred within a limited time interval.

Mineralogical analogues of foraminiferal mixing

Berger and Heath (1968) first formulated a theoretical model of vertical mixing to reconstruct the redistribution of particulate matter (mineral, faunal or floral) across boundaries marking sudden extinctions, first appearances or simple abrupt changes in abundance. This quantitative model confirmed previous observations (Bramlette and Bradley, 1941; McIntyre *et al.*, 1967) that originally abrupt boundaries of concentrated particulate matter are

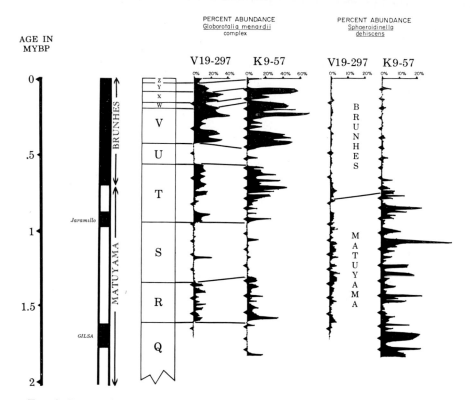

FIG. 9. Increased percentage abundance at greater water depth of the *Globorotalia menardii* d'Orbigny and *G. tumida* species complex, and of *Sphaeroidinella dehiscens* Parker and Jones in two eastern equatorial Atlantic cores spanning the entire Quaternary. (After Ruddiman, 1971.) Core V19–297 from 4122 m, core K9–57 from 4479 m.

blurred in cores by the burrowing infauna, and, as a result these particles gradually increase or decrease exponentially in abundance. The Berger–Heath model, however, contained the following critical assumptions: (*i*) cores with constant sedimentation rates; (*ii*) the existence of a thoroughly homogenised mixed layer of constant thickness, which migrates continually upward through the sediment column with the addition of new pelagic material from above and the loss of sediment to the historical layers below; and (*iii*) no penetration by burrowers to sediments beneath the hypothetical mixed-layer thickness.

Glass (1969) used sand-sized glassy spherules (microtektites) as tracers of past mixing. Working with limited numbers of spherules thought to have been deposited instantaneously at a level coincident with the Brunhes/Matuyama palaeomagnetic boundary, he traced their dispersal through zones averaging 60 cm thickness, equivalent to time spans averaging 120 000 years (Fig. 10). Such a degree of dispersal seemed to preclude the accurate resolution of Quaternary palaeoclimatic episodes of a few thousand years' duration, but the sedimentation rates in the microtektite-bearing cores averaged only 0.4 cm/1000 years, with none exceeding 1 cm/1000 years.

To test the Berger–Heath model and to estimate the effects of vertical mixing on climatic analysis in an area of higher sedimentation rates, Ruddiman and Glover (1972) made quantitative counts in North Atlantic cores with linear sedimentation rates through zones of dispersed silicic ash first noted by Bramlette and Bradley (1941). Counting only the sand-sized shards (which are closely analogous to foraminifera in 2 of 3 dimensions), they found that the ash, which is mixed through zones as shown in Fig. 11, decreases upward and downward from its peak abundance in a roughly exponential manner. They concluded that complete homogenisation of the mixed layer could not have been achieved in the North Atlantic cores, since the zones of ash concentration would show a scarp-like lower limit if mixed according to the Berger–Heath model (Fig. 12). The burrowed, but still discernible colour contacts in these and most other deep-sea sediments also indicate that complete homogenisation is not common. Ruddiman and Glover (1972) also showed that, since homogenisation in the mixed layer is incomplete, the peak abundance layer of shards must lie at the original depositional horizon, and not one mixed layer thickness below as predicted by the Berger–Heath model (Fig. 12). This result has important consequences for the use of such peak abundance layers as chronostratigraphic horizons. Counts of the mean shard size and range of sizes also revealed that the smallest shards are mixed a greater distance upward and downward from the level of original deposition than are the larger shards (Fig. 13), probably because of gut diameter restrictions in many of the burrowers. By analogy, smaller foraminifera will be further mixed than larger species, and foraminifera will be less dispersed than coccoliths or diatoms.

Ruddiman and Glover (1972) found that the ash shards are dispersed through zones averaging 50 cm thickness, approximately the same thickness as the microtektite zones of Glass (1969). However, because the North Atlantic sedimentation rates average an order of magnitude higher (4 cm/1000 years) than the sedimentation rates in Glass' Indian Ocean cores, the ash shards are dispersed through zones spanning an order of magnitude less time (10 000–15 000 years) than the microtektites. More than 90% of the ash is concentrated in less than 10 000 years of accumulated sediment. The hypothetically reconstructed, originally pure shard layers of about 0.1 cm average thickness were calculated to be equivalent in thickness (and, hence, time) to

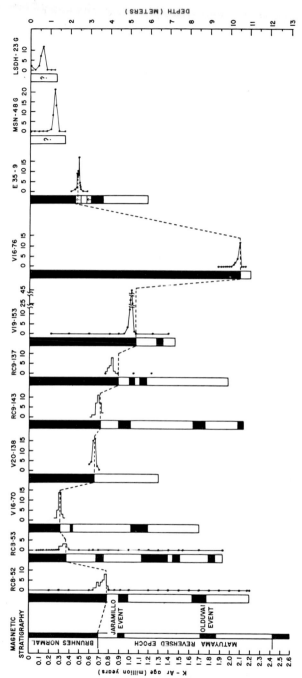

FIG. 10. Absolute abundance of glassy spherules (microtektites) from Australasian microtektite field, plotted against depth in core, showing mixing upward and downward from level of initial deposition at Brunhes/Matuyama palaeomagnetic boundary. (After Glass, 1972.)

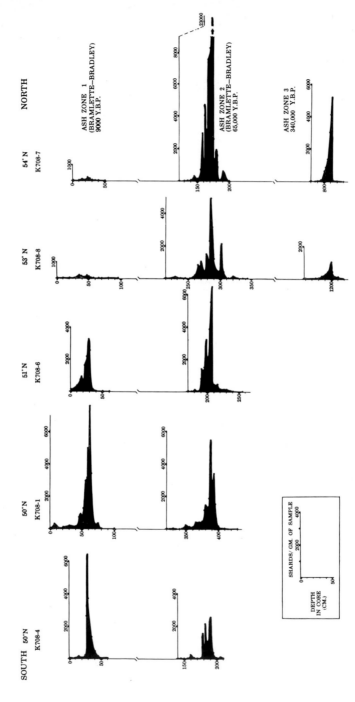

Fig. 11. Ash shard abundance plotted against depth in northeast Atlantic cores, showing mixing upward and downward from level of initial deposition, which is equivalent to horizon of present peak abundance. (From Ruddiman and Glover, 1972.)

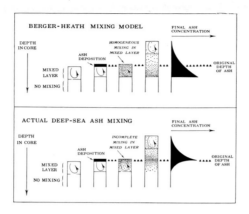

FIG. 12. Two possible vertical mixing profiles of "event" detritus: complex homogenisation of mixed-layer as predicted by Berger–Heath model, and actual profiles observed in most deep-sea sediments.

intense faunal episodes lasting 10–100 years, and to have been later reduced in concentration to roughly 1–10% of their original abundance by mixing. If such short episodes could leave even barely detectable spikes in the pelagic record, it was predicted that the peak abundances of episodes lasting more than 1000–2000 years should be left largely intact. These findings are more encouraging for palaeoclimatic resolution in high sedimentation rate areas than those inferred from Glass (1969) in slowly accumulated sediments.

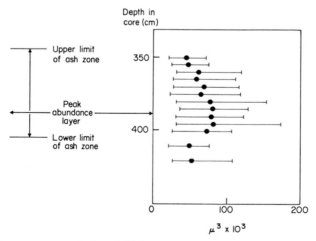

FIG. 13. Average measured size of silicic shards from a northeast Atlantic core plotted as product of maximum and median diameters in $10^3\mu^3$, ●, mean; ↔, range. (After Ruddiman and Glover, 1972.)

There is an inherent upper limit in the use of mineralogical analogues to faunal mixing. Normal ash layers more than one centimeter thick would presumably disrupt the depositional environment substantially and inhibit for a significant interval the normal feeding activities of burrowing infauna, causing underestimates of true mixing.

Mixing inferred from actual faunal data

Under optimal conditions, intense faunal episodes of shoit duration may also set limits on the maximum possible degree of vertical mixing.

McIntyre and others (1972a) and Ruddiman and McIntyre (in press) have found evidence of a very short, intensely cold climatic episode at roughly 195 000 years B.P. which they call carbonate minimum 5. The particulate material used to detect mixing around this climatic episode in two North Atlantic cores (Fig. 14) consists of the polar foraminifer *Globigerina pachyderma* Ehrenberg (left-coiling). Since there is virtually a 100% change in polar faunal abundance from almost zero at the episode margins to 90% at its peak development, it is certain that this palaeoclimatic pulse has retained at least 90% of its original peak concentration despite mixing. Sedimentation rates are known to be linear in these cores within the limits of sensitivity of the dating techniques, and the faunal shift in Fig. 14 is thus plotted against both depth in the core and against a time scale set by the prevailing sedimentation rate. The level of peak polar faunal abundance is set at a reference level of 0 years.

The shape of the original faunal episode is not known, but two extreme configurations can be postulated. The present form of the faunal curve represents one extreme based on the unlikely assumption that there has been no vertical mixing since deposition; this "unmixed" configuration takes the shape of a smooth curve spread across some 9000–10 000 years of time. To reconstruct the other extreme, it could be assumed that there was an instantaneous increase from zero to 100% polar fauna across the two sharp episode boundaries (as shown schematically in Fig. 14a). Assuming constant sedimentation rates, these hypothetical boundaries can be placed roughly at the levels for which the plotted areas $A_1 = A_2$ and $B_1 = B_2$, equivalent to compressing all cold episode material into a single interval. In the second extreme assumption, the faunal episode lasted only 4200 years in both cores (Figs 14b and 14c). From these two extreme limiting possibilities, it can be concluded that a climatic episode of between 4200 and 9500 years duration would be left in the sedimentary record after mixing at a minimum of 90% of its original peak intensity (88% and 91% in the cores in Figs 14b and 14c). It is also possible that there was no loss of peak intensity (that is, that the polar fauna originally reached only 90%).

Ericson and Wollin (1956) have detected brief changes in coiling direction of *Globorotalia truncatulinoides* d'Orbigny from right- to left-coiling in equatorial Atlantic sediments (Fig. 15). These changes are spread across 25–35 cm intervals in cores with sedimentation rates averaging 3 to 4 cm/1000

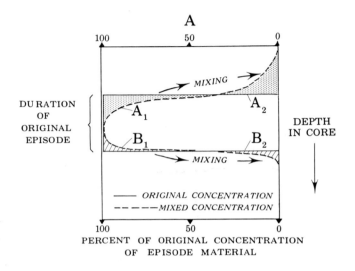

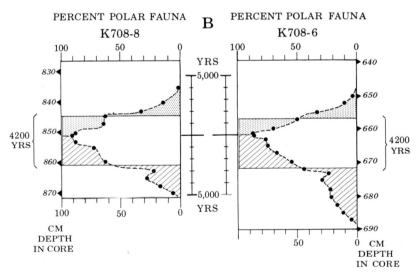

FIG. 14. Mixing of short faunal episodes in two northeast Atlantic cores. Episode duration was between 4200 years B.P. and 10 000 years, with loss of peak episode intensity of at most 10%. (From Ruddiman and McIntyre, 1976).

years, equivalent to time intervals of just under 10 000 years. Like the faunal peak in Fig. 14, they can be used as a means of setting limits on the mixing of short faunal episodes. The maximum reduction of peak abundance by vertical mixing may have been higher in the equatorial cores, since the peak left-coiling percentages average 70 % (ranging from 53–81 %); however, closer

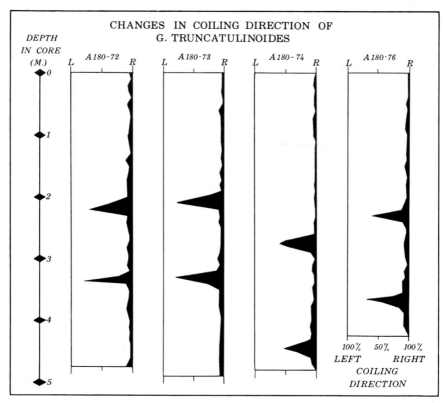

FIG. 15. Short (<10 000 year) episodes of changed coiling direction in equatorial Atlantic cores which set limits on the maximum possible loss of peak intensity of short faunal episodes. (After Ericson and Wollin, 1956.)

sampling intervals would probably result in increased peak percentages and thus reduce the estimated maximum loss of peak intensity.

From the two lines of evidence presented, it is clear that only deep-sea cores with sedimentation rates higher than 3 cm/1000 years are likely to reproduce accurately the peak intensity of palaeoclimatic episodes with a duration of as little as 5000 to 10 000 years. Although the long term (50 000 to 100 000 years) glacial/interglacial cycles can be grossly detected at much lower sedimentation rates, climatic episodes of short duration produce distinctive variation in the shape of the longer cycle curves and provide a criterion for correlating curves which are constructed for different regions. For much of the Quaternary record the stratigraphy of deep-sea cores is worked out by matching the shapes of faunal abundance curves. It is also possible that despite their brevity, intense climatic episodes of short duration recorded for deep-sea sediments may possibly correlate with events recorded in continental sediments.

For these reasons, areas of the deep sea with sedimentation rates higher than 3 cm/1000 years are mapped in Fig. 16 as regions likely to be optimal for detailed palaeoclimatic resolution. In such areas, one glacial or inter-glacial cycle should be distinguishable from another simply on the basis of shape, and even episodes lasting only 10^4 years should be accurately recorded.

Although hampered by a lesser knowledge of the effects of mixing on climatic resolution in the intermediate ranges of sedimentation rates, the author has selected 1 cm/1000 years as the probable lower limit of accumulation rates for which accurate and detailed palaeontology at the scale of 10 000-year episodes should be feasible. Areas of 1–3 cm/1000 years (accep-table) and < 1 cm/1000 years (unacceptable) are accordingly mapped in Fig. 16.

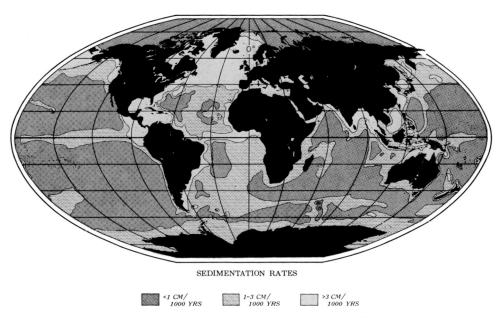

SEDIMENTATION RATES

<1 CM/ 1000 YRS   1-3 CM/ 1000 YRS   >3 CM/ 1000 YRS

FIG. 16. World map of late Quaternary sedimentation rates. (Primarily after Lisitzin, 1972.)

Most Indian and Pacific Ocean sediments are, by this criterion, unsuitable for detailed Quaternary palaeoclimatology. Some of the Indo-Pacific sediment areas mapped as having optimal or acceptable depositional rates are actually regions of rapid turbidite accumulations and contain negligible pelagic microfossil records. By comparison, much of the Atlantic and its surrounding seas is optimal, and most is suitable, for sensitive Quaternary palaeoclimatic resolution.

SEDIMENTATION PROCESSES

Downslope and across-slope sedimentation processes directly disrupt the pelagic palaeoclimatic record by both erosional and depositional interference.

Turbidites

The disruption due to downslope sedimentation can be additive (turbidite layers sandwiched between pelagic layers), subtractive (turbidites eroding underlying pelagic layers), or both at once. Channels and proximal portions of fans and abyssal plains show the most chaotic subtractive effects, whereas distal fan and abyssal plain regions tend to be largely additive, leaving the possibility that the pelagic record between turbidite layers is largely intact (Ericson *et al.*, 1961; Belderson and Laughton, 1966; Bornhold and Pilkey, 1971).

Since accurate diagnoses of the nature of turbidite effects on pelagic deposition can be made for only a few portions of the world's oceans, the author has simply mapped all abyssal plains, fans, and major channels as sites of likely alteration of the pelagic record (Fig. 17). The Ganges and Indus Cones in the Indian Ocean are the largest turbidite provinces in the world's oceans, but considerable portions of the deepest Atlantic Basins are also included. Not included are any indications of local downslope (ponded) flow; the many small regions subject to this process are usually adjoined by larger unaffected areas, so that the region as a whole is accessible to the palaeoclimatologist.

Bottom currents

Prevalent scour effects (gaps in the stratigraphic record, size sorting of fauna, accelerated solution due to long exposure time) associated with thermohaline-driven bottom-current circulation have been mapped in two large regions of the deep sea. Cores in a large area midway between Australia and Antarctica were shown to have major gaps in sedimentation because of prevalent regional erosion or nondeposition (Watkins and Kennett, 1971). Hollister and Heezen (1967) mapped an extensive region of severe scour effects in the deep passage between South America and Antarctica and in the Argentine Basin. These two areas (shown in Fig. 17) are considered marginally suitable

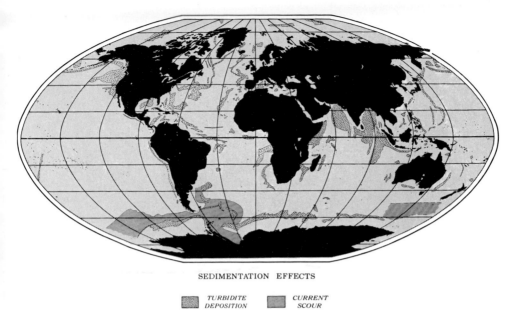

SEDIMENTATION EFFECTS

TURBIDITE DEPOSITION   *CURRENT SCOUR*

FIG. 17. World map of areas affected by turbidites and severe bottom current scour. (After Heezen and Hollister, 1971; Watkins and Kennett, 1971; and other sources.)

for pelagic analysis, since local pockets of relatively unaffected sediments may exist within the prevailing erosive regime. Other regions of the world's oceans are insufficiently studied to attempt judgments on their suitability by this criterion, but more detailed work obviously will increase the areal extent of current-altered stratigraphic records.

OPTIMAL DEPOSITIONAL REGIONS FOR
PALAEOCLIMATIC RECONSTRUCTION

By combining Figs 6, 16 and 17, we have arrived at a map of the world's oceans which roughly evaluates the regional possibilities of establishing detailed foraminiferal palaeoclimatic sequences (Fig. 18). Areas shown as optimal are those with no likely differential solution, turbidites, or bottom current scour, and with sedimentation rates in excess of 3 cm/1000 years. Unsuitable areas are those with either sedimentation rates below 1 cm/1000 years, carbonate percentages of less than 10%, turbidite deposition, or some combination of the three. Marginal to suitable areas are those which fail to meet one or more of the optimal criteria but are not unsuitable by any criterion. Optimal areas include large parts of the North Atlantic basin and

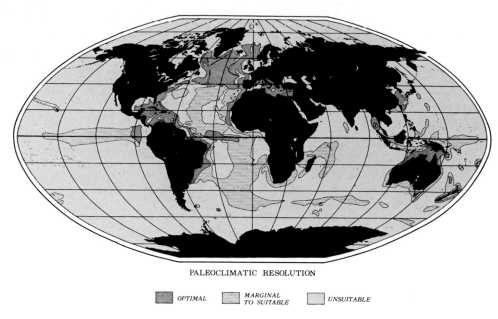

PALEOCLIMATIC RESOLUTION

OPTIMAL   MARGINAL
                 TO SUITABLE   UNSUITABLE

FIG. 18. World map of regions considered optimal, marginally suitable, and unsuitable
for accurate and detailed Quaternary foraminiferal palaeoclimatology; compiled from
Figs 6, 16, and 17.

surrounding seas, lesser parts of the South Atlantic, and local elevations in
the Indian and Pacific Oceans.

Studies of Quaternary palaeoclimatology

APPROACHES TO PALAEOCLIMATIC RECONSTRUCTION

Foraminiferal counts from Quaternary cores can be processed in several ways
for palaeoclimatic reconstruction. Plotting the estimated or actual abundance
changes with depth in a core of a species, or in the ratio of several species, is
the simplest form of data presentation, but by definition fails to take into
account the entire foraminiferal assemblage (Schott, 1935; Ericson and
Wollin, 1956; 1968; Bandy, 1960; Lidz, 1966; Frerichs, 1968; Blanc-Vernet
et al., 1969; Lisitzin, 1971). A more common presentation is to combine all
species which are typical of a modern set of environmental conditions into a
diagnostic assemblage and present in a composite plot the simultaneous
changes in abundances of all such assemblages (Todd, 1957; Parker, 1958;

Kennett, 1970; Barash, 1971; Thiede, 1971; McIntyre *et al.*, 1972a; Keany and Kennett, 1972; Kennett and Huddleston, 1972a). This procedure produces several partially interdependent curves, each of which represents a portion of the total palaeoclimatic information.

Phleger *et al.* (1953) presented single idealised palaeoclimatic curves and considered each counted sample as indicative of mid-, high-, or low- latitudes, but they gave no quantitative criteria for their choices. Ruddiman (1971) combined warm and cold foraminiferal species into a single palaeoclimatic curve, but did not make gradational distinctions within the warm and cold faunal groups among species whose temperature preferences are known to vary in the modern oceans. McIntyre *et al.* (1972a) used present-day foraminiferal (and coccolith) environmental preferences to reconstruct past geographic movements of what might be called "ecological water masses", with the latitudinal extent of the environmental shifts serving as a quantitative measure of palaeoclimatic change.

All of the above methods have the same limitation: a fundamental reliance upon the micropalaeontologist to choose diagnostic species or to ascertain significant species groupings (assemblages). As an objective means of making such decisions, the techniques of factor and vector analysis quantify the degree of correlation between species abundance trends either areally or with core depth and select the diagnostic species whose abundances vary most strikingly. This approach has been attempted in the Indian Ocean by Oba (1969), in the Tongue of the Ocean by Lynts (1971), in the South Pacific by Blackman and Somayajulu (1966), and in the Mediterranean by Ryan (1972). The greatest limitation to these methods is that while they select the diagnostic species or assemblage groupings to which other species are quantitatively related, they neither specify an ecological rationale for these choices nor provide a means of quantitatively transforming assemblages into palaeoenvironmental estimates. The environmental controls can only be surmised by subjectively comparing the known present-day temperature, salinity and productivity preferences of the species, as well as their susceptibility to differential solution, against their modern (surface-sediment) distribution.

The major problem inherent in using living foraminiferal species as standards of reference for Quaternary palaeoclimatology is the interrelationship of temperature, salinity and productivity in the oceans. All of these factors in combination have produced the Quaternary assemblages now left in the sedimentary record. Unfortunately, low salinities, low temperatures, and high productivity at present invariably occur together, except in the juxtaposition of infertile saline subtropical central waters against somewhat more productive and less saline equatorial waters. As a result of the prevalent covariation of these factors, it is difficult to extract from deep-sea faunal assemblages in surface sediments independent estimates of palaeotemperature, palaeosalinity, or palaeoproductivity.

The palaeoenvironmental technique of Imbrie and Kipp (1971) is designed to overcome this problem with a multivariate approach that ultimately reconstructs a palaeoestimate of the individual environmental parameters by using surface-sediment assemblages and modern atlas data as a predictive data base. The Imbrie–Kipp method specifies absolute values of each palaeo-estimate from any Quaternary assemblage composed of species existing at present. This technique has recently been applied by Sanchetta et al. (1972, 1973); Kellog (1973); Luz (1973); Imbrie et al. (1973); Gardner and Hays (1976); and Prell and Hays (1976).

From a theoretical standpoint, the Imbrie–Kipp method is a more comprehensive technique than any other conceived of to-date. This technique, or any later improvements or variations, will be subject to the following constraints:

(i) Until a sufficient number of deep, differentially dissolved surface sediment samples (preferentially in different stages of dissolution) are introduced into the core-top (modern assemblage) data base, differential solution cannot be considered to have been quantified as a factor.

(ii) For species which have different ecological preferences in different geographical regions of the modern oceans, various areas, if introduced into the core-top data base, might substantially alter the absolute values of the palaeoenvironmental estimates. This could be the case, for example, with several foraminifers common to both the North Atlantic and circum-Antarctic oceans; these species appear to favour much cooler waters in the Southern Ocean than in the northern hemisphere (Bé, 1969). This possible discrepancy could create an important subjective decision about whether the data base should include surface samples from throughout the world's oceans or strictly from the region being studied. This problem can be tested by using core-top faunas and accompanying environmental data from different regions or combinations of regions as a data base for projecting palaeoestimates on the down-core faunas; any differences in palaeoestimates would reflect different ecological preferences of the modern fauna in the different oceans. In addition, the communalities for the estimates on each down-core sample give some idea of the "appropriateness" of the chosen faunal-environmental model (and data base) in explaining past variations.

(iii) Environments (or, alternatively, ecologically controlling variables) which uniquely existed in the Pleistocene and which are therefore not represented in the modern data base may be impossible to reconstruct by extrapolation from the closest modern analogues; again, communalities aid in the detection of such differences. In general, there are greater dangers in extra-polating beyond environments represented in the modern data base (extremely high salinities, for example) than in interpolating new combinations within the limits of the modern data base.

(*iv*) No model can completely distinguish between a "mixed" fauna due to rapid seasonal succession of water masses, one due to a depth stratification of coexisting species assemblages in the water column, one due to short-term (10–100 year) secular climatic variations, and one due to actual bioturbation. To some extent, this problem can be attacked by comparing concordant palaeoenvironmental estimates from different faunal or floral planktonic groups which may inhabit different parts of the water column and thus would best reproduce estimates at that level.

(*v*) The modern data base contains a large element of time heterogeneity. Atlas data spanning roughly 50 years are the environmental calibration set; core-top faunal assemblages spanning 500 to 5000 years are the biotic calibra-tion set. If the mean oceanic response is grossly different across these various time intervals, the calibrations will be wrong. This is potentially the greatest single source of error. It could only be overcome by collecting vast numbers of plankton samples in wide areas, covering all seasons, and at many depth levels, accompanied by whatever environmental measurements might be deemed requisite.

Because of the above factors, it should be kept in mind that precise recon-struction of palaeotemperatures and other environmental parameters is not possible by any palaeoenvironmental technique. The palaeoestimates generated by the Imbrie–Kipp method are at present the best approxima-tions. If future studies can be quantified by the Imbrie–Kipp approach, they will continue to provide the globally consistent view of Quaternary palaeo-climatic change important to numerical models of climatic variation.

REGIONAL STUDIES

In order to simplify comparison of Quaternary palaeoclimatic changes from different regions of the world's oceans, it is first necessary to comment on the existing time scales, on the stratigraphical use of oxygen isotope curves, and on problems of nomenclature.

In the deep-sea, palaeomagnetic stratigraphy provides only a crude stratigraphical control for the Quaternary, with reversal dates ar toughly 0.69, 0.89, 0.95, 1.61 and 1.79 M.Y.B.P. (Cox, 1969). Unfortunately, there is substantial disagreement in the late Quaternary, with the two most promi-nent time scales differing by about 25% (Rona and Emiliani, 1969; Broecker and Ku, 1969). Two arguments favour the Broeckcr–Ku scale: (*i*) the diag-nostic very warm climate found in most deep-sea curves at the base of the last interglacial section falls, according to their scheme, at 120 000 years B.P., which correlates with the best-dated Quaternary high sea level stand (McIntyre *et al.*, 1972a; Ku *et al.*, 1972); (*ii*) linear extrapolation backward from the Brunhes–Matuyama boundary at 0.69 M.Y.B.P. to 0 years B.P. at the core tops supports the Broecker–Ku time scale for the last climatic cycle (Glass *et al.*, 1967; Ericson and Wollin, 1968; Shackleton and Opdyke, 1973).

Since the Broecker–Ku time scale was established at the prominent boundaries of long-term (10^5; years) climatic cycles, it is first necessary in any region to ascertain the depth-in-core of these fundamental cycle boundaries. Although the long-term cycles are often visible in faunal, floral, or lithologic data, because of local climatic effects these three parameters do not always record the critical cycles in some regions. It is thus preferable to look to a parameter that offers the possibility of synchronous global correlations. In most areas, the long-term cycles can be easily recognised from oxygen isotope curves (Broecker and van Donk, 1970).

Oxygen isotope variations are significantly controlled by preferential storage of O^{16} in continental ice sheets, leaving excess O^{18} in sea water during glaciations; this establishes cyclic variations which are in effect globally synchronous across a wide range of local oceanic environments (Shackleton, 1967). Because planktonic foraminiferal isotopic ratios are significantly affected by surface water temperature variations, Shackleton and Opdyke (1973) have recommended that benthonic species be used because of the minimal temperature-caused isotopic variations in deep waters. Thus isotopic curves can be used as a stratigraphical tool, and not just as a local palaeothermometer. Foraminiferal variations of interest to this study then can be compared and contrasted in most areas with the sequence of isotopic cycles dated by the Broecker–Ku time scale.

With a duplicated sequence of faunal curves established for each oceanic region, it is necessary to devise some kind of terminology to describe the stratigraphical succession. Duplication of form variations among cycles is necessary to overcome small-scale stratigraphical irregularities, such as the small hiatuses so common in deep-sea cores. A sequential numbering of the major deep-sea cycles is obviously advisable as a shorthand reference, but should be treated informally and cautiously until substantial duplication of all portions of the record is available to formulate standard sequences.

It has proven possible in several of the optimal to marginally suitable areas of the world's oceans shown in Fig. 18 to extract by duplication from a suite of cores a standard foraminiferal palaeoclimatic section.

Antarctic and Arctic Oceans

Kennett (1970) reported on 11 ELTANIN cores from 52–57°S latitude in the

Southern Ocean near Antarctica, and was able to define at least six major cycles of climatic warming and cooling within the Brunhes Epoch. Because of low sedimentation rates (averaging about 1 cm/1000 years) and wide sampling intervals (50 cm), the chronologic spacing of samples is roughly 50 000 years; consequently, only the approximate duration and intensity of the longest cycles can be distinguished. Keany and Kennett (1972) extended the study through the Matuyama Epoch, sampling at closer core intervals (20 cm) but effectively the same mean time interval (50 000 years). Solution, hiatuses, barren zones, and changes in sedimentation rates further hamper Antarctic palaeoclimatic reconstructions, despite the availability of numerous cores spanning the Quaternary. Considerably more detailed future study of optimal cores in circum-Antarctic seas may lead to the development of a better climatic resolution of the longer glacial cycles and of high-latitude southern hemisphere water mass movements.

Arctic Ocean foraminifera have been studied by Hermann (1969, 1970), Hunkins et al. (1971), and Clark (1969, 1971). As in the Antarctic, low sedimentation rates (averaging 0.2 cm/1000 years) and wide sample spacing in time (greater than 30 000 years) have precluded detailed resolution of time in Artic cores. Because of several taxonomic disagreements among these studies, there is still significant uncertainty about the degree of Arctic ice cover through the Quaternary. Variations in foraminiferal percentages do not bear directly on this problem, because of the overwhelming dominance of *Globigerina pachyderma* at all levels. Minor variations through time in the percentage of planktonic foraminifera as a fraction of the sediment have been attributed either to changing thicknesses of the pack-ice cover (Clark, 1971; Hunkins et al., 1971) or to alternating ice-free and ice-covered conditions (Hermann, 1970).

Mediterranean

The Mediterranean Sea is uniquely situated for monitoring certain glacial phenomena. Despite being enclosed by land, through most of the Quaternary it has supported foraminiferal assemblages similar to those of an open ocean environment. Mean sedimentation rates of 5 cm/1000 years, little or no solution at the sea floor, and abundant lithological time markers (ash falls and sapropelitic muds) offer the possibility of extraordinary climatic resolution; unfortunately, numerous small hiatuses occur. Only one Mediterranean core has been examined for oxygen isotopic variations (Emiliani, 1955b). The Mediterranean is a volatile environment for isotopic changes, since great influxes of cold meltwater from Alpine glaciers will radically alter the isotopic balance (Olausson, 1961). Long-term faunal cycles which appear to be time-correlative with those in other oceans (Ryan, 1972) are evident.

Todd (1957) and Parker (1958) found that foraminiferal assemblages in long Mediterranean cores alternated through two or three cycles of low-latitude ("warm") foraminifers and high-latitude ("cold") species. Radio-carbon dating of the upper layers suggested that these cycles lasted 10^5 years and equated with glacial-interglacial climatic changes. They noted, as had Kullenberg (1952), anomalous floods of the species *Globigerina eggeri* Rhumbler (now referred to *Globoquadrina dutertrei* d'Orbigny) at or near the levels of sapropelitic (stagnant) muds. Bradley (1938) and Kullenberg (1952) hypothesised that stagnant conditions would prevail during glacial maxima because of the stable stratification of low-salinity and low-density surface waters resulting from heavy precipitation, lessened evaporation, and generally reduced exchange with Atlantic waters across the shallower Gibraltar sill. Later studies (Mellis, 1954; Rubin and Suess, 1955) revealed, however, that the most recent stagnation began 9000 years ago late in an interval of world-wide warming from glacial to interglacial conditions.

Olausson (1961) determined from cross-correlation of numerous ALBA-TROSS cores, many of which contained perceptible hiatuses, that several sapropels were deposited shortly after major faunal cold intervals; he attributed this to incursions of Alpine glacial meltwater from the north or from the Black Sea. Ryan (1972) concluded from ALBATROSS and CONRAD cores that all stagnations are associated with the warming trends occurring just before periods of major high sea levels. This conclusion was achieved in part by successfully transferring the Broecker and Ku (1969) chronology to the Mediterranean record, and it confirms Olausson's suggestion that cold brackish meltwater from the Black Sea periodically blankets the Mediterranean, favouring apparent low-salinity species like *Globoquadrina dutertrei* at the expense of the rest of the foraminiferal population. These anomalous faunas thus define intervals of stable stratification of Mediterranean surface waters during which the lack of oxygen supply to bottom waters led to stagnation.

Any palaeoecological analysis of Quaternary Mediterranean climates will probably run into difficulty with the faunas found near and in the sapropelitic muds. This appears to be an unusually distinct case in which temperature is a less important factor than either salinity or some productivity-related effect; during times of apparent stagnation, faunas are temporarily over-whelmed by other palaeoenvironmental factors, and may have to be con-sidered separately from the rest of the record in order to achieve reasonable climate reconstructions.

Since the foraminifera during the predominantly nonstagnant sedimenta-tion resemble a normal oceanic suite, they probably contain retrievable palaeotemperature information. Ryan (1972) ran a factor analysis of the

Parker counts and isolated three end-member groups, which he inferred to be controlled respectively by temperature, salinity, and surface water mixing and productivity. He generated a family of curves showing the quantitative similarity of all samples to the reference sample inferred to represent extreme cold, using the oblique projection technique. As discussed previously, the major limitation of this kind of technique lies in the subjective decision that a particular reference sample defines a specific environmental factor. It appears, however, that over the last two or three climatic cycles the Mediterranean temperatures varied in phase with northern hemisphere glaciations.

It will be interesting to test the ability of the Imbrie–Kipp technique to extrapolate from known open-ocean environments and faunas to the different, somewhat anomalous conditions existing in the past in the enclosed Mediterranean basin.

Caribbean

Three studies have been made of Caribbean planktonic foraminiferal variations in cores with duplicated and correlative stratigraphies (Lidz, 1966; Imbrie and Kipp, 1971; Prell and Hays, 1976). Correlations between the forms of isotopic curves (Emiliani, 1966; Broecker and van Donk, 1970), *Globorotalia menardii* abundance zones (Ericson and Wollin, 1968), and coiling direction changes in *Globorotalia truncatulinoides* (Emiliani, 1966) provide the necessary stratigraphy. Rapid deposition rates (averaging 2.5 cm/1000 years) and the absence of significant solution permit the accumulation of detailed palaeoclimatic sequences.

Although there is no question that both Lidz (1966) and Imbrie and Kipp (1971) studied cores with correlative stratigraphies, there are considerable discrepancies between their basic species counts. Unfortunately, Lidz did not specify the sieve size of his samples, so it is not possible to determine whether the differences are taxonomic, mechanical, or environmental.

Imbrie and Kipp (1971) derived palaeotemperature curves for both summer and winter in VEMA core V12-122, as well as a mean annual palaeosalinity trend (Fig. 19). They noted a moderately good correlation between temperature curves and the oxygen isotope curve from the same core back to 150 000 years B.P. on the Broecker–Ku time scale, but marked divergence frequently occurred below this level, often to the point of opposite phasing. They also noted secular trends in the palaeotemperature curve, including a fundamental displacement of the curve to a cooler temperature range at levels prior to 380 000 years B.P. These long-term trends were not found in the isotopic record, which varied between roughly constant extremes. Because of the partially global control over isotopic curves (moraines give evidence of ice sheets having repeatedly achieved roughly the same maximum and minimum

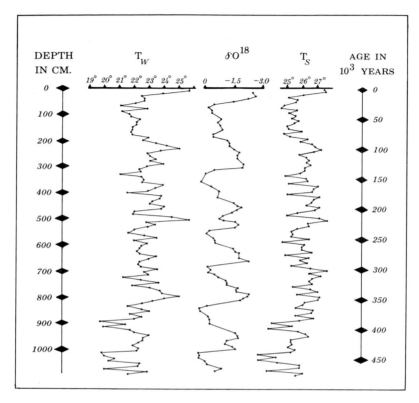

FIG. 19. Comparison of summer (T_S) and winter (T_W) palaeotemperature estimates of Imbrie and Kipp (1971) with oxygen isotope variations from Broecker and van Donk (1970), both from core V12–122.

sizes) it is not surprising that the locally controlled Caribbean palaeotemperature curves show different trends.

One interesting observation which can be made from the Imbrie–Kipp study was that the semiquantitative *Globorotalia menardii* curves in core V12–122 from Ericson and Wollin (1968) correlated better with the oxygen isotope variations than did the Imbrie–Kipp summer and winter (Ts and Tw) palaeotemperature estimates. The calculated temperature variations, however, were judged not to have been sufficiently large to explain the *G. menardii* abundance variations. By inference from North Atlantic surface sediment distributions, Imbrie and Kipp (1971) concluded that percentage abundance variations of *G. menardii* and other members of the "gyre margin" assemblage are controlled, at present, primarily by productivity and salinity. This conclusion had earlier been reached by Ruddiman (1969), who defined a "peripheral" suite of foraminifera with low concentrations in the highly saline Atlantic central waters.

Abundant *G. menardii* during isotopically defined interglacials was attributed to high productivity or moderate salinity, and its scarcity during glacials to low productivity or high salinity. It was inferred that glacial stages in the Caribbean were characterised (as are present winters) by relatively unproductive surface waters mainly of North Atlantic origin, whereas interglacial surface waters were largely of South Atlantic origin and were relatively productive.

Prell and Hays (1976) confirmed that there was a very small cooling (2°C in winter) relative to the present in the Caribbean during the last glacial maximum (18 000 years B.P.) and also supported the Imbrie–Kipp suggestion of unproductive, high-salinity glacial surface waters. They noted that the very high percentages of *Globigerinoides ruber* in the glacial Caribbean resemble the modern Sargasso fauna and suggest analogous conditions.

If the Imbrie–Kipp model of Caribbean glacial/interglacial circulation is correct, it can be surmised that although the *G. menardii* variations in the Caribbean are indirectly good indicators of the net extent of Northern Hemisphere glaciations, they ironically have little or nothing to do with actual Caribbean temperature variations. Such an explanation will apply with equal validity to the generally impressive correlation of species ratio curves of Lidz (1966) with the oxygen isotope trends of Emiliani (1966) in core P6304–8. Without exception, the species ratios have gyre margin species in the numerators and thus may primarily record salinity and productivity variations.

The palaeotemperature estimates of Imbrie and Kipp (1971); Imbrie *et al.* (1973); and Prell and Hays (1976), by contrast, appear to accurately define local Caribbean palaeotemperature variations, but give less indication of the sequence of distant northern hemisphere glaciations. The reason for the disparity between local Caribbean temperature cycles and hemispheric glaciations remains a significant problem.

Equatorial Atlantic

Foraminiferal cycles in the equatorial Atlantic have been reported by Phleger *et al.* (1953); Ericson and Wollin (1956, 1968); Ruddiman (1971); and Gardner and Hays (1976). The stratigraphic control is analogous to that used in the Caribbean, consisting of isotopic curves, *G. menardii* faunal zones and coiling direction changes of *G. truncatulinoides*. As discussed earlier, differential solution is particularly effective at depths below 4000 m, and only cores lying above that depth are optimal for palaeoclimatic work. Deposition rates at depths above 4000 m average more than 3 cm/1000 years, permitting detailed palaeoclimatic resolution.

Drawing in part upon data from Phleger *et al.* (1953), Ruddiman (1971) examined a suite of cores in the equatorial Atlantic. By lumping all species into either a "warm" or a "cold" assemblage, and combining these into a single plot, he produced nonquantitative late Quaternary palaeoclimatic curves which bear a striking resemblance (Fig. 20) to equatorial Atlantic and Caribbean oxygen isotope curves from Emiliani (1955a, 1966). There is, however, both a temperature and a salinity message in these faunal changes, as well as seasonal variations of both parameters, as shown recently by Gardner and Hays (1976).

Palaeoenvironmental estimates made by Gardner and Hays (1976) for the last glacial maximum (18 000 years B.P.) indicate that along the equator winter temperatures were only 1–2°C cooler than at present, while summer sea-surface temperatures were about 4°C cooler (Fig. 21). This predominantly summer cooling was attributed to the greater influence of the Benguela Current on the South Equatorial Current during the vigorous circulation of the southern hemisphere winter. Gardner and Hays also noted (Fig. 21) that in the region of upwelling off Africa there was a much stronger cooling during the northern hemisphere winter (10°C) than summer (1–4°C). This was attributed to more intense glacial circulation in the compressed subtropical gyre (McIntyre *et al.*, 1972a, b) with stronger advection of cool waters from the north in the Canaries Current.

Gardner and Hays note that the seasonal variations of the Intertropical Convergence Zone during the last glaciation matched the modern geographical limits closely; the primary glacial/interglacial difference they thus assumed to be one of intensity, with more vigorous glacial oceanic circulation. In particular, they noted the possibility of heightened southeast trade winds in the northern hemisphere winter.

Foraminiferal variations in the western longitudes of the equatorial Atlantic may have been mainly controlled by salinity and productivity since the glacial temperature decrease to the west was smaller. As in the Caribbean, variable northward influxes of South Atlantic water probably exerted an influential control.

Gulf of Mexico

Phleger (1951, 1955); Beard (1969); and Kennett and Huddlestun (1972a, b) have examined Quaternary foraminiferal changes in Gulf of Mexico cores. In the Gulf, volcanic ash layers, coiling direction changes, species extinctions, isotopic curves (Sackett and Rankin, 1970), and *G. menardii* abundance zones

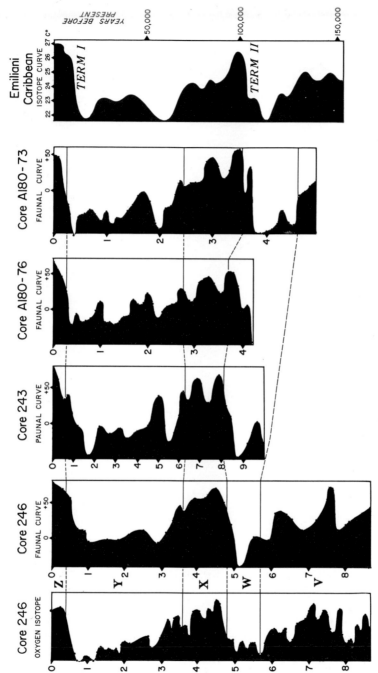

FIG. 20. Correlation of equatorial Atlantic faunal curves with oxygen isotope curves from Emiliani (1955a, 1966). (After Ruddiman, 1971.)

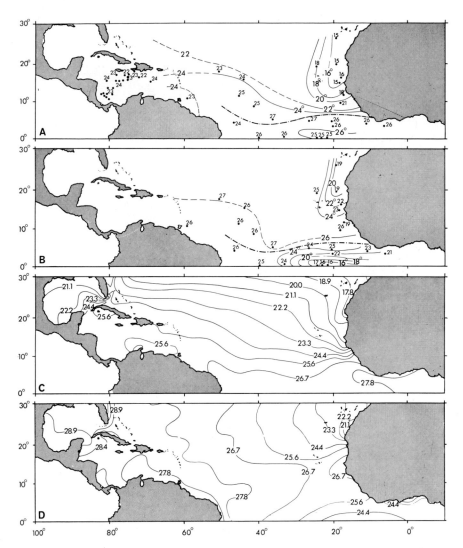

Fig. 21. Comparison of February (A) and August (B) sea-surface temperature estimates in °C for 18 000 years B.P. with present-day February (C) and August (D) observed values. (After Gardner and Hays, 1976.)

(Ewing *et al.*, 1958) provide a correlative framework. Because of the very high mean sedimentation rates (20 cm/1000 years), fine-scale resolution of very short climatic episodes is potentially feasible in the Gulf (Kennett and Huddlestun, 1972b). Using detailed stratigraphic control, Kennett and

Huddlestun (1972a) were able to examine the entire last glacial cycle with samples at time intervals of a few thousand years. They noted, however, that sedimentation rates are highly variable within individual cores because of numerous inconformities and erratic pulses of nonlinear deposition, which cause depositional rates to vary by more than an order of magnitude.

Kennett and Huddlestun (1972a) interpret Gulf of Mexico foraminiferal changes primarily in terms of palaeotemperature oscillations. Two factors suggest that palaeosalinity effects should be more seriously considered: 1, the large volumes of Laurentide glacial meltwater that drained into the Mississippi River during at least the early phases of climatic warmings; and 2, the likelihood that the Gulf of Mexico was a major source of warm moist air to feed rapidly growing glaciers, thus increasing Gulf salinity because of evaporation. Again, the Imbrie–Kipp technique should be applied to this diverse, warm water fauna to distinguish the effects of these two environmental parameters.

North Atlantic palaeoclimatology

The North Atlantic above 40°N latitude is ideal for detailed Quaternary palaeoclimatic interpretation (Bramlette and Bradley, 1941; McIntyre et al., 1972a). Its shallow depths contain sequences of virtually undissolved calcareous ooze. Because of productive surface waters and abundant nearby sources of ice-rafted detritus, the high average sedimentation rates (4–5 cm/1000 years) in the North Atlantic permit palaeoclimatic resolution of episodes as short as a few thousand years. Finally, the subpolar North Atlantic is flanked on three sides by land and water that were ice-covered during glaciations; its waters are thus an ideally situated palaeothermometer of ice fluctuations in the two major temperate-latitude ice sheets of the Quaternary (Laurentide and Scandinavian). The interlayering in North Atlantic cores of detritus eroded from the continents by ice sheets and calcareous oozes reflecting changing sea surface temperatures permits simultaneous monitoring of oceanic and continental responses to global climatic change.

In the following sections the results of three different, but related, approaches to palaeoclimatic reconstruction will be examined in this optimal region.

NORTH-EAST ATLANTIC LATITUDINAL
PALAEOCLIMATIC VARIATIONS

Following pioneering work by Bramlette and Bradley (1941) and Cushman and Henbest (1941), McIntyre et al. (1972a) traced Quaternary faunal variations and ecological water mass movements in Northeast Atlantic sediments.

Ruddiman and McIntyre (1976) have extended this study to 13 cores covering Quaternary sections of as much as 600 000 years.

Faunal assemblages

Ruddiman and McIntyre (1976) assigned foraminiferal species to five assemblages whose abundances can be used to define different ecological water masses (Table 1); counts of the five groups with depth in 13 cores in a transect across the Northeast Atlantic are shown in Fig. 22. Sample intervals average less than 5000 years, with closer control around intervals where scanning of additional samples showed rapidly changing conditions.

TABLE 1. Water mass designations of Northeast Atlantic foraminifera. s, sinistral coiling; d, dextrally coiled.

Water mass	Foraminiferal species
Polar	*Globigerina pachyderma* (s)
Subpolar	*Globigerina bulloides*
	Globigerina pachyderma (d)
	Globigerina quinqueloba
Transitional	*Globorotalia inflata*
	Globoquadrina dutertrei
Northern	*Globorotalia hirsuta*
	Globorotalia truncatulinoides
	Globorotalia scitula
Subtropical	*Globigerinoides ruber*
	Globigerinoides conglobatus
Southern	*Globigerina rubescens*
	Globorotalia crassaformis
(Cosmopolitan)	*Orbulina universa*
	Globigerinita glutinata

The faunal data from each core were plotted on a chronologic scale by alignment of the following reference points: 0 years B.P. for the core top; 13 500 years B.P. for the last deglacial warming called termination I; and 127 000 years B.P. for the next-to-last deglacial warming called termination II. Counts between these reference points were plotted by linear interpolation. Additional datum levels used to plot the longest cores are discussed by Ruddiman and McIntyre (1976). Despite minor core-to-core variations, this plotting scheme results in a north-to-south sequence of strikingly similar faunal trends.

The foraminiferal assemblage variations in Fig. 22 show eight major cycles (seven of which are complete) over the last 600 000 years. These oscillations range between the one extreme (glacial) of total dominance of polar fauna and the other (interglacial) extreme in which transitional and cold subtropical species may exceed 50% of the total. The cycles are not symmetrical, but generally show gradual increases in the percentage of polar species (coolings) followed by abrupt decreases (warmings). Recent application of the Imbrie–Kipp palaeoclimatic technique fully substantiates the use of polar foraminiferal percentages as the dominant indicator of temperature fluctuations in these cores (Sanchetta and others, 1972, 1973). The polar faunal changes in Fig. 22 can thus be referred to for palaeotemperature trends, permitting extremely detailed assessments of northern hemisphere glacial conditions.

Viewed in a lateral (geographical) sense, the most striking feature of the faunal variations in Fig. 22 is the separation of the 13 cores into two distinct groups. The nine cores taken north of 45°N contain very similar faunal records along synchronous horizons, with only slightly colder assemblages at the higher latitudes. The three cores taken south of 45°N (including two from Phleger et al., 1953) reveal similar but less extreme faunal oscillations and no dominance of the polar fauna. Core V23–84 is gradational between the two groups.

Long-term climatic cycles with lengths varying from 56 000 to 113 000 years are evident in the faunal data shown in Fig. 22. Excluding climatic pulses with durations of less than 20 000 years, we note seven complete major cycles in the last 605 000 years (Table 6) averaging 84 000 years in length. The same approximate number of major cycles has been noted in most previous studies of deep-sea sediments over comparable time intervals (Emiliani, 1966; Hays et al., 1969; Ruddiman, 1971; Kent et al., 1971; Imbrie and Kipp, 1971; Shackleton and Opdyke, 1973).

It is also possible to detect additional short but severe cold climatic pulses lasting less than 20 000 years. Most prominent among these are the brief cold pulses at roughly 195 000 years B.P. in the lower interglacial portion of major cycle C, and at 72 000 years B.P. in the middle of major cycle B (Fig. 22). There are analogous short warm climatic episodes which appear to represent significant deglaciations, particularly in the glacial portions of major cycles C and D.

If the numbering of oceanic glacial maxima is extended to encompass these shorter episodes, there were at least 11 prominent polar water advances in the last 600 000 years. Since there are significant amounts of coarse ice-rafted terrigenous debris at each of these maxima, we infer that continental glaciers

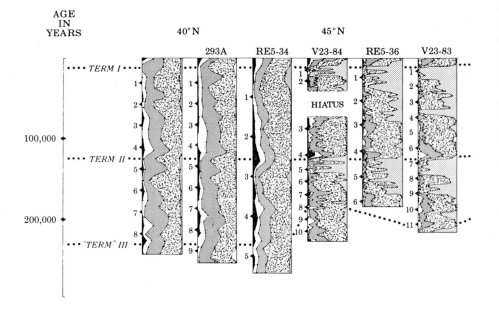

AGE IN YEARS

40°N

45°N

293A   RE5-34   V23-84   RE5-36   V23-83

TERM I

TERM II

TERM III

100,000

200,000

HIATUS

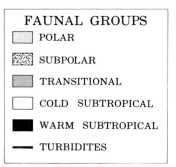

FAUNAL GROUPS

POLAR

SUBPOLAR

TRANSITIONAL

COLD SUBTROPICAL

WARM SUBTROPICAL

TURBIDITES

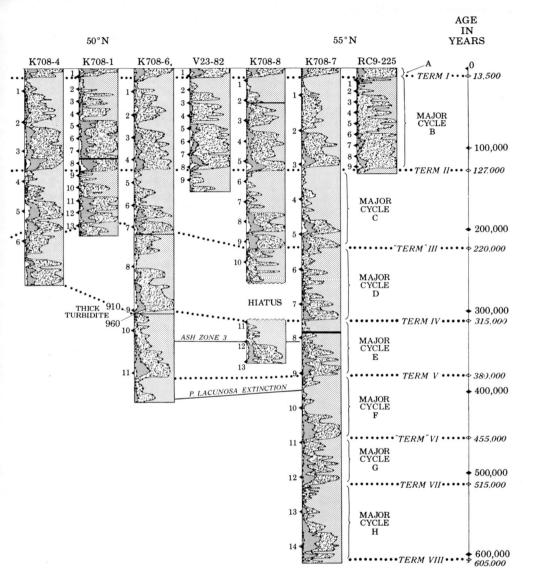

Fig. 22. Foraminiferal abundance curves for 13 cores along a north–south transect in the Northeast Atlantic. (From Ruddiman and McIntyre, 1976.)

may have reached almost full size even during the short cold climatic episodes. Within the full 1.2 M.Y. of extensive Pleistocene ice-rafting in mid-latitudes of the Northern hemisphere (Kent *et al.*, 1971), there may have been 20 or more significant continental ice sheet advances.

Latitudinal ecological water mass movements

The assemblages from the counted samples in Fig. 22 are, by analogy with present distributions, indicative of a set of past environmental conditions; such conditions can be termed an "ecological water mass". Since each sample characterises the ecological water mass overlying a given coring location for one brief interval of geological time, the combined pattern of samples down through many cores can be used to reconstruct the movement of water mass environments across the North Atlantic through the late Quaternary. Ruddiman and McIntyre (1976) assigned these counts to ecological water masses according to the following percentage limits: polar water = polar species $\geq 75\%$; subpolar water = polar species $< 75\%$; transitional species $< 35\%$; transitional water = transitional species $\geq 35\%$; cold subtropical species $< 15\%$; cold subtropical water = cold subtropical species $\geq 15\%$. In these latitudes the warm subtropical species never approach the dominance values (75%) necessary to designate any sample as a warm subtropical water mass indicator.

North–south water mass movements across the eastern North Atlantic reconstructed from foraminiferal assemblages in Fig. 22 are shown in Fig. 23. The most significant aspect of the migrations is the great lateral movement across more than 10° to 20° of latitude, as inferred by McIntyre (1967). The southernmost penetration of polar water conditions along the eastern Atlantic coast can be corroborated from the faunal counts of Thiede (1971), which show no indication of polar water at the latitudes of southern Portugal (37°N), and from Caralp (1967), who shows polar assemblages in the Bay of Biscay (44°N). Thus, polar water in the eastern North Atlantic penetrated southward to about 42°N during glacial maxima, and retreated to the 65°N latitude of Iceland during interglacials (McIntyre *et al.*, 1972a). This defines latitudinal displacements of cold faunal assemblages of more than 20° (2000 km), with the warm fauna moving more than 10° to 15° (transitional assemblages from below 40°N to above 55°N; the cold subtropical fauna from below 40°N to just above 50°N).

The rate of water mass movement can be very rapid. At latitudes between 45° and 50°, four distinct ecological water masses have frequently moved in rapid succession back and forth across the area during the Pleistocene. During the abrupt deglacial warming designated termination II, the four water masses moved successively in a cold-to-warm, south-to-north geographical retreat of 15° to 20° of latitude compressed into less than 5000 years. This implies rates of ecological water mass migration averaging more than 150 to 200 m/year, which continued uninterrupted for thousands of years; short-term rates may have been substantially larger. Measurements of continental glacier movement on land, by comparison, indicate advance and retreat rates on the order of 10 to 500 m/year (Goldthwait, 1958; Willman and Frye, 1970; Kempton and Gross, 1971).

Ecological water mass can also be used to infer the intensity of palaeo-temperature changes of North Atlantic surface waters. At present, the polar front, which defines the southern limit of polar water, is marked by the 0°C winter isotherm and the 6°C summer isotherm. Since the 45°N polar front position during full glaciations overlies the modern 12.5°C winter and 19°C summer isotherms, it can be surmised that glacial surface water temperatures were lower by 12.5°C in the winter and 13°C in the summer than at present. Using transfer function techniques on faunas in maximum Wisconsin samples, McIntyre *et al.* (1972b) suggested a 12°C glacial winter cooling in this area relative to the present.

During peak interglacial warmth at 120 000 years B.P., the polar front was displaced north at least to the present positions of the 2°C summer and −1°C winter isotherms. Such an offset implies conditions warmer than those at present by 3°C in the winter and 2°C in the summer. On a scale ranging between the two Quaternary climatic extremes of maximum glacial cold and maximum interglacial warmth, the present surface-sediment conditions (actually a mix of the last several thousand years) lie 75–80% toward the maximum warmth seen in the Quaternary record.

PALAEOCIRCULATION TRENDS AT 9300 YEARS B.P.

Since modern circulation in the subpolar North Atlantic intrudes across lines of latitude and creates east–west thermal and salinity gradients, it would be an over-simplification to analyse North Atlantic water mass migrations and palaeocirculation trends only in terms of north–south displacements. Several lines of evidence defining circulation trends at 9300 years B.P. also emphasise the significance of meridional (north–south) circulation and of zonal (east–west) migrations of ecological water masses relative to their modern positions.

Faunal assemblages at 9300 years B.P.

Ruddiman and Glover (1972) have traced two zones of silicic volcanic ash through a large area of the North Atlantic cores (Fig. 24). Ruddiman and Glover (1975) used the peak abundance of the upper ash zone as a datum plane for synchronous reconstruction of North Atlantic palaeocirculation

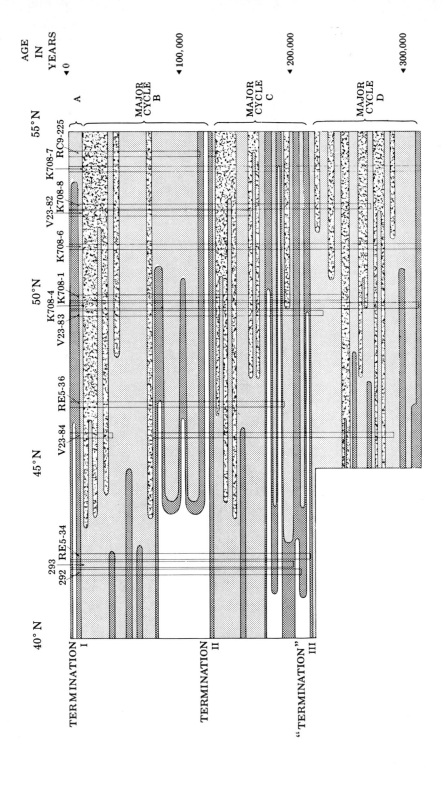

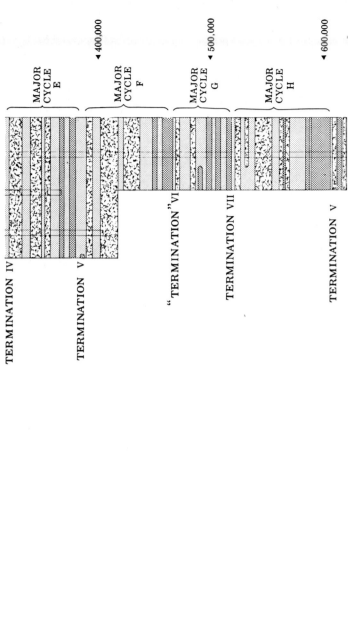

FIG. 23. Water mass movements in the Northeast Atlantic, based on faunal counts shown in Fig. 21. (From Ruddiman and McIntyre, 1976.)

trends. Deposition of the upper ash zone occurred at 9300 years B.P. (Ruddiman and McIntyre, 1973). This was a time of rapid transition from glacial to
interglacial conditions, with the temperate-latitude glaciers in rapid retreat.
The section of sediments spanning the time of ash zone deposition has been
studied in a total of 42 cores (Fig. 24).

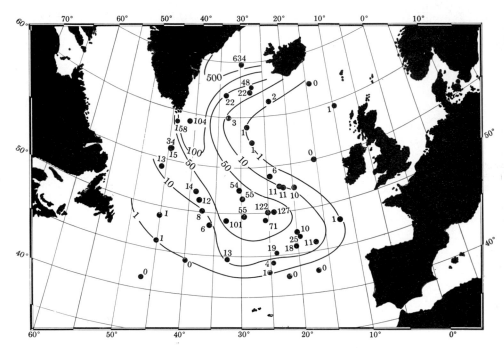

FIG. 24. Abundance contours of 9300-year ash (in 10^3 sand-sized shards). (From Ruddiman and Glover, 1975.)

Physical constraints (sedimentation rate, sample thickness and vertical
mixing) dictate that faunal assemblages counted at the 9300-year datum level
contain mixed records spanning 10^2 to 10^3 years. Thus, any trends in the
faunal maps must indicate conditions prevalent over several centuries surrounding the time of ash deposition.

At the 9300-year level, the percentage abundance of the polar fauna
(Table 1, Fig. 25) reveals two distinctive patterns: 1, a general alignment
of the 90% to 60% contours along a NNE–SSW trend that marks the
9300-year polar front; and 2, a lobate eastward protrusion of the 30% to
50% contours along latitudes 47°N to 53°N.

The warm extreme of foraminiferal species present at 9300 years B.P. is represented by the transitional fauna (Table 1, Fig. 26). The contours of percent transitional fauna at 9300 B.P. fan out from a NNE–SSW trend along Greenland and Newfoundland to an ENE–WSW trend near Great Britain.

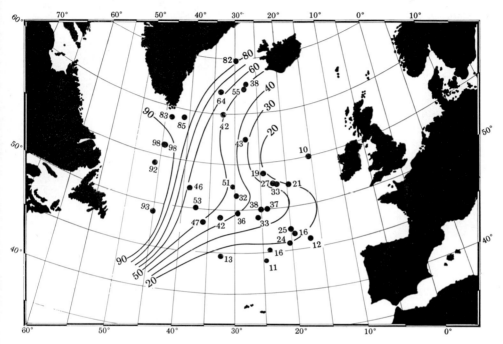

FIG. 25. Percentage abundance in each core of polar fauna at synchronous 9300-year ash datum; lobe at 50°N suggests counterclockwise-flow of Arctic water in the northern North Atlantic. (From Ruddiman and Glover, 1975.)

There is no region of diminished transitional species abundance corresponding to the lobe of enhanced polar fauna percentages. The relative increase in polar fauna in the lobe at 50°N occurs at the expense of the subpolar fauna (Table 1), particularly the right-coiling form of *G. pachyderma*.

The existence of an eastward–protruding lobe of cold water (polar) fauna can be interpreted in terms of environmental variations that are either seasonal, secular (long-term) or depth-dependent. Regardless of which interpretation is chosen, the faunal lobe suggests that eastward-moving polar water (or wind-driven upwelling of cold water along an east–west axis) was a significant feature of the mean circulation regime during the centuries surrounding 9300 years B.P.

Palaeoenvironmental estimates at 9300 *years B.P.*

The simplest interpretation that can be made of a sample assemblage which represents time intervals of several hundred years or more is that the counted faunal percentages are not just averages but that yearly values repeatedly fell near this mean, so that the mode and the mean were nearly the same. Such

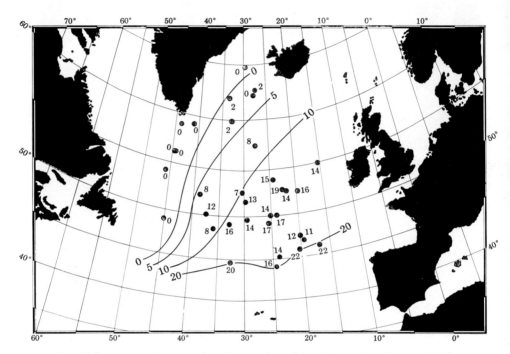

FIG. 26. Percentage abundance in each core of transitional fauna at synchronous 9300-year ash datum; transitional species inhabit North Atlantic Drift waters. (From Ruddiman and Glover, 1975.)

an assumption is inherent in the technique of Imbrie and Kipp (1971). It is also possible, however, that longer-term secular changes, possibly among quasi-stable modal climatic regimes having no resemblance to the reconstructed average annual environment, have dominated the actual palaeo-climatic variations. Unfortunately, such secular changes cannot be reconstructed, and this forces the micropalaeontologist to make the above simplifying assumption as to the relevance of mean annual climate.

Ruddiman and Glover (1975) applied Imbrie–Kipp transfer function F13 to the 9300-year foraminiferal counts. The maps of estimated mean palaeotemperature (Fig. 27) and palaeosalinity (Fig. 28) for the winter season clearly resemble the main features of the 9300-year polar fauna map (Fig. 25). The close, near-parallel spacing of the 1°C to 4°C isotherms confirms the existence of a rather steep thermal gradient along the faunally-postulated polar front, and the 5°C to 8°C isotherms show an eastward-protruding cold water lobe at 47°N to 53°N with a less intense zonal thermal front on its

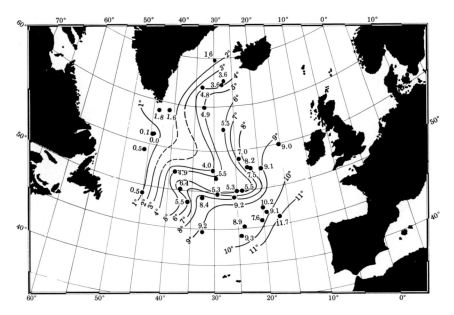

FIG. 27. Estimated mean 9300-year February surface water temperature, reconstructed using Imbrie–Kipp technique. (From Ruddiman and Glover, 1975.)

southern margin (Fig. 27). The major salinity gradients (Fig. 28) lie along the zonal front at the southern margin of the 50°N lobe.

This faunal evidence thus suggests that cyclonic flow in the subpolar North Atlantic gyre significantly affected ecological water mass boundaries and circulation patterns at 9300 years B.P. Cold, low-salinity Arctic waters flowed southward along the coasts of Greenland and Newfoundland, with a component apparently turning eastward across the Atlantic toward Europe; warm North Atlantic Drift water flowed to the north and turned to the west in the eastern North Atlantic. The 9300-year flow differs from modern circulation in the existence of a stronger eastward flow of polar water along 50°N and in the more eastward location of the meridional polar front compared with its present location along the coasts of Greenland and Newfoundland.

REGIONAL NORTH ATLANTIC WATER MASS MOVEMENTS

The 9300-year ash datum has been used to extract additional evidence of
North Atlantic water mass movements and palaeoclimatic change. Ruddiman
and McIntyre (1973) have used the ash datum as a chronostratigraphical
horizon against which the chronologic level of climatic change in cores is
compared and regionally mapped.

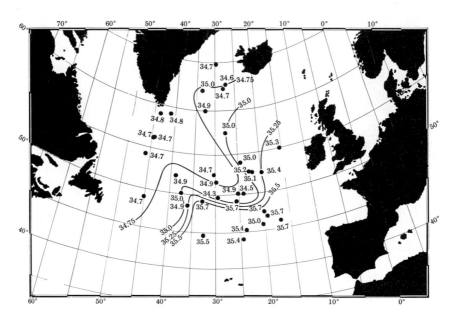

FIG. 28. Estimated mean 9300-year February surface water salinity reconstructed using
Imbrie–Kipp technique. (From Ruddiman and Glover, 1975.)

Time-transgressive faunal warming

Counts of the percentage of polar foraminifera (*G. pachyderma*, s) are plotted
against time in Fig. 29 along an east–west profile of six cores from latitudes
50° to 55°N. The level of most rapid decreases in polar fauna (and, hence,
most rapid deglacial warming) changes position relative to the 9300-year ash
level, and therefore is time-transgressive. If the assumptions of Ruddiman
and McIntyre (1973) concerning the sedimentation rates are reasonably
accurate, the faunally-defined deglacial warming transgresses at least 7000
years of time in the Atlantic above 45°N, ranging in age from 13 500 years

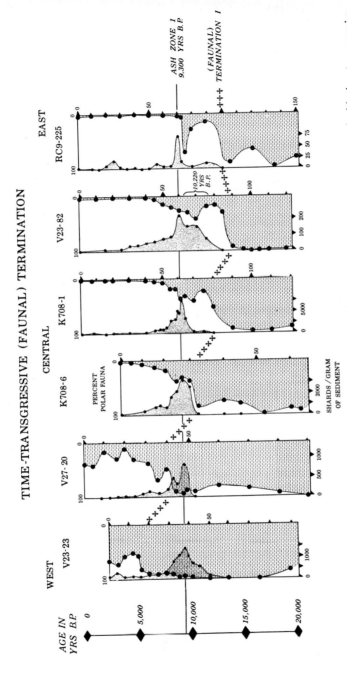

Fig. 29. Plot of chronostratigraphic peak ash abundance (stippled) in east–west core transect compared with time-transgressive change from dominantly polar to subpolar foraminiferal fauna. (From Ruddiman and McIntyre, 1973.)

B.P. in the east near Great Britain to less than 6500 years B.P. in the west near Greenland. The extrapolated ages of this abrupt warming (termination I) plotted for 28 cores (Fig. 30) form a coherent geographic pattern.

Since at these latitudes the faunally-defined warming at any location coincides with the end of major ice-rafted sand deposition (Bramlette and Bradley, 1941), the level in time marking the last deposition of glacial detritus must also be time-transgressive from east-to-west across the North Atlantic.

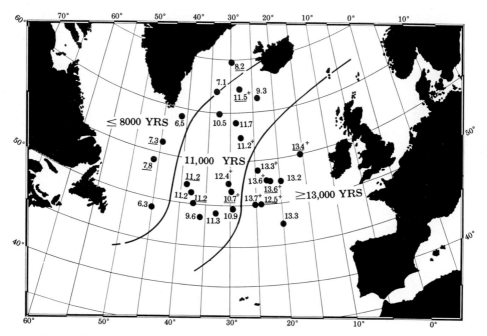

FIG. 30. Map of age of deglacial warming in 28 cores across the North Atlantic (in 10^3 years B.P.) with oceanic deglacial warming chosen as in Fig. 29. (+) indicates cores with evidence of prominent cooling (re-advance of cold water) at 10 000 years B.P. (From Ruddiman and McIntyre, 1973.)

Deglacial Polar water retreat from the North Atlantic

The southern limit of polar water constitutes the critical palaeo-oceanographic boundary in this regional sequence of deglacial warming. McIntyre *et al.* (1972a) noted that in the modern ocean this boundary separates the polar water mass, which is cold, prevalently ice-bearing, coccolith-free, and dominated by the single Arctic foraminifera, *G. pachyderma* (s), from subpolar water, which is warmer, and faunally and florally diverse, and which receives only brief seasonal incursions of sea ice. Thus the geographic sequence of warmings shown in Fig. 30 traces the east-to-west migration of the boundary between the polar and subpolar water masses across the North Atlantic (Fig. 31).

This northwestward withdrawal of polar water carrying ice-rafted glacial detritus is the most direct oceanographic analogue to ice sheet retreat during continental deglaciation. Retreat of polar water from the critical North

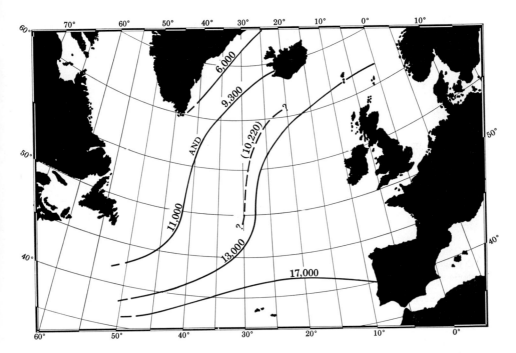

FIG. 31. Map of retreat positions of polar water in the North Atlantic from 17 000 to 6000 years B.P. (From Ruddiman and McIntyre, 1973.)

Atlantic latitudes 40° to 65°N is compiled in Fig. 32 as a plot of the decreasing percentage of ocean area covered by polar water. The water mass positions shown in Fig. 31 are the plotted control points for Fig. 32; the history of polar water retreat between these points is inferred in Fig. 32. The faunally-defined geographic pattern of water mass retreat can thus be quantified as a palaeoclimatic indicator in a manner analogous to quantitative curves of ice sheet retreat (Bloom, 1971; McDonald, 1971).

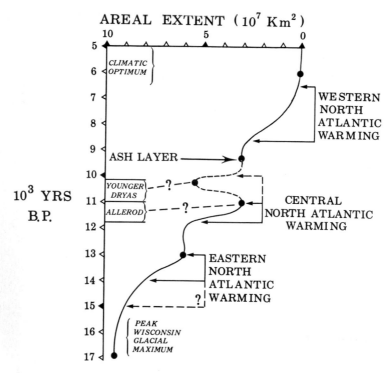

FIG. 32. Portion of North Atlantic between 40° and 65°N covered by polar water during deglaciation from 17 000 to 6000 years B.P. (From Ruddiman and McIntyre, 1973.)

SUMMARY OF GLACIAL–INTERGLACIAL WATER MASS MOVEMENTS

The preceding section shows that under optimal conditions, palaeoclimatologists can, by using foraminifera, decpiher regional circulation patterns and water mass movements which relate convincingly to flow directions in the modern circulation. The accurate definition of such patterns carries implications far beyond the immediate problem of establishing past temperature and salinity changes. Portions of the oceans can then be viewed not in the static sense of examining one core at one point in space for variations in the environment through time, but in a quasi-synoptic way, with the oceans considered as a dynamic system. Frontal boundaries and circulation trajectories can be reconstructed to infer the mean prevailing forces that drive the oceanic circulation.

For example, the sequence of deglacial retreat of polar water across the North Atlantic inferred from foraminiferal evidence (Fig. 31) conveniently summarizes several aspects of North Atlantic palaeocirculation related to the cyclonic (counterclockwise) circulation at subpolar latitudes. The prevailing surface water flow is cyclonic in response to the opposed wind stress of the polar easterlies in the north and the mid-latitude westerlies in the south, coupled with the blockage of zonal flow by continental boundaries. The relation of water mass movements to this cyclonic flow can be traced through the entire sequence of deglacial retreat.

During the last glacial maximum at 18 000 years B.P., ice-bearing polar waters completely enveloped the North Atlantic above 45°N (Weyl, 1968; McIntyre *et al.*, 1972a, b). An abrupt frontal system along 45°N separated two well-developed North Atlantic gyres during glaciations: 1, a counterclockwise subpolar gyre bounded by latitudes 65°N, 45°N, and the continental margins; and 2, a clockwise subtropical gyre positioned just to the south of the present subtropical gyre.

The zonal (latitudinally defined) orientation of this glacial North Atlantic configuration resembles the modern Antarctic Convergence; the abruptness of the isotherm and isohaline gradients (McIntyre *et al.*, 1972b) suggests that the subtropical gyre posed a significant barrier to southward penetration of polar water to latitudes below 40°N.

From this zonal glacial limit along 40°N to 45°N, the boundary of polar water began to retreat meridionally (from east-to-west) towards the 9300-year position (Fig. 31). Significantly warmer and more saline North Atlantic Drift waters intruded northward along the coasts of Europe beginning at 13 500 years B.P.

The lobate tongue at 50°N (Figs 24–28) in the 9300-year faunal patterns represents a significant distortion of the generally meridional isotherms and isohalines. Ruddiman and Glover (1975) suggest that this bulge could be explained by wind stress, since the prevailing westerlies would have intercepted, at an appreciable angle, the meridional oceanic circulation. Following 9300 years B.P., the southward polar water outflow moved farther to the west and gradually became confined as a coastal current along Greenland and Newfoundland. During the warmth of the climatic optimum at 6000 years B.P., even the southward intrusion of polar water diminished to a narrow cold flow appressed by the coriolis deflection to the east coast of Greenland.

This geometry of east–west water mass withdrawal (and, presumably, advance) is significantly different from that which might be inferred from just the north–south transect of 13 cores shown in Figs 21 and 22. That transect suggested north–south movements of polar water across 20° or more of latitude. The complete regional study (Fig. 31), however, shows that, due to cyclonic flow tendencies, polar water in fact moved along an east–west axis across the North Atlantic, and that the entire north–south core transect

(Fig. 21) was thus affected simultaneously by east–west or southeast–north-west translations and not by simple north–south water mass movements. The sustained rates of east–west migration (Fig. 31) are measured in hundreds of metres per year, similar to the range inferred from the north–south transect in Fig. 22.

Future directions of quaternary foraminiferal palaeoclimatology

For the future, four major directions of study in foraminiferal palaeo-climatology appear to offer the most promising means of developing a unified view of global palaeoclimatic change. These are:

(*i*) Additional surface-sample definition of the simultaneous variations of environmental parameters and faunal assemblages.

Quantitative techniques like that of Imbrie and Kipp (1971) interpret faunal variations in terms of the separate environmental factors which are responsible, and the larger the data base, the better the distinction among covarying parameters. Ocean-wide taxonomic standardisation is necessary for this approach to succeed, as is accurate quantification of differential solution effects for all foraminiferal assemblages.

(*ii*) Regional definition of long-term climatic deteriorations.

The fundamental Quaternary and pre-Quaternary climatic coolings lasting of the order of 10^6 years or more represent step-like baseline shifts in mean climate initiated by causes presumably different from those controlling the more repetitive (10^5-year) Quaternary climatic cycles. Such long-term changes are likely to be related to tectonic alterations of the earth's crust by orogeny and sea-floor spreading. These processes can result in climatically significant lateral and vertical movements, including: opening and closing of sills and isthmus barriers to shallows and deep oceanic circulation, uplifting of mountain ranges which may affect atmospheric circulation patterns; uplifting of broad areas to altitudes at which glaciation becomes possible, and gradual rearrangement of continents with respect to latitude and climatic belts.

Major questions about the comparative regional timing of long-term climatic deteriorations remain unanswered. Significant coolings have been observed in the North Atlantic at 3.0 m.y.B.P. (Berggren, 1972) and at 1.3–1.0 m.y.B.P. (Berggren, *et al.*, 1967; Kent *et al.*, 1971; Ruddiman, 1971). In direct contrast, Antarctic foraminiferal data suggest a significant warming during the Brunhes Epoch (0–0.7 m.y.B.P.) relative to the Matuyama Epoch (0.7–2.43 m.y.B.P.). Further definition of the worldwide phasing of these climatic shifts is essential,

(*iii*) Synchroneity of 10^5-year climatic cycles.

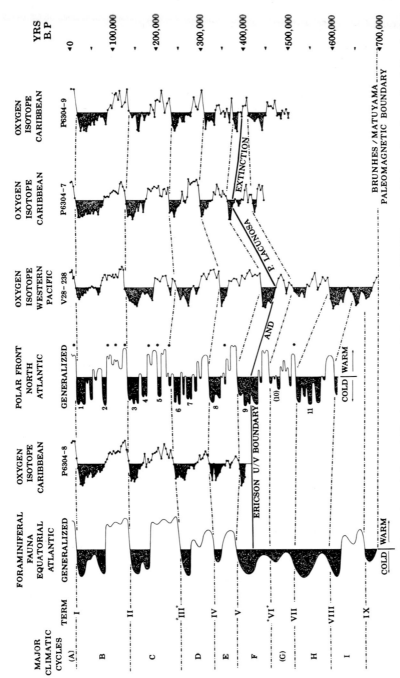

Fig. 33. Correlation of isotopic and foraminiferal parameters during the last 700 000 years. Major climatic cycles lettered A–J; glacial maxima shown in Arabic numerals; terminations shown by Roman numerals; high sea levels from Mesolella and others (1969) shown by asterisks. Equatorial Atlantic curve from Ruddiman (1971); North Atlantic curve from Ruddiman and McIntyre (1976); Caribbean isotopic curves from Emiliani (1966, 1972); Pacific isotopic curve from Shackleton and Opdyke (1973).

Basically synchronous variations of oxygen isotope ratios are established for the North Atlantic above 40°N (Emiliani, 1958), the equatorial Atlantic (Emiliani, 1955a, 1966), the Caribbean (Emiliani, 1966, 1972), the Mediterranean (Emiliani, 1955b), the equatorial Indian Ocean (Oba, 1969), and the western Equatorial Pacific (Shackleton and Opdyke, 1973).

Foraminiferal variations in the Atlantic north of 40°N and in the equatorial Atlantic south of 10°N match the synchronous isotopic curves in detail over the Brunhes Epoch (Fig. 33). Waters in those areas thus respond to climate in the same manner as the ice sheets in the Northern hemisphere. Caribbean, Mediterranean, Gulf of Mexico, and South Pacific foraminiferal changes follow the isotopic variations to a lesser degree, with a still weaker correlation in the equatorial Indian Ocean, the central North Atlantic, and the Antarctic.

Correlation of the 10^5-year cycles can also lead to definition of the extreme oceanic palaeosalinity and palaeotemperature responses to Quaternary climatic change in a given region, but this is possible only if sedimentation rates are high enough to permit accurate recording of the intensity of the oceanic responses. With sufficient dating precision, it may also be possible to establish the rates of oceanic response.

(*iv*) Datum-plane reconstructions of quasi-synoptic patterns.

Before meaningful datum-plane reconstructions are attempted for short-term (10^3- to 10^4-year) climatic episodes, it will be necessary to define more clearly the mixing effects by burrowing infauna on the faunal record of short climatic episodes. In particular, the percentage loss in original peak intensity of climatic episodes of varying duration must be ascertained in each area for a range of sedimentation rates.

The quasi-synoptic studies, where feasible, can provide two essential types of information. First, the regional water mass configurations which are derived form a basis for directly inferring mean ocean palaeocirculation trends. In ideal cases, isohaline and isotherm alignments may suggest the balance of forces affecting the circulation. Second, if the datum levels used as controls are closely spaced in time, very precise monitoring of oceanic response rates to rapid climatic shifts is possible.

Planktonic foraminifera can thus portray, in varying levels of detail, the earth's oceanic response to Quaternary climatic cycles and shifts ranging in duration from 10^3 to 10^6 years. Because atmospheric and oceanic responses to the causal mechanisms behind climatic change are complex, a global approach that reveals as much as possible about the regional responses is necessary. More than any other existing organism on the earth, planktonic foraminifera have been, and will be, the major source of data for quantitative global mapping of palaeoclimatic change.

Acknowledgments

I thank L. K. Glover for both editorial and laboratory assistance and particularly for helpful discussions during all phases of the North Atlantic work; A. McIntyre for general collaboration and for useful discussions; and G. Garner for aid in the laboratory and with the manuscript.

References

Arrhenius, G. (1952). Sediment cores from the east Pacific. *Reports Swedish Deepsea Exped.* **105**(1), 1–227.

Bandy, O. L. (1960). The geological significance of coiling ratios in the foraminifer *Globigerina pachyderma* (Ehrenberg). *J. Paleont.* **34**, 671–681.

Barash, M. S. (1971). The vertical and horizontal distribution of planktonic Foraminifera in Quaternary sediments of the Atlantic Ocean. *In* (Funnell, B. M. and Reidel, W. R., eds), *The Micropaleontology of Oceans*, 433–442. Cambridge University Press.

Bé, A. W. H. (1969). Planktonic Foraminifera, *In* Marine Invertebrates south of 35°S. latitude. Antarct. Map Folio Ser., **Folio 11**, 9–12, *Amer. Geogr. Soc.N.Y.*

Beard, J. H. (1969). Pleistocene paleotemperature record based on planktonic foraminifers, Gulf of Mexico. *Trans. Gulf Coast Assoc. of Geol. Soc.* **19**, 535–553.

Belderson, R. H. and Laughton, A. S. (1966). Correlation of some Atlantic turbidites. *Sedimentol.* **7**, 103–116.

Berger, W. H. (1967). Foraminiferal ooze: solution at depths. *Science* **156**, 383–385.

Berger, W. H. (1968). Planktonic Foraminifera: selective solution and paleoclimatic interpretation. *Deep-Sea Res.* **15**, 31–43.

Berger, W. H. (1970). Biogenous deep-sea sediments: fractionation by deep-sea circulation. *Geol. Soc. Amer. Bull.* **81**, 1385–1402.

Berger, W. H. (1971). Sedimentation of planktonic Foraminifera. *Mar. Geol.* **11**, 325–358.

Berger, W. and Heath, G. R. (1968). Vertical mixing in pelagic sediments. *J. Mar. Res.* **26**, 135–143.

Berggren, W. A. (1972). Late Pliocene-Pleistocene Glaciation. *In Initial Reports of the Deep Sea Drilling Project*, **12**, 953–963.

Berggren, W. A., Phillips, J. D., Bertels, A. and Wall, D. (1967). Late Pliocene-Pleistocene stratigraphy in deep-sea cores from the south-central North Atlantic. *Nature, Lond.* **216**, 253–254.

Blackman, A. and Somayajulu, B. L. K. (1966). Pacific Pleistocene cores: faunal analyses and geochronology. *Science* **154**, 886–889.

Blanc-Vernet, L., Chamley, H. and Froget, C. 1969. Analyses paleoclimatique d'une carotte de Méditerranée nord-occidental. Comparaison entre les résultats de trois études: foraminifères, ptéropodes, fraction sédimentaire issue du continent. *Palaeogeogr. Palaeoclimat. Palaeoecol.* **6**, 215–235.

Bloom, A. L. (1971). Glacial-eustatic and isostatic controls of sea level since the last glaciation. *In* (Turekian, K. K., ed.), *Late Cenozoic Glacial Ages*, 355–380. Yale University Press, New Haven.

Bornhold, B. D. and Pilkey, O. H. (1971). Bioclastic turbidite sedimentation in Columbus Basin, Bahamas. *Geol. Soc. Amer. Bull.* **82**, 1341–1354.

Bradley, W. H. (1938). Mediterranean sediments and Pleistocene sea levels. *Science* **88**, 376–379.

Bramlette, M. N. and Bradley, W. H. (1941). Geology and biology of North Atlantic deep-sea cores between Newfoundland and Ireland, Part 1. Lithology and geologic interpretations. *U.S. Geol. Survey Prof. Paper* **196-A**, 1–34.

Broecker, W. S. and Ku, T. L. (1969). Caribbean cores P6304–8 and P6304–9: new analysis of absolute chronology. *Science* **166**, 404–406.

Broecker, W. S. and van Donk, J. (1970). Insolation changes, ice volumes, and the O^{18} record in deep-sea cores. *Reviews of Geophys. and Space Phys.* **8**, 169–198.

Caralp, M. (1967). Les Foraminifères planctoniques d'une carotte atlantique (Golfe de Gascogne) dans la mise en évidence d'une glaciation. *C. r. hebd. Séanc. Acad. Sci. Paris* **265**, 1588–1591.

Clark, D. L. (1969). Paleoecology and Sedimentation in Part of the Arctic Basin. *Arctic* **22**, 233–245.

Clark, D. L. (1971). Arctic Ocean ice cover and its late Cenozoic history. *Geol. Soc. Amer. Bull.* **82**, 3313–3324.

Cox, A. (1969). Geomagnetic reversals. *Science* **163**, 237–245.

Cushman, J. A. and Henbest, L. G. (1941). Geology and biology of North Atlantic deep-sea cores between Newfoundland and Ireland. Part 2. Foraminifera. *U.S. Geol. Surv. Prof. Paper* **196-A**, 35–54.

Emiliani, C. (1955a). Pleistocene temperatures. *J. Geol.* **63**, 538–578.

Emiliani, C. (1955b). Pleistocene temperature variations in the Mediterranean. *Quatern.* **2**, 87–98.

Emiliani, C. (1958). Palaeotemperature analysis of core 280 and Pleistocene correlations. *J. Geol.* **66**, 264–275.

Emiliani, C. (1966). Paleotemperature analysis of the Caribbean cores P6304-8 and P6304-9 and a generalized temperature curve for the last 425,000 years. *J. Geol.* **74**, 109–126, 1966.

Emiliani, C. (1972). Quaternary paleotemperatures and the duration of the high temperature intervals. *Science* **178**, 398–401.

Ericson, D. B. and Wollin, G. (1956). Correlations of six cores from the equatorial Atlantic and the Caribbean. *Deep-Sea Res.* **8**, 104–125.

Ericson, D. B. and Wollin, G. (1968). Pleistocene climates and chronology in deep-sea sediments. *Science* **162**, 1227–1234.

Ericson, D. B., Ewing, M., Wollin, G. and Heezen, B. C. (1961). Atlantic deep-sea sediment cores. *Geol. Soc. Amer. Bull.* **72**, 193–286.

Ewing, M., Ericson, D. B. and Heezen, B. C. (1958). Sediments and topography of the Gulf of Mexico. *In* (Weeks, L. G., ed.), *Habitat of Oil*, 995–1053. Amer. Assoc. of Petr. Geol.

Frerichs, W. E. (1968). Pleistocene-Recent boundary and Wisconsin Glacial biostratigraphy in the Northern Indian Ocean. *Science* **159**, 1456–1458.

Gardner, J. V. and Hays, J. D. (1976). The Eastern Equatorial Atlantic: Sea-surface temperature and circulation responses to climatic change during the past 200,000 years. *Geol. Soc. Amer. Mem.* **145**, 221–246.

Glass, B. P. (1969). Reworking of deep-sea sediments as indicated by the vertical dispersion of the Australasian and Ivory Coast microtektite horizons. *Earth Planet. Sci. Lett.* **6**, 409–415.

Glass, B. P. (1972). Australasian microtektites in deep-sea sediments. *Ant. Res. Ser.* **19**, 335–348.

Glass, B. P., Ericson, D. B., Heezen, B. C., Opdyke, N. D. and Glass, J. A. (1967). Geomagnetic reversals and Pleistocene chronology. *Nature* **216**, 437–442.

Goldthwait, R. P. (1958). Wisconsin age forests in western Ohio. I. Age and glacial events. *Ohio Jour. Sci.* **58**, 24–27.

Hamilton, E. L. (1957). Foraminifera from an equatorial Pacific core. *Micropal cont.* **3**, 69–73.

Hays, J. D., Saito, T., Opdyke, N. D. and Burckle, L. H. (1969). Pliocene-Pleistocene sediments of the equatorial Pacific: their paleomagnetic, biostratigraphic and climatic record. *Geol. Soc. Amer. Bull.* **80**, 1481–1514.

Heezen, B. C. and Hollister, C. D. (1971). *The Face of the Deep*, Oxford University Press, 659 pp.

Hermann, Y. (1969). Arctic Ocean Quaternary microfauna and its relation to paleoclimatology. *Palaeogeogr. Palaeoclim., Palaeoecol.* **6**, 251–276.

Hermann, Y. (1970). Arctic paleo-oceanography in late Cenozoic time. *Science* **169**, 474–477.

Hollister, C. D. and Heezen, B. C. (1967). The floor of the Bellingshausen Sea. *In* (Hershey, J. B., ed.), *Deep-Sea Photography*, 117–189. Johns Hopkins Press.

Hunkins, K., Bé, A. W. H., Opdyke, N. D. and Mathieu, G. (1971). The late Cenozoic history of the Arctic Ocean. *In* (Turekian, K. K., ed.), *Late Cenozoic Glacial Ages*, 215–237. Yale University Press, New Haven.

Imbrie, J. and Kipp, N. (1971). A new micropaleontological method for quantitative paleoclimatology: application to a late Pleistocene Caribbean core. *In* (Turekian, K. K., ed.), *Late Cenozoic Glacial Ages*, 71–181. Yale University Press, New Haven.

Imbrie, J., van Donk, J. and Kipp, N. (1973). Paleoclimatic investigation of a late Pleistocene Caribbean deep-sea core: comparison of isotopic and faunal methods. *Quat. Res.* **3**, 10–38.

Keany, J. and Kennett, J. P. (1972). Pliocene-early Pleistocene paleoclimatic history recorded in Antarctic-Subantarctic deep-sea cores. *Deep-Sea Res.* **19**, 529–548.

Kellog, T. B. (1973). Climate of the Norwegian and Greenland Seas: Today and 17,000 years ago. Abstracts, Annual Meeting, *Geol. Soc. Amer.* 691.

Kempton, J. P. and Gross, D. L. (1971). Rate of advance of the Woodfordian (Late Wisconsonian) Glacial Margin in Illinois: stratigraphic and radiocarbon evidence. *Geol. Soc. Amer. Bull.* **82**, 3245–3250.

Kennett, J. P. (1966). Foraminiferal evidence of a shallow calcium carbonate solution boundary, Ross Sea, Antarctica. *Science* **153**, 191–193.

Kennett, J. P. (1970). Pleistocene paleoclimates and foraminiferal biostratigraphy in subantarctic deep-sea cores. *Deep-Sea Res.* **17**, 125–140.

Kennett, J. P. and Huddlestun, P. (1972a). Late Pleistocene paleoclimatology, foraminiferal biostratigraphy, and tephrachronology, Western Gulf of Mexico. *Quat. Res.* **2**, 38–69.

Kennett, J. P. and Huddlestun, P. (1972b). Abrupt climatic change at 90,000 Yr. B.P.: faunal evidence from Gulf of Mexico cores. *Quat. Res.* **2**, 384–395.

Kent, D., Opdyke, N. D. and Ewing, M. (1971). Climatic change in the North Pacific using ice-rafted detritus as a climatic indicator. *Geol. Soc. Amer. Bull.* **82**, 2741–2754.

Ku, T. L., Bischoff, J. L. and Boersma, A. (1972). Age studies of Mid-Atlantic Ridge sediments near 42°N and 20°N. *Deep-Sea Res.* **19**, 233–247.

Kullenberg, B. (1952). On the salinity of the water contained in marine sediments. *K. Vet. Vitt. Somh. Handl.* Ser. B. **6**, 3–37.

Lidz, L. (1966). Deep-Sea Pleistocene biostratigraphy. *Science* **154**, 1448–1451.

Lisitzin, A. P. (1971). Distribution of carbonate microfossils in suspension and in bottom sediments. *In* (Funnell, B. M. and Riedel, W. R., eds), *The Micropaleontology of Oceans*, 197–218. Cambridge University Press.

Lisitzin, A. P. (1972). Sedimentation in the world ocean. *Soc. Ec. Pal. and Min. Special Publ.* No. **17**, 218. pp.

Lynts, G. W. (1971). Analysis of the planktonic foraminiferal fauna of core 6275, Tongue of the Ocean, Bahamas. *Micropaleont.* **17**, 152–166.

Luz, B. (1973). Stratigraphic and paleoclimatic analysis of late Pleistocene tropical southeast Pacific cores. (with an appendix by N. J. Shackleton). *Quat. Res.* **3**, 56–72.

McDonald, B. C. (1971). Late Quaternary stratigraphy and deglaciation in Eastern Canada. *In* (Turekian, K. K., ed.), *Late Cenozoic Glacial Ages*, 331–354. Yale University Press, New Haven.

McIntyre, A. (1967). Coccoliths as paleoclimatic indicators of Pleistocene glaciation. *Science* **158**, 1314–1317.

McIntyre, A., Bé, A. W. H. and Preikstas, R. (1967). Coccoliths and the Plio-Pleistocene boundary. *Progress in Oceanography* **4**, 3–25.

McIntyre, A., Ruddiman, W. F. and Jantzen, R. (1972a). Southward penetrations of the North Atlantic polar front: faunal and floral evidence of large-scale surface water mass movements over the last 225,000 years. *Deep-Sea Res.* **19**, 61–77.

McIntyre, A., Bé, A., Biscaye, P., Burckle, L., Gardner, J., Geitzanauer, K., Goll, R., Kellog, T., Prell, W., Roche, M., Imbrie, J., Kipp, N., Ruddiman, W., Moore, T. and Heath, R. (1972b). The glacial North Atlantic 17,000 years ago: paleoisotherm and oceanographic maps derived from floral-faunal parameters by CLIMAP. *Geol. Soc. Amer.*, Annual Meeting, Abstr. 590–591.

Mellis, O. (1954). Volcanic ash-horizons in deep-sea sediments. *Deep-Sea Res.* **2** 89–92.

Mesolella, K. J., Matthews, R. K., Broecker, W. S. and Thurber, D. L. (1969). The astronomical theory of climatic change: Barbados data. *J. Geol.* **77**, 250–274.

Oba, T. (1969). Biostratigraphy and isotopic paleotemperature of some deep-sea cores from the Indian Ocean. *Science Repts. of the Tohuku Univ.* **41**, 129–195.

Olausson, E. (1960). Description of sediment cores from the North Atlantic. *Repts. Swedish Deep-Sea Exped.* 1947–1948, **7**, 229–386.

Olausson, E. (1961). Studies of deep-sea cores. *Repts. Swedish Deep-Sea Exped.* 1947–1948, **8**, 353–391.

Parker, F. L. (1958). Eastern Mediterranean foraminifera. *Repts. Swedish Deep-Sea Exped.* 1947–1948, **8**, 219–283.

Parker, F. L. and Berger, W. H. (1971). Faunal and solution patterns of planktonic foraminifera in surface sediments of the South Pacific. *Deep-Sea Res.* **18**, 73–107.

Phleger, F. B. (1951). Ecology of foraminifera, northwest Gulf of Mexico. Part I, Foraminifera distribution. *Geol. Soc. Amer. Mem.* **46**, 1–88.

Phleger, F. B. (1955). Foraminiferal faunas in cores offshore from the Mississippi Delta. *Deep-Sea Res. Suppl.* (papers in Mar. Biol. & Oceanogr.), **3**, 45–57.

Phleger, F. B., Parker, F. L. and Peirson, J. F. (1953). North Atlantic Foraminifera. *Repts. Swed. Deep-Sea Exped.* 1947–1948, **7**, 1–122.

Prell, W. L. and Hays, J. D. (1976). Late Pleistocene faunal and temperature patterns of the Columbia Basin, Caribbean Sea. *Geol. Soc. Amer. Mem.* **145**, 247–266.

Rona, E. and Emiliani, C. (1969). Absolute dating of Caribbean cores P6304–8 and P6304–9. *Science* **163**, 66–68.

Rubin, M. and Suess, H. E. (1955). U.S. Geological Survey radiocarbon dates, II. *Science* **121**, 481–488.

Ruddiman, W. F. (1969). Recent planktonic foraminifera: dominance and diversity in North Atlantic surface sediments. *Science* **164**, 1164–1167.

Ruddiman, W. F. (1971). Pleistocene sedimentation in the equatorial Atlantic: stratigraphy and faunal paleoclimatology. *Geol. Soc. Amer. Bull.* **82**, 283–302.

Ruddiman, W. F. and Glover, L. K. (1972). Vertical mixing of ice-rafted volcanic ash in North Atlantic sediments. *Geol. Soc. Amer. Bull.* **83**, 2817–2836.

Ruddiman, W. F. and Glover, L. K. (1975). Subpolar North Atlantic circulation at 9,300 yrs. B.P.: faunal evidence. *Quat. Res.* **5**, 361–389.

Ruddiman, W. F. and Heezen, B. C. (1967). Differential solution of planktonic foraminifera. *Deep-Sea Res.* **14**, 801–808.

Ruddiman, W. F. and McIntyre, A. (1973). Time-transgressive deglacial retreat of polar water from the North Atlantic. *Quat. Res.* **3**, 117–130.

Ruddiman, W. F. and McIntyre, A. (1975). Northeast Atlantic paleoclimatic changes over the last 600,000 years. *Geol. Soc. Amer. Mem.* **145**, 111–146.

Ryan, W. B. F. (1972). Stratigraphy of Late Quaternary sediments in the Eastern Mediterranean. *In* (Stanley, D. J. ed.), *The Mediterranean Sea: A natural sedimentation laboratory*, 149–169. Dowden, Hutchinson and Ross, Stroudsburg, Pa.

Sackett, W. M. and Rankin, J. G. (1970). Paleotemperatures for the Gulf of Mexico. *Jour. Geophys. Res.* **75**, 4557–4560.

Sanchetta, C., Imbrie, J., Kipp, N., McIntyre, A. and Ruddiman, W. F. (1972). Climatic record in North Atlantic deep-sea core V23-82: comparison of the last and present interglacials based on quantitative time series. *Quat. Res.* **2**, 363–367.

Sanchetta, C., Imbrie, J. and Kipp, N. G. (1973). Climatic record of the past 130,000 years in North Atlantic deep-sea core V23-82: comparison with the terrestrial record. *Quat. Res.* **3**, 110–116.

Schott, W. (1935). Die Foraminiferen in den aequatorialen teil des Atlantischen Ozeans, Dt. Atlantische Exped. 3, 43–135.

Shackleton, N. (1967). Oxygen isotope analyses and Pleistocene temperatures reassessed. *Nature* **215**, 15–17.

Shackleton, N. J. and Opdyke, N. D. (1973). Oxygen isotope and paleomagnetic stratigraphy of Equatorial Pacific core V28-238: oxygen isotope temperatures and ice volumes on a 10^5 year and 10^6 year scale. *Quat. Res.* **3**, 39–55.

Thiede, J. (1971). Planktonische foraminiferen in sedimenten vom Ibero-Marokkanishchen kontinentalrand. *"Meteor"* Forsch-Ergebnisee C., **7**, 15–102.

Todd, R. (1957). Foraminifera from western Mediterranean deep-sea cores. *Repts. Swed. Deep-Sea Exped.* 1947–1948, **8**, 169–217.

Valencia, M. J. (1973). Calcium carbonate and gross-size analyses of surface sediments, western Equatorial Pacific. *Pacific Science* **27**, 290–303.

Watkins, N. D. and Kennett, J. P. (1971). Antarctic Bottom Water: major change in velocity during the Late Cenozoic between Australia and Antarctica. *Science* **173**, 813–818.

Weyl, P. K. (1968). The role of the oceans in climatic change: A theory of the ice ages. *Meteorol. Mon.* **8**, 37–62.

Willman, H. B. and Frye, J. L. (1970). Pleistocene stratigraphy of Illinois: *Illinois. Geol. Survey Bull.* **94**, 204. pp.

3: Cainozoic Planktonic Foraminifera Zonation and Selective Test Solution

WILLIAM N. ORR[1] and D. GRAHAM JENKINS[2]

[1]*Department of Geology,*
University of Oregon,
Eugene 97403, U.S.A.

[2]*Department of Geology,*
University of Canterbury,
Christchurch 1, New Zealand

Difficulties experienced while attempting to use published planktonic foraminiferal zonal schemes in the Eastern Pacific were found to be related to the selective destructive solution of foraminiferal tests. Existing zonal schemes have made use of many solution-susceptible planktonic foraminiferal taxa common to continental shelf sections. Upon identifying the problems of destructive solution of calcium carbonate a new zonal scheme for deep-sea stratigraphic Upper Eocene–Recent sections was proposed by using solution-resistant taxa.

Outline

Introduction

Drilling operations on Leg 9 of the Deep Sea Drilling Project (J.O.I.D.E.S.) yielded 5045 ft of Upper Eocene–Pleistocene cores from nine sites in the eastern equatorial Pacific in depths of water of 10 159–15 085 ft (Hays *et al.*, 1972) see Fig. 1.

The cores yielded Upper Eocene to Pleistocene planktonic foraminiferal faunas and no major intraformational stratigraphic faunal breaks were recorded (Fig. 2).

An attempt was made to use previously published planktonic foraminiferal zonal schemes, but it was found that many of the zonal markers were either absent or very rare. Many of the zones established on faunas from shallower deposits were not recognisable. Examples of some zones published by Blow (1969) which were unrecognisable in the Eastern Pacific include "N4", "N13", "N14", "N15", "N16", "N21" and "N23" (Fig. 3).

The zonal scheme of Blow (1969) seems to be a modification of Bolli's (1957a, b, c) zonation of Tertiary sediments of Trinidad. Deep-sea sediments

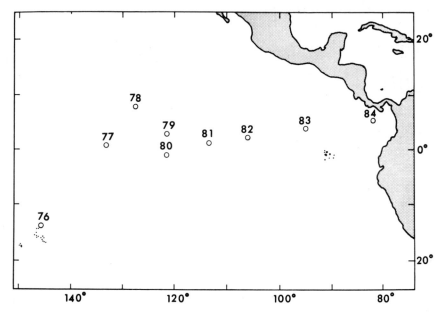

FIG. 1. Location of sites drilled on Leg 9 Deep Sea Drilling Project.

encountered on Leg IX differ from Caribbean sediments considerably with respect to the foraminiferal faunal content. Important disparities between Caribbean and Eastern Pacific sediments are to be found in the varying apparent palaeodepth and the chemical and physical characteristics of the overlying water masses. Difficulty in assigning Blow's "N" and "P" zonal numbers to the Eastern Pacific sediments was most intense in the deepest sediments on the western part of Leg 19. These sediments were encountered at the outset of the leg, and it was soon apparent that a new or modified zonation would have to be introduced in order to submit meaningful correlation data to the cruise programme. Analysis of our zonation as well as that of Bolli (1957) and Blow (1969), in the light of the growing literature on deep-sea solution of biocarbonates, has convinced us that the major shortcomings of Blow's zonal scheme in the Eastern Pacific arise because of the effects of solution on the zonal indices.

Selective solution of planktonic foraminifera

A great deal of literature has been published on the solution of calcium carbonate in the marine environment. The most authoritative work to date on the destructive solution of planktonic foraminiferal tests is that of Berger (1967, 1968, 1970a), Berger and Heath (1968) and Berger and Souter (1970).

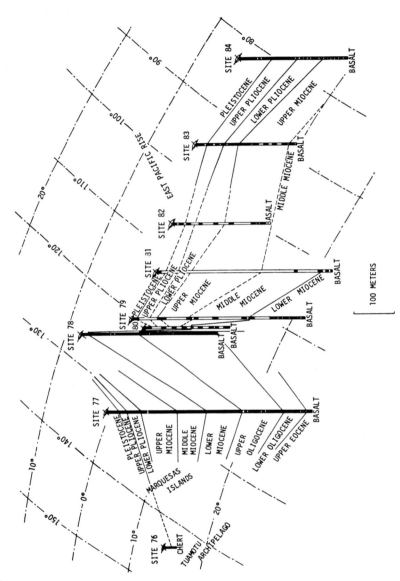

Fig. 2. Deep Sea Drilling Project Leg 9 series correlations (cored intervals in solid black).

Series Subseries	Planktonic Foraminiferal Zones	Definition of Zonal Boundaries IA = Appearance E = Extinction	Bolli (1957 a, b, c, 1966, 1970 in press Bolli & Bermudez 1966	Banner & Blow (1965) Parker (1967) Blow (1969)
Pleistocene	*Pulleniatina obliquiloculata*		*G. truncatulinoides truncatulinoides*	N-22-23
		◄ *G. fistulosus* (E)		
Upper Pliocene	*Globigerinoides fistulosus*		*G. truncatulinoides tosaensis*	N-21
		◄ *G. fistulosus* (IA)		
Lower Pliocene	*Sphaeroidinella dehiscens*		*G. exilis/G. miocenica*	N-19-20
		◄ *S. dehiscens* (IA)		
	Globorotalia tumida		*G. margaritae*	N-18
Upper Miocene		◄ *G. tumida* (IA)		
	Globorotalia plesiotumida		*G. dutertrei G. obliquus extremus*	N-17
		◄ *G. plesiotumida* (IA)		
	Globoquadrina altispira		*G. acostaensis G. fohsi lobata*	N-12-16
		◄ *G. fohsi lobata* (E)		
	Globorotalia fohsi lobata		*G. fohsi lobata*	
Middle Miocene		◄ *G. fohsi lobata* (IA)		N-10-11
	Globorotalia fohsi fohsi- Globorotalia peripheroacuta		*G. fohsi fohsi G. fohsi barisanensis*	
		◄ *G. peripheroacuta* (IA)		
	Globorotalia peripheroronda	◄ *P. glomerosa curva* (E)		N-9
	Praeorbulina glomerosa curva Subzone	◄ *P. glomerosa curva* (IA)		
	Globigerinoides bisphericus Subzone	◄ *G. bisphericus* (IA)	*G. insueta*	N-8
	Globoquadrina venezuelana	◄ *G. dissimilis* (E)		N-7
Lower Miocene	*Globigerinita dissimilis*		*C. stainforthi C. dissimilis*	N-5-6
		◄ *G. kugleri* (E)		
	Globorotalia kugleri		*G. kugleri*	N-4
		◄ *G. kugleri* (IA)		
	Globigerina angulisuturalis		*G. ciperoensis ciperoensis*	N-3
		◄ *G. opima* (E)		
	Globorotalia opima			
Upper Oligocene		◄ *C. cubensis* (E)	*G. opima opima*	N-2
	Chiloguembelina cubensis			
		◄ *C. opima* (IA)		
	Globigerina ampliapertura		*G. ampliapertura*	N-1
		◄ *P. barbadoensis* (E)		
Lower Oligocene	*Pseudohastigerina barbadoensis*		*C. chipolensis H. micra*	P-18-19
		◄ *G. insolita* (E)		
Upper Eocene	*Globorotalia insolita*		*G. cerroazulensis*	P-17

Fig. 3. Deep Sea Drilling Project Leg 9 definition and correlation of foraminiferal zones.

Berger's 1970 work deals in particular with solution of planktonic foraminifera in water masses overlying the East Pacific rise in the area traversed by
the drilling ship *Glomar Challenger* during Leg 9 of the J.O.I.D.E.S. Project.
While a few authors have considered the problems of solution of extant species
of planktonic foraminifera very little attention has been given to solution
susceptibility of extinct species. (Hays *et al.*, 1969; Cita, 1971).

Foraminiferal species have been ranked by various authors with respect
to their susceptibility to solution (see Fig. 4). The hierarchies proposed rank
species from low to high resistance and are based on various lines of evidence
including observational data from sediments (Schott, 1935), data from field
experiments (Berger, 1967) and laboratory experimental data (Berger, 1970).

Low Resistance   1 *Globigerinoides ruber*
                 2 *Orbulina universa*
                 3 *Globigerinella aequilateralis*
                 4 *Globigerina rubescens*
                 5 *Globigerinoides sacculifer*
                 6 *Globigerinoides tenellus*
                 7 *Globigerinoides conglobatus*
                 8 *Globigerina bulloides*
                 9 *Globigerina quinqueloba*
                10 *Globigerinita glutinata*
                11 *Candeina nitida*
                12 *Globororotalia hirsuta*
                13 *Globorotalia truncatulinoides*
                14 *Globorotalia inflata*
                15 *Globorotalia cultrata*
                16 *Globoquadrina dutertrei*
                17 *Globigerina pachyderma*
                18 *Pulleniatina obliquiloculata*
                19 *Globorotalia crassaformis*
                20 *Sphaeroidinella dehiscens*
                21 *Globorotalia tumida*
High Resistance 22 *Turborotalia humilis*

FIG. 4. Hierarchy of planktonic foraminiferal resistance to solution. (After Berger, 1970.)

Because little published data is available on the southeast biota with respect
to annual bio-productivity, there is no way of distinguishing absolutely
between truly rare foraminiferal species and solution susceptible species in
a given geographic area. Two papers on living foraminifera (Parker, 1960;
Bradshaw, 1959) give an approximation of the frequency of Holocene species.
Berger's (1970) (Fig. 4) ranking of Holocene species with respect to susceptibility to solution seems to approximate most closely the observations
of Pliocene and Pleistocene faunas made during Leg 9.

Departures in Leg 9 foraminiferal faunas from the solution susceptibility ranking proposed by Berger (1970) are relatively minor. *Globigerinoides ruber* (d'Orbigny) and *Orbulina universa* d'Orbigny were found to be far more common in sediments than their reported high susceptibility to solution would suggest. Both Parker (1960) and Bradshaw (1959) have, on the other hand, tabulated very high local tropical abundances for both of these species. Work in the Gulf of Mexico (Orr, 1967, 1969) and in the Northeast Pacific (Miles and Orr, 1972) has shown moreover that both *O. universa*—and *G. ruber* may precipitate a thick solution resistant cortex of calcite during their life cycle. *Globorotalia inflata* (d'Orbigny), *Globorotalia pachyderma* (Ehrenberg) and *Turborotalia humilis* (Brady) were rare in Leg 9 cores despite their recorded high resistance to solution. These three species, in addition to *Globigerina bulloides* d'Orbigny, are all characteristic of Holocene high-latitude water masses. Local occurrences of these species have been interpreted elsewhere as evidence of incursions of high latitude water masses.

A plot of the distribution of the relative abundances in Leg 9 late Pleistocene intervals for Berger's (1970) solution susceptible species (see Fig. 5) demonstrates that almost all of the species are more abundant toward the east. The most abrupt increases are between Sites 79 and 80. Also important here are the laterally discontinuous distribution patterns of some of the indices to important datums used by Blow (1969), for example, *Orbulina* sp., *Globigerinoides* spp., *Candeina* sp., etc. Because we lack experimental data, it is difficult to assign relative susceptibility to the tests of extinct species. We can suggest, however, a short list of species used elsewhere for zonal parameters that failed empirically in the southeast equatorial Pacific due to apparent high susceptibility to solution. These are: *Globigerina nepenthes* Todd, *Globigerina angulisuturalis* Bolli, *Globigerinatella insueta* Cushman and Stainforth, *Globigerinoides primordius* Banner and Blow, *Globorotalia tosaensis* Takayanagi and Saito, *Globorotalia truncatulinoides* (d'Orbigny), and *Orbulina saturalis*. Brönnimann.

Solution and diversity

Diversity of fossil faunas is the product of several complex factors of which destructive solution is only one. Primary or predepositional diversity may itself fluctuate substantially through time. Abnormally high diversity may result from undetected contamination, but the effects of contamination were easily recognised on Leg 9 faunas. Contaminants which were introduced by downhole contamination or by stratigraphical reworking were almost invariably solution-resistant species. Two of the commonest species involved in downhole contamination, for example, were *Pulleniation abliquiloculata* Parker and Jones and *Globorotalia tumida* (Brady). Both species were usually accompanied by pipe-scale in contaminated samples. *Globorotalia kugleri* Bolli was common in stratigraphically reworked faunas at Site 76.

Lipps (1970) has considered the evolution of primary planktonic faunal diversity through time. He noted persistent and co-ordinated fluctuations in the diversities of several groups of marine plankton and suggested a model based on variations in high-latitude sea surface temperatures to explain these fluctuations. Leg 9 faunas reflect Lipps' observations in as much as the lowest

				Sites			
	77	79	80	81	82	83	84
1 *Globigerinoides ruber*	+	|	+	*	+	*	●
2 *Orbulina universa*	+	+	*	*	*	*	●
3 *Globigerinella aequilateralis*	*	|	+	+	+	*	●
4 *Globigerina rubescens*	·	·			·	|	+
5 *Globigerionoides sacculifer*	+	+	*	|	*	*	*
6 *Globigerinoides conglobatus*	+	+		+	+	+	*
7 *Globigerina bulloides*	·	·	+	·		|	*
8 *Globigerina quinqueloba*	·						
9 *Globigerinita glutinata*	+	·	|	+	·	*	*
10 *Cadeina nitida*							|

Symbols: *, absent; ·, very rare; |, rare; +, scattered; *, common; ●, abundant.

[a] These values are the product of rapid counts of 50 to 100 specimens where very rare = 1 specimen, rare = 2.5 specimens, scattered = irregular high and low frequencies through the interval examined, common = up to 10% of the fauna, abundant = 10% of the fauna or more.

FIG. 5. Late Pleistocene occurrences of solution-susceptible species.

foraminiferal diversities are found in Oligocene faunas and the highest in the late Miocene and Pliocene (see Fig. 8). Keeping these background diversity fluctuations in mind, it is not difficult to distinguish between primary diversity and secondary low diversity which is brought about by solution.

In the absence of a more quantitative method, we subjectively distinguished three "states" of solution to which any given foraminiferal fauna may have been exposed. "Initial" solution is a condition recognised more by negative evidence than by position. Faunas which were subjected to this condition rarely exhibit definite evidence for solution such as partially dissolved tests, and only the spinose thin-walled forms (e.g. *Globigerina* spp. and *Globigerinoides* spp.) are totally dissolved. The "intermediate" stage of intensity in destructive solution is the easiest to recognise as the individuals which remain bear visible evidence of solution. At this stage of solution diversity is remarkably low, and all small thin-walled or juvenile individuals are absent as are most of all spinose types. In addition even thick-walled "robust" individuals which have resisted total solution usually have partially dissolved penultimate and ultimate chambers in their tests (e.g. *Globorotalia* spp., *Sphaeroidinella* spp., *Pulleniatina* spp., *Globoquadrina* spp.). "Total" solution signifies the complete removal of the foraminifera. Intervals that exhibit total solution in Leg 9 cores cannot be considered to reflect a local abatement of biologic productivity because, as Berger (1970) has observed elsewhere under parallel circumstances, the remainder of the fossil fauna and flora be it nannoplankton or Radiolaria do not exhibit a similar waning productivity.

Compensation depth and the lysocline

Berger (1967) has proposed the term lysocline for the level in a marine profile where the maximum change takes place with respect to the solution index. This level seems to be more useful than the calcium carbonate compensation depth because it is more easily recognised and measured than the latter level. The compensation depth has been defined as the level at which the solution rate depth profile and the rate of supply of calcium carbonate to the ocean floor are equal. Berger (1968) has related the lysocline to the top of the Antarctic bottom water in the Atlantic, and it seems probable that a similar condition exists in the Pacific. According to Berger (1970) the entire Pacific Ocean is in an undersaturated condition with respect to calcium carbonate except for the uppermost few hundred metres. Berger (1970) has used data supplied by Blackman (1966) to postulate the lysocline surface in the southeast Pacific. In this area the lysocline (Figs 6 and 7) assumes the shape of a quarter bowl over the East Pacific Rise and opens towards the west. By superimposing the Leg 9 drilling sites on Berger's map, we see that while every site is presently below the lysocline, Sites 79 and 80 display the greatest proximity to this level. We can see no appreciable change in diversities of Late Pleistocene faunas over the same geographic area, but faunal changes over the same area (Fig. 5) appear to corroborate the presence of the lysocline in the configuration proposed by Berger (1970).

Berger (1970) has suggested that the lysocline surface is related to the upper layer of the Antarctic bottom water. Following this, we might assume that at some time before the Pleistocene, the lyoscline should have occupied a lower level in the bathymetric column than at present. This should be displayed by foraminiferal faunas in pre-Pleistocene sediments either by an increase in species diversity or in the increased appearance of solution susceptible species. Examination of the diversity values for each site (see Fig. 8)

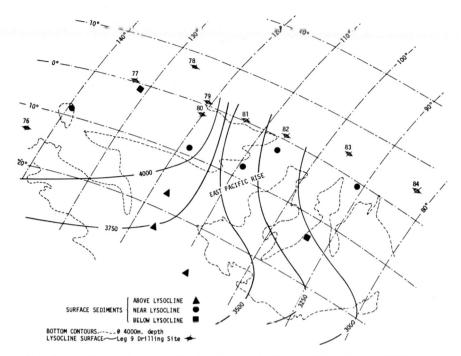

FIG. 6. Lysocline surface in the Eastern Equatorial Pacific. (After Berger, 1970.)

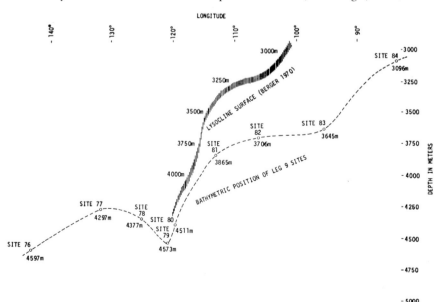

FIG. 7. Bathymetry of J.O.I.D.E.S. Leg 9 sites and the lysocline in the Eastern Equatorial Pacific.

does not show any appreciable increase below the Pleistocene. Diversity instead appears to decrease steadily below the *Sphaeroidinella dehiscens* Zone (Mid-Lower Pliocene) and drops to near zero in much of the *Globoquadrina altispira* Zone (Upper-Middle Miocene). Similarly, diversity values in much of the *Globoquadrina venezuelana* Zone (Upper-Lower Miocene) drop to near zero. These lows in diversity do not correspond to Lipps' (1970) model of primary diversity and appear to be local to this area in the southeast Pacific. The low-diversity high-solution intervals of the *G. altispira* and *G. venezuelana* zones observed on Leg 9 might be interpreted as temporary upward fluctuations in the lysocline. If we examine the faunal diversity by zones (see Fig. 8), we find that these Miocene low diversity intervals are restricted to Sites 77, 78 and 79. Further to the east in Sites 80 through 84 diversities appear to return to "normal" for both of these zones. This suggests that these low diversities are the product of the intersection of the sea-floor and the calcium carbonate compensation depth at various times during deposition. The diversity values presented here correspond remarkably well with calcium carbonate percentage curves produced for each of the 9 drilling sites for Leg 9. Intervals of low diversity are invariably intervals of low calcium carbonate percentage. Similar trends in deep-sea sediments were noted by Arrhenius (1948).

It is possible to relate the solution stratification of the marine bathymetric column to some of the solution phenomena characteristic of Leg 9 cores. It should be borne in mind that because of the ephemeral nature of the Antarctic water masses and the phenomena of sea-floor spreading, pre-Pleistocene marine profiles should bear little relation to the present lysocline. Faunas characterised here as having been exposed to "initial" solution were undoubtedly deposited in sediments near to or just above the lysocline. In these faunas only species with the thin delicate tests, the most solution-susceptible species, are removed by the undersaturated waters above the lysocline. Faunas exposed to "intermediate" stage solution where destructive solution is apparent both in faunal diversity and actual specimen appearance accumulated below the lysocline. "Total" solution faunas would imply deposition of sediments well below the lysocline and undoubtedly near, if not below, the compensation depth. By comparing diversities of faunas from Cainozoic epochs in the geographic area covered in Leg 9, we may characterise the individual epochs with respect to the state of solution of their foraminiferal faunas and the implied position of the lysocline. The marked vacillation in diversities of the Pleistocene and Pliocene faunas suggests that the lysocline lay very near the bottom (as at present) and periodically intersected the sediments. Miocene sediments appear to have been near to or above the lysocline with the exception of the *G. altispira* and *G. venezuelana* zones when the

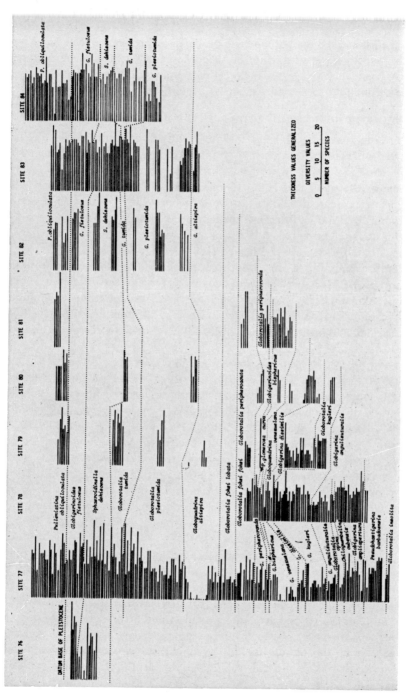

Fig. 8. Diversity values for planktonic foraminifera arranged by site D.S.D.P. Leg 9.

lysocline intersected the bathymetric profile in an area east of Site 80. Assuming the lysocline existed during the Oligocene and Upper Eocene, it appears to have been sited well below the level of sedimentation, as diversities remain stable throughout these epochs.

Planktonic foraminiferal zones

Most of the zones used on Leg 9 (Figs 9 and 10) are stratigraphically large units and normally within each zone there are several datum horizons which can be used to accurately position a sample within a named zone.

Some of the published zones (for example, Blow, 1969; Berggren, 1969) appear to represent such short time intervals that they should be relegated to subzonal rank. Many of these subzones (for example, published zones of the *G. fohsi* lineage) are useful in thick continental shelf sequences but they are of very limited use in the condensed sections from deep-sea cores.

There are a number of biostratigraphic problems which arise because certain taxa are prone to solution in deep oceanic water, and the primary aim of the proposed zonal scheme is to established the basis of a planktonic foraminiferal zonal scheme for tropical deep-sea sediments. The zonal markers which have been selected and proposed here appear to be resistant to corrosion and solution.

Existing Cainozoic planktonic foraminiferal zonal schemes for the equatorial region have been established using faunas obtained from continental shelf and slope sediments. Stratigraphically important solution-resistant taxa include species of *Globorotalia, Sphaeroidinella, Globoquadrina* and *Pulleniatina*. There are exceptions and Hays *et al.* (1969) noted the following extinct taxa as solution-susceptible species: *Globorotalia margaritae* Bolli and Bermudez, *Globigerina nepenthes* Todd, *Globoquadrina altispira* Cushman and Jarvis and *Globigerinoides fistulosus* (Schubert). While agreeing that the first two species are comparatively rare in sediments examined on Leg 9 both *G. altispira* and *G. fistulosus* were usually common and well preserved.

ZONAL CONSISTENCY AND PRECISION

The precision of Cainozoic biostratigraphic correlation by means of planktonic foraminifera is dependent on a number of qualitative methods and assumptions. An important aspect of the methods includes the accurate and consistent identification of taxa used for correlation. This is normally qualitative and subject to operator interpretation and errors. Although it is possible for one operator to be consistently right or wrong regarding an identification of a taxon throughout its stratigraphic range, it was possible on Leg 9 to have two operators and thus the taxonomic identifications were constantly checked.

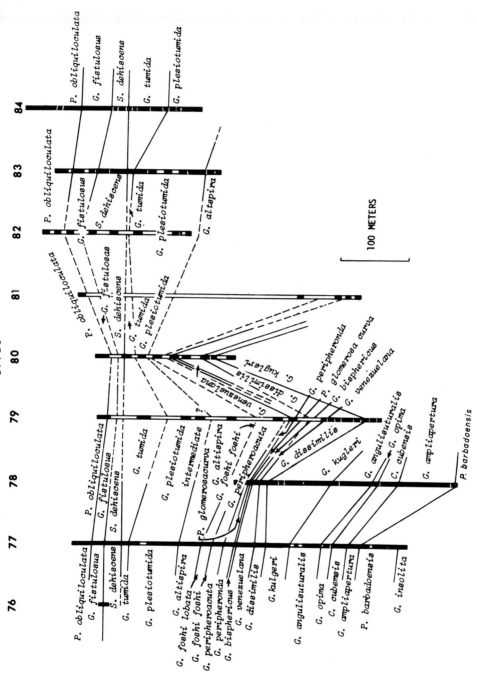

FIG. 9. Deep Sea Drilling Project Leg 9 foraminiferal zone correlations (cored intervals in solid black).

Series	Foraminifera	Nannoplankton	Radiolaria
Pleistocene	*Pulleniatina obliquiloculata*	*Gephyrocapsa "oceanica" Ceratolithus cristatus* Subzone	No zonal name
		Gephyrocapsa "oceanica" C. leptoporus macintyrei S.Z.	*Pterocanium prismatium*
Upper Pliocene	*Globigerinoides fistulosus*	*G. "oceanica"-C. carteri* S.Z.	
		Discoaster brouweri Cyclococcolithus leptoporus Subzone	*Spongaster pentas* East
Lower Pliocene	*Sphaeroidinella dehiscens*	*Ceratolithus rugosus*	
	Globorotalia tumida	*Ceratolithus tricorniculatus*	*Stichocorys peregrina*
Upper Miocene	*Globorotalia plesiotumida*		West
		Discoaster variabilis Discoaster challengeri S.Z.	*Ommatartus penultimus*
			Ommatartus antepenultimus
	Globoquadrina altispira	*D. variabilis* *D. variabilis D. hamatus* S.Z. *D. exilis D. variabilis* *C. eopelagicus* S.Z. Subzone	*Cannartus (?) petterssoni*
	Globorotalia fohsi lobata	*Discoaster exilis Discoaster kugleri* Subzone	*Cannartus laticonus*
Middle Miocene	*Globorotalia fohsi fohsi Globorotalia peripheroacuta*	*D. exilis-Cyclococcolithus neogammation* S.Z.	
		S. heteromorphus T. rugosus S.Z.	*Dorcadospyris alata*
	Globorotalia peripheroronda Praeorbulina glomerosa curva Subzone	*Sphenolithus heteromorphus Helicopontosphaera ampliaperta* S.Z.	*Calocycletta costata*
	Globigerinoides bisphericus Subzone *Globoquadrina venezuelana*	*Triquetrorhabdulus carinatus Sphenolithus heteromorphus* S.Z.	
Lower Miocene	*Globigerinita dissimilis*	*T. carinatus S. belemnos* S.Z.	*Calocycletta virginis*
		Triquetrorhabdulus carinatus Cyclococcolithus neogammation S.Z.	
	Globorotalia kugleri		*Lychnocanium bipes*
	Globigerina angulisuturalis	*T. carinatus C. bisectus var.* S.Z.	*Dorcadospyris papilio*
		C. bisectus T. carinatus S.Z.	
Upper Oligocene	*Globorotalia opima*	*C. bisectus S. distentus* Subzone	*Theocyrtis annosa*
	Chiloguembelina cubensis		
	Globigerina ampliapertura	*C. bisectus Helicopontosphaera compacta* S.Z.	*Lithocyclia angustum*
Lower Oligocene	*Pseudohastigerina barbadoensis*	*C. lucitanicas S. predistentus*	
		C. lucitanicas I. recurvus S.Z.	*Theocyrtis tuberosa*
Eocene	*Globorotalia insolita*	*Discoaster barbadiensis*	

FIG. 10. Correlations and equivalents: foraminifera, nannoplankton and radiolaria J.O.I.D.E.S. Leg 9.

Some of the assumptions used in the inter-site correlations on Leg 9 include:

The assumption that within their known stratigraphic ranges certain well-documented taxa had been widely distributed in the upper oceanic layers of the eastern equatorial Pacific. It was further assumed that the phylogenetic events used in inter-site correlation had been widespread synchronous events (Jenkins, 1965).

This uncomplicated over-simplified model had to be modified because it was complicated by the further assumption that the oceanic current pattern had remained fixed and unchanging from the Upper Eocene to Pleistocene.

If one accepts that sea-floor spreading has occurred from the Upper Eocene to the Pleistocene another complication arises because parts of the sea floor must have moved under different water masses which could have contained slightly different faunas, and also the sea floor moved under areas of varying productivity.

There is evidence that some taxa have not been uniformly distributed within the east–west equatorial belt traversed on Leg 9. This is especially true for the Pliocene–Pleistocene: for example, *Pulleniatina spectabilis* was limited to the western Site 77B, and *Globorotalia inflata* and *G. pachyderma* were only found in the eastern Sites 83 and 84 except for one record of *G. inflata* at Site 77. It is possible that known climatic changes affected the distribution of faunas during the Pliocene–Pleistocene interval.

A fundamental complication is the apparent selective action of solution on foraminiferal tests which could explain the interrupted stratigraphic record of a particular species at one or more localities.

Within the resolution of accuracy of the foraminiferal zonal scheme it was not possible to detect and differentiate Recent faunas from Pleistocene faunas in the uppermost core samples at the various sites.

ZONAL PARAMETERS AND CHARACTERISTICS

Pulleniatina obliquiloculata Zone

Definition:

Top: Not defined, but the zone includes all faunas above the extinction of *Globigerinoides fistulosus* (Schubert).

Base: Extinction of *Globigerinoides fistulosus* (Schubert). *P. obliquiloculata* (Plate 3, figs 13–14) serves as the "name bearer" for this zone particularly because changes in the coiling direction of this species is potentially useful for defining subzones.

Age: Pleistocene–Recent.

Taxa:

Extinctions within zone:

Globigerina praedigitata
Globigerinoides bolli
G. obliquus (in lower part of
    zone)

Globorotalia humerosa
    (in lower part of zone)
G. tosaensis
Pulleniatina primalis

Initial appearances within zone:

Globorotalia fimbriata
G. pachyderma

Hastigerina pelagica
H. rhumbleri

Species present and ranging throughout zone:

Candeina nitida
Globigerina bradyi
Globigerina bulloides
G. calida
G. digitata
G. falconensis
G. juvenilis
G. rubescens
*Globigerinella
    aequilateralis*
Globigerinoides conglobatus
G. ruber
G. sacculifer
G. trilobus
Globoquadrina venezuelana
Globigerinita glutinata

Globorotalia crassaformis
G. crassula
G. dutertrei
G. hirsuta
G. inflata
G. menardii
G. tumida flexuosa
G. tumida tumida
G. scitula
G. ungulata
Globorotaliodes hexagona
Orbulina universa
Pulleniatina obliquiloculata
Sphaeroidinella dehiscens
Turborotalia humilis

Species restricted to zone:
Globorotalia fimbriata
G. truncatulinoides pachytheca
G. truncatulinoides

Reference section: Site 77B: 00°28.90′N, 133°13.70′W.

Occurrence of zone: Continuously cored at Sites 77, 79, 84 and spot cored
    at Sites, 80, 81, 82 and 83 (Figs 1 and 2).

Potential subzones: Based on coiling changes of *P. obliquiloculata.*

Correlation: Bolli (1970): *G. truncatulinoides truncatulinoides* Zone.

Banner and Blow (1965), Blow (1969): Broadly equivalent to Zones N22–23.

Parker (1967): Broadly equivalent to Zones N22–23.

International series: Pleistocene–Recent.

Globigerinoides fistulosus Zone

Definition:

Top: Extinction of *Globigerinoides fistulosus* (Schubert) (Plate 1, figs 11–12).
Base: Initial evolutionary appearance of *Globigerinoides fistulosus* (Schubert).
Age: Upper Pliocene.
Taxa:
Extinctions within zone:

Globigerina apertura
Globigerina decoraperta
Globoquadrina altispira

Globorotalia exilis
G. multicamerata
Orbulina biolobata
Sphaeroidinella seminulina

Initial appearances within zone:

Globigerina digitata
Globorotalia dutertrei
G. inflata

G. tosaensis
G. ungulata

Species present and ranging throughout zone:

Candeina nitida
Globigerina bulloides
G. calida
G. juvenilis
Globigerinoides conglobatus
*Globigerinella
    aequilateralis*
G. obliquus
G. ruber
G. sacculifer
G. miocenica
G. scitula
G. tumida flexuosa
G. tumida tumida
Globorotaloides hexagona

G. trilobus
Globigerinita glutinata
Globoquadrina venezuelana
Globorotalia crassaformis
G. crassula
G. hirsuta
G. humerosa
G. inflata
G. menardii
Orbulina universa
Pulleniatina obliquiloculata
P. primalis
Sphaeroidinella dehiscens

Species restricted to zone:

Globigerinoides fistulosus

Datum planes within zone:
Extinctions of:

Globoquadrina altispira
Globorotalia multicamerata
G. exilis

Reference section: Site 77B: 00°28.90′N, 133°13.70′W.

Occurrence of zone: Continuously cored at Sites 77 and 84, and spot-cored at Sites 76, 79, 80, 82 and 83 (Figs 1 and 2).

Correlation: Bolli (1970): *G. truncatulinoides* cf. *tosaensis* Zone.

Banner and Blow (1965), Blow (1969): Broadly equivalent to Zone N21.

Parker (1967): Broadly equivalent to Zone N21. International series: Upper Pliocene.

Sphaeroidinella dehiscens Zone

Definition:

Top: Initial evolutionary appearance of *Globigerinoides fistulosus* (Schubert).

Base: Initial evolutionary appearance of *Sphaeroidinella dehiscens* (Parker and Jones), illustrated on Plate 3, figs 17–18.

Age: Lower–Upper Pliocene.

Taxa:

Extinctions within zone:

Globigerina nepenthes   *G. cibaoensis*
Globoquadrina dehiscens   *G. margaritae*
Globorotalia acostaensis   *G. pseudomiocenica*

Initial appearance within zone:

Globigerina rubescens   *Pulleniatina obliquiloculata*
Globorotalia crassaformis   *Sphaeroidinella dehiscens*
G. hirsuta

Species present and ranging throughout zone:

Candeina nitida   *G. decoraperta*
Globigerina apertura   *G. juvenilis*
G. bradyi   *Globigerinella aequilateralis*
G. bulloides   *Globigerinoides bolli*
G. calida   *G. conglobatus*
G. ruber   *G. obliquus*
G. sacculifer   *G. multicamerata*
G. trilobus   *G. pseudomiocenica*
Globigerinita glutinata   *G. scitula*
Globoquadrina altispira   *G. tumida fluexuosa*
G. venezuelana   *G. tumida tumida*
Globigerinita glutinata   *Globorotaloides hexagona*
Globorotalia exilis   *Orbulina bilobata*
G. humerosa   *Orbulina universa*
G. menardii   *Pulleniatina primalis*
G. miocenica   *Sphaeroidinella seminulina*
                                     S. subdehiscens

Species restricted to zone:
Globorotalia crassiformis viola
Pulleniatina spectabilis

Reference section: Site 77B: 00°28.90'N, 133°13.70'W.

Occurrence of zone: Continuously cored at Sites 77 and 84, and spot-cored at Sites 76, 82 and 85 (Figs 1 and 2).

Potential subzones: Based on the total range of *P. spectabilis* in the lower part of *S. dehiscens* Zone at the western Site 77B.

Correlation: Bolli (1970): *G. exilis*/*G. miocenica*

Zone: Upper *G. margaritae* Zone.

Banner and Blow (1965), Blow (1969): Broadly equivalent to Zones N19–20.

Parker (1967): Broadly equivalent to Zones N19–20.

International series: Lower–Upper Pliocene.

Globorotalia tumida Zone

Definition:

Top: Initial evolutionary appearance of *Sphaeroidinella dehiscens* (Parker and Jones).

Base: Initial evolutionary appearance of *Globorotalia tumida* (Brady), illustrated on Plate 3, figs 7–9.

Age: Upper Miocene–Lower Pliocene.

Taxa:

Extinction within zone:

G. merotumida
G. plesiotumida

Initial appearances within zone:

Globigerina apertura   *G. tumida flexuosa*
G. crassula   *Globorotalia tumida tumida*
Globorotalia margaritae

Species present and ranging throughout zone:

Candeina nitida   *G. nepenthes*
Globigerina bulloides   *G. praedigitata*
Globigerina calida   *Globigerinella aequilateralis*
G. decoraperta   *Globigerinoides conglobatus*
G. juvenilis   *G. obliquus*
G. ruber   *G. humerosa*
G. sacculifer   *G. menardii*
G. trilobus   *G. multicamerata*
Globigerinita glutinata   *G. pseudomiocenica*

Globoquadrina altispira
G. venezuelana
G. dehiscens
Globorotalia acostaensis
Globorotalia anfracta
G. cibaoensis
G. exilis

Globorotaloides hexagona
Orbulina bilobata
O. universa
Pulleniatina primalis
Sphaeroidinella seminulina
S. subdehiscens

Reference section: Site 77B: 00°28.90′N, 133°13.70′W.

Occurrence of zone: Continuously cored at Sites 77, 83 and 84, and spot-cored at Sites 79 and 80 (Figs 1 and 2).

Correlation: Bolli (1970): Broadly equivalent to the *G. margaritae* Zone.

Banner and Blow (1965), Blow (1969): Broadly equivalent to Zone N18.

Parker (1967): Broadly equivalent to Zone N18.

International series: Upper Miocene–Lower Pliocene.

Globorotalia plesiotumida Zone

Definition:

Top: Initial evolutionary appearance of *Globoortalia tumida* (Brady).

Base: Initial evolutionary appearance of *Globorotalia plesiotumida* Banner and Blow, illustrated on Plate 3, figs 4–6.

Age: Upper Miocene.

Taxa:

Extinctions within zone:

Globorotalia continuosa

Initial appearances within zone:

Candeina nitida
Globigerinoides conglobatus
Globorotalia exilis
G. humerosa

G. multicamerata
G. plesiotumida
G. pseudomiocenica
Pulleniatina primalis

Species present and ranging throughout zone:

Globigerina bulloides
G. calida
G. decoraperta
G. juvenilis
G. nepenthes
G. ruber
G. sacculifer
G. trilobus
Globigerinita glutinata

G. praedigitata
G. woodi
Globigerinella aequilateralis
Globigerinoides bolli
G. obliquus
Globorotaloides hexagona
Orbulina bilobata
Orbulina suturalis
O. universa

Globoquadrina altispira   *Sphaeroidinella seminulina*
G. dehiscens   *S. subdehiscens*
G. venezuelana
Globorotalia acostaensis
G. cibaoensis
G. menardii
G. merotumida
G. miocenica
G. scitula

Reference section: Site 77B: 00°29.90′N, 133°13.70′W.

Occurrence of zone: Continuously cored at Site 77, and spot-cured at Sites 79, 80, 82, 83 and 84 (Figs 1 and 2).

Potential subzones: Based on the initial appearance of *Pulleniatina primalis*.

Correlation: Bolli (1966, 1970), Bolli and Bermúdez (1966): Broadly equivalent to *G. dutertrei-G. obliquus extremus* Zone.

Banner and Blow (1965), Blow (1969): Broadly equivalent to Zone N17.

Parker (1967): Broadly equivalent to Zone N17.

International series: Upper Miocene.

Globoquadrina altispira Zone

Definition:

Top: Initial appearance of *Globorotalia plesiotumida* Banner and Blow.

Base: Extinction of *Globorotalia fohsi lobata*.

G. altispira serves as the name bearer for the zone because of its resistance to solution and abundance. This species is illustrated on Plate 1, figs 16–17.

Age: Middle–Upper Miocene.

Taxa:

Extinctions within zone:
Globorotalia siakensis
Orbulina suturalis

Initial appearances within zone:
Globigerina bulloides   *G. cibaoensis*
G. calida   *G. merotumida*
G. decoraperta   *Orbulina bilobata*
Globorotalia acostaensis   *Sphaeroidinella subdehiscens*

Species present and ranging throughout zone:
Globigerina bradyi   *Globigerinita glutinata*
G. praedigitata   *Globoquadrina altispira*
Globigerinella aequilateralis *G. dehiscens*

Globigerinoides bolli
G. obliquus
G. ruber
G. sacculifer
G. trilobus
O. universa

G. venezuelana
Globorotalia continuosa
G. menardii
Globorotaloides hexagona
Sphaeroidinella seminulina

Reference section: Site 77B: 00°28.90′N, 133°13.70′W.

Occurrence of zone: Continuously cored at Site 77, and spot-cured at Sites 79 and the top of the zone penetrated at Site 83 (Figs 1 and 2).

Potential subzone: Based on the initial appearance of *G. merotumida* (Plate 2, figs 13–15), although there is evidence that its initial appearance is diachronous within the examined area as for example at Sites 77 and 83.

General: The *G. altispira* Zone is a fairly large stratigraphic unit and it was not subdivided on Leg 9 because of the relatively poor faunas in the examined area.

Correlation: Bolli (1957a): *G. fohsi robusta-G. mayeri-G. menardii-G. acostaensis* Zones.

Banner and Blow (1965), Blow (1969): Broadly equivalent to Zones N12–16.

Parker (1967): Upper part of *G. altispira* Zone equivalent to Zone N16.

International series: Middle–Upper Miocene.

Globorotalia fohsi lobata Zone

Definition:

Top: Extinction of *Globorotalia fohsi lobata* Bermúdez.

Base: Initial evolutionary appearance of *G. fohsi lobata* Bermúdez (Plate 2, figs 4–6).

Age: Middle Miocene.

Taxa:

Extinctions within zone:
Globorotalia fohsi fohsi

Initial appearances within zone:
Globorotalia menardii

Species present and ranging throughout zone:

Globigerina nepenthes
G. juvenilis
Globigerinoides bolli
G. ruber
G. sacculifer
G. trilobus
Globoquadrina altispira

G. dehiscens
G. venezuelana
Globorotalia continuosa
G. siakensis
Orbulina universa
Sphaeroidinella seminulina

Species limited to the zone:
Globorotalia fohsi lobata

Reference section: Site 77B: 00°28.90′N, 133°13.70′W.
Occurrence of zone: Continuously cored at Site 77B (Figs 1 and 2)
General: *Globorotalia fohsi robusta*, which Bolli (1957a) regarded as the
terminal taxon of the *G. fohsi* lineage and which was used as zonal marker
in Trinidad, was not identified in the examined area.
Correlation: Bolli (1957a): Equivalent to the *G. fohsi lobata* Zone.
Banner and Blow (1965), Blow (1969): Broadly equivalent to the middle part
of Zone N12.
International series: Middle Miocene.

Globorotalia fohsi fohsi-Globorotalia peripheroacuta Zone

Definition:
Top: Initial evolutionary appearance of *Globorotalia fohsi lobata* Bermúdez.
Base: Initial evolutionary appearance of *Globorotalia peripheroacuta* Banner
and Blow (Plate 2, figs 19–21).
Age: Middle Miocene.
Taxa:
Extinctions within zone:

Globigerina praedigitata	*G. scitula*
Globigerinoides bolli	*Orbulina suturalis*
Globorotalia fohsi fohsi	*O. universa*

Species present and ranging through zone:

Globigerina bradyi	*Globigerinatella aequilateralis*
G. juvenilis	*Globigerinoides ruber*
G. falconensis	*G. trilobus*
G. foliata	*Globigerinita glutinata*
Globoquadrina altispira	*G. siakensis*
G. dehiscens	*Globorotaloides hexagona*
G. venezuelana	*Orbulina universa*
Globorotalia continuosa	*Sphaeroidinella seminulina*
G. praemenardii	

Species limited to zone:
Globigerina bulbosa
Globorotalia peripheroacuta
G. cf. *miozea*
G. obesa

Reference section: Site 77B: 00°28.90′N, 133°13.70′W.

Occurrence of zone: Continuously cored at Site 77B, and spot-cored at Sites 78, 79, 80 and 81 (Figs 1 and 2).

Potential subzone: Based on the extinction of *G. peripheroacuta*.

Correlation: Bolli (1957): Equivalent to the *G. fohsi fohsi* Zone.

Banner and Blow (1965), Blow (1969): Broadly equivalent to the Zones N10 and the lower part of N12.

International series: Middle Miocene.

Globorotalia peripheroronda Zone

Definition:

Top: Initial evolutionary appearance of *Globorotalia peripheroacuta* Banner and Blow.

Base: Initial evolutionary appearance of *Praerobulina glomerosa curva* (Blow).

G. peripheroronda (Plate 3, figs 1–3) serves as the name bearer for this zone because of its resistance to solution and abundance.

Age: Middle Miocene.

Praeorbulina glomerosa curva Subzone

Definition:

Top: Extinction of *Praerobulina glomerosa curva* (Plate 3, figs 10–12).

Base: Initial evolutionary appearance of *Praeorbulina glomerosa curva*.

Remarks: The *P. glomerosa curva* Subzone is positioned in the lower part of the *G. peripheroronda* Zone.

Taxa:

Extinctions within zone:

Globigerinatella insueta
Globigerinoides bisphericus
Globorotalia archaeomenardii
G. praescitula

Initial appearance within zone:

Sphaeroidinella seminulina

Species present and ranging throughout zone:

Globigerina bradyi
G. foliata
G. juvenilis
G. falconensis
Globigerinoides mitra
G. ruber
G. sacculifer
G. trilobus

Globoquadrina altispira
G. dehiscens
G. venezuelana
Globorotalia continuosa
G. cf. minutissima
G. praemenardii
G. peripheroronda
G. siakensis

Globigerinita glutinata

Globorotaloides hexagona
Hastigerinella bermudezi

Species limited to zone:
Globoquadrina langhiana
Globorotalia archaeomenardii
Praeorbulina glomerosa circularis

Datum planes within zone:
Extinction:
Globigerinatella insueta
Initial appearances:
Globorotalia praemenardii
Sphaeroidinella seminulina

Reference section: Site 78: 07°57.00′N, 127°21.35′W.

Occurrence of zone: Completely penetrated and continuously cored at Sites 77B and 78, and spot-cored at Sites 79, 80 and 81 (Figs 1 and 2).

Correlation: Bolli (1957a): Equivalent to the upper part of the *G. barisanensis* (=*G. peripheroronda*) Zone. Banner and Blow (1965), Blow (1969): Broadly equivalent to Zone N9.

International series: Middle Miocene.

Globoquadrina venezuelana Zone

Definition:

Top: Initial evolutionary appearance of *Praeorbulina glomerosa curva* (Blow).

Base: Extinction of *Globigerinita dissimilis* (Cushman and Bermúdez).

G. venezuelana (Plate 2, figs 1–3) serves as the name bearer for this zone due to its abundance and resistance to solution.

Age: Lower Miocene.

Globigerinoides bisphericus Subzone

Definition:

Top: Initial evolutionary appearance of *Praeorbulina glomerosa curva* (Blow).

Base: Initial evolutionary appearance of *Globigerinoides bisphericus* Todd (Plate 1, figs 8–10).

Remarks: The *G. bisphericus* Subzone is positioned in the upper part of the *G. venezuelana* Zone.

Taxa:

Extinctions within zone:
Cassigerinella chipolensis

Initial appearances within zone:
Globigerinatella insueta
Globigerinoides bisphericus

Globorotalia archaeomenardii
G. praemenardii

G. mitra   *Globorotaloides hexagona*
G. ruber   *Hastigerinella bermudezi*
G. sacculifer

Species present and ranging throughout zone:
Globigerina bradyi   *Globoquadrina altispira*
G. juvenilis   *G. dehiscens*
G. foilata   *G. venezuelana*
Globigerinoides obliquus   *Globorotalia continuosa*
G. ruber   *G. peripheroronda*
G. trilobus   *G. praescitula*
                                             G. siakensis

Species limited to zone:
Globorotalia cf. *bella*

Datum planes within zone:
Initial appearances:
Globigerinoides bisphericus
Hastigerinella bermudezi

Reference Section: Site 77B: 00°28.90′N, 133°13.70′W.

Occurrence of Zone: Continuously cored at Sites 77 and 78, and spot-cored at Sites 80 and 81 (Figs 1 and 2).

Correlation: Bolli (1957a): Equivalent to the *G. insueta* Zone except for the upper part *G. insueta* Zone above the appearance of *Praeorbulina glomerosa curva*. Banner and Blow (1965), Blow (1969): Broadly equivalent to Zones N7–8.

International series: Upper Aquitanian to Lower Burdigalain: Lower Miocene.

Globigerinita dissimilis Zone

Definition:

Top: Extinction of *Globigerinita dissimilis* (Cushman and Bermúdez), this species is illustrated on Plate 1, figs 13–15.

Base: Extinction of *Globorotalia kugleri* Bolli.

Age: Lower Miocene.

Taxa:

Extinctions within zone:
Globigerina angustiumbilicata
Globigerinita dissimilis
Globorotaloides suteri

Initial appearances within zone:
Globigerina foliata

Globigerinoides obliquus
Globorotalia praescitula

Species present and ranging throughout zone:

Cassigerinella chipolensis *Globoquadrina altispira*
Globigerina bradyi *G. dehiscens*
G. juvenilis *G. tripartita*
G. angustiumbilicata *G. venezuelana*
Globigerinoides altiperturus *Globorotalia continuosa*
G. trilobus *G.* cf. *minutissima*
Globigerinita glutinata *G. peripheroronda*
  G. siakensis

Species limited to zone:
Globorotaloides stainforthi

Reference section: Site 79: 02°33.02′N, 121°34.00′W.

Occurrence of zone: Continuously cored at Sites 77 and 78, and spot-cored at Sites 79 and 80 (Figs 1 and 2).

General: Relatively thin development at both Sites 77 and 78 with apparently thicker sequences at both Sites 79 and 80.

Correlation: Bolli (1957a): Equivalent to the combined *G. dissimilis* and *G. stainforthi* Zones.

Banner and Blow (1965), Blow (1969): Approximately equivalent to the Zones N5–N6.

International series: Lower Miocene.

Globorotalia kugleri Zone

Definition:

Top: Extinction of *Globorotalia kugleri* Bolli (Plate 2, figs 10–12).

Base: Initial appearance of *G. kugleri*.

Age: Lower Miocene.

Taxa:

Extinctions within zone:
Globorotalia nana

Initial appearances within zone:

Globigerina juvenilis *G. dehiscens*
Globigerinoides trilobus *Globorotalia peripheroronda*
Globigerinita glutinata
Globoquadrina altispira

Species present and ranging throughout zone:

Cassigerinella chipolensis *Globorotalia continuosa*

Globigerina angustiumbilicata
G. bradyi
Globigerinita dissimilis
Globoquadrina tripartita
G. venezuelana

G. cf. *minutissima*
G. siakensis
Globorotaloides suteri
G. unicava

Species limited to zone:
Globigerinoides primordius
Globorotalia kugleri
G. mendacis

Datum planes within zone:
Initial appearance:
Globigerinoides trilobus

Reference section: Site 78: 07°57.00′N, 127°21.35′W.
Occurrence of zone: Continuously cored at Sites 77 and 78, and spot-cored at Sites 79 and 80 (Figs 1 and 2).
Potential subzones:

Zone	Subzones
G. kugleri	*G. kugleri-G. trilobus*
	G. kugleri-G. tripartita

General: Compared with the other zones, the *G. kugleri* Zone has a comparatively large number of initial appearances, which amount to eight in the examined area as compared with seven recorded by Bolli in Trinidad (1957a).
Correlation: Bolli (1957a): Equivalent to the *G. kugleri* Zone.
Banner and Blow (1965), Blow (1969): Equivalent to the upper part of Zone N3 and Zone N4.
International series: Lower Miocene: Aquitanian Stage (see Jenkins, 1966b).

Globigerina angulisuturalis Zone

Definition:
Top: Initial appearance of *Globorotalia kugleri* Bolli.
Base: Extinction of *Globorotalia opima* Bolli.
G. angulisuturalis (Plate 1, figs 5–7) is used here as the name bearer due to its resistance to solution and abundance within the zone.
Age: Upper Oligocene.
Taxa:
Extinctions within zone:
Globigerina bradyi
Globorotalia continuosa

Species present and ranging throughout zone:

Cassigerinella chipolensis	*Globoquadrina venezuelana*
Globigerina angustiumbilicata	*Globorotalia* cf. *minutissima*
G. euapertura	*G. nana*
Globigerinita dissimilis	*G. pseudokugleri*
Globoquadrina tripartita	*G. siakensis*
	Globorotaloides suteri

Reference section: Site 78: 07°57.00′N, 127°21.35′W.

Occurrence of zone: Continuously cored at Sites 77 and 78 (Figs 1 and 2).

General: At Site 78, *G. angulisuturalis* ranges from the *C. cubensis* Zone to the upper part of the *G. kugleri* Zone, and its highest frequency is in the *G. angulisuturalis* Zone although it was not found in four of the samples.

Correlation: Bolli (1957a): *Globigerina ciperoensis* Zone.

Banner and Blow (1965), Blow (1969): Broadly equivalent to Zone N3.

International series: Upper Oligocene.

Globorotalia opima Zone

Definition:

Top: Extinction of *Globorotalia opima* Bolli (Plate 2, figs 16–18).

Base: Extinction of *Chiloguembelina cubensis* (Palmer).

Age: Upper Oligocene.

Taxa:

Extinctions within zone:

Globorotalia opima

Initial appearances within zone:

Globorotalia cf. *minutissima*

Species present and ranging throughout zone:

Cassigerinella chipolensis	*Globigerinita dissimilis*
Globigerina angustiumbilicata	*Globoquadrina tripartita*
G. ciperoensis	*G. venezuelana*
G. euapertura	*Globorotalia nana*
	Globorotaloides unicava

Reference section: Site 77B: 00°28.90′N, 133°13.70′W.

Occurrence of zone: Continuously cored at Sites 77 and 78 (Figs 1 and 2).

Correlations: Bolli (1957a): Upper part of the *G. opima* Zone.

Banner and Blow (1965), Blow (1969): Broadly equivalent to the upper part of Zone N2.

International series: Upper Oligocene.

Chiloguembelina cubensis Zone

Definition:

Top: Extinction of *Chiloguembelina cubensis* (Palmer), illustrated on Plate 1, fig. 1.

Base: Initial appearance of *Globorotalia opima* Bolli.

Age: Upper Oligocene.

Taxa:

Extinctions within zone:

Globigerina ampliapertura

G. selli

Globorotalia gemma

G. cf. *siakensis*

Initial appearances within zone:

Globorotalia opima

G. siakensis

Species present and ranging throughout zone:

Cassigerinela chipolensis	*Globigerina ouachitaensis*
Chiloguembelina cubensis	*Globigerinita dissimilis*
Globigerina angulisuturalis	*Globoquadrina tripartita*
G. angustiumbilicata	*G. venezuelana*
G. euapertura	*Globorotalia nana*
G. gortani	*G.* cf. *siakensis*
	Globorotaloides unicava

Reference section: Site 77B: 00°28.90′N, 133°13.70′W.

Occurrence of zone: Continuously cored at Sites 77 and 78 (Figs 1 and 2).

General: A differently defined *C. cubensis* Zone has previously been used by Lindsay (1969) in South Australia; the lower boundary is defined by the extinction of *Globigerina linaperta* Finlay and the upper boundary is defined by the extinction of *C. cubensis*.

Correlation: Bolli (1957a): *G. opima opima* Zone.

Banner and Blow (1965), Blow (1969): Broadly equivalent to the lower part of N2.

International series: Upper Oligocene.

Globigerina ampliapertura Zone

Definition:

Top: Initial appearance of *Globorotalia opima* Bolli.

Base: Extinction of *Pseudohastigerina barbadoensis* Blow.

G. ampliapertura (Plate 1, figs 2–4) is used here as the name bearer due to its resistance to solution and abundance.

Age: Upper Oligocene.
Taxa:
Extinctions within zone:
Globigerina cf. *angiporoides*
G. tapuriensis

Initial appearances within zone:
Globigerina angulisuturalis

Species present and ranging throughout zone:

Cassigerinella chipolensis	*G. selli*
Chiloguembelina cubensis	*Globigerinita dissimilis*
Globigerina ampliapertura	*Globoquadrina tripartita*
G. angulisuturalis	*Globorotalia gemma*
G. angustiumbilicata	*Globorotalia nana*
G. euapertura	*Globorotalia* cf. *siakensis*
G. gortanii	*Globorotaloides suteri*

Datum planes within zone:
Extinction:
G. tapuriensis

Reference section: Site 78: 07°57.00′N, 127°21.35′W.
Occurrence of zone: Continuously cored at Sites 77 and 78 (Figs 1 and 2).
Potential subzone: Based on the extinction of *G. tapuriensis*.
Correlation: Bolli (1957a): *G. ampliapertura* Zone.
Banner and Blow (1965), Blow (1969): Broadly equivalent to Zone N1.
International series: Upper Oligocene.

Pseudohastigerina barbadoensis Zone

Definition:
Top: Extinction of *Pseudohastigerina barbadoensis* Blow (Plate 3, figs 15–16).
Base: Extinction of *Globorotalia insolita* Jenkins.
Age: Lower Oligocene.
Taxa:
Extinctions within zone:
Globigerina linaperta n. subsp.
Pseudohastigerina micra

Initial appearances within zone:

Cassigerinella chipolensis	*G. tapuriensis*
Globigerina cf. *angiporoides*	*Globoquadrina tripartita*
G. euapertura	*G. venezuelana*

G. gortanii   *Globorotalia* cf. *siakensis*
G. selli

Species present and ranging throughout zone:
Chiloguembelina cubensis   *Globigerinita dissimilis*
Globigerina ampliapertura   *Globorotalia gemma*
G. *angustiumbilicata*   G. *nana*
                                     Globorotaloides suteri

Species limited to zone:
Pseudohastigerina barbadoensis

Potential datum planes within zone:
Extinction:
P. *micra*   Initial appearances:
                                     Cassigerinella chipolensis
                                     Globigerina euapertura
                                     G. *gortanii*
                                     G. *selli*
                                     G. *tapuriensis*

Reference section: Site 77B: 00°28.90′N, 133°13.70′W.

Occurrence of zone: Continuously cored at Site 77B, and spot-cored at Site 78 (Figs 1 and 2).

Potential subzone: Based on initial appearance of G. *euapertura*.

General: P. *barbadoensis* was found to be fairly common within its zone, although in some samples it appeared to be missing.

Correlation: Bolli (1957a): Upper G. *cerroazulensis* Zone–lower G. *ampliapertura* Zone.

Banner and Blow (1965), Blow (1969): Broadly equivalent to Zone P19.

International series: Upper Eocene–Lower Oligocene.

Globorotalia insolita Zone

Definition:

Top: Extinction of *Globorotalia insolita* Jenkins (Plate 2, figs 7–9).

Base: Not defined and it is considered that at Site 77B only the uppermost part of the zone was penetrated.

Age: Upper Eocene.

Taxa:

Species present and ranging throughout upper part of zone:
Chiloguembelina cubensis   G. *insolita*
Globigerina ampliapertura   G. *nana*
G. *linaperta* n. subsp.   *Globorotaloides suteri*
Globigerinita dissimilis   G. *unicava*
Globorotalia gemma   *Pseudohastigerina micra*

Reference section: Site 77B: 00°28.90'N, 133°13.70'N.

Occurrence of zone: Spot cored at Site 77B (Figs 1 and 2).

General: The zonal marker was first described from the upper half of the Upper Eocene *Globigerina linaperta* Zone in New Zealand (Jenkins, 1966a). The planktonic foraminiferal fauna found in the *G. insolita* Zone is limited to only ten taxa and the explanation for such a relatively low diversity is not obvious.

Correlation: Bolli (1957a): Broadly equivalent to the *G. cerroazulensis-G. semiinvoluta* Zones (in part).

Banner and Blow (1965), Blow (1969): Broadly equivalent to Zones P17(?)–18.

International series: Upper Eocene.

Acknowledgements

We wish to thank James D. Hays and The Leg 9 scientific staff as well as the entire J.O.I.D.E.S. staff for co-operation and assistance in producing this zonal scheme for publication. Dr. H. M. Bolli is responsible for the excellent scanning electron micrographs. Finally, the professional staffs in the geological departments at Eugene and Christchurch are sincerely acknowledged for assistance with the manuscript, plates and figures.

References

Arrhenius, G. (1948). *In Rept. Swed. Deep-Sea Exped.* (1947–1948), **5**, 1–89.

Banner, F. T. and Blow, W. H. (1965). Progress in the planktonic foraminiferal biostratigraphy of the Neogene. *Nature* **208**, 1164–1166.

Berger, W. H. (1967). Foraminiferal ooze: solution at depths. *Science* **156**, 383–385.

Berger, W. H. (1968a). Planktonic foraminifera: selective solution and paleoclimatic interpretation. *Deep-Sea Res.* **15**, 31–43.

Berger, W. H. (1970a). Planktonic foraminifera: selective solution and the lysocline. *Mar. geol.* **8**, 111–138.

Berger, W. H. and Heath, G. R. (1968b). Vertical mixing in pelagic sediments. *J. marine Res.* **26**, 134–143.

Berger, W. H. and Soutar, A. (1970). Preservation of plankton shells in an anaerobic basin off California. *Bull. geol. Soc. Am.* **81**, 275–282.

Berggren, W. A. (1969). Rates of evolution in some Cenozoic planktonic foraminifera. *Micropaleont.* **15**, 351–365.

Blackman, A. (1966). *Pleistocene Stratigraphy of Cores from the Southeast Pacific Ocean.* (Thesis), University of California, San Diego, California.

Blow, W. H. (1969). Late middle Eocene to Recent planktonic foraminiferal biostratigraphy. *In* (Brönnimann, P. and Renz, H. H., eds), *Proc. Int. Plankt. Conf.* **2**, 199–421, pls 1–54. E. J. Brill, Leiden.

Bolli, H. M. (1957a). Planktonic foraminifera from the Oligocene-Miocene Cipero and Lengua Formations of Trinidad, B. W. I. *U.S. Nat. mus. Bull.* **215**, 97–123, pls 22–29.

Bolli, H. M. (1957b). Planktonic foraminifera from the Eocene Navet and San Fernando Formations of Trinidad, B. W. I. *U.S. Nat. mus. Bull.* **215**, 155–172, pls 35–39.

Bolli, H. M. (1957c). The genera *Globigerina* and *Globorotalia* in the Paleocene-lower Eocene Lizard Springs formation of Trinidad, B. W. I. *U.S. Nat. mus. Bull.* **215**, 61–81.

Bolli, H. M. (1970). The foraminifera of Sites 23–31. *In* (Bader, R. G. *et al.*) 1970. *Initial Reports of the Deep Sea Drilling Project*, **4**, U.S. Government Printing Office, Washington.

Bolli, H. M. and Bermúdez, P. J. (1966). Zonation based on planktonic foraminifera of Middle Miocene to Pliocene warm water sediments. *Boln. int. Assoc. Venez. Geol. Mineralog. y petrogr.* **8**, 119–149.

Bradshaw, J. S. (1959). Ecology of living planktonic foraminifera in the North and Equatorial Pacific. *Contr. Cushman Fdn. Foram. Res.* **10**, 25–64, pls 6–8.

Cita, M. B. (1971). Paleoenvironmental aspects of DSDP Legs. I–IV. *In* (Farinacci, A., ed.), *Proc. II Plankt. Conf.*, 251–285, pls 1–5. Tecnoscienza, Roma.

Hays, J. D., Saito, T., Opdyke, N. D. and Burckle, L. H. (1969). Pliocene-Pleistocene sediments of the Equatorial Pacific: their paleomagnetic, biostratigraphic, and climatic record. *Bull. geol. Soc. Am.* **80**, 1481–1514.

Hays, J. D. *et al.* (1972). *Initial Reports of the Deep Sea Drilling Project* **9**, U.S. Government Printing Office, Washington.

Jenkins, D. G. (1965). Planktonic Foraminifera and Tertiary intercontinental correlations. *Micropaleont.* **11**, 265–277, pls 1–2.

Jenkins, D. G. (1966a). Planktonic foraminiferal zones and new taxa from the Danian to Lower Miocene of New Zealand. *N.Z. J. geol. Geophys.* **8**, 1088–1126.

Jenkins, D. G. (1966b). Planktonic foraminifera from the type Aquitanian-Burdigalian of France. *Contr. Cushman Fdn. Foram. Res.* **17**, 1–15, pls 1–3.

Lindsay, L. M. (1969). Cenozoic foraminifera and stratigraphy of the Adelaide Plains Sub-basin, South Australia. *Bull. Geol. Surv. South Australia* **42**, 1–60, pls 1–2.

Lipps, J. H. (1970). Plankton evolution. *Evolution* **24**(1), 1–22.

Miles, G. A. and Orr, W. N. (1972). Holocene pelagic Foraminifera in the Northeast Pacific. *Abstract, geol. Soc. Am.* **4**, 3.

Nyaudu, Y. R. (1964). Carbonate deposits and paleoclimatic implications in the northeast Pacific Ocean. *Science* **146**(3643), 515–517.

Orr, W. N. (1967). Secondary Calcification in the Foraminiferal Genus *Globorotalia*. *Science* **157**, 1554–1555.

Orr, W. N. (1969). Variation and distribution of *Globigerinoides ruber* in the Gulf of Mexico. *Micropaleont.* **15**, 373–379, pl. 1.

Parker, F. L. (1960). Living planktonic foraminifera from the equatorial and southeast Pacific. *Sci. Rep. Tohuku Univ.* **2**, 71–82.

Parker, F. L. (1967). Late Tertiary biostratigraphy (planktonic foraminifera) of tropical Indo-Pacific deep-sea-cores. *Bull. Am. Paleont.* **52**, 115–208.

Saidova, H. M. (1965). Sediment stratigraphy and paleography of the Pacific Ocean by benthonic foraminifera during the Quaternary. *In* (Sears, M., ed.), *Progress in Oceanography*, **4**, 143–151. Pergamon Press, London.

Schott, W. (1935). Die Foraminiferen in dem äquatorialen Teil des Atlantischen Ozeans, Deutsch Atlantic Expedition "Meteor", 1925–1927. *Wiss. Ergebn.* **3**, 43–134.

PLATES

PLATE 1

fig. 1. *Chiloguembelina cubensis* Palmer. Side view (×149). Site 77B, Core 51, core-catcher; *P. barbadoensis* Zone, Lower Oligocene.

figs 2–4. *Globigerina ampliapertura* Bolli. (2) Spiral view (×75). Site 77B, Core 41, core-catcher; *C. cubensis* Zone, Oligocene; (3) umbilical view (×70); (4) side view (×70). Site 78, Core 32, core-catcher; *G. ampliapertura* Zone, Oligocene.

figs 5–7. *Globigerina angulisuturalis* Bolli. (5) Spiral view (×125). Site 78, Core 25, core-catcher: *C. cubensis* Zone, Upper Oligocene; (6) side view (×125); (7) umbilical view (×125).

figs 8–10. *Globigerinoides bisphericus* Todd. (8) Side view (×45). Site 78, Core 3, core-catcher; (9) umbilical view (×43); (10) spiral-side (×65).

figs 8–10. (10) Site 78, Core 3, Section 3, top; *G. bisphericus* Subzone, Lower Miocene.

figs 11–12. *Globigerinoides fistulosus* Schubert. (11) Spiral view (×63). Site 83A, Core 5, core-catcher; *G. fistulosus* Zone, Upper Pliocene; (12) umbilical view (×40). Site 76, Core 1, core-catcher; *G. fistulosus* Zone, Upper Pliocene.

figs 13–15. *Globigerinita dissimilis* Cushman and Bermúdez. (13) Side view (×50); (14) umbilical view (×50); (15) spiral view (×45). Site 77B, Core 30, core-catcher; *G. dissimilis* Zone, Lower Miocene.

figs 16–17. *Globoquadrina altispira* Cushman and Jarvis. (16) Umbilical view (×43); (17) side view (×35). Site 78, Core 3, Section 1, 50–52 cm; *G. peripheroronda* Zone, Middle Miocene.

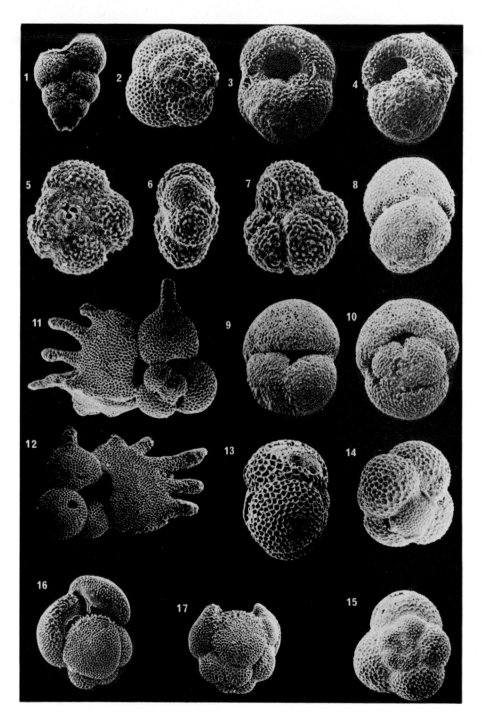

PLATE 2

figs 1–3. *Globoquadrina venezuelana* Hedberg. (1) Spiral view (×37). Site 77A, Core 1, core-catcher; *P. obliquiloculata* Zone, Pleistocene; (2) side view (×37); (3) umbilical view (×25).
figs 4–6. *Globorotalia fohsi lobata* Bermúdez. (4) Side view (×45); (5) umbilical view (×40); (6) spiral view (×45). Site 77B, Core 21, core-catcher; *G. fohsi lobata* Zone, Middle Miocene.
figs 7–9. *Globorotalia insolita* Jenkins. (7) Side view (×125); (8) umbilical view (×149); (9) spiral view (×225). Site 77B, Core 53, core-catcher; *G. insolita* Zone, Upper Eocene.
figs 10–12. *Globorotalia kugleri* Bolli. (10) Spiral view (×55). Site 79, Core 13, Section 4, top; *G. kugleri* Zone, Lower Miocene; (11) side view (×90); (12) umbilical view (×80).
figs 13–15. *Globorotalia merotumida* Banner and Blow. (13) Side view (×75); (14) spiral view (×75). Site 77B, Core 13, Section 6, 16–18 cm; *G. plesiotumida* Zone, Upper Miocene; (15) umbilical view (×75).
figs 16–18. *Globorotalia opima* Bolli. (16) Side view (×50); (17) spiral view (×50). Site 77B, Core 39, core-catcher; *G. opima* Zone, Oligocene.; (18) umbilical view (×50).
figs 19–21. *Globorotalia peripheroacuta* Blow and Banner. (19) Spiral view (×123). Site 77B, Core 23, Section 3, top; *G. fohsi fohsi-G. peripheroacuta* Zone, Middle Miocene; (20) side view (×69); (21) umbilical view (×69).

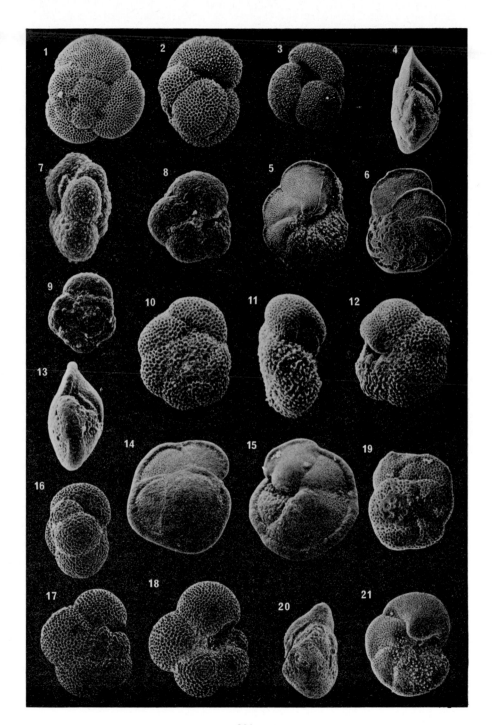

PLATE 3

figs 1–3. *Globorotalia peripheroronda* Blow and Banner. (1) Spiral view (×100). Site 79A, Core 4, core-catcher; *G. venezuelana* Zone, Lower Miocene; (2) side view (×100); (3) umbilical view (×90).

figs 4–6. *Globorotalia plesiotumida* Banner and Blow. (4) Umbilical view (×45); (5) spiral view (×45). Site 77B, Core 9, Section 6, 3–5 cm; *G. plesiotumida* Zone, Upper Miocene; (6) side view (×69).

figs 7–9. *Globorotalia tumida tumida* Brady. (7) Umbilical view (×20); (8) side view (×20); (9) spiral view (×20). Site 76, Core 1, core-catcher; *G. fistulosus* Zone, Upper Pliocene.

figs 10–12. *Praeorbulina glomerosa curva* (Blow). (10) Side view (×55); (11) side view (×55); (12) spiral view (×55). Site 81, Core 3, Section 2, top; *G. peripheroronda* Zone, Middle Miocene.

figs 13–14. *Pulleniatina obliquiloculata* Parker and Jones. (13) Side view (*P. finalis*) (×40); (14) umbilical view (×45). Site 77B, Core 1, Section 2, 39–41 cm; *P. obliquiloculata* Zone, Pleistocene.

figs 15–16. *Pseudohastigerina barbadoensis* Blow. (15) Peripheral view (×123). Site 77B, Core 51, core-catcher; *P. barbadoensis* Zone, Lower Oligocene; (16) side view (×123).

figs 17–18. *Sphaeroidinella dehiscens* Parker and Jones. (17) Umbilical view (×35); (18) side view (×25). Site 77B, Core 2, Section 6, 12–14 cm: *P. obliquiloculata* Zone, Pleistocene.

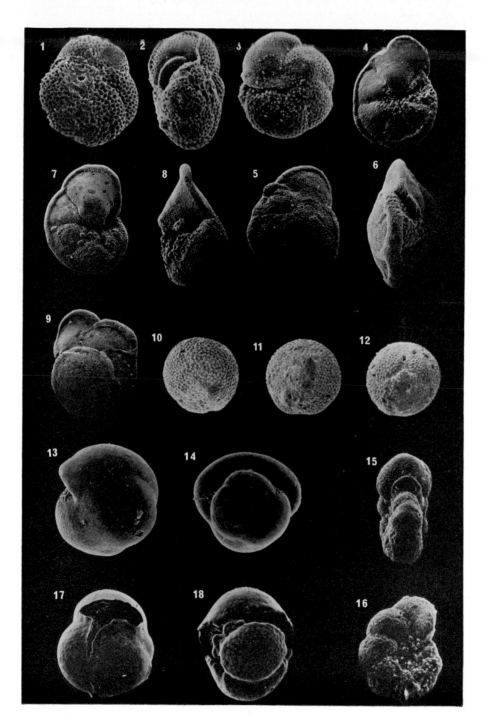

4: Atlas of Palaeogene Planktonic Foraminifera

Some species of the genera, *Subbotina, Planorotalites, Morozovella, Acarinina* and *Truncorotaloides*

W. A. BERGGREN

Department of Geology and Geophysics,
Woods Hole Oceanographic Institution,
Woods Hole,
Massachusetts 02543, U.S.A.

The taxonomy, nomenclature, phylogenetic relationships, distribution and stratigraphic range of over 50 species of Palaeogene planktonic foraminifers are discussed. The holotype of each and several (interpreted) valid illustrated references to these species are presented in a series of charts. Of these species 3 are assigned to *Subbotina*, 26 to *Morozovella*, 7 to *Planorotalites*, 12 to *Acarinina* and 3 to *Truncorotaloides*.

As a result of the author's recent investigations on type and comparative material as well as those of colleagues around the world, several significant taxonomic changes have been made in order to ensure stability of nomenclature and consistency of identification. The more significant taxonomic changes are listed below for each of the genera.

In the genus *Planorotalites* the following taxonomic changes are made:
Globorotalia haunsbergensis Gohrbandt, 1963 is placed in the synonymy of *Globorotalia ehrenbergi* Bolli, 1957; *Globorotalia pseudoscitula* Glaessner, 1937 is made a senior synonym of *Globorotalia renzi* Bolli, 1957. *Globorotalia palmerae* is a true planktonic form and not a pararotaliid and is placed in the genus *Planorotalites*. The following are placed in *Subbotina* on phylogenetic grounds: *Globigerina spiralis* Bolli, 1957, *Globigerina inconstans* Subbotina, 1953; *Acarinina praecursoria* Morozova, 1957 (=*Globorotalia trinidadensis* Bolli, 1957 = *Acarinina shachdagica* Morozova, 1961). *G. inconstans* is a senior synonym of *Acarinina indolensis* Morozova, 1957. The following changes are made in *Morozovella*: *Globorotalia pusilla pusilla* Bolli, 1957 is a senior synonym of *Planorotalites tauricus* Morozova, 1961; *Globorotalia albeari* Cushman and Bermúdez, 1949 is a senior synonym of *Globorotalia pusilla laevigata* Bolli, 1957; *Globorotalia conicotruncata* Subbotina, 1953 is a senior synonym of *Globorotalia angulata abundocamerata* Bolli, 1957; *Globorotalia subbotinae* Morozova, 1939 is a senior synonym of *Globorotalia rex* Martin, 1943; *Pulvinulina crassata* Cushman, 1925 is a senior synonym of *Globorotalia spinulosa* Cushman, 1927.

In the genus *Acarinina* the following taxonomic changes are made:
Globigerina coalingensis Cushman and Hanna, 1927, is a senior synonym of *Globoquadrina primitiva* Finlay, 1947 and *Acarinina triplex* Subbotina, 1953; *Globigerina esnaensis* Le Roy, 1953 is a senior synonym of *Acarinina intermedia* Subbotina, 1953, *Globigerina stonei* White, 1955 and *Globorotalia irrorata* Loeblich and Tappan, 1957 (although placed in the genus *Acarinina* by this writer, a recent illustration by Luterbacher, 1964, fig. 18 of a form probably referable to this species from the *G. pseudomenardii* Zone of El Quss Abu Said, Farafra Oasis, Egypt, has relict sutural apertures on the spiral side indicating a relationship with *Truncorotaloides*); *Globigerina nitida* Martin, 1943, is a senior synonym of *Acarinina acarinata* Subbotina, 1953; *Globigerina mckannai* White, 1928, is a senior synonym of

Globigerina subsphaerica Subbotina, 1947 and *Globigerina spiralis* Loeblich and Tappan, 1957 *partim* (*non* Bolli, 1957); *Globorotalia pentacamerata* Subbotina, 1947 (*non* 1936) is a senior synonym of *Globigerina gravelli* Brönnimann, 1952, *Globigerina dubia* var. *lakiensis* Haque, 1956, *Acarinina pentacamerata* var. *camerata* Khalilov, 1956 and *A. pentacamerata* var. *acceleratoria* Khalilov, 1956; *Globigerina soldadoensis* Brönnimann, 1952 is a senior synonym of *Acarinina interposita* Subbotina, 1953 and *Acarinina clara* Khalilov, 1956; *Pulvinulina crassata* var. *densa* Cushman, 1925, is a senior synonym of *Globorotalia crassaformis* Subbotina, 1947 *et seq.* (*non* Galloway and Wissler, 1927) and *Globorotalia bullbrooki* Bolli, 1957.

In the genus *Truncorotaloides* the following taxonomic changes are made:

Truncorotaloides rohri Brönnimann and Bermudez, 1953 is a senior synonym of *Acarinina rugosoaculeata* Subbotina, 1953; *Globorotalia collactea* Finlay, 1939 is a senior synonym of *Acarinina rotundimarginata* Subbotina, 1953 and *Globorotalia spinuloinflata* Bolli, 1957 (*non* Bandy, 1949).

Outline

Introduction

The usefulness of planktonic foraminifera in biostratigraphic studies is now well appreciated through the detailed investigations by workers in various parts of the world. Yet in spite of the numerous publications in this field it is all too frequently observed that the concept of a particular planktonic foraminiferal species varies considerably from one investigator to another and in some cases bears little relationship to the original concept of the author of the species. Another common occurrence is the fact that although general agreement as to the concept of various planktonic species may exist among specialists in other parts of the world, these conspecific forms are given different specific, and even generic names. This has been nowhere more evident than in a comparison of studies of western and Soviet specialists. As a result of this duplication and overlap, distinct difficulties have been encountered in attempts to correlate various marine stratigraphic sequences in different parts of the world.

Investigations by the author (1960, 1966) and Luterbacher (1964), among others, have attempted to reconcile some of these discrepancies. On the basis of studies of museum type and comparative collections in the Soviet Union, Europe and the United States, the author has accumulated data which has enabled at least a partial resolution of the taxonomic chaos which has been impending. This illustrated atlas of some stratigraphically important Palaeogene planktonic foraminifera has been prepared in the hope that it will provide a guide for the accurate recognition and identification of these forms, aid in the formulation of a concept of the variability of a given species, provide criteria for an understanding of the phylogenetic relationships between the various specific forms and some of the problems of homeomorphy associated with them, and an understanding and appreciation of their restricted stratigraphic range.

Discussion

Our knowledge of Palaeogene planktonic foraminiferal faunas has grown steadily during the past few decades. If the publications of Glaessner and Subbotina in the decade 1930–1940 marked the beginning of studies in Palaeogene planktonic foraminifera (see Berggren, 1960), then the comprehensive studies by Subbotina (1953), Bolli (1957) and Loeblich and Tappan (1957) ushered in a new era of modern treatment of planktonic foraminifera and their use in regional stratigraphic correlation. During the past 15 years there have been numerous monographs on various aspects of Palaeogene planktonic foraminiferal faunas from different parts of the world. The data from these sources indicate that Palaeocene faunas are remarkably uniformly distributed around the world. Faunal provincialisation is seen to develop in the Eocene, particularly in the middle Eocene, and this trend continues throughout the remainder of the Cainozoic as a reflection of the progressive provincialisation of climate (Berggren and Hollister, 1974). In this chapter I have attempted to present a summary of the stratigraphic and geographic distribution of various Palaeogene planktonic foraminifera as well as some of the taxonomic problems attendant upon their recognition. The charts include illustrations which I believe are representative of the species discussed in the descriptive section. It is in this sense that this work differs from the exhaustive compilation of Ellis *et al.* (1969): the synonymic compilation is subjective rather than objective. The reader is invited to eliminate those references (and illustrations) which do not encompass his own concept of a particular species.

Palaeogene planktonic foraminiferal studies have been reviewed by Berggren (1960, 1966), McGowran (1968), and more recently by Krasheninnikov (1969) and Jenkins (1971) among others. By the time this volume is published, the long awaited monograph on Palaeogene planktonic foraminifera by the late Dr. W. H. Blow will be nearing publication.

A brief review is presented of the two recent papers cited above because of their interesting contribution to our knowledge concerning regional geographic distribution patterns among Palaeogene forms.

Krasheninnikov (1969) has presented a detailed summary of the geographic and stratigraphic distribution of Palaeogene planktonic foraminifera in tropical and subtropical areas of the world. He notes the similarities and differences between faunal assemblages in the Crimean–Caucasus region and the Mediterranean. The differences are such (greater development of acarininids coupled with the rarity of *Morozovella velascoensis* and related forms and *Planorotalites pseudomenardii* in the Crimea–Caucasus region) that he suggests that the Crimean–Caucasus region be considered a distinct "subprovince" or "palaeobiogeographic region". This provincialisation is not seen during Danian time but develops gradually during late Palaeocene–early Eocene time. However, it is in the middle Eocene that a marked faunal differentiation between the two regions is seen. In the Caucasus–Crimea, *Orbulinoides*, *Globigerinatheka*, *Clavigerinella* are absent, and *Hantkenina* and *Cribrohantkenina* are rare. Such species as *Morozovella spinulosa*, *M. lehneri*, *Subbotina pseudomayeri*, *S. bolivariana*, *S. senni*, *Globigerinita echinata*, *G. unicava*, *Hantkenina aragonensis*, *Globigerina rohri*, and *G. ampliapertura*, *i. al.*, have not been recorded in the Crimean–Caucasus region. This faunal provincialisation within the middle Eocene is similar to that described by Jenkins (1965, 1971) in New Zealand. This provincialisation of planktonic faunal assemblages was probably a result of a developing climatic provincialisation which, in turn, was related to changing palaeogeographic relationships between continents and oceans and concomitant changes in circulation patterns (Berggren and Hollister, 1974).

A note of caution may be appropriate at this point. The question of provincial faunal assemblages or endemic elements is, however, fraught with problems. Krasheninnikov (1969), for instance, cites the occurrence of several elements in the late Palaeocene–early Eocene fauna of California which he says do not occur in the Crimea–Caucasus area. And yet, a comparison of the faunas of these two regions reveals that, in some instances, the same forms have been recorded under different names in these regions (i.e. *Acarinina triplex* = *A. coalingensis*, *A. acarinata* = *A. nitida*).

It is interesting to note that Krasheninnikov (1971a, b) has observed that Palaeogene planktonic foraminiferal faunas of the western Pacific are diverse and contain many of the forms originally described from the Soviet Union. Far from being provincial, or endemic, they provide the basis for worldwide stratigraphic correlation. As the author has repeatedly stressed in previous publications, the problems in biostratigraphy and correlation are often taxonomic and lie with the palaeontologist rather than with the fossils.

The Cainozoic planktonic foraminiferal faunas of New Zealand have been studied in detail by Jenkins (1965, 1971). The Palaeogene planktonic foraminiferal faunas are, in general, comparable to those described elsewhere in low-mid latitude regions of the world. A certain amount of indigeneity in the fauna is suggested by the (apparent) presence of certain taxa not recorded outside of New Zealand. On the other hand certain taxa which were thought to be indigenous to New Zealand–Australia, for instance *Subbotina angiporoides* (Jenkins, 1971) is a common component of late Eocene–early Oligocene assemblages in deep-sea cores of the North and South Atlantic, have a wider geographic distribution than first suggested. The reason is probably differences in taxonomic concept among workers in various parts of the world.

The Palaeocene planktonic foraminiferal faunal succession recorded from New Zealand (Jenkins, 1971) is markedly similar to those recorded from other low–mid latitude areas of the world. In particular, the angulo-conical keeled morozovellids, spinose acarininids, and compressed planorotaliids, indicate a broad latitudinal distribution of these forms.

It is during the Eocene (54–37 my) that the first indication of a provincialisation of the planktonic foraminiferal fauna is seen (e.g. absence in New Zealand of typically tropical forms as *Orbulinoides beckmanni*, *Cribrohantkenina*, *Clavigerinella*, *Turborotalia cerroazulensis*, *Morozovella spinulosa*, *M. lehneri*, and *M. aragonensis* (Jenkins, 1971). In the case of the latter species, however, the fact that both *M. lensiformis*, which was recorded as *G. dolabrata* by Jenkins, *op. cit.*, and *G. crater = caucasica*, antecedent and descendant species of *M. aragonensis*, respectively, are present in New Zealand sediments leads me to suggest that *M. aragonensis* will be recorded from the New Zealand region. Indeed, in New Zealand the angulo-conical morozovellid lineage ends at the Heretaungan/Porangan boundary which is approximately equivalent to the early/middle Eocene boundary. The hantkeninids are rare and sporadic and have a shorter total stratigraphic range in New Zealand than in low latitude areas. On the other hand the extension of *Truncorotaloides collactea* into the late Eocene in New Zealand mirrors its stratigraphic distribution in Denmark and the North Sea (Berggren, 1969).

This provincialisation of New Zealand planktonic foraminiferal faunas is probably due to the combined effects of sea-floor spreading (Margolis and Kennett, 1971) and oceanic current patterns. The general WNW movement of New Zealand has probably resulted in only a few degrees latitudinal translation since the early Cainozoic. At the same time New Zealand has probably moved through the belt of West Wind Drift. It would seem that New Zealand has occupied a position intermediate between the tropical and Antarctic water masses during the Cainozoic and this is reflected in the nature of the faunas. Climatic belts, and concomitantly biogeographic provinces were much wider in the early Cainozoic.

In the Oligocene the continued provincialisation of the planktonic foraminiferal fauna of New Zealand is evident. The so-called "large globigerinid" (*G. tapuriensis, G. tripartita, G. sellii, G. binaiensis, i. al.* (Berggren, 1971)) faunas typical of tropical regions, have not been recorded from the Australia–New Zealand area. The Oligocene faunas consist predominantly of the relatively nondescript globigerinid and non-keeled globorotaliid faunas characteristic of mid–high latitudes of northern Europe and the North Atlantic (Berggren, 1969, 1972). An interesting example of the widespread distribution of a seemingly small and insignificant form is the occurrence of the short ranging *Globorotalia munda* in the Whaingaroan–Duntroonian stages of New Zealand and the Rupelian of Germany and in sediments of equivalent age in the North Atlantic. This is one of the most useful forms for recognising and correlating sediments of middle Oligocene age in high latitudes and we may expect to see it recorded in intermediate low latitude localities.

Remarks on classification

Five genera of Palaeogene planktonic foraminifera are considered in this paper: *Subbotina, Morozovella, Planorotalites, Acarinina* and *Truncorotaloides*. Within these genera various phyletic lineages and branches may be discerned; as this paper is primarily taxonomic in content they are only briefly mentioned here. A more detailed discussion of the phylogenetics of Palaeogene planktonic foraminifera may be found in other publications by this author (Berggren, 1965, 1966, 1969).

The genus *Morozovella* as used in this paper embraces both keeled and non-keeled forms in agreement with McGowran (1968).

The genus *Acarinina* Subbotina, 1953, is accepted and employed here for a group of species which are distinguished from *Subbotina* and *Morozovella* by developing a strongly spinose wall texture. Two, and possibly three, branches within the acarininid appear lineage to lead independently, though parallel trends, to the development of the polytypic, but monophyletic, genus *Truncorotaloides*. In terms of normal criteria of divergence (gaps) and monophyly there would seem to be justification for recognition of the acarininids as a genus, providing the definition of the genus is emended to include primarily spinose forms such as those mentioned in this paper.

The genus *Truncorotaloides* is also accepted, although available data appears to indicate that there is a certain overlap in the criteria currently used to distinguish the two genera. In several cases it has been found difficult to assign a given species to one or the other of the two genera.

The author is in agreement with most of McGowran's (1968) perceptive insight and persuasive arguments for a reclassification of Palaeogene "globorotaliids". Two points of minor disagreement are discussed below, however.

McGowran (1968) has expanded the concept of *Truncorotaloides* to include *Acarinina* and *Morozovella* as subgenera and placed them in the subfamily Truncorotaloidinae. He includes such species as *inconstans* Subbotina, *praecursoria* Morozova and *uncinata* Bolli in *Acarinina* and thus derives the subgenus *Morozovella* directly from *Acarinina* (McGowran, 1968). But *inconstans* and *praecursoria* have a reticulate wall structure and are more properly placed in *Subbotina*. On the other hand *uncinata*, with its low angulo-conical test and curved sutures on the spiral side of the test represents the initiation of the angulo-conical morozovellid lineage. The earliest typical acarininids (*esnaensis–acarinata–coalingensis, mckannai*) appear in the *Planorotalites pseudomenardii* Zone (middle Palaeocene). Both *Morozovella* and *Acarinina* have evolved independently from *Subbotina* (Berggren, 1966).

I have outlined the reasons for distinguishing *Acarinina* and *Truncorotaloides* at the generic level on the basis of divergence (gaps) and monophyly (Berggren, 1966), and McGowran (1968) has independently arrived at the same conclusion (cf., however, Steineck, 1971, who considers *Acarinina* a junior synonym of *Truncorotaloides*). Thus the difference in interpretation lies mainly in the origin of *Morozovella* which is based, in turn, on the generic assignment of several early Palaeocene transitional forms. I would prefer to ascribe generic status to *Truncorotaloides*, *Acarinina* and *Morozovella* rather than treating the latter two as subgenera of the former.

The second point of disagreement deals with the group of compressed, keeled and non-keeled, smooth-walled "globorotaliids". McGowran (1968) includes these forms (his Group 6, Fig. 1) in *Planorotalites* Morozova 1957 and uses *Globanomalina* Haque, 1956, for the planispiral–pseudoplanispiral forms which the author has assigned to *Pseudohastigerina* Banner and Blow, 1969. *Globanomalina* Haque, 1956, has been shown (Berggren *et al.*, 1967) to be based primarily upon the concept of a low trochoid spiral test. It would seem preferable to retain the name *Pseudohastigerina* for a clearly defined group (Banner and Blow, 1959; see emended diagnosis in Berggren *et al.*, 1967) which is based upon a planispiral mode of coiling. If we accept this concept, then the names *Planorotalites* Morozova, 1957 and *Globanomalina* Haque, 1956 are available for McGowran's (1968) Group 6 (including *imitata* in Group 7). Because the type species of *Globanomalina* (*G. ovalis* Haque) is morphologically and phyletically very close to the earliest species of *Pseudohastigerina* (*P. wilcoxensis = P. pseudoiota*), the name *Globanomalina* is the less desirable of the two. The type species of *Planorotalites* (*P. pseudoscitula*) is both morphologically and phyletically distinct from *Pseudohastigerina* and represents the last member of a distinct lineage.

The taxonomic and phylogenetic problems associated with the genus *Pseudohastigerina* have been studied by Berggren *et al.* (1967) and Cordey *et al.* (1970) and are reviewed here for the purpose of historical perspective. The holotype specimen of *Globanomalina ovalis* was not deposited in the collections of the British Museum (Natural History). The illustration of this specimen suggests a planispiral test. Of the two paratypes of this species deposited in the British Museum (Natural History), one is clearly trochoid, the other doubtfully so (due to poor preservation, unequivocal distinction was not possible). Haque himself (1956) indicated that the genus *Globano-malina* was defined to include forms where chambers were arranged "in a trochoid spiral, usually very low". He pointed out that "all the chambers are visible from the dorsal side, which may be flattened or convex; ventral side completely involute, umbilicate". He also stated that the aperture is "peripheral, sometimes extending to the umbilicus on the ventral side, usually with a short lip". In his description of the type species, *G. ovalis*, Haque (1956) stated that the aperture is "an elongate slit at the base of the last chamber of the outer whorl extending from the periphery to the umbilicus with a narrow lip". Thus, the basis for the genus *Globanomalina* is a low trochoid spire and an extra-umbilical aperture which extended in some forms to the spiral side, not a planispiral test (cf. Loeblich and Tappan, 1964). On the other hand Haque apparently included both low trochoid and asymmetrically coiled "planispiral" tests in his concept of *Globanomalina*. The type level of his *G. ovalis* is at the Palaeocene/Eocene boundary where the evolutionary transition from low trochoid to planispiral coiling occurs so that it is understandable that a clear differentiation between these types of coiling was not made. If we base the concept of *Globanomalina* rigidly on the illustration of a single holotype specimen which appears to be planispiral, then *Globanomalina* would be the correct name for the planispiral forms of the Eocene–middle Oligocene. If, on the other hand, we base our concept of *Globanomalina* on the paratype material at the British Museum (Natural History) to the exclusion of the figured holotype, *Globanomalina* might serve as the name of the smooth globorotaliids. If we combine the holotype and paratype material in our concept we would probably include both planispiral and trochoid forms in the same genus. Coiling mode, however, is normally given a suprageneric rank in the hierarchy of classification.

A distribution plot of greatest breadth of final chamber vs. greatest test diameter has suggested some interesting possible relationships within the pseudohastigerinids (Cordey *et al.*, 1970). It was shown that *pseudoiota* has a distribution pattern which falls largely outside that of *wilcoxensis* and appears closer to *sharkriverensis*. The holotype of *ovalis* is also closer to the *sharkriverensis* group based on size alone. The latter is a senior synonym (by page priority) of *Pseudohastigerina wilcoxensis globulosa* Gohrbandt, 1967. At the same time it was suggested that *eocenica* may be more appropriately included in the *sharkriverensis* group rather than with the smaller *wilcoxensis*. These differences may be those of phyletic size increase (the type level of *wilcoxensis* is in the lower part of the *Morozovella subbotina* Zone, that of *eocenica* near the *formosa/aragonensis* boundary and that of *sharkriverensis* in the early middle Eocene).

We need not belabour the obvious point that the holotype is but a single specimen. Populations, not individuals, are classified and types serve merely as the legislative requirement of taxonomy. The holotype should be representative of the populations from which it was chosen and, similarly, the genotype should typify the genus of which it is representative. *Globanomalina ovalis* is at the marginal end of the spectrum of smooth-walled "globorotaliids" and represents actually a transitional phase to the planispiral tests of the Eocene members. Inasmuch as it is difficult to determine whether *Globanomalina ovalis* is "fish or fowl" and inasmuch as it is certainly not typical of the group of species in which it is included, I would suggest abandonment of the *Globanomalina* in favour of *Planorotalites* for the group of keeled and non-keeled compressed, smooth-walled "globorotaliids". The name *Planorotalites* has now been used by McGowran (1968) and Jenkins (1971) so that its retention will add stability to an otherwise insecure Palaeogene planktonic foraminiferal classification.

A suggested classification of Palaeogene planktonic foraminifera is shown in Fig. 1a and the approximate duration (in my) of several genera whose range is restricted to the Palaeogene is shown in Fig. 1b.

The use of lineage zones—phylozones—in achieving higher resolution in planktonic foraminiferal stratigraphy was discussed by Berggren (1971). Several phylozonation schemes for the Cenozoic were suggested. Data dealing with the morozovellids (1971, Fig. 2), acarininids–truncorotaloidids (1971, Fig. 3), planorotaliids and hantkeninids (Fig. 4) were combined into a comparative phylozonation scheme (1971, Table 5).

Use of the Charts

There are twelve charts which accompany this chapter. Species of the genera *Subbotina*, *Morozovella*, *Planorotalites*, *Acarinina* and *Truncorotaloides* are illustrated. The holotype illustration of each species (with the exception of *Pulvinulina crassata* var. *densa* Cushman which has not been illustrated to date) is placed at the top by itself. In some instances a recent reillustration of a poorly figured holotype is placed beside the original illustration (see *Globigerina coalingensis* Cushman and Hanna, 1927; *Globorotalia collactea* Finlay, 1939). The illustrated references are presented in a group sequence

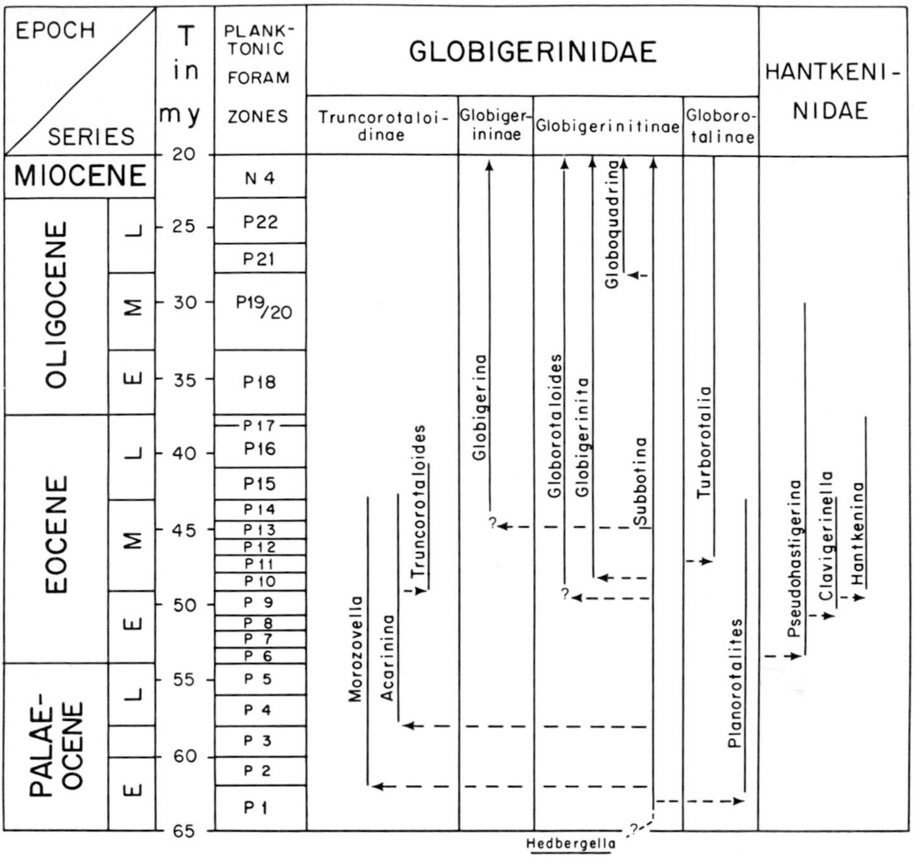

GENUS	T in my
Acarinina	15
Clavigerinella	7
Hantkeninia	11
Morozovella	19
Planorotalites	20
Pseudohastig-erina	24
Truncorotal-oides	8

FIG. 1a (above). A classification of Paleogene planktonic foraminifera.

FIG. 1b. Range (in my) of some Paleogene planktonic foraminiferal genera.

by general relationship for the sake of continuity and ease of comparison. Thus illustrations of forms belonging to the *Planorotalites compressa*–*P. ehrenbergi*–*P. chapmani* and *P. pseudomenardii* bioseries can be readily compared on a single chart. Arranged beneath in general chronological order are the author's interpretations of valid references to the species. The name given to the particular species is given below this illustration; beneath this in parenthesis are given the author and date of the publication. Reference to the synonymy will then readily reveal the source of the illustrated reference and further data can be obtained by checking the reference. In some instances pertinent stratigraphic data has been included with the illustration. In this way the reader can readily compare illustrations of a given species and either accept, reject or modify the interpretation of the species and its taxonomic status.

These charts, it should be emphasised, are to be used together with the descriptive discussion presented for each species. Aspects of the taxonomy, nomenclature, occurrence, and interpreted phylogenetic relationships where justified by data and stratigraphical range are presented for each species.

The ranges of the planktonic species discussed in this paper are given in terms of a zonation which the author has developed over the past five years. It is quite close to that devised by Bolli (1957) in actual practice. Comparison with the zonations in use in the Soviet Union, as well as modifications of these schemes made by various authors in recent works is made wherever pertinent.

Descriptive sections

Genus S U B B O T I N A Brotzen and Pozaryska, 1961

Subbotina praecursoria (Morozova)

Acarinina praecursoria Morozova, 1957, *Akad. Nauk. U.S.S.R. Doklady* **114**, 1111, text-figs 1a–c.

Globigerina shachdagica Khalilov, 1956, *Tr. Inst. Geol N.A.U.K., Azerb. S.S.R.* **17**, 246, pl. 1, figs 3a–c.

Globorotalia trinidadensis Bolli, 1957, *U.S. Nat. Mus. Bull.* **215**, 73, pl. 16, figs 19–23—Bolli and Cita, 1960, *Riv. Ital. Paleont.* **66**, 29, pl. 33, fig. 1a–c.—Gohrbant, 1963, *Geol. Ges. Wien, Mitt.* **55**, 45, pl. 1, figs 13–15.—El Naggar, 1966, *Bull. Brit. Mus.* (Nat. Hist.), ꞁ Geol. Suppl. 2, 236, pl. 18, figs 7a–c.

Globorotalia shachdagica (Khalilov), Morozova, 1959, *Dokl. Akad. Nauk. S.S.R.* **124**, 1116, fig. 5.—Luterbacher, 1964, *Eclog. geol. Helv.* **57**, 664, fig. 24.

Globorotalia praecursoria (Morozova). Luterbacher, 1964, *Eclog. geol. Helv.* **57**, 652, fig. 25.–Berggren, 1965, *Micropaleont.* **11**, p. 293, pl. 1, figs 1–4.

Subbotina praecursoria (Morozova). Berggren, 1971, *Proc. II Plankt. Conf.* pl. **4**, figs 6, 7.

Remarks: The author's concept of this species is illustrated on Chart I. Unfortunately Morozova only illustrated a single specimen of her new species. This is a large robust form which (as this author would interpret it) lies towards one end of the range of infraspecific variability exhibited by this species. The writer would include specimens like Bolli's (1957, pl. 16, figs 19–21) holotype of *G. trinidadensis* at the other end of the variability curve.

In the author's opinion *S. praecursoria* evolved from *S. pseudobulloides* from which it is distinguished by its larger size, the larger number of chambers per whorl (5-7), the slower rate of increase in chamber size, and the development of spinose ornament on the peripheral margin of the chambers of the final whorl. It is difficult to separate the two forms in the lower Danian where the large robust type illustrated by Morozova (1957) has not been observed by the author. The increase in size and morphological variation shown in Morozova's (1957) and Bolli's (1957) illustrations is only attained in the upper Danian.

Stratigraphical range and distribution: Bolli (1957, fig. 12) gave the stratigraphical range of *S. praecurzoria* (*G. trinidadensis*) as *S. praecursoria* Zone (Danian) to the *Morozovella* (*Globorotalia*) *uncinata* zone. Morozova (1959) considered *S. praecursoria* as a guide form of "Zone 4: cancellate globigerinids and chiloguembelinids" (=Montian), which the author would correlate with the *Morozovella* (*Globorotalia*) *uncinata* and *Subbotina* (*Globigerina*) *inconstans* Zones. Gohrbandt (1963) recorded *S. praecursoria* from his Zone A and the lower two thirds of his zone B (equivalent to the *S. praecursoria* and *M. uncinata* zones. Similar observations have been made by the author in North Africa. In the Mexia Clay Member of the Wills Point Formation, Texas, *S. praecursoria* is a common associate of *S. pseudobulloides* and *S. inconstans*, and is not observed in younger strata. It also occurs in the Porters Creek of Mississippi and Alabama.

Subbotina spiralis Bolli, 1957

Globigerina spiralis Bolli, 1957, *U.S. Nat. Mus. Bull.* **215**, 70, pl. 16, figs 16–18.—Bolli and Cita, 1960, *Riv. Ital. Paleont.* **66**, 12, pl. 32, fig. 2.—Gohrbandt, 1963, *Mitt. geol. Ges. Wien* **56, H.** 1, 46, pl. 1, figs 10–12.—Hillebrandt, 1962, *Bayer, Akad. Wissen. Math.-Naturw. Kl. Abh. N.F.* **108**, 122, pl. 11, fig. 20 (?).

Remarks: This species is characterised by its relatively small size, tight coiling and medium to high trochospire. The tendency for the last chamber to develop an extraumbilical–umbilical aperture and the strongly curved sutures on the spiral side of the test reveal its affinities with *Morozovella pusilla pusilla.* It has probably evolved from *S. pseudobulloides* by way of forms closely related to *P. imitata.* It is here included in *Subbotina* for reasons similar to those discussed in including *Globigerina inconstans* in *Subbotina.*

Stratigraphical Range: This species has a restricted range—*M. uncinata* Zone—and as such is a useful guide form in the lower part of the Palaeocene. It has been reported in younger strata in the Palaeocene (Loeblich and Tappan, 1957; Jenkins, 1971), but these records are believed to be of *A. mckannai* (see discussion under that species).

Subbotina inconstans (Subbotina) 1953

Globigerina inconstans Subbotina, 1952, *Trudy, Vses. Neft. Nauk.-Issled. Geol.-Razved. Inst.* (V.N.G.R.I.) *n.s.* **76,** 58, pl. 3, figs 1–2.—Subbotina, 1960, *Akad. Nauk. U.S.S.R.* text-fig. 1.

Globorotalia (*Globorotalia*) *inconstans* (Subbotina).—Hillebrandt, 1962, *Bayer, Akad. Wiss. Math.-Naturw. Kl. Abh.* **108,** 130, pl. 12, figs 7–8.

Gl. (*Acarinina*) *inconstans* (Subbotina).—Leonov and Alimarina, 1961, (*pars*), *Moscow, Univ. Geol. Fac. Sbornik Trudov* pl. 3, figs 1–6, 8 (*non* 7a–c).

"Transitional form between *Globorotalia pseudobulloides* (Plummer) and *Globorotalia uncinata* Bolli, New species.—Bolli, 1957, *U.S. Nat. Mus. Bull.* **215,** pl. 17, figs 16–18.

Globorotalia varianta (Subbotina).—Loeblich and Tappan, 1957, *U.S. Nat. Mus. Bull.* **215,** 196, p. 44, figs 1–2 (*non* pl. 45, figs 4a–c).

Acarinina indolensis Morozova, 1959, *Akad. Nauk. U.S.S.R. Doklady* **124,** no. 5, 1116, text-figs 1g, d, e.

Globorotalia inconstans (Subbotina). Luterbacher, 1964, *Eclog. geol. Helv.* **57,** 650, figs 19–23.—Berggren, 1965, *Micropaleont.* **11,** 291, text-fig. 9, figs 3a–4c.

Transitional stage between *Globorotalia trinidadensis* Bolli and *Globorotalia uncinata uncinata* Bolli. El Naggar, 1966, *Bull. Brit. Mus.* (*Nat. Hist.*), Geol. Suppl. 2, 241, pl. 18, figs 5a–c.

Remarks: The author's concept of *Subbotina inconstans* is shown on Chart I, and by the forms which have been assigned to this species in the synonymy.

The author has examined the holotype (no. 3992) and paratype (no. 3993) in the collections of the Micropalaeontological Laboratory of V.N.I.G.R.I. in Leningrad. The five chambers of the last whorl, the slightly incised sutures, sub-angular peripheral margin, and weak overlap of earlier chambers has been observed as diagnostic. These forms agree most closely with Bolli's (1957) concept of the transition between *G. pseudobulloides* and *G. uncinata,* and not with *Globorotalia trinidadensis*, as suggested by Lenov and Alimarina (1961).

The species is included here in *Subbotina* because of its close relationship to *S. pseudobulloides*. The species actually lies in an intermediate position between *S. pseudobulloides* and *Morozovella uncinata* Chart I. This writer would disagree with Hillebrandt's (1962) conclusion that *S. (Globorotalia) inconstans* is a senior synonym of *Morozovella (Globorotalia) uncinata*, and although both species are closely related they are retained as separate species here. The author also disagrees with McGowran's (1968) suggestion that the boundary between *Subbotina* and *Acarinina* should be placed between *S. pseudobulloides* and *S. inconstans*.

Stratigraphical range and distribution: Subbotina inconstans evolved from *S. pseudobulloides* (Plummer) in strata which are equivalent to the Upper Danian *S. Str.* Forms which can be assigned to *S. inconstans* continue above the *Morozova uncinata* Zone, where they gradually lose their identity. The form is abundant in the Mexia Member of the Wills Point formation at various localities in Texas, and in the Porters Creek Clay in Mississippi and Alabama.

Genus PLANOROTALITES Morozova, 1957

Planorotalites compressa (Plummer) 1926
Globigerina compressa Plummer, 1926, *Univ. Texas Bull.* **2644,** 135, pl. 8, figs 11a–c (fig. 11a is holotype by subsequent designation of Brönnimann, 1952, p. 25).—Glaessner, 1937, *Pub. Lab. Pal. Moscow Univ. Prob. Pal.* **2–3,** pl. 4, figs 32a–c.—Cushman and Todd, 1942, *Cushman Lab. Foram. Res., Contr.* **18,** 44, pl. 8, figs 5, 6.—Troelsen, 1957, *U.S. Nat. Mus. Bull.* **215,** 129, pl. 30, figs 1a–c.—Hofker, 1960, *Contr. Cushman Found. Foram. Res.* **11,** pt. 3, no. 210, 78, figs 35a–c.
Globorotalia compressa (Plummer). Brönnimann, *Bull. Am. Paleontology* **34,** 25, pl. 2, figs 19–24.—Dalbiez and Glintzboekel (*in* Cuvillier *et al.*), 1955, *Proc. IV World Petrol. Congr. sect.* **1/D,** pl. 1, figs 3a–4c.—Bolli, 1957, *U.S. Nat. Mus. Bull.* **215,** 77, pl. 20, figs 21–23.—Loeblich and Tappan, 1957, *Ibid.* 188, pl. 40, figs 5a–c; pl. 41, figs 5a–c; pl. 42, figs 5a–c; pl. 44, figs 9a–10c.—Olsson, 1960, *J. Paleont.* **34,** no. 1, 45, pl. 8, figs 20–22.—Bolli and Cita, 1960, *Riv. Ital. Paleont.* **66,** no. 3, 20, pl. 32, figs 3a–c.—Gohrbandt, 1963, *Mitt. geol. Ges. Wien* **56,** 50, pl. 6, figs, 7–9.—El Naggar, 1966, *Bull. Brit. Mus. (Nat. Hist.), Geol. Suppl.* **2,** 203, pl. 17, figs 1a–3c.
Globorotalia membranacea (Ehrenberg). Subbotina, 1953 (*pars*), *Trudy, Vses. Neft. Nauk.-Issled. Geol.-Razved. Inst.* (V.N.I.G.R.I.) *n.s.* **76,** 205, pl. 16, figs 7a–c, 10a–c (*non* 8a–9c; 11a–13c) (*non* Ehrenberg).

Globigerina compressa Plumer var. *caucasica* Khalilov, 1956, *Akad. Nauk Azerb. S.S.R. Inst. Geol. Baku, Trudy* **17**, 237, pl. 1, figs 2a–c.

    Globorotalia (Turborotalia) compressa (Plummer). Berggren, 1962, *Stockholm Contr. Geol.* **11**, 94, pl. 14, figs 5a–c; text-figs 13 (1–6).—Jenkins, 1971, *N.Z. Geol. Surv. Pal. Bull.* **42**, p. 113, pl. 9, figs 227–229.

    Globorotalia kilabiyaensis. El Naggar, 1966, *Bull. Brit. Mus. (Nat. Hist.), Geol. Suppl.* **2**, 218, pl. 17, figs 4a–c.

    Planorotalites compressus (Plummer). McGowran, 1968, *Micropaleont.* **14**, no. 2, pl. 4, figs 10–11.

Remarks: This species is distinguished from the associated *S. pseudobulloides,* from which it is probably descended, by its more acute periphery, and the smaller, more densely distributed, pores. The wall lacks the reticulate ("honeycombed") pattern of *S. pseudobulloides.*

    P. compressa, which appears in the younger part of the type Danian, is an early member of one evolutionary lineage which leads to compressed, smooth-walled, keeled and non-keeled planorotaliids in the Palaeocene. It evolved, by gradual compression of the test, into *P. ehrenbergi.*

    Globigerina compressa var. *caucasica* was said to differ from the typical form in "the smaller number of chambers, including the final whorl" (Khalilov, 1956). *Planorotalites compressa* has 4–5 chambers in the final whorl. The stability of nomenclature is little served by distinctions such as this and the illustration appears assignable to *P. compressa* (Plummer) as it has generally been interpreted.

Stratigraphical range: Base of the *Subbotina trinidadensis* Sub-zone to lower part of *Morozovella pusilla pusilla* Zone.

Planorotalites ehrenbergi Bolli, 1957

    Globorotalia membranacea (Ehrenberg). White, 1928, *J. Paleont.* **2**, 280, pl. 38, fig. 1.—Cushman and Bermudez, 1949, *Contr. Cushman Lab. Foram. Res.* **25**, 34, pl. 6, figs 16–18 (see synonymic list).—Subbotina, 1953 (*pars*), 1953, *Trudy, Vses. Neft. Nauk.-Issled. Geol.-Razved. Inst. (V.N.I.G.R.I.), n.s.* **76**, 205, pl. 16, figs 8a–9c; 11a–12c (*non* 7a–c; 10a–c; 13a–c) (*non* Ehrenberg).

    Globorotalia ehrenbergi Bolli, 1957, *U.S. Nat. Mus. Bull.* **215**, 77, pl. 20, figs 18–20.—Bolli and Cita, 1960, *Riv. Ital. Paleont.* **66**, 21, pl. 33, figs 4a–c.— El Naggar, 1966, *Bull. Brit. Mus. (Nat. Hist.), Geol. Suppl.* **2**, 207, pl. 17, figs 5a–c.

    Globorotalia pseudomenardii, Bolli, Loeblich and Tappan, 1957, *U.S. Nat. Mus. Bull.* **215**, 193, pl. 45, figs 10a–c.

    Globorotalia haunsbergensis Gohrbandt, 1963, *Mitt. geol. Ges. Wien* **56**, 53, pl. 6, figs 10–12.

Planorotalites chapmani ehrenbergi (Bolli). McGowran, 1968, *Micropaleont.* **14,** no. 2, pl. 4, fig. 12.

Remarks: The holotype of this species contains a faint keel in the last two chambers indicating its relationship to the stratigraphically younger *P. pseudomenardii.*

This species, as well as *P. pseudomenardii,* has appeared in the literature under the name *Globorotalia membranacea* (Ehrenberg)—a Pliocene form of dubious affinities. *Globorotalia haunsbergensis* Gohrbandt is here included in the synonymy of this species. Differences between these forms appear too slight to warrant separation as distinct species.

Stratigraphical range: Lower part of *Morozovella pusilla* Zone to lower part of *P. pseudomenardii* Zone.

Planorotalites pseudomenardii Bolli, 1957

Globorotalia membranacea (Ehrenberg). Toulman, 1941, *J. Paleont.* **15,** no. 6, 608, pl. 82, figs 4, 5.—Subbotina (*pars*), 1953, *Trudy Vses. Nauk.-Issled. Geol.-Razved. Inst.* (*V.N.I.G.R.I.*) *n.s.* **76,** 205, pl. 16, figs 13a–c (*non* 7a–12c). —Aubert, 1962, *Notes Serv. géol. Maroc.* **21,** 56, pl. 2, fig. 6.

Globorotalia cf. *membranacea* (Ehrenberg). Hofker, 1955, *Rep. McLean Foram. Lab.* no. **2,** 14, pl. 4.

Globorotalia pseudomenardii Bolli, 1957, *U.S. Nat. Mus. Bull.* **215,** 77, pl. 20, figs 14–17.—Loeblich and Tappan, 1957, *Ibid.,* 193, pl. 45, figs 10a–c; pl. 47, figs 4a–c; pl. 49, figs 6a–c; pl. 54, figs 10a–13c; pl. 59, figs 3a–c; pl. 60, figs 1a–c.—Bolli and Cita, 1960, *Riv. Ital. Paleont.* **66,** 26, pl. 33, figs 2a–c.— Olsson, 1960, *J. Paleont.* **34,** 47, pl. 9, figs 10–12.—Bermúdez, 1961, *Memoria del III Congresso Geologico Venezolano,* **III,** *Boletin de Geologia, Publicacion especial* 3, Caracas, Venezuela, 1298, pl. 15, fig. 9.—Gartner and Hay, 1962, *Eclog. geol. Helv.* **55,** 566, pl. 1, fig. 5 (see synonymic list).—Gohrbandt, 1963, *Mitt. geol. Ges. Wien* **56,** 52, pl. 6, figs 16–18.—El Naggar, 1966, *Bull. Brit. Mus.* (*Nat. Hist.*)*, Geol. Suppl.* **2,** 227, pl. 17, figs 7a–8c.

Planorotalites psuedomenardii (Bolli). McGowran, 1968, *Micropaleont.* **14,** no. 2, pl. 4, figs 5–9.

Globorotalia (*Planorotalites*) *pseudomenardii* Bolli. Jenkins, 1971, *N.Z. Geol. Surv. Pal. Bull.* **42,** 109, pl. 9, figs 217–220.

Remarks: This is one of the most distinctive and easily recognisable species in the Palaeocene. The strongly compressed test with moderately inflated chambers, lobulate peripheral margin and keeled periphery sets it apart from all other species. Its restricted stratigraphic range makes it an excellent guide form for the middle part of the Palaeocene (see discussion in Gartner and Hay, 1962).

In addition to the references above which give an idea of its geographic distribution, this species occurs at latitude 30°S, west of the Mid-Atlantic Ridge (DSDP, Leg 3).

Stratigraphical range: This species is restricted to the *Planorotalites pseudomenardii* Zone. Despite reports to the contrary, *P. pseudomenardii* is not found in type Montian deposits (Berggren, 1962; Gohrbandt, 1963). The *P. pseudomenardii* Zone can be correlated with part of the type Thanetian (Gohrbandt, 1963; Berggren, 1965); the upper Montian, or Montian *s. str.* is older than the *P. pseudomenardii* Zone, and is equivalent, at least in part, with the *M. uncinata* Zone.

Planorotalites chapmani Parr, 1938

Globorotalia chapmani Parr, 1938, *J. Roy. Soc. West. Australia* **24,** (1937–1938), 87, pl. 3, figs 8, 9a, b.—McGowran, 1964, *Ibid.,* vol. 42, pt. 3, 85, text-figs 1–9.

Planorotalites chapmani (Parr). McGowran, 1968, *Micropaleont.* **14,** pl. 4, figs 13–18, 21.

Globorotalia compressa (Plummer). Jennings, 1936, *Bull. Am. Paleont.* **23,** 193, pl. 4, fig. 8.—Toulmin, 1938, *J. Paleont.* **15,** no. 6, 607, pl. 82, figs 1, 2 (*non* Plummer).

Globorotalia membranacea (Ehrenberg). Glaessner, 1937, *Probl. Paleontology, Moscow Univ. Lab. Paleontology* **2–3,** 385, pl. 4, fig. 38.—Subbotina, 1953 (*pars*), *Trudy, Vses. Neft. Nauk.Issled. Geol.-Razved. Inst.* (*V.N.I.G.R.I.*) n.s. **76,** 205, pl. 16, fig. 12 (*non* pl. 16, figs 7a–11b; 13) (*non* Ehrenberg).

Globorotalia luxorensis Nakkady, 1951, *J. Paleont.* **24,** 691, pl. 90, figs 39–41.—Le Roy, 1953, *Geol. Soc. Am. Mem.* **54,** 18, pl. 3, figs 5–7.—Nakkady, 1959, *Micropaleont.* **5,** 465, pl. 5, fig. 1.

Globanomalina ovalis Haque var. *lakiensis* Haque, 1956, *Pakistan Geol. Survey, Mem. Pal. Pakistanica, Quetta* **1,** 149, pl. 14, figs 2a–c.

Globanomalina simplex Haque, 1956, *Ibid.,* 149, pl. 30, figs 2a–c.

Globanomalina simplex Haque var. *orbicularis, Ibid.,* 149, pl. 27, figs 1a–c.

Globorotalia elongata Glaessner. Bolli, 1957, *U.S. Nat. Mus. Bull.* **215,** 77, pl 20, figs 11–13.—Loeblich and Tappan, 1957, *Ibid.,* 189, pl. 45, figs 5a–c; pl. 46, figs 5a–c; pl. 48, figs 5a–c; pl. 49, figs 7a–c; pl. 54, figs 1a–5; pl. 59, figs 4a–c; pl. 60, figs 9a–c; pl. 63, figs 2a–c.—Olsson, 1960, *J. Paleont.* **34,** 45, pl. 9, figs 4–6 (*non* Glaessner).

Globorotalia troelseni Loeblich and Tappan, 1957, *U.S. Nat. Mus. Bull.* **215,** 196, pl. 60, figs 4a–c; pl. 63, figs 5a–c; Berggren, 1960, *Rept. XXI Int. Geol. Congr.* pt. **6,** 53, pl. 1, figs 21a–c.—Gohrbandt, 1963, *Mitt. geol. Ges. Wien* **56,** 51, pl. 6, figs 13–15.—El Naggar, 1966, *Bull. Brit. Mus.* (*Nat. Hist.*), *Geol. Suppl.* **2,** 238, pl. 17, figs 10a–c.

Globorotalia (*Globorotalia*) *elongata* Glaessner. Hillebrandt, 1962, *Bayer, Akad. Wissen. Math.-Naturw. Kl. Abh. N.F.* **108,** 127, pl. 12, figs 9a–c (*non* Glaessner).

Globorotalia emileis El Naggar, 1966, *Bull. Brit. Mus.* (*Nat. Hist.*), *Geol. Suppl.* **2,** 208, pl. 17, figs 9a–c.

Remarks: In this author's interpretation *Planorotalites chapmani* is a relatively long-ranging species with a wide range of intraspecific variation (see synonymic list).

The taxonomic affinities of the present species were clarified in the recent study by McGowran (1964) of Parr's original material from the Kings Park Shale (Perth Basin, W. Australia). He concluded that *Globorotalia chapmani* is the correct name for Palaeocene forms formerly identified as *G. elongata.* McGowran's findings are substantiated by the author's comparison of specimens of this species from the Boongerooda Greensand of the Carnarvon Basin, N.W. Australia, which were supplied by McGowran, with hypotypes of *G. elongata* designated by Bolli, Loeblich and Tappan in the U.S. National Museum, Washington, D.C.

Stratigraphical range and distribution: Planorotalites chapmani ranges from the *P. pseudomenardii* Zone to the *Morozovella subbotinae* Zone. The author has observed this species in the lower part of the *M. subbotinae* Zone of the Bashi Marl member of the Hatchetigbee Formation (Wilcox Group) in Alabama (which is within the uppermost part of the *Discoaster multiradiatus* Zone) in samples from the lower part of the *Discoaster tribrachiatus* Zone in the Lodo Formation in California (Bramlette and Sullivan, 1961, tab. 1), in the lower part of the *M. rex* Zone of Libya and Egypt (North Africa) and in the *Globorotalia subbotinae* Zone (equivalent to the *Globorotalia marginodentata* Subzone of Subbotina, 1953) of the North–Central Caucasus in the Soviet Union.

Planorotalites planoconica Subbotina, 1953

Globorotalia planoconica Subbotina, 1953, *Trudy, Vses. Neft. Nauk.-Issled. Geol.-Razved. Inst.* (*V.N.I.G.R.I.*) n.s. **76,** 210, pl. 17, figs 4a–6c.—Said and Sabry, 1964, *Micropaleont.* **10,** 384, pl. 3, figs 12a–c.

Remarks: This species remains little known to micropalaeontologists outside the Soviet Union. The holotype (V.N.I.G.R.I. no. 4081) and two paratypes (V.N.I.G.R.I. nos 4082, 4083) from the Zone of conical globorotaliids, Foraminiferal beds, Suite F_1, in the vicinity of the town of Nal'chik, Khieu River, northern Caucasus, Soviet Union, are characterised by their small size, compressed test and acute periphery. Subbotina (1953) mentions that the peripheral margin has a keel in some instances; none were observed by this writer on the three specimens studied in the collections of V.N.I.G.R.I.

This species bears a close resemblance to *P. pseudomenardii* in some respects. Its origin remains an uncertainty; it may have evolved from forms close to the *P. pseudomenardii-P. chapmani* complex. It may well have been the ancestor of *Planorotalites pseudoscitula* Glaessner (=*G. renzi* Bolli) (see Berggren, 1960).

Stratigraphical range and distribution: Planorotalites pseudomenardii Zone to *Morozovella aragonensis* Zone. The range attributed to *P. planoconica* by Subbotina (1953, tab. 3, fig. 7; 1960, fig. 4) appears excessive: Zone of rotaliid-like globorotaliids through the Zone of conical globorotaliids. It is likely that more than one species is involved in this case.

This writer has observed *P. planoconica* in samples from the Palaeocene (*P. pseudomenardii* Zone) of New Zealand, kindly sent by Dr. Graham Jenkins. Elsewhere he has observed this species in the *M. subbotinae* Zone of Alabama (Bashi Marl Formation), the Lodo Formation of California and the lower Eocene *M. rex* to *M. aragonensis* Zones of the Esna Shale Formation, Egypt.

Planorotalites imitata Subbotina, 1953

Globorotalia imitata Subbotina, 1953, *Trudy, Vses. Neft. Nauk.-Issled. Geol.-Razved. Inst. (V.N.I.G.R.I.) n.s.* **76**, 206, pl. 16, figs 14a–16c.—Loeblich and Tappan, 1957, *U.S. Nat. Mus. Bull.* **215**, 190, pl. 44, figs 3a–c; pl. 45, figs 6a–c; pl. 59, figs 5a–c; pl. 63, figs 3a–c.—Subbotina, 1960, *Izdat. Akad. Nauk., S.S.S.R.* fig. 4.—Berggren, 1965, *Micropaleont.* **11**, 290, text-figs 8a–f.

Planorotalites imitatus (Subbotina). McGowran, 1968, *Micropaleont.* **14**, no. 2, pl. 4, figs 1–3.

Remarks: This distinctive little form is distinguished by its small size, four to five finely perforate, inflated chambers in the final whorl, curved, depressed sutures and umbilical–extra-umbilical aperture. The last chamber, in edge view, is typically subrounded to truncate.

This species has probably evolved from forms intermediate between *S. pseudobulloides* and *S. spiralis* (see Berggren, 1966).

Stratigraphical range and distribution: Morozovella uncinata Zone through *P. pseudomenardii* Zone.

Loeblich and Tappan (1957) recorded this species from various formations of the Gulf and Atlantic Coastal Plain which range from the *M. uncinata* Zone (see Berggren, 1965) to the *P. pseudomenardii* Zone. A similar range has been observed in North Africa.

Planorotalites pseudoscitula Glaessner, 1937

Globorotalia pseudoscitula Glaessner, 1937, *Studies in Micropaleontology*, *Publ. Lab. Paleontology, Moscow Univ.* 32, text-figs 3a–c.—Subbotina, 1953 (*pars*), *Trudy, Vses. Neft. Nauk.-Issled. Geol.-Razved. Inst. (V.N.I.G.R.I.) n.s.* **76,** 208, pl. 17, figs 1a–c (*non* pl. 16, figs 17a–c; pl. 18, figs 18a–c (?)).— Shutskaya, 1956, *Akad. Nauk, S.S.S.R., Trudy Inst. geol. Nauk.* **164,** *geol. ser.* (no. 71), 95, pl. 4, figs 5a–c.—Pokorny, 1960, *Rev. de l'Inst. fran. du Pétrole et Ann. des Combust Liq.* **15,** 1126, pl. 5, figs 3a–c.

Globorotalia renzi Bolli, 1957, *U.S. Nat. Mus. Bull.* **215,** 168, pl. 38, figs 3a–c.—Berggren, 1960, *XXI Int. Geol. Congr. Proc. Sect. 6,* 53, pl. 1, figs 16a, b.

Globorotalia (Planorotalites) renzi Bolli. Jenkins, 1971, *N.Z. Geol. Surv. Bull.* **42,** 110, pl. 9, figs 224–226.

Remarks: The holotypes of *P. pseudoscitula* and *P. pseudoscitula* var. *elongata* were unfortunately lost during the Second World War. However, a satisfactory concept of this species is contained in the work of Subbotina (1953) and Shutskya (1956).

Bolli's (1957, p. 168) *Globorotalia renzi* clearly shows the characters of this species. Although superficially similar to *Planorotalites pseudomenardii,* it is consistently smaller, the test is more compressed and the outline is generally more circular. The form described by Glaessner as *G. pseudoscitula* var. *elongata* is regarded here as a morphotypic variant of the species.

Planorotalites pseudoscitula appears to be the end member of a lineage which began with *Subbotina pseudobulloides* and led through *Planorotalites compressa* to the compressed smooth planorotaliids (see Berggren, 1966; McGowran, 1968).

Stratigraphical range and distribution: Upper part of the *Morozovella aragonensis* Zone to the top of the *Truncorataloides rohri* Zone.

P. pseudoscitula has been recorded from various Middle Eocene sections in the S.W. Soviet Union, Caribbean, Mediterranean (Berggren, personal collections) and New Zealand. It is also a relatively common component of Middle Eocene assemblages in mid-low latitudes in the Atlantic and Indian Ocean.

Planorotalites palmerae (Cushman and Bermúdez), 1937

Globorotalia palmerae Cushman and Bermúdez, 1937, *Contr. Cushman Lab. Foram. Res.* **13,** 26, pl. 2, figs 51–53.—Bermúdez, 1937, *Mem. Soc. Cubana Hist. Nat.* **11,** 167; 1938, **12,** 11.—Cushman and Bermúdez, 1949, *Contr. Cushman Lab. Foram. Res.* **25,** pt. 2, 31, 32, pl. 6, figs 4–6.—Bolli, 1957, *U.S. Nat. Mus. Bull.* **215,** 166, pl. 38, figs 2a–c.—Schmidt and Raju,

1973, *Koninkl. Nederl. Akademie Wetensch, Proc. ser. B* **76,** no. 2, 177, pl. 1, figs 5a–c; pl. 2, figs 5, 6; 9, 10.

Remarks: Although the author has elsewhere considered this species to belong to the genus *Pararotalia* and thus not planktonic (Berggren, 1966), Schmidt and Raju (1973) have recently shown conclusively that this form is, indeed, planktonic. They suggest that it has evolved from *P. pseudoscitula* by the radial prolongation of the chambers at the axial periphery which develop a radially elongate shape terminating in a spine. It is the last in sequence of appearance of the phylogenetic lineage *planoconica–pseudo-scitula–palmerae*, although it is succeded by its longer-ranging antecedent *pseudoscitula* in the middle Eocene.

Stratigraphical range: Restricted to the zone of which it is the nominate species; because of its restricted geographic distribution this zone is here replaced by the *Acarinina densa* Zone.

Genus MOROZOVELLA McGowran in Luterbacher, 1964

Morozovella uncinata Bolli, 1957

Globorotalia uncinata Bolli, 1957, *U.S. Nat. Mus. Bull.* **215,** 74, pl. 17, figs 13–15.—Bolli and Cita, 1960, *Riv. Ital. Paleont.* **66,** 391, pl. 32, figs 5a–c, 7a–c.—Berggren, 1965, *Micropaleont.* **11,** 294, text-fig. 9, figs 5a–c.—El Naggar, 1966, *Bull. Brit. Mus. (Nat. Hist.), Geol. Suppl.* **2,** 240, pl. 18, figs 1a–c; pl. 19, figs 2a–c.
Acarinina indolensis Morozova, 1959, *Dokl. Acad. Nauk. S.S.R.* **124,** 1114, figs 1d–f.
Globorotalia (Acarinina) inconstans (Subbotina).—Leonov and Alimarina, 1961 (*part*), *Moscow, Univ. Geol. Fac. Sbornik Trudov* pl. 3, figs 7a–c (*non* figs 1–6, 8).
Globorotalia indolensis (Morozova) Luterbacher, 1964, *Eclog. geol. Helv.* **57,** 656, figs 33, 34.
Globigerina? uncinata (Bolli. Gohrbandt, 1963, *Geol. Ges. Wien. Mitt.* **55,** 49, pl. 3, figs 7–9.
Globorotalia perclara Loeblich and Tappan, El Naggar, 1966, *Bull. Brit. Mus. (Nat. Hist.), Geol. Suppl.* **2,** 263, pl. 21, figs 2a–c (*non* Loeblich and Tappan).

Remarks: This species has developed from *Subbotina pseudobulloides* via *S. inconstans.* The distinction between *Subbotina* and *Morozovella* is made on the basis of the development in *M. uncinata,* of a low angulo-conical test, flat spiral side, lateral truncation of the chambers and sharply curved sutures on the spiral side.

M. uncinata is part of a gradually evolving lineage which leads to the first angulo-conical, keeled morozovellid *M. angulo* (White).

Stratigraphical range and distribution: The appearance of *Morozovella uncinata* defines the base of the *M. uncinata* Zone. This species ranges into the *M. pusilla pusilla—M. angulata* Zone where it is gradually replaced by *M. angulata.*

The species has a narrow geographic distribution and probably does not occur further north than 45°N (Bandy, 1960). Data for the southern hemisphere are scarce or incomplete; *M. uncinata* has, however, been recorded at 30°S in the South Atlantic. *M. uncinata* is recorded from the Caribbean, Mediterranean and N.W. Caucasus.

The *M. uncinata* Zone is an important biostratigraphic interval which because of its limited vertical extent has either been included in the Danian (*s. str.*) or Thanetian, or has been overlooked altogether.

Morozovella pusilla Bolli, 1957

Globorotalia pusilla pusilla Bolli, 1957, *U.S. Nat. Mus. Bull.* **215,** 78, pl. 20, figs 8–10.—Bolli and Cita, 1960, *Riv. Ital. Paleont.* **66,** 28, pl. 32, figs 4a–c.—Said and Sabry, 1964, *Micropaleont.* **10,** 385, pl. 1, figs 11a–c.—El Naggar, 1966, *Bull. Brit. Mus.* (*Nat. Hist.*), *Geol. Suppl.* **2,** 232, pl. 17, figs 11–ac. *Planorotalites tauricus* Morozova, 1961, *Paleontol. Zhurnal Akad. Nauk S.S.S.R. no.* **2,** 16, pl. 2, fig. 3.

Remarks: This stratigraphically important species may be distinguished from *Subbotina spiralis* by its more acute periphery and the strongly curved, depressed spiral sutures. With the development of a more circular outline, a more acute axial periphery, development of a keel, and flush spiral sutures, it evolved into the distinct *Morozovella albeari.*

This species appears to have been described from the Crimea under the name of *Planorotalites tauricus* Morozova (1961). Examination of the holotype of this species failed to reveal significant differences with *M. pusilla* (see further remarks in Berggren, 1966).

Stratigraphical range: Morozovella pusilla Zone.

Morozovella albeari (Cushman and Bermúdez) 1949

Globorotalia albeari Cushman and Bermúdez, 1949, *Contr. Cushman Lab. Foram. Res.* **25,** 33, pl. 6, figs 13–15.
Globorotalia pusilla laevigata Bolli, 1957, *U.S. Nat. Mus. Bull.* **215,** 78, pl. 20, figs 5–7.—Bolli and Cita, 1960, *Riv. Ital. Paleont.* **66,** no. 3, 27, pl. 32, figs 6a–c.—Said and Sabry, 1964, *Micropaleont.* **10,** 385, pl. 3, figs 7a–c.—El

Naggar, 1966, *Bull. Brit. Mus. (Nat. Hist.)*, *Geol. Suppl.* **2**, 229, pl. 17, figs 12a–c.

Globorotalia pseudoscitula Glaessner. Loeblich and Tappan, 1957 (*pars*), *U.S. Nat. Mus. Bull.* **215**, 193, pl. 48, figs 3a–c; pl. 59, figs 2a–c; pl. 62, figs 6a–c (*non* pl. 46, figs 4a–c).

Globorotalia (Globorotalia?) pusilla laevigata Bolli. Hillebrandt, 1962, *Bayer, Akad. Wissen. Math.-Natur. w. Kl. Abh. N.F.* **108**, 128, pl. 11, figs 17a–c.

Globorotalia pusilla aff. *laevigata* Bolli. Gohrbandt, 1963, *Mitt. geol. Ges. Wien* **56**, 54, pl. 3, figs 19–21.

Remarks: This species was originally described from the Madruga Formation of Cuba (Cushman and Bermúdez, 1949). It was recorded as *G. pusilla laevigata* by Bolli (1957) from the *G. pseudomenardii* Zone, Lower Lizard Springs Formation of Trinidad and appears to have been simultaneously recorded by Loeblich and Tappan (1957) as *Globorotalia pseudoscitula*. The latter species evolves in the upper part of the Lower Eocene and ranges through the Middle Eocene; it is a senior synonym of *Planorotalia renzi* Bolli and is unrelated—relatively speaking—to *M. albeari* (scc Berggren, 1966).

This writer has suggested (*op. cit.*) derivation of *Morozovella convexa* from *M. pusilla laevigata* in the *P. pseudomenardii* Zone.

Stratigraphical range: Planorotalites pseudomenardii Zone.

Morozovella convexa Subbotina, 1953

Globorotalia convexa Subbotina, 1953, *Trudy, Vses. Neft. Nauk.-Issled. Geol.-Razved. Inst. (V.N.I.G.R.I.) n.s.* **76**, 209, pl. 17, figs 2a–c.—Loeblich and Tappan, 1957, *U.S. Nat. Mus. Bull.* **215**, 188, pl. 48, figs 4a–c; pl. 50, figs 7a–c; pl. 53, figs 6a–8c; pl. 57, figs 5a–c; pl. 61, figs 4a–c; pl. 63, figs 4a–c.—Olsson, 1960, *J. Paleont.* **34**, 45, pl. 9, figs 13–15.—Berggren, 1960, *Stockholm Contr. Geol.* **5**, 91, pl. 11, figs 1a–c.—Gartner and Hay, 1962, *Eclog. geol. Helv.* **55**, 562, pl. 1, figs 4a–c.—Gohrbandt, 1963, *Mitt. geol. Ges. Wien* **56**, 54, pl. 3, figs 4–6, 10–12.

Turborotalia (Acarinina) convexa Subbotina. Pokorny, 1960, *Rev. Inst. Fran. Pétrole et Ann. Comb. Liquides* **15**, 126, pl. 5, figs 2a–c.

Truncorotaloides (Morozovella) convexus (Subbotina). McGowran, 1968, *Micropaleont.* **14**, pl. 2, figs 11–14.

Remarks: This species is interpreted here as the ancestor of *Morozovella broedermanni* (Bolli, 1957).

Stratigraphical range: Planorotalites pseudomenardii Zone to *Morozovella aragonensis* Zone. The range given by Subbotina (1953) is considered as somewhat excessive by this author.

Morozovella broedermanni Cushman and Bermúdez, 1949

Globorotalia (*Truncorotalia*) *broedermanni* Cushman and Bermúdez, 1949, *Contr. Cushman Lab. Foram. Res.* **25**, 40, pl. 7, figs 22–24.

Globorotalia broedermanni Cushman and Bermúdez. Bolli, 1957 (*pars*), *U.S. Nat. Mus. Bull.* **215**, 80, pl. 19, figs 13–15.

Remarks: This species appears to be related to *M. convexa* from which it differs in its generally larger size, larger umbilicus and less restricted aperture.

Stratigraphical range: Morozovella subbotinae Zone to the *M. aragonensis* Zone. This species, as conceived here, has a shorter range than that given by Bolli (1957).

Morozovella lodoensis (Mallory) 1959

?*Globorotalia broedermanni* Cushman and Bermúdez. Bolli, 1957 (*pars*), *U.S. Nat. Mus. Bull.* **215**, 80, pl. 37, figs 13a–c (*non* pl. 19, figs 13–15).

Globorotalia broedermanni Cushman and Bermúdez var. *lodoensis* Mallory, 1959, Lower Tertiary Biostratigraphy of the California Coast Ranges. *Am. Assoc. Petrol. Geol. Spec. Publ.* 253, pl. 23, figs 3a–c.

Globorotalia mattseensis Gohrbandt, 1967, *Micropaleont.* **13**, 322, pl. 1, figs 25–30.

Remarks: Mallory (1959) distinguished this species by means of its larger umbilical opening, straighter radial ventral sutures and more lobulate periphery.

Further work is desirable in order to determine whether Bolli's illustration of *G. broedermanni* is, indeed, the same as Mallory's form. An examination of topotype specimens of *Globorotalia mattseensis* Gohrbandt indicates that it is conspecific with *M. lodoensis* (Mallory).

Stratigraphical range: Acarinina densa Zone to *Globigerapsis Kugleri* Zone.

Morozovella aequa (Cushman and Renz) 1942

Globorotalia angulata (White). Glaessner, 1937 (*partim*), *Probl. Paleont.* **2–3**, 383, pl. 4, figs 36a–c (*non* figs 35a–c, 37a–c).

Globorotalia crassata (Cushman) var. *aequa* Cushman and Renz, 1942, *Contr. Cushman Lab. Foram. Res.* **18**, pl. 1, 12, pl. 3, figs 3a–c.—Weiss, 1955, *J. Paleont.* **29**, 19, pl. 6, figs 4–6.—Sacal and Debourle, 1957, *Mém. Géol. Soc. France no.* **78**, 64, pl. 29, figs 10–12.—Aubert, 1962, *Notes Serv. géol. Maroc.* **21**, 55, pl. 2, fig. 5.

Globorotalia crassata (Cushman). Subbotina, 1947, *Mikrofauna Kavkaza, Emby i Srednei Azii, Lengostoptekhizdat* 119–121, pl. 5, figs 31–33 (*non* pl. 9, figs 15–17).—Subbotina, 1953 (*partim*), *Trudy, Vses. Neft. Nauk.-Issled.*

Geol.-Razved. Inst. (*V.N.I.G.R.I.*) *n.s.* **76**, 211, pl. 17, figs 11a–12c (*non* figs 7a–10c, 13a–c).

Truncorotalia crassata var. *aequa* (Cushman and Renz). Dalbiez and Glintzboeckel, 1955, *in* Cuvillier *et al., Proc. IV World Petrol Congr.* sect. 1/D, 535, pl. 2, fig. 9.

Globorotalia praenartanensis Shutskaya, 1956, *Akad. Nauk. S.S.S.R., Trudy Inst. Geol. Nauk.* **164**, 98, pl. 3, fig. 5.

Globorotalia aequa Cushman and Renz. Bolli, 1957, *U.S. Nat. Mus. Bull.* **215**, 74, pl. 17, figs 1–3; pl. 18, figs 13–15.—Loeblich and Tappan, 1957 (*partim*), *Ibid.,* 186, pl. 50, fig. 6; pl. 55, fig. 8; pl. 59, fig. 6; pl. 60, fig. 3; pl. 64, fig. 4 (*non* pl. 46, figs 7a–8c).—Bolli and Cita, 1960, *Riv. Ital. Paleont.* **66**, 17, pl. 31, fig. 51.—Gartner and Hay, 1962, *Eclog. geol. Helv.* **55**, 560, pl. 2, figs 1a–2b.—Said and Sabry, 1964, *Micropaleont.* **10**, no. 3, 381, pl. 2, fig. 8.—Luterbacher, 1964, *Eclog. geol. Helv.* **57**, 670, figs 63–71.—El Naggar, 1966, *Bull. Brit. Mus.* (*Nat. Hist.*), *Geol. Suppl.* **2**, 190, pl. 21, figs 6a–c.

Globorotalia aff. *crassata aequa* Cushman and Renz. Hornibrook, 1958, *Micropaleont.* **4**, pl. 1, figs 1–3.

Globorotalia (*Truncorotalia*) *aequa aequa* Cushman and Renz. Hillebrandt, 1962, *Bayer, Akad. Wissen. Math.-Naturw. K. Abh. N.F.* **108**, 133, pl. 13, figs 1–4.

Truncorotalia aequa (Cushman and Renz). Gohrbandt, 1963, *Mitt. geol. Ges. Wien* **56**, 58, pl. 4, figs 10–12.

Truncorotaloides (*Morozovella*) *aequus* (Cushman and Renz). McGowran, 1968, *Micropaleont.* **14**, pl. 2, fig. 15.

Globorotalia aequa bullata Jenkins, 1965, *N.Z. Geol. Geophys.* **8**, 1110, fig. 10 (87–91).

Globorotalia (*Morozovella*) *bullata* Jenkins, 1971, *N.Z. Geol. Surv. Bull.* **42**, 100, pl. 7, figs 172–176.

Globorotalia (*Morozovella*) *aequa aequa* Cushman and Renz. Jenkins, 1971, *N.Z. Geol. Surv. Bull.* **42**, 100, pl. 7, figs 167–171.

Morozovella aequa (Cushman and Renz). Berggren, 1971, *Proc. II Plankt. Conf.* pl. 5, fig. 6.

Remarks: This species is distinguished from associated forms in the Palaeocene and lower Eocene in having a $3\frac{1}{2}$–$4\frac{1}{2}$ chamber in the final whorl, strongly convex umbilical side, narrow and deep umbilicus and a peripheral margin which ranges from subacute to weakly keeled. It probably evolved in the lower part of the *Planorotalites pseudomenardii* Zone from *M. angulata*.

Although Bolli and Cita (1960) have suggested that *M. aequa* is closely related to *G. lensiformis*, this writer believes *M. aequa* is the ancestor of *Morozovella subbotinae* (=*M. rex*), and that *M. lensiformis* is intermediate between *M. subbotinae* and *M. aragonensis* (see Berggren, 1966).

Stratigraphical range: Planorotalites pseudomenardii Zone through *Morozovella subbotinae* Zone.

Morozovella angulata (White) 1928

Globigerina angulata White, 1928, *J. Paleont.* **2,** no. 3, 191, pl. 27, fig. 13.
Globorotalia angulata (White). Glaessner, 1937b, *Probl. Paleont.* **2–3,** 383, pl. 4, figs 35a–c (*non* 36a–37c).—Bolli, 1957, *U.S. Nat. Mus. Bull.* **215,** 74, pl. 17, figs 7–9.—Loeblich and Tappan, 1957, *Ibid.*, 187, pl. 45, fig. 7; pl. 48, fig. 2; pl. 50, fig. 4; pl. 55, figs 2, 6, 7; pl. 58, fig. 2; pl. 64, fig. 5.—Berggren, 1960, *Int. Geol. Congress 21st Sess. Copenhagen* pt. **6,** *Proc.* 53, pl. 1, figs 11a–14.—Olsson, 1960, *J. Paleont.* **34,** p. 44, pl. 8, figs 14–16.—Bolli and Cita, 1960, *Riv. Ital. Paleont.* **66,** 18, pl. 33, fig. 8.—Gartner and Hay, 1962, *Eclog. geol. Helv.* **55,** no. 2, 559, pl. 1, fig. 6.—Said and Sabry, 1964, *Micropaleont.* **10,** 392, pl. 1, fig. 3.—Luterbacher, 1964, *Eclog. geol. Helv.* **57,** no. 2, 658, figs 37–39.—El Naggar, 1966, *Bull. Brit. Mus. (Nat. Hist.), Geol. Suppl.* **2,** 197, pl. 22, figs 1a–c.
Acarinina conicotruncata (Subbotina). Subbotina, 1953 (*pars*), *Trudy, Vses. Neft. Nauk.-Issled. Geol.-Razved. Inst. (V.N.I.G.R.I.),* n.s. **76,** 220, pl. 20, figs 9a–12c, (*non* figs 5a–8c).
Globorotalia apanthesma Loeblich and Tappan, 1957, *U.S. Nat. Mus. Bull.* **215,** 187, pl. 68, fig. 1; pl. 55, fig. 1; pl. 59, fig. 1.
Globorotalia (Morozovella) apanthesma Loeblich and Tappan. Jenkins, 1971, *N.Z. Geol. Surv. Bull.* **42,** 102, pl. 8, figs 186–188.
Globorotalia quadrata Nakkady and Talaat, 1959, *Micropaleont.* **5,** no. 4, 462, pl. 7, figs 3a–c.
Globorotalia (Truncorotalia) angulata (White). Leonov and Alimarina, 1961, *Sb. Trudov Geol. Fak. (k XXI Mezhd. Geol. Kongr.) Izd. Mosk. Univ.* 53, pl. 4, figs 1–4; pl. 5, figs 1–10; pl. 7, figs 5, 8, 9.—Hillebrandt, 1962, *Bayer, Akad. Wissen. Math.-Naturw. Kl. Abh. N.F.* **108,** 131, pl. 13, figs 14, 15.
Globorotalia (Acarinina) conicotruncata Subbotina. Leonov and Alimarina, 1961, *Ibid.,* 53, pl. 4, figs 5–9.
Truncorotalia angulata angulata (White). Gohrbandt, 1963, *Mitt. geol. Ges. Wien.* **56,** 57, pl. 4, figs 4–6.
Globorotalia (Morozovella) angulata (White). Jenkins, 1971, *N.Z. Geol. Surv., Bull.* **42,** 102, pl. 8, figs 183–185.

Remarks: This is one of the first Palaeogene angulo-conical morozovellids, and it is a common, and often the dominant species, in middle Palaeocene faunas. It exhibits a relatively wide range of morphological variation (Chart V).

M. angulata evolved from *M. uncinata* in the upper part of the *M. uncinata*

Zone; it evolved into *M. aequa* in the lower part of the *Planorotalites pseudomenardii* Zone. *Morozovella conicotruncata* evolved from *M. angulata* within the *M. pusilla* Zone and this evolutionary sequence can be used to subdivide the *M. pusilla–M. angulata* Zone.

Stratigraphical range: Uppermost part of the *M. uncinata* Zone or base of the *M. pusilla pusilla* Zone to the lower part of the *P. pseudomenardii* Zone.

Morozovella conicotruncata (Subbotina) 1947

Globorotalia angulata (White). Glaessner, 1937 (*partim*), *Probl. Paleont. Publ. Lab. Pal. Moscow Univ.* **2–3**, 383, pl. 4, fig. 37 (*non* figs 35a–35c).

Globorotalia conicotruncata Subbotina, 1947 (*partim*), *Mikrofauna Kavkaza, Emby i Srednei Azii, Lengostoptekhizdat* 115, pl. 4, figs 11–13; pl. 9, figs 9–11.— Subbotina, 1950, *Mikrofauna S.S.S.R., sb. 4, Trudy, Vses. Neft. Nauk.-Issled. Geol.-Razved. Inst.* (*V.N.I.G.R.I.*) **51**, 107, pl. 5, figs 25–29.—Luterbacher, 1964, *Eclog. geol. Helv.* **57**, no. 2, 660, figs 40–42, 46–51.

Acarinina conicotruncata (Subbotina). Subbotina, 1953 (*partim*), *Trudy, Vses. Neft. Nauk.-Issled. Geol.-Razved. Inst.* (*V.N.I.G.R.I.*) n.s. **76**, p. 220, pl. 20, figs 5a–8c (*non* pl. 20, figs 9a–12c).

Globorotalia angulata White var. *kubanensis* Shutskaya, 1956, *Akad. Nauk S.S.S.R. Trudy Inst. Geol. Nauk* **164**, 93, pl. 3, fig. 3.

Globorotalia angulata abundocamerata Bolli, 1957, *U.S. Nat. Mus. Bull.* **215**, 74, pl. 17, figs 4–6.—Bolli and Cita, 1960, *Riv. Ital. Paleont.* **66**, 19, pl. 33, fig. 6.—Said and Sabry, 1964, *Micropaleont.* **10**, 382, pl. 1, fig. 2.—El Naggar, 1966, *Bull. Brit. Mus.* (*Nat. Hist.*), *Geol. Suppl.* **2**, 194 pl 22, figs 2a–c.

?Globorotalia hispidicidaris Loeblich and Tappan, 1957, *U.S. Nat. Mus. Bull.* **215**, 190, pl. 58, figs 1a–c.

Globorotalia trichotrocha Loeblich and Tappan, 1957, *U.S. Nat. Mus. Bull.* **215**, 195, pl. 50, fig. 5; pl. 57, fig. 57, figs 1, 2.—Olsson, 1960, *J. Paleont.* **34**, 49, pl. 10, figs 1–3.

Globorotalia convexa Subbotina. Said and Kerdany, 1961, *Micropaleont.* **7**, 329, pl. 1, fig. 1 (*non* Subbotina).

Globorotalia (Acarinina) conicotruncata Subbotina. Leonov and Alimarina. 1961, *Sb. Trudov Geol. Fak.* (*k XXI Mezhd. Geol. Kongr.*) *Izd. Mosk. Univ.* 53, pl. 6, figs 1–3, 5, 7.

Truncorotalia angulata abundocamerata (Bolli). Gohrbandt, 1963, *Mitt. geol. Ges. Wien* **56**, 58, pl. 4, figs 7–9.

Morozovella conicotruncata (Subbotina). Berggren, 1971. *Proc. II Plankt. Conf.* pl. 4, figs 8–14.

Remarks: From an examination of specimens in this writer's collection identified by Subbotina in 1958, as well as Subbotina's discussions and illustrations of this species, it would seem that the following forms can be included under *M. conicotruncata: G. angulata kubanensus* Shutskaya, *G. angulata abundocamerata* Bolli, *G. trichotrocha* Loeblich and Tappan, and *G. hispidicidaris* Loeblich and Tappan. Further, some of Subbotina's (1953) illustrations of *M. conicotruncata* would seem to refer more properly to *M. angulata*. The illustrations of *G. trichotrocha* are similar to small forms identified as *G. conicotruncata* by Subbotina in this writer's collections. The author has found it impossible to differentiate consistently between these various forms, nor to determine a distinct stratigraphic range for any one of them. Accordingly they are here included in the synonymy of *M. conicotruncata* Subbotina.

The reader is referred to studies by Leonov and Alimarina (1961), Alimarina (1963), Gohrbandt (1963), and Luterbacher (1964) for additional discussion of the affinities and stratigraphic distribution of this species.

Stratigraphical range: Middle part of *Morozovella pusilla* Zone to upper part of *P. pseudomenardii* Zone.

Morozovella velascoensis (Cushman), 1925

Pulvinulina velascoensis Cushman, 1925, *Contr. Cushman Lab. Foram. Res.* **1,** 18, pl. 3, figs 5a–c.
Globorotalia velascoensis (Cushman). White, 1928, *J. Paleont.* **2,** 281, pl. 38, figs 1a–c.—Applin and Jordan, 1945, *J. Paleont.* **19,** 146, pl. 19, fig. 8.—Le Roy, *Geol. Soc. Am. Mem.* **54,** 33, pl. 3, figs 1–3.—Sacal and Debourle, 1957, *Mém. Soc. Géol. France* **78,** 64, pl. 29, figs 7–9.—Bolli, 1957, *U.S. Nat. Mus. Bull.* **215,** 76, pl. 20, figs 1–4.—Loeblich and Tappan, 1957, *Ibid.,* 196, pl. 64, figs 1a–2c.—Nakkady, 1959, *Micropaleont.* **5,** 462, pl. 4, figs 4a–c.—Bolli and Cita, 1960, *Riv. Ital. Paleont.* **66,** 31, pl. 33, figs 7a–c.—Said and Kerdany, 1961, *Micropaleont.* **7,** no. 3, 330, pl. 1, figs 10a–c.—Aubert, 1962, *Notes Serv. géol. Maroc.* **21,** 53, pl. 1, fig. 1.—Said and Sabry, 1964, *Ibid.,* **10,** no. 3, 386, pl. 2, fig. 9.—Luterbacher, 1964, *Eclog. geol. Helv.* **57,** 681, figs 92–94, 98, 99.—El Naggar, 1966, *Bull. Brit. Mus. (Nat. Hist.), Geol. Suppl.* **2,** 246, pl. 20, figs 3a–d; pl. 21, fig. 3.
Globorotalia (Truncorotalia) velascoensis (Cushman).—Cushman and Bermúdez, 1949, *Contr. Cushman Lab. Foram. Res.* **25,** 41, pl. 8, figs 4–6.—Hillebrandt, 1962, *Bayer, Akad. Wissen. Math.-Naturw. Kl. Abh. N.F.* **108,** 139, pl. 13, figs 16–21.
Truncorotalia velascoensis (Cushman). Dalbiez and Glintboeckel, *in* Cuvillier *et al.,* 1955, *Proc. IV World Petrol. Congr. sect.* **1/D,** 535, pl. 2, fig. 8.
Pseudogloborotalia velascoensis (Cushman). Bermúdez, 1961, *Mem. III Congr. Geol. Venez.* **3,** *Bol. de. Geol. Publ. Espec.* **3,** 1394, pl. 16, figs 11a–b.
Truncorotalia velascoensis velascoensis (Cushman). Gohrbandt, 1963, *Mitt. geol. Ges. Wien.* **56,** 59, pl. 5, figs 7–9.

Truncorotaloides (*Morozovella*) *velascoensis* (Cushman). McGowran, 1968, *Micropaleont.* **14**, pl. 2, fig. 1.

Truncorotaloides (*Morozovella*) sp. aff. *T.* (*M.*) *velascoensis* (Cushman). McGowran, 1971, *N.Z. Geol. Surv. Pal. Bull.* **42**, pl. 2, figs 2–4.

Globorotalia (*Morozovella*) *velascoensis velascoensis* (Cushman). Jenkins, 1971, *N.Z. Geol. Surv. Bull.* **42**, 107, pl. 9, figs 214–216.

Morozovella velascoensis (Cushman). Berggren, 1971, *Proc. II Plankt. Conf.* pl. 5, fig. 12.

Remarks: Morozovella velascoensis is a characteristic species in the middle and upper Palaeocene. The high conical test, 5–8 chambers in the final whorl and the thick circum–umbilical collar distinguish it from associated species. There appears to be an allometric relationship between the size of the final chamber and the number of chambers in the final whorl; in those individuals in which the last whorl contains 6 or more chambers, the final chamber is smaller than the penultimate chamber in most instances.

Although some authors have preferred to include *Morozovella acuta* in the synonymy of *M. velascoensis,* they are distinguished here at the specific level. While linked by intermediate forms, it is generally possible to distinguish between the two forms. In addition, the more robust, more highly ornamented *M. velascoensis* is absent from samples containing forms referable to *M. acuta* in some areas.

The present species has been confused with *M. caucasica* by several workers, including the author, in the past. Although clearly homeomorphic, these two species have distinctly different stratigraphic ranges. The transition from *M. aragonensis* to *M. caucasica* can be observed in the lower Eocene *M. aragonensis* Zone. *M. velascoensis* may have developed from *M. conicotruncata* near the top of the *M. pusilla* Zone (cf. Bolli, 1957, fig. 12).

The range of variation in this species has been discussed recently by Hillebrandt (1962, 1963) and Luterbacher (1964). *Morozovella pasionensis* Bermúdez, if not a flattened morphotypic variant of *M. velascoensis,* is very closely related to it (see discussion under that species).

Stratigraphical range: Upper part of *M. pusilla* Zone to basal part of *M. subbotina* Zone.

The author has observed *M. velascoensis* in association with *M. acuta,* *M. rex* (=*M. subbotinae*), *Acarinina wilcoxensis* and *Pseudohastigerina wilcoxensis* in the Bashi Marl Member of the Hatchetigbee Formation, Alabama, in the Lodo Formation, Fresno County, California and in the lower levels of the Esna Shale exposed at Luxor, Egypt—all of which he would place in the lowermost Eocene *M. subbotinae* Zone. The base of the Eocene is drawn by this writer, in the absence of other criteria, at the first occurrence of *Pseudohastigerina wilcoxensis* (Cushman and Ponton), a plani-spiral planktonic species.

Morozovella acuta (Toulmin) 1941

Globorotalia wilcoxensis Cushman and Ponton var. *acuta* Toulmin, 1941, *J. Paleont.* **15**, 608, pl. 82, figs 6–8.—Cushman and Renz, 1942, *Contr. Cushman Lab. Foram. Res.* **18**, 12, pl. 3, fig. 2.—Cushman, 1944, *Am. J. Sci.* **242**, 15, pl. 2, figs 16, 17.—*Contr. Cushman. Lab. Foram. Res.* **20**, 48, pl. 8, fig. 5.—Shifflet, 1948, *Maryland Dept. Geol. Mines and Water Resources Bull.* **3**, 73, pl. 4, figs 23a–c.—Cushman and Bermúdez, 1949, *Contr. Cushman Found. Foram. Res.* **25**, p. 39, pl. 7, figs 19–21.

Globorotalia (Truncorotalia) lacerti Cushman and Renz.—Hofker, 1955, *Rept. McLean Foram. Lab.* no. **2**, 14, pl. 1.

Globorotalia acuta Toulmin. Loeblich and Tappan, 1957, *U.S. Nat. Mus. Bull.* **215**, 185, pl. 47, figs 5a–c; pl. 55, figs 4a–5c; pl. 58, figs 5a–c.—Berggren, 1960, *Int. Geol. Congr. 21st Sess. Copenhagen* pt. **6,** Proc. 53, pl. 1, figs 15a–c.—Luterbacher, 1964, *Eclog. geol. Helv.* **57**, 686, figs 101–104.—Aubert, 1962, *Notes Serv. géol. Maroc.* **21**, 54, pl. 1, fig. 3.

Globorotalia aequa Cushman and Renz. Loeblich and Tappan, 1957 (*pars*), *U.S. Nat. Mus. Bull.* **215**, 185, pl. 46, figs 7a–8c (*non* pl. 50, figs 6a–c; pl. 55, figs 8a–c; pl. 59, figs 6a–c; pl. 60, figs 3a–c; pl. 64, figs 1a–c).

Truncorotalia velascoensis acuta (Toulmin). Gohrbandt, 1963, *Mitt. geol. Ges. Wien.* **56**, 61, pl. 4, figs 13–15.

Globorotalia (Morozovella) velascoensis acuta Toulmin. Jenkins, 1971, *N.Z. Geol. Surv. Pal. Bull.* **42**, 106, figs 205–207.

Remarks: The characteristics which serve to distinguish this species from *Morozovella velascoensis* have been discussed by Loeblich and Tappan (1957). In particular, *M. acuta* has fewer chambers than *M. velascoensis* and a less pronounced ornament on the test. It is possible that *M. velascoensis*—by its robust and heavily ornamented test—is an indicator of relatively deeper water than *M. acuta*.

This species shows a marked morphologic similarity to *M. angulata* from which it may have evolved near the top of the *M. pusilla* Zone.

Stratigraphical range: Uppermost part of *M. pusilla* Zone to basal part of *M. subbotina* Zone.

Morozovella occlusa (Loeblich and Tappan) 1957

?*Discorbina simulatilis* Schwager, 1883, *Palaeontographica* **30**, 120, pl. 29, fig. 15.

Truncorotalia crassata aequa (Cushman and Renz). Said and Kenaway, 1956, *Micropaleont.* **2,** 151, pl. 6, fig. 8 (*non* Cushman and Renz).

Globorotalia occlusa Loeblich and Tappan, 1957, *U.S. Nat. Mus. Bull.* **215,** 191, pl. 55, figs 3a–c; pl. 64, figs 3a–c.—Luterbacher, 1964, *Eclog. geol. Helv.* **57,** 690, figs 112–114.—El Naggar, 1966, *Bull. Brit. Mus.* (*Nat. Hist.*) *Geol. Suppl.* **2,** 221, pl. 22, figs 4a–c (*non* pl. 20, figs 2a–d).

Globorotalia crosswicksensis Olsson, 1960, *J. Paleont.* **34,** 47, pl. 10, figs 7–9.

Globorotalia simulatilis (Schwager). Said and Kerdany, 1961, *Micropaleont.* **7,** 329, pl. 1, fig. 8.—Said and Sabry, 1964, *Ibid.,* **10,** 385, pl. 1, fig. 1.

Globorotalia (*Truncorotalia*) *velascoensis occlusa* Loeblich and Tappan.— Hillebrandt, 1962, *Bayer, Akad. Wissen. Math.-Naturw. Kl. Abh. N.F.* **108,** 139, pl. 13, figs 22–26.

Truncorotalia velascoensis occlusa (Loeblich and Tappan). Gohrbandt, 1963, *Mitt. geol. Ges. Wien.* **56,** 60, pl. 4, figs 16–18; pl. 5, figs 1–3.

Globorotalia (*Morozovella*) *velascoensis occlusa* Loeblich and Tappan. Jenkins, *N.Z. Geol. Surv. Pal. Bull.* **42,** 106, pl. 9, figs 208–210.

Remarks: This species is distinguished from associated forms in the Palaeocene by its low conical test, narrow deep umbilicus and flat to low trochospire. A somewhat different interpretation of this species has been presented by Hillebrandt (1962) and Luterbacher (1964). This writer believes, however, that the concepts of these authors overlap with that of Loeblich and Tappan (1957) and that the forms they have illustrated can be included in the present species. *G. crosswicksensis* Olsson is also interpreted here as a synonym of *M. occlusa* (cf. Luterbacher, *loc. cit.*).

Specimens identified as *Globorotalia simulatilis* (Schwager) from Farafrah Oasis, Egypt, were sent to this writer by Professor Rushdi Said, Cairo University. They agree in most repsects with *Morozovella occlusa* and *G. crosswicksensis* (cf. El Naggar, 1966). Hillebrandt (1964) has recently resurrected *Globorotalia simulatilis* based on a study of material from El Quss Abu Said, Farafrah Oasis, Egypt. According to him (*loc. cit.*) *G. simulatilis* is related to *G. conicotruncata.*

Stratigraphical range: Upper part of *P. pseudomenardii* Zone to top of *M. velascoensis* Zone.

Morozovella pasionensis (Bermúdez) 1961

Pseudogloborotalia pasionensis Bermúdez, 1961, *Mem. del III Congr. Geol. Venez. Publ. esp.* **3,** 1346, pl. 16, figs 8a, b.

Globorotalia pasionensis (Bermúdez). Luterbacher, 1964, *Eclog. geol. Helv.* **57,** 690, figs 108–110.

Remarks: This species differs from typical *Morozovella velascoensis* (Cushman) in having a flatter test, more lobulate periphery, more rounded umbilical shoulders and a broader umbilicus; on some individuals the chambers on the spiral side overlap in a weakly developed imbricate pattern. Further work may reveal that this form is but an extreme morphotypic variant of *Morozovella velascoensis* (Cushman).

Stratigraphical range: Bermúdez (1961) described this species from the Lower Eocene of Guatemala. Luterbacher (1964) recorded it from levels G-71 and G-70a in the section at Gubbio, Central Apennines, Italy, i.e. in the upper part of his *Globorotalia aequa* Zone. The *Globorotalia aequa* Zone of Luterbacher corresponds to the uppermost Palaeocene–lowermost Eocene. Luterbacher (*loc. cit.*) also recorded this species from the Velasco Formation of Mexico at Ebane (*Globorotalia velascoensis* Zone).

Morozovella simulatilis (Schwager) cf. Luterbacher, 1964

Discorbina simulatilis, Schwager, 1883, *Palaeontographica* **30,** 120, pl. 29, figs 15a–d.
?Globorotalia simulatilis (Schwager), 1883. Luterbacher, 1964, *Eclog. geol. Helv.* **57,** 665, figs 53–60.

Remarks: This species, originally described by Schwager (1883) from El Quss Abu Said, Farafrah Oasis, Egypt, has been seldom used in palaeontologic literature. On the basis of a study of material from the type locality, Luterbacher (1964) has resurrected the species.

The author has not recognised *G. simulatilis* in his work for the following reasons: (*a*) the illustration of the holotype of this species is hardly "quite clear" as Luterbacher states; it is exceedingly small and does not exhibit the morphologic details necessary to form a judgement and base a species concept upon, (*b*) forms identified by Professor Rushdi Said as *Globorotalia simulatilis* agree well with *G. occlusa* and *G. crosswicksensis,* Luterbacher apparently considers Said and Kerdany's (1961) record of this species as valid). Luterbacher's illustrations of *G. simulatilis* are presented here for purposes of comparison with other associated and related forms.

Stratigraphical range: M. pusilla Zone—*P. pseudomenardii* Zone (Luterbacher, 1964, fig. 133).

Morozovella parva Rey, 1955

?Truncatulina colligera Schwager, 1883, *Palaeontographica* **30,** 126, pl. 29, fig. 14.
Globorotalia simulatilis (Schwager). Le Roy, 1953, *Geol. Soc. Am. Mem.* **54,** 32, pl. 9, figs 1–3.—Nakkady, 1959, *Micropaleont.* **5,** 462, pl. 4, figs 4a–c.

Globorotalia velascoensis (Cushman). Hamilton, 1953, *J. Paleont.* **27**, no. 2, 231, pl. 31, figs 24, 28, 29.

Globorotalia velascoensis (Cushman) var. *parva* Rey, 1955, *Bull. Soc. Geol. France sér.* **6, 4,** 209, pl. 12, fig. 1.—Bolli and Cita, 1960, *Riv. Ital. Paleont.* **66,** 32, pl. 33, figs 5a–c.—Gartner and Hay, 1962, *Eclog. geol. Helv.* **55,** 565, pl. 2, figs 5a–c.—Aubert, 1962, *Notes Serv. géol. Maroc.* **21,** 54, pl. 1, figs 2a–c.—Luterbacher, 1964, *Eclog. geol. Helv.* **57,** 678, figs 91a–c.—El Naggar, 1966, *Bull. Brit. Mus. (Nat. Hist.), Geol. Suppl.* **2,** 264, pl. 20, figs 4a–d.

Truncorotalia simulatilis (Schwager). Said and Kenaway, 1956, *Micropaleont.* **2,** 151, pl. 6, fig. 6.

Globorotalia rex Martin. Loeblich and Tappan, 1957, *U.S. Nat. Mus. Bull.* **215,** 195, pl. 60, figs 1a–c.

Globorotalia colligera (Schwager). Said and Kerdany, 1961, *Micropaleont.* **7,** 328, pl. 1, fig. 14.—Said and Sabry, 1964, *Ibid.* **10,** 382, pl. 1, fig. 14.

Truncorotalia velascoensis parva (Rey). Gohrbandt, 1963, *Mitt. geol. Ges. Wien* **56,** 61, pl. 5, figs 4–6.

Globorotalia (Morozovella) velascoensis parva Rey. Jenkins, 1971, *N.Z. Geol. Surv. Pal. Bull.* **42,** 106, pl. 9, figs 211–213.

Remarks: This species is characterised by having 4 chambers (which differ little in size) in the final whorl and slightly raised, beaded sutures on the spiral side. Luterbacher (1964) re-examined topotype material and concluded that: (*a*) the type level of *G. velascoensis parva* is younger than the *G. velascoensis* Zone, although he did not give reasons for this conclusion; (*b*) the form described as *G. velascoensis parva* from the *G. pseudomenardii* Zone by Gartner and Hay (1962), Bolli and Cita (1962), Gohrbandt (1963) and Aubert (1963) is not the same as *G. velascoensis parva* Rey. Although this may eventually prove to be true, this writer prefers to retain the name *G. velascoensis parva* for the present, rather than rename the forms described from the *G. pseudomenardii* Zone.

Specimens identified as *Globorotalia colligera* (Schwager) from Farafrah Oasis were recently sent to the author by Professor Rushdi Said of Cairo University. They are similar in most respects to the forms mentioned above, and agree well with the illustration of Loeblich and Tappan (1957, pl. 60, fig. 1) of "*Globorotalia rex*" from the Nanafalia Formation of Alabama. The illustration of Loeblich and Tappan (1957) is very similar to that of Luterbacher (1964) of *G. velascoensis parva* and the two forms are judged conspecific here.

This species may be related to, and descended from, *Morozovella acuta*.

Stratigraphical range: Planorotalites pseudomenardii Zone to *M. velascoensis* Zone.

Morozovella tadjikistanensis (Bykova) 1953

Globorotalia tadjikistanensis Bykova, 1953, *Mikrofauna S.S.S.R., sb.* 6, *Trudy Vses. Neft. Nauk.-Issled. Geol.-Razved. Inst. n.s.* **69**, 86, pl. 3, fig. 5.—Luterbacher, 1964, *Eclog. geol. Helv.* **57**, 663, fig. 52.

Globorotalia (Truncorotalia) tadjikistanensis (Bykova). Leonov and Alimarina, 1961, *Sb. Trudov Geol. Fak. (k XXI Mezhd. Geol. Kongr.) Izd. Mos. Univ.* 53, pl. 7, figs 1–4, 6, 7.

Globorotalia pusilla mediterranica El Naggar, *Bull. Brit. Mus. (Nat. Hist.), Geol. Suppl.* **2**, 230, pl. 19, figs 3a–c.

Remarks: The holotype (V.N.I.G.R.I. no. 1794) of this species was described from the Tadjik Depression, Ak-Tay, Suzaksian Stage, *Globorotalia tadjikistanensis* Zone. The species was said to have developed from *Globorotalia angulata*, from which it differs in being somewhat smaller, lower and in having a greater number of chambers, a more rounded and conically arched spiral side and more strongly curved sutures. The same distinction was subsequently made (Bykova *in* Subbotina, 1953, p. 223) to distinguish it from *Acarinina conicotruncata*. Perhaps the most distinctive feature of this form is its biconvex test.

Globorotalia tadjikistanensis djanensis Shutskaya (ms) is not separated here from the typical form (see also Luterbacher, 1964) for want of sufficient data on this form.

This form is generally quite rare in samples and it may be that additional data will reveal it to be but a morphotypic variant of *M. conicotruncata*.

Stratigraphical range: Upper part of *M. pusilla* to upper part of *P. pseudomenardii* Zone (see also Luterbacher, 1964).

Morozovella acutispira (Bolli and Cita) 1960

Globorotalia acutispira Bolli and Cita, 1960, *Riv. Ital. Pal. Strat.* **66**, 375, pl. 35, figs 3a–c.—Cita and Bolli *in* Bolli, Cita and Schaub, 1961, *Riv. Ital. Pal. Strat.* **64**, 386, fig. 2.—Luterbacher, 1964, *Eclog. geol. Helv.* **57**, p. 673, figs 72a–c.

Globorotalia occlusa Loeblich and Tappan. El Naggar, 1966, *Bull. Brit. Mus. (Nat. Hist.), Geol. Suppl.* **2**, 221, pl. 20, figs 2a–d (*non* pl. 22, figs 4a–c; *non* Loeblich and Tappan).

Truncorotaloides (Morozovella) sp. aff. (*M.*) *acutispira* (Bolli and Cita). McGowran, 1968, *Micropaleont.* **14**, pl. 2, figs 6–10.

Remarks: The holotype of this species, recently refigured by Luterbacher (1964) and illustrated here, is from the *Planorotalites pseudomenardii* Zone, 'aderno d'Adda, Northern Italy. It is characterised by a flattened test, thick peripheral keel and the apiculate spire around which the early chambers are coiled.

This species bears a superficial resemblance to *M. marginodentata* from the lower Eocene. The latter may be distinguished by its generally flatter test, more ornate ornament, thicker keel and flat to low-convex umbilical side.

Stratigraphical range: Planorotalites pseudomenardii Zone.

Morozovella kolchidica Morozova, 1961

Globorotalia kolchidica Morozova, 1961, *Paleont. Zhurnal Acad. Nauk. S.S.S.R.* **1/2**, 17, pl. 2, fig. 2.
Globorotalia sp. aff. *G. kolchidica* Morozova. Luterbacher, 1964, *Eclog. geol. Helv.* **57**, 668, figs 61, 62.

Remarks: The holotype of this species was described from the "Montian" beds of the Khokodze River, Crimea and said to occur in the North Caucasus and Kopet Dag as well. I have examined the holotype of this form. It is a 5-chambered form with a thick, spinose keel and thickened umbilical shoulder. It resembles, in some respects, *Morozovella marginodentata* from the lower Eocene.

I have not recognised this species in my stratigraphic studies. *M. kolchidica* may have branched off from *M. angulata* with which it is associated along with *M. conicotruncata* (*fide* Morozova, 1962). Forms probably related to this species which were recorded by Luterbacher (1964) from the Gubbio section, Central Apennines, Italy, are also figured here.

Stratigraphical range: Morozovella pusilla Zone.

Morozovella subbotinae (Morozova) 1939

Globorotalia subbotinae Morozova, 1939, *Byul. Mosc. Obshschestva Isp. Prirody, otd. geol.* **17**, 80, pl. 1, figs 16, 17.—Shutskaya, 1956, *Akad. Nauk. S.S.S.R. Trudy Geol. Inst. Nauk.* **164**, 4, pl. 4, fig. 3 (?), 4.—Luterbacher, 1964, *Eclog. geol. Helv.* **57**, no. 2, p. 676, figs 85–90.
Globorotalia rex Martin, 1943, *Stanford Univ. Publ. Univ. Ser. Geol. Sci.* **3**, 117, pl. 8, figs 2a–c.—Bolli, 1957, *U.S. Nat. Mus. Bull.* **215**, 75, pl. 18, figs 10–12.—Said and Sabry, 1964, *Micropaleont.* **10**, 385, pl. 2, figs 3a–c.
Globorotalia marginodentata Subbotina, 1953 (*partim*), *Trudy, Vses. Neft. Nauk.-Issled. Geol.-Razved. Inst. (V.N.I.G.R.I.) n.s.* **76**, 212, pl. 18, figs 3a–c (transitional form to *G. lensiformis* Subbotina) (*non* pl. 17, figs 14a–16c; pl. 18, figs 1a–2c).
Truncorotalia marginodentata marginodentata (Subbotina). Gohrbandt, 1963, *Mitt. geol. Ges. Wien* **56**, 62, pl. 6, figs 4–6.

Truncorotalia marginodentata aperta Gohrbandt, 1963. *Mitt. geol. Ges. Wien* **56,** 63, pl. 5, figs 10–15.

Globorotalia velascoensis (Cushman) var. *acuta* Toulmin. Graham and Clausen, 1955, *Contr. Cushman Found. Foram. Res.* **6,** 29, pl. 5, figs 8a–9c.

Truncorotalia spinulosa (Cushman). Said and Kenaway, 1956, *Micropaleont.* **2,** 151, pl. 6, figs 3a–c.

Globorotalia (Truncorotalia) aequa simulatilis (Schwager). Hillebrandt, 1962, *Bayer, Akad. Wissen. Math.-Naturw. Kl. Abh. N.F.* **108,** 134, pl. 13, figs 6–8 (*non* Schwager).

Globorotalia bollii El Naggar, 1966, *Bull. Brit. Mus. (Nat. Hist.) Geol. Suppl.* **2,** (*pars*), 202, pl. 22, figs 5a–c; ?6a–d.

Globorotalia (Morozovella) aequa rex Martin. Jenkins, 1971, *N.Z. Geol. Surv. Pal. Bull.* **42,** 101, pl. 7, figs 180–182.

Morozovella subbotinae (Morozova). Berggren, 1971, *Proc. II Plankt. Conf.* pl. 5, figs 10, 11.

Remarks: This robust form is a characteristic species in the lower Eocene. It probably evolved from *Morozovella aequa* from which it is distinguished by its generally larger size, thicker and more distinct keel and more evolute spire.

Globorotalia rex Martin has generally been used in the literature in the West for this distinct lower Eocene form. In the Soviet Union the name *Globorotalia subbotinae* Morozova has been used for the same form. The author would disagree with Luterbacher (1964) that the form illustrated by Bolli (1957, pl. 18, figs 10–12) is to be interpreted as a variety of *Globorotalia aequa* and not as *G. rex* Martin. The minor difference cited by Luterbacher (*loc. cit.*) to support this opinion *are those of degree not kind*; the illustration of Bolli (*loc. cit.*) is certainly more similar to the typical *G. rex* and *G. subbotinae* as figured by various authors than to *G. aequa* and is here included in the synonymy of *G. subbotinae*. This is borne out by a recent examination of the holotype of *Globorotalia rex* Martin (no. 7404) from sample S-2 + 48, L.S.J.U., Loc. M-74, Lodo Formation, California.

This species is very similar to forms which occur in the Palaeocene and have been recorded under the name *Globorotalia velascoensis parva* (see discussion under that species). Indeed, this similarity led Loeblich and Tappan (1957) and later the author (Berggren, 1960) to record *Globorotalia rex* from the *Globorotalia pseudomenardii* Zone of the Gulf Coast and Nigeria respectively.

Studies by this writer suggest that *Morozovella subbotinae* is ancestral to *Morozovella lensiformis* (Subbotina) which, in turn, evolved into *M. aragonensis* (Nuttall) (see also Bolli, 1957, figs 12, 13).

For further discussion of this species the reader is referred to publications by Gohrbandt (1963), Luterbacher (1964) and Berggren (1965).

Stratigraphical range: Morozovella subbotinae Zone—Morozovella formosa Zone.

Morozovella marginodentata (Subbotina) 1953

Globorotalia crassata (Cushman). Glaessner, 1937, *Etyudy po mikropaleontologii* **1**, 31, pl. 1, figs 7a–c.—Subbotina, 1947 (*partim*), *Trudy, Vses. Neft. Nauk.-Issled. Geol.-Razved. Inst.* (*V.N.I.G.R.I.*), *n.s.* **76**, 211, pl. 17, fig. 13 (*non* figs 7–12).

Globorotalia marginodentata Subbotina, 1953, *Trudy Vses. Neft. Nauk.-Issled. Geol.-Razved. Inst.* (*V.N.I.G.R.I.*) *n.s.* **76**, 212, pl. 17, figs 14a–16c; pl. 18, figs 1a–2c (3a–c is interpreted by Subbotina as a transitional form to *G. lensiformis*).—Luterbacher, 1964, *Eclog. geol. Helv.* **57**, 673, figs 75–84.

?Globorotalia bolli El Naggar, 1966, *Bull. Brit. Mus.* (*Nat. Hist.*), *Geol. Suppl.* **2**, (*pars*), 202, pl. 22, figs 6a–d (*non* pl. 22, figs 5a–c).

Globorotalia (*Truncorotalia*) *aequa marginodentata* Subbotina. Hillebrandt, 1962, *Bayer, Akad. Wissen. Math.-Naturw. Kl. Abh. N.F.* **108**, 135, pl. 13, figs 9a–11.

Globorotalia (*Morozovella*) *aequa marginodentata* Subbotina. Jenkins, 1971, *N.Z. Geol. Surv. Pal. Bull.* **42**, 101, pl. 7, figs 177–179.

Morozovella marginodentata (Subbotinae). Berggren, 1971, *Proc. II Plankt. Conf.* pl. 5, fig. 9.

Remarks: This form is distinguished from associated species by its flattened test, raised and pointed umbilical chamber tips, and broad, thick keel. The ornament of the test is highly variable and gradation into typical *G. subbotinae* can be observed in some material. The geographic distribution of these two forms is not identical, however, and *M. marginodentata* is here retained as a species distinct from *M. subbotinae*.

Forms bearing a marked resemblance to the present species are observed in the Palaeocene, primarily within the *Planorotalites pseudomenardii* Zone. These are interpreted here as probable intraspecific morphotypic variants of one, or two, Palaeocene morozovellid species; i.e. the relationship would be considered homeomorphic.

Stratigraphical range: Morozovella subbotinae–M. formosa Zones.

Morozovella gracilis (Bolli) 1957

Globorotalia formosa gracilis Bolli, 1957, *U.S. Nat. Mus. Bull.* **215**, 75, pl. 18, figs 4–6.—Luterbacher, 1964, *Eclog. geol. Helv.* **57**, 692, figs 115–117, see also figs 105–107.

Globorotalia (Morozovella) gracilis Bolli. Jenkins, 1971, *N.Z. Geol. Surv. Pal. Bull.* **42**, 105, pl. 9, figs 202–204.

Morozovella gracilis (Bolli). Berggren, 1971, *Proc. II Plankt. Conf.* pl. 5, figs 7, 8.

Remarks: This lower Eocene form is characterised by its lobulate periphery, acute and spinose keel and 5–6 chambers in the last whorl. It is similar to *M. marginodentata* (Subbotina) but the latter has a broader and thicker keel and less conical test.

Stratigraphical range and distribution: Morozovella subbotinae Zone–*M. formosa* Zone.

This species is associated with *Morozovella subbotinae, M. marginodentata* and *M. aequa* in the *M. subbotinae* Zone in various sections in the world. It was originally described from the *Morozovella rex* Zone of Trinidad and was shown to range through the *M. formosa* Zone.

Morozovella formosa (Bolli) 1957

Globorotalia formosa formosa Bolli, 1957, *U.S. Nat. Bull.* 76, pl. 18, figs 1–3.—Luterbacher, 1964, *Eclog. geol. Helv.* **57**, 694, figs 118–120.

Morozovella formosa (Bolli). Berggren, 1971, *Proc. II Plankt. Conf.* 1970, pl. 5, figs 15, 16.

Remarks: This species is distinguished from the related *M. gracilis* by its less lobulate periphery and in having 6–8 chambers in the final whorl. Hillebrandt (1962) placed *M. formosa formosa* in the synonymy of *M. caucasica* without giving any specific reason for this. Luterbacher (1964) has observed that in the lower Eocene of the Central Apennines the youngest representative of this species are intermediate to *Morozovella aragonensis.* The author has suggested (Berggren, 1966, fig. 3) that *M. aragonensis* has evolved from *M. lensiformis* within the *M. formosa* Zone and that *M. caucasica* is a descendent of *M. aragonensis.* It may be that *M. caucasica* is more closely related to *M. formosa formosa* from which it differs in having a wider and deeper umbilicus and thick, everted umbilical collar.

Stratigraphical range and distribution: Morozovella formosa Zone and *M. aragonensis* Zone.

Morozovella formosa is a common form in Lower Eocene sediments west of the Mid-Atlantic Ridge, Lat. 30°S (Berggren, 1971).

Morozovella lensiformis Subbotina, 1953

Globorotalia lensiformis Subbotina, 1953, *Trudy, Vses. Neft. Nauk.-Issled. Geol.-Razved. Inst. (V.N.I.G.R.I.) n.s.* **76,** 214, pl. 18, figs 4a–5c.—Luterbacher, 1964, *Eclog. geol. Helv.* **57,** 671, figs 74a–c.

Globorotalia californica Smith, 1957, *Univ. Calif. Publ. Geol. Sci.* **32,** 190, pl. 28, figs 22a–23c.—Mallory, 1959, Lower Tertiary biostratigraphy of the California Coast Ranges. *Am. Assoc. Petrol. Geol. Spec. Publ.* 253, pl. 38, fig. 4. (The name *Globorotalia californica* Smith is a junior homonym of *Globorotalia californica* Cushman and Todd, 1948, *Cushman Lab. Foram. Res. Contr.* **24,** pl. 16, figs 22, 23, from the "Lower Cretaceous?, Franciscan group . . .", Santa Clara County, California.)

Globorotalia (*Truncorotalia*) *lensiformis* Subbotina, Hillebrandt, 1962, *Bayer, Akad. Wissen. Math.-Naturw. Kl. Abh.* N.F. **108,** 136, pl. 13, figs 12a–13c.

Globorotalia dolabrata Jenkins, 1965, *N.Z. J. Geol. Geophysics*, **8,** 1113, fig. 12 (104–112).

Globorotalia (*Morozovella*) *dolabrata* Jenkins. Jenkins, 1971, *N.Z. Geol. Surv. Pal. Bull.* **42,** 104, pl. 10, figs 233–241.

Morozovella lensiformis (Subbotina). Berggren, 1971, *Proc. II Plankt. Conf.* pl. 5, figs 18–20.

Remarks: This species is characterised by its rather granular texture (particularly on the umbilical side) and the nearly flat spiral side and flush sutures.

Although recognising the close affinities between this species and *M. aragonensis*, Subbotina (1953) considered that both species evolved independently from *M. marginodentata* at the base of the Zone of conical globorotaliids. Bolli and Cita (1960) and Luterbacher (1964) consider *G. lensiformis* to be closely related to *Globorotalia aequa*. The author would interpret *M. lensiformis* as a descendent of *M. subbotinae* from which it differs in its more granular test, narrower and less ornamented keel, flatter spiral side and simpler sutures (on the spiral side).

The forms described by Smith (1957) as *Globorotalia californica* from the lower Eocene of California appear to be synonymous with the present species. The name *G. californica* Smith, 1957 is preoccupied by *Globorotalia californica* Cushman and Todd, 1948 (which may be a praeglobotruncanid).

An examination of topotype material sent by Jenkins reveals that his *Globorotalia dolabrata* is a junior synonym of *M. lensiformis*.

Stratigraphical range and distribution: Middle part of *M. formosa* Zone through *M. aragonensis* Zone.

This species appears at stratigraphic levels which may be correlated with the *M. formosa* Zone; it evolved relatively rapidly into *M. aragonensis* and does not range above the *M. aragonensis* Zone. It is a common form in early Eocene assemblages in the Caucasus but has not been recorded in significant numbers elsewhere to the author's knowledge.

Morozovella aragonensis (Nuttall) 1930

Globorotalia aragonensis Nuttall, 1930, *J. Paleont.* **4**, 3, 288, pl. 24, figs 6–8, 10, 11.—Glaessner, 1937, *Etyudy Mikropal.* **1**, 30, pl. 1, fig. 5.—Subbotina, 1953, *Trudy, Vses. Neft. Nauk.-Issled. Geol.-Razved. Inst.* (*V.N.I.G.R.I.*) *n.s.* **76**, 215, pl. 18, figs 6a–7c.—Shutskaya, 1956, *Akad. Nauk S.S.S.R., Trudy Inst. Geol. Nauk.* **164**, 100, pl. 5, fig. 2 (?), 3 (?).—Bolli, 1957, *U.S. Nat. Mus. Bull.* **215**, 75, pl. 18, figs 7–9; 167, pl. 38, fig. 1.—Mallory, 1959, Lower Tertiary biostratigraphy of the California Coast Ranges, *Am. Assoc. Petrol Geol.* Spec. Publ. 252, pl. 35, fig. 1.—Pessagno, 1961, *Micropaleont.* **7**, no. 3, 356, pl. 1, figs 14–16.—Aubert, 1962, *Notes Serv. géol. Maroc.*, **21**, 56 pl. 21, fig. 1.—Luterbacher, 1964, *Eclog. geol. Helv.* **57**, 696, figs 121–126.

Globorotalia marksi Martin, 1943, *Stanford Univ. Publ. Univ. Serv. Geol. Sci.* **3**, 25, pl. 8, figs 1a–c.

Globorotalia velascoensis (Cushman). Subbotina, 1947 (*partim*), *Mikrofauna Kavkaza, Emby i Srednei Azii, Lengostoptekhizdat* 123, pl. 7, figs 9–11 (*non* pl. 9, figs 21–23).

Globorotalia (*Truncorotalia*) *aragonensis* Nuttall. Cushman and Bermúdez, 1949, *Contr. Cushman Lab. Foram. Res. Spec. Publ.* **25**, 38, pl. 7, figs 13–15.

Remarks: The circular outline of the test, the granular rugose surface, the indistinct sutures on the spiral side and the deep, narrow umbilicus serve to distinguish this species from associated forms.

The holotype of *Globorotalia marksi* Martin (Stanford Univ. Coll. 7402) from sample S − 6 + 11, Lodo Formation, Fresno Co., California, has been examined and found to be a typical *aragonensis*. The tightly coiled test and narrow umbilicus, high conical umbilical side, and thick keel are visible on the specimen. The coarsely papillate surface, although apparent, is not well-preserved. The type level of this species is in the lower Eocene, within the *Morozovella aragonensis* Zone.

Examination of the holotype of *Globorotalia naussi* Martin (Stanford Univ. Coll. 7403) from sample S − 4 + 8 suggests that this form is also *aragonensis* and transitional to *caucasica*. The final chamber(s) is/are broken off and it is difficult to form a judgment.

Stratigraphical range and distribution: Morozovella aragonensis Zone through *Globigerapsis kugleri* Zone.

This species is one of the most distinct components of early Eocene assemblages (although it ranges into the middle Eocene). It is particularly abundant in early Eocene assemblages in the Caucasus, in various parts of the Mediterranean and Caribbean and in the California (Lodo Formation) sequence.

Morozovella caucasica (Glaessner) 1937

Globorotalia velascoensis (Cushman) var. *aragonensis* Nuttall. Subbotina, 1936, *Trudy, Neft. Geol.-Razved. Inst.* (*V.N.I.G.R.I.*) *ser.* **A, 96,** pl. 3, figs 1–3.

Globorotalia aragonensis Nuttall var. *caucasica* Glaessner, 1937, *Etyudy Mikropaleont.* **1,** 31, pl. 1, fig. 6.—Shutskaya, 1956, *Akad. Nauk, S.S.S.R., Trudy Inst. Geol. Nauk.* **164,** 100, pl. 5, figs 1a–c.

Globorotalia crater Finlay, 1939, *Roy. Soc. New Zealand, Trans.* **69,** 125 (*nomen nudum*).

non *Globorotalia crater* Finlay. Finlay, 1939, *Roy. Soc. New Zealand, Trans.* **69,** pt. 3 pl. 29, figs 157, 162, 163.

Globorotalia velascoensis (Cushman). Subbotina, 1947 (*partim*), *Mikrofauna Kavkaza, Emby i Srednei Azii, Lengostoptekhizdat* 123, pl. 9, figs 21–23 (*non* pl. 7, figs 9–11).—Subbotina, 1953, *Trudy, Vses. Neft. Nauk.-Issled. Geol.-Razved Inst.* (*V.N.I.G.R.I.*) *n.s.* **76,** 216, pl. 19, figs 1, 2 (3 and 4 are transitional forms from *G. aragonensis* to *G. aragonensis caucasica*).—Said, 1960, *Micropaleont.* **6,** 284, pl. 1, fig. 2.

Globorotalia (*Truncorotalia*) *crater,* Hornibrook, 1958, *Micropaleont.* **4,** no. 1, 33, pl. 1, figs 3–5.

Globorotalia aragonensis Nuttall *twisselmanni* Mallory, 1959, Lower Tertiary biostratigraphy of the California Coast Ranges, *Am. Assoc. Petrol. Geol.* Spec. Publ. 252, pl. 23, fig. 1.

Truncorotalia caucasica (Glaessner). Kraeva *in* Kaptarenko-Chernoussova, Golyak, Zernetskii, Kraeva and Lipnik, 1963, *Akad. Nauk Ukrainskii S.S.R. Inst. Geol. Nauk ser. strat. i paleont.* **45,** 150, pl. 31, figs 4a–c.

Globorotalia caucasica Glaessner. Luterbacher, 1964, *Eclog. geol. Helv.* **57,** 684, pl. 97, figs 97a–c.

Globorotalia (*Morozovella*) *crater crater* Finlay. Jenkins, 1971, *N.Z. Geol. Surv. Pal. Bull.* **42,** 103, pl. 8, figs 192–197.

Globorotalia (*Morozovella*) *crater caucasica* Glaessner. Jenkins, 1971, *N.Z. Geol. Surv. Pal. Bull.* **42,** 103, pl. 8, figs 189–191.

Remarks: The identification of this species with *G. velascoensis* (Cushman)—a Palaeocene form—has probably caused more problems in stratigraphic correlation than any other Palaeogene planktonic species. The difference in stratigraphic ranges of the two species as well as the transition from *M. aragonensis* to *M. caucasica* which can be observed in lower Eocene sections illustrates their genetic independence. Here is a good illustration of the practicality of the phylogenetic approach in the identification of planktonic foraminifers.

In erecting a new species, *Globorotalia crater,* Finlay (1939a) gave a brief description but no illustration. Thus, *G. crater* is a *nomen nudum*. Specimens figured by Finlay (1939b, pl. 29, figs 157, 162, 163) are from older strata than the type level and are of a different species (Hornibrook, 1958). The first valid reference to this species is that of Hornibrook (1958, p. 33, pl. 1, figs 3–5) in which a holotype specimen is illustrated. It would seem, then, that Hornibrook (1958) is to be credited with authorship of the species *crater* not Finlay (1939).

Jenkins (1971) distinguishes 5-chambered *crater* from 6–8-chambered *caucasica* in New Zealand sections. The author would consider the structural and ornamental similarity in these forms as more important and group these together under the name *caucasica*. The author would disagree with Jenkins' (1971) suggestion that *formosa* and *crater* are synonymous. These forms are morphologically quite distinct and have decidedly different stratigraphic ranges.

G. aragonensis twisselmanni, while exhibiting some features intermediate between *M. aragonensis* and *M. caucasica*, is here included also in its synonymy.

Stratigraphical range: Morozovella aragonensis Zone through *Hantkenina aragonensis* Zone.

Morozovella crassata (Cushman) 1925

Pulvinulina crassata Cushman, 1925, *Am. Assoc. Petrol. Geol. Bull.* **9**, 300, pl. 7, fig. 4.

Globorotalia spinulosa Cushman, *Contr. Cushman Lab. Foram. Res.* 1927, **3**, 114, pl. 23, fig. 4.—Cole, 1927, *Bull. Am. Palaeont.* **14**, 34, pl. 2, fig. 9.— Nuttall, 1930, *J. Palaeont.* **4**, 276, 288.—Howe, 1939, *Geol. Bull.* **14**, *Louisiana Geol. Survey* 85, pl. 12, figs 10–12.—Cushman, 1939, *Contr. Cushman Lab. Foram. Res.* **15**, 75, pl. 12, fig. 21.—Franklin, 1944, *J. Paleont.* **18**, 318, pl. 48, fig. 8.—Beckmann, 1953, *Eclog. geol. Helv.* **46**, 397, pl. 26, fig. 13.—Bolli, 1957, *U.S. Nat. Mus. Bull.* **215**, 168, pl. 38, figs 6a–7c.—Pessagno, 1961, *Micropaleont.* **7**, no. 3, 356, pl. 2, figs 11–13.—Saito, *Trans. proc. Paleont. Soc. Japan n.s.* **45**, 215, pl. 33, figs 9a–c.—Aubert, 1962, *Notes Serv. géol. Maroc.* **21**, 61, pl. 4, fig. 4.

Globorotalia crassata (Cushman). Cushman and Barksdale, 1930, *Contr. Dept. Geol. Stanford Univ.* **1**, 68, pl. 12, fig. 8.—Bandy, 1964, *Contr. Cushman Found Foram. Res.* **15**, pl. 1, no. 279, 34, fig. 1.

Globorotalia (Truncorotalia) spinulosa Cushman. Cushman and Bermúdez, 1949, *Contr. Cushman Lab. Foram. Res.* **25**, 40, pl. 8, figs 1–3.

Globorotalia hadii Aubert, 1962, *Notes Serv. géol. Maroc.* **21**, 62, pl. 4, figs 1–3.

Pseudogloborotalia spinulosa (Cushman) Bermúdez, 1961, *Mem. III Congr. Geol. Venez.* **3**, *Bol. Geol. Publ. espec.* 1347, pl. 17, figs 2a, b.

Remarks: The original description of this species (Cushman, 1925) was brief and the figure (spiral view only) poor with the result that the name has been used for forms which differ considerably from the recently designated lectotype by Bandy (1964, fig. 1). As Banner and Blow (1965) have pointed out, this lectotype is very similar to Bolli's (1957, figs 6, 7) hypotypes of *Globorotalia spinulosa* Cushman as well as to Cushman's illustration of the holotype itself (1927, pl. 23, fig. 4). They have suggested that *G. crassata* (Cushman) should be considered a prior synonym of *G. spinulosa* Cushman; this procedure is adopted herein. An examination of the type material in the collections of the U.S. National Museum by Dr. Blow and the author in 1967 has confirmed this. Blow (1969) points out that Bolli's (1957, pl. 38, figs 6, 7) hypotypes of *G. spinulosa* differ from the holotype of *P. crassata* and *P. spinulosa*. He notes that "the taxon *G. (G.) spinulosa* of Bolli and the present author, is being renamed in a forthcoming paper by Blow and Berggren". We chose the name "*Globorotalia coronata*" for this taxon: however, this paper was never written. This form will be described, however, in the monograph on Palaeogene planktonic foraminifera which Blow was working on at the time of his death, and which is currently being edited for publication.

Unfortunately, the stratigraphical range (Upper Palaeocene into Middle Eocene) and relationships (intermediate between *G. angulata* and *G. rex*) suggested for this species by Bandy (1964) are incorrect. This worker appears to have been misled by earlier, incorrect references to *G. crassata*.

The holotype of *G. spinulosa* was described by Cushman (1927) from the Middle Eocene Alazan Clay Formation of Vera Cruz, Mexico.

In the upper part of its stratigraphic range *M. aragonensis* has a tendency to become less conical, less rugose and thinner walled. The number of chambers is, as a rule, 4–5 and the increase in size is more rapid than in typical *M. aragonensis* (see also Luterbacher, 1964). For these reasons this writer would regard *M. aragonensis* as the ancestor of *M. crassata*. The transition between the two species is illustrated in such forms as that illustrated by Saito (1962, pl 33, fig 9) as *Globorotalia spinulosa* from the Middle Eocene of Haha-Jima (Hillsborough Island), Japan.

The author would include in the synonymy of this species *Globorotalia hadii* Aubert, described from the "Bartonian" of Jebel Si-Ameur-el-Hadi, valley of Basra, Morocco, North Africa. Topotypes kindly sent by Mme. Aubert do not differ significantly from typical *G. spinulosa* and she herself notes its close relationship with *G. spinulosa*. Confirmation of this has been received in a communication (September, 1965) from Mme. Aubert.

Stratigraphical range: Hantkenina aragonensis Zone to *Orbulinoides beckmanni* Zone.

Morozovella lehneri Cushman and Jarvis, 1929

Globorotalia lehneri Cushman and Jarvis, 1929, *Contr. Cushman Lab. Foram. Res.* **5**, 17, pl. 3, fig. 16.—Cushman and Renz, 1948, *Cushman Lab. Foram. Res. Spec. Publ.* **24**, 40, pl. 8, figs 3, 4.—Cushman and Bermúdez, 1949, *Contr. Cushman Lab. Foram. Res.* **25**, 32, pl. 6, figs 7–9.—Bolli, 1957, *U.S. Nat. Mus. Bull.* **215**, 169, pl. 38, figs 9a–13.—Pessagno, 1961, *Micropaleont.* **7**, 358, pl. 1, figs 5–7.—Saito, 1962, *Trans. Proc. Paleont. Soc. Japan* n.s. **45**, 215, pl. 32, figs 11a–c.—Aubert, 1962, *Notes Serv. géol. Maroc.* **21**, 62, pl. 51, fig. 1.

Pseudogloborotalia lehneri (Cushman and Jarvis) Bermúdez, *Mem. III Congr. Geol. Venez.* **3**, *Bol. Geol. Publ. espec.* **3**, 1345, pl. 16, fig. 9.

Remarks: This species is distinguished from the related *M. crassata*, from which it probably evolved, by its strongly lobulate, densely spinose keeled periphery, and the smooth and distinctly compressed test. It is one of the most distinctive species in the middle and upper part of the Middle Eocene. Its extinction heralds the end of the lineage of angulo-conical, keeled morozovellids.

Stratigraphical range and distribution: Upper part of *Globigerapsis kugleri* Zone through *Truncorotaloides rohri* Zone.

In addition to the distribution cited in the synonymic list above, this species occurs as a relatively common component of low and mid-latitude assemblages of the Atlantic Ocean.

Genus ACARININA Subbotina, 1953

Acarinina nitida (Martin) 1943

Globigerina nitida Martin, 1943. *Stanford Univ. Publ. Univ. Ser. Geol. Sci.* **3**, 25, pl. 7, figs 1a–c.—Graham and Classen, 1955, *Cushman Found. Foram. Res. Contr.* **6**, 28, pl. 4, figs 20a, b, 21a–c.—Mallory, 1959, Lower Tertiary biostratigraphy of the California Coast Ranges, *Am. Assoc. Petrol. Geol. Spec. Publ.* 248, pl. 30, figs 5a–c.

Acarinina acarinata Subbotina, 1953, *Trudy, Vses. Neft. Nauk -Issled. Geol.-Razved. Inst. (V.N.I.G.R.I.)* n.s. **76**, 229, pl. 22, figs 4–10.

Globigerina cf. *G. soldadoensis* Brönniman. Loeblich and Tappan, 1957, *U.S. Nat. Mus. Bull.* **215**, 182, pl. 53, fig. 4.

Turborotalia acarinata (Subbotina). Gohrbandt, 1963, *Mitt. geol. Ges. Wien* **56**, 66, pl. 2, figs 13–15.

Remarks: The holotype has been compared, by the author, with topotype material of *A. acarinita*, the type species of the genus *Acarinina*, from the

Caucasus. As Martin (1943) perceptively observed over twenty years ago, the species to which *A. nitida* is most closely related is probably *Globigerina coalingensis* Cushman and Hanna (= *Acarinina coalingensis* (Cushman and Hanna)) which is a senior synonym of *Acarinina triplex* Subbotina. *A. coalingensis* is generally a larger form, with more inflated chambers, and a densely spinose or papillate test surface.

Stratigraphical range: P. pseudomenardii Zone to *M. formosa* Zone.

Acarinina esnaensis (Le Roy) 1953

Globigerina esnaensis Le Roy, 1953, *Geol. Soc. Am. Mem.* **54**, 31, pl. 6, figs 8–10 (27 February).—Gohrbandt, 1963, *Mitt. geol. Ges. Wien* **56**, 49, pl. 2, figs 19–21.

Acarinina intermedia Subbotina, 1953, *Trudy, Vses. Neft. Nauk.-Issled. Geol.-Razved. Inst. (V.N.I.G.R.I.)* n.s. **76**, 227, pl. 20, figs 14–16, ? 1–4 (19 November).

Truncorotalia esnaensis (Le Roy). Said and Kenaway, 1956, *Micropaleont.* **2**, 151, pl. 6, figs 7a, b.

Globorotalia esnaensis (Le Roy). Loeblich and Tappan, 1957, *U.S. Nat. Mus. Bull.* **215**, 189, pl. 57, fig. 7 (?); pl. 61, figs 1, 2, 9.—Nakkady, 1959, *Micropaleont.* **5**, 461, pl. 3, figs 2a–c.—Berggren, 1960, *Stockholm Contr. Geol.* **5**, 92, pl. 5, figs 3a–c; pl. 6, figs 1a–c; pl. 10, figs 3a–c.—Said and Kerdany, 1961, *Micropaleont.* **7**, 328, pl. 1, fig. 6.—Gartner and Hay, 1962, *Eclog. geol. Helv.* **55**, 563, pl. 2, figs 4a–c.—Said and Sabry, 1964, *Micropaleont.* **10**, no. 3, 383, pl. 1, fig. 5.—El Naggar, 1966, *Bull. Brit. Mus. (Nat. Hist.), Geol. Suppl.* **2**, 210, pl. 21, figs 6a–c.

Globigerina stonei Weiss, 1955, *J. Paleont.* **29**, 18, pl. 5, figs 16–21.

Globorotalia irrorata Loeblich and Tappan, 1957, *U.S. Nat. Mus. Bull.* **215**, 191, pl. 46, figs 2a–c; pl. 61, figs 5a–c.

Globorotalia sp. Luterbacher, 1964, *Eclog. geol. Helv.* **57**, 643, figs 18a–c.

Truncorotaloides (Acarinina) sp. aff. *T. (A.) esnaensis* (Le Roy). McGowran, 1968, *Micropaleont.* **14**, pl. 3, figs 3–6.

Globorotalia (Acarinina) esnaensis (Le Roy). Jenkins, 1971, *N.Z. Geol. Surv. Pal. Bull.* **42**, 82, pl. 3, figs 84–88.

Remarks: This species was described in the same year, 1953, under two different names from the Palaeocene of Egypt and the North Caucasus, Soviet Union. The name *Globigerina esnaensis* Le Roy has precedence by date of publication over *Acarinina intermedia* Subbotina (see synonymy above). The species is characterised by its almost quadrate outline, rounded periphery and roughly textured (papillate or bluntly spinose) wall. It is probably the ancestor of *Acarinina pseudotopilensis* Subbotina.

Stratigraphical range: Planorotalites pseudomenardii Zone to *Morozovella subbotinae* Zone.

A. esnaensis is a common species in the Palaeocene–Lower Eocene whose stratigraphic value has been hampered by the variety of names under which it has been recognised.

Acarinina pseudotopilensis Subbotina, 1953

Acarinina pseudotopilensis Subbotina, 1953, *Trudy, Vses. Neft. Nauk.-Issled. Geol.-Razved. Inst. (V.N.I.G.R.I.) n.s.* **76,** 227, pl. 21, figs 8a–9c; pl. 22, figs 1a–3c.

Globorotalia pseudotopilensis (Subbotina). Loeblich and Tappan, 1957, *U.S. Nat. Mus. Bull.* **215,** 194, pl. 60, fig. 2.—Reyment, 1959, *Rept. Geol. Surv. Nigeria* (1957), 81, pl. 15, figs 14a–c, 15, 16a, b; pl. 16, figs 1a, b.— Berggren, 1960, *Stockholm Contr. Geol.* **5,** 94, pl. 11, fig. 4; pl. 12, fig. 1;— Dieci, *Soc. Pal. Ital. Boll.* **4,** p. 17, pl. 1, figs 3a–c.

Globorotalia tortiva Bolli (*nom. nov, pro Globigerina velascoensis*, var. *compressa* White 1928), 1957, *U.S. Nat. Mus. Bull.* **215,** 78, pl. 19, figs 19–21.

Turborotalia (*Acarinina*) cf. *pseudotopilensis* (Subbotina 1953). Pokorny, 1960, *Rev. Inst. Franc. Pétrole et Ann. Comb. Liquides* **15,** pl. 5, figs 4a, b.

Globorotalia (*Acarinina*) *pseudotopilensis* (Subbotina). Hillebrandt, 1962, *Bayer, Akad. Wissen. Math.-Naturw. Kl. Abh. N.F.* **108,** 143, pl. 14, figs 1a–c.

Turborotalia pseudotopilensis (Subbotina). Gohrbandt, 1963, *Mitt. geol. Ges. Wien* **56,** 66, pl. 3, figs 13–15.

Globorotalia whitei Weiss. El Naggar, 1966, *Bull. Brit. Mus.* (*Nat. Hist.*), *Geol. Suppl.* **2,** 249, pl. 23, figs 3a–c.

Truncorotaloides pseudotopilensis (Subbotina). Jenkins, 1971, *N.Z. Geol. Surv. Pal. Bull.* **42,** 135, pl. 13, figs 382–387.

Remarks: A. pseudotopilensis has probably evolved from *A. esnaensis* from which it is distinguished by its subtruncate periphery, the stronger separation of the chambers, the trapezoidal shape of the final chamber and the generally more spinose wall surface.

Stratigraphical range and distribution: Planorotalites pseudomenardii Zone to lower part of *Morozovella aragonensis* Zone.

A. pseudotopilensis is particularly common in Early Eocene faunas of the Caucasus region but it occurs in both the Caribbean and Mediterranean–Alpine region as well.

Acarinina wilcoxensis (Cushman and Ponton) 1932

Globorotalia wilcoxensis Cushman and Ponton, 1932, *Contr. Cushman Lab. Foram. Res.* **8,** 71, pl. 9, figs 10a–c.—Hamilton, 1953, *J. Paleont.* **17,** 231, pl. 32, fig. 7.—Weiss, 1955, *J. Paleont.* **29**, 19, pl. 6, figs 7–9.—Bolli, 1957, *U.S. Nat. Mus. Bull.* **215**, 79, pl. 19, figs 17–19.—Berggren, 1960, *Stockholm Contr. Geol.* **5**, 97, pl. 13, figs 3a–4c.—El Naggar, 1966, *Bull. Brit. Mus. (Nat. Hist.), Geol. Suppl.* **2**, 250, pl. 23, figs 5a–c.

Globorotalia whitei Weiss, 1955, *J. Paleontol.* **29**, 18, pl. 6, figs 1–3.

Globorotalia cf. *wilcoxensis* Cushman and Ponton, Asano, 1958, *Sci. Repts. Tohoku Univ. ser.* 2, **29**, 46, pl. 12, figs 4a–c.

Truncorotalia ? wilcoxensis (Cushman and Ponton). Gohrbandt, 1963, *Mitt. geol. Ges. Wien* **56**, 64, pl. 4, figs 1–3.

Truncorotaloides (*Acarinina*) *wilcoxensis* (Cushman and Ponton). McGowran, 1968, *Micropaleont.* **14**, pl. 3, fig. 1.

Acarinina wilcoxensis (Cushman and Ponton). Berggren, 1971, *Proc. II Plankt. Conf.* pl. 5, figs 4, 5.

Remarks: *A. wilcoxensis* probably evolved from *A. esnaensis* from which it is distinguished by its larger size and more acute axial periphery. *A. wilcoxensis* is probably the ancestor of *A. quetra*. Valid references to this species in the literature would seem to indicate that this species is restricted to the *Morozovella subbotinae* Zone.

Stratigraphical range: Morozovella subbotinae Zone.

Acarinina quetra (Bolli), 1957

Acarinina crassaformis (Galloway and Wissler). Subbotina, 1953 (*partim*), *Trudy, Vses. Neft. Nauk.-Issled. Geol.-Razved. Inst.* (*V.N.I.G.R.I.*) *n.s.* **76**, 223, pl. 21, fig. 5 (*non* 1–4, 6, 7).

Globorotalia quetra Bolli, 1957, *U.S. Nat. Mus. Bull.* **215**, 79, pl. 19, figs 1–6.

Remarks: This species is distinguished from *A. wilcoxensis* by the acute shape of the test and the development of a distinctly keeled periphery.

The author would interpret the evolution of *A. esnaensis–A. wilcoxensis–A. quetra* as a separate and distinct branch of the acarininids (see Berggren, 1966). *A. quetra*, which is interpreted as the end member of this branch, is thus probably not related in an ancestral-descendant manner to *A. densa* (=*A. bullbrooki*). The latter may have evolved from *A. soldadoensis angulosa*.

Stratigraphical range: Morozovella subbotinae Zone (upper half) to *G. aragonensis* Zone; ? *Acarinina densa* Zone.

Acarinina coalingensis (Cushman and Hanna) 1927

Globigerina coalingensis Cushman and Hanna, 1927, *San Diego Soc. Nat. Hist. Trans.* **5**, 219, pl. 14, figs 4a–b.—Mallory, 1959, *Am. Assoc. Petrol. Geol. Spec. Publ.* 249, pl. 34, figs 6a–c.

Globoquadrina primitiva Finlay, 1947, *New Zealand J. Sci. Tech.* **28**, 291, pl. 8, figs 129–134.—Jenkins, 1965a, *Micropaleont.* **11**, 269, pl. 1, figs 3a–c (outline drawings of holotype of Finlay).

Acarinina triplex Subbotina, 1953, *Trudy, Vses. Neft. Nauk.-Issled. Geol.-Razved. Inst.* (*V.N.I.G.R I.*) *n.s.* **76**, 230, pl. 23, figs 1–5.

Globigerina primitiva Finlay. Brönnimann, 1952, *Bull. Am. Paleont.* **34**, 11, pl. 1, figs 10–12.—Bolli, 1957, *U.S. Nat. Mus. Bull.* **215**, 71, pl. 15, figs 6–8.

Globigerina inaequispira Subbotina. Loeblich and Tappan, 1957 (*pars*), *U.S. Nat. Mus. Bull.* **215**, 181, pl. 61, figs 3a–c (*non* pl. 49, figs 2a–c; pl. 52, figs 1a–2c; pl. 56, figs 7a–c; pl. 62, figs 2a–c) (*non* Subbotina).

Globigerina triplex (Subbotina). Berggren, 1960, *Stockholm Contr. Geol.* **5**, 71, pl. 6, figs 2a–3c; pl. 13, figs 1a–2c.

Globorotalia (*Acarinina*) *primitiva* (Finlay). Hillebrandt, 1962, *Bayer, Akad. Wissen. Math.-Naturw. Kl. Abh. N.F.* **108**, 141, pl. 14, figs 2a, b; 4a, c.

Turborotalia primitiva (Finlay). Gohrbandt, 1963, *Mitt. geol. Ges. Wien* **56**, 67, pl. 1, figs 19–21.

Pseudogloboquadrina primitiva (Finlay). Jenkins, 1965b, *N.Z. J. Geol. Geophys.* **8**, 1122, fig. 9 (81–86).—Jenkins, 1971, *N.Z. Geol. Surv. Pal. Bull.* **42**, 170, pl. 18, figs 551–561.

Remarks: This species has generally been identified under the name of *Globigerina primitiva* Finlay or *Acarinina triplex* Subbotina (see synonymic list). However, receipt in 1963 by Dr. K. Gohrbandt, then with Mobil Oil Co., Tripoli, Libya, of a reillustration of the holotype of *Globigerina coalingensis* Cushman and Hanna, which is deposited in the palaeontological collections of the California Academy of Sciences, San Francisco, California, as well as topotypic material, has permitted a restudy of this form and its determination as the correct name for this species. The type level is lower Eocene, not as stated in Cushman and Hanna (1927), upper Eocene.

Globoquadrina primitiva was originally described by Finlay (1947) from the middle Eocene of New Zealand. It was included in his new genus *Globo-quadrina* because of its supposed relationship with the type species *Globo-quadrina dehiscens* (Chapman, Parr and Collins). Various authors have since referred Finlay's *primitiva* to *Globigerina*, *Globorotalia*, *Turborotalia* or *Acarinina*. Jenkins (1965a, pl. 1, figs 3a–c; 1965b, fig. 9 (81–83)) recently reillustrated the holotype of *G. primitiva* Finlay, and a paratype (1965b, fig. 9 (84–86) and erected the new genus *Pseudogloboquadrina*.

Pseudogloboquadrina, however, is a synonym of *Acarinina*. The flattened umbilical face above the aperture seen in some individuals of *A. primitiva* from New Zealand is also seen in *A. triplex* from the North Caucasus and is reminiscent of a similar development in *Globigerina sellii* and *G. binaiensis* in the late Oligocene.

The holotype of *Acarinina triplex* Subbotina, (V.N.I.G.R.I.) no. 4135, was described from the foraminiferal beds, Suite F_1, Zone of compressed globorotaliids, *Globorotalia marginodentata* Subzone, Khieu River, vicinity of town of Nal'chik, North Caucasus, Soviet Union. This level is approximately equivalent to the *Morozovella subbotinae* Zone or lower part of the *M. formosa* Zone. Three additional specimens were illustrated from the zone of conical globorotaliids in the same stratigraphic section. This zone is approximately equivalent to the upper part of the *M. formosa* Zone and the *M. aragonensis* Zone.

Acarinina coalingensis (Cushman and Hanna) has probably evolved from *Acarinina nitida* (Martin) from which it is distinguished by its larger size, more inflated chambers, and densely spinose to papillate test. Separation of the two species has been difficult in well preserved material, in the author's collection from the *P. pseudomenardii* Zone of Nigeria (West Africa).

Stratigraphical range: Upper part of *Planorotalites pseudomenardii* Zone to within the Middle Eocene (upper limit not determined exactly).

Acarinina mckannai (White) 1928

Globigerina mckannai White, 1928, *J. Paleont.* **2**, 194, pl. 27, fig. 16.— Loeblich and Tappan, 1957 (*partim*), *U.S. Nat. Mus. Bull.* **215**, 181, pl. 47, fig. 7; pl. 53, figs 1, 2; pl. 57, fig. 8; pl. 62, figs 5–7.—Gohrbandt, 1963, *Mitt. geol. Ges. Wien* **56**, 47, pl. 2, figs 4–6.

Globigerina subsphaerica Subbotina, 1947, *Mikrofauna Kavkaza, Emby i Srednei Azii, Lengostoptekhizdat* 108, pl. 5, figs 23–28.—Subbotina, 1953, *Trudy, Vses. Neft. Nauk.-Issled. Geol.-Razved. Inst. (V.N.I.G.R.I.)* n.s. **76**, 59, pl. 2, fig. 15.—Shutskaya, 1956, *Akad. Nauk. S.S.S.R., Tr. Inst. Geol. Nauk.* **164**, geol. ser. 91, pl. 3 fig. 1.

Globigerina cretacea var. *esnehensis* Nakkady, 1950, *J. Paleont.* **24**, 689, pl. 90, figs 14–16.

Globorotalia mckannai (White). Bolli, 1957, *U.S. Nat. Mus. Bull.* **215**, 79, pl. 19, figs 14–16.—Bolli and Cita, 1960, *Riv. Ital. Paleont.* **66**, 23, pl. 31, fig. 6.—Gartner and Hay, 1962, *Eclog. geol. Helv.* **55**, p. 564, pl. 1, figs 1a–c.

Globigerina spiralis Bolli. Loeblich and Tappan, (*pars*), 1957, *U.S. Nat. Mus. Bull.* **215**, 182, pl. 47, fig. 3; pl. 49, fig. 3; pl. 51, figs 6 (?), 7–9; (*non* pl. 53, fig. 3) (*non* Bolli).

Acarinina subsphaerica (Subbotina). Shutskaya, 1958, *Akad. Nauk. SSSR, Voprosy Mikropaleontologii* 2, 89, pl. 2, figs 6–14; pl. 3, figs 1–21.—Shutskaya, 1960, *Trudy, Vses. Nauk.-Issled. Geol.-Razved. Neft. Inst. (V.N.I.G.R.I.)* **16**, *Paleont. Sb.* **3**, 249, pl. 2, fig. 8.

Globorotalia (*Acarinina*) *mckannai* (White). Hillebrandt, 1962, *Bayer, Akad. Wissen. Math.-Naturw. Kl. Abh. N.F.* **108**, 140, pl. 14, figs 8a–10c.
Globigerina alanwoodi El Naggar, 1966, *Bull. Brit. Mus.* (*Nat. Hist.*), *Geol. suppl.* **2**, p. 156, pl. 16, figs 16a–c.

Remarks: A. mckannai is characterised by its nearly circular outline, 5–7 inflated chambers in the last whorl, and relatively large variation in the size of the umbilicus. The author has previously expressed the opinion that *Acarinina pentacamerata* (Subbotina) is a junior synonym of *A. mckannai* (Berggren, 1960), but Bolli and Cita (1960) and Gohrbandt (1963) have presented arguments for separating these two forms. The taxonomic questions relating to *A. pentacamerata* are complex (and discussed below under that species). Although the two forms are very similar in morphology it would appear that they do have different stratigraphic ranges. Whereas *A. mckannai* appears to have developed from *A. nitida* in the upper part of the *P. pseudomenardii* Zone, *A. pentacamerata* probably developed from *A. soldadoensis* within the *M. subbotinae* Zone.

Stratigraphical range: Upper part of *Planorotalites pseudomenardii* Zone through lower half of *Morozovella velascoensis* Zone.

Acarinina pentacamerata (Subbotina) 1947

Globorotalia pentacamerata Subbotina, 1947 (*partim*), *Mikrofauna Kavkaza, Emby i Srednei Azii, Lengostoptekhizdat*, 128, pl. 7, figs 12–17 (*non* pl. 9, figs 24–26).
Acarinina pentacamerata (Subbotina). Subbotina, 1953 (*partim*), *Trudy, Vses. Neft. Nauk.-Issled. Geol.-Razved. Inst.* (*V.N.I.G.R.I.*) *n.s.* **76**, 233, pl. 23, fig. 8; pl. 24, figs 1–6 (*non* 7–8; 9(?)).—Shutskaya, 1956, *Akad. Nauk. S.S.S.R. Trudy Inst. Geol. Nauk* **164**, geol. ser. 103, pl. 3, figs 6a–c.—Kraeva *in* Kaptarenko-Chernoussova, Golyak, Zernetskii, Kraeva and Lipnik, 1963, *Akad. Nauk Ukrain. S.S.R. Inst. Geol. Nauk. ser. strat. i paleont.* **45**, 152, pl. 31, figs 6a, b.
Globigerina lgravelli Brönnimann, 1952, *Bull. Am. Paleont.* **34**, 12, pl. 1 (11), figs 16–18.—Bolli, 1957, *U.S. Nat. Mus. Bull.* **215**, 72, pl. 16, figs 1–3.— Hornibrook, 1958, *N.Z. J. Geol.* and *Geophys.* **1**, no. 4, 665, pl. 2, figs 23, 15.—Gohrbandt, 1963, *Mitt. geol. Ges. Wien* **56**, 48, pl. 2, figs 10–12.
Acarinina pentacamerata (Subbotina) var. camerata Khalilov, 1956, *Akad. Nauk Azerb. S.S.R. Inst. Geol. Baku, Trudy* **17**, 252, pl. 5, figs 6a–c.
Acarinina pentacamerata (Subbotina) var. *acceleratoria* Khalilov, 1956, *Ibid.*, 253, pl. 7, figs 7a–c.
Globigerina dubia var. *lakiensis* Haque, 1956, *Pakistan Geol. Surv. Mem. Pal. Pakistanica* **1**, 174, pl. 4, figs 2a–c.

Globigerina mckannai White. Berggren, 1960 (*partim*), *Stockholm Contr. Geol.* **5,** 68, pl. 9, figs 2, 3 (*non* pl. 1, fig. 4; pl. 9, fig. 4; pl. 10, fig. 1, text-fig. 7). *Globorotalia* (*Acarinina*) *pentacamerata* Subbotinae. Hillebrandt, 1962, *Bayer, Akad. Wissen. Math.-Naturw. Kl. Abh. N.F.* **108,** 142, pl. 14, figs 7a–c.

Remarks: The taxonomy of *Acarinina pentacamerata* (Subbotina) is complex. In 1936 Subbotina mentioned and illustrated (pl. 3, figs 7–9) a new species *Globorotalia crassa* (d'Orbigny) var. *pentacamerata* Subbotina. This species, however, was invalid at the time according to the International Rules of Zoological Nomenclature, art. 25 (c), 1–2. The illustration of this specimen agrees well with what has later become the commonly accepted concept of *Globigerina soldadoensis* Brönnimann.

In 1947 Subbotina formally described and illustrated *Globorotalia pentacamerata* Subbotina for the first time. Three specimens were figured. The holotype (pl. 7, figs 15–17) was recorded from the *Globorotalia crassaformis* Zone (which Subbotina considered middle Eocene in age); the two paratypes (pl. 7, figs 12–14 and 24–26) were from the "*Globorotalia velascoensis* Zone" of lower Eocene age (the latter Zone was subsequently, in 1953, termed the Zone of conical globorotaliids and is characterised by *Morozovella aragonensis* Nuttall and *M. caucasica* Glaessner).

It is doubtful whether the three illustrated specimens mentioned above belong to the same species. The form figured on pl. 7, figs 24–26 by Subbotina (1947) shows the distinct lateral separation of chambers peripherally which is characteristic of Bolli's (1957) *Acarinina angulosa.* Indeed, Subbotina's specimen is almost identical to that figured by Bolli (1957, pl. 35, figs 8a–c) from the *Globorotalia palmerae* Zone, Navet Formation, Trinidad. There may be some question whether Bolli's specimen from the *G. palmerae* Zone is conspecific with that figured (*op. cit.,* pl. 16, figs 4–6) from the type locality of the *G. formosa formosa* Zone of Trinidad. They are considered conspecific here. The specimen figured by Subbotina (1947) as figs 12–14 on pl. 7 shows the characters which have subsequently come to be associated with *A. pentacamerata* as elucidated by Subbotina in 1953. It is probably conspecific with Brönnimann's *Globigerina gravelli* (see also Bolli, 1957; Hornibrook, 1958). This leaves the holotype. The figured specimen shows six chambers in the final whorl (Subbotina mentions five), a tightly coiled test and a relatively narrow umbilicus.

Bandy (1964) has suggested that *G. pentacamerata* is a junior synonym of *Globigerina aspensis* Colom. Judging from the holotype of *A. pentacamerata* alone this is hard to determine. Colom's species is not well-known and further work is needed on this form to elucidate its characters. The holotype and paratype specimens of *A. pentacamerata* were not found by this writer in the

collections of V.N.I.G.R.I. on the occasion of several visits there in 1962 and
1963. It is possible that they were lost or destroyed during the siege of
Leningrad (1941–1943). On the other hand this writer has examined all the
specimens figured by Subbotina (1953) in her subsequent study of a *A. pentacamerata*. It is upon these specimens that the general concept of the species
should be formed. The specimens figured from the zone of conical globorotaliids in the North Caucasus (pl. 24, figs 1a–5c) are typical representatives
of this species. The 5–7 rounded chambers, relatively wide umbilicus and
weakly developed circum-umbilical shoulder distinguish this species from
associated forms in the lower Eocene. It is in this sense that the concept
of *A. pentacamerata* should be used (to the exclusion of forms illustrated by
Subbotina (1953, pl. 24, figs 6a–8c) from the *G. crassata* Subzone of the Zone
of compressed globorotaliids (upper Palaeocene). The suggestion that
A. pentacamerata be considered a junior synonym of *G. aspensis* does not
seem a satisfactory solution. It is based on the assumption that the holotype
of *A. pentacamerata* (which is apparently lost) belongs to *A. aspensis* which
is difficult to prove from available evidence. It also requires the rejection
of the concept of the species as subsequently elucidated by Subbotina (1953)
and the inclusion of these forms in some other species—the lower Eocene
ones probably in *A. gravelli*, the Palaeocene ones in *A. mckannai*. Inasmuch
as the species *Globorotalia pentacamerata* Subbotina (=*Acarinina pentacamerata* (Subbotina)) was validly described and illustrated in 1947, and
inasmuch as it would seem possible to formulate a valid concept of this
species based on its authors subsequent work, this writer would suggest
retention of the name *A. pentacamerata* for the sake of stability of nomenclature.

    Acarinina pentacamerata var. *camerata* and *A. pentacamerata* var. *accelera-
toria*, described by Khalilov (1956, pp. 252, 253) from the lower Eocene of
northeastern Azerbaidzhan, Soviet Union are placed in the synonymy
of *A. pentacamerata* here. *A. pentacamerata* var. *camerata* was said (*op. cit.*
p. 252) to differ from the typical form "in the larger number of chambers in
the first whorl, and in the considerable inflation of the dorsal side". *A. pentacamerata* var. *acceleratoria* was said (*op. cit.* p. 253) to differ from the typical
form "in the lesser number of chambers in the early whorls". These minor
differences are, in themselves, deemed insufficient to distinguish two subspecies of *A. pentacamerata*; the illustrated holotypes of both "varieties"
show the typical characters associated with *A. pentacamerata*.

    Subbotina (1947) described the holotype of *G. pentacamerata* from the
Globorotalia crassaformis Zone. This level would be above the top of the
G. aragonensis Zone, probably in the *G. palmerae* Zone. She subsequently
(1953, p. 234, tab. 3, p. 29) showed it to range from the *G. crassata* Subzone

of the zone of compressed globorotaliids through the zone of conical globo-rotaliids, but observed that the species reached its acme of development in the zone of conical globorotaliids. However, in her chart showing the suggested phylogeny of the acarinids (1953, fig. 8, p. 153) Subbotina shows *A. pentacamerata* evolving from *A. interposita* Subbotina (=*A. soldadoensis* (Brönnimann)) at the base of the *G. marginodentata* Subzone (approximately equivalent to the *M. subbotinae* Zone) of the zone of compressed globo-rotaliids and ranging to the top of the zone of conical globorotaliids. She mentioned (*op. cit.* p. 234) that the forms from the Palaeocene of Mangyshlak (illustrated as figs 6a–8c, pl. 24) differed from the lower Eocene forms in their significantly smaller size, indicating that she probably maintained some reservations on the conspecificity of these two groups.

Brönnimann (1952, p. 13) gave the following stratigraphic range for *Globigerina gravelli* in Trinidad: both zones of the Lizard Springs Formation, rare to common; Ramdat Marl, rare. Bolli (1957, p. 72) gave a stratigraphic range of *Globorotalia rex* Zone to *G. aragonensis* Zone.

The author would suggest that *A. pentacamerata* may have evolved into *A. aspensis* in the upper part of the lower Eocene (Berggren, 1966). Bolli (1957, p. 162) suggested that *G. soldadoensis angulosa* might have been the ancestor of *G. aspensis*.

Stratigraphical range: Morozovella subbotinae Zone through *M. aragonensis* Zone; ? *A. densa* Zone.

Acarinina soldadoensis (Brönnimann) 1952

Globorotalia crassa (d'Orbigny) var. *pentacamerata* Subbotina, 1936, *Trudy, Neft. Geol.-Razved. Inst.* (*V.N.I.G.R.I.*) ser. **A. 96,** 11, 14, 16, pl. 3, figs 7–9 (invalid acc. to Intern. Rules of Zoological Nomenclature, Art. 25(c), 1–2).

Globigerina soldadoensis Brönnimann, 1952, *Bull. Am. Paleont.* **34,** 9, pl. 1 (11), figs 1–9.—Bolli, 1957, *U.S. Nat. Mus. Bull.* **215,** 71, pl. 16, figs 7–9, 162, pl. 35, fig. 9(?).—Berggren, 1960, *Proc. Int. Geol. Congr.* 21st Sess, Copenhagen, Sect **6,** pl. 1, figs 17a, b.—Gartner and Hay, 1962, *Eclog. geol. Helv.* **55,** 569, pl. 1, figs 3a–c.

Acarinina interposita Subbotina, 1953, *Trudy, Vses. Neft. Nauk.-Issled. Geol.-Razved. Inst.* (*V.N.I.G.R.I.*) n.s. **76,** 231, pl. 23, figs 6, 7.—Kraeva *in* Kaptarenko-Chernoussova, Golyak, Zernetskii, Kraeva and Lipnik, 1963, *Akad. Nauk. Ukrainskii S.S.R. Inst. Geol. Nauk., ser. strat. i paleont.* **45,** 152, pl. 31, figs 7a–c.

Acarinina clara Khalilov, 1956, *Akad. Nauk Azerb. S.S.R. Inst. Geol. Baku, Trudy* **17,** 250, pl. 5, figs 4a–c.

Globigerina mckannai White. Berggren, 1960 (*partim*), *Stockholm Contr. Geol.* **5,** 68, pl. 1, fig. 4; pl. 9, fig. 4; pl. 10, 1, text-fig. 7 (*non* pl. 9, figs 2, 3).

Globorotalia interposita (Subbotina), Said, 1960, *Micropaleont.* **6**, 283, pl. 1, fig. 9 (?).
Turborotalia soldadoensis (Brönnimann). Gohrbandt, *Mitt. geol. Ges. Wien* **56**, 65, pl. 2, figs 7–9, 16–18.
Globigerina mckannai White. El Naggar, 1966, *Bull. Brit. Mus. (Nat. Hist.) Geol. Suppl.* **2**, 120, pl. 16, figs 5a–c.
Globorotalia (Acarinina) soldadoensis (Brönnimann). Jenkins, 1971, *N.Z. Geol. Surv. Pal. Bull*, **42**, 83, pl. 4, figs 94–98,

Remarks: Although Subbotina's (1936) illustration of *Globorotalia crassa* has no official status (see discussion of *A. pentacamerata*) it is nevertheless included here in the synonymy of *A. soldadoensis*.

Stratigraphical range: Morozovella velascoensis Zone through *Acarinina densa* Zone; ? *Hautkenina aragonensis* Zone.

The author has observed specimens as high as the *H. aragonensis* Zone which may be referable to this species. It may be that the observations at these stratigraphically higher levels involve an overlap in species concept with one or the other closely related forms in this group such as *A. aspensis* or *A. pentacamerata*.

Acarinina angulosa (Bolli) 1957

Globorotalia pentacamerata Subbotina, 1947 (*pars*), *Trudy, Vses. Neft. Nauk.-Issled. Geol.-Razved. Inst. (V.N.I.G.R.I.) n.s.* **76**, 128, pl. 9, figs 24–26 (*non* pl. 7, figs 12–17). (Note that the specimen illustrated by Subbotina, 1936, pl. 3, figs 7–9 is *not* the same as the one referred to here.)
Globigerina soldadoensis angulosa Bolli, 1957, *U.S. Nat. Mus. Bull.* **215**, 71, pl. 16, figs 4–6; p. 162, pl. 35, figs 8a–c.

Remarks: This species was erected as a subspecies by Bolli (1957) for forms which differed from *A. soldadoensis* in having more angular chambers and a somewhat different stratigraphic range: *G. formosa formosa* Zone to *G. palmerae* Zone. Because of the angular disposition of the chambers, the specimen illustrated by Subbotina (1947, pl. 9, figs 24–26) from the *G. velascoensis* Zone (=Zone of conical globorotaliids of Subbotina, 1953) on the Kuban River, North Caucasus, Soviet Union, is included here in the synonymy of this species.

Stratigraphical range: Morozovella formosa Zone through *A. densa* Zone.

Acarinina aspensis (Colom) 1954

Globigerina aspensis Colom, 1954, *Bol. Inst. Geol. y Min. España* **66**, 151–154, pl. 3, figs 1–35; pl. 4, figs 1–31.

Globorotalia aspensis (Colom). Bolli, 1957, *U.S. Nat. Mus. Bull.* **215,** 166, pl. 37, figs 18a–c.

Remarks: As Bolli (1957) has pointed out, the illustrations by Colom of this species show a considerable amount of variation in test size, chamber shapes, and number of chambers in last whorl. Although it seems possible to distinguish these forms as a distinct species from the allied *A. pentacamerata* and *A. soldadoensis,* further work is needed on this species before its use in stratigraphic studies can be evaluated. This writer has mentioned in his discussion of *A. pentacamerata* his reasons for rejecting Bandy's (1964) suggestion that *A. pentacamerata* is a junior synonym of *A. aspensis.*

Bolli (*loc. cit.*) suggests that *A. aspensis* evolved from *A. angulosa;* this writer has suggested that *A. pentacamerata* may have been its ancestor (Berggren, 1964).

Stratigraphical range: Acarinina densa Zone to *Globigerapsis kugleri* Zone (Bolli, 1957).

Acarinina densa (Cushman) 1925

Pulvinulina crassata Cushman, var. *densa* Cushman, 1925, *Bull. Amer. Ass. Petrol. Geol.* **9,** 301.

Globorotalia crassata (Cushman) var. *densa* (Cushman). Cushman and Barksdale, 1930, *Contr. Dept. Geol., Stanford Univ.* **1.**—Cushman, 1939, *Contr. Cushman Lab. Foram. Res.* **15,** 74, pl. 12, fig. 20.—Cushman and Renz, 1948, *Cushman Lab. Foram. Res. Spec. Publ.* **24,** 40, pl. 8, figs 7, 8.—Cushman, 1948, *Bull., Maryland Dept. Geol. Mines and Water Resources* **2,** 243.—Saito, 1962, *Trans. Proc. Palaeont. Soc. Japan n.s.* **45,** 214, pl. 33, figs 5a–6b, 10a–11c. —Aubert, 1962, *Notes Serv. géol. Maroc.* **21,** no. 156, 57, pl. 21, fig. 4.

Globorotalia crassaformis (Galloway and Wissler). Subbotina, 1947, *Mikrofauna Kavkaza, Emby i Srednei Azii, Lengostoptekhizdat* 129, pl. 8, figs 17–19; pl. 9, figs 27–32 (*non* Galloway and Wissler).

Globorotalia (Truncorotalia) crassata (Cushman) var. *densa* (Cushman). Cushman and Renz, 1949, *Contr. Cushman Lab. Foram. Res.* **25,** 38, pl. 7, figs 10–12.

Globigerina spinuloinflata Bandy, 1949. *Bull. Am. Paleont.* **32,** 122, pl. 23, figs 1a–c.

Acarinina crassaformis (Galloway and Wissler). Subbotina, 1953 (*pars*), *Trudy, Vses. Neft. Nauk.-Issled. Geol.-Razved. Inst.* (*V.N.I.G.R.I.*) *n.s.* **76,** 223, pl. 21, figs 2a–4c, 7a–c; ? figs 1a–c; 5a–6c.

Globorotalia bullbrooki Bolli, 1957, *U.S. Nat. Mus. Bull.* **215,** 167, pl. 38, figs 4a–5c.

Globorotalia densa (Cushman). Pessagno, 1960, *Micropaleont.* **6**, 99, pl. 5, no. 3.—Pessagno, 1961, *Ibid.*, **7**, no. 3, 356, pl. 1, figs 1–3.
  Turborotalia (*Acarinina*) cf. *bullbrooki* (Bolli, 1957). Pokorný, 1960, *Rev. Inst. Franc. Petrole et Ann. Comb. Liquides* **15**, pl. 5, figs 1a–c.
  Globorotalia crassata (Cushman). Aubert, 1962, *Notes Serv. géol. Maroc.* **21**, 58, pl. 2, figs 3a–c; ? 2a–c.

Remarks: This characteristic late early Eocene–early middle Eocene species is interpreted as an acarininid here, not a morozovellid. It is part of an evolutionary sequence within the acarininid group which, as in the case of one, and perhaps two other branches, leads to the development of species of *Truncorotaloides* (see Berggren, 1966). Indeed Bolli (1957) has suggested that *A. bullbrooki* itself may be a truncorotaloid. Convergent evolution has led, in the case of *A. densa*, to a superficial resemblance to forms normally placed in *Morozovella*.

Stratigraphical range: Acarinina densa Zone—*Globigerapsis kugleri* Zone. The range attributed to this species here is greater than that attributed to *G. bullbrooki* (=*A. densa*) by Bolli (1957) but less than the range attributed to *A. crassaformis* (=*A. densa*) by Subbotina (1953) which is considered excessive.

Genus TRUNCOROTALOIDES Brönnimann and Bermúdez, 1953

Truncorotaloides rohri Brönnimann and Bermúdez, 1953

Truncorotaloides rohri Brönnimann and Bermúdez, 1953, *J. Paleont.* **27**, 818, pl. 87, figs 7–9.—Beckmann, 1953, *Eclog. geol. Helv.* **46**, 396, pl. 26, figs 10, 11. Bolli, Loeblich and Tappan, 1957, *U.S. Nat. Mus. Bull.* **215**, 42, pl. 10, figs 5a–c.—Bolli, 1957, *Ibid.*, 170, pl. 39, figs 8a–12c.—Berggren, 1960, *Proc. 21st Int. Geol. Congr., Copenhagen, sect.* **6**, 53, pl. 1, figs 18a, b.—Bermúdez, 1961, *Mem. III Congr. Geol. Venez.* **3**, *Bol. Geol. Publ. espec.* **3**, 1352, pl. 17, figs 3a, b, pl. 20, figs 10a–c.
  Truncorotaloides rohri Brönnimann and Bermúdez var. *guaracaraensis* Brönnimann and Bermúdez, 1953, *J. Paleont.* **27**, 818, pl. 87, figs 1–3.
  Truncorotaloides rohri Brönnimann and Bermúdez var. *piparoensis* Brönnimann and Bermúdez, 1953, *J. Paleont.* **27**, 818, pl. 87, figs 4–6.
  Truncorotaloides rohri Brönnimann and Bermúdez var. *mayoensis* Brönnimann and Bermúdez, 1953, *J. Paleont.* **27**, p. 819, pl. 87, figs 7–9.
  Acarinina rugosoaculeata Subbotina, 1953, *Trudy, Vses. Neft. Nauk.-Issled. Geol.-Razved. Inst.* (*V.N.I.G.R.I.*) *n.s.* **76**, 235, pl. 25, figs 4a–6c.

Remarks: Truncorotaloides rohri Brönnimann and Bermúdez is the type species of the genus *Truncorotaloides*. It is a characteristic form of the middle

Eocene and would appear to be related to, and probably a descendant of, *T. collactea* (Finlay).

Bandy (1964) has stated that Bolli's (1957) illustrations of *T. rohri* are of a species previously described by him (Bandy, 1949) as *Globigerinoides pseudodubia*. However, an examination of the holotype of *Globigerinoides pseudodubia* (No. 4911) from the Tallahatta Formation, Claiborne Group at Little Stave Creek, Alabama reveals that this is not *T. rohri* Brönnimann and Bermúdez. This *"Globigerinoides" pseudodubia* is considerably larger (diameter 0.45 mm), and has a coarsely perforate wall in contrast to the densely hispid–spinose wall and finely perforate test in *T. rohri*.

Acarinina rugosoaculeata Subbotina is placed in the synonymy of this species. It appears to possess the characteristics of this species. The stratigraphic range attributed to it by Subbotina (1953, p. 236, tab. 3, p. 29, fig. 8, p. 153) of middle Eocene–middle Oligocene is, however, difficult to interpret. Subbotina (1953, tab. 3, p. 29) indicates that *A. rugosoaculeata* is most abundant in the middle Eocene Zone of acarininids and rare in younger strata. There may be more than one species involved here.

Stratigraphical range and distribution: Globigerapsis kugleri Zone—*Truncorotaloides rohri* Zone, middle Eocene.

This species is common and widespread in deep-sea sediments of middle Eocene age within the low to mid-latitude regions.

Truncorotaloides collactea (Finlay) 1939

Globorotalia collactea Finlay, 1939, *Trans. Proc. Roy. Soc. New Zealand*, **69**, 37, pl. 29, figs 164–165.

?*Globigerina collactea* (Finlay). Brönnimann, 1952, *Bull. Am. Paleont.* **34**, 13, pl. 1, figs 13–15.

Acarinina rotundimarginata Subbotina, 1953, *Trudy, Vses. Neft. Nauk.-Issled. Geol.-Razved. Inst. (V.N.I.G.R.I.)* n.s. **75**, 234, pl. 25, figs 1a–3c.

Globorotalia spinuloinflata Bolli, 1957, *U.S. Nat. Mus. Bull.* **215**, 168, pl. 38, figs 8a–c (*non* Brady).

Truncorotaloides rotundimarginata (Subbotina). Berggren, 1960, *Proc. 21st Int. Geol. Congr. Copenhagen sect.* **6**, 53, pl. 1, figs 19a, b.

Truncorotaloides collactea (Finlay). Jenkins, 1965, *N.Z. J. Geol. Geophys.* **8**, 843, figs 1–27.—Jenkins, 1971, *N.Z. Geol. Surv. Pal. Bull.* **42**, 134, pl. 14, figs 402–407.

?*Globorotalia (Acarinina) spinoinflata (sic.)* (Bandy). Jenkins, 1971, *N.Z. Geol. Surv. Pal. Bull.* **42**, 83, pl. 4, figs 99–101.

?*Globorotalia (Acarinina)* cf. *spinoinflata (sic.)* (Bandy). Jenkins, 1971, *N.Z. Geol. Surv. Pal. Bull.* **42**, 83, pl. 6, figs 117–122.

Remarks: The affinities of this species have been clarified by Jenkins (1965).

Originally described as a species of *Globorotalia* by Finlay (1939) from the Bortonian Stage (Middle Eocene) of Hampden Beach, North Otago, New Zealand, Jenkins has observed that supplementary apertures occur on the spiral side on some individuals of this species and has placed it in *Truncorotaloides*. This species is characterised by having 5 hispid to weakly spinose chambers in the final whorl and a subacute periphery in edge view. The holotype and four paratypes of *T. collactea* have been illustrated by Jenkins (1965) and these figures are included on Chart XII.

Jenkins (1971) indicates that *Acarinina rotundimarginata* Subbotina is a junior synonym of *Truncorotaloides collactea* (Finlay) with which the author agrees. However, he considers *Globorotalia* (*Acarinina*) *spinoinflata* (*sic.*) Bandy to be the same as the form figured as *G. spinuloinflata* by Bolli (1957, pl. 38, figs 8a–c) and distinct from *T. collactea* (Jenkins, 1971). However, comparison of the holotype (pl. 25, figs 1a–c) and paratype (pl. 25, figs 2a–3c) illustrations of *Acarinina rotundimarginata* Subbotina (1953) with that of Bolli (1957, pl. 38, figs 8a–c) shows no significant differences. The author would interpret Bolli's (1957) *G. spinuloinflata* as a subangular variant of the more typically rounded *T. collactea*. *Globorotalia spinuloinflata* Bandy, 1949, is conspecific with, and a senior synonym of *Globorotalia bullbrooki* Bolli, 1957. However, both are predated by *Pulvinulina densa* Cushman, 1925, which is apparently the earliest valid name for this species. For a fuller account of the taxonomic problems associated with these forms see Berggren (1965, 1966). *Acarinina rotundimarginata* Subbotina from the Middle Eocene of the North Caucasus, Soviet Union and *Globorotalia spinuloinflata* Bolli (*non* Bandy) from the Middle Eocene of Trinidad are placed in the synonymy of this species.

T. collactea may have evolved from *Acarinina nitida* in the *A. densa* Zone.

Stratigraphical range and distribution: Acarinina densa Zone—*?Globigerapsis mexicana* Zone.

It appears that this species continues into late Eocene time in high latitudes (Berggren, 1969; Jenkins, 1965, 1971) but may have become extinct at the top of the middle Eocene in low latitudes. An examination of the stratigraphic level from which *T. collactea* has been recorded in Denmark (Moesgaard Strand) by K. Perch-Nielsen indicates that it is in the *Isthmolithus recurvus* Zone which is within the lower part of the upper Eocene. Thus, *T. collactea* may not extend to the top of the Eocene as suggested previously by this writer (Berggren, 1969).

Truncorotaloides topilensis (Cushman) 1925

Globigerina topilensis Cushman, 1925, *Contr. Cushman Lab. Foram. Res.* **1,** 7, pl. 1, figs 9a–c.

Truncorotaloides topilensis (Cushman). Bolli, 1957, *U.S. Nat. Mus. Bull.* **215**, 170, pl. 39, figs 13–16b.—Saito, 1962, *Trans. Proc. Paleont. Soc. Japan n.s.* **45**, 215, pl. 33, figs 8a–c.

Pseudogloborotalia topilensis (Cushman) Bermúdez, 1961, *Mem. III, Congr. Geol. Venez.* **3**, *Bol. Geol. Publ. espec.* **3**, 1348, pl. 16, fig. 12.

Globorotalia topilensis (Cushman). Aubert, 1962, *Notes Serv. géol. Maroc.* **21**, 60, pl. 4, fig. 5.

Remarks: This distinct form is distinguished by its angular periphery, wedge-shaped chambers, strongly spinose test and discrete supplementary apertures on the spiral side.

This species is interpreted here as a descendant of *Acarinina densa* ($= A.$ *bullbrooki*). *Truncorotaloides rohri* var. *mayoensis*, if not conspecific with this species, is probably closely related to it.

The short stratigraphic range and distinct morphology of this species makes this a very useful form in identifying the mid-part of the middle Eocene and in regional correlation.

Stratigraphical range: Globigerapsis kugleri Zone to *Orbulinoides beckmanni* Zone, middle Eocene.

Acknowledgements

The ideas expressed in this chapter are the fruit of more than a decade of experience in working with Palaeogene planktonic foraminifera. I should like to acknowledge, in particular, the co-operation and stimulating discussions of several colleagues which have played a significant role in the opinions expressed here: Dr. V. A. Krasheninnikov and Dr. V. G. Morozova, Geological Institute, Academy of Sciences, Moscow; Dr. V. P. Alimarina, Faculty of Geology, Moscow State University, Moscow; Dr. N. N. Subbotina, All-Union Petroleum Research Institute (V.N.I.G.R.I.), Leningrad; Dr. R. K. Olsson, Rutgers University, New Brunswick, and Dr. W. H. Blow (deceased), Sunbury-on-Thames. This paper has profited from an initial review by two of my colleagues at Woods Hole, Dr. David Wall and Dr. Bilal Ul Haq. This investigation has been supported by the Oceanography Section of the National Science Foundation, Grant GA-30723.

References

Bandy, O. (1960). Planktonic foraminiferal criteria for paleoclimatic zonation. *Tohoku Univ., Sci. Repts. ser* 2 (*Geol.*), spec. vol. **4**, 1–8.

Bandy, O. (1964). Cenozoic planktonic foraminiferal zonation. *Micropaleont.* **10**, 1–17.

Banner, F. T. and Blow, W H. (1959). Classification and stratigraphic distribution of the Globigerinaceae. *Palaeont.* **2**, 1–27, pls. 1–3.

Berggren, W. A. (1960). Paleogene Biostratigraphy and planktonic Foraminifera of the SW Soviet Union: An analysis of recent Soviet investigations. *Stockholm Contr. Geol.* **6**, 63–125, 8 tabs.

Berggren, W. A. (1965). Some problems of Paleocene–Lower Eocene planktonic foraminiferal correlations. *Micropaleont.* **11**, 278–300, 1 pl.

Berggren, W. A. (1966). Phylogenetic and taxonomic problems of some Tertiary planktonic foraminiferal lineages. *Voprosy Mikropal.* **11**, 309–332, 4 tabs (in Russian).

Berggren, W. A. (1969). Rates of evolution in some Cenozoic planktonic foraminifera: *Micropaleont.* **1**, 351–365, 13 text-figs.

Berggren, W. A. (1971). Multiple phylogenetic zonations of the Cenozoic based on planktonic foraminifera. *In* (Farinacci, A., ed.), *Proc. II Plankt. Conf.*, 1970, **I**, 41–56. Tecnoscienza, Roma.

Berggren, W. A. (1972). Cenozoic biostratigraphy and paleobiogeography in North Atlantic. *In* (Laughton, A. S. *et al.*), *Initial Reports of the Deep Sea Drilling Project*, **12**, 965–1001, 13 pls. U.S. Government Printing Office, Washington.

Berggren, W. A., Olsson, R. K. and Reyment, R. A. (1967). Origin and development of the foraminiferal genus *Pseudohastigerina* Banner and Blow, 1959. *Micropaleont.* **13**, 265–288, 1 pl. 12 text-figs.

Berggren, W. A. and Hollister, C. D. (1974). Paleogeography, Paleobiogeography and the history of circulation in the Atlantic Ocean. *In* (Hay, W. W., ed.), *Studies in paleo-oceanography*. Tulsa, Oklahoma, Soc. Econ. Paleontologists and Mineralogists Special Publication 20, 126–186.

Blow, W. H. (1969). Late Middle Eocene to Recent planktonic Foraminiferal Biostratigraphy. *In* (Brönnimann, P. and Renz, H. H., eds), *Proc. I. Plankt. Conf.*, Geneva, 1967, **1**, 199–422, 54 pls., 43 figs. E. J. Brill, Leiden.

Bolli, H. M. (1957). The genera *Globigerina* and *Globorotalia* in the Paleocene–Lower Eocene Lizard Springs Formation of Trinidad, B.W.I. *U.S. Nat. Mus. Bull.* **215**, 61–81.

Cordey, W. G., Berggren, W. A. and Olsson, R. K. (1970). Phylogenetic trends in the planktonic foraminiferal genus *Pseudohastigerina* Banner and Blow, 1959. *Micropaleont.* **16**, 235–242.

Haque, A. F. N. (1956). The foraminifera of the Ranikot and the Laki of the Nammal Gorge, Salt Range. *Pakistan, Geol. Survey, Pal. Pakistanica*, **1**, 1–300, pls. 1–34.

Ellis, B. F., Messina, A. R., Charmatz, R. and Ronai, L. E. (1969). Catalogue of Index Smaller Foraminifera, **2**, Tertiary planktonic Foraminifera. Spec. Publ. Amer. Mus. Natural History, New York.

Jenkins, D. G. (1965). Planktonic foraminiferal zones and new taxa from the Danian to Lower Miocene of New Zealand. *N.Z. Geol. J. Geol. Geophys.* **8**, 1088–1126, 15 text-figs.

Jenkins, D. G. (1971). New Zealand Cenozoic planktonic foraminifera. *N.Z. Geol. Surv. Pal. Bull.* **42**, 278, 23 pls.

Krasheninnikov, V. A. (1969). Geographical and stratigraphical distribution of planktonic foraminifers in Paleogene deposits of tropical and subtropical areas. *Acad. Sciences U.S.S.R., Geol. Inst. Publ. Office "NAUKA".* 1–188, Moscow (in Russian).

Krasheninnikov, V. A. (1971a). Cenozoic Foraminifera. *In* Initial Reports of the Deep Sea Drilling Project, **6**, 1055–1068. U.S. Government Printing Office, Washington.

Krasheninnikov, V. A. (1971b). Stratigraphy and foraminifera of Cenozoic pelagic sediments of the northwest part of the Pacific Ocean. *Vopros. Mikropal.* **14**, 140–199 (in Russian).

Loeblich, A. R., Jr. and Tappan, H. (1957). Planktonic foraminifera of Paleocene and early Eocene age from the Gulf and Atlantic Coastal Plain. *U.S. Nat. Mus. Bull.* **215**, 173–197, pls. 40–64.

Loeblich, A. R. and Tappan, H. (1964). *Treatise on Invertebrate Paleontology*, Part C, Protistata 2, Sarcodina chiefly "Thecamoebians" and Foraminiferida, **1** and **2.** Geol. Soc. Amer. University of Kansas Press.

Luterbacher, H. P. (1964). Studies in some *Globorotalia* from the Paleocene and Lower Eocene of the Central Apennines. *Eclog. geol. Helv.* **57**, 631–730, 134 figs.

Margolis, S. B. and Kennett, J. P. (1971). Cenozoic paleoglacial history of Antarctica recorded in subantarctic deep-sea cores. *Am. Jour. Sci.* **271**, 1–36.

McGowran, G. (1968). Reclassification of early Tertiary *Globorotalia. Micropaleont.* **14**, no. 2, 179–198, pls. 1–4.

Steineck, P. L. (1971). Phylogenetic reclassification of Paleocene planktonic foraminifera. *Texus J. Sci.* **23**, 167–178.

Subbotina, N. N. (1953). Globigerinidae, Hantkeninidae, and Globorotaliidae, Fossil Foraminifera of the U.S.S.R. *Trudy, Vses. Neft. Nauk-Issled. Geol.-Razved. Inst.* (V.N.I.G.R.I.) *n.s.* **76**, no. 9, 1–296, 41 pls. Leningrad (in Russian).

a   b   c

Subbotina praecursoria (Morozova,1957)

19   20

21   22   23

Globorotalia trinidadensis (Bolli, 1957)

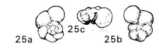

5a   5b   5c

Globorotalia schachdagica (Morozova, 1959)

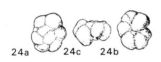

13   14   15

Globigerina trinidadensis (Gohrbandt, 1963)

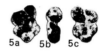

25a   25c   25b

Globorotalia praecursoria (Luterbacher, 1964)
Globoconusa daubjergensis / *Acarinina indolensis*
Zone (Dn2 III) Tarkhankut, NW Crimea

24a   24c   24b

Globorotalia schachdagica
(Luterbacher, 1964)
Middle part of G. inconstans Zone
western Turkmenia, USSR

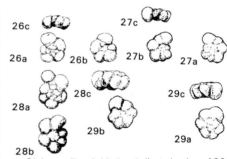

26c   27c

26a   26b   27b   27a

28c   29c

28a   29b   29a

28b

Globorotalia trinidadensis (Luterbacher, 1964)
26, 27: topotypes, type sample of *G. trinidadensis*
Zone (TLL 192632), Lower Lizard Springs
Fm., Trinidad; 28, 29: Gubbio section, level G-86

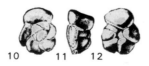

10   11   12

transition *G. uncinata* − *G. angulata* (Bolli, 1957)

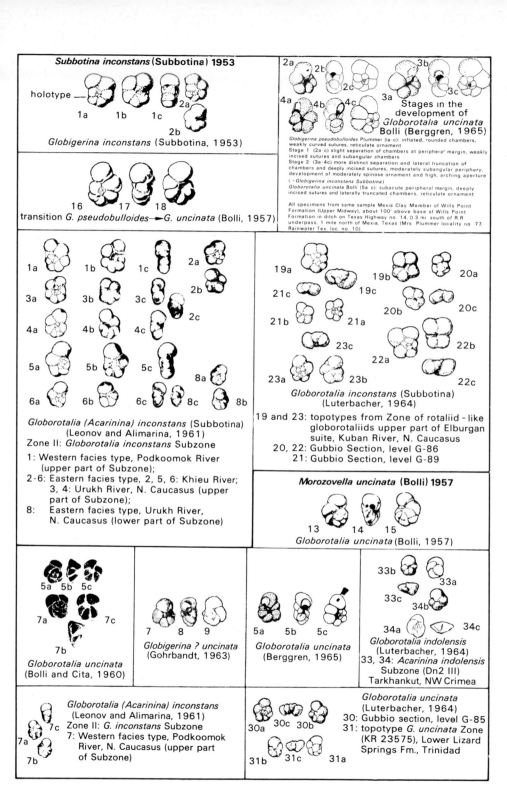

Subbotina inconstans (Subbotina) **1953**

holotype —

1a 1b 1c 2a 2b

Globigerina inconstans (Subbotina, 1953)

16 17 18

transition *G. pseudobulloides* → *G. uncinata* (Bolli, 1957)

2a 2b 2c 3b 3c 3a

4a 4b 4c

Stages in the development of *Globorotalia uncinata* Bolli (Berggren, 1965)

Globigerina pseudobulloides Plummer (1a-c): inflated, rounded chambers, weakly curved sutures, reticulate ornament
Stage 1 (2a-c): slight separation of chambers at peripheral margin, weakly incised sutures and subangular chambers
Stage 2: (3a-4c) more distinct separation and lateral truncation of chambers and deeply incised sutures, moderately subangular periphery, development of moderately spinose ornament and high, arching aperture
(= *Globigerina inconstans* Subbotina)
Globorotalia uncinata Bolli (5a-c): subacute peripheral margin, deeply incised sutures and laterally truncated chambers, reticulate ornament

All specimens from same sample Mexia Clay Member of Wills Point Formation (Upper Midway), about 100' above base of Wills Point Formation in ditch on Texas Highway no. 14, 0.3 mi south of R.R underpass, 1 mile north of Mexia, Texas (Mrs Plummer locality no 77, Rainwater Tex. loc. no. 10)

1a 1b 1c 2a 2b 2c

3a 3b 3c

4a 4b 4c

5a 5b 5c

6a 6b 6c 8c 8a 8b

Globorotalia (Acarinina) inconstans (Subbotina)
(Leonov and Alimarina, 1961)
Zone II: *Globorotalia inconstans* Subzone

1: Western facies type, Podkoomok River (upper part of Subzone);
2-6: Eastern facies type, 2, 5, 6: Khieu River; 3, 4: Urukh River, N. Caucasus (upper part of Subzone);
8: Eastern facies type, Urukh River, N. Caucasus (lower part of Subzone)

19a 19b 20a

21c 19c

21b 21a 20b 20c

23c 22b

22a

23a 23b 22c

Globorotalia inconstans (Subbotina)
(Luterbacher, 1964)
19 and 23: topotypes from Zone of rotaliid-like globorotaliids upper part of Elburgan suite, Kuban River, N. Caucasus
20, 22: Gubbio Section, level G-86
21: Gubbio Section, level G-89

Morozovella uncinata (Bolli) **1957**

13 14 15

Globorotalia uncinata (Bolli, 1957)

5a 5b 5c

7a 7c

7b

Globorotalia uncinata
(Bolli and Cita, 1960)

7 8 9

Globigerina ? uncinata
(Gohrbandt, 1963)

5a 5b 5c

Globorotalia uncinata
(Berggren, 1965)

33b 33a

33c 34b

34a 34c

Globorotalia indolensis
(Luterbacher, 1964)
33, 34: *Acarinina indolensis* Subzone (Dn2 III) Tarkhankut, NW Crimea

7c

7a

7b

Globorotalia (Acarinina) inconstans
(Leonov and Alimarina, 1961)
Zone II: *G. inconstans* Subzone
7: Western facies type, Podkoomok River, N. Caucasus (upper part of Subzone)

30a 30c 30b

31b 31c 31a

Globorotalia uncinata
(Luterbacher, 1964)
30: Gubbio section, level G-85
31: topotype *G. uncinata* Zone (KR 23575), Lower Lizard Springs Fm., Trinidad

Planorotalites compressa (Plummer) 1926

II

11a 11b 11c

Globigerina compressa (Plummer) 1926

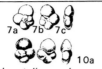

7a 7b 7c

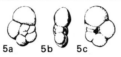

10a

Globorotalia membranacea
(Subbotina, 1953)

2a 2c 2b

Globigerina compressa (Plummer) var.
caucasica Khalilov, 1956

5a 5b 5c

Globigerina compressa
(Troelsen, 1957)

3a 3b 3c

Globorotalia compressa
(Bolli and Cita, 1960)

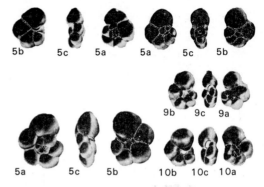

5b 5c 5a 5a 5c 5b

9b 9c 9a

5a 5c 5b 10b 10c 10a

Globorotalia compressa (Loeblich and Tappan, 1957)

7 8 9

Globorotalia compressa (Gohrbandt, 1963)

Planorotalites ehrenbergi (Bolli) 1957

18 19 20

Globorotalia ehrenbergi Bolli, 1957

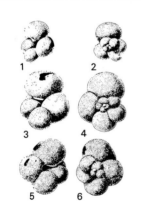

1 2

3 4

5 6

Intraspecific variation in *Globorotalia*
(Turborotalia) compressa (Plummer)

5a 5b 5c

(Berggren, 1962)

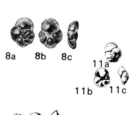

8a 8b 8c

11a

11b 11c

9a 9b 9c 12a 12b 12c

Globorotalia membranacea
(Subbotina, 1953)

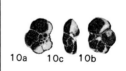

10a 10c 10b

Globorotalia pseudomenardii
(Loeblich and Tappan, 1957)

12 11 10

Globorotalia haunsbergensis
(Gohrbandt, 1963)

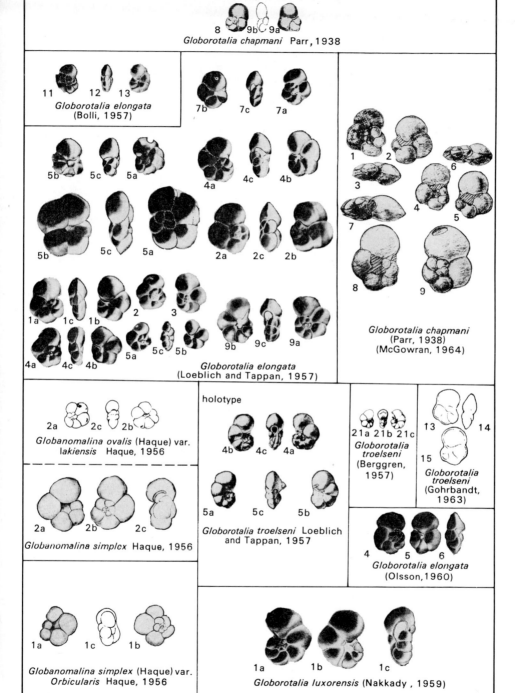

Planorotalites chapmani (Parr) 1938

8 9b 9a
Globorotalia chapmani Parr, 1938

11 12 13
Globorotalia elongata
(Bolli, 1957)

7b 7c 7a

5b 5c 5a
4a 4c 4b

5b 5c 5a
2a 2c 2b

1a 1c 1b 2 3
5a 5c 5b 9b 9c 9a
4a 4c 4b
Globorotalia elongata
(Loeblich and Tappan, 1957)

1 2 6
3
4 5
7
8 9

Globorotalia chapmani
(Parr, 1938)
(McGowran, 1964)

2a 2c 2b
Globanomalina ovalis (Haque) var.
lakiensis Haque, 1956

2a 2b 2c
Globanomalina simplex Haque, 1956

holotype

4b 4c 4a

5a 5c 5b
Globorotalia troelseni Loeblich
and Tappan, 1957

21a 21b 21c
Globorotalia troelseni
(Berggren, 1957)

13 14
15
Globorotalia troelseni
(Gohrbandt, 1963)

4 5 6
Globorotalia elongata
(Olsson, 1960)

1a 1c 1b
Globanomalina simplex (Haque) var.
Orbicularis Haque, 1956

1a 1b 1c
Globorotalia luxorensis (Nakkady, 1959)

Planorotalites pseudomenardii (Bolli) 1957

holotype ——

14 15 16

17

Globorotalia pseudomenardii Bolli, 1957

13a 13b 13c

Globorotalia membranacea (Subbotina, 1953)

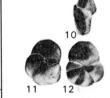

10

11 12

Globorotalia pseudomenardii
(Olsson, 1960)

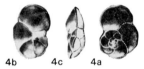

4b   4c   4a   3b   3c   3a

5

Globorotalia pseudomenardii
(Gartner and Hay, 1962)

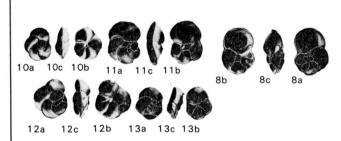

10a 10c 10b   11a 11c 11b

8b   8c   8a

12a 12c 12b   13a 13c 13b

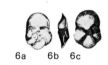

6a   6b   6c

Globorotalia membranacea
(Aubert, 1962)

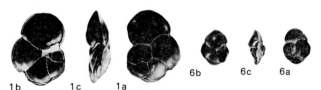

1b   1c   1a   6b   6c   6a

Globorotalia pseudomenardii
(Loeblich and Tappan, 1957)

16 17 18

Globorotalia pseudomenardii
(Gohrbandt, 1963)

Planorotalites planoconica (Subbotina) 1953

holotype ——

4a 4b 4c 5a 5b 5c 6a 6b 6c

Globorotalia planoconica Subbotina, 1953

Planorotalites imitata (Subbotina) 1953

16a 16b 16c

holotype ——

14a 14b 14c 15a 15b 15c

Globorotalia imitata Subbotina, 1953

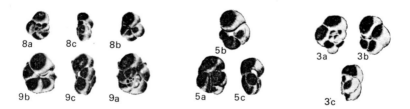

3b   3c   3a   6a   6c   6b

8a   8c   8b   5b   3a   3b

9b   9c   9a   5a   5c   3c

Globorotalia imitata (Loeblich and Tappan, 1957)

Subbotina spiralis (Bolli) 1957

16   17   18

Globigerina spiralis (Bolli, 1957)

Morozovella broedermanni
(Cushman and Bermudez) 1949   III

22   23   24

Globorotalia (Truncorotalia) broedermanni
Cushman and Bermúdez, 1949

2a   2b   2c   10   11   12

Globigerina spiralis
(Bolli and Cita, 1960)

Globigerina spiralis
(Gohrbandt, 1965)

13   14   15

Globorotalia broedermanni (Bolli, 1957)

Morozovella pusilla Bolli, 1957

8   9   10

3a

3c

4a   4b   4c

3b

Planorotalites tauricus
Morozova, 1961

Globorotalia pusilla pusilla
(Bolli and Cita, 1960)

6a   6b   6c

Pseudogloborotalia broedermanni (Bermudez, 1961)

Morozovella lodoensis (Mallory) 1959

3a   3b   3c

Globorotalia broedermanni (Cushman and Bermúdez)
var. *lodoensis* (Mallory, 1959)

Globorotalia albeari (Cushman and Renz) 1949

13   14   15

13b   13c

13a

Globorotalia broedermanni (Bolli, 1957)

3a   3b   3c

5a   5b   5c

2b   2c   2a

6a   6c   6b

Globorotalia pseudoscitula
(Loeblich and Tappan, 1957)

6a   6b   6c

Globorotalia pusilla laevigata
(Bolli and Cita, 1960)

19   20

21

Globorotalia pusilla aff.
laevigata (Gohrbandt, 1963)

5   6   7

Globorotalia pusilla Bolli
subsp. *laevigata*
Bolli, 1957

Planorotalites pseudoscitula (Glaessner) 1937

tf.3a tf.3b tf.3c

Globorotalia pseudoscitula Glaessner, 1937

tf. 3d tf. 3e tf. 3f

Globorotalia pseudoscitula (Glaessner) var. *elongata* Glaessner, 1937

18a 18b 18c

1a 1b 1c

Globorotalia pseudoscitula (Subbotina, 1953)

5a 5b 5c

Globorotalia pseudoscitula (Shutskaya, 1956)

3a 3c 3b

Globorotalia renzi Bolli, 1957

3a 3b 3c

Globorotalia pseudoscitula (Pokorný, 1960)

16a

16b

Globorotalia renzi (Berggren, 1960)

Morozovella convexa (Subbotina) 1953

2a 2b 2c 3a 3b 3c

Globorotalia convexa Subbotina, 1953

7b 7c 7a

6a 6c 6b

7a 7c 7b 8a 8c 8b

5a 5c 5b 6b 6c 6a 4c 4b 4a 4a 4c 4b

Globorotalia convexa (Loeblich and Tappan, 1957)

13 14 15

Globorotalia convexa (Olsson, 1960)

2a 2b 2c

Turborotalia (Acarinina) convexa (Porkorný 1960)

4a 4b 4c

Globorotalia convexa (Gartner and Hay, 1962)

4 5 6 12 11 10

Globorotalia ? convexa (Gohrbandt, 1963)

Morozovella aequa (Cushman and Renz) 1942

3a   3c   3b

Globorotalia crassata (Cushman) var. *aequa*
(Cushman and Renz, 1942)

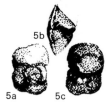

5b

5a   5c

Globorotalia aequa   (Aubert, 1962)

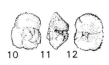

10   11   12

Truncorotalia aequa (Gohrbandt, 1963)

5a   5b   5c

Globorotalia praenartanensis
(Shutskaya, 1956)

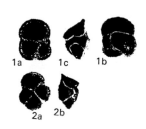

1a   1c   1b

2a   2b

Globorotalia aequa
(Gartner and Hay, 1962)

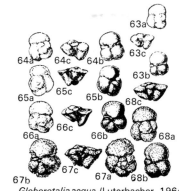

63a

64a   64c   64b   63c

63b

65a   65c   65b   68c

66c

66a   66b   68a

67c   67a   68b

67b

Globorotalia aequa (Luterbacher, 1964)
61, 62: Glubbio section, level G-81
63, 66: Topotypes, Soldado Formation, Soldado Rock, Trinidad
67, 68: *G. velascoensis* Zone, Velasco Fm., Ebano, E. Mexico

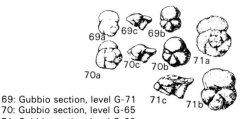

69a   69c   69b

70a   70c   70b   71a

71c   71b

69: Gubbio section, level G-71
70: Gubbio section, level G-65
71: Gubbio section, level G-60

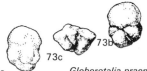

73b

73c

73a

Globorotalia praenartanensis
(Luterbacher, 1964)
73: Topotype, *Acarinina acarinata* Zone
Kuban River, N. Caucasus

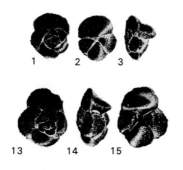

1 2 3

13 14 15

Globorotalia aequa (Bolli, 1957)

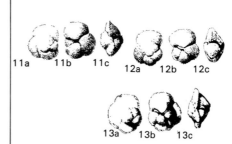

11a 11b 11c 12a 12b 12c

13a 13b 13c

Globorotalia crassata (Subbotina, 1953)

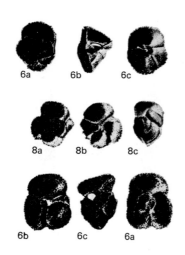

6a 6b 6c

8a 8b 8c

6b 6c 6a

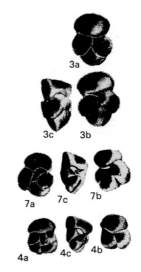

3a

3c 3b

7a 7c 7b

4a 4c 4b

Globorotalia aequa (Loeblich and Tappan, 1957)

1 2

Globorotalia crassata aequa (Hornibrook, 1958)

Morozovella angulata (White) 1927

 13a   13b   13c

Globigerina angulata White,1927

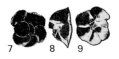
7 8 9

 2a   2b   2c

Globorotalia angulata (Bolli,1957)

Globorotalia angulata (Shutskaya,1956)

 6a   6b   6c

Globorotalia angulata
(Gartner and Hay,1962)

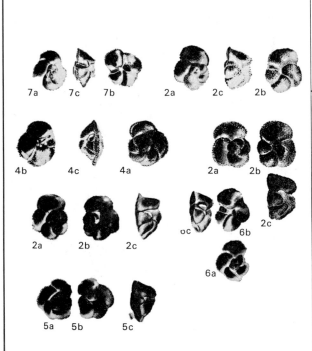

7a 7c 7b 2a 2c 2b

4b 4c 4a 2a 2b

2a 2b 2c oc 6b 2c

6a

5a 5b 5c

Globorotalia angulata
(Loeblich and Tappan, 1957)

 3a   3b 3c

Globorotalia angulata
Nakkady and Talaat, 1959

4 5 6

Truncorotalia angulata
(Gohrbandt, 1963)

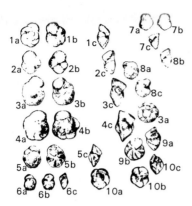

Globorotalia (Truncorotalia) angulata
(Leonov and Alimarina, 1961)
Zone II: *G. angulata* Subzone
Eastern facies type, Urukh River, N. Caucasus
1, 2, 4, 6: upper part of subzone
3, 5, 7-10: lower part of subzone

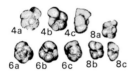

Globorotalia (Acarinina) conicotruncata
(Leonov and Alimarina, 1961)
Zone III: *G. tadjikistanensis* Subzone
4, 6, 8: Eastern facies type; Urukh River,
lower part of subzone

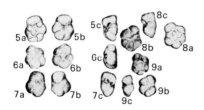

Globorotalia (Acarinina) conicotruncata
(Leonov and Alimarina, 1961)
Zone II: *G. angulata* Subzone
Eastern facies type: 5, 8-9, upper part of
subzone, 6, 7 middle part of subzone

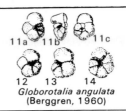

Globorotalia angulata
(Berggren, 1960)

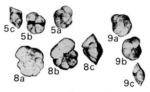

Globorotalia (Truncorotalia) angulata
(Leonov and Alimarina, 1961)
Zone III: *G. tadjikistanensis* Subzone
5: transitional form to *G. (Truncorotalia)* aff. *crassata*
Cushman; eastern facies type, Urukh River (upper
part of subzone)
8: Eastern facies type, Urukh River (middle part of
subzone)
9: transitional form to *G. (Truncorotalia)
marginodentata* Subbotina, eastern facies type
Urukh River (upper part of subzone)

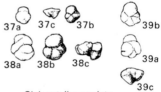

Globorotalia angulata
(Luterbacher, 1964)
37: Gubbio section, level G-78
38: Gubbio section, level G-81
39: Gubbio section, level G-82

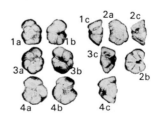

Globorotalia (Truncorotalia) angulata
(Leonov and Alimarina, 1961)
Zone II: *Globorotalia angulata* Subzone
1-4: Eastern facies type, Urukh River,
N. Caucasus (1, 2: lower part of subzone;
3, 4 upper part of subzone)

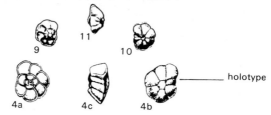

Morozovella conicotruncata (Subbotina) 1947

9   11   10

4a   4c   4b   —— holotype

Globorotalia conicotruncata Subbotina, 1947

4a   4b   4c

Globorotalia angulata var. *kubanensis*
Shutskaya, 1956

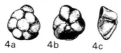

4   5   6

Globorotalia angulata abundocamerata
Bolli, 1957

pl.48, fig. 1a   pl.48, fig. 1c   pl.48, fig. 1b

pl.55, fig. 1b   pl.55, fig. 1c   pl.55, fig. 1a

4a   4c   4b

1b   1c   1a

2b   2c   2a

5a   5c   5b

Globorotalia trichotrocha
Loeblich and Tappan, 1957

pl.59, fig. 1a   pl.59, fig. 1c   pl.59, fig. 1b

Globorotalia apanthesma
Loeblich and Tappan, 1957

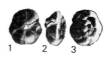

1   2   3

Globorotalia trichotrocha
(Olsson, 1960)

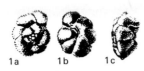

Globorotalia hispidicidaris
Loeblich and Tappan, 1957

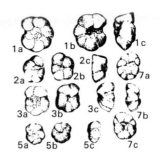

Globorotalia (Acarinina) conicotruncata
(Leonov and Alimarina, 1961)
        Zone III: *G. tadjikistanensis* Subzone
1, 3, 5, 7: Eastern facies type, Urukh River
1, 3, 5:   lower part of subzone
7:           upper part of subzone
2:           Eastern facies type Khieu River,
            lower part of subzone)

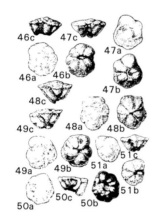

Globorotalia conicotruncata
(Luterbacher, 1964)
46, 48, 50: Gubbio section, level G-78
47:       Gubbio section, level G-77
51:       Gubbio section, level G-79

Globorotalia conicotruncata
(Luterbacher, 1964)
40: Zone of rotaliid-like globorotaliids,
     Khieu River, N. Caucasus

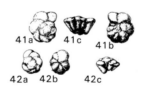

Globorotalia angulata abundocamerata
(Luterbacher, 1964)
41, 42: topotypes, G. pusilla pusilla Zone
(TLL 232705), Lower Lizard Springs Fm., Trinidad

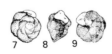

Truncorotalia angulata abundocamerata
(Gohrbandt, 1963)

Globorotalia kubanensis
(Luterbacher, 1964)
42: topotype, Acarinina
     conicotruncata Zone
     Kuban River, N. Caucasus

Morozovella velascoensis (Cushman) 1925

5a   5b   5c

Pulvinulina velascoensis Cushman, 1925

1   2   3   4

Globorotalia velascoensis (Bolli, 1957)

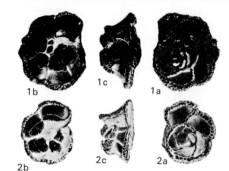

1b   1c   1a

2b   2c   2a

Globorotalia velascoensis
(Loeblich and Tappan, 1957)

11a   11b

Pseudogloborotalia velascoensis
(Bermúdez, 1961)

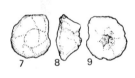

7   8   9

Truncorotalia velascoensis
velascoensis (Gohrbandt, 1963)

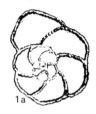 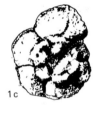 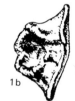

1a   1c   1b

Globorotalia velascoensis (Aubert, 1962)

98b   98a   98c

99b   99a   99c

100c   100a   100b

Globorotalia velascoensis (Cushman)
(Luterbacher, 1964)
98: *G. velascoensis* Zone, Velasco Formation,
Ebano, Eastern Mexico
99: Gubbio section, level G-74
100: *G. aff. velascoensis*; Gubbio section,
level G-71

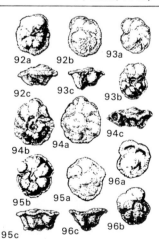

92a   92b   93a

92c   93c   93b

94c

94b   94a

95b   95a   96a

95c   96c   96b

Globorotalia velascoensis (Cushman)
92-94: *G. velacoensis* Zone, Velasco Formation,
Ebano, Eastern Mexico
95-96: *G. aff. velascoensis*; G velascoensis Zone,
Valesco Formation, Ebano, Eastern Mexico

VI

Morozovella acuta (Toulmin) 1941

6   7   8

Globorotalia wilcoxensis Cushman and Ponton var. *acuta* Toulmin, 1941

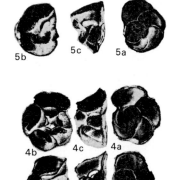

5b   5c   5a

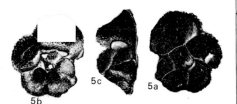

4b   4c   4a

5b   5c   5a

5c   5a

5b

Globorotalia acuta
(Loeblich and Tappan, 1957)

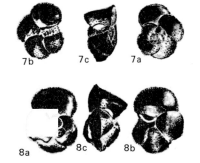

7b   7c   7a

8a   8c   8b

Globorotalia aequa
(Loeblich and Tappan, 1957)

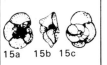

15a   15b   15c

Globorotalia acuta
(Berggren, 1960)

15   14   13

*Truncorotalia velascoensis
acuta* (Gohrbandt, 1963)

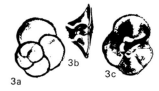

3a   3b   3c

Globorotalia acuta
(Aubert, 1962)

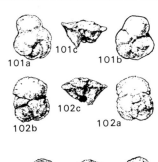

101a   101c   101b

102b   102c   102a

103a   103c   103b

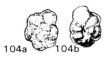

104a   104b

Globorotalia acuta Toulmin
(Luterbacher 1964)
101: *G. pseudomenardii* Zone, El Quss
Abu Said, Farafrah Oasis, Egypt
102,104: *G. velascoensis* Zone, Velasco Formation,
Ebano, Eastern Mexico

VII

Morozovella occlusa (Loeblich and Tappan) 1957

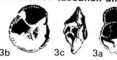

3b   3c   3a

3a   3c   3b

Globorotalia occlusa Loeblich and Tappan, 1957

7   8   9

***Globorotalia
crosswicksensis***
Olsson, 1960

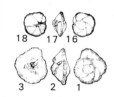

18   17   16

3   2   1

*Truncorotalia velascoensis
occlusa* (Gohrbandt,
1963)

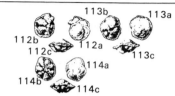

113b   113a
112b   112a
112c   113c
114a
114b   114c

Globorotalia occlusa (Luterbacher, 1964)
112, 113: *G. velascoensis* Zone, Velasco Fm.,
Ebano, Eastern Mexico
114: Gubbio section, level G-74

Morozovella pasionensis (Bermúdez, 1961)

8a

8b

Pseudogloborotalia pasionensis Bermúdez, 1961

Morozovella simulatilis (Schwager, 1883)

15a   15c   15b

Discorbina simulatilis Schwager 1883

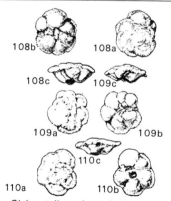

108b   108a
108c   109c
109a   109b
110c
110a   110b

Globorotalia pasionensis (Bermúdez)
(Luterbacher, 1964)
108: topotype Lower Eocene, Rio de la Pasion,
El Petén, Guatemala
109: Gubbio section, level G-71 – 110:
Gubbio section, level G-70a

54a 53a 53c 53b
55a 55c 55b
54c 56c 56b
54b 57c 57a 59a 56a
57b 59c 59b
58c 60c
58a 58b 60a 60b

Globorotalia simulatilis (Schwager)
(Luterbacher, 1964)
53-55: *G. pseudomenardii* Zone,
El Quss Abu Said, Farafrah Oasis, Egypt
56: Gubbio section, level G-81
57, 59: Gubbio section, level G-78
58: Gubbio section, level G-82
60: *G.* sp. aff. *simulatilis*, Gubbio section, G-78

111a   111c   111b

Globorotalia sp. aff. *pasionensis* (Bermúdez)
(Luterbacher, 1964)
111: *G. velascoensis* Zone, Velasco Formation,
Ebano, Eastern Mexico

Morozovella parva Rey, 1955

14a   14c   14b

Truncatulina colligera
Schwager, 1883

holotype

14a

Globorotalia velascoensis
(Cushman) var. *parva*
Rey, 1955

?holotype

91a   91c   91b

Globorotalia velascoensis parva
(Luterbacher, 1964)
Koudiat bou Khelif, Northern Morocco

1   2   3

Globorotalia simulatilis (Le Roy, 1953)

1a   1c   1b

Globorotalia rex (Loeblich and Tappan, 1957)

5a   5b   5c

Globorotalia velascoensis parva
(Gartner and Hay, 1962)

2a   2b   2c

*Globorotalia velascoensis
parva* (Aubert, 1962)

4   5   6

Truncorotalia velascoensis parva
(Gohrbandt, 1963)

Morozovella tadjikistanensis (Bykova, 1953)

5a   5b   5c

Globorotalia tadjikistanensis (Bykova, 1953)

Morozovella kolchidica (Mocozova, 1961)

2a   2b   2c

Globorotalia kolchidica (Morozova, 1961)

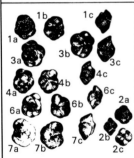

1b   1c
1a
3a   3b
   3c
   4c
4a   4b   6c   2a
6a   6b
      7c   2b
7a   7b   2c

Globorotalia (Truncorotalia) tadjikistanensis
(Leonov and Alimarina, 1961)
Zone III: *G. tadjikistanensis* Subzone
1-4, 6, 7: Eastern facies type, Urukh River
1, 2, 7: upper part of subzone;
3, 4, 6: lower part of subzone, N. Caucasus

52a   52c   52b

Globorotalia tadjikistanensis
(Luterbacher, 1964)
52: Gubbio section, level G-79

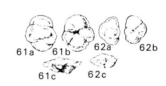

61a   61b   62a   62b
61c   62c

Globorotalia sp. aff. *kolchidica*
(Luterbacher, 1964)
61, 62: Gubbio section, level G-81

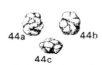

44a   44b
44c

Globorotalia tadjikistanensis djanensis
44: *Acarinina tadjikistanensis djanensis* Zone,
eastern Caucasus

Morozovella acutispira
Bolli and Cita, 1960

72c

72b   72a

Globorotalia acutispira
(Luterbacher, 1964)
72: holotype, *G. pseudomenardii* Zone,
Paderno d'Addo, Northern Italy

Morozovella subbotinae (Morozova) 1939

VIII

16   17

Globorotalia subbotinae Morozova, 1939

Globorotalia rex (Martin, 1943)

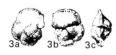
3a   3b   4

Globorotalia subbotinae (Shutskaya, 1956)

10   11   12

Globorotalia rex (Bolli, 1957)

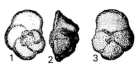 
3a   3b   3c

Globorotalia marginodentata (Subbotina, 1953)

1   2   3

Truncorotalia cf. rex (Gohrbandt, 1963)

Morozovella gracilis (Bolli, 1957)

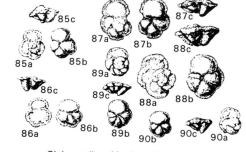
85c   87c
85a   85b   87a   87b   88c
86c   89a   89c   88b
86a   86b   89b   90b   90c   90a
88a

Globorotalia subbotinae Morozova
(Luterbacher, 1964)
85: Gubbio section, level G-73;
86: Gubbio section, level G-71;
87: *G. subbotinae* Zone, E. Caucasus;
88: Gubbio section, level G-58;
89: Gubbio section, level G-71;
90: Gubbio section, level G-73;

4   5   6

Globorotalia formosa Bolli, subsp. *gracilis* Bolli, 1957

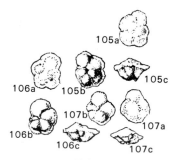
105a
105c
106a   105b
106b   107b   107a
106c   107c

Globorotalia sp. aff. *formosa gracilis* Bolli
(Luterbacher, 1964)
105: *G. velascoensis* Zone, Velasco Formation,
Ebano, Eastern Mexico
106, 107: Gubbio section, level G-74

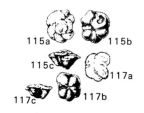
115a   115b
115c
117a
117c   117b

Globorotalia formosa gracilis Bolli
(Luterbacher, 1964)
115: topotype, *G. rex* Zone (TLL 232994)
Upper Lizard Springs Formation,
Trinidad
117: Gubbio section, level G-58

Morozovella marginodentata (Subbotina) 1953

original
(after Glaessner)

14a 14b 14c

holotype
(Subbotina, 1953)

15a 15b 15c

16a 16b 16c

1a 1b 1c 2a 2b 2c

Globorotalia marginodentata (Subbotina, 1953)

4 5 6

Truncorotalia marginodentata marginodentata
(Gohrbandt, 1963)

10 11 12

13 14 15

Truncorotalia marginodentata
apertura (Gohrbandt, 1963)

75a 75b 76a

75c 77c

76b

77b 77a

76c

79a

78b 78a

79b 79c 78c

Globorotalia marginodentata Subbotina
(Luterbacher, 1964)
75, 76: Zone of compressed globorotaliids,
Khieu River, N. Caucasus
77, 78: G. velascoensis Zone, Velasco
Formation, Ebano, Mexico
79: Gubbio section, level G-75

80a

80b

80c 81c

81b

82a 82c

82b 81a

83c

83a 84b

83b 84a

84c

Globorotalia marginodentata (Luterbacher, 1964)
80: Gubbio section level G-74
81, 82: Gubbio section level G-71
83: Gubbio section level G-71
84: Gubbio section level G-58

118a 118b 119a
119c

118c

119b

120b 120a 120c

Globorotalia formosa formosa Bolli
(Luterbacher. 1964)
118: topotype, *G. formosa formosa* Zone,
Upper Lizard Springs Formation, Trinidad
119: Gubbio section, level G-70
120: Gubbio section, level G-52

Morozovella formosa (Bolli, 1957)

1 2 3

Globorotalia formosa Bolli subsp.
formosa Bolli, 1957

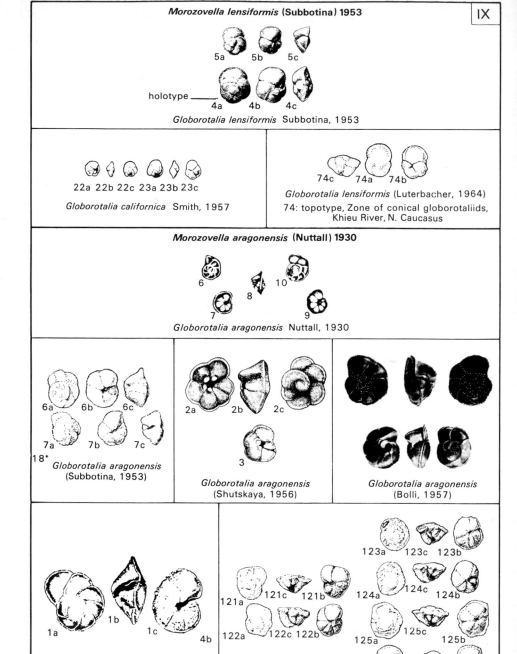

IX

Morozovella lensiformis (Subbotina) 1953

5a 5b 5c

holotype ——— 4a 4b 4c

Globorotalia lensiformis Subbotina, 1953

22a 22b 22c 23a 23b 23c

Globorotalia californica Smith, 1957

74c 74a 74b

Globorotalia lensiformis (Luterbacher, 1964)
74: topotype, Zone of conical globorotaliids,
Khieu River, N. Caucasus

Morozovella aragonensis (Nuttall) 1930

6 8 10

7 9

Globorotalia aragonensis Nuttall, 1930

6a 6b 6c

7a 7b 7c

18* *Globorotalia aragonensis*
(Subbotina, 1953)

2a 2b 2c

3

Globorotalia aragonensis
(Shutskaya, 1956)

Globorotalia aragonensis
(Bolli, 1957)

1a

1b

1c

4b

Globorotalia aragonensis
(Aubert, 1962)

123a 123c 123b

121a 121c 121b 124a 124c 124b

122a 122c 122b 125a 125c 125b

126a 126c 126b

Globorotalia aragonensis
(Luterbacher, 1964)
121: Gubbio Section, level G-52 *Globorotalia aragonensis*
122: Gubbio Section, level G-49 (Nuttall, 1930)

Morozovella caucasica (Glaessner) **1937**

6a   6c   6b

Globorotalia aragonensis (Nuttall) var. *caucasica*   Glaessner, 1937

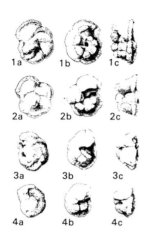

1a   1b   1c

2a   2b   2c

3a   3b   3c

4a   4b   4c

Globorotalia aragonensis (Subbotina, 1953)

1a   1b   1c

Globorotalia aragonensis caucasica
(Shutskaya, 1956)

3   4   5

Globorotalia crater (Hornibrook, 1958)

1a   1b   1c

Globorotalia aragonensis   Nuttall   var.
twisselmanni   Mallory, 1959

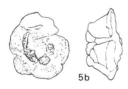

5b

Globorotalia crater (Pokorný, 1960)

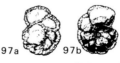

97a   97b

97c

Globorotalia caucasica
(Luterbacher, 1964)

97: Zone of conical globorotaliids,
Khieu River, N. Caucasus

Morozovella lehneri (Cushman and Jarvis) 1929

16a   16b   16c

Globorotalia lehneri Cushman and Jarvis, 1929

9

Pseudogloborotalia lehneri (Bermúdez, 1961)

11a   11b   11c

Globorotalia lehneri (Saito, 1962)

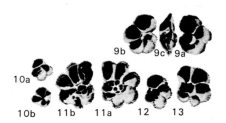

9b   9c 9a

10a

10b   11b   11a   12   13

Globorotalia lehneri (Bolli, 1957)

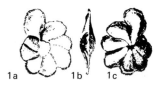

1a   1b   1c

Globorotalia lehneri (Aubert, 1962)

Morozovella crassata (Cushman) 1925

Re-illustration of holotype by Bandy (1964)

Pulvinulina crassata Cushman, 1925

  — holotype

4a   4c   4b

Globorotalia spinulosa (Cushman, 1927)

6a 6c 6b

7a 7c 7b

Globorotalia spinulosa (Bolli, 1957)

2a   2b

Pseudogloborotalia spinulosa (Bermúdez, 1961)

9a   9b   9c

Globorotalia spinulosa (Saito, 1962)

4a

4b

4c

Glolorotalia spinulosa (Aubert, 1962)

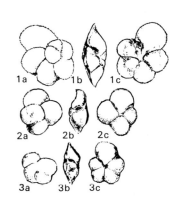

1a   1b   1c

2a   2b   2c

3a   3b   3c

Globorotalia hadii   Aubert, 1962

Acarinina esnaensis (Le Roy) 1953

X

8   10   9

Globigerina esnaensis Le Roy, 1953

15a   15b   15c

16   17   18   19   20   21

Globigerina stonei Weiss, 1955

14a 14b 14c 16a 16b 16c

Acarinina intermedia Subbotina, 1953

10   11   12

Globorotalia whitei (Bolli, 1957)

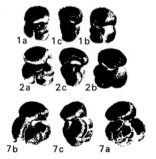

1a   1c   1b

2a   2c   2b

7b   7c   7a

Globorotalia esnaensis
(Loeblich and Tappan, 1957)

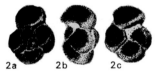

2a   2b   2c

Globigerina esnaensis (Nakkady, 1959)

4a   4b   4c

Globigerina esnaensis (Gartner and Hay, 1962)

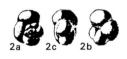

2a   2c   2b

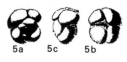

5a   5c   5b

Globorotalia irrorata
(Loeblich and Tappan, 1957)

19   20   21

Globigerina esnaensis (Gohrbandt, 1963)

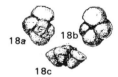

18a   18b

18c

Globorotalia (Luterbacher, 1964)

Acarinina nitida (Martin) 1943

1a   1b   1c

Globigerina nitida (Martin, 1943)

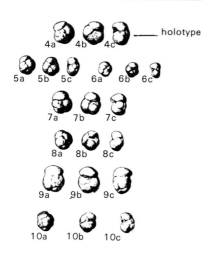

4a   4b   4c ——— holotype

5a   5b   5c   6a   6b   6c

7a   7b   7c

8a   8b   8c

9a   9b   9c

10a   10b   10c

Acarinina acarinata Subbotina, 1953

5a   5b   5c

Globigerina nitida (Mallary, 1959)

13   14   15

Turborotalia acarinata (Gohrbandt, 1963)

Acarinina pseudotopilensis Subbotina, 1953

8a   8b   8c ——— holotype

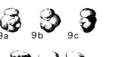

9a   9b   9c

1a   1b   1c

2a   2b   2c

3a   3b   3c

Acarinina pseudotopilensis (Subbotina, 1953)

2a   2c   2b

Globorotalia pseudotopilensis
(Loeblich and Tappan, 1957)

19   20   21

Globorotalia tortiva Bolli, 1957, new name

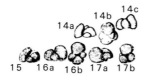

14a   14b   14c

15   16a   16b   17a   17b

Globorotalia pseudotopilensis (Reyment, 1959)

4a

4b

Turborotalia (Acarinina)
pseudotopilensis
(Pokorný, 1960)

13   14   15

Turborotalia
pseudotopilensis
(Gohrbandt, 1963)

Acarinina wilcoxensis (Cushman and Ponton)1932

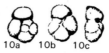

10a   10b   10c

Globorotalia wilcoxensis Cushman and Ponton, 1932

1   3   2

Globorotalia whitei Weiss, 1955

7   8   9

Globorotalia wilcoxensis (Bolli, 1957)

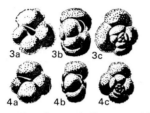

3a   3b   3c

4a   4b   4c

Globorotalia wilcoxensis (Berggren, 1960)

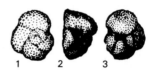

1   2   3

Turborotalia ? wilcoxensis (Gohrbandt, 1963)

Acarinina quetra (Bolli) 1957

1   2   3   ——— holotype

Globorotalia quetra Bolli, 1957

5a

5b   5c

Acarinina crassaformis (Subbotina, 1953)

Acarinina coalingensis (Cushman and Hanna) 1927

4a   4b

re-illustration
of holotype
(from Gohrbandt,
unpublished)

Globigerina coalingensis Cushman and Hanna, 1927 , holotype 2548

holotype —

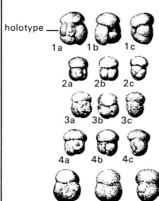

1a   1b   1c

2a   2b   2c

3a   3b   3c

4a   4b   4c

5a   5b   5c

Acarinina triplex Subbotina, 1953

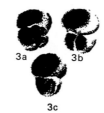

3a   3b

3c

Globigerina inaequispira
(Loeblich and Tappan, 1957)

6a   6b   6c

*Globigerina
coalingensis*
(Mallory, 1959)

4a   4c   4b

Globigerina cf. *soldadoensis*
(Loeblich and Tappan, 1957)

19   20

21

Turborotalia primitiva
(Gohrbandt, 1963)

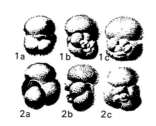

1a   1b   1c

2a   2b   2c

Globigerina triplex (Berggren. 1960)

outline drawing of holotype

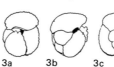

3a   3b   3c

Globoquadrina primitiva Finlay, 1939
(Jenkins, 1965)

Acarinina mckannai (White) 1928

XI

16a 16b 16c

Globigerina mckannai White, 1928

16

17 18
Globorotalia mckannai (Bolli, 1957)

23 24 25

26 27 28
Globigerina subsphaerica
Subbotina, 1947

1a 1b 1c
Globigerina subsphaerica
(Shutskaya, 1956)

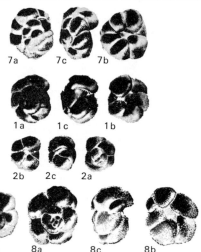
7a 7c 7b

1a 1c 1b

2b 2c 2a

5a

8a 8c 8b

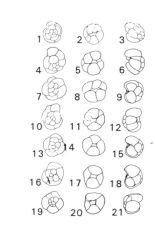
1 2 3

4 5 6

7 8 9

10 11 12

13 14 15

16 17 18

19 20 21

Acarinina subsphaerica (Shutskaya, 1958)

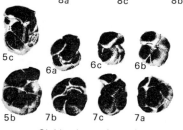
5c

6a 6c 6b

5b 7b 7c 7a

Globigerina mckannai
(Loeblich and Tappan, 1957)

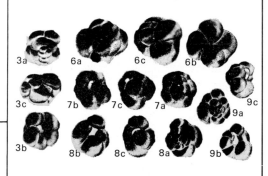
3a 6a 6c 6b

3c 7b 7c 7a 9c

9a

3b 8b 8c 8a 9b

9 10
Globigerina
mckannai
(Berggren, 1960)

1a 1b 1c
Globorotalia mckannai
(Gartner and Hay, 1962)

6 5 4
Globigerina mckannai
(Gohrbandt, 1963)

3a 3c 3b

Globigerina spirallis
(Loeblich and Tappan, 1957)

Acarinina soldadoensis (Brönnimann) 1952

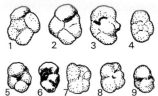

Globigerina soldadoensis Brönnimann, 1952

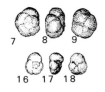

Globigerina soldadoensis (Bolli, 1957)

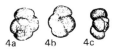

Globorotalia crassa (d'Orbigny) var. pentacamerata Subbotina, 1936 (invalid)

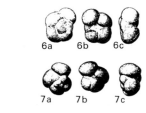

Acarinina interposita Subbotina, 1953

Globigerina soldadoensis (Berggren, 1960)

Globorotalia soldadoensis (Gartner and Hay, 1962)

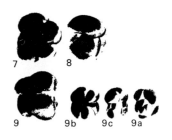

Turborotalia soldadoensis (Gohrbandt, 1963)

Acarinina clara Khalilov, 1956

Transition *A. soldadoensis - A. gravelli* (Bolli, 1957)

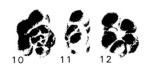

Acarinina angulosa (Bolli) 1957

Globigerina soldadoensis (Brönnimann) subsp. *angulosa* (Bolli, 1957)

8a   8c   8b

Globigerina soldadoensis angulosa (Bolli, 1957)

24   25   26

Globorotalia pentacamerata Subbotina, 1957

Acarinina aspensis (Colom) 1954

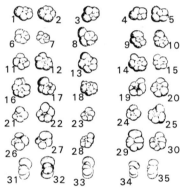

Pl. 3, figs. 1-35

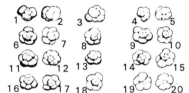

Pl.4, figs. 1-20

Globigerina aspensis Colom, 1954

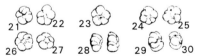

Pl.4, figs. 21-31

Globigerina aspensis (Colom, 1954)

Globorotalia aspensis
(Bolli, 1957)

18a   18c   18b

Acarinina pentacamerata (Subbotina) 1947

holotype (15-17) —
15 17 16

12 14 13

Acarinina pentacamerata (Subbotina) 1947

17 18 16

Globigerina gravelli (Brönnimann, 1952)

2a 2b 2c

Globigerina dubia (Egger) var.
lakiensis (Haque, 1956)

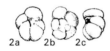
6a 6b 6c

Globorotalia pentacamerata
(Shutskaya, 1956)

1 2 3

Globigerina gravelli
(Bolli, 1957)

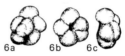
6a 6b 6c

Acarinina pentacamerata (Subbotina) var.
camerata Chalilov, 1956

10 11 12

Globigerina gravelli
(Gohrbandt, 1963)

7a 7b 7c

Acarinina pentacamerata (Subbotina) var.
acceleratoria Chalilov, 1956

Truncorotaloides collactea (Finlay) 1939

holotype —

164   165

Globorotalia collactea Finlay, 1939

1   2   3

holotype re-illustrated
by Jenkins (1-3)

4   5   6

4-15 paratypes

7   8   9

10   11   12

13   14   15

Truncorotaloides collactea
(Finlay) 1939 (Jenkins, 1965)

8a   8c   8b

Globorotalia spinuloinflata (Bolli, 1957)

1a 1b 1c

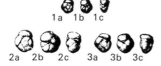

2a   2b   2c   3a   3b   3c

Acarinina rotundimarginata
Subbotina, 1953

19a   19b

*Truncorotaloides
rotundimarginata*
(Berggren 1960)

Acarinina densa (Cushman) 1925
(holotype not illustrated)

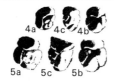

1a   1b   1c

Globigerina spinuloinflata
Bandy, 1949

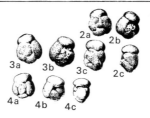

2a   2b

3a   3b   3c   2c

4a   4b   4c

Acarinina crassaformis
(Subbotina, 1953)

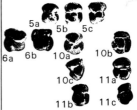

5a   5b   5c

6a   6b   10a   10b

10c   11a

11b   11c

Globorotalia crassata densa
(Saito, 1962)

4a   4c   4b

5a   5c   5b

Globorotalia bullbrooki
(Bolli, 1957)

1a   1b   1c

Turborotalia (Acarinina) cf. *bullbrooki* (Porkorný, 1960)

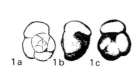

4a   4b

4c

Globorotalia crassata densa
(Aubert, 1962)

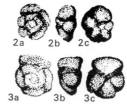

2a   2b   2c

3a   3b   3c

Globorotalia crassata
(Aubert, 1962)

Truncorotaloides topilensis (Cushman) 1925

9a   9b   9c

Globigerina topilensis (Cushman, 1925)

13 14 15   16b   16c 16a

Truncorotaloides topilensis (Bolli, 1957)

5a   5b   5c

Truncorotaloides topilensis
(Aubert, 1962)

5a   5b   5c

Truncorotaloides topilensis
(Saito, 1962)

12

*Truncorotaloides
topilensis*
(Bermúdez, 1961)

7   8   9

Truncorotaloides rohri Brönnimann and Bermúdez, 1953

10   11   12

Truncorotaloides rohri
Bronnimann and Bermúdez var.
mayoensis Brönnimann and
Bermudez, 1953

1   3   2

Truncorotaloides rohri
Bronniman and Bermúdez var.
guaracaraensis Brönnimann and
Bermudez, 1953

4   6   5

Truncorotaloides rohri
Bronnimann and Bermúdez
var. *piparoensis* Brönnimann
and Bermudez, 1953

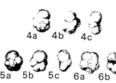

4a   4b   4c

5a   5b   5c   6a   6b

Acarinina rugosoaculeata
(Subbotina, 1953)

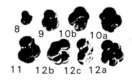

8   9   10b   10a

11   12b   12c 12a

Truncorotaloides rohri
(Bolli, 1957)

3a

3b

Truncorotaloides rohri
(Bermúdez, 1961)

18a   18b

Truncorotaloides rohri
(Berggren, 1960)

5: Mesozoic Planktonic Foraminifera

A world-wide review and analysis

BRUCE A. MASTERS

Amoco Production Company,
Research Center,
P.O. Box 591,
Tulsa, Oklahoma 74102, U.S.A.

With increasing frequency, students of foraminifera are directing their talents toward endeavours other than the more traditional taxonomical or biostratigraphical studies. Yet, few groups of foraminifers have been adequately studied and compared. All studies, including biostratigraphical, are an outgrowth of descriptive taxonomy. It is, therefore, imperative that taxonomy be placed upon as sound a foundation as possible.

To establish such a stable base for the Mesozoic planktonic foraminifers, all taxa were re-evaluated from primary types, topotypes and numerous suites. Less than 24% of the described taxa represent senior synonyms. The most significant reason for this is the failure to allow for and to recognise intraspecific variants. Data obtained from studies of living species have permitted a more realistic interpretation of extinct taxa.

Improved taxonomy has led to a modified phylogenetic proposal, which necessitated the establishment of the new subfamily Ticinellinae. Because of the taxonomic changes, the traditional zonal schemes are no longer applicable. Although no substitute is offered at this time, the stratigraphical ranges, geographical distributions and accompanying descriptions should serve as a basis for the development of an improved scheme.

Outline

Introduction

During the last decade, there has been an extraordinary increase in the study of planktonic foraminifers, particularly those of the Mesozoic. Almost without exception, publications resulting from these studies have been descriptive, biostratigraphical or both. There has been an increase in the number of attempts to recognise phylogenetic relationships, and palaeoecological and palaeogeographical distributions of these organisms. The literature, then, has followed a trend from a purely descriptive phase to an interpretive one.

However, until taxonomic stability among the taxa which have emerged, and are still emerging, from descriptive works is achieved, any interpretations will be, at best, only partially valid. Thus taxonomic evaluation has to be the first step in an interpretative study. The literature is filled with synonymous taxa, and failure to recognise these results in a loss of valuable data upon which all subsequent interpretations must be made.

In order to make a meaningful taxonomic diagnosis, the question of what constitutes a fossil species must be answered. Then, and only then, will it be possible to derive the fullest information from the available data. The major portion of this chapter is concerned with a critical analysis of this problem. This analysis is subjective, and the conclusions derived from it should not be construed as being the final, unalterable solution. As new evidence becomes available, continual reassessment must follow to avoid a self-defeating dogmatic position.

To effectively evaluate the validity of a taxon, it is necessary to examine primary types and topotypic suites. The first is important because too often type descriptions and illustrations are inadequate and misleading. The examination of topotypic suites is equally essential to determine the range of morphologic variability. One should also compare other suites to that of the topotype, so that local environmental influences upon test morphology may be recognised. Finally a comparison must be made with other taxa, after following the same procedure with them.

Unfortunately, the above steps are infrequently followed prior to the introduction of a new species. The potential author of a new species must be thoroughly familiar with the worldwide literature and faunas, if multiple names for a single organism are to be avoided. It is equally critical that he have an understanding of the variability of a species. This variability takes four forms as recognised in the ontogenetic development of the individual, in the variation within a population due to external influences, in the phylogenetic development of the lineage and in genetic differences.

In fossil organisms, none of these variables can be tested directly and the interpretation of each is subjective. Nevertheless, all of these are known to influence test morphology, and must be considered in the determination of what constitutes a species.

Presented here is an attempt to deal with all described Mesozoic planktonic foraminifers. Of the more than 600 species and subspecies, 144 (less than 24%) are judged valid. To document species validity, over 200 primary types (holotypes, cotypes, etc.) were studied. Supplementing these, several hundred topotypes and topotypic suites were analysed in addition to innumerable non-topotypic suites. An attempt has been made to examine all literature in which a new taxon has been described, and in which Mesozoic planktonic foraminifers are illustrated. Papers which included only lists or reports of species' occurrences have not been used. Such references have essentially no value, and can be very misleading. Without illustrations, it is not possible to determine whether the author's identifications and interpretations are correct. The coverage is nearly exhaustive since only approximately one per cent of the known illustrated publications is not available in the United States.

The species concept

Foraminiferologists have usually relied solely upon test morphology in order to define a species. This approach has caused an unwarranted splitting of taxa. From studies of living species, Arnold (1964, 1968), Bé (1965) and Bé and Hemleben (1970), among others, have demonstrated a high degree of test variability within a given species. It seems incongruous that, in a group as large and important as the Foraminiferida, so little is known about the living organism. Yet, even the minimal information available has had a significant impact upon species interpretation.

The test is, and will remain, the primary key to the diagnosis of fossil species. Analytical studies of the test are being conducted along several lines using new instruments (e.g. the scanning electron microscope) to aid in the refinement of species categorisation.

The Linnean system of nomenclature is an attempt to express the natural relationship of one individual to another or group of others. It is a rigid system imposed upon a non-rigid set of organisms. At the very least, such a system can be only partially successful. It does not easily accommodate intra- or inter-specific variability. Sigal (1966b) recognised this deficiency, and proposed the concept of the "spectre".

Our present system of zoological nomenclature, although not wholly satisfactory, is workable. McKerrow (1956) was correct in his observation that "if chaos is to be avoided no drastic changes [in the system] can be tolerated", nor is it "practical to alter the system of nomenclature to correspond to each new conception". What can be done is to be more critical in the analysis of data leading to the establishment of a new species and to apply this same judicious evaluation to previously named species. Reiss (1971) emphasised the need to use all observable characters and features, and that this "whole range of character combinations must be analysed against the background of geographical and stratigraphical distribution, as well as that of knowledge on biology and ecology of living species".

The trinomial

The International Code of Zoological Nomenclature (I.C.Z.N.) serves as a guide to the procedure of establishing a new taxon. As emphasised above, these rules and recommendations are necessary for nomenclatorial stability. Although major deviation is not recommended, non-disruptive modification is certainly in order.

One such modification advocated here is the discontinuance of the trinomen category. The trinomen, or subspecies, implies a close relationship to the species, *sensu stricto*. The difference, although sufficiently distinct to be recognised, is thought not significant enough for the establishment of a new species. It would seem desirable to be able to demonstrate phyletic relationships in such a manner, but very often the purported relationship is not supported by later evidence and interpretation. This can be both confusing and misleading, and clearly impedes the development of a phylogenetic classification.

In the taxa studied here, the majority of the subspecies were of two types. One differs so little from the species, *sensu stricto*, as to be an intraspecific variant. The other departs from the species, *sensu stricto*, so completely as to be a valid new species. In the latter case the subspecies are commonly interpreted as not even being in the same lineage as the restricted species. Because of this situation, the subspecies encountered during this study were either suppressed as intraspecific variants or elevated to specific rank.

It is recommended that the use of the trinomen be discontinued. If two forms are as close morphologically as the trinomen implies, they most likely belong to the same population.

Holotype syndrome

Many workers have a too narrow definition for taxa. This has created a literature replete with synonyms. Neontologists allow great latitude for the morphologic variability of a living taxon. This is frequently not done with fossil organisms, particularly the foraminifers.

This "splitting," based upon minor genetically and non-genetically induced morphologic variation, has two underlying causes: lack of sufficient biological knowledge on the part of palaeontologists and the need for continuous biostratigraphic refinement.

The first is centred around what could be called the "holotype syndrome". According to the Commission on Zoological Nomenclature (Recommendation 73A), to establish a new species one should select a holotype. This specimen is supposed to best represent the new species, and is necessary for taxonomic stability. But what so often happens is that the holotype becomes the *only* guide to that species. Paratypes are usually selected not to show species variability, which may well not be fully known at the time, but to conform as closely as possible to the holotype. This is like saying that to be classified as *Homo sapiens*, all men must look exactly alike. While we all bear major characteristics in common, there exists an almost infinite genetic diversity at a lower level to allow for individual recognition below the specific or subspecific levels. In other words, this is intraspecific variation. It exists for all organisms, a fact frequently not recognised by palaeontologists.

The second cause for "splitting" of taxa has been encouraged, directly or indirectly, by the need to facilitate intraregional, interregional and intercontinental correlation. Butler (1963) noted that the typological species concept is attractive in its simplicity. But this approach has been, to a large degree, self-defeating. The literature is clogged with so many new names that it has become difficult for most workers to keep abreast with them, much less to analyse their validity. As a result much potentially valuable biostratigraphical, environmental, phylogenetical and biogeographical data has been non-retrievable.

To avoid the narrow definition of a species resulting from this syndrome, the author of a new taxon should make explicitly clear the extent of intraspecific variation. If sufficient material or experience is lacking to determine the degree of variability, then the taxon is best left unnamed. When these criteria can be met, then the variation should be described, illustrated and represented in the selected paratypes.

INTRASPECIFIC VARIABILITY

Historically, few workers have attempted to establish variability ranges for a species. Recognition of these minor differences is not always possible in a single area of study. Most new species have not been proposed from monographic or interregional investigations, but from local areas consisting of one or few exposures of strata.

A second hindrance in the recognition of intraspecific variation is a lack of experience on the part of palaeontologists with living organisms. This is not to suggest that the neontologist has only to observe or directly test an organism in order to determine whether he is dealing with a separate species or a variant. Such is not always the case. Subjectivity enters into his analyses as well, but to a lesser degree than those of the palaeontologist. For example, the neontologist cannot bring two organisms or allopatric populations together to test the ability to interbreed without disrupting natural balances. However, he is able to test this potential by controlled laboratory or field studies. This testing will not be conclusive proof that two organisms are or are not the same species, but certainly the degree of subjectivity can be reduced. The palaeontologist does not have this opportunity, but the results of the neontological examination can be used to interpret fossil organisms.

One of the most successful pioneers in the laboratory culturing of foraminifers is Zach M. Arnold. For many years his studies have pointed out the need for a careful re-evaluation of the palaeontological species concept and the recognition of intraspecific variation within a population. Arnold (1964) has demonstrated that certain morphologic characteristics which were once believed to be of significance at the generic level are in fact mere variants of the same species.

Unfortunately this data is very slowly revealed because so few biologists and palaeontologists are working with so few of the living foraminifers. Nevertheless, the key word from these studies is "caution". In fossil organisms, the extent of the morphologic diversity must be evaluated carefully with respect to possible causes for these differences before a new species is proposed. One must look for alternative explanations for morphologic variations.

1. *Genetic variation*

(*a*) *Sexual dimorphism.* The presence of two sexes, the most readily discernible genetic variation of many organisms, has not been recognised in the foraminifers, although Le Calvez (1950) reported indirect evidence for its existence. Under controlled laboratory conditions, he found that only certain individuals of *Glabratella patelliformis* would join together in plastogamy. This at least suggests sexual differentiation in this species, even though it is not apparent in the morphology of the test.

(*b*) *Alternation of generations.* Morphologic dimorphism is known to occur in numerous living and extinct species. This variation results as reproduction changes between sexual and asexual, commonly referred to as alternation of generations. Two morphotypes resulting from this alternation are a megalospheric form, which is typically associated with the gamont generation, and a microspheric form resulting from the schizont generation. However, in certain species the associations may be reversed.

(*c*) *Coiling.* The direction of coiling in some species is apparently controlled by the method of reproduction. Asexual microspheric individuals are dextrally coiled, while the sexual megalospheric forms are sinistrally coiled. Coiling direction has been an important correlative tool for the Pleistocene. During glacial advances and thus, cooler temperatures, coiling direction is found to be predominantly sinistral in some species (Ericson *et al.*, 1963). During interglacial periods, the reverse is true. There has been some attempt to apply this concept to species as old as Cretaceous. Bolli (1971) summarised the current knowledge of coiling directions for Cretaceous to Recent planktonics.

(d) *Spire height*. Another pair of morphotypes, seldom recognised in the fossil record but known from living species, are the high and low spired trochoid forms of a single species. Loeblich and Tappan (1964) cited the example of *Glabratella patelliformis* (Brady) in which the schizont was larger and lower spired than the gamont. There are numerous fossil forms in which the only difference is spire height. An attempt has been made in this study to recognise these variants.

(e) *Growth pattern*. Arnold (1964, 1968) in examination of laboratory-maintained cultures, has discovered conclusive evidence of multiple growth forms in certain benthonic foraminifers. For example, *Spiroloculina hyalina* Schulze may be represented by quinqueloculine, triloculine, spiroloculine and uncoiled individuals from the same asexual brood. The paucity of such studies does not allow for any generalisations. However, one conclusion is readily apparent. There is a great need to examine as many foraminiferal life cycles as possible, including those of the planktonics. As yet it is not possible to analyse the Mesozoic planktonics in this sense, but such studies should cause workers to be cautious in their interpretations and their proposals of new taxa.

2. Ontogenetic variation

(a) *Life history*. The progressive stages of maturity (nepionic through gerontic) of foraminifera are not well-known and many palaeontologists have erected separate species and even new genera on the basis of variations between different stages in the life cycle of a species. Interpretations of this kind can be avoided through population analysis. Ontogenesis, as illustrated by the test morphology of a population, is completely gradational. This may not, however, be discernable from a single sample owing to the influence of environmental factors on development or the selective modification of a population by processes which occur after death. Thus the investigation of all size fractions from many samples is essential in order to obtain meaningful results from population studies.

(b) *Growth*. As individual forms mature new chambers are added to the test. Typically these chambers gradually enlarge throughout the life of the individual, presumably to accommodate the increasing volume of cytoplasm.

There are exceptions to this generalisation. In *Heterohelix americana* (Ehrenberg) and *Guembelitra cretacea* Cushman, one or several chambers which follow the proloculus may be smaller than the initial chamber. In some forms chambers formed during the ephebic stage of growth remain essentially unchanged in size. In species where such characteristics are constant the cause is genetic, but where it fluctuates the environment is probably responsible.

(c) *Chamber arrangement*. Another important ontogenetic development is the modification in the pattern of chamber addition. Perhaps the most notable is the change from an initially planispiral coil to a biserial growth form in many heterohelicid species.

The number of chambers per whorl was once thought to remain constant throughout the ontogeny of trochospiral forms. Now it is known that in some species a greater number of chambers per whorl in the neanic stage gives way to fewer chambers in the ephebic and gerontic stages.

(d) *Ornamentation.* The secondary surface deposits termed ornamentation serve no known function. These deposits have been shown in the Cretaceous Globigerinacea (Pessagno and Miyano, 1968) to consist of ultragranular hyaline calcite.

Variation in the coarseness of ornamentation has been observed in all Mesozoic planktonics. Ornamentation is lacking in the nepionic chambers although it may be added over a portion of these chambers during a post neanic stage. During the neanic stage ornamentation is poorly developed, and is represented by a few randomly scattered pustules or spines, very fine costae or weak costellae. In the ephebic and gerontic stages additional ornament develops and the coarseness of all ornament increases with the continuous deposition of new calcite. The ultimate chamber, however, is less coarsely ornamented than the penultimate. Consequently the same individual will increase in the coarseness of ornamentation from the nepionic to neanic stages, followed in the later stages by a gradual decrease to the ultimate chamber.

(e) *Tegilla.* Species which develop a tegillum do so during or after the neanic stage. Initially the tegillum is little more than an extension of the apertural lips (portici). As the individual matures, the tegillum becomes more complex by the fusion of each new extension to the previous ones. This creates numerous openings along the margins of each extension.

Bolli *et al.* (1957) defined two types of openings related to the tegillum. One type, which they termed infralaminal accessory apertures, was defined as located along the margins of the tegilla, adjacent to the chambers. The second variety was said to pierce the tegilla, and was called "intralaminal". Subsequent examination has shown no difference between these "types", which are best referred to as tegillar apertures.

With each new chamber, a new extension is formed. That portion toward the already existing tegillum may or may not be in contact and cemented along its full length. If not, the so called "pierced" variety occurs. Although this type is most common, the non-pierced form is also found. Some species and individuals are predominantly one or the other, but both types have been noted at different ontogenetic stages of the same species and even of a given individual. This variation has no taxonomic significance, but the presence or absence of a tegilla in mature individuals does.

(*f*) *Pore density*. A rather minor ontogenetic variation between the nepionic and later stages can be noted. Not only does the nepionic stage generally lack ornamentation, it also has a significantly reduced pore density. The proloculus of all Mesozoic planktonic species observed thus far is imperforate except for one or two apertural openings. The succeeding nepionic chambers have an increasing pore density, but even here a reduction factor of ten to a hundred or more may exist as compared to later chambers.

(*g*) *Wall structure*. Reiss (1957b) recognised a bilamellar wall structure in most of the planktonic groups, fossil and living. He stated that the chamber walls were double, "formed by an outer lamella, one per instar [new chamber] covering the whole test, and by an inner one, lining each chamber and confined to it . . .". With the use of scanning electron microscopy, Bé and Hemleben (1970) discovered that the outer and inner lamellar units are composed of many more layers than recognised by Reiss. This has been confirmed for certain Cretaceous planktonic foraminifers also.

Contrary to Reiss' observations, Bé and Hemleben suggest that chamber calcification occurs continually and independently of a newly forming chamber. Furthermore, they believe that a lamina, rather than being a continuous sheet over the last whorl, is deposited in patches.

By whichever means calcification is accomplished, the end result is the same. The wall thickness of the earlier chambers in the last whorl becomes thicker than that of the later ones. Such species as *Globigerina washitensis* Carsey display this phenomenon well. This particular species also has a rather thick primary organic membrane separating the inner and outer lamellae.

Bé (1965) discovered what he and Hemleben (1970) later referred to as two phases of wall calcification in Recent planktonic foraminifers. The first phase consists of the "bilamellar" growth to normal adult size, as discussed above. The second phase is the superposition of a "calcite crust" over the adult test. Such a crust would soon conceal surface ornamentation. The development of a "calcite crust" has not been recognised so far in Mesozoic planktonic species.

3. *Phylogenetic variation*

Phylogenetic variation can be studied at any taxonomic level down to that of the species. Yet it is the most elusive and subjective of the four primary types of variation discussed in this section. There are several reasons for this: (1) incompleteness of the fossil record, (2) failure to recognise variability in the individuals composing populations, which leads to (3) the great number of synonyms in the literature, (4) insufficient knowledge of certain groups and (5) the determination of which morphologic characteristics have phylogenetic significance.

In spite of these difficulties, a number of phylogenetic variations have been established within species and higher taxonomic categories. As recognition of variation allows a more complete interpretation of a species, so it permits a more accurate induction of phylogenetic lineages.

The variations discussed below are those which are noted in the phylogeny of certain species, and should not necessarily be construed to be generally applicable to other species. The adaptive significance of these changes is not understood.

(a) *Size*. A gradual, steady increase in the size of some species, such as *Heterohelix globulosa* (Ehrenberg), can be observed through time. In the determination of a phylogenetic change in size, sufficiently large samples must be examined to distinguish between genetic and phylogenetic changes. The trend toward greater size has been recognised in many organisms, but usually at the supraspecific levels.

(b) *Ornamentation*. Several species are known to exhibit a phylogenetic increase in coarseness of ornamentation. Two species displaying this well-documented variation are *Heterohelix globulosa* (Ehrenberg) and *Rugoglobigerina rugosa* (Plummer). In each case the height and width of the ornamentation, costae and costellae respectively, gradually becomes greater with time.

(c) *Spire height*. An increase in the spiroconvexity of trochospiral forms has been recognised only in *Globotruncana fornicata* Plummer. The phyletic lineage between *G. fornicata* and *G. contusa* (Cushman) passes through a series of gradational forms with increasingly greater spire heights. Although some workers have regarded these changes as representing new species or subspecies and others as intraspecific characters, most accept them as representing an intergradational series.

4. *Non-genetic variation*

The tests of planktonic foraminifers are frequently modified by a variety of external influences. The size, chamber arrangement, ornamentation and wall structure of the test of some individuals have been sufficiently altered so that they have been described as new species. These deviations from the typical test usually occur during the lifetime of the organism, but may also result from post-mortem and post-burial agents. Such modifications can be recognised if a large enough suite of specimens is examined, or suites from different regions are compared.

(a) *Environmental influences on chamber size*. Examples of an abrupt increase in chamber size has been noted in most genera. New species have been based upon this phenomenon, e.g. *Heterohelix ultimatumida* (White) [= *H. globulosa* (Ehrenberg)]. Presumably the addition of a new chamber is in response to a volumetric increase of the cytoplasm. This growth, as recognised by chamber size, may be slow or rapid, but is usually gradational.

Within genetically determined limits, growth is controlled by such factors as nutrient supply, temperature and salinity. If, during either vertical or lateral migration, an individual is transported into an environment better suited to growth, an abnormally large chamber would likely be secreted. These chambers may occur singly or in a series of two to four. Those subsequent to the first large chamber usually do not continue to increase in size, but maintain dimensions approximately equivalent to the first. These chambers may be followed by ones which are reduced in size.

A reduction in chamber size is also a common occurrence and may be a result of adverse environmental conditions. The abrupt reduction in the size of the last one or two chambers is termed "kummerform growth". Hecht and Savin (1972) showed that for three modern species kummerform growth "appears to be related to test formation under stressed conditions". These occurred when water temperature was too high or too low for a given species.

Either an increase or decrease in chamber size will alter the number of chambers per whorl in coiled forms. Past workers have established new species with as little as a half of a chamber per whorl difference from an already existing species. Thus, the need to establish the mean and the full range of variability for each species becomes apparent.

Test size. The ultimate size attained by an individual is limited by the time of its death or of its reproduction. The former needs no further explanation. In those individuals or species where all of the protoplasm is utilised during reproduction, the parent test is discarded, thereby ending growth.

Bradshaw (1957) discovered that both temperature and salinity affect the rate of growth and reproductive activity. Although his and other studies dealt only with benthonic species, the findings are equally applicable to the planktonics.

Unlike most other organisms, foraminifers may become larger under less favourable environmental conditions. If a given set of environmental parameters delays reproductive activity the individual will continue to grow until those parameters become more suitable for reproduction or kill it outright. Thus, the size of the individual is not necessarily directly related to reproductive maturity.

Spire height. Yvonne Herman, in a paper presented to the Twenty-fourth International Geological Congress (Montreal, 1972), stated that she has found evidence which suggests that variations in the spire height of the Quaternary *Globorotalia truncatulinoides* can be attributed to seasonal changes in temperature.

If further study on this and other genera support this observation, then a re-examination of many trochospirally coiled Mesozoic species is in order. Such a variation may also be due to temperature changes with latitude or water depth.

Test porosity. Bé (1968) and Frerichs *et al.* (1972) have been able to relate porosity and pore densities directly to surface water temperatures. As yet these studies have not been applied to the Mesozoic planktonic foraminifers. Nevertheless, latitudinal variation in pore diameters and densities of a given species should be recognisable providing there has been no diagenesis. Pflaumann (1971) has shown that pore diameters vary on an individual specimen. Presumably, this is attributable to diagenesis (see remarks under *Post-burial–Inorganic precipitation*).

Injuries. A common but usually overlooked morphologic variant results because of damage to the test. Such injuries may be divisible into organic (predator) and inorganic (mechanical) categories.

Sliter (1971) reported nematode predation on certain benthonic foraminifers. Nematodes enter the tests by boring. Sliter noted that planktonics were rarely bored. His conclusion is borne out in this study. Borings can be distinguished from pores by their larger size, and from apertures by their random occurrence.

Specimens have been found which have repaired chambers. The remnants of the original chamber are typically ragged and angular. This is assumed to be mechanical damage at the time the chamber in question was in the ultimate position. A new chamber, slightly reduced in size, was added beneath the remnant.

In species with globigerinid-shaped chambers, a form of aberrantism related to injury occurs. Whenever a chamber is broached, a new chamber may be added over this hole as though it were a normal aperture (see Pl. 53, fig. 5). No more than one such chamber has ever been observed for a given injury. The aberrant chamber is of "normal" size for the position of the injury. The cause of the hole is not known. However, the degree of frequency of these aberrant chambers is approximately the same as that for borings.

(*b*) *Post-mortem influences.* Any factor which affects the test during the interval of time from the death of the organism until burial is for convenience categorised as post-mortem. During this period a test may be altered by many agents, most of which probably cannot be determined. There are, however, two effects which can be recognised: solution and breakage.

Solution. Modern oceans are generally undersaturated with calcium carbonate below a depth of 3.7 kilometers. Inasmuch as planktonic foraminiferal tests are predominantly calcium carbonate, they will dissolve in such an environment. Those tests which settle near the saturated–unsaturated horizon may be only partially dissolved before burial. Shallower water microenvironments with a calcium carbonate deficiency may also exist. Again partial or complete solution may occur.

Solution will affect first the more delicate structures, such as spines and surface ornamentation. New species of Heterohelicidae, which are speciated in large part upon ornamentation, have been based upon smooth, non-ornamented surfaces, e.g. *Heterohelix washitensis* (Tappan) and *Ventilabrella multicamerata* de Klasz. It is not possible to distinguish post-mortem solution and post-burial decalcification. It is likely that the above mentioned species are the result of the latter conditions.

Breakage. Once the tests have settled to the bottom, mechanical breakage may, and often does, occur. They may be ingested and passed through the digestive tract of a sediment feeder. The surface sediment can be reworked by currents or waves. In either situation the more delicate structures will again be most affected.

The ultimate chamber, being thinnest, is frequently missing. Apertural lips, tegilla and spines are commonly damaged or removed altogether. New species have been erected based upon the unrecognised lack of such structures. *Hedbergella beegumensis* Marianos and Zingula lacks the thin ultimate chamber, and was thus characterised by "having the last few chambers thicker . . .".

(*c*) *Post-burial influences.* The preservation of the foraminiferal test is the determining factor in identification. Many diagenetic agents affect the test after burial. The test can be altered or completely destroyed by these agents. It is this post-burial stage which is probably the most critical of the non-genetic changes.

Breakage. During the lithification of the enclosing sediments, the tests may be broken by compaction. The degree of compaction, and hence breakage, is dependent upon the nature of the sediments, e.g. sorting, grain size and shape, and the amount of interstitial fluids. The entire test may be shattered or only those features which lack sufficient structural strength to withstand minimal compression, such as the tegilla or the ultimate chamber. Such test damage may also occur during laboratory preparation. The post-mortem breakage is usually not distinguishable from that of post-burial.

Inorganic precipitation. Foraminiferal tests are frequently filled with a single crystal of calcite, and the recrystallised test wall may be in optical continuity with the internal mould. Typically only the gross surface features are retained during this process.

In studies on a Recent *Neogloboquadrina*, Pflaumann (1971) discovered that pore diameters and porosity do not correlate with the spiral geometry of the test. Mapping of the diameters and concentrations of pores disclosed a random pattern on an individual and between individuals. The author attributed this to diagenesis. Pores may be closed off and their number significantly reduced by inorganic precipitation.

Secondary overgrowth can alter or mask the organically deposited surface features. Papillose or vermicular structures result. These have been incorrectly interpreted as primary (organic) structures.

Decalcification. When using light microscopy, care must be taken in recognising internal moulds which arise through decalcification.

Internal moulds can be recognised by the following traits: lack of aperture(s); lack of pores; presence of a transparent test; lack of expected ornamentation or primary structures such as a keel. Any of the preceding traits can have other causes. Thus, it is important to use as many of these as possible in the determination.

Early planktonic forms and the origin of planktonic foraminifers

Few attempts have been made to determine the ancestral stock of the planktonic foraminifers. A number of reasons may be responsible for this situation: a lack of strata of the appropriate age, environmentally unsuitable strata, a scarcity of early planktonic species, an inadequate knowledge of their morphology or poor preservation.

Certainly, before one can assess potential ancestors of the planktonics an analysis of the pre-Cretaceous species, their morphology and time of appearance is necessary.

1. TRIASSIC

Oberhauser (1960) recorded *Globigerina* from middle Ladinian strata and assigned two forms, *G. mesotriassica* and *G. ladinica* to this genus. Kristan–Tollman (1964) described forms which are preserved as casts from the Rhaetian of Austria as *Globigerina rhaetica*. All these forms have morphological characteristics in common with members of the subfamily Discorbinae. These are: indentations along the ventral (*G. mesotriassica*) or dorsal and ventral sides (*G. rhaetica*) which may correspond with accessory apertures, the early coiling is discorbid (*G. mesotriassica. G. ladinica*) and chamber inflation is slight. For these reasons none of these forms are believed to be planktonic. Fuchs (1967a) has reassigned these forms to his new genera *Oberhauserella* (*G. mesotriassica* and *G. rhaetica*) and *Kallmanita* (*G. ladinica*) which he considers (Fuchs, 1960) as members of a new family Oberhauserellidae.

2. JURASSIC

Terquem and Berthelin (1875) were the first workers to assign a Jurassic species to the genus *Globigerina*. Later workers added another fourteen Jurassic species to this genus. In my opinion of the fifteen Jurassic species of *Globigerina* described only two distinct forms exist, the high-spired *G. jurassica* Hofman, and the low-spired *G. hoterivica* Subbotina. The others which are discussed below are either synonymous of these species or are not planktonic.

G. liasiana Terquem and Berthelin, *G. lobata* Terquem and *G. Oolithica* Terquem lack the tightly coiled form with three or four subspherical chambers per whorl of the early *Globigerina*. *G. helvetojurassica* Haeusler, though morphologically similar to early *Globigerina*, cannot be accurately assigned to this genus because of the poor state of preservation of the type specimens. Forms attributed by Balakhmatova (1953) to *Globigerina* ex. gr. *bulloides* are figured as line drawings; this prevents corroboration of their planktonic nature. Three species *G. avarica*, *G. dagestanica* and *G. gaurdakensis* which were described by Morozova and Moskalenko (1961) are here considered as benthonic forms or indeterminable species. Bars and Ohm (1968) erected *Globigerina spuriensis* for Upper Dogger forms which appears to possess a bulla-like ultimate chamber and a slit-like aperture along the ventral suture. Further study is required to resolve whether this form can be considered as planktonic.

The following forms are considered to be synonyms of *G. hoterivica:*

(*i*) *G. oxfordiana* Grigelis and forms assigned to this species by Bignot and Guyader (1966, 1971) and Premoli Silva (1966).

(*ii*) *G. balakhmatovae* Morozova and forms assigned to "*G.*" *balakhmatovae* by Brönnimann and Wernli (1971).

(*iii*) *G. terquemi* Iovčeva and Trifonova (1961).

(*iv*) *G. bathonica* Pazdrowa (1969), only the low-spired variety.

The following forms are considered to be synonyms of *Globigerina jurassica*:

(*i*) The forms attributed to "*G.*" *avarica* Morozova by Brönnimann and Wernli (1971) which though poorly preserved are strongly similar to *G. jurassica*.

(*ii*) *G. conica* Iovčeva and Trifonova and the high-spired variety of *G. bathonica* Pazdrowa.

In the section on intraspecific variation it was suggested that spire height may be either an environmentally or genetically controlled character. Thus it is possible that *G. hoterivica* and *G. jurassica* are one and the same species. The two are retained here as separate species because of the differences in their stratigraphical range.

3. THE PRECURSOR OF PLANKTONIC FORAMINIFERS

The earliest record of true planktonic forms is from the Bathonian (M. Jurassic). Thus the ancestor of the first planktonic species must be looked for in benthonic species of pre-Bathonian strata.

The earliest planktonics are very small members of the genus *Globigerina*. The associated fauna, when reported (e.g. Iovčeva and Trifonova, 1961; Bignot and Guyader, 1966; Brönnimann and Wernli, 1971), is characteristic of the neritic zone and typically includes members of the Discorbinae, Nodosariacea, and assorted agglutinated forms and miliolids.

The fact that the early planktonic species are found in what has been interpreted as a neritic environment should not be disturbing, in fact just the opposite. The neritic zone has greater species diversity and environmental stress than any other marine habitat, which greatly increase the probability that this was the ancestral environment. Living planktonic species are not as frequent in shallow water as in deeper. Possibly the early planktonics had not acquired the trait of diurnal migration, or perhaps because of the very small size of the adult, progressive sinking in the water column did not occur with ontogeny. Nevertheless, an environment in proximity to that of the ancestor should be expected.

The discorbids have long been suggested as the possible ancestor of the planktonics (Galloway, 1933; Cushman, 1948b; Pokorný, 1963). This was based upon only the gross test morphology, however. Loeblich and Tappan (1964) described the Discorbidae as being monolamellar. Although according to an earlier study (Reiss, 1963b), they are bilamellar as are the Globigerinacea. The evolutionary step would have been too great to consider either the miliolids or agglutinated species as potential ancestors, and the Nodosariacea exhibit neither the appropriate apertural type nor coiling.

Palaeoecology

Palaeoecological investigations of Mesozoic planktonic foraminifers are based on concepts and methods derived from intensive and continuing ecological studies of living planktonics. These include studies of the areal distribution of Mesozoic forms, and investigations of morphological features for which environmental control is recorded in the Recent.

1. MORPHOLOGY

By matching his findings with a previously established palaeotemperature curve, Frerichs (1971) demonstrated that certain morphological features of the Cretaceous planktonic foraminifers were temperature dependent. Keels, accessory apertures, clavate chambers, tubular spines and planispiral coiling were all found to develop during warming trends and disappear during cooling. He showed that during cold intervals new phenotypes rarely developed and that the planktonic population was reduced to its more conservative members. Major taxonomic radiation occurs during warm periods.

Data obtained for living taxa show that dextrally coiled trochospiral forms characterise warm water, and sinistrally coiled forms cooler water. Coiling data for keeled Cretaceous species compiled by Bolli (1971) do not, however, conform with Cretaceous palaeotemperature curves. No shift to sinistral coiling occurs during the Cenomanian–Turonian and Maastrichtian cooling. Since very few data are available for coiling directions of Mesozoic species the interpretation of climate based on this parameter is uncertain.

In certain extant species pore density has been shown to decrease with distance from the equator and to be related to temperature (Frerichs *et al.*, 1972). Since this parameter can be influenced by diagenetic changes in the wall structure of fossil species its use as a temperature indicator is limited in ancient sediments.

2. LATITUDINAL AND VERTICAL DISTRIBUTION

Several workers (Bartlett, 1969; Scheibnerová, 1971a; Sliter, 1972a, and others) have begun to determine latitudinal distribution patterns for specific taxa. This work is still in a state of flux as general agreement has still to be reached on the number of palaeobiogeographic provinces and their boundaries.

Eicher and Worstell (1970b) suggested that *Heterohelix* is a form which frequented shallower levels of the water column. They determined that *Heterohelix* was first among planktonics to enter a newly transgressed region and last to be forced out during regression.

Sliter (1972a, b) in a study of Cretaceous foraminifera regarded the globotruncanids as bathypelagic, and the heterohelicids and hedbergellids as largely epipelagic but extending to the bathypelagic realm.

Palaeobiogeography

A number of distributional patterns have been established for living marine organisms. McGowan (1971) for instance established a twelve-fold division of regions and subregions which are primarily delimited by latitude and water currents. Such a complex pattern is, as yet, not applicable to the geologic past. The current state of knowledge permits the establishment of only rather broad distributional patterns for Mesozoic foraminifers, whose precision is dependent on sampling density and recognition of taxonomic synonyms.

During the last decade a number of studies have been concerned with the dispersal patterns of Mesozoic planktonics (Bandy, 1960, 1967; Kent, 1969; Sliter, 1968; Bergquist, 1971; Scheibnerová, 1970, 1971a, 1971b; Dilley, 1971; Douglas, 1972). All these studies suffer in varying degrees from nomenclatorial fragmentation at the generic and especially the specific levels. Nevertheless, certain generalisations can be made from them. Already, latitudinal distribution patterns, agreed by all workers to have been temperature dependent, have emerged from these studies. Faunal provinces have also been defined. Using the terminology of Scheibnerová (1970, 1971a, 1971b) these are: Tropical province characterised by large numbers of species representing most genera, but particularly the more complex keeled varieties; the Boreal and Austral provinces have fewer species which are primarily heterohelicids and simple globigerine forms. As one progresses from the Tropical province poleward, evidence suggests that variety becomes less. Intermediate and high-latitude provinces are recognised more on the absence of certain genera and species rather than from endemic taxa as the simpler forms seem to range throughout all provinces (Sliter, 1972a). This corresponds to the conditions which exist in modern oceans, and all workers concur to the striking similarity and parallelism between the Mesozoic and modern provinces.

Included in the Systematic Descriptions section (p. 329) of this paper are global distribution maps of the majority of the Mesozoic taxa. The distribution is a composite for the entire range of a species. The pattern circumscribes the maximum geographic range of an area. This should not be interpreted as meaning that the taxon is necessarily found at every place within that area. Denoted on each map is an approximation of the land representing maximum transgression during the Mesozoic. The land–water contact is an amalgamation of data taken from Douglas (1972) and many other sources.

Phylogeny

Numerous phylogenetic schemes have been proposed for the Mesozoic Globigerinacea (e.g. Banner and Blow, 1959; Pessagno, 1967 and El-Naggar, 1971a), and each have been slightly different. No attempt is made here to comment upon earlier schemes because this would require a detailed analysis of the generic-level interpretations given by the authors of these proposals. The plan given here is organised upon the latest findings of other authors as well as the author's observations.

Table 1 indicates the first development of each genus. This is not meant to exclude the possibility that different members of a genus evolved at different times from the same ancestral stock. This is recognised in several genera and is treated in the systematics section.

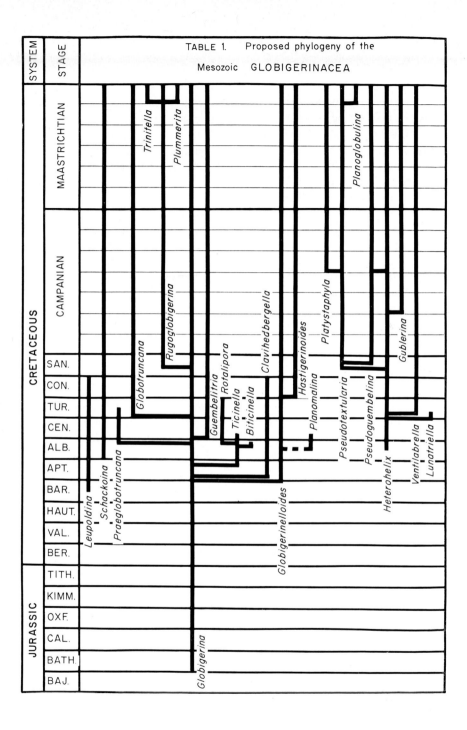

TABLE 1.   Proposed phylogeny of the

Mesozoic GLOBIGERINACEA

The ancestors of *Leupoldina* and *Schackoina* have not been recognised. It is unlikely that *Schackoina* evolved from *Leupoldina*. Some species exhibit an initial trochospire which might suggest affinities with *Globigerina*, but this is deemed unlikely at this time, because of rather important morphologic differences.

Other than the genera noted above, there are two major phylogenetic stocks. One, the earliest, arises from the genus *Globigerina*, the other from *Heterohelix*. The ancestors of these genera have not been recognised with any degree of certainty. *Globigerina* may have evolved from the Discorbinae but the ancestor of *Heterohelix* remains unknown. It is suspected that the two stocks developed independently from unrelated benthonic foraminifers.

The ancestors of the very earliest species of *Globigerinelloides* have not been discerned. However, another early species, *G. ultramicra* (Subbotina), is transitional with *Globigerina planispira* Tappan. Presumably *Planomalina* evolved from a *Globigerinelloides* by developing a keel, but this is unconfirmed. The *Hastigerinoides* species arose nearly simultaneously from two species of *Globigerinelloides*. Both *H. alexanderi* (Cushman) and *H. subdigitata* (Carman) developed from *G. ultramicra* (Subbotina), while *H. watersi* (Cushman) is transitional with *G. escheri* (Kaufmann).

Clavihedbergella bizonae (Chevalier), the oldest member of the genus, probably evolved through *Globigerina planispira* Tappan. Other species evolved from *Globigerina* or from other members of *Clavihedbergella*.

Ticinella is believed to have developed through *Globigerina trocoidea* (Gandolfi) which appears transitional to *T. roberti* (Gandolfi). *Ticinella roberti* grades into *Biticinella breggiensis* (Gandolfi). *Rotalipora ticinensis*, the first *Rotalipora*, descended from *Ticinella praeticinensis* Sigal.

Globigerina delrioensis Carsey probably gave rise to the *Praeglobotruncana* through *P. delrioensis* (Plummer).

The predecessor of *Guembelitria* is not known, but its similarity to the early *Globigerina* is unmistakeable.

Globotruncana cretacea (d'Orbigny) is thought to be the earliest member of the genus because of its globigerine characteristics. The only difference between this species and *Globigerina delrioensis* Carsey is the development of a keel and a tegillum.

Rugoglobigerina rugosa (Plummer) also descended from *Globigerina delrioensis* Carsey. Except for the tegillum on *R. rugosa*, the two are virtually indistinguishable. *Trinitella scotti* Brönnimann may have evolved from either *R. rugosa* or *R. hexacamerata* Brönnimann. *Plummerita hantkeninoides* Brönnimann probably developed from *R. rugosa*.

Heterohelix gave rise directly or indirectly to seven genera. *Lunatriella spinifera* Eicher and Worstell probably evolved from *H. pulchra* (Brotzen).

The first *Ventilabrella, V. austinana* Cushman, is gradational with *H. globulosa* (Ehrenberg). *Pseudoguembelina* is polyphyletic with various members developing through species of both *Heterohelix* and *Pseudotextularia. Gublerina* evolved from *H. pseudotessera* (Cushman) forming *G. reniformis* (Marie).

By developing chambers with a greater thickness than height or width *Heterohelix globulosa* (Ehrenberg) evolved into the first *Pseudotextularia, P. browni* Masters. *Platystaphyla* descended from a yet undetermined *Pseudotextularia. Planoglobulina varians* (Rzehak) is gradational with *Pseudotextularia elegans* (Rzehak).

Biostratigraphy

Using traditional zonal methods, we have reached a plateau beyond which we are unable to increase our biostratigraphic resolution. A thorough review and discussion of various Cretaceous planktonic zonal schemes and a contrasting of the American Code on Stratigraphic Nomenclature with the Stratigraphical Code of the Geological Society of London has been previously presented (Masters, in press). At that time the author introduced the use of "datum levels" for the upper Cretaceous planktonic foraminifers. The advantages of this method over the traditional were discussed and demonstrated.

To develop a system of datum levels, it is necessary to have as chronologically long and continuous sequence of strata as possible. The reason for this is that accurate first and last occurrences of each species must be obtained. Many of the topotype and hypotype specimens used in this study have come from isolated worldwide localities. For this reason, it has not been possible to propose additional datum levels for the remainder of the Mesozoic. Because of the taxonomic re-evaluation presented here, the previously established and widely used traditional zonal schemes are no longer valid. Consequently, a temporary zonal vacuum is created until such time as datum levels are proposed for the rest of the Mesozoic or, at least, the old schemes are reappraised.

The species ranges presented (see Tables 2–7) do not necessarily represent total ranges because only illustrated publications were used for this project. Furthermore, the use of papers illustrated by thin-sections is not always possible. As pointed out by many authors (e.g. Sartoni, 1965), most species of planktonic foraminifers, described from surface features, cannot be recognised in thin sections. Of those few which can be identified in this manner, often one must have at least a general, if not specific knowledge, of the age of the enclosing strata before identification can be made. This is the reverse of what is usually sought, i.e. the age of the strata.

The ranges of species illustrated on Tables 2–7 should serve as a base from which to reconstruct the total range of these species. Some may be extended significantly, others not at all. The Santonian to Maastrichtian segment of datum levels should require at most only minor modifications.

Worthy of note is a recent comment by Scheibnerová (1972) that the standard zonal schemes cannot be applied on a worldwide basis because of the existence of faunal provinces. Increased research on Mesozoic palaeobiogeography has indicated the presence of two major latitudinally controlled faunal provinces: one in the high latitudes, the other tropical with intervening transitional provinces. A characteristic fauna is present in each of the two main provinces. As a result, Scheibnerová proposed the creation of a dual zonal scheme with the fauna of the transitional province permitting an integration of the two schemes.

Glossary

Below is a short list of definitions for some of the terms used in this paper. For a more complete list the reader is referred to Loeblich and Tappan (1964).

Allele(s). One of alternate states of the same gene which leads to a different phylogenetic end product.

Allopatic populations. Those which inhabit areas separated by physical barriers, and are thus geographically isolated.

Alternation of generations. The periodic, usually irregular, change between asexually and sexually produced progeny.

Antipenultimate. The third from the last chamber.

Apertural flap. A thin, wide, imperforate extension bordering the aperture.

Apertural lip. Analogous to the apertural flap, but narrower.

Aperture, accessory. Subordinate to the primary aperture in the sense of never supporting secondary chambers.

Aperture, primary. Major opening through the chamber.

Aperture, relict. Umbilical portion of primary aperture not covered by successive chambers.

Aperture, supplementary. Additional to the primary aperture in the sense of supporting secondary chambers.

Aperture, umbilical. One which opens only into the umbilicus in trochospiral forms.

Aperture, umbilical–equatorial. One which extends from one umbilicus to the periphery and into the other umbilicus in planispiral forms.

Aperture, umbilical–extraumbilical. One which extends from the umbilicus to the periphery in trochospiral forms. Synonymous with extraumbilical–umbilical.

Carina(e). An imperforate ridge formed by merging pustules.

Convolute. Planispiral coiling in which successive whorls are in contact with but not covering earlier ones.

Costa(e). A raised imperforate ridge as surface ornamentation.

Costellae. Raised imperforate ridges meridionally arranged as surface ornamentation.

Cytoplasm. Protoplasm outside of the nuclear membrane.

Dorsal. Spiral side of trochospirally coiled forms.

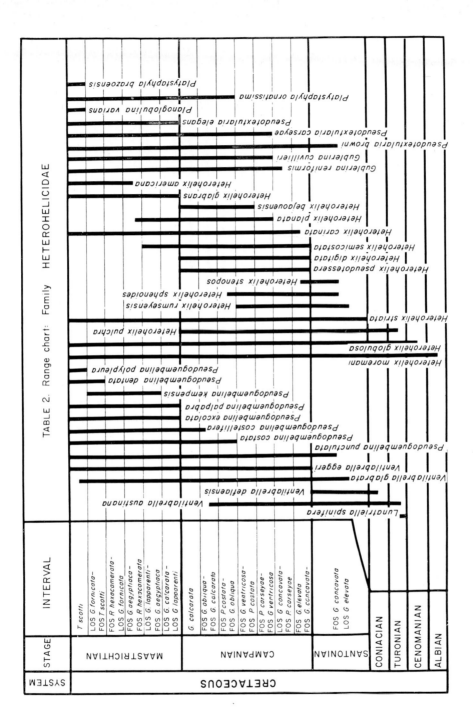

TABLE 2. Range chart: Family HETEROHELICIDAE

TABLE 3. Range chart: Family PLANOMALINIDAE

TABLE 4. Range chart: Family SCHACKOINIDAE

SYSTEM-SERIES	STAGE	INTERVAL
CRETACEOUS	MAASTRICHTIAN	T. scotti
		LOS G. fornicata – FOS T scotti
		FOS R. hexacamerata – LOS G. fornicata
		FOS G. aegyptiaca – FOS R. hexacamerata
		LOS G. lapparenti – FOS G. aegyptiaca
		LOS G. calcarata – LOS G. lapparenti
	CAMPANIAN	G. calcarata
		FOS G. obliqua – FOS G. calcarata
		FOS P. costata – FOS G. obliqua
		FOS G. ventricosa – FOS P. costata
		FOS P. carseyae – FOS G. ventricosa
		LOS G. concavata – FOS P. carseyae
		FOS G. elevata – LOS G. concavata
	SANTONIAN	FOS G. concavata – LOS G. elevata
	CONIACIAN	
	TURONIAN	
	CENOMANIAN	
	ALBIAN	
	APTIAN	
	BARREMIAN	
	HAUTERIVIAN	
	VALANGINIAN	
	BERRIASIAN	

Range taxa:
Leupoldina cabri, Leupoldina pentagonalis, Schackoina cenomana, Schackoina masellae, Schackoina bicornis, Schackoina alberti, Schackoina utriculus, Schackoina moliniensis, Schackoina primitiva, Schackoina tappanae, Schackoina sellaeforma, Schackoina multispinata

TABLE 5. Range chart
Family GLOBIGERINIDAE

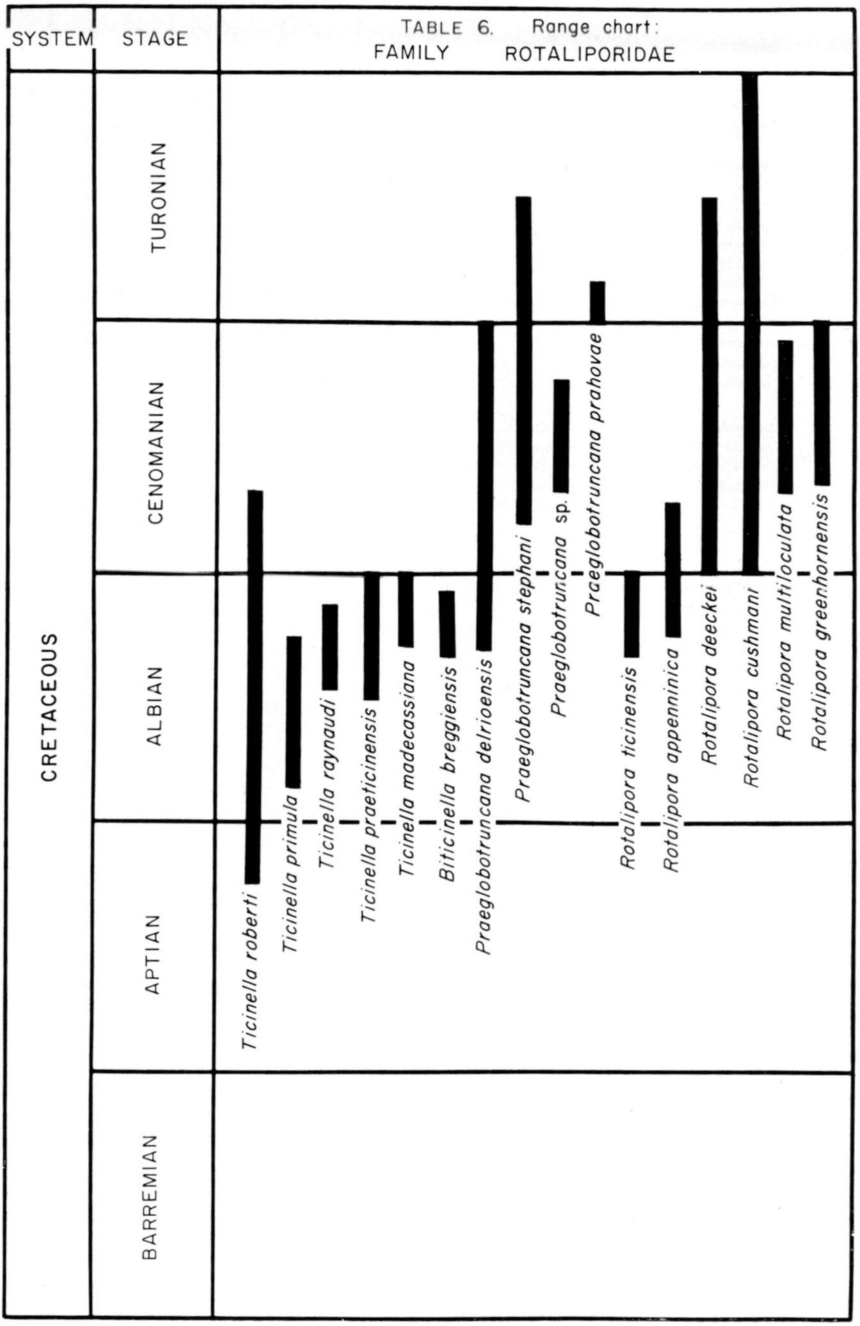

TABLE 6. Range chart: FAMILY ROTALIPORIDAE

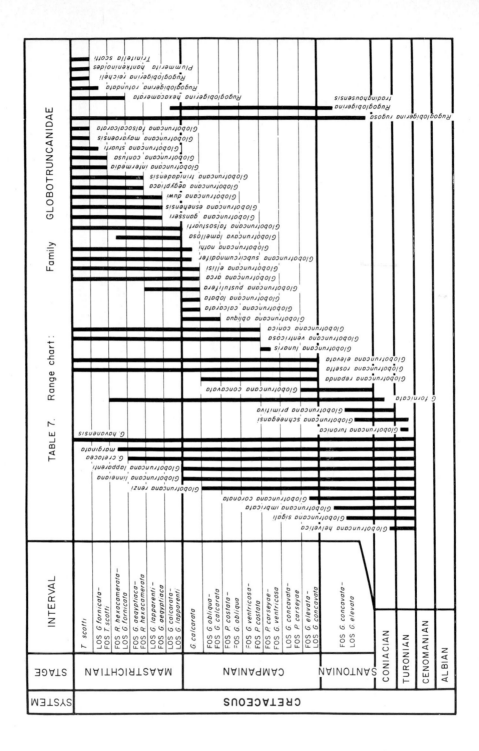

TABLE 7. Range chart: Family GLOBOTRUNCANIDAE

Ephebic. Adult stage of ontogeny.
Evolute. Coiled form with all chambers visible.
Gamont. Sexually reproducing generation.
Gerontic. Senile stage of ontogeny.
Hispid. Numerous, fine, short spines.
Imbricate. Overlapping.
Involute. Coiled form in which only the chambers of the final whorl are visible.
Kummerform. Reduced final chamber size.
Limbate. Thickened (widened) suture.
Megalospheric. Form with large proloculus, usually associated with sexually reproducing generation.
Microspheric. Form with small proloculus, usually associated with asexually reproducing generation.
Mutation. A fundamental change in the genetic factor.
Neanic. Juvenile stage of ontogeny.
Nepionic. Embryonic stage of ontogeny.
Ontogeny. The course of development of an individual.
Ornamentation. Secondary, imperforate, superficial structures.
Palingenesis. Being produced again.
Penultimate. Next to last chamber added.
Phenotype. Organism defined solely by appearance.
Plastogamy. Fusion of tests by their umbilical surfaces during sexual reproduction.
Population. A body of similar individuals.
Porticus (*portici*). Wide, asymmetrical apertural flaps or extensions.
Proloculus. Initial chamber of the test.
Protoplasm. Living matter consisting of cytoplasm and nucleus.
Pustule. Short, thick, conical, imperforate surface elevation.
Schizont. Asexually reproducing generation.
Tegillum (*tegilla*). Apertural flaps irregularly fusing into an umbilical cover plate.
Tubulospine. A hollow, spine-like extension of the chamber.
Ventral. Umbilical side of trochospiral forms.

Systematic descriptions

The classification presented is a modified version of that used by Loeblich and Tappan (1964). This in-depth investigation of the Mesozoic planktonic foraminifers revealed certain inconsistencies in the classification scheme proposed by these authors. The changes made are explained in the appropriate sections.

It should be emphasised that any classification tends to be artificial and highly subjective. Naturally the subjective interpretations are based upon as much objective observation as possible. Nevertheless, these observations can be variously interpreted depending upon the priorities established by a given author. In other words, one classification may not be any more "correct" than another. The important characteristics of any scheme of classification are its logical progression and its workability.

In the following classification, each species was considered on the basis of not only its morphology and stratigraphic and geographic occurrences, but also of the analysis of populations in an attempt to recognise intraspecific variants. These organisms were once living, and must be interpreted as such. Hence, each individual within a species is different, however slight. These differences must be recognised before we can gain the full value of our investigative efforts into other facets of these organisms.

The types are either deposited in the collections of the U.S. National Museum, Washington, D.C. or housed in my personal collection.

Phylum PROTOZOA
Subphyllum SARCODINA
Class RETICULAREA
Subclass GRANULORETICULOSIA
Order FORAMINIFERIDA
Superfamily GLOBIGERINACEA
Family HETEROHELICIDAE Cushman, 1927c

Definition: Neanic stage either biserial or with a small planispiral coil, later stages entirely biserial or with a diverging biseries, supplementary chambers, or dumb-bell-shaped primary chambers; supplementary apertures may be present along the biserial axis or at the base of the back of the chambers; bilamellar walls; always costate.

Generic Key
I. Chamber thickness does not exceed height.
A. Without accessory or supplementary apertures.
1. Simple biserial test in all ontogenetic stages . . . . . . *Heterohelix*
2. Vertically elongate chambers with lateral spine-like projections . . . . *Lunatriella*
3. Divergent biseries with partially septate interarea . . . . *Gublerina*
B. With accessory or supplementary apertures.
1. Paired accessory apertures along median line of test . *Pseudoguembelina*
2. Supplementary apertures and chambers in biserial plane. *Ventilabrella*
II. Chamber thickness exceeds height in early ontogenetic stages.
A. Without supplementary apertures.
1. Simple biserial test in all ontogenetic stages . . . *Pseudotextularia*
B. With supplementary apertures.
1. Reduced supplementary chambers in biserial plane . . *Platystaphyla*
2. Dumb-bell-shaped gerontic chambers . . . . . *Planoglobulina*

Remarks: The retention of *Guembelitria*, *Guembelitriella*, *Gubkinella*, *Chiloguembelina*, *Woodringina* and *Bifarina* in the Heterohelicidae is not justified, a conclusion also reached by Brown (1969). *Guembelitria* and its gerontic synonym *Guembelitriella* are herein transferred to the Globigerinidae. The type species of *Gubkinella* may be a benthonic form, but other species attributed to it are typical species of *Globigerina*. Brown agreed with Reiss (1963b) in the conclusion that *Chiloguembelina*, *Bifarina* and probably *Woodringina* have a single-layered wall and are probably benthonic. The Family Heterohelicidae is restricted to the Cretaceous. Supposed post-Cretaceous forms are either misidentified or reworked from the Cretaceous.

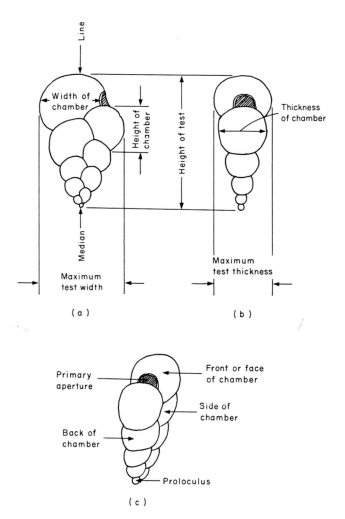

FIG. 1. A simple heterohelicid. (a) side view, (b) edge view, (c) oblique view.

The Heterohelicidae have never taken their rightful place among other planktonics in biostratigraphical studies. Brown (1969) pointed out that there are complex taxonomic and morphologic reasons for this. However, if given as careful consideration as other groups, the Heterohelicidae are equally useful for correlation.

The morphologic terminology of Brown (1969) is used here. An adaptation of his diagrams is given in Figs 1 and 2.

Range: Albian–Maastrichtian (*Trinitella scotti* Interval).

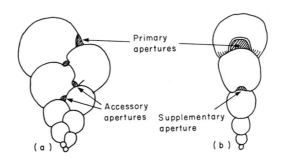

FIG. 2. Aperture types. (a) accessory apertures along median line of test, (b) supplementary aperture at the base of the back of the chamber.

Genus GUBLERINA Kikoïne, 1948

Gublerina Kikoïne, 1948, p. 26.

Type species: *Gublerina cuvillieri* Kikoïne, 1948.

Description: Test laterally compressed, flabelliform; nepionic stage a small planispiral coil, neanic stage biserially arranged primary chambers, in later stages biserial chambers become progressively more divergent, with insertion of secondary chambers or a partially septate interarea between divergent primary chambers. Chambers of nepionic and neanic stages inflated, becoming gradually larger in size, height and thickness of chambers approximately equal but width greatest, set obliquely to biserial axis; later primary chambers laterally more compressed, width increases but no longer biserially alternating; secondary chambers small, blister-like, irregularly shaped, introduced into a thinner, non-septate area between primary chambers. Primary and supplementary apertures low, interiomarginal arches, with variably developed lips and flaps. Sutures depressed to flush, oblique, straight to curved. Surface longitudinally costate, perforate.

Remarks: The divergent bioseries with secondary chambers in the partially septate interarea distinguish this genus from all other heterohelicids.

Brown (1969) envisioned the laterally depressed interarea to have originated from greatly extended apertural lips, which become perforate. These lips or panels, according to Brown, "bridge the gap between the divergent primary chambers" until the divergence becomes so great that supplementary chambers must also be introduced in the interarea in order to bridge the gap. At best this is an oversimplification. In no other planktonic species do apertural lips or flaps become perforate, no matter the degree of extension. These panels are interpreted to be modified, non-inflated, parallel-sided chambers. In the gerontic stage, the primary chambers may become reduced and indistinguishable from the irregularly shaped and positioned secondary chambers. These latter chambers become more numerous in the gerontic stage, often to the exclusion of the panel-like chambers.

Loeblich and Tappan (1964) regarded *Sigalia* as a junior synonym of *Gublerina*. However, *Sigalia*, determined to be a synonym of *Heterohelix* (Masters, in press), lacks the divergent bioseries typical of *Gublerina*.

De Klasz *et al.* (1969) proposed the new species *Gublerina prima* from the Turonian of Gabon. The illustration of the holotype is a line drawing which imparts little information. This specimen is more likely a *Ventilabrella austinana* Cushman. However, because of the poor quality of the type figure and the lack of topotypic material for this study, this species is not treated here.

Range: Campanian (LOS *Globotruncana concavata*—FOS *Pseudotextularia carseyae* Interval)—Maastrichtian (*Trinitella scotti* Interval).

Gublerina cuvillieri Kikoïne

Ventilabrella carseyae Plummer and van Wessem, 1943, p. 45, pl. 1, figs 41, 42.

Gublerina cuvillieri Kikoïne, 1948, p. 26, pl. 2, figs 10a–c; de Klasz, 1953b, p. 245, 246, pl. 8, figs 1a, b; Bettenstaedt and Wicher, 1956, p. 502, pl. 2, fig. 15; Sacal and Debourle, 1957, p. 13, pl. 3, fig. 6; Wille-Janoschek, 1966, p. 117, pl. 8, fig. 11; Lehmann, 1966a, p. 315, pl. 2, fig. 5; *non* Dupeuble, 1969, p. 159, p. 4, figs 17, 18; Brown, 1969, p. 57–59, pl. 2, fig. 7; pl. 3, figs 5a, b; pl. 4, figs 5a, b; Hanzlíková, 1972, p. 93, 94, pl. 24, figs 13 (?), 14.

Gublerina glaessneri Brönnimann and Brown, 1953, p. 155, 156, text figs 13a–14b; Montanaro Gallitelli, 1957, p. 140, 141, pl. 32, fig. 7; Salaj and Samuel, 1963, p. 107, pl. 8, figs 3a, b; Salaj and Samuel, 1966, p. 228, pl. 26, figs 3a, b.

Gublerina ornatissima (Cushman and Church). Montanaro Gallitelli, 1957 p. 140, 141, pl. 32, figs 1, 3 (non 2, 4–6b).

Description: Test flabelliform, moderately to rapidly increasing in width; small, initial planispiral coil, often obscured by ornamentation, followed by a brief biserial stage which quickly becomes divergent, secondary chambers inserted between divergent primary chambers. Chambers of primary series wider than thick or high with width of divergent portion being first subequal then reduced, set obliquely to biserial axis. Secondary chambers initially inserted between the divergent primary series, with reduced thickness, width and often height; with varying proportions of essentially parallel-sided, incompletely septate chambers and irregularly positioned and shaped chambers with minimal inflation, the latter being the exclusive component of the gerontic stage. Apertures low, interiomarginal arches. Sutures depressed to flush, oblique to curved. Surface coarsely costate in the coiled and biserial portions becoming reduced in later primary chambers.

Remarks: *Gublerina cuvillieri* is the only *Gublerina* with coarse costae developed on the pre-ephebic portion of the test. The ephebic and gerontic stages vary from individual to individual because of the irregular nature of the secondary chambers. This species probably developed from *Gublerina reniformis* (Marie).

Kikoïne's type figure gave no indication as to the presence of costae. De Klasz (1953b) was not able to locate the holotype of *Gublerina cuvillieri*, but did examine topotypes. He illustrated a coarsely costate hypotype, which has become a standard of reference.

Brönnimann and Brown (1953), believing *Gublerina cuvillieri* to lack coarse costae as the type figure indicated, established *G. glaessneri* for this form. Their species is now considered a synonym of Kikoïne's.

Loeblich and Tappan (1964) incorrectly believed *Ventilabrella ornatissima* Cushman and Church to be conspecific with *Gublerina cuvillieri*. Examination of the holotype of *V. ornatissima* has shown it to have an entirely different chamber arrangement. It is a valid species belonging to the genus *Platystaphyla* Masters.

Distribution: This is shown in Fig. 3.

Range: Campanian (FOS *Pseudotextularia carseyae*—FOS *Globotruncana ventricosa* Interval)—Maastrichtian (*Trinitella scotti* Interval).

Gublerina reniformis (Marie)

Pseudotextularia (Guembelina) acervulinoides Egger. Liebus, 1927, p. 375, 376, pl. 14, figs 3a, b (non 2).

Ventilabrella reniformis Marie, 1941, p. 185, pl. 28, figs 277a–c.

?*Ventilabrella compressa* van der Sluis, 1950, p. 20, 21, pl. 1, figs 1a–c.

Gublerina acuta de Klasz, 1953b, p. 246, 247, pl. 8, figs 3a, b; Perlmutter and Todd, 1965, p. 14, 15, pl. 2, fig. 18; Todd, 1970, p. 151, pl. 5, fig. 4.

Gublerina acuta robusta de Klasz, 1953b, p. 247, pl. 8, figs 4a–5b; Montanaro Gallitelli, p. 140, 141, pl. 32, fig. 9; Salaj and Samuel, 1963, p. 106, 107, pl. 8, figs 5a–6; Ansary and Tewfik, 1968, p. 44, pl. 3, figs 14a, b.

Gublerina hedbergi Brönnimann and Brown, 1953, p. 155, text figs 11a–12b.

Gublerina ornatissima (Cushman and Church). Montanaro Gallitelli, 1957, p. 140, 141, pl. 32, figs 2 (?) 4–6b; Graham, 1962, p. 105, pl. 19, figs 16Ca, b (*non* 16A–Bc); Brönnimann and Rigassi, 1963, pl. 17, fig. 4; Bandy, 1967, p. 25, text-fig. 11 (3).

Gublerina cuvillieri Kikoïne. Dupeuble, 1969, p. 159, pl. 4, figs 17, 18; Weaver *et al.*, 1969, pl. 3, fig. 1.

Gublerina reniformis (Marie). Brown, 1969, p. 59, pl. 2, fig. 6; pl. 4, figs 3a–4b; Cita and Gartner, 1971, pl. 5, fig. 2.

Description: Test flabelliform, moderately to rapidly increasing in width; with a small, initial planispiral coil succeeded by a short biserial stage, which later becomes divergent with secondary chambers between the divergent series. Chambers of the primary type gradually increasing size until the gerontic stage where the trend is reversed, width greater than other dimensions; set obliquely to biserial axis; becoming divergent in ephebic and gerontic stages. Secondary chambers may be thin, parallel-sided, partially septate or irregularly shaped and positioned with minimal inflation; the latter comprise the entire gerontic stage. Both primary

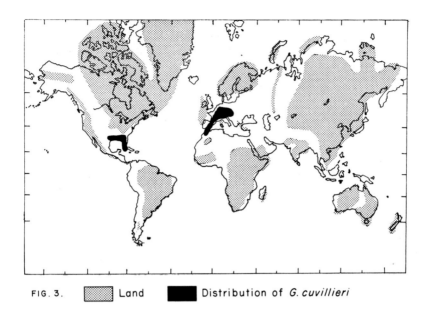

FIG. 3. ▨ Land ■ Distribution of *G. cuvillieri*

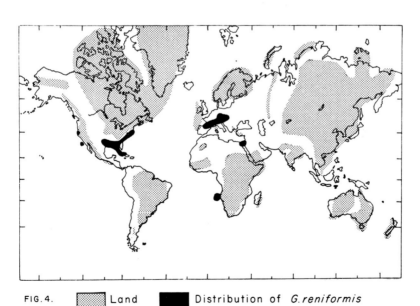

FIG. 4. ▨ Land ■ Distribution of *G. reniformis*

and secondary chambers tend to become reniform. Apertures low, interiomarginal arches. Sutures depressed to flush, oblique to curved. Surface with finely developed costae throughout.

Remarks: This species is finely costate throughout, whereas *Gublerina cuvillieri* Kikoïne has coarse costae in the early portion. The secondary chambers vary greatly, giving a wide range of test variants.

Brown (1969) envisioned a *Heterohelix pulchra* (Brotzen)—*H. pseudotessera* (Cushman)—*Gublerina reniformis* lineage. Although *Heterohelix pulchra* is not believed to be part of this lineage because of the presence of pore-mounds, *H. pseudotessera* is the most likely progenitor of *G. reniformis*. No transitional forms have been recognised, but with the presence of reniform gerontic chambers and depressed panels, little modification would be necessary.

Brown (1969) was the first to regard Marie's *Ventilabrella reniformis* as a *Gublerina*, and senior synonym of *G. acuta* de Klasz. The type figure is merely a line drawing. Although the figure does not preclude the possibility of it being a *Gublerina*, neither is it conclusive. Brown's conclusion is accepted until such time as an examination of the types of *V. reniformis* should prove otherwise.

In 1950 van der Sluis erected the new species *Ventilabrella compressa*. He described this taxon as having "a wedge-shaped plate between the two rows" of biserial chambers. This along with the type figure strongly suggests that this species is a *Gublerina*. It is tentatively referred to *G. reniformis* because it seems to lack the coarse ornamentation typical of *G. cuvillieri* Kikoïne.

Distribution: This is shown in Fig. 4.

Range: Campanian (LOS *Globotruncana concavata*—FOS *Pseudotextularia carseyae* Interval)—Maastrichtian (*Trinitella scotti* Interval).

Genus HETEROHELIX Ehrenberg, 1843

Heterohelix Ehrenberg, 1843, p. 429.

Spiroplecta Ehrenberg, 1844, p. 75.

Guembelina Egger, 1899, p. 31.

Type species: *Spiroplecta americana* Ehrenberg, 1844.

Description: Test small, initially with a minute planispiral coil followed by a biserial stage or entirely biserial; laterally compressed; both edge and side view lobate; thickness gradually increases, width gradually to rapidly. Chambers of coil appressed, margin entire, 3–12 in number, very gradually increasing in size, rectangular except spherical proloculus, which is imperforate with 2 subequal apertures, one of which remains open in completely biserial forms; biserially alternating chambers initially appressed, rectangular in side view, becoming subspherical or reniform later, up to 14 in number. Aperture a low to high interiomarginal, symmetrical arch, usually bordered by an imperforate lip or flap. Sutures oblique to biserial axis, straight to curved; median sutures zigzag. Costae of varying strengths always present; finely to coarsely perforate.

Remarks: *Heterohelix* is the most primitive of the Heterohelicidae, lacking supplementary and accessory apertures, supplementary and divergent chambers and the rapid increase in chamber thickness characteristic of other genera in this family.

Loeblich (1951) reviewed the taxonomic history of *Heterohelix*. The genus was originally described without naming a type species. From Ehrenberg's later works (1844, 1854), Loeblich determined that the name *Spiroplecta* was invalidly substituted for *Heterohelix*, and that *S. americana*, the type species of that genus, becomes the type species for *Heterohelix*.

Until Montanaro Gallitelli (1957) restudied *Heterohelix*, forms now assigned to the genus with the exception of *H. americana*, were assigned to *Guembelina*. Loeblich and Tappan (1961b) reaffirmed the conclusion that *Guembelina*, in addition to being a junior homonym of *Guembelina* Kuntz (1895), is a junior synonym of *Heterohelix*.

Stenestad (1968a) presented an excellent description and discussion of the genus *Heterohelix*. In referring to the costae, he stated that the "character of the shell surface by itself is hardly suitable as a criterion for distinguishing between the species, since striation may be present in 'smooth' species and 'striate' species may be only slightly striate or partly smooth." The author agrees that it is generally unwise to speciate this or any other genus on a single feature. The nature of the costae in this genus are as valuable as any other criterion of speciation, but must be used in conjunction with all other distinctive features.

The concept of unornamented species of *Heterohelix* has arisen because of the inadequate resolution of light microscopy. All *Heterohelix* are costate, except *H. pulchra* (Brotzen) and *H. glabrans* (Cushman) which possess pore-mound surface structures. It is recommended that the use of the term "striae" or any derivative thereof be discontinued and replaced by appropriate forms of the term "costae". "Striae" in the sense of Stenestad (1968a) and others refers to very fine costae. However, by definition striae are narrow grooves. These structures are ridges, not grooves.

The type description and figures of *Heterohelix* (*Chiloguembelina*) *pseudotessera* subsp. *directa* Aliyulla (1965) are inadequate for recognition. However, *Chiloguembelina* is neither a subgenus of *Heterohelix* nor found in Cretaceous strata.

The remark by El-Naggar (1971b) that "chamber proliferation" is "considered of no taxonomic importance" is totally unacceptable. Such a remark stems from a misunderstanding of the origin and the development of supplementary chambers and their shapes. His treatment of *Pseudotextularia*, *Planoglobulina* and *Racemiguembelina* as synonyms of *Heterohelix* is rejected. It is interesting, though somewhat puzzling, that he retains both *Gublerina* and *Ventilabrella*.

Range: Albian–Maastrichtian (*Trinitella scotti* Interval).

Heterohelix americana (Ehrenberg), Plate 1, figs 1–3
Spiroplecta americana Ehrenberg, 1844, p. 75; Ehrenberg, p. 854, p. 24, pl. 32I,
figs 13, 14; pl. 32II, fig. 25; Ehrenberg, 1856, p. 175, pl. 7, fig. 6; Egger, 1899, p. 30,
pl. 14, fig. 23.
Guembelina striata Ehrenberg. Egger, 1899, p. 33, pl. 14, figs 5–7 (?), 39 (*non* 10,
11, 37, 38).
Heterohelix americana (Ehrenberg), Cushman, 1927a, p. 214, pl. 34, figs 13, 14;
pl. 36, fig. 25; Cushman, 1928b, p. 229, pl. 33, fig. 1; pl. 34, figs 1a, b; Cushman,
1933a, pl. 26, fig. 1; Cushman, 1933c, pl. 21, figs 1a, b; Cushman 1946, p. 101, pl.
44, fig. 3; Loeblich, 1951, p. 107, text-fig. 1; Hofker, 1957, p. 423, text-figs 479d, e;
Loeblich and Tappan, 1964, p. C652–654, text-fig. 523: 5a, b.
Heterohelix americanus (Ehrenberg). Bykova and Subotina, 1959, p. 338, text-figs
850a, b.
Heterohelix pulchra (Brotzen). Olsson, 1960, p. 27, pl. 4, fig. 6.
Heterohelix navarroensis Loeblich 1951, p. 107, 108, pl. 12, figs 1–36, text-fig. 2;
non Olsson, 1960, p. 27; pl. 4, fig. 5; Pessagno, 1967, p. 261, pl. 89, figs 8, 9; *non*
Rasheed and Govindan, 1968, p. 79, text-figs 3, 4; Hanzlíková, 1972, p. 40, pl. 7,
figs 11a–12, 23; Brown, 1969, p. 33, 34, pl. 1, figs 1a–6b; Bertels, 1970, p. 31, pl. 2,
figs 1a–2; Hanzlíková, p. 91, 92, pl. 23, figs 12a, (?) b, 13.
Description: Test small with an initial planispiral coil, gradually increasing in
size; periphery in side view mildly lobate, periphery of coil entire. Chambers of coil
3–12, commonly 6, in number, closely packed, non-costae, imperforate, with an
acute but non-keeled periphery. Chambers of biserial portion appressed, rectangular
in side view, gradually increasing in size; 6–12, commonly 10, in number. Aperture
a low, interiomarginal arch bordered by a lip which is narrow at the crest of the
arch becoming wider at its attachment to previous chamber. Sutures depressed,
straight to gently curved, oblique to biserial axis. Surface finely perforate between
moderately coarse, discontinuous costae curving away from the aperture in a
radiating fashion; costae tend to merge at the back of the chamber, but never form
a true keel.
Variation: The rate of test width expansion varies resulting in very slender side
views or more flaring ones. The greatest variable is the initial coil, which may be
so small as to appear non-coiled or to be large, imparting a rather blunt appearance
to the initial end of the test.
Remarks: The combination of the typically large coil, appressed chambers and
typically slender test distinguish this species from all other species assigned to
Heterohelix. The rate of expansion of the width of the test varies and results in very
slender or more flaring side views. Greatest variation occurs in the initial coil which
may be so small that it appears non-coiled, or so large that the early part of the test
has a blunt appearance. A non-coiled generation has not yet been recognised in
H. americana. Although this generation may be represented by *H. carinata* (Cushman) these two taxa are retained here because of their different statigraphic ranges.
Heterohelix americana strongly resembles *H. carinata* which is considered to be
ancestral. The main differences are the lack of keels and development of a coil in
H. americana.
Brown (1969) maintained that the name *Heterohelix navarroensis* Loeblich is
the senior synonym for this taxon on the basis that "*H.* [*Spiroplecta*] *americana*
(Ehrenberg), 1844, is a junior secondary homonym of *H.* [*Textilaria*] *americana*
(Ehrenberg), 1843". No type figure was given by Ehrenberg for his *Textilaria*

americana, and its type description is inadequate for recognition. Furthermore, because it has not been reported within the last fifty years, it should be considered a nomen oblitum. This would leave *H*. [*Spiroplecta*] *americana* as the valid senior synonym.

Distribution: This is shown in Fig. 5.

Range: Maastrichtian (FOS *Globotruncana aegyptiaca*—FOS *Rugoglobigerina hexacamerata* Interval to *Trinitella scotti* Interval).

Heterohelix bejaouensis Salaj and Maamouri

Sigalia bejaouensis Salaj and Maamouri, 1971, p. 75, 76, text-figs 5a, b.

Description: Test biserial, with laterally compressed subparallel sides; rapidly expanding in side view, often as wide as high; profile slightly lobate. Chambers 10–15 in number; initially rectangular in side view later becoming subcircular; sides parallel or with very slight inflation; rapidly increasing in width. Aperture a moderate to high, interiomarginal arch bordered by a narrow imperforate lip. Numerous discontinuous, moderately coarse costae conform to chamber curvature as seen in side view. Walls densely perforated by fine to moderate diameter pores, irregularly arranged between costae. Sutures depressed; in early chambers, straight, oblique to biserial axis; later curved convexly in direction of growth; median suture formed by relict apertural lips, wide, depressed.

Variation: There is slight variation in the rate of increase in the test width. It is typically as wide as it is high.

Remarks: *Heterohelix bejaouensis* differs from its congeners by its subparallel sides along with rapid increase in test width. There is only a slight variation in the rate of increase in test width and the species is as high as it is wide. This species evolved from *Heterohelix globulosa* (Ehrenberg) by maintaining essentially a constant chamber and test thickness.

Although neither primary types nor topotypes of *Heterohelix bejaouensis* have been examined, its distinctive characters seem to be identical to those of *H. sphenoides* Masters (1976).

Distribution: This is shown in Fig. 6.

Range: Campanian (? FOS *Globotruncana ventricosa*—FOS *Pseudoguembelina costata* Interval to *Globotruncana calcarata* Interval).

Heterohelix carinata (Cushman), Plate 1, figs 4, 5

Guembelina carinata Cushman, 1938a, pl. 3, figs 10a, b; Cushman, 1944b, p. 92, pl. 14, figs 6a, b; Cushman, 1946b, p. 105, pl. 45, figs 8a, b; (?) Shaw, 1953, pl. 1, fig. 22; Frizzell, 1954, p. 108, pl. 15, figs 13a, b.

Non *Heterohelix carinata* (Cushman). Montanaro Gallitelli, 1957, p. 137–139, pl. 31, figs 16, 17.

Heterohelix pachymarginata, Stenestad, 1968a, p. 66, 67, pl. 1, figs 1, 2; pl. 2, figs 4–6; Stenestad, 1969, p. 659, 660, pl. 1, figs 15, 16; text-figs 14a–c.

Description: Test small, biserial throughout, laterally compressed; with a thickened peripheral keel along the chamber edges, periphery typically entire in all but last 1–2 chambers. Chambers 14–16 in the adult, with greatest inflation near but not at the biserial axis, tapering to keeled edge; moderately increasing in width; secondary calcification typical over vertex of chamber near biserial axis. Costae moderately coarse, continuous in calcified areas, fine, discontinuous elsewhere. Proloculus with a single open aperture or primary pore at initial end. Aperture a low, broad, interiomarginal arch bordered by a moderately wide imperforate lip.

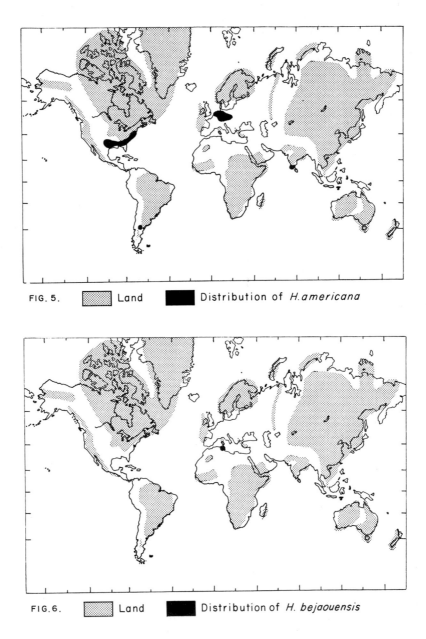

FIG. 5.　▨ Land　■ Distribution of *H. americana*

FIG. 6.　▨ Land　■ Distribution of *H. bejaouensis*

Wall irregularly and finely to coarsely perforate; proloculus and first 1–2 chambers imperforate. Sutures depressed, curving downward at keeled periphery.

Remarks: The keeled edges, degree of lateral compression and small size distinguish this species from its congeners. All characteristic features of this species appear to be rather stable.

Heterohelix carinata may have originated from *H. semicostata* (Cushman) by developing a completely costate test and a keeled periphery or from *H. bejaouensis* (Salaj and Maamouri) by a reduction in size and the development of a keel. No intermediate forms are yet recognised. *Heterohelix semicostata* is superficially more similar.

As previously noted (Masters, in press), the contour of the chambers and the median suture combine to occasionally appear in light microscopy as accessory apertures. Examination of numerous topotypes by scanning electron microscopy have established that they are not present in this species.

Although the photographs of the holotype of *Heterohelix pachymarginata* Stenestad (1968a) are not clear, there are no features illustrated or discussed which preclude it being a junior synonym of *H. carinata* (Cushman). Stenestad suggested that *H. carinata* probably has a true carina rather than the thickened margin of his species. Masters (in press) referred to the margin of *H. carinata* as being keeled. This keel, as revealed by scanning electron micrographs, is not single but a broad thickened one as described for *H. pachymarginata*. This is merely a matter of semantics. In this respect the two appear identical.

Distribution: This is illustrated in Fig. 7.

Range: Campanian (FOS *Globotruncana elevata*—LOS *Globotruncana concavata* Interval)—Maastrichtian (*Trinitella scotti* Interval).

Heterohelix digitata Masella

Heterohelix digitata Masella, 1959, p. 15, 17, pl. figs 1a–10.

Description: Test biserial, slowly to moderately increasing in width; edge rounded in earlier portion, digitate in gerontic stage. Chambers subspherical in all but last 2, which become elongate parallel to the biserial axis and form a digitate extension on the backs of the chambers; these extensions may be perpendicular to the chamber wall or more typically oblique, sloping back toward the older chambers; 7–10 in number; gradually increasing in size. Aperture a low, interiomarginal arch. Sutures depressed, straight to gently convex adaperturally.

Remarks: *Heterohelix digitata* differs from all other species assigned to this genus by possessing digitate chamber extensions. (See also discussion under *Lunatriella spinifera* Eicher and Worstell, 1970a). The height and degree of digitation of the ultimate and penultimate chambers and the test width are the main variables of this species.

The test width of some of the specimens illustrated by Masella suggest a close relationship to *Heterohelix moremani* (Cushman). Others, including the holotype, more closely resemble *H. globulosa* (Ehrenberg). Ancestral forms will most likely be found in one or the other of these species.

Masella (1959) described *Heterohelix digitata* as having an initial planispiral coil. In each of the thin-section sketches, she shaded the "proloculus". This chamber has a diameter slightly smaller than or almost the same as the chamber which forms the tip of the test. It is not uncommon for the second chamber to be smaller than the proloculus in many species, both planktonic and benthonic. I believe that this is the case with *H. digitata* and that no coil is present.

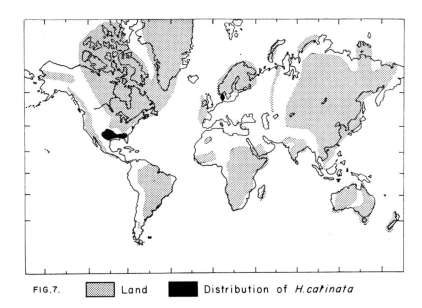

FIG.7.   [Land]   [Distribution] of *H. carinata*

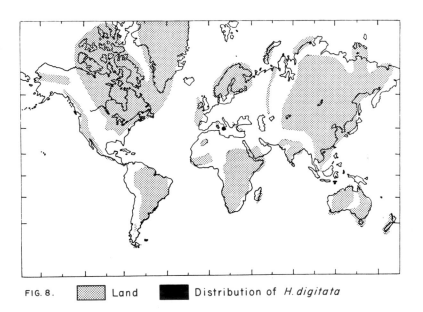

FIG. 8.   [Land]   [Distribution] of *H. digitata*

This species was also originally described as being unornamented. This is unlikely as all other heterohelicids have costae.

Ehrenberg (1854) described and illustrated under the name of *Textilaria missouriensis* a form which also possesses chamber extensions. Because of Ehrenberg's stylized, transmitted light illustrations, it is not possible to determine whether his and Masella's species are synonymous.

Distribution: This is shown in Fig. 8.

Range: Campanian (FOS *Globotruncana elevata*—LOS *Globotruncana concavata* Interval to *Globotruncana calcarata* Interval).

Heterohelix glabrans (Cushman), Plate 1, figs 6, 7

Guembelina glabrans Cushman, 1938a, p. 15, pl. 3, figs 1a–2; Cushman, 1946b, p. 109, pl. 46, figs 17a–18; Kikoïne, 1948, p. 17, pl. 1, figs 2a–c.

Heterohelix glabrans (Cushman). Olsson, 1960, p. 26, 27, pl. 4, fig. 4; Pessagno, 1967, p. 259, pl. 88, figs 1, 2, 10, 11; Ansary and Tewfik, 1968, p. 39, pl. 3 figs 12a, b; *non* Stenestad, 1968b, pl. 2, fig. 9; Stenestad, 1969, p. 656, 657, pl. 1, figs 5–7; text-figs 10a–e; Hanzlíková, 1969, p. 38, pl. 7, figs 1a–2, 18; Todd, 1970, p. 148, pl. 5, fig. 2; Hanzlíková, 1972, p. 90, pl. 23, figs 1a, b; Govindan, 1972, p. 169, pl. 1, figs 11, 12.

Heterohelix pulchra (Brotzen). Stenestad, 1969, p. 654, 656, pl. 1, fig. 3 (*non* 1, 2); text-figs 9a–c; Neagu, 1970, p. 60, pl. 14, figs 6–8; Govindan, 1972, p. 168, 169, pl. 1, figs 9, 10.

Description: Test small, biserial, highly compressed laterally, rapidly increasing in width; edge acute but not keeled, lobate. Chamber-backs form a vertically acute edge, greatest inflation near midpoint of side or slightly toward biserial axis; rapidly but regularly increasing in width, width nearly double thickness; 14–15 in number; early chambers rectangular in side view, later becoming circular. Aperture a high, interiomarginal arch bordered by a wide imperforate lip. Sutures depressed, straight to slightly curved, oblique to biserial axis. Surface finely perforate, with pore mounds longitudinally aligned.

Remarks: *Heterohelix glabrans* differs from *H. carinata* (Cushman), the only other species with an acute periphery, by having a lobate, non-keeled edge. This species is one of two *Heterohelix* species which show pore mound development. The features of this distinctive species remain more or less constant.

The ancestor of *Heterohelix glabrans* has not yet been recognised.

This species is one of two *Heterohelix* with a pore mound development. No keel is developed as claimed by Cushman (1938a).

Distribution: This is shown in Fig. 9.

Range: Maastrichtian (LOS *Globotruncana calcarata*—LOS *Globotruncana lapparenti* Interval to *Trinitella scotti* Interval).

Heterohelix globulosa (Ehrenberg)*, Plate 1, figs 8, 9

Textularia globulosa Ehrenberg, 1840, p. 135, pl. 4, figs 2, 4, 5, 7, 8.

Textilaria globulosa (Ehrenberg). Ehrenberg, 1854, pl. 23, figs 3–6; pl. 24, figs 12–14; pl. 25, figs A8, 9, 11, 13; pl. 26, fig. 10 (*non* 9); pl. 27, fig. 6; pl. 28, figs 9, 10; pl. 29, figs 17a, b; pl. 30, figs 3a, b; pl. 32:1, fig. 8?; pl. 32:2, fig. 12.

Textilaria globifera Reuss, 1860, p. 232, pl. 13, fig. 8 (*non* 7a, b).

(?) *Oligostegina laevigata* Kaufmann, 1865, in Heer, p. 197, text-figs 108a–c.

(?) *Textularia brevicona* Perner, 1892, p. 25, pl. 9, figs 12a, b.

* (A complete synonymy list is available from the author upon request.)

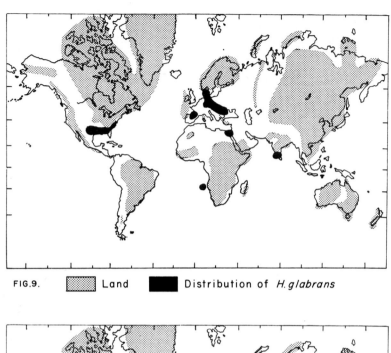

FIG.9. ▨ Land ■ Distribution of *H.glabrans*

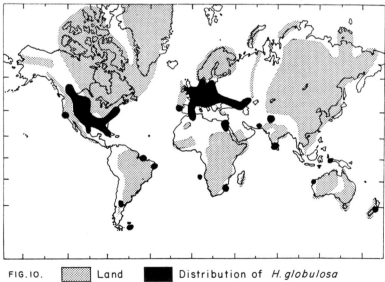

FIG.10. ▨ Land ■ Distribution of *H.globulosa*

Textularia decurrens Chapman, 1892, p. 515, pl. 15, fig. 6.
Guembelina globulosa (Ehrenberg). Egger, 1899, p. 32, pl. 14, fig. 43.
Guembelina ultimatumida White, 1929, p. 39, pl. 4, figs 13a, b.
Pseudotextularia globulosa (Ehrenberg). Macfadyen, 1933, p. 10, text-fig. 1 (K, L).
Guembelina tessera (Ehrenberg). Voorwijk, 1937, p. 194, pl. 1, figs 3, 4.
Guembelina globocarinata Cushman, 1938a, p. 10, 11, pl. 2, figs 4a–5.
Guembelina reussi Cushman, 1938a, pl. 11, pl. 2, figs 6a–9b.
Guembelina complanata Marie, 1941, p. 184, pl. 28, figs 276a–c.
(?) *Pseudotextularia trilocula* Marie, 1941, p. 186, pl. 28, figs 278a–d.
Guembelina globulosa var. *striatula* Marie, 1941, p. 182, 183, pl. 28, figs 273a–275b.
Guembelina ventilabrelliformis van der Sluis, 1950, p. 21, pl. 1, figs 2a–c.
Heterohelix globulosa (Ehrenberg). Montanaro Gallitelli, 1957, p. 137–139, pl. 31, figs 12–15.
Heterohelix orientalis Kavary, 1963, p. 61, 62, pl. 12, figs 27–29.
"*Pseudoguembelina*" *frizzelli* Kavary, 1963, p. 66, pl. 13, figs 19, 20.
Heterohelix (*Pseudoguembelina*) *striata* subsp. *anfracta* Aliyulla, 1965, p. 233, pl. figs 4a, b.
Heterohelix (*Pseudoguembelina*) *porosa* Aliyulla, 1965, p. 224, pl. figs 2a–3b.
Ventilabrella (?) *austinana* Cushman. Martin, 1972, p. 84, 85, pl. 1, figs 1, 2.

Description: Test with an initial coil of 4–5 chambers or biserial throughout, moderately to rapidly increasing width and thickness, edge rounded, periphery lobate. Chamber width, height and thickness subequal; inflated; 9–12 in number; early chambers rectangular to square in side view, becoming circular. Aperture a moderately high to high, interiomarginal arch bordered by a narrow imperforate lip. Wall with narrow to wide diameter pores, irregularly spaced; with fine to moderately coarse, discontinuous costae, radiating from the aperture following the curvature of the chamber. Proloculus and initial coil imperforate. Sutures depressed, straight, oblique to biserial axis.

Variation: An abrupt increase in the rate of chamber growth may occur. An occasional supplementary chamber may form at the base of the back of a chamber. The costae become more coarse through younger strata (Masters, in press).

Remarks: *Heterohelix globulosa* is distinguished from *H. moremani* (Cushman) by having fewer chambers in the adult test and a more rapidly increasing test width. It differs from *H. striata* (Ehrenberg) by possessing more numerous and less coarse costae. The following variations have been recorded: an abrupt increase in the rate of chamber growth, the formation of occasional supplementary chambers on the base of the back of a chamber, and a coarsening of the costae through younger strata (Masters, in press).

Heterohelix moremani (Cushman) is thought to be ancestral to *H. globulosa*.

Heterohelix globulosa is the most frequently cited Mesozoic species, probably because it usually outnumbers all other species in samples which span its stratigraphical range. Owing to very generalised illustrations and descriptions it is difficult to evaluate some of the earlier citations.

Pessagno (1967) designated fig. 5β of Ehrenberg (1840) as representative of the lectotype of *Heterohelix globulosa*. Unfortunately, any such designation is premature and fruitless without concurrent examination of Ehrenberg's specimens.

Kaufman (1865) described *Oligostegina laevigata* from thin-section. Subsequently the genus has become a catchall for several unrelated organisms; e.g. *Pithonella* and *Calcisphaerula* (Incertae Sedis), and *Orbulina* (Foraminiferida). *Oligostegina laevi-*

gata as interpreted by Kaufman, are nothing more than oblique sections of *Heterohelix*, which show only one to three chambers. It is uncertain as to which *Heterohelix* species is involved, because three-dimensional specimens are necessary for species recognition. *Heterohelix globulosa* is a likely candidate because it occurs in such prodigious numbers.

Marie's (1941) *Guembelina globulosa* var. *striatula* and *G. complanata* appear to be conspecific with *Heterohelix globulosa*. *Pseudotextularia trilocula* Marie, 1941, is an aberrant *H. globulosa*, which has an additional chamber contrary to the normal biserial growth pattern. Such forms, although not common, can be recognised in any population.

Guembelina ventilabrelliformis van der Sluis, 1950, was misinterpreted by its author as becoming triserial in the final growth stage. The ultimate chamber of the holotype is reduced in size and positioned more or less axially. Such a phenomenon is common in the heterohelicids. This typically represents the final growth stage, after which no other chambers are added. It is still a biserial pattern. Van der Sluis' species is a synonym of *Heterohelix globulosa*.

The poorly preserved holotype of *Heterohelix orientalis* Kavary, 1963, is identical with *H. globulosa*. The holotype of "*Pseudoguembelina*" *frizzelli* Kavary, 1963, had large, inflated final chambers which collapsed during preservation, a fact not recognised by its author. This species, also, is a synonym of *H. globulosa*.

From their type descriptions and their distributions, the specimens illustrated and named by Aliyulla (1965) as *Heterohelix* (*Pseudoguembelina*) *striata* subsp. *anfracta* and *H.* (*Pseudoguembelina*) *porosa* are regarded as junior synonyms of *H. globulosa*.

The author has previously concluded (in press) that *Heterohelix globulosa* is the senior synonym of *Guembelina reussi* Cushman, *G. ultimatumida* White and *G. globocarinata* Cushman.

Distribution: This is shown in Fig. 10.

Range: Cenomanian—Maastrichtian (*Trinitella scotti* Interval).

Heterohelix moremani (Cushman), Plate 2, fig. 1

Textilaria globulosa Reuss (*non* Ehrenberg), 1846, p. 39, pl. 12, fig. 23.

Textilaria globifera Reuss, 1860, p. 232, pl. 13, figs 7a, b (*non* 8).

Textularia pygmaea d'Orbigny, (?) Calvin, 1895, p. 226, 227, pl. 19, fig. 7.

Guembelina globifera (Reuss), Egger, 1899, p. 33, pl. 14, figs 35, 36, 53, 55; Carman, 1829, p. 311, 312, pl. 34, fig. 3; Loetterle, 1937, p. 34, pl. 5, fig. 3.

Guembelina moremani Cushman, 1938a, p. 10, pl. 2, figs 1a–3; Cushman, 1944b, p. 90, pl. 14, figs 1a, b; Cushman, 1946b, p. 103, pl. 44, figs 15a–17; Kikoïne, 1948, p. 18, pl. 1, figs 4a–c; Bolin, 1952, p. 38, 39, pl. 2, figs 17a, b; Shaw, 1953, pl. 1, figs 24, 25; Frizzell, 1954, p. 109, pl. 15, figs 28–30b; Bolin, 1956, p. 290, pl. 38, figs 15a, b.

Guembelina washitensis, Tappan, 1940, p. 115, pl. 19, figs 1a, b.

Guembelina globulosa (Ehrenberg), Cushman, 1946b, p. 105, 106, pl. 45, fig. 9 (*non* 10–15b); Young, 1951, p. 63, pl. 14, fig. 23 (*non* 12, 24–26); Frizzell, 1954, p. 109, pl. 15, fig. 24 (*non* 25a–27b); Ansary, Andrawis and Fahmy, 1962, pl. 4 (part).

Heterohelix moremani (Cushman), Petri, 1962, p. 88, 89, pl. 11, figs 1, 2; Pessagno, 1967, p. 260, 261, pl. 48, figs 10, 11; pl. 89, figs 1, 2; Michael, 1967, p. 208, pl. 6, fig. 12; Bandy, 1967, p. 22, text-fig. 11 (1), 12 (1B); Barr, 1968, pl. 1, figs 14, 15; Brown, 1969, p. 35, 36, pl. 1, figs 8a, b; *non* Neagu, 1970b, p. 60, pl. 14, figs 9, 10.

Heterohelix washitensis (Tappan), Bandy, 1967, p. 22, text-fig. 12 (1A); Brown, 1969, p. 34, 35, pl. 1, figs 7a, b; (?) Gower-Biedowa, 1972, p. 61, 62, pl. 5, figs 3a, b.

Heterohelix globulosa (Ehtenberg) Eicher and Worstell, 1970, p. 296, pl. 8, figs 3, 4 (? *non* 5, 6).

Guembelina boliviniformis Agalarova, 1951, in Dzhafarov, Agalarova, and Khalilov, p. 88, pl. 13, figs 9, 10.

Description: Test slender, biserial, edge rounded, periphery moderately lobate. Chambers subequal in width, height and thickness or with width slightly greater; 13–20 in number; very slowly increasing in size; appressed, square to rectangular in side view. Aperture a low to moderately high, interiomarginal arch bordered by a narrow, imperforate lip; an occasional single supplementary aperture and chamber may occur at the base of the back side of the primary chamber. Walls finely perforate; with closely spaced, thin, discontinuous costae. Sutures depressed, straight, abaperturally oblique to the biserial axis.

Remarks: The very slowly increasing test width and the large number of chambers, 13–20, distinguish *Heterohelix moremani* from all its congeners except *H. stenopos*, which is more laterally compressed and has coarser costae. The morphological features of this species are constant; only the costae tend to coarsen slightly through time.

This species is the oldest known *Heterohelix*, and its progenitor has not been recognised.

Heterohelix washitensis (Tappan), often regarded as non-costae (e.g. Brown, 1969), has been shown to possess very thin costae, and to be a junior synonym of *H. moremani* (Masters, in press).

The holotype, refigured by Pessagno (1967), appears to be a calcite cast or internal mold.

Distribution: This is shown in Fig. 11.

Range: Mid-Albian—Maastrichtian (*Trinitella scotti* Interval).

Heterohelix planata (Cushman), Plate 2, fig. 3

Guembelina planata Cushman, 1938a, p. 12, 13, pl. 2, figs 13a–14; Cushman, 1946b, p. 105, pl. 45, figs 6a, b (*non* 7); Frizzell, 1954, p. 109, pl. 15, figs 32a, b.

Heterohelix planata (Cushman), *non* Montanaro Gallitelli, 1958, p. 144, pl. 2, figs 9, 10; Pessagno, 1967, p. 261, 262, pl. 86, figs 3, 4; (?) Bandy, 1967, p. 25, text-fig. 12 (8); *non* Neagu, 1970b, p. 60, pl. 13, figs 34–36.

Heterohelix pulchra (Brotzen), Weaver *et al.*, 1969, pl. 3, figs 2a, b.

Description: Test biserial, width moderately increasing, edge rounded, periphery lobate. Chamber height and thickness subequal with the width being approximately 50% greater; 12–14 chambers, moderately increasing in size; early chambers rectangular in side view, later becoming more tear-drop shaped with the apex positioned at the median suture; the ephebic tear-drop shaped chamber gradually becomes less inflated toward the median sutures and merges into the apertural flap. Proloculus imperforate. Aperture a moderately high, interiomarginal arch; bordered by a wide, imperforate flap which tends to flare open, the narrowest width is at the crest of the apertural arch. Sutures depressed, straight, slanted back toward older chambers. Surface finely perforate; with fine to moderately coarse, discontinuous, closely spaced costae which radiate from the aperture following the curvature of the chamber.

Remarks: *Heterohelix planata* lacks the depressed panels and well-developed reniform chambers typical of *H. pseudotessera* (Cushman). It differs from *H. globulosa* (Ehrenberg) by having ephebic chambers with greater overlap. There is some variation in the width of the apertural flap within members of this species.

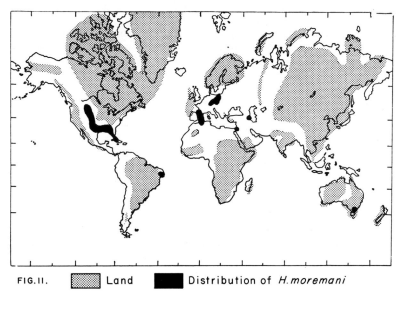

FIG.II.   ▒ Land   ■ Distribution of *H. moremani*

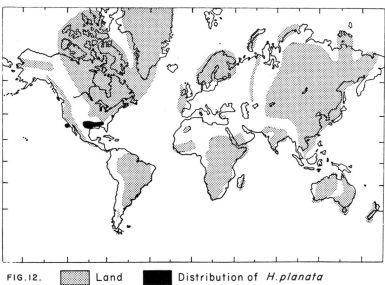

FIG.12.   ▒ Land   ■ Distribution of *H. planata*

Heterohelix planata evolved from *H. globulosa* (Ehrenberg) by developing an accentuated apertural flap and a marked increase in the degree of overlap in the ephebic chambers.

The holotype of *Heterohelix planata* does not have a keel as presumed by Cushman (1938a). Nor are there "depressed triangular areas" as given in the type description; these are merely portions of the chambers which are gradually less inflated, but never depressed. The holotype was refigured by Pessagno (1967).

Distribution: This is shown in Fig. 12.

Range: Campanian (FOS *Pseudotextularia carseyae*—FOS *Globotruncana ventricosa* Interval)—Maastrichtian (FOS *Globotruncana aegyptica*—FOS *Rugoglobigerina hexacamerata* Interval).

Heterohelix pseudotessera (Cushman), Plate 1, figs 10, 11

Guembelina pseudotessera Cushman, 1938a, p. 14, pl. 2, figs 19, 20 (*non* 21); Cushman, 1946b, p. 106, 107, pl. 45, figs 16, 17 (*non* 18); Frizzell, 1954, p. 109, pl. 15, figs 34a, b (*non* 33); *non* Ansary and Fakhr, 1958, p. 121, pl. 1, figs 27a, b.

Heterohelix pseudotessera (Cushman), Barr, 1968, pl. 1, fig. 12 (*non* 13).

Description: Test biserial, moderately to rapidly increasing width, edge rounded, periphery moderately lobate. Chambers of early portion equidimensional or gradually increasing in width; last 1–3 become increasingly reniform, partially separated by medially located, depressed, flat, subtriangular panels; 12 16 chambers, regularly increasing in size. Aperture a moderately high, interiomarginal arch. The panels appear to be a single wall, but probably are broad, parallel, closely spaced, apertural flaps. Wall finely and irregularly perforate, panels generally imperforate. Surface with fine, closely spaced, discontinuous costae following the chamber curvature from the aperture. Sutures depressed, straight, angled abaperturally.

Remarks: *Heterohelix pseudotessera* differs from *H. plabata* Cushman by possessing depressed panels and reniform chambers. It is separated from *H. pulchra* (Brotzen), the only other species with reniform chambers, by the presence of panels and the absence of pore mounds. The most variable character is panel size, the increase of which accentuates the reniform chamber shape.

The phylogeny of *H. pseudotessera* is unconfirmed though this species probably arose from *Heterohelix globulosa* (Ehrenberg), which it closely resembles in all but the final reniform chambers.

The holotype and paratypes were examined during this study. Cushman's fig. 20 is a *Heterohelix pulchra* (Brotzen).

Distribution: This is shown in Fig. 13.

Range: Campanian (FOS *Globotruncana elevata*—LOS *Globotruncana concavata* Interval—*Globotruncana calcarata* Interval).

Heterohelix pulchra (Brotzen), Plate 2, fig. 2

Guembelina tessera (Ehrenberg), Cushman, 1932, p. 338, pl. 51, figs 4, (?) 5a, b; Loetterle, 1937, p. 34, pl. 5, fig. 4; Kalinin, 1937, p. 37, pl. 4, figs 52, 53.

Guembelina pulchra, Brotzen, 1936, p. 121, pl. 9, figs 3a, b (*non* 2a, b); Cushman, 1938a, p. 12, pl. 2, figs 12a, b; Hofker, 1956f, p. 77, pl. 9, fig. 69.

Guembelina pseudotessera Cushman, 1938a, p. 14, 15, pl. 2, fig. 21 (*non* 19a, b, 20); Cushman, 1944b, p. 91, pl. 14, figs 5a, b; Cushman, 1946b, p. 106, 107, pl. 45, figs 18–20a (*non* 16a–17), Schijfsma, 1946, p. 76, 77, pl. 4, fig. 7; Hamilton, 1953, p. 234, pl. 30, fig. 14; Frizzell, 1954, p. 109, pl.15, fig. 33 (*non* 34a, b).

Guembelina dagmarae Suleimanov, 1955, p. 624, text-figs 2a, b.

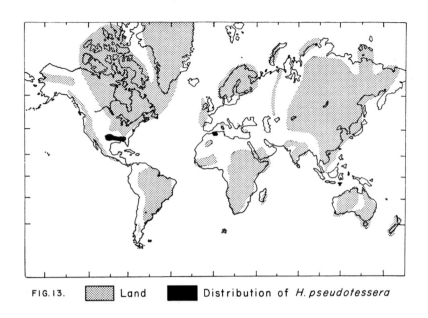

FIG.13. [Land]   [■]   Distribution of *H. pseudotessera*

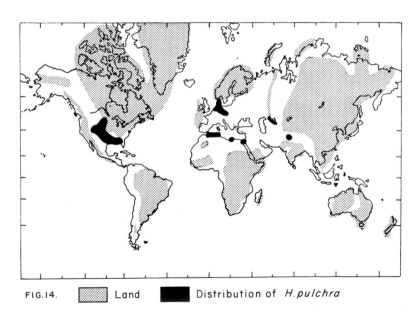

FIG.14. [Land]   [■]   Distribution of *H. pulchra*

Heterohelix pulchra (Brotzen), *non* Montanaro Gallitelli, 1958, p. 144, 145, pl. 2, figs 11, 12; *non* Pessagno, 1962, p. 358, pl. 1, fig. 3; *non* Takayanagi, 1965, p. 197, 198, pl. 20, figs 3a, b; Kent, 1967, p. 1446, pl. 183, fig. 9; (?) Bandy, 1967, p. 23, text-fig. 12 (7); Pessagno, 1967, p. 262, pl. 87, fig. 4; *non* Sliter, 1968, p. 95, 96, pl. 14, figs 4a–6, 9a, b; Ansary and Tewfik, 1968, p. 40, pl. 3, figs 8a, b; *non* Weaver *et al.*, 1969, pl. 3, figs 2a, b; Stenestad, 1969, p. 654, 656, pl. 1, figs 1, 2 (*non* 3; text-fig. 9a–c); Eicher, 1969, text-fig. 3 (part); (?) Douglas and Rankin, 1969, p. 190, 191, text-figs 4A, B; Mello, 1969, p. 71, 72, pl. 8, figs 4a, b; *non* Douglas, 1969, p. 158, pl. 11, figs 3a, b, 14a, b; Eicher and Worstell, 1970b, p. 296, pl. 8, figs 9, 10; *non* Neagu, 1970b, p. 60, pl. 14, figs 6–8; (?) Rahhali, 1970, p. 68, text-figs 7i–k; Eicher and Worstell, 1970a, pl. 1, figs 1–3; *non* Morris, 1971, p. 280, pl. 7, figs 2a, b; *non* Mello, 1971, p. 35, pl. 5, figs 10a, b; *non* Hanzlíková, 1972, p. 92, pl. 23, figs 10a–11; *non* Govindan, 1972, p. 168, 169, pl. 1, figs 9, 10.

Bolivinoides senonicus Dain. Balakhamatova, 1960, in Glazunova *et al.*, p. 98, pl. 17, figs 2a, b.

(?)*Heterohelix dentata* Stenestad, 1968a, p. 67, 68, pl. 1, figs 3–6, 8, 9; pl. 2, figs 1–3.

Heterohelix pseudotessera (Cushman), Barr, 1968, pl. 1, fig. 13 (*non* 12).

Heterohelix lata (Egger), Hanzlíková, 1969, p. 40, pl. 7, figs 8–10, 20–22; Hanzlíková, 1972, p. 91, pl. 23, figs 5, 6 (*non* 7a, b).

Description: Test with an initial planispiral coil or biserial throughout, gradually to rapidly increasing in width; edge rounded; early periphery slightly lobate, later deeply lobate. Chambers of neanic portion subspherical, ephebic portion rectangular, finally developing elongate reniform shape; final chambers often extend from edge to edge; initial coil of 4–7 chambers, 12–14 in biserial portion; usually rapidly increasing in width, which becomes the greatest dimension. Primary aperture an interiomarginal arch on pregerontic chambers and an exteriomarginal arch on the very wide gerontic chambers, arch essentially the full height of the gerontic chambers; with a very wide, imperforate apertural flap flaring out and down onto the previous chamber; paired relict apertures appear at the abapertural end of successive reniform chambers. Wall finely perforate with each pore penetrating the centre of a low mound (pore mound); noncostate; proloculus and initial coil imperforate. Sutures depressed; straight and abaperturally oblique to the biserial axis in pregerontic portion, broadly convex in the direction of growth in gerontic portion.

Remarks: *Heterohelix pulchra* is the only *Heterohelix* with both pore mounds and well-developed reniform chambers. No other species is known to possess relict apertures. It differs from *H. glabrans* (Cushman), the only other species bearing pore mounds, in its reniform chambers. Morphological variation includes planispiral coil which is not present in all specimens. The width of the reniform chambers increases rapidly during ontogeny. Thus, depending upon the stage of growth, the appearance of the test varies.

The origin of *Heterohelix pulchra* is not known.

Brown (1969) designated the specimen represented by fig. 3 on pl. 9 of Brotzen (1936) as the lectotype of *Heterohelix pulchra*. A paralectotype, on deposit at the U.S. National Museum, does not have as well developed reniform chambers as more advanced forms.

There is little doubt that *Guembelina dagmarae* Suleimanov, 1955, is conspecific with *Heterohelix pulchra*, since it has the same size, reniform chambers and stratigraphic range.

Contrary to the statement of Stenestad (1968a), the ephebic and gerontic stages of *Heterohelix pulchra* do typically possess a "dentate margin" in side view. It would seem that he is distinguishing between ontogenetic stages of *H. pulchra*. Stenestad's (1968a) *H. dentata* is believed to be a junior synonym of *H. pulchra*. Recognition of the presence of pore mounds is needed to confirm this assignment, however.

Distribution: This is shown in Fig. 14.

Range: Turonian—Campanian (*Globotruncana calcarata* Interval).

Heterohelix rumseyensis Douglas

Heterohelix rumseyensis Douglas, 1969, p. 159, pl. 11, figs (?) 9a, b, 10a, b (*non* 11a, b).

Description: "Test large, biserial, rapidly tapering in early portion, gently tapering to straight-sided in adult; periphery lobate throughout. Chambers subglobular, wider than high, inflated in early portion; later pairs quadrate. Sutures depressed, distinct, straight. Wall calcareous, perforate; surface costate to smooth; costae fine, longitudinal. Aperture a low arch on inner margin of last chamber, with apertural flaps along outside of arch; flaps of previous chambers observable along medial suture of last pair of chambers." (Type description).

Remarks: *Heterohelix rumseyensis* differs from *H. moremani* (Cushman) by being more laterally compressed and having a slightly more rapid increase in width. The variation within this species is unknown.

This species may have arisen from *Heterohelix moremani* (Cushman), a form which has a similar number of chambers.

The taxon *rumseyensis* does appear to be new. However, the holotype shows traces of what may represent accessory apertures along the midline of the test, in which case it should be transferred to the genus *Pseudoguembelina*. These structures are not apparent from the type figure. Electron microscopy is needed to determine their true nature. Until such time, transfer is not justified.

Distribution: This is shown in Fig. 15.

Range: Santonian (FOS *Globotruncana concavata*—FOS *Globotruncana elevata* Interval)—Campanian (FOS *Globotruncana ventricosa*—FOS *Pseudoguembelina costata* Interval).

Heterohelix semicostata (Cushman), Plate 2, figs 4, 5; plate 3, figs 1, 2

Guembelina semicostata. Cushman, 1938a, p. 16, pl. 3, figs 6a, b; Frizzell, 1945, p. 110, pl. 15, figs 41a–44; Cushman, 1946b, p. 107, pl. 46, figs 1a–5; Kikoïne, 1948, p. 19, pl. 1, figs 6a–c.

Heterohelix semicostata (Cushman). Eternod Olvera, 1959, p. 70, 71, pl. 1, figs 14, 15.

Gublerina semicostata (Cushman). Barr, pl. 1, fig. 5.

Gublerina ornatissima (Cushman and Church). Neagu, 1970b, p. 60, pl. 14, fig. 11.

Gublerina rajagopalani Govindan, 1972, p. 170, pl. 2, figs 1–5.

Description: Test biserial, laterally compressed, rapidly increasing in width, edge rounded, periphery mildly lobate. Chambers wider than high and thick; rectangular in side view in neanic portion, later becoming arched in the direction of growth; sides become subparallel in adult or thickness may even decrease; 14–16 chambers, rapidly increasing in width. Aperture an interiomarginal arch essentially the full height of the chamber, bordered by a narrow, imperforate lip. Sutures of early portion obscured by costae, later becoming limbate, raised, convexly curved in the direction of growth, not appearing to overlap in gerontic stage. Surface with coarse costae covering all of the neanic portion, restricted to the chamber backs and only

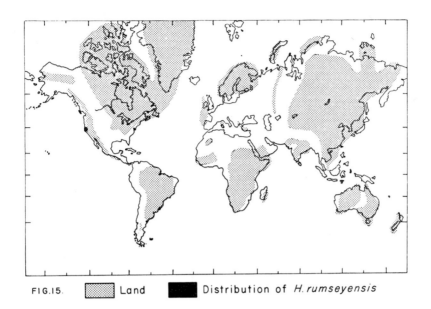

FIG.15. ▨ Land ■ Distribution of *H. rumseyensis*

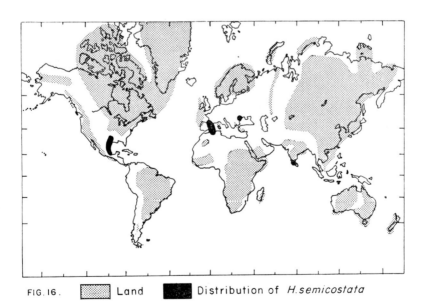

FIG.16. ▨ Land ■ Distribution of *H. semicostata*

the immediately adjacent areas of the sides in later portions; costae follow peripheral curvature of chambers; 3/4 or more of chamber sides lack costae.

Remarks: *Heterohelix semicostata* bears uniquely positioned costae which permit easy recognition of this taxon.

The incomplete overlap of the sutures in the last two to four chambers creates the illusion of a non-septate area between the biseries. As a result, Neagu (1970b) attributed such a specimen to *Gublerina ornatissima* (Cushman and Church). Govindan (1972) proposed the new name *G. rajagopalani* for specimens from India. Although neither the specimens of Neagu nor Govindan were examined, the distribution of the costae as illustrated by them are known only from *Heterohelix semicostata*.

Distribution: This is shown on Fig. 16.

Range: Campanian (FOS *Globotruncana elevata*—LOS *Globotruncana concavata* Interval)—Maastrichtian (LOS *Globotruncana lapparenti*—FOS *Globotruncana aegyptiaca* Interval).

Heterohelix sphenoides Masters 1976, p. 318, pl. 1, figs. 1–3

Description: "Test biserial, laterally compressed; rapidly expanding in side view, often as wide as long; profile slightly lobate. Chambers typically 12–15 in number with a circular outline as viewed from the side, but walls subparallel in edge view with only slight inflation. Numerous irregularly positioned pores with moderate diameters becoming finer in later chambers, but in some specimens finely perforate throughout. Discontinuous costae of moderate width and height on early chambers gradually becoming finer in later chambers; costae conform to the curvature of the chamber as seen in side view. Aperture a high, narrow arch bordered by a narrow, imperforate lip. Chamber sutures imperforate, depressed, convex toward later chambers; zig-zag median suture twice the width of the other sutures." (Type description).

Remarks: *Heterohelix sphenoides* is distinguished from all other species of *Heterohelix* except *H. bejaouensis* (Salaj and Maamouri) by its subparallel sides and rapid increase in width. It may be conspecific with *H. bejaouensis*. There is a slight variation in the rate at which the width of the test increases. Other features are stable.

This species evolved from *Heterohelix globulosa* (Ehrenberg) through a decrease in the chamber thickness and an increase in their width.

Salaj and Maamouri (1971) established the new species *Sigalia bejaouensis* from the *Globotruncana calcarata* subzone in Tunisia. Type material of this species was not examined during this study. Based solely upon the type figure and description, a marked resemblance is noted between this species and *Heterohelix sphenoides*. If proven conspecific, Salaj and Maamouri's species, here considered a *Heterohelix*, would have priority and extend its stratigraphic range.

Distribution: This is shown in Fig. 17.

Range: Santonian (FOS *Globotruncana concavata*—FOS *Globotruncana elevata* Interval)—Campanian (FOS *Pseudoguembelina costata*—FOS *Globotruncana obliqua* Interval).

Heterohelix stenopos Masters 1976, p. 319, pl. 1, figs 4, 5

Description: Test biserial, elongate; laterally compressed, but unkeeled; slightly lobate in profile; minimal increase in width throughout. Chambers 16–17 in number; only slightly inflated; each succeeding chamber but little larger than the previous one, wider than thick. Pores of fine to moderate diameter; in rows between costae.

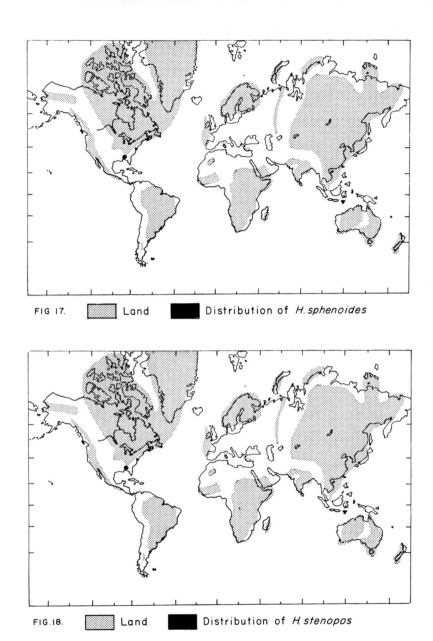

FIG 17. Land Distribution of *H. sphenoides*

FIG.18. Land Distribution of *H. stenopos*

Costae discontinuous, of moderate width and height; conforming to the curvature of the chamber as viewed from the side. Aperture a low, narrow arch with an imperforate lip, which increases in width toward contact with previous chamber. The sides of the relict apertural lips left as depressed triangular areas between adjacent chambers. Sutures straight, depressed, normal to edge, imperforate.

Remarks: *Heterohelix stenopos* differs from *H. rumseyensis* Douglas by having a narrower test width and coarser costae. It is distinguished from *H. moremani* (Cushman) in its greater lateral compression and coarser costae. Little variation has been recognised in the morphological characteristics of this species.

Heterohelix moremani (Cushman) was ancestral to this short ranging species.

Distribution: This is shown in Fig. 18.

Range: *Santonian* (FOS *Globotruncana concavata*—FOS *Globotruncana elevata* Interval)—Campanian (FOS *Globotruncana elevata*—LOS *Globotruncana concavata* Interval).

Heterohelix striata (Ehrenberg), Plate 3, figs 2, 3

Textularia striata Ehrenberg, 1840, p. 135, pl. 4, figs 1, 1¹, 2, 3, (?9); (?) Tutovskii, 1887, p. 348, 349, pl. 3, figs 1a–e; pl. 7, fig. a; (?) Egger, 1899, p. 19, pl. 5, fig. 13.

Textilaria striata (Ehrenberg). Ehrenberg, 1854, pl. 25, fig. Cl; pl. 27, fig. 3; pl. 28, fig. 6; pl. 31, fig. 9; pl. 32:1, fig. 4b; pl. 32:2, figs 11, 14; (?) Ehrenberg, 1856, p. 175, pl. 7, fig. 3.

Textilaria sulcata, Ehrenberg, 1854, p. 25, pl. 27, fig. 4; pl. 29, figs 16, 216; pl. 30, figs 4a–c, pl. 31, fig. 10.

Textilaria americana Ehrenberg. (?) Ehrenberg, 1856, p. 175, pl. 7, figs 1, 2.

(?) *Textularia striatopunctata* Egger, 1857, p. 294, pl. 12, figs 27–29.

Guembelina striata (Ehrenberg). Egger, 1899, p. 33, pl. 14, figs 10, 11 (5–7?, 37, 38?, *non* 39); Cushman, 1931c, p. 43, pl. 7, figs 6a, b (?7); Cushman, 1933a, pl. 26, fig. 8; (?) Jedlitschka, 1935, p. 41, pl. figs 11, 12; *non* Cushman, 1936, p. 418, pl. 1, figs 11a, b; *non* Brotzen, 1936, pl. 9, figs 1a, b; text-figs 1 (5), 2A–E; *non* Voorwijk, 1937, p. 194, pl. 1, figs 9, 10; van Wessem, 1943, p. 45, pl. 1, figs 39, 40; Cushman, 1946b, p. 104, 105, pl. 45, figs 4, 5; Cushman, 1948, p. 253, Key pl. 26, fig. 8; Kikoïne, 1948, p. 19, 20, pl. 1, figs 7a–c; Cushman, 1949, p. 7, pl. 3, fig. 24; *non* Voorthuysen, 1951, pl. 2, fig. 22; Hagn, 1953, p. 73, pl. 6, figs 14, 15; (?) Shaw, 1953, pl. 1, fig. 23; Hamilton, 1953, p. 235, pl. 30, fig. 13; *non* Maslakova, 1959, p. 117, pl. 15, fig. 3; (?) Hofker, 1961c, text-figs 58, 59; (?) Haake, 1962, p. 46, pl. 4, fig. 14; Hofker, 1962a, p. 1063, text-fig. 8c, p. 1067, text-fig. 12c; *non* Khan, 1970, p. 31, 32, pl. 1, fig. 10.

Guembelina paucistriata Albritton, 1937, p. 22, pl. 4, figs 8–10.

Pseudoguembelina striata (Ehrenberg). Brönnimann and Brown, 1953, p. 154, text-fig. 6; Belford, 1960, p. 60, 61, pl. 15, figs 12, 13; (?) Neagu, 1970b, p. 61, pl. 14, figs 4, 5.

Guembelina cf. *G. striata*. Bolin, 1956, p. 291, pl. 38, figs, 12, 17.

Guembelina decurrens Chapman. Hofker, 1956f, p. 77, pl. 9, fig. 66.

Guembelina striata f. *turonica* Hofker, 1957, p. 419, 420, Abb, 477:1d, e (?a–c), (?2a–f), 3c, e, (?a, b, d), 4a, c–f (?b).

Heterohelix striata (Ehrenberg). Eternod Olvera, 1959, p. 71, 72, pl. 2, figs 4, 8; Berggren, 1962a, p. 20, 21, pl. 6, figs 1a–5B; Pessagno, 1962, p. 358, pl. 1, fig. 5; Skinner, 1962, p. 39, pl. 5, fig. 10; *non* Cati, 1964, p. 260, 261, pl. 42, figs 6a, b; Takayanagi, 1965, p. 198, 199, pl. 20, figs 4a, b; Aliyulla, 1965, p. 219, text-figs 3a, b; Lehmann, 1966, p. 315, pl. 2, fig. 8; Pessagno, 1967, p. 264, pl. 78, figs 4, 5; pl. 88, figs 3–7; Bandy, 1967, p. 23, 24, text-figs 11(2), 12(9); Ansary and Tewfik, 1968, p. 41, pl. 3, figs 3a, b; Sliter, 1968, p. 96, pl. 13, figs 13a, b; Douglas, 1969, p. 159, 160, pl. 11, figs 4a, b, 7a–8b; Hanzlíková, 1969, p. 40, 41, pl. 7, figs 13–17, 24–28; (?) Funnell, Friend and Ramsay, 1969, p. 21, pl. 1, figs 3, 4; text-figs 2a, b; Bertels, 1970, p. 31, 32, pl. 2, figs 7a–8b; *non* El-Naggar, 1971a, pl. 7, fig. i; (?) Hanzlíková, 1972, p. 93, pl. 23, figs 14–18; Govindan, 1972, p. 168, pl. 1, figs 13, 14.

Heterohelix navarroensis Loeblich. Pessagno, 1962, p. 358, pl. 1, fig. 4; Bertels, 1970, p. 31, pl. 2, figs 1a–2.

Pseudoguembelina excolata (Cushman). Lehmann, 1966, p. 316, pl. 2, fig. 7.

Heterohelix globulosa var. *striatula* (Marie). Ansary and Tewfik, 1968, p. 39, 40, pl. 3, figs 7a, b.

Heterohelix striata var. *aegyptica*, Ansary and Tewfik, 1968, p. 41, pl. 3, figs 2a, b.

Heterohelix globulosa (Ehrenberg). Douglas, 1969, p. 157, 158, pl. 11, figs 12a, b.

Pseudoguembelina costulata, Douglas, 1969, p. 160, pl. 11, figs 6a, b; (?) Govindan, 1972, p. 169, pl. 1, figs 15, 16.

Heterohelix sp. cf. *H. planata* (Cushman). Bertels, 1970, p. 31, 32, pl. 2, figs 3a, b.

Description: Test biserial, regularly and moderately increasing in width, edge rounded, periphery moderately lobate. Chambers equidimensional, appearing square in side view, ultimate and penultimate subcircular; 11–15 chambers moderately increasing in size; with only a slight overlap on the previous chamber. Aperture moderately high, interiomarginal bordered by a narrow imperforate lip, which becomes wider at the contact with the previous chamber. Wall perforated by narrow to moderate diameter pores situated between continuous, coarse costae, which follow the curvature of the chamber radiating from the aperture; proloculus imperforate. Sutures depressed, straight, normal to the edge.

Remarks: *Heterohelix striata* possess the coarsest costae of any species assigned to the genus *Heterohelix*. It is the only *Heterohelix* in which the costae are continuous over each chamber. There is also a tendency for fewer costae per chamber. The rate of increase in the test width is somewhat variable. The costae, particularly on the earlier portions of the test, tend to be uninterrupted from chamber to chamber.

This species descended from *Heterohelix globulosa* (Ehrenberg) through a decrease in the number of costae and the development of continuous costae over the entire test.

Ehrenberg's (1840) 2α was designated by Pessagno (1967) as representative of the lectotype of *Heterohelix striata*. This designation accomplishes nothing without accompanying redescription and reillustration, and may create problems if this specimen is not in keeping with the concept of this species.

Egger (1857) proposed the new species *Textularia striato-punctata* from the Miocene of Lower Bavaria. He remarked on the similarity of his species with Ehrenberg's *Textilaria striata* and others. Egger's species is most probably a reworked *Heterohelix striata*.

Heterohelix striata var. *aegyptiaca* Ansary and Tewfik, 1968, differs, according to the authors, from *H. striata* by having fewer and more rapidly expanding chambers. It is regarded as an intraspecific variant. According to Article 45e (ii) of the I.C.Z.N., the use of the term "variety" relegates this form to infrasubspecific rank.

The specimen illustrated by Govindan (1972) as *Pseudoguembelina costulata* (Cushman) is believed to be a *Heterohelix striata*. The test and chamber shape is typical for *H. striata*, and not for a *Pseudoguembelina*. The "faintly developed" sutural openings mentioned by Govindan may be the result of damage to the test. In any event, these openings, as drawn, are neither the right shape nor in the correct position for *Pseudoguembelina*.

Distribution: This is shown in Fig. 19.

Range: Santonian (FOS *Globotruncana concavata*—FOS *Globotruncana elevata* Interval)—Maastrichtian (*Trinitella scotti* Interval).

Genus LUNATRIELLA Eicher and Worstell, 1970a

Lunatriella Eicher and Worstell, 1970a, p. 117.

Type species: *Lunatriella spinifera* Eicher and Worstell, 1970a.

Description: "Test elongate; initial portion of nearly equidimensional chambers arranged biserially like *Heterohelix*; later portion composed of vertically elongated chambers which form an irregularly uniserial pattern; sutures depressed; later chambers with or without spinelike lateral projections; aperture in early biserial chambers a high arch with lateral apertural lips; aperture in elongate later chambers terminal, the lateral apertural lips merging below the aperture into a troughlike projection which is attached to the preceding chamber; wall of radially arranged calcite, finely perforate." (Type diagnosis.)

Remarks: *Lunatriella* differs from *Heterohelix* by possessing gerontic chambers which are elongate parallel to the biserial axis, and which develop spinose projections along the chamber backs and troughlike extensions of the apertural lips. *Lunatriella* does not develop a uniserial growth pattern in the true sense. It is more of a loosely biserial arrangement.

Range: Lower Turonian.

Lunatriella spinifera Eicher and Worstell, Plate 3, figs 5, 6

New genus (unnamed) Eicher, 1969, text-fig. 3 (part).

Lunatriella spinifera Eicher and Worstell, 1970a, p. 118, 119, pl. 1, figs 5–17; Eicher and Worstell, 1970b, p. 296, 297, pl. 8, figs 7, 8, 12.

Description: "Test elongate, compressed; initial biserial portion consisting of up to four pairs of subglobular chambers; later chambers becoming increasingly elongate vertically, with curved axes convex outward; final one to three elongate chambers in fully developed specimens irregularly uniserial, the chambers arcing alternately right and left; commonly the last chamber and rarely the last two chambers of the uniserial portion having a large aboral spine-like elongation protruding at an angle that varies greatly among specimens; sutures distinct, depressed, strongly curved in later chambers of biserial portion; aperture in early chambers a high arch bounded by lateral lips, which become increasingly elongate as chambers increase in height. In elongate but still biserially arranged chambers the lateral lips join below the aperture to form a troughlike flange (here termed the 'apertural trough'), which extends downward to the top of the chamber below. In uniserially arranged chambers the apertural trough commonly forms a delicate flying buttress separated from the main body of the test by a window; wall of radially built calcite, finely perforate." (Type description.)

Remarks: A monotypic genus in which the degree of chamber elongation, the

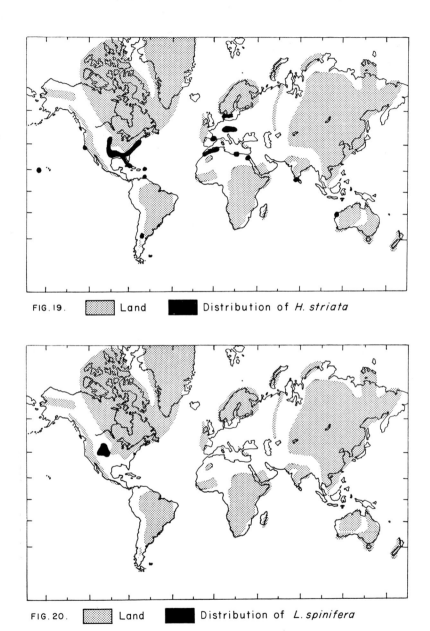

FIG.19. ▨ Land ■ Distribution of *H. striata*

FIG.20. ▨ Land ■ Distribution of *L. spinifera*

position of the spinose projections and the preservation of the apertural trough flying buttresses singly or in combination vary considerably.

Eicher and Worstell (1970a) infer the evolution of *Lunatriella spinifera* from *Heterohelix pulchra* (Brotzen). Gradational specimens were recognised, but unfortunately preservation prevents determination of the presence of the pore mounds characteristic of *H. pulchra*.

Earlier named taxa show similarity to *Lunatriella spinifera*. *Textilaria missouriensis* Ehrenberg, 1854, apparently possesses some degree of chamber elongation. Because of the schematic type figure, drawn using transmitted light, and the fact that it has not been recognised since its original description, the taxonomic status of this form remains unclear. Cushman (1938a) suggested that *T. missouriensis* was probably synonymous with *T. (=Heterohelix) americana* (Ehrenberg), but this is unlikely.

Masella (1959) described the new species *Heterohelix digitata* from Sicily. Eicher and Worstell (1970a) distinguish *Lunatriella spinifera* from Masella's species by the the presence of apertural lips and troughs, and the more elongate spine-bearing chambers. In the suite of specimens illustrated by Eicher and Worstell, only those which develop the loosely biserial stage have apertural troughs. It is possible that the Sicilian specimens simply have not reached this ontogenetic stage of development. Most of the *H. digitata* illustrated by Masella do not exhibit chamber elongation to the extent of *L. spinifera*. However, figs 2 and 6 of Masella show a degree of elongation comparable to figs 4 and 5 of Eicher and Worstell. *Heterohelix digitata* must be re-examined as the original stylised drawings are inadequate. Because *H. digitata* appears to lack the distinctive reniform chambers typical of the earlier stages of *L. spinifera*, both are accepted as separate but convergent species.

It should be noted that where Eicher and Worstell (1970a) refer to "vertically elongate" chambers, they mean elongation of the chambers in the direction of test growth, parallel to the biserial axis. The reference to "vertical" might mistakenly be interpreted as relating to a living posture in the water column. The term "loosely biserial" also seems more appropriate than "uniserial" for the terminal growth pattern.

Distribution: This is shown in Fig. 20.

Range: Lower Turonian.

Genus PLANOGLOBULINA Cushman, 1927f

Planoglobulina Cushman, 1927f, p. 77.

Racemiguembelina Montanaro Gallitelli, 1957, p. 142.

Type species: *Guembelina acervulinoides* Egger, 1899, (part) [= *Pseudotextularia varians* Rzehak, 1895, (part)].

Description: Early stage may be planispiral, later developing typical, ephebic stage, biserial chambers which have a greater thickness than width or height. A supplementary chamber may be added at the base of the back of the last 1 or 2 biserial stage chambers. In the gerontic stage, the primary chambers become increasingly restricted by the progressively enlarging supplementary aperture, which results in dumb-bell-shaped chambers. Supplementary chambers are added with each new primary chamber. The resulting outline as viewed from the apertural end is sub-circular. The primary aperture is a broad, interiomarginal arch. The primary aperture is exceeded in size by the later supplementary apertures, which ultimately divide the final 1 or 2 chambers into 2 equal chamberlets. These chamberlets occupy a peripheral position. The test is ornamented by strong, discontinuous, longitudinal costae, with intercostate pores.

Remarks: *Planoglobulina* differs from all other Heterohelicidae by possessing dumbbell-shaped gerontic chambers which impart a subcircular crosssectional outline to the test.

Martin (1972) attempted to resolve the difficult taxonomic entanglements and structural misinterpretations surrounding *Planoglobulina* and *Ventilabrella*. She correctly recognised the validity of both genera, but misinterpreted the morphology of *Planoglobulina*, thus perpetuating the confusion.

The author has already given an in-depth evaluation of the more complex Heterohelicidae, including *Planoglobulina* (Masters, 1976), and the reader is referred to that paper for detail. In summary it was demonstrated that *Planoglobulina* is restricted to forms with the subcircular cross-section and typical pseudotextularian biserial chambers which later develop into dumbbellshaped chambers. The specimens referred to this genus by Martin (1972) belong to *Platystaphyla*. Her specimens have an early pseudotextularian portion to which later supplementary chambers are added in the plane of biseriality.

Range: Maastrichtian (*Trinitella scotti* Interval).

Planoglobulina varians (Rzehak), Plate 4, figs 1, 2
Pseudotextularia varians, Rzehak, 1895, p. 2, pl. 2, figs 2, 3 (*non* 1a, b); Cushman, 1926b, p. 17, pl. 2, figs 4a, b; Cushman, 1927e, p. 157, pl. 27, figs 2a, b; Cushman, 1928b, p. 231, pl. 33, figs 5, 6; pl. 34, figs 4a, b; White, 1929, p. 40, pl. 4, figs 15a, b; Voorwijk, 1937, p. 194, pl. 1, figs 14, 15; Cushman, 1938a, p. 21, 22, pl. 4, figs 1a–4; Keller, 1939, pl. 2, fig. 2; Cushman and Todd, 1943, p. 65, pl. 11, fig. 17; Cushman, 1946b, p. 110, pl. 47, figs 4a–7b; (?) Cita, 1948a, p. 125, 126, pl. 2, fig. 7; Kikoïne, 1948, p. 23, pl. 2, figs 4a–c; Cushman, 1949, p. 8, pl. 3, fig. 26; Noth, 1951, p. 62, pl. 7, figs 20, 21; Itzhaki, 1952, p. 187–189, text-figs 1–8; Huss, 1957, pl. 6, fig. 2(9); pl. 7, fig. 1 (7); Bukowy and Geroch, 1957, p. 317, pl. 28, figs 12a–13b; Maslakova, 1959, p. 118, pl. 15, fig. 7; Seiglie, 1959, p. 76, 77, pl. 5, fig. 6; pl. 7, figs 6a, b (*non* 5a, b); Vinogradov, 1960a, pl. 6, figs 34a–35c; *non* Hofker, 1960e, text-fig. 11; Vasilenko, 1961, p. 207, 208, pl. 41, figs 10a, b; Jurkiewicz, 1961, pl. 23, figs 26a–b; Pokorný, 1963, p. 390, fig. 441 (part); Wille-Janoschek, 1966, p. 121, 122, pl. 8, figs 9a, b.
Guembelina acervulinoides Egger, 1899, p. 35, pl. 14, fig. 20 (*non* 14–18), 21, 22.
Guembelina fructicosa Egger, 1899, p. 35, pl. 14, figs 8, 9, 24 (*non* 25, 26).
Pseudotextularia fructicosa (Egger). Cushman, 1928b, p. 231, pl. 33, figs 7a, b; Cushman, 1933a, pl. 26, figs 16a, b; Cushman, 1948, p. 256, Key pl. 26, figs 16a–c; Seiglie, 1959, p. 56, pl. 2, figs 3a, b, 5a, b; Pessagno and Brown, 1969, p. 116, pl. 1, figs 1–4; Hanzlíková, 1972, p. 96, pl. 24, figs 9a, b.
Planoglobulina acervulinoides (Egger). Cushman, 1928b, p. 231, pl. 33, fig. 9 (*non* 8; pl. 34, fig. 5); Cushman, 1933a, pl. 26, fig. 17; Jedlitschka, 1935, p. 41, pl. figs 15, 16 (?); Cushman, 1948, p. 257, Key pl. 26, fig. 17.

Pseudotextularia varians var. *mendezensis* White, 1929, p. 41, figs 16a, b; Voorwijk, 1937, p. 194, pl. 1, figs 18, 24; Cushman, 1946b, pl. 47, figs 9a, b.

Pseudotextularia varians var. *textulariformis* White, 1929, p. 41, figs 17a, b; Voorwijk, 1937, p. 195, pl. 1, figs 16, 17; Cushman, 1946b, pl. 47, figs 8a, b; Kikoïne, 1948, p. 23, 24, pl. 2, figs 5a–c.

Pseudotextularia elegans var. *varians* Rzehak. Glaessner, 1936, p. 101, 102, pl. 1, figs 3–5, text-fig. 1c; Glaessner, 1948, p. 152, pl. 10, fig. 15e.

Transitional *Pseudotextularia elegans* Rzehak to *Pseudotextularia varians*. Rzehak. Noth, 1951, pl. 7, figs 18, 19.

Pseudotextularia intermedia de Klasz, 1953, p. 231, 232, pl. 5, figs 2a–c; Danilova, 1958, p. 225, pl. 23, fig. 2; Pessagno, 1967, p. 269, 270, pl. 86, fig. 11.

Pseudotextularia elegans (Rzehak). Frizzell, 1954, p. 111, pl. 16, figs 5a–7; Naggappa, 1959, pl. 7, figs 7a–8; Hiltermann and Koch, 1962, p. 337, pl. 46, fig. 12 (*non* 11); tab. 19 (part); [incorrectly attributed to Grzbowski] Scheibnerová, 1969, tab. 6, figs 7, 8 (*non* tab. 7, figs 1, 2).

Pseudotextularia varians (elegans) Rzehak. Hofker, 1957, p. 422–424, text-figs 478a, b, d, e (*non* c, f-m).

Racemiguembelina fructicosa (Egger). Montanaro Gallitelli, 1957, p. 142, 143, pl. 32, figs 14a–15b; Enternod Olvera, 1959, p. 78, 79, pl. 2, figs 5–7, 11; Said and Kerdany, 1961, p. 334, pl. 2, fig. 17; Loeblich and Tappan, 1964, p. C656, fig. 525:8a, b; Saavedra, 1965, p. 344, text-fig. 93; Pessagno, 1967, p. 270, 271, pl. 90, figs 14, 15; Neagu, 1968, text-fig. 3, fig. 58; Sliter, 1968, p. 98, pl. 14, figs 16a, b; Ansary and Tewfik, 1968, p. 43, pl. 3, figs 16a, b; Dupeuble, 1969, p. 158, pl. 4, figs 12a, b; Funnell, Friend and Ramsay, 1969, p. 25, 26, pl. 2, figs 3, 4; text-figs 8a, b; Hanzlíková, 1969, p. 44, pl. 8, figs 14a, b; Bate and Bayliss, 1969, pl. 4, figs 1a–c; Neagu, 1970, pl. 42, fig. 58; Todd, 1970, p. 152, pl. 5, figs 7a, b; Cita, 1970, pl. 4, fig. 6; Govindan, 1972, p. 171, pl. 1, figs 7, 8; pl. 2, figs 8, 9.

Pseudotextularia plummerae (Loetterle). Seiglie, 1959, p. 77–79, pl. 5, figs 4a, b (*non* 5; pl. 7, figs 2a, b).

Racemiguembelina textuliformis (White). Salaj and Samuel, 1966, p. 233, tab. 37 (12).

Racemiguembelina varians (Rzehak). Salaj and Samuel, 1966, p. 233, 234, tab. 37, fig. 13; Cita and Gartner, 1971, pl. 3, fig. 2.

Pseudotextularia elegans fructicosa (Egger). Bandy, 1967, p. 24, 25, text-fig. 12 (13).

Description: Test biserial, conical, with supplementary chambers. Ephebic primary chambers thicker than high or wide. Gerontic primary chambers constricted into 2 subspherical chamberlets connected by a narrow imperforate bridge, which is lost in the final growth stage. Supplementary apertures at central, basal back of the ephebic chambers, initially moderately arched later becoming highly arched. Supplementary chambers broad at onset, later subspherical. Primary aperture a broad, low interiomarginal arch bordered by an imperforate lip. Few coarse costae, generally continuous on each chamber; with 1–3 rows of narrow to moderate diameter pores in the intercostal area.

Remarks: A monotypic genus. The number of supplementary chambers is variable, but the primary chambers of the gerontic stage show the most variation. The number of dumb-bell chambers, the extent of their separation and width of their connecting bridge, and the number of final chamberlets are the most diverse structures of this species.

Planoglobulina varians (Rzehak) exhibits a gradational bioseries with the ancestral form *Pseudotextularia elegans* (Rzehak).

According to the I.C.Z.N., Article 75c (4), to be valid a neotype must be "consistent with what is known of the original type-material, from its description and from other sources . . ". Martin's (1972) designation of a neotype for *Planoglobulina acervulinoides* (Egger) (= *P. varians*) does not meet these requirements, and is, therefore, rejected.

A complete review of the chaotic taxonomy of *Planoglobulina* has already been presented (Masters, 1976), but summarisation follows. Egger (1899) illustrated eight syntypic specimens for his new species *Guembelina acervulinoides*. Of these syntypes, fig. 15, 16 and possibly 14 are representative of *Pseudotextularia elegans* (Rzehak), 1891, as emended by Masters (1976) and generally understood by most workers. Figures 17 and 18 were re-illustrated (although not designated) by Cushman (1928a) as the type species for *Ventilabrella*, i.e. *V. eggeri*. Figures 21 and 22, the former being a transmitted light sketch, represent an unknown species of *Ventilabrella*, as is indicated by the numerous, small, subequal chambers, plus the fact that it had to have been very thin to have been so illustrated.

All of Egger's syntypes have, then, been accounted for except that represented by fig. 20. As was observed by Brown (1969) and noted by Martin (1972), the largest, centrally located chamber of this figure has a circular outline and shading which indicate that this chamber does not lie in the biserial plane. In fact this chamber is characteristic of a typically oriented dumb-bell-shaped chamber, which is sufficiently elongated normal to the biserial plane so as to stand out in strong relief. Martin (1972) recognised this fact when she stated that "In the writer's opinion Egger's figure 20 most likely represents a *Racemiguembelina*". *Racemiguembelina* has been documented by the author (1976) as a junior synonym of *Planoglobulina*. Moreover, fig. 20 was selected as the type species of *Planoglobulina* by Cushman (1928a), and thus, the lectotype of *Guembelina acervulinoides* by subsequent designation.

Thus, Martin's (1972) neotype is not consistent with what is known about Egger's syntypes. The form she selected as the neotype is now referred to *Platystaphyla brazoensis* (Martin).

Interestingly, Martin (1972) does not cite any of Egger's syntypes in her synonymy of *Planoglobulina acervulinoides*. Thus, her selection of a neotype is also invalid because the name "*acervulinoides*" remains available [I.C.Z.N., Art. 17(1)], and cannot be used for an altogether different form.

Distribution: This is shown in Fig. 21.

Range: Maastrichtian (*Trinitella scotti* Interval).

Genus PLATYSTAPHYLA Masters, 1976

Planoglobulina Cushman. Montanaro Gallitelli, 1957, p. 141, 142 (part); Loeblich and Tappan, 1964, p. C655, C656 (part); Pessagno, 1967, p. 271 (part).

Heterohelix Ehrenberg. El-Naggar, 1971, p. 446 (part).

Type species: *Plano-globulina brazoensis* Martin, 1972.

Description: The test is flabelliform, rapidly increasing in width. The initial 8–12 chambers are pseudotextularian, being biserial and laterally compressed. The remaining biserial chambers become subglobular to slightly irregular, but lack the lateral compression of the earlier portion. Also rapidly

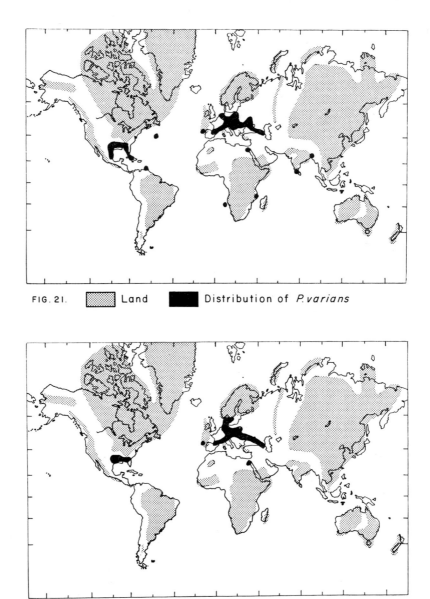

FIG. 21. ▦ Land   ■ Distribution of *P. varians*

FIG. 22. ▦ Land   ■ Distribution of *P. brazoensis*

becoming subglobular to irregular are the supplementary chambers, which in turn support additional supplementary chambers. All chambers lie in the plane of biseriality. The surface ornamentation consists of strong discontinuous, longitudinal costae, which follow the contours of the chamber. The apertures of the biserial chambers are low interiomarginal arches. Supplementary apertures occur on the back of the base of each subglobular biserial chamber. Each supplementary chamber may have one or two apertures.

Remarks: *Platystaphyla* differs from *Pseudotextularia* by possessing supplementary chambers, and from *Planoglobulina* by having these chambers added in the biserial plane.

A detailed discussion of the growth pattern of *Platystaphyla* has been presented (Masters, 1976). The pattern is regular and predictable, not a random proliferation as frequently reported.

Martin (1972) introduced two new terms which require comment—"cameral flange" and "cameral ridge". As defined, a cameral flange is a perforate, lateral extension on either side of each aperture in the final series of chambers, and these may merge between adjacent, separated chambers. A cameral ridge was defined as an imperforate rim extending continuously from flange to top of adjacent chambers. These features have no taxonomic significance, and as seen below, are merely remnant structures. Although it may be convenient to have a name for such structures, it seems to be an unnecessary proliferation of terms.

Most frequently when *Platystaphyla* reaches the gerontic stage, a final, single, very thin-walled, atypical chamber will be added across the entire top of the test. The cameral ridges and flanges of Martin are nothing more than remnants of this final chamber. The imperforate ridge represents the sutural contact of this atypical chamber. The flange is the structurally stronger, more protected portion of the wall. The same features may also be found in *Ventilabrella*.

Range: Campanian (FOS *Pseudoguembelina costata*–FOS *Globotruncana obliqua* Interval)–Maastrichtian (*Trinitella scotti* Interval).

Platystaphyla brazoensis (Martin), Plate 4, figs. 3, 4
Pseudotextularia acervulinoides (Egger). Cushman, 1926b, p. 17, 18, pl. 2, fig. 5; Wille-Janoschek, p. 119, pl. 8, figs 8a, b; Sturm, 1970, p. 113, pl. 8, figs 3a–c.
Planoglobulina acervulinoides (Egger). Cushman, 1927e, p. 158, pl. 27, fig. 3; Cushman, 1938a, p. 23, pl. 4, figs 5a, b, 7a, b, 8a, b (*non* 6a, b); Kline, 1943, p. 45, pl. 7, fig. 1; Cushman, 1946b, p. 111, pl. 47, figs 12a–13b, 15a, b (*non* 14a, b); *non* Kikoïne, 1948, p. 24, pl. 2, figs 6a, b; *non* Cita, 1948a, p. 126, 127, pl. 2, figs 8a, b; Cushman, 1949, p. 8, pl. 3, fig. 27; (?) Itzhaki, 1952, p. 187–189, text-figs 9–11; Said and Kenawy, 1956, p. 140, pl. 3, fig. 45; Maslakova, 1959, p. 119, pl. 15, figs 8a, b; Said and Kerdany, 1961, p. 334, pl. 2, fig. 15; Saavedra, 1965, p. 344, text-fig. 91;

Salaj and Samuel, 1966, p. 234, 235, pl. 26, figs 1, 2; tab. 37 (14); Pessagno, 1967, p. 271, pl. 87, fig. 14; Ansary and Tewfik, 1968, p. 43, 44, pl. 3, figs 15a, b; Dupeuble, 1969, p. 158, pl. 4, fig. 16; Funnell, Friend and Ramsay, 1969, p. 22, 23, pl. 1, figs 7, 8, text-figs 4a, b; Hanzlíková, 1969, p. 41, pl. 8, figs 2a, b (*non* 1); *non* Rahhali, 1970, p. 68, text-figs. 7L–O; (?) Al-Shaibani, 1971, p. 112, pl. 3, fig. 3; *non* Cita and Gartner, 1971, pl. 5, fig. 3; Martin, 1972, p. 81, 82, pl. 3, figs 3–6.

   Planoglobulina elegans var. *acervulinoides* (Egger). Glaessner, p. 102, 103, pl. 1, figs 6a–9; Glaessner, 1948, p. 152, pl. 10, figs 15c, d.

   (?) *Pseudotextularia varians* Rzehak. Hofker, 1960e, text-fig. 11.

   Pseudotextularia (*Racemiguembelina*) *fructicosa* (Egger). Berggren, 1962a, p. 22–24, pl. 6, figs 6a, b.

   Pseudotextularia elegans (Rzehak). Bettenstaedt and Wicher, 1956, p. 502, pl. 1, fig. 6 (part); Hofker, 1956f, p. 77, pl. 9, fig. 78; Witwicka, 1958; p. 195, 196, pl. 8, fig. 6, 7 (?); Brown, 1969, p. 47–54, pl. 2, figs 4a, b; pl. 3, figs 2, 3; text-figs 13, 14 (*non* 9, 10); (incorrectly attributed to Grzybowski) Scheibnerová, 1969, tab. 7, fig. 1, (*non* 2; tab. 6, fig. 7, 8.

   Ventilabrella eggeri Cushman. Said and Kenawy, 1956, p. 140, pl. 3, fig. 38; Hanzlíková, 1972, p. 96, 97, pl. 23, figs 19a, b (*non* 23).

   Ventilabrella glabrata Cushman. Said and Kenawy, 1956, p. 140, pl. 3, fig. 46.

   Planoglobulina acervulinoides acervulinoides (Egger). Bandy, 1967, p. 25, text-fig. 11 (6).

   Racemiguembelina fructicosa (Egger). Neagu, 1970b, p. 61, 62, pl. 14, fig. 3.

   Planoglobulina brazoensis Martin, 1972, p. 82, 83, pl. 3, fig. 7; pl. 4, figs 1a–2.

   Description: Test biserial with supplementary chambers added in the biserial plane; rapidly increasing in width; periphery in side view lobate; in edge view, sides subparallel or gently biconvex with greatest thickness at midpoint. Primary biserial chambers thicker than wide or high, after the initial 6–10 chambers the width remains essentially constant. Supplementary chambers typically smaller and more equidimensional than the primary chambers; becoming generally less thick and more irregular in shape and size in the gerontic stage. Costae thick, high, discontinuous, longitudinal; being finer and more numerous on the final supplementary gerontic chambers. Primary apertures narrow, moderately high arches with narrow lips; all apertures of supplementary chambers moderately high, often irregularly shaped and positioned. Surface perforated by numerous small to moderate diameter pores.

   Remarks: This species has slightly less coarse and more numerous costae than does *Platystaphyla ornatissima* (Cushman and Church). The number and shape of the supplementary chambers is highly variable. The final one or two chambers when preserved, often cover several or all of the previous ones.

   Platystaphyla brazoensis descended from *Pseudotextularia elegans* (Rzehak), and is identical with this form in the biserial stage.

   In discussing her new species *Planoglobulina brazoensis*, Martin (1972) stated that it "differs from *P. acervulinoides* (Egger), in its less symmetrical, more inflated test, narrower costae, and fewer number of chambers (15 to 18)". The specimens she attributed to Egger's species are intraspecific variants of her species, which is here transferred to *Platystaphyla*.

   Test symmetry has no taxonomic value with this group, but is dependent solely upon the number and position of supplementary chambers. The narrowness of the costae Martin mentioned is due to a better state of preservation. Her specimens of "*Planoglobulina acervulinoides* (Egger)" are considerably abraded, with costae having been worn away on some chambers. Costae have an approximately triangular cross-section; such that as abrasion occurs, the crest of the costae becomes wider. The number of chambers is directly proportional to the age of the individual. Chamber inflation is an intraspecific variant, most probably environmentally controlled.

Distribution: This is shown in Fig. 22.

Range: Maastrichtian (*Trinitella scotti* Interval).

Platystaphylla ornatissima (Cushman and Church)

Ventilabrella ornatissima Cushman and Church, 1929, pl. 39, figs 12–15; Cushman, 1938a, p. 27, pl. 4, fig. 11; *non* Kikoïne, 1948, p. 25, 26, pl. 2, figs 8a–c; Brown, 1969, p. 41, 42, pl. 3, figs 6a–c (*non* 7a, b).

Gublerina ornatissima (Cushman and Church). *non* Montanaro Gallitelli, 1957, p. 140, 141, pl. 32, figs 1–6b; Graham, 1962, p. 105, pl. 19, figs 16A–Bc (*non* Ca, b); Graham and Church, 1963, p. 61, pl. 7, figs 10a, b; (?) Martin, 1964, p. 86, pl. 11, figs 3a–c; *non* Takayanagi, 1965, p. 200, 201, pl. 20, figs 6a–8b; *non* Saavedra, 1965, p. 343, text-fig. 89; *non* Bandy, 1967, p. 25, pl. 11 (3); *non* Neagu, 1970b, p. 60, pl. 14, fig. 11.

Planoglobulina ornatissima (Cushman and Church). Douglas, 1969 p. 160, 161, pl. 11, fig. 2 (*non* 1a, b).

Description: Test basically biserial, rapidly increasing in width; periphery in side view mildy lobate. Chambers of pre-gerontic stage typically pseudotextularian with thickness exceeding height and equal to or greater than width, later becoming less wide; supplementary chambers small, subspherical; biserial portion slowly increasing in size in side view or may maintain the same dimensions after the initial 4–5 chambers. Primary aperture a very low, wide, interiomarginal arch; supplementary apertures occur at the base of the back of gerontic chambers. Costae discontinuous, wide and high, typically 4–6 per chamber in the biserial portion as seen in side view; later chambers with finer, more numerous costae. Sutures straight, depressed or slightly raised in this depression.

Remarks: *Platystaphyla ornatissima* is distinguished from *P. brazoensis* (Martin) by having fewer and coarser costae in the biserial portion. The rate of expansion of the biserial portion and the number of supplementary chambers comprise the major variables of this species.

This species certainly evolved from a *Pseudotextularia*, but transitional forms have not yet been recorded.

Martin (1964) illustrated three views of a specimen attributed to *Gublerina ornatissima* (Cushman and Church) in which the characteristics of the costae of this species are only shown in the edge view. In his remarks, he noted the presence of longitudinal costae in the earlier chambers. Therefore, it is assumed that the side view is incorrectly drawn, and on this basis his specimen is tentatively referred to *Platystaphyla ornatissima*.

Distribution: This is shown in Fig. 23.

Range: Campanian (FOS *Pseudoguembelina costata*—FOS *Globotruncana obliqua* Interval)—Maastrichtian (*Trinitella scotti* Interval).

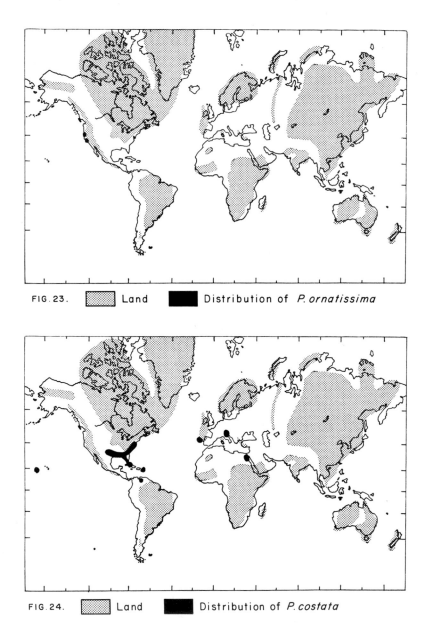

FIG. 23.   [Land]   ▉ Distribution of *P. ornatissima*

FIG. 24.   [Land]   ▉ Distribution of *P. costata*

Genus PSEUDOGUEMBELINA Brönnimann and Brown, 1953

Pseudoguembelina Brönnimann and Brown, 1953, p. 150–153.

Type species: *Guembelina excolata* Cushman, 1926b.

Description: Test small to large, biserial throughout or with initial plani-spiral coil; laterally compressed; lobate in both edge and side views; thickness gradually increases up to the last one or two chambers in which the thickness decreases, often markedly; width regularly increases until the final chamber or two where the trend is reversed. Chambers of the nepionic stage sub-spherical, imperforate except for a terminal pore in the proloculus; later chambers appressed, square to rectangular in side view; the ultimate and occasionally the penultimate chambers reduced in size, and atypically shaped. Primary aperture a low to moderate, interiomarginal arch bordered by a narrow imperforate lip. Accessory apertures paired, one per side per chamber; along median line of test at lower interior margin of later chambers; may extend below level of associated chamber via a tube-like projection or may be covered by a short lip-like extension, each of which is partially or completely imperforate. Sutures depressed, straight to curved, oblique to biserial axis; median suture zig-zag, depressed. Costae may be fine or coarse; longitudinal, conforming to chamber curvature. Finely to coarsely perforate.

Remarks: *Pseudoguembelina* is the only member of the Heterohelicidae to possess paired accessory apertures along the median line of the test.

Because *Pseudoguembelina* does not support accessory apertures on pre-ephebic chambers, it may not always be possible to distinguish its species from those of *Heterohelix*.

The restriction by Pessagno (1967) of *Pseudoguembelina* to those forms which possess only "tubelike apertural flaps" over the accessory apertures is unwarranted. Such a limitation places generic weight on an intraspecifically variable structure. This "flap" or projection may be a single simple tube-like extension with a single opening, or it may be split by a thin bridge into two openings, or the tube itself may bifurcate. Frequently, the tube is absent, and in its place is a narrow, eyelid-like structure. Any combination or all of these varieties may be found on a single specimen. Contrary to Pessagno's opinion, it is the presence of paired accessory apertures, not the nature of the "flap" that distinguishes this genus from *Heterohelix*.

Range: Upper Santonian (FOS *Globotruncana concavata*–FOS *Globotruncana elevata* Interval)–Maastrichtian (*Trinitella scotti* Interval).

Pseudoguembelina costata (Carsey), Plate 4, fig. 6
Guembelina decurrens Chapman. (?) Egger, 1899, p. 33, pl. 14, fig. 1 (*non* 2, 3).
Textularia costata Carsey, 1926, p. 26, pl. 1, fig. 4.
Guembelina costulata Cushman, 1938a, p. 16, 17, pl. 3, figs 7a–9; Cushman and Todd, 1943, p. 64, pl. 11, fig. 13; van Wessem, 1943, p. 43, pl. 1, figs 29, 30; Cushman, 1946b, p. 108, pl. 46, figs 11, 12 (*non* 10a, b); Cushman, 1948, p. 258, pl. 24, fig. 4; Cushman, 1949, p. 7, pl. 3, fig. 25; Sellier de Civrieux, 1952, p. 271, pl. 6, figs. 15a, b;

Hamilton, 1953, p. 234, pl. 30, fig. 12; Frizzell, 1954, p. 108, pl. 15, figs 17a–19; Ansary and Fakhr, 1958, p. 119, 120, pl. 1, figs 23a, b.

Pseudoguembelina costulata (Cushman). Brönnimann and Brown, 1953, p. 153, 154, text-fig. 5; Montanaro Gallitelli, 1957, p. 139, 140, pl. 31, figs 21, 22; Pessagno, 1962, p. 358, pl. 1, fig. 6; Skinner, 1962, p. 40, pl. 5, fig. 12; Said and Sabry, 1964, p. 392, 394, pl. 3, fig. 19; Cati, 1964, p. 261, 262, pl. 42, figs 9a, b; Perlmutter and Todd, 1965, p. 14, pl. 2, fig. 14; Pessagno, 1967, p. 266, pl. 79, fig. 1; pl. 88, figs 8, 9; pl. 90, fig. 3; Sliter, 1968, p. 97, 98, pl. 114, figs 11a–12b; Funnell, Friend and Ramsay, 1969, p. 24, pl. 1, figs 11, 12; text-figs 1a, b (incorrectly cited as 6a, b); *non* Todd, 1970, p. 151, pl. 5, fig. 3; *non* Govindan, 1972, p. 169, pl. 1, figs 15, 16.

Heterohelix costulata (Cushman). *non* Barr, 1969, pl. 1, fig. 7.

Description: Test of small to moderate size; slender to moderate width; margin of neanic portion may be entire or slightly lobate, increasing to moderate lobation in later chambers. Biserially alternating chambers equidimensional except for the penultimate and/or ultimate, which may be reduced in size and be irregular in shape; all others square as viewed from the side; 13–16 regularly enlarging chambers. Primary aperture a low, interiomarginal arch bordered by a narrow, imperforate lip which may be absent in the gerontic chambers. Accessory apertures paired, along the median line of each side of the chamber directed toward the proloculus; typically with a variably shaped tube-like extension bearing a single opening. Sutures depressed, straight, occasionally slightly convex in the direction of growth; oblique to the median axis, slanting toward the initial end. Costae of moderate width and height, continuous over each chamber, often continuous between neanic chambers. One to three irregular rows of small diameter pores between the costae.

Remarks: The costae of *Pseudoguembelina costata* are more numerous, by a factor of two, and less coarse than those of *P. excolata* (Cushman). Whereas, those of *P. costellifera* Masters are much finer and more numerous, in addition to being discontinuous. The rate of increase in the width of the test is the most variable feature of this species. The atypical final chambers also vary in shape as do those of all species of this genus.

Heterohelix striata (Ehrenberg) is now believed to be the ancestor to *Pseudoguembelina costata*. Each possesses the same distinctive continuous costae.

Examination of the holotypes and of topotypes (Masters, in press) has shown *Pseudoguembelina costata* to be identical with *P. costulata* (Cushman), rather than *P. excolata* (Cushman) as many have thought.

Distribution: This is shown in Fig. 24.

Range: Mid-Campanian (FOS *Pseudoguembelina costata*—FOS *Globotruncana obliqua* Interval)—Maastrichtian (*Trinitella scotti* Interval).

Pseudoguembelina costellifera Masters, Plate 4, fig. 5

Pseudoguembelina excolata excolata (Cushman). Bandy, 1967, p. 24, text-fig. 12 (11).

(?) *Heterohelix dentata* Stenestad. Stenestad, 1969, p. 658, pl. 1, figs 9, 10, 14; text-figs 12a–c.

Pseudotextularia costellifera, Masters, 1976, p. 319, pl. 1, figs 6–8.

Description: Test small to moderate in size, regularly and gradually to rapidly increasing in width; periphery slightly lobate in all but neanic portion which is entire. Chambers 14–16 in number, inflated; rectangular in side view, with a tendency toward increasing width and the development of ultimate and penultimate reniform chambers. Primary aperture a low to moderately high, interiomarginal arch bordered by an imperforate, narrow lip. Accessory apertures on short to long tube-like chamber extensions in the interio-sutural position on each side of the ephebic chambers; tube-like extensions tend to become longer and more complex with successive chambers. Sutures depressed, initially straight, normal to edge; later convex in the direction of growth. Costae thin, very numerous, closely packed, generally but irregularly conforming to the chamber curvature, discontinuous. Small diameter pores in single intercostate rows in the post-nepionic stages.

Remarks: *Pseudoguembelina costellifera* is similar to *P. kempensis*, but lacks the acute, keeled neanic periphery of that species. It can be distinguished from *P. costata* (Carsey) and *P. excolata* (Cushman) by its much finer, more numerous costae, and from *P. polypleura* Masters, which has a much larger test with only eyelid-like accessory apertural extensions. The shape of the accessory apertural projections is the most diverse structure of this species. Eyelid-like projections, short to long tubes, bifurcating and doubly apertate types have been observed, often on the same specimen.

Heterohelix carinata (Cushman) is thought to have given rise to *Pseudoguembelina costellifera* through the loss of the peripheral keel and development of accessory apertures.

Distribution: This is shown in Fig. 25.

Range: Campanian (FOS *Globotruncana obliqua*—FOS *Globotruncana calcarata* Interval)—Maastrichtian (*Trinitella scotti* Interval).

Pseudoguembelina excolata (Cushman), Plate 4, figs 7, 8

Guembelina sulcata Ehrenberg. (?) Egger, 1902, p. 33, pl. 14, fig. 30.

Guembelina excolata Cushman, 1926b, p. 20, pl. 2, fig. 9; Cushman, 1927e, p. 157, pl. 28, fig. 13; White, 1929, p. 34, pl. 4, figs 7a, b; (?) Voorwijk, 1937, p. 194, pl. 1, figs 7, 8; Cole, 1938, pl. 3, fig. 4; Cushman, 1938a, p. 17, pl. 3, figs 11a, b; Cushman and Hedberg, 1941, p. 92, pl. 22, fig. 14; Cushman and Todd, 1943, p. 64, pl. 11, figs 15a, b; Cushman, 1946b, p. 108, 109, pl. 46, figs 16a, b; Cushman and Renz, 1947, p. 44, pl 11, figs 13, 14; Hamilton, 1953, p 234, pl. 30, fig. 11; LeRoy, 1953, p. 34, pl. 7, figs 24, 25; Frizzell, 1954, p. 108, pl. 15, figs 20a, b; Vinogradov, 1960a, pl. 6, figs 33a, b.

Guembelina costulata Cushman. Cushman, 1946b, p. 108, pl. 46, figs. 10a, b (*non* 11, 12).

Guembelina cf. *Guembelina excolata* Cushman. Sellier de Civrieux, 1952, p. 271, pl. 6, figs 16a, b.

Pseudoguembelina excolata (Cushman). Brönnimann and Brown, 1953, p. 153, text-figs 1–4; Said and Kenawy, 1956, p. 139, pl. 3, fig. 36; *non* Montanaro Gallitelli, 1957, p. 139, 140, pl. 31, fig. 23; Eternod Olvera, 1959, p. 75, 76, pl. 2, figs 2, 3; Olsson, 1960, p. 28; pl. 4, fig. 11; Said and Kerdany, 1961, p. 332, pl. 2, fig. 11; Loeblich and Tappan, 1964, p. C656, fig. 525:6 (*non* fig. 5); Said and Sabry, 1964, p. 394, pl. 3, fig. 20; Saavedra, 1965, p. 344, text-fig. 92; Pessagno, 1967, p. 266, 267, pl. 68, figs 4, 5; pl. 90, fig. 5; Ansary and Tewfik, 1968, p. 42, 43, pl. 3, figs 10a, b; Brown, 1969, p. 37, 38, pl. 1, figs 11–12b; *non* Funnell, Friend and Ramsay, 1969, p. 24, 25, pl. 2, figs 1, 2; (?) text-figs 7a, b; *non* Hanzlíková, 1969, p. 43, pl. 8, figs 9a–10b; Cita, 1970, pl. 4, fig. 8; Cita and Gartner, 1971, pl. 3, fig. 4; *non* Hanzlíková, 1972, p. 94, pl. 24, figs. 3a–4b.

Pseudoguembelina excolata excolata (Cushman). *non* Bandy, 1967, p. 24, text-fig. 12 (11).

Pseudoguembelina excolata costulata (Cushman). Bandy, 1967, p. 24, text-fig. 12 (10).

Pseudoguembelina costulata (Cushman). Todd, 1970, p. 151, pl. 5, fig. 3.

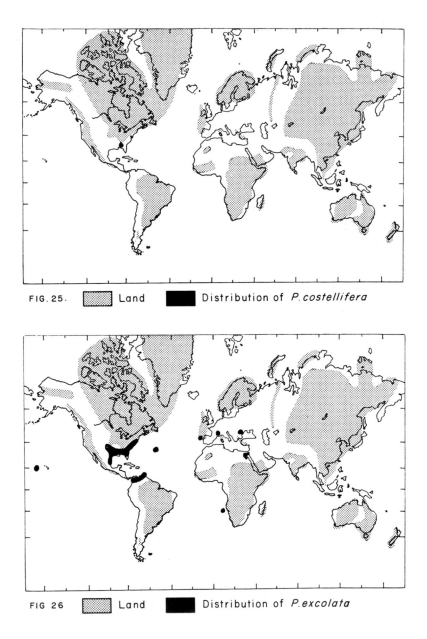

FIG. 25.   [ ] Land   ■ Distribution of *P. costellifera*

FIG 26   [ ] Land   ■ Distribution of *P. excolata*

Description: Test small, biserial; width gradually increasing; post-neanic thickness unchanging; periphery slightly lobate in all but the neanic stage which is entire Chambers of all but last 1–2 equidimensional, but these 2 tend to have a reduced thickness, greater width and be crescent-shaped; closely appressed; 14–16, rarely 20, in number. Primary aperture a low to moderately high, interiomarginal arch relative to chamber height, with an imperforate, narrow lip. Accessory apertures paired at the lower interiomarginal position of both sides of each post-neanic chamber; covered by progressively enlarging flaps or short tube-like structures. Sutures depressed, straight and normal to edge in neanic stage, later convex in the direction of growth. Costae coarse, 14–18 per ephebic chamber, continuous over each chamber, often continuous between neanic chambers, with tendency to converge at the rear of the chamber, otherwise conforming to chamber curvature. Nepionic portion non-costate, imperforate; remainder of test with two or more rows of small diameter pores between the costae.

Remarks: *Pseudoguembelina excolata* is easily differentiated from all other species, except for *P. palpabra* Brönnimann and Brown, on the basis of its fewer and coarser costae. It is distinguished from *P. palpabra* by its smaller size, the position and orientation of its accessory apertures and its chamber shape. There is slight variation in the costae on the ultimate and, occasionally, the penultimate chambers where they become finer and more numerous. The accessory apertures vary through ontogeny from minute openings with almost no flap to openings which are large tube-like extensions.

This species evolved from *Pseudoguembelina costata* (Carsey) through a reduction in the number of costae per chamber and an increase in their coarseness.

Distribution: This is shown in Fig. 26.

Range: Maastrichtian (LOS *Globotruncana calcarata*—LOS *Globotruncana lapparenti* Interval to *Trinitella scotti* Interval).

Pseudoguembelina kempensis Esker

Pseudoguembelina kempensis Esker, 1968a, p. 168, text-figs 1–5.

Description: Test small, biserial or with an initial planispiral coil; gradually increasing in width; neanic and early ephebic stages with entire margin, later becoming slightly lobate; neanic and early ephebic portions with a relatively thick, rounded keel along the edge. Chambers of post-neanic portions thicker than wide or high, except the final pair in which the width decreases; 13–16 in number; the nepionic stage lacks the strong inflation of the later chambers, which develop this along the test median line and taper toward the rear of these chambers. Primary aperture a low to high, interiomarginal arch. Accessory apertures on either side of the last 2–4 chambers along the median position of the test; covered by short to long, tube-like extensions. Sutures of post-neanic chambers deeply depressed, accentuating the chamber inflation; gently convex in the direction of growth. Costae very fine, discontinuous, longitudinal. Pore diameter very small.

Remarks: The combination of the fine longitudinal costae and the thick peripheral keel separate this species from its congeners. Insufficient data is available to determine the extent of intraspecific variation within *P. kempensis*.

There may be a phyletic relationship between *Pseudoguembelina kempensis* and *P. costellifera* Masters.

Distribution: This is shown in Fig. 27.

Range: Maastrichtian (LOS *Globotruncana lapparenti*—FOS *Globotruncana aegyptiaca* Interval to LOS *Globotruncana fornicata*—FOS *Trinitella scotti* Interval).

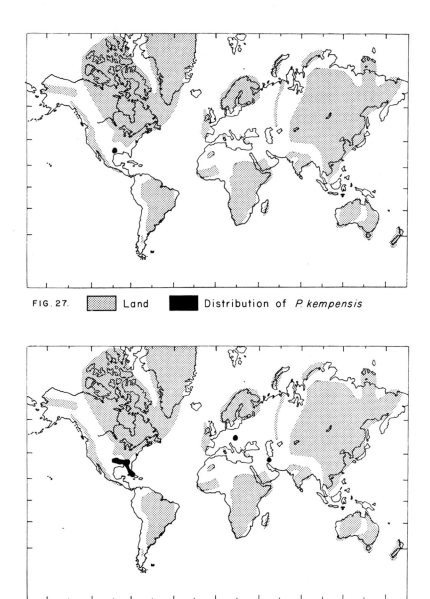

FIG. 27.　▨ Land　■ Distribution of *P. kempensis*

FIG. 28.　▨ Land　■ Distribution of *P. palpabra*

Pseudoguembelina palpabra Brönnimann and Brown, Plate 4, fig. 9

Pseudoguembelina palpabra Brönnimann and Brown, 1953, p. 155, text-figs 9a–10b; Pessagno, 1967, p. 267, pl. 78, figs 1, 2 (3?; *non* pl. 89, figs 3, 4); Brown, 1969, p. 38, pl. 1, figs 9, 10.

Pseudoguembelina cornuta Seiglie, 1959, p. 60, 61, pl. 4, figs 1a–7.

Pseudoguembelina leroyi Kavary, 1963, p. 66, pl. 13, figs 17, 18.

Pseudoguembelina excolata (Cushman). Hanzlíková, 1969, p. 43, pl. 8, figs 9a–10b; Hanzlíková, 1972, p. 94, pl. 24, figs 3a–4b.

Description: Test of moderate size, biserial; rapidly increasing in width in neanic portion, later the width may remain nearly constant; margin of neanic portion entire, later stages moderately lobate; post neanic thickness constant except for the final 1–2 chambers which may be 1/3 the thickness of the others. Chambers initially rectangular in side view, later becoming arched convexly in the direction of growth and strongly inflated; 10–16 in number. Primary aperture a low, interiomarginal arch. Paired accessory apertures on the last 2–4 chambers, covered by wide, short flaps directed away from the test midline along the sutures. Sutures of the neanic stage flush; later deeply depressed, convex in the direction of growth. Costae very coarse, conforming to the chamber contours; area immediately above the accessory apertures non-costate as is the neanic stage. Pores of wide diameters; distributed in 2, occasionally more, irregular, intercostate rows; apertural flaps and neanic chambers imperforate.

Remarks: Only one other species, *Pseudoguembelina polypleura* Masters, has accessory apertures which are located at the crest of the preceding chamber, rather than along the test midline, and which are directed away from the midline. *Pseudoguembelina palpabra* is easily distinguished from *P. polypleura* by its fewer and coarser costae. There is some variation in the degree of reduction and lateral compression of the final pair of chambers. Otherwise, the features of this species are constant.

The phylogeny of this species is unknown. However, the shape of the ephebic chambers and the coarseness of the costae suggest a close tie with *Pseudotextularia elegans* (Rzehak) or its immediate predecessor, *P. browni* Masters.

Pseudoguembelina palpabra has coarser costae than any of its congeners.

Of the four syntypes of *Pseudoguembelina leroyi* Kavary, 1963, only one exhibits an accessory aperture. The species is judged to be a junior synonym of *P. palpabra* Brönnimann and Brown because of the coarse costae and the position of the accessory aperture.

Distribution: This is shown in Fig. 28.

Range: Maastrichtian (LOS *Globotruncana calcarata*—LOS *Globotruncana lapparenti* Interval to *Trinitella scotti* Interval).

Pseudoguembelina polypleura Masters, Plate 5, fig. 1

Pseudoguembelina polypleura Masters, 1976, p. 319, pl. 1, fig. 9.

Description: Test large, biserial throughout, width rapidly increasing, periphery lobate. Chambers of earlier portion rectangular in side view, later becoming subspherical; numbering 13–16; rapidly increasing in size; inflated. Primary aperture a low, interiomarginal arch. Paired accessory apertures on each of the last 2–4 chambers; covered by wide, flaring, imperforate flaps; located above the previous chamber, adjacent to but directed away from the test midline. Sutures depressed; straight, normal to edge initially, later convex in the direction of growth. Costae numerous, fine to moderately wide, discontinuous, conforming to the chamber contours. Small diameter pores restricted to single, irregular, intercostal rows.

Remarks: *Pseudoguembelina polypleura* is distinguished from other members of the genus by its large size, the position and shape of its accessory apertures and the presence of numerous discontinuous costae. The accessory apertures are paired, developed on each of the last 2–4 chambers, and located above the previous chamber, adjacent to but directed away from the midline of the test. Major variations have not been recognised in this species.

Pseudoguembelina costellifera Masters is suspected to be ancestral to *P. polypleura*. Both species have similar costal strength, and show similar rates of increase in the width of the test.

Distribution: This is shown in Fig. 29.

Range: Maastrichtian (*Trinitella scotti* Interval).

Pseudoguembelina punctulata (Cushman), Plate 5, fig. 2

Guembelina sp. (?) Voorwijk, 1937, p. 194, pl. 1, figs 11, 13.

Guembelina punctulata Cushman, 1938a, p. 13, pl. 2, figs 15–16b; Cushman, 1946b, p. 108, pl. 46, figs 13–14b; Frizzell, 1954, p. 109, pl. 15, figs 37a, b.

Pseudoguembelina punctulata (Cushman). Brönnimann and Brown, 1953, p. 154, text-figs 7, 8.

Heterohelix (*Heterohelix*) *planeobtusa* Aliyulla, 1965, p. 222, 223, pl. figs 5a, b.

Heterohelix punctulata (Cushman). Bandy, 1967, p. 23, text-fig. 12(4); Pessagno 1967, p. 262, 263, pl. 86, figs 7–10; Sliter, 1968, p. 96, pl. 14, figs 7a, b; (?) Ansary and Tewfik, 1968, p. 40, pl. 3, figs 8a, b.

Heterohelix robusta, Stenestad, 1968a, p. 68, 69, pl. 1, figs 12–14; pl. 3, figs 1–3; (?) Stenestad, 1969, pl. 658, 659, pl. 1, figs 17–19; text-figs 13a–c.

Description: Test large, biserial throughout; initial end rapidly increasing in width, markedly slowing in later stages; thickness essentially unchanging after neanic chambers; periphery moderately lobate in later stages. Chambers in side view appear rectangular becoming more square in ephebic stage, where the thickness approximates the width and height; 15–18 in number; post-neanic stage highly inflated. Primary aperture a low to moderately high arch with a narrow imperforate lip. Accessory apertures on either side of the final 2–3 chambers, openings are small with only a narrow bordering rim rather than a flap or tubular extension; located at the junction of the 2 preceding chambers. Sutures flush in the neanic portion; later depressed, straight, normal to edge. Costae fine, numerous, discontinuous, conforming to chamber contours. Pores large in diameter, numerous, irregularly situated.

Remarks: The large size, highly inflated chambers, coarse perforation and small accessory apertures are distinctive of this species. Variation occurs in the uninflated neanic chambers which may number from seven to twelve, and in the diameter of the pores which increases toward the initial end.

Previously (in press) the author suggested a phyletic relationship between *Pseudoguembelina punctulata* and *Pseudotextularia browni* Masters. This *Pseudoguembelina* is unlike any other in several ways (as noted above), but does share certain of these features with *Pseudotextularia browni*. The chamber width, although never becoming truly pseudotextularian, exceeds that of its congeners. In addition, the test size and shape, and the nature of the costae are similar to *Pseudotextularia browni*.

The holotype clearly exhibits accessory apertures, which were illustrated by Pessagno (1967) on the redrawn figure. It is also coarsely porous which is not shown on Pessagno's figures.

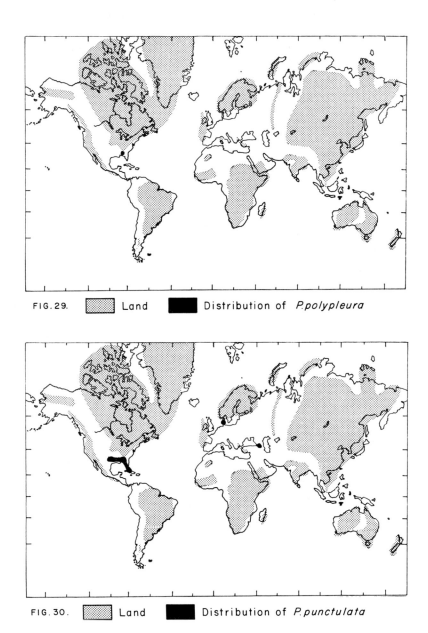

FIG.29. ▨ Land ■ Distribution of *P.polypleura*

FIG.30. ▨ Land ■ Distribution of *P.punctulata*

Although no mention of accessory apertures was made in the type description of *Heterohelix planeobtusa* Aliyulla, 1965, nor shown on the type figure, their existence is suspected. This taxon appears to share with *Pseudoguembelina punctulata* all other morphologic characteristics, as well as being found at essentially identical stratigraphic levels. For these reasons *H. planeobtusa* is regarded as a junior synonym of *P. punctulata*.

Stenestad (1968a) proposed the name *Heterohelix robusta* for a form which from the type figures and description seems to be conspecific with *Pseudoguembelina punctulata*. He did not observe the presence of accessory apertures, though these are easily overlooked because of their inordinately small size. The author has shown (in press) *P. punctulata* to be costate throughout and to lack an initial marginal keel, criteria which Stenestad used to define *H. robusta*. His species is tentatively judged to be a junior synonym of *P. punctulata*.

Distribution: This is shown in Fig. 30.

Range: Santonian (FOS *Globotruncana concavata*—FOS *Globotruncana elevata* Interval)—Maastrichtian (*Trinitella scotti* Interval).

Genus PSEUDOTEXTULARIA Rzehak, 1891

Pseudotextularia Rzehak, 1886, (nomen nudum), p. 8; Rzehak, 1891 (part) p. 2.

Bronnibrownia Montanaro Gallitelli, 1955a (nomen nudum), p. 215, 220, 222.

Bronnimannella Montanaro Gallitelli, 1956, p. 35.

Type species: *Cuneolina elegans* Rzehak, 1891.

Description: Test of moderate to large size; biserial with or without a small, initial planispiral coil; width rapidly increasing, but may remain constant or even decrease in the late ephebic and gerontic stages; thickness rapidly increases, often being the largest dimension. Chambers of the neanic stage subspherical; all later chambers thicker than wide or high except for the ultimate and occasionally the penultimate which may be aberrant; width may increase steadily or remain unchanged during the later ephebic stage and may be reduced in the gerontic portion. Primary aperture a wide, low to moderately high, interiomarginal arch bordered by a narrow imperforate lip or face. An occasional supplementary aperture may be found on the back of the final 1–2 chambers, occupying a central–basal position, with its height and width typically equidimensional; rarely covered by a small supplementary aperture. Sutures depressed, straight, normal to edge. Costae fine to coarse, conforming to curvature of chamber. Pores range from small to moderately wide in diameter.

Remarks: In *Pseudotextularia* the supplementary chambers are rare, usually not exceeding two, and are strictly of secondary importance to the biserial arrangement. This distinguishes it from either *Planoglobulina* or *Platystaphyla*. It differs from all other Heterohelicidae by possessing chambers which are thicker than they are wide or high.

Misinterpretations of *Pseudotextularia* and its allied genera, *Planoglobulina*, *Platystaphyla*, and *Ventilabrella*, have led to untenable taxonomic and morphologic diagnoses. Only an exacting examination of original publications and of type specimens has enabled this confusion to be resolved. A detailed analysis has previously been presented (Masters, 1976), necessitating only a summation here.

Rzehak (1891) originally regarded two different growth forms as belonging to *Cuneolina elegans* Rzehak (= *Pseudotextularia elegans*); four years later he restricted his concept of the species to the simple biserial form. This designation restricts the growth form having "chamber proliferation with a circular cross-section" to the species *varians* Rzehak now treated as a *Planoglobulina*. A third growth form having "proliferation of chambers in a single plane" is placed in the genus *Platystaphyla*. *Ventilabrella* possesses a growth form similar to that of *Platystaphyla*, but it lacks the pseudotextularian chamber shape. Even though all four forms have been variously treated as synonyms, each is a valid taxon.

Brown (1969) recognised the origin of the more complex pseudotextularian growth forms, but failed to correctly discern between their ontogeny and phylogeny. He envisioned the forms with supplementary apertures as an ontogenetic development of *Pseudotextularia*, when in fact their differing stratigraphic positions support a phylogenetic cause. This in turn lends credence to the arguement for separate taxa.

Pseudotextularia, then, is limited to those forms which are biserial. Although they may develop one or two supplementary apertures and chambers, these are subordinate to the biserial chamber arrangement. This restriction is not without its problems. Mature *Pseudotextularia* with a supplementary aperture could be confused with an immature *Platystaphyla* inasmuch as their stratigraphic ranges overlap. If mature *Platystaphyla* are not present, then one can say with reasonable certainty that the specimens belong to *Pseudotextularia*.

Keller (1946) described *Pseudotextularia koslovi* from thin-section. It is, however, not possible to speciate the Heterohelicidae from thin-sections, and is recommended that *P. koslovi* be disregarded.

The three syntypes of *Pseudotextularia* (?) *reissi* Kavary, 1963, were examined. None of these represent a new species, nor do they belong to *Pseudotextularia*. One syntype (pl. 13, fig. 7) is the last three chambers of a *Heterohelix*, probably *H. globulosa* (Ehrenberg). The two remaining syntypes (pl. 13, fig. 8 and one unfigured) had their last two chambers considerably inflated, but which were axially flattened during preservation. These to most likely belong to *H. globulosa*.

Range: Late Santonian (FOS *Globotruncana concavata*–FOS *Globotruncana elevata* Interval)–Maastrichtian (*Trinitella scotti* Interval).

Pseudotextularia browni Masters, Plate 5, figs 3, 4

Textularia globulosa Ehrenberg. Carsey, 1926, p. 25, pl. 5, figs 2a, b.

Guembelina striata (Ehrenberg). Kalinin, 1937, p. 37, 38, pl. 4, fig. 56; Cushman, 1938a, p. 8–10, pl. 1, figs 4a, b (*non* figs 37–39b); Cushman, 1944b, p. 91, pl. 14, figs 4a, b.

(?) *Guembelina nuttalli* Voorwijk, 1937, p. 192, pl. 2, figs 1–9.

Guembelina plummerae Loetterle, 1937, p. 33, 34, pl. 5, figs 1, 2; Cushman, 1938a, p. 15, 16, pl. 3, figs 3a–5b; Cushman, 1944b, p. 90, 91, pl. 14, figs 3a, b; Cushman, 1946b, p. 104, pl. 45, figs 1a–3; *non* Kikoïne, 1948, p. 18, pl. 1, figs 5a–c; *non* Hamilton, 1953, p. 234, pl. 30, fig. 10; Frizzell, 1954, p. 109, pl. 15, figs 36a, b.

(?) *Guembelina santonica* var. *pseudoobtusa* Agalarova, 1951, in Dzhafarov, Agalarova and Khalilov, p. 99, pl. 14, figs 3, 4.

(?) *Guembelina santonica* Agalarova, 1951, in Dzhafarov, Agalarova, Khalilov, p. 100, pl. 14, figs 7, 8; Geodakchan and Aliyulla, 1959, p. 56, pl. 1, fig. 1.

Bronnimannella plummerae (Loetterle). Montanaro Gallitelli, 1956, p. 35, pl. 7, figs 1a–2c.

Pseudotextularia elegans (Rzehak). Montanaro Gallitelli, 1957, p. 138, 139, pl. 33, figs 6a–c; Cati, 1964, p. 261, pl. 42, figs 10a, b; Loeblich and Tappan, 1964, p. C656, figs 525:7a–c; (?) Hanzlíková, p. 43, 44, pl. 8, figs 11a, b, 13a, b (*non* 12a, b); Latif, 1970, p. 38, pl. 5, figs 9, 10.

Pseudotextularia plummerae (Loetterle). *non* Seiglie, 1958, p. 77–79, pl. 5, figs 4a–5b; pl. 7, figs 2a, b.

(?) *Guembelina subplummerae* Aliyulla, 1959, in Geodakchan and Aliyulla, p. 57, pl. 1, fig. 5.

Heterohelix striata (Ehrenberg). Cati, 1964, p. 260, 261, pl. 42, figs 6a, b.

Heterohelix ultimatumida (White). Funnell, Friend and Ramsay, 1969, p. 21, 22, pl. 1, figs 5, 6; text-figs 3a, b.

Planoglobulina (?) *carseyae* (Plummer). Martin, 1972, p. 83, pl. 4, figs 4a–c (*non* 5–7b).

Pseudotextularia browni Masters, 1976, p. 321, pl. 1, figs 10–12.

Description: Test biserial; width and thickness rapidly increasing from initial end, but width may decrease in the gerontic stage; neanic stage with entire margin, later portions moderately lobate. Chambers rectangular in side view, but thickness exceeds the other dimensions; inflated; numbering 10–14. Aperture a low, broad, interiomarginal arch bordered by a narrow imperforate face or lip; an occasional supplementary aperture on the final 1–2 chambers, located in the central-basal position of the back of the chamber, rarely with small supplementary chambers. Nepionic sutures flush, later sutures depressed, straight, normal to edge; sutures on edge may be straight or convex in the direction of growth. Costae of fine to moderate strength, numerous, generally discontinuous. Small diameter pores, generally restricted to single intercostal rows.

Remarks: The numerous, fine to moderate, discontinuous costae serve to distinguish this species from its congeners. The rate of increase in chamber thickness is usually uniform, at approximately 50% from chamber to chamber, though this may change to as much as a 100% increase. Supplementary apertures may not be present, but any stage from dimpling to various sizes of apertures may be observed. The width of these apertures is typically equivalent to their height, but always narrower than the primary aperture.

A completely gradational series between *Heterohelix globulosa* (Ehrenberg) and *Pseudotextularia browni* has been observed (Masters, 1976).

During the examination of the cotypes of *Guembelina plummerae* Loetterle, which are deposited at the University of Nebraska State Museum, it was deemed necessary to select a lectotype to represent this taxon. The specimen (UNSM IP-10034) represented by Loetterle's fig. 1 is selected as the lectotype of *G. plummerae*, while the one (UNSM IP-10035) illustrated by his fig. 2 is designated as the paralectotype. Contrary to Loetterle's description, no initial coil is present on either of his specimens.

The author proposed the name *Pseudotextularia browni* (Masters, 1976) to replace that of *Guembelina plummerae* Loetterle, 1937, a junior secondary homonym of *Ventilabrella plummerae* Sandidge, 1932. The submission of this name may have been premature. It is possible that one of four other species, *Guembelina nuttalli* Voorwijk, 1937, *G. santonica* Agalarova, 1951, *G. santonica pseudoobtusa* Agalarova, 1951, or *G. subplummerae* Aliyulla, 1959, may prove to be the senior synonym.

The type figures of *Guembelina nuttalli* Voorwijk are little more than outline sketches, and no mention of the coarseness of the costae is made. On the basis of a comparison of these figures with those of other species given by Voorwijk, it is suspected that the costae are very fine. Brown (1969) provisionally regarded Aliyulla's and both of Agalarova's species as synonyms of *Pseudotextularia plummerae* Loetterle (= *P. browni*), but noted that the costae as illustrated appear coarser than those of Loetterle's species. The taxonomic status of these four potential synonyms can be resolved only after examination of the original material.

Distribution: This is shown in Fig. 31.

Range: Late Santonian (FOS *Globotruncana concavata*—FOS *Globotruncana e.evata* Interval)—Maastrichtian (*Trinitella scotti* Interval).

Pseudotextularia carseyae (Plummer), plate 6, figs 1, 2

Ventilabrella carseyae Plummer, 1931, p. 178, 179, pl. 9, figs 7a–9c (10?); Cushman, 1938a, p. 26, 27, pl. 4, figs 20–22 (*non* 23a–24b); Cushman, 1946b, p. 112, pl. 48, figs 1a–4b (*non* 5a, b); Loeblich, 1951, p. 109, pl. 12, figs 6–8; Sellier de Civrieux, 1952, p. 271, pl. 6, figs 18a, b; *non* Salaj and Samuel, 1966, p. 230, 231, tab. 37 (8).

Ventilabrella plummerae Sandidge, 1932a, p. 195, 196, pl. 19, figs 5, 6.

Pseudotextularia carsyae (Plummer). Rauzer-Charnousova and Fursenko, p. 40, 41, text-figs 17I–III; (?) Hanzlíková, 1972, p. 94, pl. 23, figs 20–22.

Guembelina plummerae Loetterle. Cushman, 1938a, p. 15, pl. 3, figs 3–5; Sellier de Civrieux, 1952, p. 270, pl. 6, figs 11a–12.

Guembelina carseyae (Plummer). Frizzell, 1954, p. 108, pl. 15, figs 14–16.

Planoglobulina carseyae (Plummer). Montanaro Gallitelli, 1957, p. 141, 142, pl. 32, fig. 13; Pessagno, 1967, p. 271, 272, pl. 87, figs 10, 15, 16; (?) Hanzlíková, 1969, p. 42, pl. 7, figs 29, 30; pl. 8, figs 6–8; *non* Govindan, 1972, p. 171, 172, pl. 1, fig. 17.

Heterohelix striata (Ehrenberg). Graham and Church, 1963, p. 62, 63, pl. 7, figs 12a, b.

Pseudotextularia elegans (Rzehak). Pessagno, 1967, p. 268, 269, pl. 75, figs 12–17; pl. 85, figs 10, 11; pl. 88, figs 14–16; pl. 89, figs 10, 11.

Pseudotextularia eggeri Rzehak. Sliter, 1968, p. 98, pl. 14, fig. 15 (*non* 13a–14b).

Pseudotextularia cushmani, Brown, 1969, p. 55, 56, pl. 2, figs 2a–3b; pl. 3, figs 4a, b.

Description: Test biserial throughout or with an initial small, planispiral coil; width and thickness rapidly increasing; neanic stage margin entire, later lobate.

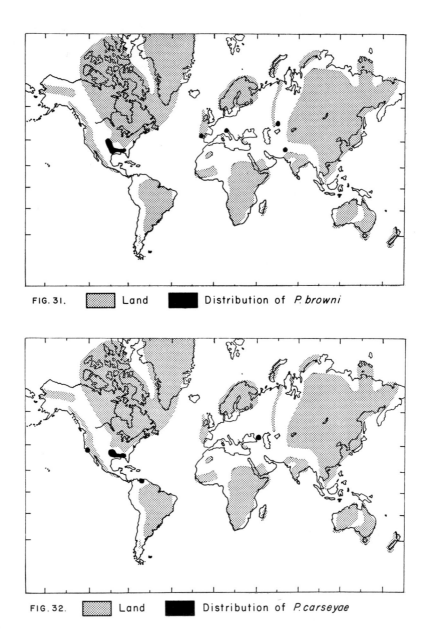

FIG. 31. [ ] Land ■ Distribution of *P. browni*

FIG. 32. [ ] Land ■ Distribution of *P. carseyae*

Chambers of coil appressed, 3–5 in number; later chambers inflated, 12–14 in number, thickness exceeding both width and height which are equidimensional or with the width slightly greater. Aperture a low to moderately high, broad, interiomarginal arch, bordered by a narrow imperforate face or lip; an occasional narrow to moderately wide supplementary aperture or chamber is present on the central-basal position of the ultimate chamber. Sutures flush in neanic stage; depressed, straight, normal to edge in later stages; sutures on edge straight or convex in the direction of growth due to presence of a dimple or supplementary aperture. Costae moderately coarse, generally continuous with a single, secondary, discontinuous costae between the primary ones. Pores of moderate diameter, generally aligned in 2 rows between the primary costae.

Remarks: This species is the only representative of the *Pseudotextularia* possessing a bimodal costal strength. The strength of its costae (primary) lies between that of *P. browni* Masters and *P. elegans* (Rzehak), and they are more continuous than those of any other species. The number of chambers in the initial coil, if present, varies. The supplementary apertures may vary considerably in size when present. Other features are stable.

The continuous costae of *Pseudotextularia carseyae* more nearly approach those of *Heterohelix striata* (Ehrenberg) than any other heterohelicid. There is little likelihood that this species is a member of the proposed *P. browni–P. carseyae–P. elegans* phyletic lineage of Brown (1969). This hypothesis is possible, but requires a change from discontinuous costae to continuous and back again. For *H. striata* to be ancestral requires only a change from heterohelicid- to pseudotextularian-shaped chambers, a phenomenon already observed from *H. globulosa* (Ehrenberg) to *P. browni*.

The holotype (= *Guembelina plummerae* of Cushman) of *Pseudotextularia cushmani* Brown, 1959, possesses continuous, evenly spaced costae as does *P. carseyae*. Brown remarked that they were difficult to distinguish. *Pseudotextularia cushmani* is regarded as a junior synonym of *P. carseyae*.

Distribution: This is shown in Fig. 32.

Range: Campanian (FOS *Pseudotextularia carseyae*—FOS *Globotruncana ventricosa* Interval)—Maastrichtian (*Trinitella scotti* Interval).

Pseudotextularia elegans (Rzehak), Plate 6, figs 3, 4

Cuneolina elegans Rzehak, 1891, p. 4.
Pseudotextularia varians Rzehak, 1895, p. 217, pl. 7, figs 1a, b (*non* 2, 3); Seiglie, 1959, p. 76, 77, pl. 7, figs 5a, b (*non* 6a, b; pl. 5, fig. 6); Pokorný, 1963, p. 390, fig. 441 (part).
Guembelina acervulinoides Egger, 1899, p. 36, pl. 14, figs 14?, 15, 16 (*non* 17, 18, 20–22).
Guembelina fructicosa Egger, 1899, p. 35, pl. 14, figs 25, 26 (*non* 8, 9, 24).
Textularia biarritzensis Halkyard, 1918, p. 34, pl. 2, fig. 6.
Guembelina striata (Ehrenberg). Cushman, 1936, p. 418, pl. 1, figs 11a–c; Voorwijk, 1937, p. 194, pl. 1, figs 9, 10.
Pseudotextularia elegans (Rzehak). Glaessner, 1936, p. 99–101, text-figs, 1a, b, pl. 1, figs 1a–2; Glaessner, 1948, p. 152, pl. 10, figs 15a, b; Noth, 1951, p. 61, 62, pl.7, figs 15–17; *non* Frizzell, 1954, p. 111, pl. 16, figs 5–7; Bettenstaedt and Wicher, 1956, p. 502, pl. 1, fig. 6 (part); *non* Witwicka, 1958, p. 195, 196, pl, 8, figs 6, 7; Eternod Olvera, 1959, p. 73, 74, pl. 1, figs 11–13; Sieglie, 1959, p. 55, 56, pl. 1, figs 1a, b, 3a, b; *non* Nagappa, 1959, pl. 7, figs 7a–8b; Olsson, 1960, p. 28, pl. 4, figs 9,

10; Said and Kerdany, 1961, p. 332, pl. 2, figs 9a, b; Skinner, 1962, p. 40, 41, pl. 5, fig. 17; Hiltermann and Koch, 1962, p. 337, pl. 46, fig. 11 (*non* 12; tab. 19); Kavary and Frizzell, 1963, p. 64, pl. 13, figs 9–10; *non* Cati, 1964, p. 261, pl. 42, figs 10a, b; Said and Sabry, 1964, p. 392, pl. 3, figs 29a, b; Perlmutter and Todd, 1965, p. 114, pl. 2, figs 17a, b; Saavedra, 1965, p. 344, text-fig. 90; Lehmann, 1966, p. 316, pl. 2, fig. 10; Salaj and Samuel 1966, p. 232, tab. 37, fig. 11; Wille-Janoschek, 1966, p. 120, 121, pl. 8, figs 10a, b; Pessagno, 1967, p. 268, 269, pl. 75 figs 12–17; pl. 85, figs 10, 11; pl. 88, figs 14–16; pl. 89, figs 10, 11; Sliter, 1968, p. 98, pl. 14, figs 13a–14b (*non* 15); Ansary and Tewfik, 1968, p. 43, pl. 3, figs 11a, b; (?) Barr, 1968, pl. 1, fig. 10; Bate and Bayliss, 1969, pl. 4, figs 3a–c; Hanzlíková, 1969, p. 43, 44, pl. 8, figs 12a, b (*non* 11a, b, 13a, b); Funnell, Friend and Ramsay, 1969, p. 23, pl. 1, figs 9, 10; text-figs 5a, b; [incorrectly attributed to Grzybowski] Scheibnerová, 1969, tab. 7, fig. (?) 2 (*non* fig. 1; tab. 6, figs 7, 8); Brown, 1969, p. 47–54, text-figs 9a–10b (*non* 13, 14; pl. 2, figs 4a, b; pl. 3, figs 2, 3); Dupeuble, 1969, p. 158, pl. 4, figs 13a, b; (?) Sturm, 1970, p. 113, pl. 8, figs 2a, b; Bertels, 1970, p. 33, pl. 2, figs 4a–5b; Neagu, 1970b; p. 61, pl. 14, fig. 1; pl. 42, fig. 60; Todd, 1970, p. 151, 152, pl. 5, figs 5a, b; Cita, 1970, pl. 4, fig. 7; Cita and Gartner, 1971, pl. 3, fig. 3; Govindan, 1972, p. 170, 171, pl. 1, figs 5, 6; pl. 2, figs 6, 7; Hanzlíková, 1972, p. 95, pl. 24, figs 10a, b (*non* 8a, 11a–12b; pl. 23, figs 24a–26).

Guembelina plummerae Loetterle. Cole, 1938, pl. 3, fig. 9; Kikoïne, 1948, p. 18, pl. 1, figs 5a–c; Said and Kenawy, 1956, p. 139, pl. 3, figs 33a, b; Vinogradov, 1960a, pl. 6, figs 31a, b.

Ventilabrella carseyae Plummer. Cushman, 1947, p. 14, pl. 4, figs 9, 10.

Guembelina striata var. *deformis* Kikoïne, 1948, p. 20, pl. 1, figs 8a–c.

Guembelina striata var. *compressa* Nakkady, 1950, p. 686, pl. 89, fig. 19.

Pseudotextularia varians (elegans) Rzehak. Hofker, 1957, p. 422–424, text-figs 1, m (*non* a–k).

Guembelina elegans (Rzehak). Maslakova, 1959, p. 117, 118, pl. 15, figs 4a, b.

Pseudotextularia bronnimanni Seiglie, 1959, p. 57, 58, pl. 1, figs 5a–8b.

Pseudotextularia elongata Seiglie, 1959, p. 58, 59, pl. 1, figs 2a, b, 4a, b; pl. 2, figs 1a–2b, 4a, b, 6a, b (*non* pl. 3, figs 1a, b).

Pseudotextularia nuttalli (Voorwijk). Eternod Olvera, 1959, p. 74, 75, pl. 2, fig. 1.

Pseudotextularia plummerae (Loetterle). Seiglie, 1959, p. 77–79; pl. 5, figs 5a, b (*non* 4a, b); pl. 7, figs 2a, b.

Planoglobulina carseyae (Plummer). Eternod Olvera, 1959, p. 76, 77, pl. 1, figs 5, 8, 9.

Pseudotextularia pecki Kavary, 1963, p. 63, pl. 13, figs 5, 6.

Racemiguembelina textulariformis (White). Salaj and Samuel, 1966, p. 233, tab 37, fig. 12.

Pseudotextularia deformis (Kikoïne). Pessagno, 1967, p. 269, pl. 90, fig. 16; pl. 92, figs 19–21.

Pseudoguembelina palpabra Brönnimann and Brown. Pessagno, 1967, p. 267, pl. 89, figs 3, 4; aff. *palpabra*, Pessagno, 1967, p. 267, pl. 89, figs 5, 12–14.

Planoglobulina acervulinoides (Egger). Cushman, 1938a, p. 23, pl. 4, figs 6a, b (*non* 5a, b, 7a–8b); Cushman, 1946b, p. 111, pl. 47, figs 14a, b (*non* 12a–13b, 15a, b); Said and Sabry, 1964, p. 394, pl. 3, fig. 30.

Pseudotextularia elegans elegans (Rzehak). Bandy, 1967, p. 24, text-fig. 12 (12).

Pseudoguembelina excolata (Cushman). Funnell, Friend and Ramsay, 1969, p. 24, 25, pl. 2, figs 1, 2 (*non* text-figs 7a, b).

Heterohelix elegans (Rzehak). El-Naggar, 1971a, pl. 7, figs a, b, d, e.

Pseudoguembelina punctulata (Cushman). El-Naggar, 1971a, pl. 7, fig. c.

Description: Test biserial throughout; width and thickness rapidly increasing; margin of nepionic stage entire, later stages lobate. Chamber width may slightly exceed height with thickness double the width; numbering 14–18; inflated. Aperture a low, broad, interiomarginal arch, bordered by a narrow imperforate face and lip. Occasional supplementary apertures, or rarely supplementary chambers, on the broad edge in a central-basal position. Sutures of neanic stage flush; in others depressed, straight and normal to edge; those on the edge are straight or convex in the direction of growth. Costae coarse, continuous to irregularly discontinuous over each chamber. Pores of moderate diameter, occasionally merging to form elongate pits; 1–4 intercostal rows.

Remarks: No other *Pseudotextularia* possesses costae as coarse as those of *P. elegans*. There is no variation in costal coarseness but their density may vary from typically four up to eight costae per unit area. The latter approaches that of *Pseudotextularia browni* Masters. Test width is also variable.

This species is believed to have descended from *Pseudotextularia browni* Masters which it resembles in all but costal strength.

Brown (1969) employed the discrimination grid devised by Mayr, Linsley and Usinger (1953) to determine whether *Pseudotextularia elegans* and *Racemiguembelina fructicosa* (Egger) [= *Planoglobulina varians* (Rzehak)] are members of a single species or of two. Brown concluded that since they occupied essentially the same geographic areas and were synchronous, then it was more likely that these two are but one species.

The first supposition Brown (1969) made was that these two forms apparently did interbreed since they were not reproductively isolated. This line of reasoning may be useful, but is nevertheless conjecture and could be misleading. Other species of *Pseudotextularia*, living concurrently with *P. elegans*, occupied the same geographical and environmental (i.e. not reproductively isolated) areas, and yet retained their morphologic integrity. As noted by Newell (1956), in synpatric forms, hybridisation is *prima facie* evidence that interbreeding occurred. No such evidence can be observed here, in fact, hybridisation would be exceedingly difficult if not impossible to recognise in such morphologically similar forms. Thus, no case can be made in support of more than the potential of interbreeding.

Brown's (1969) second assumption is erroneous. The two species in question are not synchronous over their total stratigraphic range. *Planoglobulina varians* (Rzehak) occurs only during the very youngest portion of the total range of *Pseudotextularia elegans*. This alone is reason to suspect that two species are present. More important is the development of a different growth pattern in *P. varians*. Thus, these two organisms must be regarded as distinct but phyletically related species.

Textularia biarritzensis Halkyard, 1918, is a *Pseudotextularia elegans* which was most likely reworked into the Eocene Blue Marl at Biarritz, France.

The degree of reduction in the test width of *Pseudotextularia* is highly variable, and easily encompasses that shown in Kikoïne's (1948) *Guembelina striata* var. *deformis*.

Pseudotextularia pecki Kavary, 1963, is small but as coarsely costate as is *P. elegans*, to which it is referred.

Gohrbandt (1967) stated that *Pseudotextularia elegans* was described by Rzehak (1891) from Late Paleocene sediments containing elements of a reworked Cretaceous fauna.

Distribution: This is shown in Fig. 33.

Range: Maastrichtian (LOS *Globotruncana calcarata*—LOS *Globotruncana lapparenti* Interval to *Trinitella scotti* Interval).

Genus VENTILABRELLA Cushman, 1928a

Ventilabrella Cushman, 1928a, p. 2.

Sigalia Reiss, 1957a, p. 243.

Type species: *Ventilabrella eggeri* Cushman, 1928a.

Description: "Early stage coiled in the microspheric form, followed by a biserial arrangement of subglobular chambers. In early biserial stage supplementary apertures develop at the base of the edge of the chambers. Supplementary chambers are added, each of which may have one or two apertures. From this point the biserial arrangement is superceded by numerous, less orderly, supplementary chambers, imparting a distinct fan-shape to the test. Individual chambers of the fan are subglobular to slightly elongate longitudinally, with occasional highly irregular shapes in gerontic specimens. All chambers are distinct, separate units. The surface ornamentation consists of discontinuous, longitudinal costae of varying height and width. Pores lie between the costae. The primary biserial aperture is an interiomarginal arch. The secondary apertures are randomly oriented and variable in shape." (Emended description, Masters, 1976.)

Remarks: *Ventilabrella* differs from the only other fan-shaped genus *Platystaphyla* by having subglobular heterohelicid chambers in the biserial part rather than the thicker pseudotextularian type.

Most workers have followed Montanaro Gallitelli's (1957) conclusion that *Ventilabrella* was a junior synonym of *Planoglobulina*. Brown (1969) correctly concluded that this synonymy was erroneous and re-employed the name *Ventilabrella*. This decision was later supported by Martin (1972). However, neither Brown nor Martin have fully grasped the taxonomic and morphologic intricacies of *Ventilabrella*. I have presented a detailed discussion (Masters, 1976) of the complexly intertwined problems of morphology and taxonomy for the genera *Ventilabrella*, *Planoglobulina*, *Platystaphyla* and *Pseudotextularia*; each genus was found to be valid.

Martin's (1972) emended description of *Ventilabrella* places primary emphasis on the vermicular ornamentation, the cameral flanges and cameral ridges, and the test shape in general. She proposed the term "vermicular" for an irregular type of surface structure found on some *Ventilabrella*. She considered this structure as ornamentation with taxonomic significance at the specific level. Although not previously recognised in the heterohelicids, this structure has been recorded in other groups of planktonics.

The vermicular surface may be comparable to the crustal calcification noted by Bé and Hemleben (1970) in the gerontic stage of a Recent planktonic

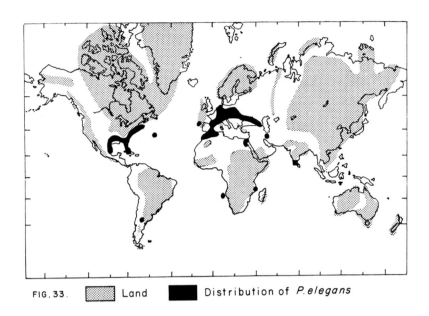

FIG. 33. ▦ Land ■ Distribution of *P. elegans*

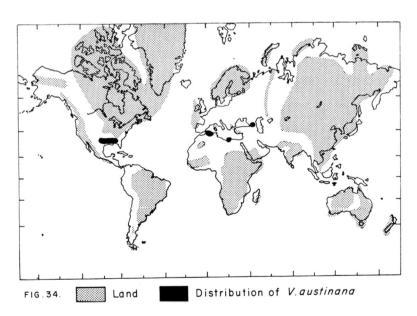

FIG. 34. ▦ Land ■ Distribution of *V. austinana*

foraminifer. Pflaumann (1971) discovered an identical vermicular structure in another Recent species, and attributed it to diagenesis. Though there is some question of diagenesis in Pflaumann's specimens, the vermicular structure is not an uncommon diagenetic product.

Martin (1972) illustrated equally mature specimens of the same species, both with and without a vermicular surface. Furthermore, some of her specimens which exhibit this structure are very poorly preserved. This suggests an inorganic (diagenetic) origin for the vermicular structure in the case of the *Ventilabrella*.

As in all planktonic foraminifers, the final chambers of *Ventilabrella* are the thinnest and, therefore, most susceptible to breakage. In fact it is rare to find the final row of chambers unbroken. The remnants of these chambers, as pointed out for those in *Platystaphyla* (see Remarks under this genus), correspond to the "cameral ridges" and "cameral flanges" of Martin (1972). No taxonomic significance can be attached to these remnants.

In defining genera and species, Martin (1972) used "the ultimate shape and form of the adult test, which is determined by the arrangement of 'primary' and 'supplementary' chambers alike". In *Ventilabrella* the ultimate shape of the test is determined by the number and position of the chambers. In turn the number of chambers is directly related to the time of the organism's death, whether by predation, senescence or reproduction, where the test is no longer occupied. Therefore, this criteria proposed by Martin is inherently misleading, and ignores the ontogenetic development of the organism as reflected in its test.

Martin's (1972) attempt to mathematically quantify *Ventilabrella eggeri* Cushman, the type species of the genus, by measuring the angle of the slope of the "final" chambers cannot be accepted on this or any other species of Heterohelicidae. With *V. eggeri*, the specimens illustrated are incomplete, broken individuals (see further comments under this species), not the complete "adult test" which she earlier emphasised as critical to species determination. The measurement of this angle is meaningless because the angle is controlled by three highly variable factors: the number of chambers (primary and supplementary), their distribution and their size. Each alone will significantly alter such an angle; together the possible variations are innumerable. Mathematical quantification is a valuable tool, but great care must be given to the feature(s) selected for quantification. Such a selection should be made only after a lengthy study of the extent and potential cause(s) of the degree of variation which may exist for a given feature.

The genus *Sigalia* was proposed by Reiss (1957a) for the species *Guembelina* (*Guembelina*, *Ventilabrella*) *deflaensis* Sigal, primarily on the basis of the raised, beaded limbate sutures. He noted that this species had a "rare tendency to build more than one (maximum two) chambers in a series (row) in very late stages". This ventilabrellid stage is neither as rare nor limited to as few chambers as Reiss indicated. Bettenstaedt and Wicher (1955) had earlier transferred Sigal's species to *Ventilabrella*.

In determining the taxonomic status of *G. deflaensis* Sigal, one is faced with two problems: (1) the significance of the raised, limbate sutures, and (2) the exact nature of its growth form.

Brown (1969) interpreted this species as having a "two-keeled imperforate band along the top of its later chambers," which form "the limbate sutures of the test" as successive chambers are added. Scanning electron microscopy has revealed that the oral surface between these "keels" is perforate. Thus, the importance typically awarded to keel structure does not apply here, as these are no more than raised, limbate sutures. The question which then arises is what is the taxonomic significance of sutures? If one is to be consistent within the Heterohelicidae, then the sutures of all the species must be considered of equal rank with depressed sutures in one group, flush in another and raised in a third. No one would agree to this. Even in the taxon *deflaensis*, the sutures run the full gamut from raised to depressed. I do not believe that the nature of the sutures is taxonomically valuable above the species level. *Sigalia*, then must become a junior synonym of either *Heterohelix* or *Ventilabrella*.

Reiss (1957a) recognised a stratigraphic difference between the two morphotypes of *Guembelina deflaensis* Sigal, with the ventilabrellid form appearing later than but accompanied by the biserial form. The two grade imperceptibly. The stratigraphic evidence submitted by Reiss is subjective. Until a detailed study of a continuous stratal sequence containing these two morphotypes is presented, *deflaensis* must be placed somewhat arbitrarily into either *Heterohelix* or *Ventilabrella*. The latter is chosen because a stratigraphic variance between the morphotypes is not discernible, and the ventilabrellid stage is interpreted as the gerontic stage of this species. Thus, *Sigalia* becomes synonymous with *Ventilabrella*.

Range: Turonian–Maastrichtian (*Trinitella scotti* Interval).

Ventilabrella austinana Cushman, Plate 6, fig. 5
Ventilabrella austinana Cushman, 1938a, p. 26, pl. 4, figs 19a, b; Cushman, 1946b, p. 111, pl. 47, figs 16a, b; *non* Kikoïne, 1948, p. 25, pl. 2, figs 7a–c; Frizzell, 1954, p. 111, pl. 16, figs 9a, b.
Planoglobulina glabrata (Cushman). Montanaro Gallitelli, 1957, p. 141, 142, pl. 32, figs 11, 12; Barr, 1968, pl. 1, figs 6, 9.
Planoglobulina austinana (Cushman). Skinner, 1962, p. 42, pl. 5, fig. 16; Bandy, 1967, p. 25, text-fig. 11 (5).
Planoglobulina transcaucasica Aliyulla, 1965, p. 226, pl. 1, figs 9a–10b.
Planoglobulina multicamerata (de Klasz). (?) Pessagno, 1967, p. 272, 273, pl. 89, fig 15.

Planoglobulina (?) *carseyae* (Plummer). Martin, 1972, p. 83, pl. 4, figs 5–7b.
Ventilabrella glabrata Cushman. Martin, 1972, p. 86, 87, pl. 1, figs 8–9b.
Ventilabrella riograndensis Martin, 1972, p. 88, 89, pl. 2, figs 1–4b.
Ventilabrella eggeri Cushman. (?) Hanzlíková, p. 96, 97, pl. 23, fig. 23 (*non* 19a, b).

Description: Test initially biserial, later developing supplementary chambers in the plane of symmetry; rapidly expanding, fan-shaped; edge rounded, periphery lobate. Chambers of biserial stage with equidimensional height, width, and thickness, gradually increasing in size, up to 14 in number, inflated; supplementary chambers more irregularly shaped with either height or width being dominant, may be smaller than, equal to or larger than associated primary chamber, inflated to laterally compressed; all covered by fine, generally discontinuous costae which conform to the curvature of the chamber. Primary aperture a low to moderately high, interiomarginal arch bordered by an imperforate lip. A single secondary aperture appears at the lower rear edge of each ephebic primary chamber; secondary chambers may have 1–2 apertures, always one in the interiomarginal position, the second on the lower, rear edge. Sutures depressed, straight in the neanic stage; depressed, straight to irregular later. Finely and densely perforate.

Remarks: This species is recognised by its numerous, fine, discontinuous costae. As with all members of *Ventilabrella*, the number of chambers in both the neanic biserial stage and the later ventilabrellid stage is highly variable. Also typically, the final supplementary chambers may be irregularly shaped, either drawn out in the direction of test growth or expanded across the terminal edge of the test.

Heterohelix globulosa (Ehrenberg) grades into *Ventilabrella austinana*, and is considered to be the ancestral form.

Aliyulla (1965) introduced the new species *Planoglobulina transcaucasica*, which was found in lower Santonian strata. In the type description, Aliyulla stated that its growth becomes triserial in the latter part. This is a misinterpretation of chamber addition, and there is no triserial portion in this specimen. From the type figures, although rather schematic, and type description, it is evident that this species belongs to *Ventilabrella* rather than *Planoglobulina*. It is considered a junior synonym of *V. austinana*, which is also known from the Santonian.

Three of the specimens attributed by Martin (1972) to *Planoglobulina* (?) *carseyae* (Plummer) appear to lack pseudotextularian biserial chambers, and should, therefore, be placed in *Ventilabrella*. The forms she assigned to *V.* (?) *austinana* are transitional with *Heterohelix globulosa* (Ehrenberg). Inasmuch as the ventilabrellid stage is not dominant, these are referred to the latter species.

The specimens, which Martin (1972) incorrectly identified as *Ventilabrella glabrata* Cushman, possess moderately fine costae of uniform strength over the entire test. This type of ornament is characteristic of *V. austinana*.

The holotype of *Ventilabrella riograndensis* Martin, 1972, also possesses fine costae throughout except where partially obliterated by secondary vermicular deposits. This species is conspecific with *V. austinana*.

Distribution: This is shown in Fig. 34.

Range: Turonian–Campanian (FOS *Globotruncana obliqua*—FOS *Globotruncana calcarata* Interval).

Ventilabrella deflaensis (Sigal), Plate 7, figs 1, 2
Guembelina (*Guembelina, Ventilabrella*) *deflaensis* Sigal, 1952, p. 36, 37, text-fig. 41.

Ventilabrella decoratissima de Klasz, 1953a, p. 228, pl. 4, figs 5a, b; Bettenstaedt and Wicher, 1956, pl. 1, fig. 3.

Ventilabrella deflaensis Sigal. Bettenstaedt and Wicher, 1956, p. 503, pl. 1, figs 1–2; Wicher and Bettenstaedt, 1957, p. 30–38, figs. 3a, b; Küpper, 1964, p. 635, 636, pl. 1, figs 14a, b.

Ventilabrella alpina de Klasz. Bettenstaedt and Wicher, 1956, pl. 1, fig. 4.

Gublerina decoratissima (de Klasz). Montanaro Gallitelli, 1957, p. 140, pl. 32, fig. 8; Salaj and Samuel, 1963, p. 106, pl. 7, figs 4–5b; pl. 8, fig. 1; Salaj and Samuel, 1966, p. 229, tab. 37 (4, 5); Bandy, 1967, p. 25, text-fig. 11 (4); Esker, 1969, p. 212, 213, pl. 1, figs 7, 8.

Sigalia deflaensis (Sigal). Reiss, 1957b, p. 243–245; Reiss, 1958b, p. 5, text-figs 1a–e; Samuel, 1962, p. 194, pl. 13, figs 2a–3b; Salaj and Samuel, 1963, p. 105, text-figs A–C (*non* pl. 7, figs 1a, b); (?) Salaj and Samuel, 1966, p. 227, tab. 37 (1); Wille-Janoschek, 1966, p. 123, 124, pl. 8, figs. 5–6; Brown, 1969, p. 42, 43, pl. 2, fig. 1; pl. 3, figs 8a–9b; text-fig. 6 (?); Porthault, 1970, in Donze *et al.*, p. 63, 64, pl. 9, figs 4, 5.

Sigalia cf. *deflaensis* (Sigal). Hanzlíková, 1972, p. 98, pl. 24, fig. 15.

Guembelina conjakica Geodakchan, 1959, in Geodakchan and Aliyulla, p. 58, pl. 1, figs 7, 8.

Guembelina malocaucasica Aliyulla, 1959, in Geodakchan and Aliyulla, p. 59, pl. 1, fig. 9.

Sigalia carpatica Salaj and Samuel, 1963, p. 105, 106, pl. 7, figs 2a–3, text-fig. C; Salaj and Samuel, 1966, p. 226, 227, tab. 2 (2).

Planoglobulina deflaensis (Sigal). van Hinte, 1965a, p. 57, text-figs 1a–d.

Gublerina primitiva Aliyulla, 1965, p. 225, 226, pl. figs 6a–8b.

Gublerina deflaensis (Sigal). Scheibnerová, 1967, p. 265, 266, text-figs 4A–G, 5(1a–2b); (?) Barr, 1968, pl. 1, fig. 16; Scheibnerová, 1969, p. 55, 56, tab. 6, figs 5a–6b; *non* Esker, 1969, p. 213, pl. 2, figs 4, 5.

Sigalia decoratissima (de Klasz). Porthault, 1970, in Donze *et al.*, p. 63, pl. 9, figs 6, 7.

Description: Test biserial throughout or occasionally developing a gerontic stage ventilabrellid chamber arrangement; laterally compressed, subparallel sides; periphery slightly lobate or irregular; slowly to moderately expanding in width. Chambers 14–16 in biserial stage; wider than high, slightly arched in direction of test growth; thickness remains essentially constant after first 2–3 chambers, slight if any lateral inflation; gradually increasing in size; supplementary chambers may be added along the backs of the primary chambers; secondary calcification may cover portions of earlier chambers imparting rough surface to the test. Proloculus imperforate with a terminal pore or aperture. Primary aperture a low, interiomarginal arch bordered by a moderately wide imperforate lip; supplementary apertures may form at the base of the back of gerontic stage primary chambers. Sutures distinctive, convex in direction of growth, limbate, raised in earlier chambers to flush or slightly depressed later; continue along backs of later chambers to form a double keel-like structure which converges at the base of each chamber; intervening surface perforate. Wall minutely perforate; with numerous, fine, nearly vertical costae which become coarse over the secondarily thickened and raised sutures.

Remarks: The near parallel sides of the test along with the limbate, raised sutures which continue as a keel-like structure at the backs of the chambers distinguish this species from all other *Heterohelix*. The features of the biserial stage are constant. Only the presence and extent of the ventilabrellid stage is variable.

The slender test of *Heterohelix deflaensis* occurs in only one other species, *H. moremani* (Cushman). Although transitional forms have not been recognised, Cushman's species seems to be the most likely ancestor.

The taxon *deflaensis* has been attributed to five different genera at one time or another. This trend was initiated by the original author (Sigal, 1952), who was undecided as to whether his species belonged to *Guembelina* or *Ventilabrella*. Reiss (1957b) proposed the genus *Sigalia* for this species based upon the test shape and its raised costate sutures, characters which should not be given supraspecific rank (see Remarks under *Ventilabrella*). Nevertheless, the genus *Sigalia* has been used most frequently for *deflaensis*, but here treated as a synonym of *Ventilabrella*.

Montanaro Gallitelli (1957) reported that this species may develop an incipient gublerinid final stage, a morphotype not recognised here for this taxon. Although the simple biserial form is most common, morphotypes with one or more supplementary chambers are occasionally found. Because of this gerontic development, *deflaensis* is assigned to the genus *Ventilabrella*. The species appears to be a phylogenetic end member, and all morphotypes seem to have an identical stratigraphic range.

Salaj and Samuel (1963) placed specimens which possess a beginning ventilabrellid stage, into *Ventilabrella decoratissima* de Klasz. The type figure of de Klasz's species does not clearly position the longitudinal costae with respect to the sutures. However, Montanaro Gallitelli (1957) illustrated a paratype of *V. decoratissima* which showed the costae to be most coarsely developed on the sutures. *Ventilabrella decoratissima*, and hence the specimens of Salaj and Samuel, are considered synonymous with *V. deflaensis*.

The type description and figures of *Gublerina primitiva* Aliyulla, 1965, establish this Santonian taxon as being conspecific to *Ventilabrella deflaensis*.

Brown (1969), without comment, listed *Guembelina conjakica* Geodakchan and *G. malocaucasica* Aliyulla as synonyms of *Sigalia* (= *Ventilabrella*) *deflaensis*. The publication in which these species were described was not obtainable during this study. They are, however, listed for completeness, but not necessarily indicating agreement with the author with this synonymy.

Distribution: This is shown in Fig. 35.

Range: Upper Coniacian–Santonian (FOS *Globotruncana concavata*—FOS *Globotruncana elevata*).

Ventilabrella eggeri Cushman, Plate 7, figs 3, 4

Ventilabrella eggeri Cushman, 1928a, p. 2, 3, pl. 1, figs 10a–11 (*non* 12); Cushman, 1928b, p. 231, pl. 33, figs 10a, b; Cushman, 1933a, pl. 26, figs 14, 15; Cushman, 1938a, p. 25, pl. 4, figs 12–14; Cushman, 1944b, p. 92, pl. 14, fig. 8; Cushman, 1946b, p. 111, pl. 47, figs 17–19; Cushman, 1948, p. 259, pl. 24, fig. 13; Huss, 1957, pl. 6, fig. 2 (10); pl. 7, fig. 1 (6); (?) Maslakova, 1959, p. 119, pl. 15, fig. 9; Vinogradov, 1960a, pl. 6, figs 36, 37; (?) Küpper, 1964, p. 636, 637, pl. 1, figs 13a, b; *non* Salaj and Samuel, 1966, p. 231, tab. 37 (6); Martin, 1972, p. 85, 86, pl. 1, figs 6a–7; *non* Hanzlíková, 1972, p. 96, 97, pl. 23, figs 19a, b, 23.

Ventilabrella austinana Cushman. Kikoïne, 1948, p. 25, pl. 2, figs 7a–c.

Ventibrella eggeri var. *eggeri* Cushman. Frizzell, 1954, p. 111, pl. 16, fig. 10.

Gublerina ornatissima (Cushman and Church). Martin, 1964, p. 86, pl. 11, fig. 3; Brown, 1969, p. 41, 42, pl. 3, figs 7a, b (*non* 6a–c).

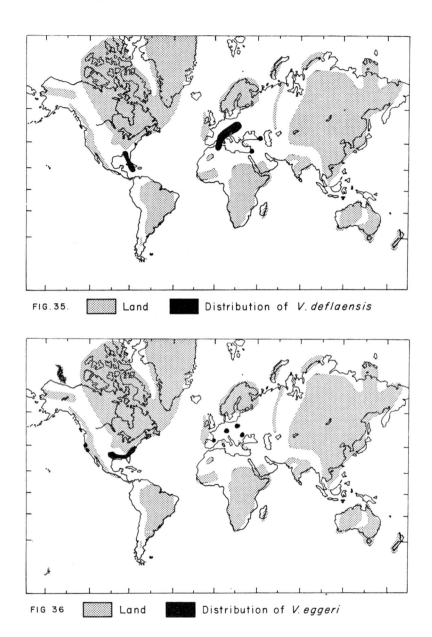

FIG. 35. ▨ Land ■ Distribution of *V. deflaensis*

FIG 36 ▨ Land ■ Distribution of *V. eggeri*

Ventilabrella ornatissima Cushman and Church.
Planoglobulina ornatissima (Cushman and Church). Douglas, 1969, p. 160, 161, pl. 11, figs 1a, b (*non* 2).
   Planoglobulina glabrata (Cushman). Neagu, 1970, p. 60, 61, pl. 14, fig. 2.
   Planoglobulina acervulinoides (Egger). (?) Rahhali, 1970, p. 68, text-figs 7L–O.
   Description: Test flabelliform, initially biserial, later developing numerous supplementary chambers in the plane of symmetry; width rapidly increasing; edge rounded, periphery generally lobate. Chambers of the biserial stage numbering 9–14, inflated, each approximately equidimensional, increasing regularly but moderately in size; supplementary may be irregularly shaped with a tendency for either the height or width to become the greatest dimension, may be smaller to larger than the primary chambers, width generally constant and slightly less than that of the primary chambers and slightly less lateral inflation; numerous. All chambers with discontinuous costae; after the initial nepionic stage the costae of the biserial portion are coarser and fewer per chamber than those of the supplementary chambers, although the difference is not great and often there is no difference; nearly longitudinal but with slight conformation to the surface contours. Primary aperture a low to moderately high, interiomarginal arch bordered by a narrow, imperforate lip. Supplementary apertures occur along the lower outside margins of the primary ephebic chambers, one per chamber; and the supplementary chambers may bear one, corresponding in position to the primary aperture, or two, the second corresponding to the supplementary apertural position for the primary chambers. Sutures depressed, straight, normal to edge in the biserial part; depressed to flush, straight to curved in the later portion. Finely and densely perforate.
   Remarks: No other *Ventilabrella* has costae as coarse as those of *V. eggeri*. Variations occur in chamber number and shape.
   Ventilabrella glabrata Cushman is believed to be the progenitor of *V. eggeri* because specimens which appear transitional between the two are recorded.
   In 1972 Martin selected a neotype for and emended the description of *Ventilabrella eggeri*. Both actions were unfortunate, and will likely create rather than resolve confusion. As noted by Brown (1969) and others, a neotype is needed. The specimen selected by Martin, however, is both incomplete and broken. Both her neotype and hypotype specimens show the remnants (the so-called "cameral ridges") of additional chambers. It seems that she attempted to find specimens which exhibit precisely the same outline and number of primary and supplementary chambers as the specimen Cushman (1928a, pl. 1, figs 10a, b) apparently intended to be the holotype.
   Brown (1969) outlined the circumstances surrounding Cushman's naming of *V. eggeri*. He supported the notion that the reason no holotype of this species is to be found in the Cushman collection is that Cushman had selected the specimen represented by figs 17 and 18 of *Guembelina acervulinoides* Egger as the type species of *Ventilabrella*. This specimen, of course, was housed in the Egger collection. I fully agree with Brown's interpretation, but on the evidence he presented, it remains as conjecture, and, therefore, has no taxonomic validity. However, a search of Cushman's publications has revealed additional information, which leaves no doubt of Cushman's intent. One month after publication of the new genus *Ventilabrella* and its type species *V. eggeri*, Cushman released the revised edition of his *Foraminifera, Their Classification and Economic Use* (1928b). Here the majority of specimens illustrated are type species (= genoholotype of Cushman). Cushman used the same illustrations on plate 33, figs 10a, b as two of those presented originally.

Moreover, he noted on the plate explanation that the figures were taken from Egger. This, then, was the holotype, later destroyed during World War II.

The first good, original illustrations of *Ventilabrella eggeri* given by Cushman (1938a) have been followed since by nearly all workers. Martin's (1972) neotype may fall within this accepted concept of the species, but because the specimen is incomplete doubt will remain.

According to Martin (1972), the neotype "was selected from the original type level . . . the Campanian 'Lower Taylor Marl' . . . McLennan County, Texas". Cushman made no such designation, either of the type locality or type level, noting only that it could be found "in certain horizons of the Taylor Marl of Texas". To this he later added (1938a) that *Ventilabrella eggeri* is found in "upper Bavaria" and in the "upper part of the Austin chalk," again without citing specific localities. It was not until eight years later (1946b) that Cushman listed for this species the localities, twenty-six of which are from the lower part of the Taylor marl of Texas. The type locality is, in fact, in the upper Bavarian Alps. In selecting the neotypes, Martin did not follow the requirement [I.C.Z.N., Art. 75c (5)] that the neotype must be selected from the type locality. Her neotype is rejected as invalid.

Martin (1972) based the emended description and distinction of *Ventilabrella eggeri* upon the angle of slope of the last "row" of chambers. The measured angle was thirty-two degrees, and the specimens were selected to meet this requirement, ignoring otherwise identical and complete specimens. Such extraneous parameters are entirely unreliable and totally without meaning taxonomically. Her emended description is rejected.

Distribution: This is shown in Fig. 36.

Range: Campanian (FOS *Globotruncana elevata*—LOS *Globotruncana concavata* Interval)—Maastrichtian (*Trinitella scotti* Interval).

Ventilabrella glabrata Cushman, Plate 8, figs 1, 2

Pseudotextularia (Guembelina) acervulinoides Egger. (?) Liebus, 1927, p. 375, 376, pl. 14, fig. 2 (*non* 3a, b).

Planoglobulina acervulinoides (Egger). White, 1929, p. 33, pl. 4, fig. 6; Noth, 1951, p. 62, 63, pl. 7, fig. 1.

Ventilabrella eggeri var. *glabrata* Cushman, 1938a, p. 26, pl. 4, figs 15–17b; Cushman, 1946b; p. 111, 112, pl. 47, figs 20–22; Frizzell, 1954, p. 111, pl. 16, figs 11, 12.

Ventilabrella cf. *eggeri* (Cushman). (?) Hagn, 1953, p. 74, pl. 6, fig. 18.

Ventilabrella multicamerata de Klasz, 1953a, p. 230, pl. 5, figs 1a, b; Martin, 1972, p. 88, pl. 3, figs 1a–2; (?) Hanzlíková, 1972, p. 97, pl. 24, figs 5, 6.

Ventilabrella bipartita de Klasz, 1953a, p. 230, pl. 4, figs 7a, b; Salaj and Samuel, 1966, p. 230, tab. 37 (9).

Ventilabrella alpina de Klasz, 1953a, p. 229, 230, pl. 4, figs 6a, b.

Planoglobulina glabrata (Cushman). Montanaro Gallitelli, 1957, p. 141, 142, pl. 32, fig. 10 (*non* 11, 12); Eternod Olvera, 1959, p. 77, 78, pl. 2, fig. 9; Skinner, 1962, p. 42, pl. 5, fig. 15; Pessagno, 1967, p. 272, pl. 88 figs 12, 13, 17; *non* Barr, 1968, pl. 1, figs 6, 9; (?) Hanzlíková, 1969, p. 42, pl. 8, figs 3a, b; (?) Esker, 1969, p. 213, pl. 2, fig. 10; Govindan, 1972, p. 172, pl. 2, fig. 10.

Planoglobulina meyerhoffi Seiglie, 1960, p. 122–124, text-figs 2a–3.

Ventilabrella eggeri Cushman. (?) Salaj and Samuel, 1966, p. 231, tab. 37, fig. 6; Wille-Janoschek, 1966, p. 125, 126, pl. 8, figs 7a, b.

Planoglobulina multicamerata (de Klasz). Pessagno, 1967, p. 272, 273, pl. 89, fig. 15; Cita and Gartner, 1971, pl. 3, fig. 1.

Planoglobulina acervulinoides glabrata (Cushman). (?) Bandy, 1967, p. 25, text-fig. 11 (7).

Heterohelix glabrata (Cushman). El-Naggar, 1971a, pl. 7, fig. f.

Ventilabrella browni Martin, 1972, p. 85, pl. 1, figs 3, 4.

Ventilabrella glabrata Cushman. *non* Martin, 1972, p. 86, 87, pl. 1, figs 8–9b; *non* Hanzlíková, 1972, p. 97, pl. 24, figs 2a, b, 7a, b.

Ventilabrella manuelensis Martin, 1972, p. 87, 88, pl. 2, figs 5–8.

Planoglobulina carseyae (Plummer). Govindan, 1972, p. 171, 172, pl. 1, fig. 17.

Description: Test biserial with supplementary chambers being added later in the biserial plane of symmetry; width rapidly increasing as supplementary chambers are added; edge rounded, periphery slightly lobate. Chambers of the initial biserial portion numbering 9–14, inflated, increasing gradually in size, subspherical; supplementary chambers numerous, laterally more compressed than the primary chambers, irregular in size, with a tendency for the chambers to be higher than wide in the gerontic stage finally becoming wide and extending over more than one earlier chamber. Costae present on all chambers; moderately coarse over the biserial stage, becoming finer and more numerous on the supplementary chambers; all generally discontinuous, longitudinal with slight conformation to the chamber contour. Primary aperture a low, interiomarginal arch, visible only on broken specimens. A single supplementary aperture on the lower outside edge of each primary ephebic chamber; each supplementary chamber with an aperture in the interiomarginal position and often on the lower outside margin. Sutures depressed, straight, normal to edge in the biserial stage; later depressed becoming flush in the gerontic stage, straight to gently curved. Pores of biserial portion irregular in position and diameter, later numerous with a small diameter.

Remarks: This species has costae whose degree of coarseness lies between those of *Ventilabrella austinana* Cushman and *V. eggeri* Cushman. The variation in morphological character is typical for the genus.

Ventilabrella glabrata may have evolved from either *V. austinana* Cushman or its ancestor, *Heterohelix globulosa* (Ehrenberg).

The costae of *Ventilabrella glabrata* are less coarse than those of *V. eggeri* Cushman. Both species may have coarser costae on the biserial stage, but those of *V. eggeri* are frequently equally coarse throughout. There is also little difference in their ranges. Further examination may prove them conspecific.

On the bases of the type figure description and topotypes, *Ventilabrella multicamerata* de Klasz, 1953a, is determined to be a junior synonym of *V. glabrata* Cushman. De Klasz's species has more numerous supplementary chambers than most illustrated specimens of *V. glabrata*. This is not regarded, however, as sufficient grounds for the erection of a new taxon, but rather represents the continued growth of an individual. Although no type material was available for examination, the two species *Ventilabrella alpina* and *V. bipartita* described by de Klasz (1953a) are thought to be identical and conspecific with *V. glabrata*, following Brown's (1969) synonymy.

Pessagno (1967) distinguished *Planoglobulina* (= *Ventilabrella*) *multicamerata* (de Klasz) from *P.* (= *Ventilabrella*) *glabrata* (Cushman) by its "not having as pointed an apical extremity in the early part of its test and by having a less lobulate, larger test with more numerous chambers". Interestingly the specimen of *V. multicamerata* illustrated by Pessagno has an initial portion more pointed than his illustrations of *V. glabrata*. The size and lobulateness of the test is directly proportional to the number of chambers in this species. Finally, he noted that the surface of *V. multi-camerata* is papillose rather than costate. In this case either the optical system he used lacked the necessary resolution to see the costae or the specimen was poorly preserved. All Heterohelicidae are known to bear costae except *Heterohelix pulchra* (Brotzen) and *H. glabrans* (Cushman) which have pore-mounds.

Although the holotype of *Planoglobulina meyerhoffi* Seiglie, 1960, was not examined during this study, it is considered a junior synonym of *Ventilabrella glabrata* for reasons noted above in the discussion of *V. multicamerata* de Klasz.

Martin (1972) stated that *Ventilabrella glabrata* "is the only species of *Ventilabrella* with fine, relatively even costae covering the entire surface of the test . . .". Examination of the holotype and topotypes reveals that the costae in the early portion are moderately coarse becoming fine on later chambers (Masters, in press). This is identical to the costae exhibited on *V. browni* Martin, which is, thus, considered a junior synonym of *V. glabrata*. Martin's *V. glabrata* is referred to *V. austinana* Cushman.

Ventilabrella manuelensis Martin, 1972, was distinguished by its "abruptly flaring, subcircular test, diminished biserial portion, larger number of chambers, and heavy vermicular ornamentation". All of the above, except the "ornamentation," are tied directly to the point at which the first supplementary chambers were added and the age of the individual. There is a progressive reduction of the biserial portion through time within a given lineage. While this is a recognisable characteristic, it is not taxonomically significant. As already noted (see Remarks under VENTILABRELLA) the vermicular "ornamentation" may well be a secondary, inorganic deposit and not the result of a physiological function. If this is the case, and it seems likely, the problem of recognition of Martin's species becomes apparent because the vermicular deposits mask the original coarseness of the costae on the earlier portion of the test.

On the best preserved specimen of *Ventilabrella manuelensis* Martin (fig. 5), which is not the holotype, fine costae can be observed on the latest chambers. Only two species of *Ventilabrella* have costae of this strength: *V. austinana* Cushman and *V. glabrata* Cushman, the former throughout and the latter only on the latest chambers. *Ventilabrata austinana*, however, has a longer biserial stage than either *V. glabrata* or *V. manuelensis*. Therefore, Martin's species is referred as a junior synonym to *V. glabrata*.

The specimens assigned to *Ventilabrella multicamerata* de Klasz by Martin (1972) are so poorly preserved few surface features can be recognised. It is doubtful if any original shell material remains. However, on the basis of de Klasz's original figure, Martin's specimens are believed to be the same and, thus are synonyms of *V. glabrata* Cushman.

Distribution: This is shown in Fig. 37.

Range: Mid-Santonian (FOS *Globotruncana concavata*—FOS *Globotruncana elevata* Interval)—Maastrichtian (*Trinitella scotti* Interval).

Family PLANOMALINIDAE Bolli, Loeblich and Tappan, 1957

Definition: Test may be initially a low trochospiral or planispiral throughout; with or without an imperforate peripheral keel; primary aperture an interiomarginal, umbilical–equatorial arch, the umbilical extensions of which

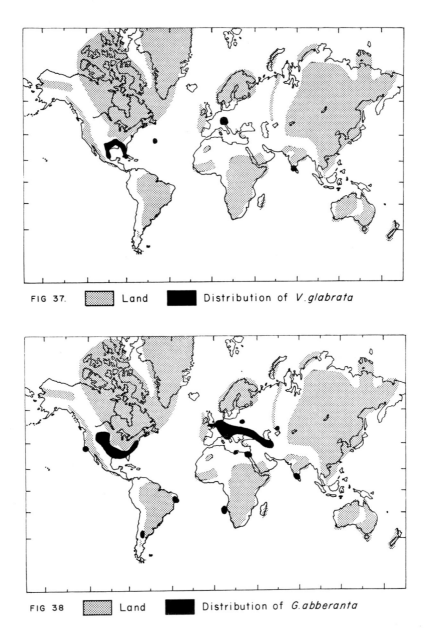

FIG 37.　▨ Land　■ Distribution of *V. glabrata*

FIG 38　▨ Land　■ Distribution of *G. abberanta*

form relict apertures; paired supplementary apertures and chambers may be present.

Generic Key
  I. With a peripheral keel . . . . *Planomalina*
  II. With paired supplementary apertures . . . . *Globigerinelloides*
  III. With radially elongate chambers . . . . . . *Hastigerinoides*

Remarks: El-Naggar (1971a) treats this group as the Subfamily Plano-malininae within the Family Globotruncanidae. Because of the great diversity in structure this appears to be an undesirable combination. The characteristics separating the globotruncanids and the planomalinids are regarded as being equivalent in taxonomic importance. Thus the classification of Loeblich and Tappan (1964) is followed in this regard.
Range: Aptian–Maastrichtian (*Trinitella scotti* Interval).

Genus GLOBIGERINELLOIDES Cushman and ten Dam, 1948
Globigerinelloides Cushman and ten Dam, 1948, p. 42.
Biglobigerinella Lalicker, 1948, p. 624.
Blowiella Krechmar and Gorbachik, 1971, p. 135.
Type species: *Globigerinelloides algeriana* Cushman and ten Dam, 1948.
Description: Test may be a low trochospire in initial whorl later becoming planispiral, biumbilicate, involute to partially evolute with a lobate, generally imperforate periphery. Chambers inflated, spherical to rhomboidal. Wall perforated usually by small diameter pores, commonly pustulose. Primary aperture an equatorial arch extending to the umbilical areas of each side. These umbilical extensions remain open as relict apertures with subsequent chamber addition. Primary aperture splits into two supplementary apertures in gerontic individuals. Paired supplementary chambers can be added over the supplementary apertures. Apertural flap present. Apertural face imperforate.
Remarks: *Globigerinelloides* lacks the peripheral keel of *Planomalina* and the radially elongate chambers of *Hastigerinoides*. It is the only genus in this family to develop paired supplementary apertures and chambers. Dain (in Subbotina, 1953) recognised numerous modifications of the primary aperture within the same species of *Globigerinelloides*. These varied from the single peripheral to the paired apertures. Berggren (1962a) correctly suggested that the paired apertural or "*Biglobigerinella*" stage was merely a late ontogenetic development. This was conclusively documented by Masters (in press) for nearly all recognised species of *Globigerinelloides* including *G. multispina* Lalicker [= *G. abberanta* (Netskaya)], the type species for *Biglobigerinella*.

Krechmar and Gorbachik (in Gorbachik, 1971) created the new genus *Blowiella* with *Planomalina blowi* Bolli as the type species. They also placed *P. saundersi* Bolli into their new genus. At the same time they envisioned *Blowiella* as belonging to the Schackoinidae, but offered no explanation for this. These authors compared *Blowiella* to *Globigerinelloides*, noting that their new genus lacked the layered, secondarily thickened walls, the septal layering, and possibly that it has a different primary wall structure. In their diagnosis of the genus Krechmar and Gorbachik stated that the wall, possibly is first single-layered, then hardly noticably thickened. The bases for their determinations are not known, but the type species *P. blowi* as well as *P. saundersi* are true *Globigerinelloides*, as shown by examination of holotypes and topotypes. The genus *Blowiella* is a junior synonym of *Globigerinelloides*.

A variety of forms have been referred to the species *Globigerinelloides aspera* (Ehrenberg). Although believed by many workers to be a *Globigerinelloides* its exact status remains unclear. In an attempt to stabilise taxonomic interpretation of the species, Pessagno (1967) selected a lectotype for *Phanerostomum asperum* Ehrenberg, 1854 (figs 26a, b). Unfortunately, Pessagno's designation does not resolve the status of this species because it did not accompany a redescription of Ehrenberg's material. Ehrenberg's figures are stylised and were illustrated using transmitted light. Is the specimen represented by fig. 26 a low trochospiral or a truely planispiral form? Is the aperture peripheral or does it extend into one or both umbilici? These questions cannot be answered without reviewing the specimen itself. Thus, all specimens referred to *G. aspera* are for the time placed into synonymy with later described species.

In 1952 Sigal described under the benthonic genus *Planulina* the new species *cheniourensis*. It was considered a planktonic by Banner and Blow (1959), who transferred it to the genus *Planomalina*. Sigal (1966b) established the new genus *Labroglobigerinella*, also planktonic, for this species. One of the primary criteria by which fossil foraminifers are determined to have been planktonic is their widespread geographic distribution. The supposed planktonic habitat of *P. cheniourensis* must be seriously questioned, because even after more than two decades, it has not been recognised outside of Algeria. This is not necessarily a valid reason for not considering this species as a planktonic form. *Planulina cheniourensis* bears a strong resemblance to the *Globigerinelloides* species. The taxon has not been treated here because type material was not available during this study.

Brönnimann (1952a) questionably assigned his new species *tururensis* to *Globigerinella*. The holotype was examined, and appears to be an internal calcitic mold. Its ultimate chamber is damaged and partially missing. Because of the poor preservation and the fact that it has not since been recorded in the literature, very little is known about this taxon. Consequently, no attempt has been made to treat this species.

It is suspected that Moullade's (1966) "*Globigerinelloides*" *gyroidinaeformis* is not a planktonic. It has a completely involute, subspherical test unlike that of any other *Globigerinelloides* or Mesozoic planktonic for that matter.

The holotype of *Globigerinella escheri* subsp. *clavata* Brönnimann was examined, and is believed to be an immature *Hastigerinoides watersi* (Cushman).

Range: Aptian–Maastrichtian (*Trinitella scotti* Interval).

Globigerinelloides abberanta (Netskaya), Plate 8, figs 3–5; plate 9, figs 1, 2
Globigerinella aspera (Ehrenberg). Brotzen, 1942, p. 25, 27, text-fig. 8C; Bukowy and Geroch, 1957, p. 317, pl. 28, figs 1a–2b; Maslakova, 1959, p. 107, pl. 10, figs 5a, b; Balakhmatova, 1960, in Glazunova *et al.*, p. 117, pl. 22, figs 1a–4.
Globigerinella ex gr. *aspera* (Ehrenberg). Ehremeeva and Belousova, 1961, p. 99, pl. 31, figs 6a–7b.
Globigerinella abberanta Netskaya, 1948, p. 220, pl. 2, figs 3a, b.
Biglobigerinella multispina Lalicker, 1948, p. 624, pl. 92, figs 1a–3c; Bermúdez, 1952, p. 115, pl. 21, figs 11–13; Frizzell, 1954, p. 127, pl. 20, figs 10a, b; Bolli, Loeblich and Tappan, 1957, p. 25, pl. 1, figs 11–12b; Rauzer-Chernousova and Fursenko, 1959, p. 297, 298, text-figs 663a, b; Herm, 1962, p. 53, pl. 3, fig. 8; Pokorný, 1963, p. 375, fig. 416; Loeblich and Tappan, 1964, p. C656, figs 4–5b; Subbotina, 1964, p. 253, 254, pl. 54, figs 10a–11c, 14a–15d; pl. 50, figs 1a–8d; Saavedra, 1965, p. 322, text-fig. 4.
Biglobigerinella algeriana ten Dam and Sigal, 1949, p. 235, text-figs 1a–d, 2, 3.
Biglobigerinella cf. *algeriana* ten Dam and Sigal. Bukowy and Geroch, 1957, p. 318, pl. 28, figs 4a–10.
Globigerinella biforaminata (Hofker). Hofker, 1956b, fig. 20; Hofker, 1956d, p. 53, text-figs 2, 5; Bukowy and Geroch, 1957, p. 317, 318, pl. 28, figs 3a, b.
Globigerina (Globigerinella) aspera (Ehrenberg). Hofker, 1959, text-figs 7a, b.
Biglobigerinella biforaminata (Hofker.) Olsson, 1960, p. 44, pl. 8, figs 7, 8; Mello, 1969, p. 95, 96, pl. 2, figs 3a–5; Morris, 1971, p. 280, 281, pl. 6, figs 13a, b.
Globigerina biforaminata Hofker, 1956f, p. 76, 77, pl. 9, figs 68a–c; Hofker, 1960e, text-figs 19a–c; Hofker, 1962, p. 1062, text-fig. 7C; p. 1063, text-fig 8D; p. 1067, text-fig. 12D; Perlmutter and Todd, 1965, p. 118, 119, pl. 5, fig. 3 (*non* figs 2a, b).
Globigerina (Biglobigerinella) biforaminata Hofker. Todd, 1970, p. 152, pl. 5, figs 10a, b.
Planomalina aspera (Ehrenberg). Barr, 1961, p. 561, 563, pl. 69, figs 4a, b.
Planomalina (Globigerinelloides) aspera aspera (Ehrenberg). van Hinte, 1965b, p. 85, 86, pl. 1, figs 2a, b.
Planomalina multispina (Lalicker). Barr, 1961, p. 563, 564, pl. 69, figs 5a, b.
Globigerinella voluta voluta (White). Herm, 1962, p. 51, 52, pl. 3, fig. 7.
Planomalina mauryae Petri, 1962, p. 116–119, pl. 16, figs 1a, b (*non* 2a–4).
Planomalina messinae biforaminata (Hofker). van Hinte, 1963, p. 102, pl. 12, figs 4a, b.
Globigerinelloides messinae (Brönnimann). Olsson, 1964, p. 174, 176, pl. 7, figs 6a–8b; Kalantari, 1969, p. 198, pl. 25, figs 6a–c.

Biglobigerinella abberanta (Netskaya). Subbotina, 1964, p. 252, 253, pl. 54, figs 6a–9c, 12a–13b.

Globigerinelloides asper (Ehrenberg). Takayanagi, 1965, p. 201, 202, pl. 20, figs 9a–c.

Planomalina (*Globigerinelloides*) *biforaminata* (Hofker). Salaj and Samuel, 1966, p. 160, 161, pl. 7, figs 1a, b.

Globigerinelloides multispina (Lalicker). Pessagno, 1967, p. 276, 277, pl. 70, figs 1, 2; pl. 82, figs 10, 11; pl. 91, figs 1, 2; Bertels, 1970, p. 34, 35, pl. 3, figs 1a–2b; text-figs 8–10; Hanzlíková, 1972, p. 99, pl. 25, figs 2a, b (*non* 3a–4b).

Globigerinelloides prairiehillensis Pessagno, 1967, p. 277, 278, pl. 60, figs 2, 3; pl. 83, fig. 1; pl. 90, figs 1, 2, 4; Mello, 1971, p. 26, 27, pl. 2, figs 5a, b.

Globigerinelloides aspera aspera (Ehrenberg). Bandy, 1967, p. 12, text-fig. 5 (8).

Globigerinelloides aspera (Ehrenberg). Barr, 1968, p. 313, 314, pl. 37, figs 6a, b (*non* 3a, b); Rasheed and Govindan, 1968, p 80, 82, pl. 8, figs 4–6; Neagu, 1970b, p. 63, pl. 25, figs 22, 23.

Globigerinelloides eaglefordensis (Moreman). Ansary and Tewfik, 1968, p. 45, pl. 4, figs 10a, b.

Biglobigerinella sp. cf. *B. algeriana* ten Dam and Sigal. Neagu, 1970b, p. 62, pl. 28, figs 15–18.

Globigerinelloides biforaminata (Hofker). Neagu, 1970b, p. 62, pl. 28, figs 19, 20.

Globigerinelloides asperus (Ehrenberg). El-Naggar, 1971a, pl. 3, figs i–k.

Description: Test planispiral, biumbilicate; equatorial periphery lobate, axial periphery rounded; with an imperforate peripheral band in all but the last 1–3 primary chambers and never in the supplementary chambers. Primary chambers subspherical, well inflated, numbering 6–7 in the last whorl, moderately to rapidly increasing in size; final 1–2 chambers of gerontic stage tend to markedly increase in thickness. Secondary chambers usually paired, offset laterally out of the plane of coiling, parallel to the axis of coiling, extending slightly if at all into the umbilicus; subspherical to irregularly shaped, usually smaller than previous primary chamber. Primary aperture in juvenile and early ephebic stages a moderately high, umbilical–equatorial arch bordered by a wide, imperforate lip; becoming broader and lower, finally splitting into 2 smaller, umbilical–extraumbilical to extraumbilical apertures, bordered by wide lips; each supplementary chamber will have 1–2 supplementary apertures, also bordered by lips. Paired relict apertures appear along the umbilical margin, one on either side of each chamber. Umbilicus shallow, relatively narrow to moderately wide. Sutures depressed, straight. Surface randomly perforated by small diameter pores; pustulose.

Remarks: Only two species occur over the range of *Globigerinelloides abberanta* which might be confused with it. *Globigerinelloides alvarezi* (Eternod Olvera) is more loosely coiled and has more chambers per whorl than *G. abberanta*, while *G. escheri* (Kaufmann) is smaller with fewer chambers per whorl. *G. abberanta* shows the typical neanic to gerontic change from single to paired apertures and paired supplementary chambers. The coiling tends to become tighter through the ontogenetic development. The final few primary chambers may become nearly double the thickness of that which would be expected by projecting the previous rate of size increase.

The phylogeny of this species is unknown.

The type description and figures, and the range of *Globigerinella abberanta* Netskaya do not differ from those of *Biglobigerinella multispina* Lalicker. Although no publication date other than the year is given for Netskaya's article, the issue was sent to press on 29 December 1947. Thus, it is reasonable to assume that it was released early in 1948, which gives *G. abberanta* priority over the September 1948 publication of *G. multispina*. It should be noted that *"abberanta"* is spelled *"aberranta"* in the plate explanation. The former spelling, appearing first in the text, is accepted as correct.

The taxon *Biglobigerinella algeriana* ten Dam and Sigal (1949) was stated to differ from *B*. (= *Globigerinelloides*) *multispina* by having a diameter/width ratio of 10/7 as opposed to 10/9. Furthermore, *G. multispina* was said to have small spines rather than a lightly rugose surface. The tiny short spines mentioned by Lalicker (1948) for *G. multispina* correspond to the rugosities mentioned by ten Dam and Sigal. The diameter/width ratio is within the range of limits exhibited by topotypes of *G. multispina*, which has already been determined as a synonym of *G. abberanta*.

Distribution: This is shown in Fig. 38.

Range: Mid-Santonian (FOS *Globotruncana concavata*—FOS *Globotruncana elevata* Interval)—Maastrichtian (*Trinitella scotti* Interval).

Globigerinelloides algeriana Cushman and ten Dam

Globigerinelloides algeriana Cushman and ten Dam, 1948, p. 43, pl. 8, figs 4–6; Bermudez, 1952, p. 115, 116, pl. 21, figs 15a–c; Glintzboeckel and Magné, 1955, p. 153–155, text-figs 1–3; Bolli, Loeblich and Tappan, 1957, p. 21–23, pl. 1, figs 1a, b; Solange and Sigal, 1958, text-fig. 1; Ruggieri, 1963, p. 77, 78, text-fig. 3; Gorbachik, 1964, p. 33–36, text-figs 1, 2, 6; pl. 2, figs 1–8; Loeblich and Tappan, 1964, p. C656–C658, fig. 526 (6a, b); Bandy, 1967, p. 12, text-fig. 5 (4).

Biglobigerinella barri Bolli, Loeblich and Tappan. Bartenstein, Bettenstaedt and Bolli, 1966, p. 164, pl. 4, fig. 388 (*non* 386, 387, 389–397).

Globigerinelloides algerianus Cushman and ten Dam. Moullade, 1966, p. 124, 125, pl. 9, fig. 15; Sidó, 1970, p. 388–391, pl. 1, figs 1–16; pl. 2, figs 1, 2, 3 (part); Gorbachik, 1971, pl. 29, figs 6a, b; Kuhry, 1971, p. 227, 228, pl. 1, fig. 1; Risch, 1971, p. 54, 55, pl. 6, figs 13, 14.

Planomalina (*Globigerinelloides*) *algeriana* (Cushman and ten Dam). Salaj and Samuel, 1966, p. 159, 160, pl. 6, figs 4a, b; tab. 33 (13).

Labroglobigerinella (*Globigerinelloides*) *spectrum algerianum* (Cushman and ten Dam). Signal, 1966b, p. 20, pl. 2, figs 1a–3b.

Globigerinelloides ferreolensis (Moullade). Moullade, 1966, p. 122–124, pl. 9, figs 1, 2 (*non* 3); Moullade, 1969, p. 463, 464, pl. 1, figs 8, 9 (*non* 7); Risch, 1971, p. 54, pl. 6, figs 15, 16.

Description: Test planispiral with a tendency to become loosely coiled or even evolute; biumbilicate; equatorial periphery slightly lobate, axial periphery rounded. Chambers of last whorl regularly but rapidly increase in height while width and thickness change very slightly; numbering 7–13. Aperture a low, interiomarginal, umbilical–equatorial arch bordered by an imperforate lip. Relict apertures present along the umbilical margins. Umbilicus shallow, wide; earlier whorls visible. Sutures on first chambers of last whorl straight to slightly curved, later curved to sigmoidal; all depressed.

Remarks: *Globigerinelloides alvarezi* (Eternod Olvera) is the only other described species whose chambers tend to become as evolute as those of *G. algeriana*. The chambers of *G. algeriana* tend to be rather high; whereas, those of *G. alvarezi* are subspherical. Sidó (1970) photographed eight specimens which showed variation in size (0.350–0.528 mm), in the number of chambers in the final whorl (7–13) and in the looseness of the coiling.

The phylogeny of this species is unknown.

In a topotypic suite of *Globigerinelloides barri* Bolli, Loeblich and Tappan, 1957 [= *G. cushmani* (Tappan)], specimens which grade into an evolute form (pl. 13, fig. 4) are found. These do not appear to differ in any way from *G. algeriana*. Although not recorded from the lower Aptian, *G. barri* may be present throughout the range of *G. algeriana*. If so one must examine the possibility either that *G. algeriana* is merely a morphotype of *G. barri* or that it evolved from the latter.

Distribution: This is shown in Fig. 39.

Range: Aptian.

Globigerinelloides alvarezi (Eternod Olvera), Plate 9, fig. 3; Plate 10, fig. 1.

Planomalina alvarezi Eternod Olvera, 1959, p. 91, 92, pl. 4, figs 5–7; Martin, 1964, p. 84, pl. 10, figs 8–9.

Planomalina ehrenbergi Barr, 1961, p. 563, pl. 69, figs 1a, b.

Globigerinelloides eaglefordensis (Moreman). Ayala-Castañares, 1966, p. 15, 16, pl. 1, figs 2a–c; pl. 6 (?) figs 2a–3b; Moullade, 1966, p. 125 126 pl. 9, fig. 7 (*non* 6, 8); Moullade, 1969, p. 463, 464, pl. 1, fig. 11 (*non* 10, 12).

Globigerinelloides alvarezi (Eternod Olvera). Sliter, 1968, p. 98, 99, pl. 15, figs 1a–2b.

Description: Test planispiral, biumbilicate, loosely coiled; equatorial periphery lobate, axial periphery rounded. Chambers subspherical, well inflated, gradually increasing in size; 7–8 in the final whorl; early chambers with an imperforate peripheral band. Aperture a low, interiomarginal, umbilical–equatorial arch bordered by a narrow, imperforate lip which widens into the umbilicus. Relict apertures visible along the margins of each umbilicus. Umbilicus wide, shallow; earlier whorls visible. Sutures depressed, straight. Surface with random, small diameter pores; pustulose.

Remarks: No other species over its range is as loosely coiled as is *Globigerinelloides alvarezi*. The older *G. algeriana* Cushman and ten Dam has rather high chambers versus equidimensional for *G. alvarezi*. No significant morphologic variation has been recognised for this species.

The phylogeny of this species is unknown.

Distribution: This is shown in Fig. 40.

Range: Mid-Albian–Campanian (*Globotruncana calcarata* Interval).

Globigerinelloides bentonensis (Morrow), Plate 10, figs 2, 3

Anomalina bentonensis Morrow, 1934, p. 201, pl. 30, figs 4a, b; Cushman, 1940, p. 28, pl. 5, figs 3a, b; Cushman, 1946b, p. 154, pl. 63, figs 7a, b; Sigal, 1952, p. 23, text-fig. 21.

Non *Anomalina bentonensis* Morrow. var. Voorwijk, 1937, p. 197, pl. 1, figs 30, 31.

Globigerinelloides bentonensis (Morrow). Loeblich and Tappan, 1961, p. 267, 268, pl. 2, figs 8–10; Todd and Low, 1964, p. 400, 401, pl. 1, figs 3a–c; Bandy, 1967, p. 12, text-fig. 5 (7); *non* Pessagno, 1967, p. 275, pl. 76, figs 10, 11; Eicher, 1969, text-fig. 3 (part); *non* Eicher and Worstell, 1970b, p. 297, pl. 8, figs 17a, b, 19a, b; pl. 9, figs 3a, b; Kuhry, 1971, p. 228, pl. 1, figs 2A, B; *non* Risch, 1971, p. 55, 56, pl. 6, figs 17, 18; Gawor-Biedowa, 1972, p. 63, 64, pl. 6, figs 7a–c; Michael, 1973, p. 208, pl. 1, figs 4–6.

Planomalina pulchella Todd and Low, 1964, p. 401, pl. 1, figs 9a, b.

Description: Test planispiral, biumbilicate; equatorial periphery slightly lobate; axial periphery rounded, with an imperforate peripheral band in the chambers of

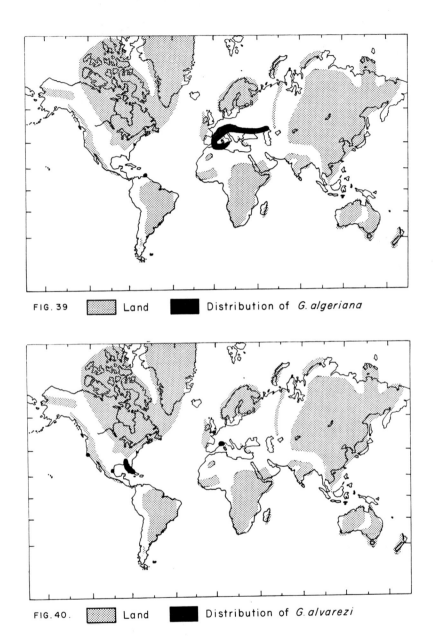

FIG. 39   ▨ Land   ■ Distribution of *G. algeriana*

FIG. 40.   ▨ Land   ■ Distribution of *G. alvarezi*

the last whorl. Chambers appressed, 7–8 in the last whorl, slowly increasing in size, inflated, appearing trapezoidal or triangular in side view, maximum dimensions of each chamber nearly equal. Aperture a low, interiomarginal, umbilical–equatorial arch; imperforate bordering lip narrow at the periphery, flaring out into the umbilicus to impart an apparent greater height to the chambers. Relict apertures present along the margins of each umbilicus. Umbilicus narrow if the apertural lips are preserved, moderately wide if not, shallow. Sutures depressed, slightly curved. Wall with randomly dispersed small and moderate diameter pores.

Remarks: *Globigerinelloides bentonensis* is distinguished by its closely packed chambers and slight lobateness of the periphery. It is very close morphologically to *G. cushmani* (Tappan), but the chambers of the latter are not as appressed and, as a result, the test is more lobate. The intraspecific variation is not yet established for this species.

The phylogeny of this species is unknown.

Examination of the holotype of *Anomalina bentonensis* Morrow revealed that the test is symmetrical in edge view rather than the somewhat asymmetrical interpretation given in the type figure. The holotype possibly has paired supplementary apertures, and relict apertures are visible in the umbilici.

The holotype of *Planomalina pulchella* Todd and Low, 1964, was examined, and is believed to be identical to *Globigerinelloides bentonensis*.

Distribution: This is shown in Fig. 41.

Range: Albian.

Globigerinelloides blowi (Bolli), Plate 11, fig. 3

Planomalina blowi Bolli, 1959, p. 260, pl. 20, figs 2a, b, 3; Hermes, 1966, pl. figs 6–8.

Planomalina maridalensis Bolli, 1959, p. 261, pl. 20, figs 4–6b; Bartenstein, Bettenstaedt and Bolli, 1966, p. 163, pl. 4, figs 371–380; Hermes, 1966, pl. figs 9–11.

Globigerinella duboisi Chevalier, 1961, p. 33, pl. 1, figs 14a–18.

Globigerinella gottisi Chevalier, 1961, p. 33, 34, pl. 1, figs 9a–11, 13 (?), (*non* 12).

Praeglobotruncana (*Globigerinelloides*) *messinae* (Brönnimann). Berggren, 1962a, p. 44–49, pl. 8, figs 6a–c (*non* 4a–5c, 7a–8c).

Globigerinelloides blowi (Bolli). Hermes, 1966, p. 161, 162, pl. figs 1, 2; Moullade, 1966, p. 119–121, pl. 8, figs 24–26; Bandy, 1967, p. 12, text-fig. 5 (1); *non* Kuhry, 1971, p. 228, pl. 1, figs 3A, B; *non* Risch, 1971, p. 53, 54, pl. 6, figs 21, 22.

Globigerinelloides maridalensis (Bolli). Bandy, 1967, p. 12, text-fig. 5 (2).

Blowiella blowi (Bolli). Gorbachik, 1971, pl. 30, figs 2a, b.

(?) *Globigerinelloides texomaensis* Michael, 1973, p. 209, pl. 6, figs 9–11.

Description: Test small, planispiral, nearly involute but with the early whorls partially visible, biumbilicate; equatorial periphery deeply lobate; axial periphery rounded, with an imperforate peripheral band in the early chambers. Chambers inflated subspherical, but width may become slightly greater imparting an ovoidal shape; 4–5 in the last whorl, size increase moderate to rather rapid. Aperture a low, interiomarginal, umbilical–equatorial arch bordered by a narrow, imperforate lip. Relict apertures paired on either side of the chambers at their umbilical margins. Umbilicus narrow, shallow. Sutures depressed, straight to slightly curved.

Remarks: Only two species are within the size range of *Globigerinelloides blowi* and coexist with it. One of these, *G. saundersi* (Bolli), has distinctively high chambers and a laterally compressed test. The other, *G. ultramicra* (Subbotina), has twice the number of chambers per whorl. Morphological features are generally stable, but

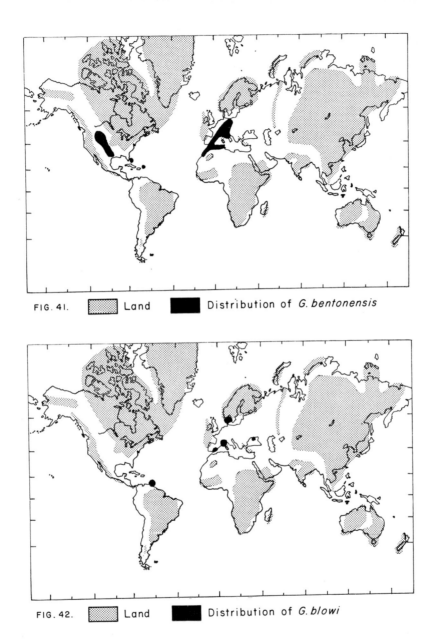

FIG. 41. Land Distribution of *G. bentonensis*

FIG. 42. Land Distribution of *G. blowi*

minor variation occurs in the degree of involuteness of the coiling and in the number of chambers in the final whorl.

The phylogeny of this form is unknown.

The ontogenetic morphologic series common to most other species has not yet been recognised in *Globigerinelloides blowi*.

Bolli (1959) distinguished *Globigerinelloides blowi* from his *G. maridalensis* on the basis of the spherical chambers of the former. Those of *G. maridalensis* were stated to be laterally compressed in the last whorl. Neither the type figure nor the actual holotype reveals the compression in *G. maridalensis*. In fact the holotypes of the two species are virtually indistinguishable. *Globigerinelloides blowi* becomes the senior synonym by earlier appearance in the same paper.

Globigerinelloides duboisi, named by Chevalier (1961), appears to be morphologically identical to *G. blowi*. The two also occur in the same stratigraphic interval. Chevalier's (1961) *Globigerinella gottisi* is regarded as an intraspecific variant of *Globigerinelloides blowi*.

In 1973 Michael proposed the new species *Globigerinelloides texomaensis* from the upper Albian. He differentiated it largely upon the laterally compressed and umbilically protruding final chamber. This type of feature is common in the gerontic stage of many species, and has no taxonomic significance. This exact structure has been observed in several species of *Globigerinelloides*. Once this feature is disregarded, the remaining five-chambered test is similar to *G. blowi*. Because the holotype of Michael's species was not examined, it is for now only tentatively assigned to *G. blowi*.

Distribution: This is shown in Fig. 42.

Range: Upper Aptian, Upper Albian (?).

Globigerinelloides cushmani (Tappan), Plate 10, fig. 4; Plate 11, figs 1, 2
Globigerinella cushmani Tappan, 1943, p. 513, pl. 83, figs 5a, b; Frizzell, 1954, p. 127, pl. 20, figs 8a, b.
Globigerinella aissana, Sigal, 1952, p. 28, text-fig. 30.
Planomalina caseyi Bolli, Loeblich and Tappan, 1957, p. 24, pl. 1, figs 4a–5b.
Planomalina (Globigerinelloides) caseyi Bolli, Loeblich and Tappan. Salaj and Samuel, 1966, p. 161, pl. 6, figs 1a, b; tab. 33 (1).
Biglobigerinella barri Bolli, Loeblich and Tappan, 1957, p. 25, pl. 1, figs 13–18b; Bartenstein, Bettenstaedt and Bolli, 1966, p. 164, pl. 4, figs 386, 387, 389–397 (*non* 388); Salaj and Samuel, 1966, p. 164, 165, pl. 6, figs 7a, b; tab. 33 (10).
"*Biglobigerinella*" sp. Solange and Sigal, 1958, text-fig. 2.
"*Globigerina*" sp. Solange and Sigal, 1958, text-fig. 3.
"*Globigerinella*" sp. Solange and Sigal, 1958, text-fig. 4.
Biglobigerinella sigali Chevalier, 1961, p. 33, 34, pl. 1, figs 19–23.
Globigerinelloides eaglefordensis (Moreman). Loeblich and Tappan, 1961, p. 268, 269, pl. 2, figs 3–7; Ayala-Castañares, 1962, p. 15, 16, pl. 1, figs 2a–c; pl. 6, figs 2a–3b; Borsetti, 1962, p. 29, pl. 2, figs 1, 2, text-figs 20–24; Loeblich and Tappan, 1964, p. C656–C658, fig. 526 (7a, b); Saavedra, 1965, p. 321, 322, text-fig. 2; Moullade, 1966, p. 125, 126, pl. 9, figs 6, 8 (*non* 7); *non* Ansary and Tewfik, 1968, p. 45, pl. 4, figs 10a, b; Barr, 1968, pl. 2, figs 3a–4b; Moullade, 1969, p. 463, 464, pl. 1, figs 10, 12 (*non* 11); Sidó, 1971, p. 46, pl. 1, fig. 3 (part).
Globigerina aspera (Ehrenberg). Hofker, 1962, p. 96, text-figs 18a–c.
Globigerinelloides caseyi (Bolli, Loeblich and Tappan). Low, 1964, p. 123;

Burckle, Saito and Ewing, 1967, text-fig. 2 (3a, b); Pessagno, 1967, p. 276, pl. 49, figs 2–5; Pessagno, 1969a, pl. 10, fig. D; *non* Eicher and Worstell, 1970b, p. 297, 298, pl. 8, figs 11, 15a–16; Belford and Scheibnerová 1971, pl. 3, figs 17–19; Douglas, 1971, pl. 2, figs 4–6; El-Naggar, 1971a, pl. 3, figs l–o; *non* Michael, 1973, p. 208, pl. 1, figs 7–9.

Biglobigerinella sp. Salaj and Samuel, 1966, p. 164, pl. 6, figs 8a, b; tab. 33 (9).

Labroglobigerinella (*Labroglobigerinella*) spectrum *algerianum* (Cushman and ten Dam). Sigal, 1966, p. 18–20, pl. 1, figs 1a–2b, 4a, b, 7a–10b; pl. 7, figs 6a, b.

Globigerinelloides ferreolensis (Moullade). Moullade, 1966, p. 122–124, pl. 9, fig. 3 (*non* 1, 2); Moullade, 1969, p. 463, pl. 1, fig. 7 (*non* 8, 9).

Globigerinelloides bentonensis (Morrow). Pessagno, 1967, p. 275, pl. 76, figs 10, 11; Eicher and Worstell, 1970b, p. 297, pl. 8, figs 17a, b, 19a, b; pl. 9, figs 3a, b.

Description: Test planispiral, biumbilicate, generally involute but with rare evolute individuals; equatorial periphery moderately lobate; axial periphery rounded with an imperforate peripheral band of varying width on the early chambers and occasionally into the last whorl. Chambers inflated, trapezoidal in side view with the ultimate chamber circular; 7–9 in the final whorl; gradually increasing in size; final 1–2 chambers rapidly increase in thickness; paired supplementary chambers are added in the gerontic stage, positioned laterally just out of the plane of coiling but usually mutually contiguous, parallel to the axis of coiling. Aperture a low interiomarginal, umbilical–equatorial arch through the early ephebic stage; in the late ephebic stage, smaller paired umbilical–extraumbilical to extraumbilical apertures form; a single, small supplementary aperture forms on each supplementary chamber; a narrow imperforate lip occurs over each aperture; in the primary apertures, these are narrow over the periphery then flare out into the umbilicus marking the position in each umbilicus of the relict apertures. Umbilicus shallow, narrow. Sutures depressed, straight to slightly curved. Surface with small to moderate diameter pores.

Remarks: *Globigerinelloides cushmani* is larger than and not as tightly coiled as *G. ultramicra* (Subbotina). It is similar to *G. bentonensis* (Morrow), but has a more lobate equatorial periphery. The single to paired apertures and chambers typical of the genus is expressed in the ontogeny of this species. Rare individuals exhibiting one or two evolute chambers have been observed. These never form the paired apertures and chambers.

The phylogeny of this species is unknown.

The holotypes and topotypes of *Planomalina caseyi* and of *Biglobigerinella barri*, both named by Bolli, Loeblich and Tappan (1957), were examined and compared with those of *Globigerinella cushmani* Tappan. The minor variations that exist are considered to be encompassed by the intraspecific variation of *G. cushmani*.

Solange and Sigal (1958) illustrated under various names (i.e. "*Biglobigerinella*," "*Globigerinella*" and "*Globigerina*") specimens which appear identical to the varieties of *Globigerinelloides cushmani*.

From the description and figure of the holotype of *Biglobigerinella sigali* Chevalier, 1961, the species is believed to be conspecific with *Globigerinelloides cushmani*.

Distribution: This is shown in Fig. 43.

Range: Upper Aptian–Cenomanian.

Globigerinelloides escheri (Kaufmann), Plate 11, figs 4, 5
Nonionina escheri Kaufmann, 1865, p. 198; Hermes, 1966, pl. figs 25–27.

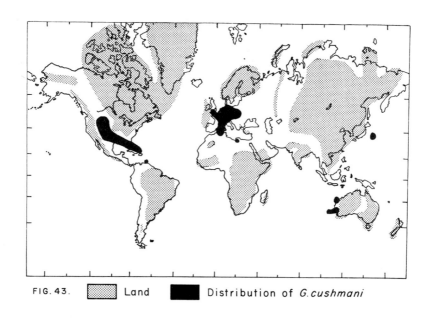

FIG. 43.   [Land]   [■]   Distribution of *G.cushmani*

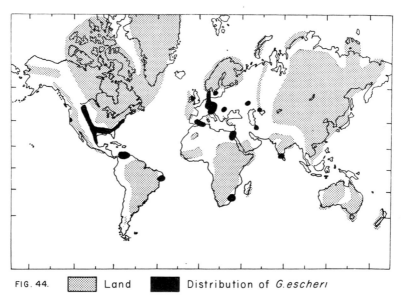

FIG. 44.   [Land]   [■]   Distribution of *G.escheri*

Globigerina voluta White, 1928a, p. 197, 198, pl. 28, figs 5a, b; Drooger, 1951, p. 69, 71, figs 8a, b.

Globigerinella aspera (Ehrenberg). Cushman, 1931c, p. 59, pl. 11, figs 5a, b; Brotzen, 1936, p. 170, pl. 13, figs 2a–c; Bandy, 1951, p. 508, 509, pl. 75, figs 3a–c; Hagn, 1953, p. 92, pl. 8, fig. 7; Subbotina, 1953, p. 107, 108, pl. 13, figs 2a–11b (*non* 12a–c); Smitter, 1957, p. 199, 201, text-figs 20a, b; Cati, 1964, p. 251, 252, pl. 41, figs 3a, b.

Globigerinella voluta (White). Sandidge, 1932d, p. 284, pl. 44, figs 1, 2; Hagn, 1953, p. 92, pl. 8, fig. 6; Hamilton, 1953, p. 227, pl. 30, fig. 6; Subbotina, 1953, p. 108–110, pl. 13, figs 13a, b (*non* 14a–15b); Ansary and Fakhr, 1958, p. 132, 133, pl. 2, figs 12a, b.

Globigerinella aequilateralis (Brady). Noth, 1951, p. 74, pl. 9, figs 7a, b.

Globigerinella messinae messinae Brönnimann, 1952a, p. 42–44, pl. 1, figs 6, 7, text-figs 20a–q; Herm, 1962, p. 51, pl. 3, fig. 5.

Globotruncana beldingi beldingi Gandolfi, 1955a, p. 31, 32, pl. 1, figs 8a–c.

Globotruncana (*Rugoglobigerina*) *beldingi* subsp. *subbeldingi* Gandolfi, 1955a, p. 32, pl. 1, figs 7a–c; text-figs 7 (3a–c).

Planomalina escheri (Kaufmann). *non* Bolli, 1959, p. 260, 261, pl. 20, flgs 7a–8; *non* Hermes, 1966, pl. figs 39–41.

Praeglobotruncana (*Globigerinelloides*) *messinae*(Brönnimann). Berggren, 1962a, p. 44–49, pl. 8, figs 4a–5c, 8a–c (*non* 6a–7c), text-figs 6 (1a–6c), 7 (1a–8c).

Planomalina mauryae Petri, 1962, p. 116–119, pl. 16, figs 3a–b (*non* 1a–2b, 4).

Planomalina messinae messinae (Brönnimann). van Hinte, 1963, p. 100, 101, pl. 12, figs 6a, b.

Planomalina aspera (Ehrenberg). Martin, 1964, p. 84, pl. 10, figs 7a–c.

Hastigerina aspera forma *digitata* Subbotina, 1964, p. 250, 251, pl. 54, figs 4a–c (*non* 1a–3c, (?) 5a–c).

Globigerina biforaminata Hofker. Perlmutter and Todd, 1965, p. I18, I19, pl. 5, figs 2a, b (*non* 3).

Globigerinelloides aspera (Ehrenberg). Barr, 1966, p. 503, 504, pl. 78, figs 4a, b; Hanzlíková, 1972, p. 98, pl. 25, figs 1a, b.

Globigerinelloides escheri (Kaufmann). Marianos and Zingula, 1966, p. 334, 335, pl. 37, figs 3a, b; Bandy, 1967, p. 12, text-fig. 5 (6); Ansary and Tewfik, 1968, p. 45, pl. 4, figs 9a, b; *non* Eicher, 1969, text-fig. 3 (part).

(?) *Planomalina* (*Globigerinelloides*) *aspera* (Ehrenberg). Salaj and Samuel, 1966, p. 160, pl. 7, figs 3a, b.

? *Planomalina* (*Globigerinelloides*) sp. Salaj and Samuel, 1966, p. 162, pl. 7, figs 4a, b.

Globigerinelloides asperus (Ehrenberg). Pessagno, 1967, p. 274, 275, pl. 60, figs 4, 5.

Globigerinelloides volutus (White). Pessagno, 1967, p. 278, 279, pl. 62, figs 10, 11.

Globigerinelloides aspera voluta (White). Bandy, 1967, p. 13, text-fig. 5 (11).

Globigerinelloides messinae messinae Brönnimann. Bandy 1967, p. 12, text-fig. 5 (9).

Globigerinella escheri (Kaufmann). Mantovani-Uguzoni and Pirini-Radrizzani, 1967, p. 1227, pl. 92, figs 9a–10b.

Globigerinelloides voluta (White). Sliter, 1968, p. 99, pl. 15, figs 3a, b, 5a, b.

Globigerinelloides messinae (Brönnimann). Ansary and Tewfik, 1968, p. 45, pl. 4, figs, 9a, b; Hanzlíková, 1969, p. 45, pl. 12, figs 1a–3b; *non* Kalantari, 1969, p. 198, pl. 25, figs 6a–c; *non* Weaver *et al.*, 1969, pl. 4, figs 2a–c; North and Caldwell, 1970, p. 49, 50, pl. 4, figs 4a–c; Govindan, 1972, p. 172, pl. 2, figs 20–23.

Globigerinelloides multispinatus (Lalicker). Douglas, 1969, p. 161, 162, pl. 9, figs 6a–c.

Globigerinelloides escheri escheri (Kaufmann). Neagu, 1970b, p. 62, pl. 25, figs 16–21; pl. 28, figs 12–14.

Globigerinelloides multispina (Lalicker). Hanzlíková, 1972, p. 99, pl. 25, figs 3a–4b (*non* 2a, b).

(?) *Globigerinelloides tumidus* Michael, 1973, p. 209, pl. 1, figs 10–12.

Description: Test small, planispiral, biumbilicate, involute to just loosely coiled enough to see inner whorls; equatorial periphery deeply lobate; axial periphery rounded, with an imperforate peripheral band on the early chambers of the last whorl. Chambers subspherical, inflated, rapidly increasing in size, 5–5½ in the last whorl. Supplementary chambers occur in pairs laterally just out of the plane of coiling; smaller than the final primary chamber. Primary aperture a low to moderately high, interiomarginal, umbilical–equatorial arch bordered by a narrow imperforate lip which flares outward into the umbilicus. Secondary apertures paired, small, usually as high as wide, extraumbilical or umbilical–extraumbilical, each with a bordering lip. Paired relict apertures on each chamber at the umbilical margins. Umbilicus narrow, shallow. Sutures depressed, straight. Small diameter pores evenly distributed over pustulose surface.

Remarks: *Globigerinelloides escheri* might be confused with only two other species. It differs from *G. subarinata* (Brönnimann) by having a rounded periphery rather than acute, and from *G. ultramicra* (Subbotina) by having fewer chambers which increase in size more rapidly. This species exhibits the typical ontogenetic progression to tighter coiling and paired apertures and chambers. Extreme increase in the final primary chamber thickness has not been recognised.

This species evolved from a flat-spired *Globigerina*. *Globigerina planispira* Tappan may be the ancestor (Masters, in press), although it has more chambers per whorl. *Globigerina delrioensis*, with five chambers per whorl is the most likely candidate.

The holotype of Gandolfi's (1955) *Globotruncana beldingi subbeldingi* has no trace of a keel nor imperforate peripheral band. Although the area of the aperture is damaged, it appears to be umbilical–equatorial. The supposed dorsal side is partially filled with sediment, but the test is apparently biumbilical. The species is judged to be a junior synonym of *Globigerinelloides escheri*.

Michael (1973) described the new species *Globigerinelloides tumidus* from the upper Albian of Texas. Its type figure and description could be substituted for those of *G. escheri*. The latter, heretofore, has been recorded only from Turonian and later strata. Michael neither illustrates nor discusses the variability of his new species, and for these reasons, *G. tumidus* is questionably assigned to *G. escheri*.

Distribution: This is shown in Fig. 44.

Range: Upper Albian (?), Turonian–Maastrichtian (*Trinitella scotti* Interval).

Globigerinelloides saundersi (Bolli), Plate 11, fig. 6

Planomalina saundersi Bolli, 1959, p. 262, pl. 20, figs 9–11; Bartenstein, Bettenstaedt and Bolli, 1966, p. 163, pl. 4, figs 384, 385, 398–402.

Globigerinelloides saundersi (Bolli). Bandy, 1967, p. 12, text-fig. 5 (3).

Blowiella saundersi (Bolli). Gorbachik, 1971, pl. 30, figs 5a, b.

Description: Test small, planispiral with the early whorls partially visible, biumbilicate; laterally compressed; equatorial periphery stellate; axial periphery rounded to bluntly pointed, with an imperforate peripheral band on the earlier subspherical chambers. Chambers initially subspherical, becoming high and bluntly tapered in the last whorl, both width and height exceed thickness; inflated, rapidly increasing in size; numbering 5–6 in the last whorl with the 3–4 last chambers being radially elongate. Aperture a low, interiomarginal, umbilical–equatorial arch bordered by a narrow, imperforate lip. Relict apertures in pairs on each chamber marginal to the umbilici. Umbilicus relatively wide, shallow. Sutures depressed, straight.

Remarks: This species is the only stellate, laterally compressed *Globigerinelloides* in the lower Cretaceous. Its morphological characteristics are generally stable. The phylogeny of this species is unknown.

Distribution: This is shown in Fig. 45.

Range: Aptian—lower Albian.

Globigerinelloides subcarinata (Brönnimann), Plate 12, figs 1, 2

Globigerinella messinae subsp. *subcarinata* Brönnimann, 1952a, p. 44, 45, pl. 1, figs 10, 11, text-figs 21a–m; Herm, 1962, p. 51, pl. 2, fig. 1; *non* Cati, 1964, p. 252, pl. 41, figs 4a, b.

Hastigerina aspera forma *digitata* Subbotina, 1964, p. 250, 251, pl. 54, figs 1a–3c, (?) 5a–c (*non* 4a–c).

Globigerinelloides subcarinatus (Brönnimann). Pessagno, 1967, p. 278, pl. 62, figs 12, 13; *non* Ansary and Tewfik, 1968, p. 46, pl. 4, figs 6a, b.

Globigerinelloides messinae subcarinata (Brönnimann). Bandy, 1967, p. 12, text-fig. 5 (10).

Globigerinelloides messinae (Brönnimann). Weaver *et al.*, 1969, pl. 4, figs 2a–c.

Description: Test small, planispiral involute, biumbilicate; equatorial periphery deeply lobate; axial periphery acutely rounded with an imperforate peripheral band generally throughout. Chambers inflated, circular in side view, laterally compressed, greatest thickness at midpoint of chamber; $4\frac{1}{2}$–5 in the last whorl, rapidly increasing in size. Aperture a low to moderately high, interiomarginal, umbilical–equatorial arch bordered by a moderately wide, imperforate lip. Relict apertures may be visible along the umbilical margins. Umbilicus narrow, shallow. Sutures depressed, straight. Surface perforated by small to moderate diameter pores; pustulose.

Remarks: No other species of *Globigerinelloides* has an acute axial margin. Other than the minor variation in chamber count, the features of this species appear stable.

Globigerinelloides subcarinata evolved from *G. escheri* (Brönnimann) by becoming laterally compressed.

In 1964 Subbotina introduced the new form *Hastigerina aspera* forma *digitata*. The use of this type of nomenclature should be avoided. "Forma" is of an infra-subspecific rank [I.C.Z.N., Art. 45e (ii)]. Although "*digitata*" was not validly proposed, the specimens themselves are typical *Globigerinelloides subcarinata*.

Distribution: This is shown in Fig. 46.

Range: Campanian (*Globotruncana calcarata* Interval)—Maastrichtian (*Trinitella scotti* Interval).

Globigerinelloides ultramicra (Subbotina), Plate 12, figs 3, 4, 5

Globigerinella aspera (Ehrenberg). Dain, 1934, p. 42, 43, pl. 4, figs 46a, b; Hamilton, 1953, p. 226, pl. 30, fig. 5; Belford, 1960, p. 91, pl. 25, figs 4, 5; Herm, 1962, p. 49, 50, pl. 3, fig. 6; Graham and Church, 1963, p. 64, 65, pl. 7, figs 17a–c.

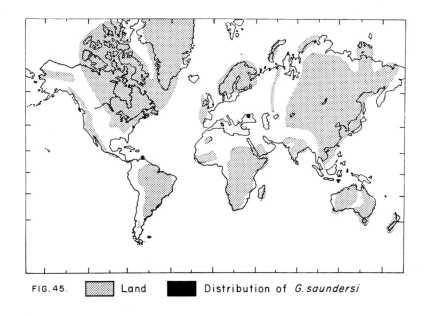

FIG. 45. ▨ Land   ■ Distribution of *G. saundersi*

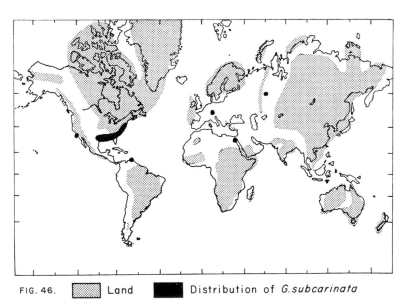

FIG. 46. ▨ Land   ■ Distribution of *G. subcarinata*

Globigerinella ultramicra Subbotina, 1949, p. 33–35, pl. 2, figs 17, 18; Subbotina, 1953, p. 106, 107, pl. 13, figs 1a, b; Maslakova, 1959, p. 107, pl. 10, figs 4a, b; Samuel, 1962, p. 180, p. 7, figs 2a, b; Cati, 1964, p. 252, pl. 41, figs 2a, b.

Globigerina aspera (Ehrenberg). Shaw, 1953, pl. 2, figs 16, 17.

Globigerinella aequilateralis (Brady). Tollman, 1960, p. 191, pl. 20, fig. 6.

"*Globigerinella*" *japonica* Takayanagi, 1960, p. 131, 132, pl. 9, figs 12a, b.

Planomalina yaucoensis Pessagno, 1960, p. 98, pl. 2, figs 14–15; Pessagno, 1962, p. 358, pl. 1, figs 1, 2.

Biticinella ferreolensis Moullade, 1961, p. 214, pl. 1, figs 1–5; Hermes, 1966, pl. figs 42, 43.

Planomalina (*Globigerinelloides*) *messinae*. Brönnimann. Berggren, 1962a, p. 44–49, pl. 8, figs 7a–c (*non* 4a–6c, 8a–c).

Planomalina mauryae Petri, 1962, p. 116–119, pl. 16, fig. 4 (*non* 1a–3b).

Planomalina aspera aspera (Ehrenberg). van Hinte, 1963, p. 97–100, pl. 12, figs 2a–3, text-figs. 15a–f.

Planomalina (*Globigerinelloides*) *ultramicra* (Subbotina). (?) Salaj and Samuel, 1966, p. 162, pl. 7, figs 2a, b.

Labroglobigerinella (*Labroglobigerinella*) spectrum-a Sigal, 1966b, p. 23, pl. 2, figs 5a, b, 7a, b.

Labroglobigerinella (*Globigerinelloides*) spectrum-a Sigal, 1966b, p. 23, pl. 2, figs 4a, b, 6a, b.

Labroglobigerinella (*Labroglobigerina*) spectrum-a Sigal, 1966b, p. 23, pl. 2, figs 8a, b, 10a, b, 12a, b.

Labroglobigerinella (*Labrobiglobigerinella*) spectrum-a Sigal, p. 23, pl. 2, figs 9a, b, 11a, b.

Globigerinelloides bolli Pessagno, 1967, p. 275, 276, pl. 62, fig. 5; pl. 81, figs 7, 8.

Globigerinelloides yaucoensis (Pessagno). Pessagno, 1967, p. 279, pl. 75, figs 9, 10, 27; Ansary and Tewfik, 1968, p. 46, pl. 4, figs 5a, b.

Globigerinelloides aspera (Ehrenberg). Barr, 1968a, p. 313, 314, pl. 37, figs 3a, b (*non* 6a, b); Barr, 1968b, pl. 2, figs 2a, b.

Globigerinelloides escheri (Kaufmann). Bandy, 1967, p. 12, text-fig. 5 (6); Eicher, 1969, text-fig. 3 (part).

Globigerinelloides caseyi (Bolli, Loeblich and Tappan). Eicher and Worstell, 1970b, p. 297, 298, pl. 8, figs 11, 15a–16; Michael, 1973, p. 208, pl. 1, figs 7–9.

Globigerinelloides escheri escheri (Kaufmann). Neagu, 1970b, p. 62, pl. 25, figs 16–18 (*non* 19–21); pl. 28, figs 12–14.

Globigerinelloides ferreolensis Moullade. Gorbachik, 1971, pl. 29, figs 5a–c; (?) Kuhry, 1971, p. 228, 229, pl. 1, figs 4A, B.

Description: Test small, planispiral with the inner whorls just visible, biumbilicate; equatorial periphery lobate; axial periphery round with an imperforate peripheral band in the early chambers of the last whorl. Primary chambers subspherical to slightly higher than wide, numbering 7–8 in the final whorl, slowly increasing in size. Secondary chambers paired, at either side of the primary chamber position; reduced in size. Primary aperture a low to moderately high, interiomarginal, umbilical–equatorial arch bordered by a narrow, imperforate lip which widens as it reaches the umbilicus. Secondary apertures form in the late ephebic stage; paired, umbilical–extraumbilical, each with a lip comparable to the primary lip. Relict apertures occur along the umbilical margins, one on either side of each chamber. Umbilicus relatively wide, shallow. Sutures depressed, straight to curved. Surface with small diameter pores; pustulose.

Remarks: The small size of the test combined with the number of slowly enlarging chambers distinguish *Globigerinelloides ultramicra* from its congeners. All typical ephebic to gerontic morphological stages are present in this species.

This species is believed to have descended from *Globigerina planispira* Tappan near the time of the first appearance of the latter.

Moullade (1961) introduced as a new species *Biticinella ferreolensis* which is considered identical to *Globigerinelloides ultramicra* on the basis of the type figures and description.

Distribution: This is shown in Fig. 47.

Range: Mid-Aptian–Maastrichtian (*Trinitella scotti* Interval).

Genus HASTIGERINOIDES Brönnimann, 1952a

Hastigerinoides Brönnimann, 1952a, p. 52.

Eohastigerinella Morozova, 1957, p. 1112.

Type species: *Hastigerinella alexanderi* Cushman, 1931c.

Description: Test planispiral, biumbilicate; equatorial periphery deeply lobate in neanic stage, becoming stellate in ephebic and gerontic stages; axial periphery rounded to pointed. Chambers of neanic stage subspherical with an imperforate peripheral band which is lost in later chambers, gradually to rapidly increasing in size; ephebic chambers rapidly becoming radially elongate, gerontic chambers become clavate or pointed, ultimate length exceeds maximum width by 3–3½ times. Primary aperture a low, interiomarginal, umbilical–equatorial arch bordered by an imperforate lip. Secondary relict apertures visible in both umbilici. Umbilici shallow, narrow. Sutures distinct, depressed, radial. Surface densely and coarsely perforate, smooth to pustulose.

Remarks: *Hastigerinoides* is the only member of the Planomalinidae with radially elongate chambers. It lacks the well developed single keel of *Planomalina* and is characteristically more coarsely perforate than *Globigerinelloides*.

The genus *Hastigerinoides* has been interpreted as being polyphyletic (Masters, in press), evolving from various species of *Globigerinelloides*. The chamber elongation cannot be regarded as an ontogenetic stage in the species of the latter genus because the species of *Hastigerinoides* typically have different and shorter stratigraphic ranges than their progenitors.

Range: Coniacian–Maastrichtian (*Trinitella scotti* Interval).

Hastigerinoides alexanderi (Cushman), Plate 12, figs 6, 7; Plate 13, fig. 1

Hastigerinella alexanderi Cushman, 1931c, p. 87, 88, pl. 11, figs 6a–9; Cushman, 1933a, pl. 34, figs 15a, b; Cushman, 1946b, p. 148, pl. 61, figs 4a–7; Cushman, 1948b, p. 324, Key pl. 34, figs 15a, b; Frizzell, 1954, p. 127, pl. 20, figs 15a–16.

Hastigerinoides alexanderi (Cushman). Bolli, Loeblich and Tappan, 1957, p. 24, 25, pl. 1, figs 7a–10b; Loeblich and Tappan, 1964, p. C658, figs 2a, b; Saavedra, 1965, p. 322, text-fig. 3; Pessagno, 1967, p. 273, 274, pl. 60, fig. 6; Bandy, 1967, p. 14, text-fig. 6 (part); Douglas, 1969, p. 162, pl. 6, fig. 6.

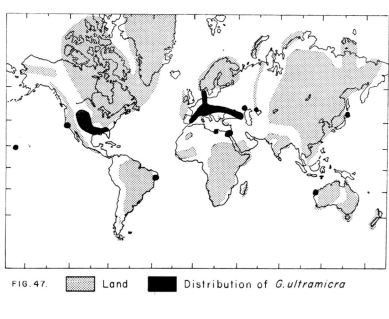

FIG. 47. ░░░ Land ██ Distribution of *G.ultramicra*

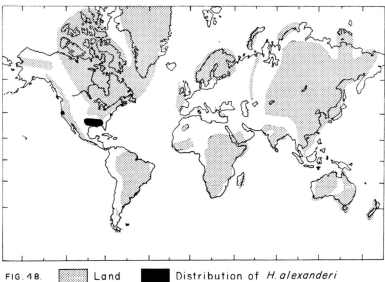

FIG. 48. ░░░ Land ██ Distribution of *H.alexanderi*

Hastigerinella (Hastigerinoides) alexanderi Cushman. Pokorný, 1963, p. 377, fig. 421.

Description: Test planispiral, involute, biumbilicate; equatorial periphery stellate and axial periphery pointed in post-neanic stages. Chambers of neanic stage sub-spherical with an imperforate peripheral band, with approximately 6 chambers per whorl; later chambers perforate, rapidly becoming radially elongate, slowly tapering over first half followed by a rather abrupt reduction in diameter which even more gradually tapers to a blunt point; 5–7 chambers in the final whorl all but the first 1–3 elongate; gradually to rapidly increasing in size; greatest diameter approximately one third of the length from the base. Aperture a low, interiomarginal, umbilical–equatorial arch bordered by a narrow to wide lip; relict apertures remain in the shallow, narrow umbilici. Sutures deeply depressed, radial. Surface densely perforate, smooth to pustulose.

Remarks: The number and the length of the radially elongate chambers alters the appearance of this species. Occasionally the ultimate chamber bifurcates at approximately one half of its length into two tapering portions oriented parallel to the axis of coiling.

Globigerinelloides ultramicra (Subbotina) is thought to be the most likely ancestor of this species. The neanic stages of each are essentially indistinguishable.

The more narrow tapering portion of the elongate chamber should probably not be referred to as a tubulospine. True tubulospines are imperforate on other genera, while the entire radially elongate chamber of *Hastigerinoides alexanderi* is perforate.

Distribution: This is shown in Fig. 48.

Range: Coniacian–Santonian (FOS *Globotruncana concavata*—FOS *Globotruncana elevata* Interval).

Hastigerinoides subdigitata (Carman)

Globigerina subdigitata Carman, 1929, p. 315, pl. 34, figs 4, 5.

Planomalina mendezensis Eternod Olvera, 1959, p. 92, 93, pl. 4, figs 9–11.

Clavihedbergella subdigitata subdigitata (Carman). Bandy, 1967, p. 13, 14, text-fig. 6 (part).

Clavihedbergella subdigitata (Carman). *non* Eicher and Worstell, 1970b, p. 306, pl. 10, figs 3a, b, 8a, b; *non* Caron, 1971, p. 150, text-figs 9a–c.

Description: Test planispiral, convolute, biumbilicate, equatorial periphery increasingly lobate, axial periphery rounded. Chambers initially subspherical to spherical, gradually increasing in size, last 1–3 becoming elongate in the plane of coiling with the length of the final chamber 2–$2\frac{1}{2}$ times its width; neanic chambers with an imperforate peripheral band, later becoming perforate; axis of elongate chambers tangential to the proloculus; 6–8 chambers in the final whorl. Aperture a low, interiomarginal, umbilical–equatorial arch bordered by a narrow lip, relict apertures visible in the umbilici. Umbilicus shallow, wide for the size of the test. Sutures depressed, radial. Surface densely perforate, pustulose particularly on the imperforate periphery.

Remarks: *Hastigerinoides subdigitata* has more chambers per whorl and fewer elongate chambers which never become clavate than does *H. watersi* (Cushman). The number of radially elongate chambers is the main variable character in this species.

This species appears to be identical to *Globigerinelloides ultramicra* (Subbotina) except in the development of the radially elongate chambers. Thus, *G. ultramicra* is suspected to be ancestral.

Globigerina subdigitata has always been assumed to be trochospiral. Examination of the holotype has shown it to be planispiral, hence, its transfer to *Hastigerinoides*. The holotype is preserved as an internal calcitic mold. A total of thirteen chambers with seven in the final whorl are visible.

Eternod Olvera (1959) described a form, *Planomalina mendezenensis*, which appears to be conspecific with *Hastigerinoides subdigitata*.

Distribution: This is shown in Fig. 49.

Range: Coniacian–Maastrichtian (*Trinitella scotti* Interval).

Hastigerinoides watersi (Cushman), Plate 13, figs 2, 3

Hastigerinella watersi Cushman, 1931c, p. 86, 87, pl. 11, figs 4–5c; Cushman, 1933a, pl. 34, figs 14a–c; Loetterle, 1937, p. 46, 47, pl. 7, figs 6a, b; Cushman, 1946b, p. 148, pl. 61, figs 8a–9; Cushman, 1948b, p. 324, Key pl. 34, figs 14a–c; Frizzell, 1954, p. 128, pl. 20, figs 17a, b.

Hastigerinella simplex Morrow. Loetterle, 1937, p. 46, 47, pl. 7, figs 6a, b.

Hastigerinoides watersi (Cushman). Bolli, Loeblich and Tappan, 1957, p. 24, 25, pl. 1, figs 6a, b; Loeblich and Tappan, 1964, p. C658, figs 3a, b; Pessagno, 1967, p. 274, pl. 51, fig. 5.

? *Hastigerinoides alpina* Sigal, 1959, p. 74, 76, pl. figs 50–53.

Globigerinelloides subcarinatus (Brönnimann). Ansary and Tewfik, 1968, p. 46, pl. 4, figs 6a, b.

Description: Test planispiral, involute, biumbilicate; equatorial periphery deeply lobate in neanic stage, later stellate; axial periphery rounded. Chambers of neanic portion spherical, rapidly increasing in size, with a wide, imperforate peripheral band; becoming cylindrically elongate in the ephebic stage to clavate in the gerontic stage with the constriction occurring approximately $\frac{1}{3}$ of the length from the bulbous tip, subsequent chambers abruptly larger than the last neanic chamber; 5–6 chambers in the final whorl. Primary aperture a low, interiomarginal, umbilical–equatorial arch bordered by a narrow imperforate lip; relict apertures visible in the shallow, narrow umbilici. Sutures deeply depressed, radial. Surface densely and coarsely perforate, pustules develop on the imperforate band.

Remarks: The number of chambers per whorl and the clavate gerontic chambers distinguish *Hastigerinoides watersi* from its congeners. The radially elongate chambers exhibit the greatest morphologic diversity in this species. Their length, number and the presence or absence of a clavate tip markedly alters the test appearance.

As previously noted (Masters, in press), juveniles of *Hastigerinoides watersi* are essentially indistinguishable from *Globigerinelloides escheri* (Kaufmann), except in the larger pore diameter of the former.

The illustrations of Sigal's (1959) new species, ? *Hastigerinoides alpina*, are identical to scanning electron micrographs (Masters, in press) of neanic topotypes of *H. watersi*, which is regarded as the senior synonym.

Distribution: This is shown in Fig. 50.

Range: Coniacian–Campanian (FOS *Pseudotextularia carseyae*—FOS *Globotruncana ventricosa* Interval).

Genus PLANOMALINA Loeblich and Tappan, 1946

Planomalina Loeblich and Tappan, 1946, p. 257.

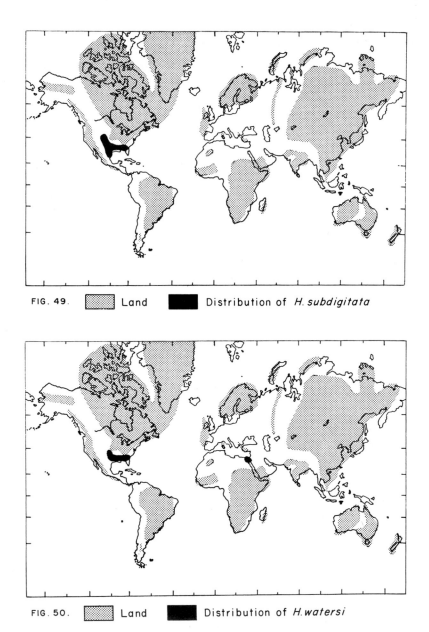

FIG. 49. ▨ Land   ■ Distribution of *H. subdigitata*

FIG. 50. ▨ Land   ■ Distribution of *H. watersi*

Type species: *Planomalina apsidostroba* Loeblich and Tappan, 1946 [= *Planulina buxtorfi* Gandolfi, 1942].

Description: Test planispiral, biumbilicate; equatorial periphery entire in earlier portion of final whorl, later becoming lobate; axial periphery acute. Chambers rather flat, crescent-shaped in side view, with a single keel. Sutures raised to depressed, limbate. Primary aperture an interiomarginal, umbilical–equatorial arch bordered by a lip. Relict apertures visible along umbilical margin.

Remarks: *Planomalina* is distinguished from *Globigerinelloides* by its peripheral keel and absence of paired supplementary apertures and chambers in the gerontic stage.

Thus far the genus is represented by a single species.

Range: Mid-Albian–lower Cenomanian.

Planomalina buxtorfi (Gandolfi), Plate 13, fig. 4; Plate 14, fig. 1

Planulina buxtorfi Gandolfi, 1942, p. 103, 104, pl. 3, figs 7a–c; pl. 5, fig. 4 (part); pl. 6, figs 1 (part), 2 (part); pl. 9, fig. 2; pl. 12, fig. 2 (part); pl. 13, figs 13, 15 (part); text-fig. 35 (1–11); Cita, 1948a, p. 15, pl. 2, fig. 13; Sigal, 1952, p. 23, fig. 22; Samuel, 1962, p. 179, pl. 7, figs 1a–c.

Planomalina apsidostroba Loeblich and Tappan, 1946, p. 258, pl. 37, figs 22–23b; Cushman, 1948b, p. 333, Key pl. 54, figs 4–5b; Bermudez, 1952, p. 92, pl. 16, figs 4–6; Frizzell, 1954, p. 131, pl. 21, figs 11a, b; Bolli, Loeblich and Tappan, 1957, p. 23, 24, pl. 1, figs 2a–3b; Bykova *et al.*, 1959, p. 287, text-figs 595, 596a, b.

Planomalina cf. *apsidostroba* Loeblich and Tappan. *non* Bolli, 1959, p. 259, 260, pl. 20, fig. 1.

Planomalina (?) *almadenensis* Cushman and Todd, 1948, p. 98, pl. 16, figs 25a, b.

Planomalina buxtorfi (Gandolfi). Reichel, 1950, p. 616; Küpper, 1955, p. 117, pl. 18, figs 8a, b; Brönnimann and Brown, 1956, text-fig. 15 (o); Sigal, 1956b, p. 211, 212, text-fig. 1; Gandolfi, 1957, p. 64, pl. 8, figs 6a, b; Klaus, 1959, p. 829, 830, pl. 8, figs 5a–c; Loeblich and Tappan, 1961, p. 269, 270, pl. 2, figs 1a–2b; Ayala-Castañares, 1962, p. 16–18, pl. 1, figs 3a–c; pl. 7, figs 1a, b; Postuma, 1962, 8 figs; Flandrin *et al.*, 1962, p. 224, 225, pl. 1, figs 12, 13; Graham, 1962, p. 105, pl. 20, figs 22Aa–Bb; Boccaletti and Sagri, 1965, p. 473, pl. 2, fig. 2; pl. 3, fig. 1 (part); pl. 6, fig. 1 (part); Broglio, Loriga and Mantovani 1965, pl. III, fig. 17 (*non* 8); Saavedra, 1965, p. 321, text-fig. 1; Marianos and Zingula, 1966, p. 334, pl. 37, figs 1a–c; Lehmann, 1966b, p. 156, pl. 1, fig. 7; text-figs 1h, i, o, p; Bandy, 1967, p. 12, text-fig. 5 (5); Barr, 1968, pl. 2, figs 1a, b; pl. 3, figs 2a, b; Scheibnerová, 1969, p. 56, tab. 7, figs 3a, b [incorrectly cited as *P. aspera* (Ehrenberg) in the plate explanation]; Caron and Luterbacher, 1969, p. 25, pl. 8, figs 5a–c; Bate and Bayliss, 1969, pl. 2, figs 7a, b; Broglio Loriga and Mantovani 1970, p. 224, 225, chart (50) ;Gorbachik and Krechmar, 1971, p. 23, pl. 4, figs 6, 7; Gorbachik, 1971, pl. 29, figs 7a, b; Sidó, 1971, p. 45, 46, pl. 1, fig. 1 (part); Brönnimann and Rigassi, 1971, pl. 19, figs 2a–c; Risch, 1971, p. 56, 57, pl. 6, figs 23, 24; Luterbacher, 1972, p. 586, pl. 5, fig. 3.

Labroglobigerinella (*Planomalina*) *spectrum buxtorfi* (Gandolfi). Signal, 1966b, p. 26, pl. 4, figs 1a, b, 4a, b.

Labroglobigerinella (? nov. morphogen. 1; ? nov. morphogen. 2) *spectrum buxtorfi* (Gandolfi). Sigal, 1966b, p. 26, pl. 4, figs 2a, b, 3a, b.

Planomalina (*Planomalina*) *buxtorfi* (Gandolfi). Salaj and Samuel, 1966, p. 163, pl. 6, figs 5a, b; tab. 33 (15).

Description: Test planispiral, biumbilicate, involute with a slight tendency to become evolute in the gerontic stage; equatorial periphery entire in the early portion of the final whorl, later becoming slightly lobate; axial periphery acute; sides subparallel. Chambers slope flatly from periphery to umbilical shoulder where they reach their greatest thickness; 11–15 crescent-shaped chambers in the final whorl, convex adaperturally; with a single peripheral keel, which is wider in the earlier chambers of the final whorl due to secondary calcification. Primary aperture a low, interiomarginal, umbilical–equatorial arch with a narrow lip which widens toward the umbilicus. Relict apertures visible in the umbilicus as successively overlapping apertural lips. Sutures limbate, raised in earlier part of final whorl, flush to depressed later, convex adaperturally, merging into the keel. Umbilicus moderately narrow, deep. Secondary calcification may occur on the chamber walls and sutures of earlier portions of the final whorl, often forming a carina on the umbilical shoulder.

Remarks: A monotypic genus in which the main features are stable. Slight variation in the degree of evoluteness of the gerontic chambers occurs. The development of a true evolute stage is never reached.

The ancestor of *Planomalina buxtorfi* is not known. *Planulina cheniourens* Sigal from the Aptian of Algeria is very similar to *Planomalina buxtorfi*, differing primarily by the absence of a keel.

The holotype of *Planomalina* ? *almadenensis* Cushman and Todd has a peripheral keel, limbate and raised sutures, a deep umbilicus and relict apertures. It is a junior synonym of *P. buxtorfi*.

The author concurs with Loeblich and Tappan (1964) that their *Planomalina apsidostroba* is conspecific with *P. buxtorfi*.

Distribution: This is shown in Fig. 51.

Range: Mid-Albian–lower Cenomanian.

Family SHACKOINIDAE Pokorný, 1958

Definition: Test may be initially a flat trochospire, later planispirally coiled; final one or one and a half whorls with hollow tubulospines or bulbous chamber extensions; aperture an equatorial to umbilical–equatorial arch bordered by a lip; surface perforated by minute pores.

Generic Key

  I. With tubulospines.   .   .   .   .   *Schackoina*
  II. With bulbous chambers .   .   .   .   *Leupoldina*

Remarks: *Schackoina* and *Leupoldina* may not be phylogenetically related. They are placed together more as a matter of convenience because of their similarity in coiling and chamber shape.

Range: Mid-Barremian–Maastrichtian (*Trinitella scotti* Interval).

Genus LEUPOLDINA Bolli, 1958

Leupoldina Bolli, 1958, p. 275.

Type species: *Leupoldina protuberans* Bolli, 1958 (= *Schackoina cabri* Sigal, 1952).

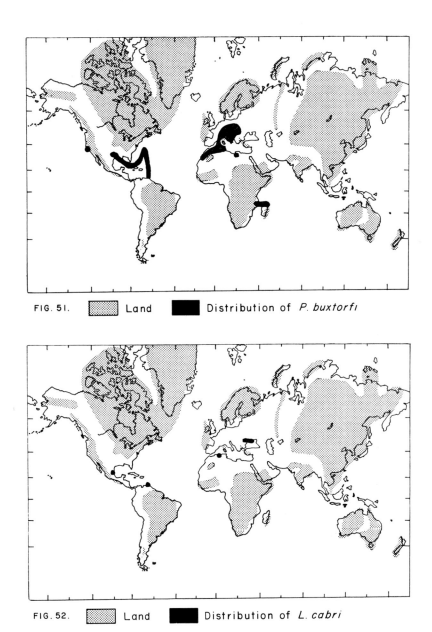

FIG. 51. ▦ Land ■ Distribution of *P. buxtorfı*

FIG. 52. ▦ Land ■ Distribution of *L. cabri*

Description: Test a flat trochospire in the first 1–1½ whorls, later becoming planispiral and biumbilicate; adult stellate in side view; non-keeled. Chambers of trochospiral portion subspherical, later becoming elongate in plane of coiling and developing 1–2 bulbous extensions in those of the last whorl. Aperture an interiomarginal, equatorial arch develops paired, umbilical, supplementary apertures when ultimate chamber is added over and around the first elongate chamber of the last whorl. Sutures depressed, radial. Surface perforate.

Remarks: *Leupoldina* possesses chambers which have bulbous extremities; whereas *Schackoina* all have tapering tubulospines.

Bolli (1958) characterised his new genus, *Leupoldina*, as having two or more bulb-shaped chamber extensions and paired umbilical apertures. This definition is regarded as too narrow. These definitive features are characteristic of only the ephebic and gerontic stages in the ontogenetic development of a single species.

The intial whorl or two have simple subspherical chambers representing the neanic stage. This is followed by the development of single tapered and then single bulbous chamber extensions in the ephebic stage. Finally, paired bulb-shaped extensions in the ultimate and occasionally in the penultimate chamber are typical of the gerontic stage. The paired apertures must out of structural necessity, occur after the first revolution of extended chambers, because these would block the "normal" single, peripheral aperture of subsequent chambers.

Forms, which are like those describe above, but with only a single peripheral aperture and bulbous extension per chamber were regarded by Bolli (1958) as species of *Schackoina*, e.g. *S. pustulans* Bolli. They are treated here as *Leupoldina*. Insufficient material was available to determine whether these forms also develop the paired apertures and bulbous extensions. At best these characters are specific traits and may even be intraspecific for all *Leupoldina*.

Bolli (1958) suspected that *Schackoina cabri* Sigal was conspecific with his *Leupoldina protuberans*. This opinion is adopted here, and thus *S. cabri* becomes the senior synonym and the type species for *Leupoldina*.

Range: Mid-Barremian–Cenomanian.

Leupoldina cabri (Sigal), Plate 14, fig. 4; Plate 15, fig. 1

Schackoina cabri Sigal, 1952, p. 20, 21, text-fig. 18; Sigal, 1959, p. 72, 74, figs 19–46; (?) Flandrin *et al.*, 1962, p. 224, pl. 3, fig. 28; (?) Salaj and Samuel, 1966, p. 165, pl. 7, figs 5a–c.

Schackoina aff. *S. cabri* Sigal. Eternod Olvera, 1959, p. 151, 152, pl. 5, figs 1a, b.

Leupoldina protuberans Bolli, 1958, p. 277, pl. II, figs 1–13a; Bolli, 1959, p. 264, pl. 20, figs 20a, b; Loeblich and Tappan, 1964, p. C658, C659, fig. 526, (10a, b); Saavedra, 1965, p. 322, 323, text-fig. 6; Neagu, 1970a, fig. 1 (3); Gorbachik, 1971, pl. 30, figs 7a, b.

Leupoldina Bolli. Bandy, 1967, p. 15, text-fig. 6 (part).

Description: Test initially a flat trochospire developing a true planispire in the last whorl; becomes biumbilicate only in last whorl; stellate in post-neanic stages. Chambers of neanic portion subspherical; later chambers become elongate in the plane of coiling, then developing a constriction at approximately mid-length resulting in a clavate or bulb-shaped extremity; the ultimate and occasionally the penultimate chamber may possess paired bulbous protuberances, arranged parallel to the coiling axis; inflated except where constricted; 4–5 in the final whorl. Primary aperture a low, interiomarginal, equatorial arch; paired supplementary apertures, low, narrow, umbilical arches; each bordered by a narrow, imperforate lip. Sutures depressed, radial. Surface finely perforate.

Remarks: *Leupoldina cabri* is distinguished from *L. pentagonalis* (Reichel) by being more tightly coiled and having shorter chambers in the final whorl. The major variation occurs in the number of bulbous protuberances and the number of chambers with multiple protuberances. The ultimate chamber commonly has two. Bolli (1958) recorded some forms with three protuberances. It is not clear from his illustrations, however, whether the third bulb is caused by that of the chamber in the previous whorl or is intrinsic to the ultimate chamber. The penultimate chamber typically has one, but may have two bulb-shaped extensions. The paired extensions may be adjacent or well separated, and are frequently broken off.

The phylogeny of the species is unknown.

Bolli (1958) implies that *Leupoldina* evolved from *Schackoina*. This has not been confirmed. *Leupoldina* is retained in the Family Schackoinidae because of their similar chamber arrangement.

The differences as noted above are regarded as intraspecific variants rather than potential subspecies as suggested by Bolli (1958).

Distribution: This is shown in Fig. 52.

Range: Mid-Barremian–Cenomanian.

Leupoldina pentagonalis (Reichel), Plate 14, figs 2, 3

Schackoina pentagonalis Reichel, 1948, p. 395, 397, text-figs 1a–f, 6 (1), 7 (1); Montanaro-Gallitelli, 1955b, tab. 1 (part).

Schackoina pentagonalis aperta Reichel, 1948, p. 397, text-figs 2a–d, 6 (2), 7 (2); Montanaro-Gallitelli, 1955b, tab. 1 (part).

Hastigerinoides rohri Brönnimann, 1952a, p. 55, 56, pl. 1, figs 8, 9; text-figs 29a–f.

Schackoina pustulans pustulans Bolli, 1958, p. 274, pl. I, figs 1–4.

Schackoina pustulans quinquecamerata Bolli, 1958, p. 274, 275, pl. I, figs 6a, 7a.

Schackoina reicheli Bolli, 1958, p. 275, pl. I, figs 8–10a; Sigal, 1959, p. 74, figs 48, 49; Bandy, 1967, p. 15, text-fig. 6 (part).

Schackoina pustulans Bolli. Sigal, 1959, p. 70, 72, pl. figs 1–18; Todd and Low, 1964, p. 407, pl. 1, fig. 7; (?) Bartenstein *et al.*, 1966, p. 163, 164, pl. 4, figs 381–383; Bandy, 1967, p. 15, text-fig. 6 (part).

Schackoina gandolfii Reichel. Bolli, 1959, p. 263, pl. 20, fig. 17 (*non* 12–16, 18).

Clavihedbergella pentagonalis (Reichel). Luterbacher and Premoli Silva, 1962, p. 269, pl. 22A, figs 1, 2.

Clavihedbergella pentagonalis aperta (Reichel). Luterbacher and Premoli Silva, 1962, p. 269, pl. 22A, figs 3, 4.

Leupoldina pustulans (Bolli). Gorbachik and Krechmar, 1971, pl. 6, fig. 1; Gorbachik, 1971, pl. 30, figs 6a, b.

Leupoldina reicheli (Bolli). Gorbachik, 1971, pl. 30, figs 8a, b.

Description: Test initially a flat trochospire which becomes planispiral and biumbilicate in the final whorl; stellate in side view. Chambers of neanic portion subspherical, slowly increasing in size; post-neanic chambers rapidly increase in size, becoming radially elongate in the plane of coiling; early chambers of last whorl may taper to point but soon develop a constriction and a bulbous extremity; 4, usually 5 in the last whorl. Aperture a low, interiomarginal, equatorial arch in the planispiral portion; bordered by a narrow, imperforate lip. Sutures depressed, radial. Surface finely perforate.

Remarks: *Leupoldina pentagonalis* is less tightly coiled and possesses more elongate final chambers than *L. cabri* (Sigal). It is not yet known to develop the paired apertures and protuberances of *L. cabri*. The bulb-shaped protuberances are often broken away leaving seemingly pointed chambers. The length of the chambers of the final whorl is quite variable. The constriction may occur at approximately mid-length or three-fourths of the distance to the tip. The longer the chamber, the farther from the axis of coiling is the constriction. The number of chambers in the final planispiral stage is also variable.

The phylogeny of this species is unknown.

Luterbacher and Premoli Silva (1962) restudied *Schackoina pentagonalis* Reichel and discovered that it possesses bulbous chamber protuberances and a low trocho-spiral coil. For these reasons they transferred it to the genus *Clavihedbergella*. Although the apertural position is not shown on their illustrations, these forms are interpreted as ones which have not yet developed the final planispiral stage. Other-wise, these are identical to the various forms illustrated by Bolli (1958) and discussed below. Thus, the taxon *pentagonalis* is herein transferred to the genus *Leupoldina*.

Schackoina pentagonalis aperta Reichel is judged to be conspecific with and a junior synonym of *Leupoldina pentagonalis*.

The holotype of *Hastigerinella* (*Hastigerinoides*) *rohri* Brönnimann, 1952a, lacks the ultimate chamber and the tips of the radially elongate chambers are broken. Because these chambers do not taper, they most likely were clavate. There is the suggestion of this on the penultimate and antipenultimate chambers of the holotype. Brönnimann's species is regarded as a junior synonym of *Leupoldina pentagonalis*.

Paratypes of Bolli's (1958) *Schackoina pustulans pustulans*, *S. pustulans quin-quecamerata* and *S. reicheli* were examined, and are believed to represent intra-specific variants of *Leupoldina pentagonalis*.

Distribution: This is shown in Fig. 53.

Range: Aptian–Cenomanian.

Genus SCHACKOINA Thalmann, 1932

Hantkenina (*Schackoina*) Thalmann, 1932, p. 288.

Schackoina Thalmann. Cushman, 1933a, p. 267.

Schackoina (*Schackoina*) Thalmann. Banner and Blow, 1959, p. 9.

Type species: *Siderolina cenomana* Schacko, 1897.

Description: Test small, a flat trochospire may be present initially, the last whorl is planispiral, biumbilicate; equatorial periphery broadly but deeply lobate, rarely entire; axial periphery rounded; tubulospinose. Chambers generally subspherical but may be laterally compressed and elongate in either the plane of coiling or the direction of coiling. Tubulospines absent in early stages later developing one or more per chamber; hollow, perforate, with only slight tapering to a blunt tip. Aperture a low interio-marginal, umbilical–equatorial arch bordered by a narrow to wide lip. Relict

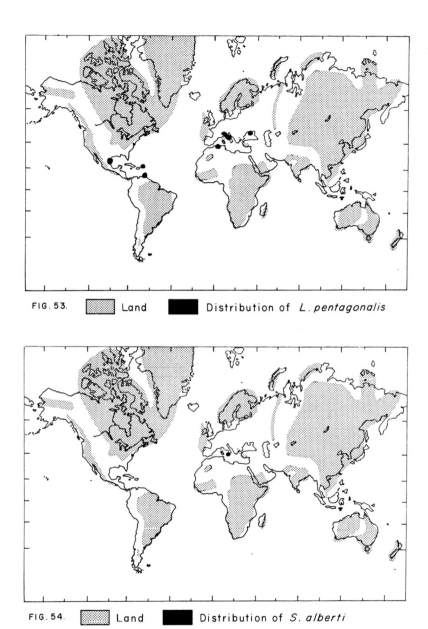

FIG. 53.　　◻ Land　　■ Distribution of *L. pentagonalis*

FIG. 54.　　◻ Land　　■ Distribution of *S. alberti*

apertures may be visible in either umbilicus. Umbilici narrow, shallow. Sutures straight, depressed, rarely flush. Surface with extremely minute pores; slightly hispid.

Remarks: *Schackoina* lacks the bulbous chambers of *Leupoldina*.

The *Schackoina* are probably one of the most over-subdivided taxa of the Mesozoic planktonic foraminifers. Thirteen or more species and subspecies have been named. Unfortunately no truly comprehensive study of this genus has been undertaken, and only three important investigations (Reichel, 1948; Masella, 1960; and Boccaletti and Pirini, 1965) have been published. Moreover, there is no uniformity of opinion among these authors.

Most of the species of *Schackoina* have been based upon the number of tubulospines per chamber and the number of chambers per whorl. Certainly many of these differences are gradational, but whether they are ontogenetic or phylogenetic has not been satisfactorily resolved. Reichel (1948) presented stratigraphic evidence for the phyletic side. Others (e.g. Masella, 1960) offered evidence in support of the ontogenetic.

Among the earliest to recognise ontogenetic change, Cushman and Wickenden (1930) proposed four growth stages (nepionic, neanic, ephebic and gerontic) to account for forms with varying numbers of tubulospines. For their *Schackoina multispinata*, they suggested that those with a single tubulospine per chamber were neanic individuals and those with several represented ephebic and gerontic individuals. This remarkable insight has for all practical purposes been ignored since its introduction. The *Schackoina* must be re-examined in this light.

Because of a lack of sufficient topotypic material during this investigation, an exhaustive study could not be undertaken. For this reason, many of the species are regarded tentatively as being valid, if there appears to be a potential stratigraphic variance. This approach is unsatisfactory, and a detailed account of their morphology and their first and last occurrences must be made before their stratigraphic usefulness can be determined.

El-Naggar (1971a) split the genus *Schackoina* into three subgenera, *S.* (*Hastigerinoides*), *S.* (*Eohastigerinella*) and *S.* (*Schackoina*). *Hastigerinoides* and its junior synonym *Eohastigerinella* are morphologically unlike and phylogenetically unrelated to *Schackoina*. They bear only superficial resemblance because of the radial elongation of the adult chambers. The small *Schackoina* have perforate peripheries and minute pores on the order of 250 mμ for those measured, features none of which are shared with *Hastigerinoides*.

Range: Albian–Maastrichtian.

Schackoina alberti Masella, Plate 15, fig. 2

Schackoina cenomana alberti Masella, 1960, p. 26, pl. 5, figs 11–16; pl. 6, figs 1–9; pl. 10, fig. 5.

Description: Test small, planispiral involute, biumbilicate; equatorial periphery deeply lobate, stellate; axial periphery initially pointed, becoming truncated and finally tripartite. Chambers inflated; 4 in the final whorl, the first is radially elongate with a single tubulospine; the second may be as the first but typically has 2 tubulospines positioned at approximately 45° from and on either side of the plane of coiling; the third also bears 2 tubulospines but the chamber itself tends to be drawn out to form the spines; the ultimate chamber is drawn into 3 prominent conical extensions which become tubulospines, with the laterally positioned cones at 45°–90° from the plane of coiling and smaller than the central cone in the plane of coiling; the radial axes are not necessarily mutually perpendicular, particularly that of the ultimate chamber. Aperture a low to moderately high, interiomarginal, umbilical–equatorial arch bordered by a narrow lip. Sutures depressed, straight. Umbilicus narrow, shallow.

Remarks: *Schackoina alberti* is the only species which has a tripartite conical construction in the last chamber. The degree of the tripartite extensions and their relative positions are variable. The second chamber of the last whorl may have one or two tubulospines.

The phylogeny of this form is unknown.

It must be emphasised that this species may represent merely the gerontic stage of another species.

Distribution: This is shown in Fig. 54.

Range: Cenomanian.

Schackoina bicornis Reichel, Plate 15, figs 3, 4

Schackoina cenomana bicornis Reichel, 1948, p. 401, 402, text-figs 4a–g, 6 (4), 7 (4), 8b, 9a–g, 10 (6, 8); Aurouze and de Klasz, 1954, p. 99, text-fig. 1B; Zanzucchi, 1955, p. 373, pl. 28, figs b, 5–7; Küpper, 1956, p. 44, 45, pl. 8, figs 5a–c; Samuel, 1962, p. 181, 182, pl. 8, figs 5, 6; Boccaletti and Pirini, 1965, p. 56, pl. 1, figs 12–18; text-fig. d; Gawor-Biedowa, 1972, p. 65, 66, pl. 6, fig. 2.

Schackoina bicornis Reichel. Bykova *et al.*, 1959, p. 300, text-figs 675A, B; Porthault, 1969, p. 529, pl. 1, fig. 1.

Schackoina cenomana cenomana (Schacko). Masella, 1960, p. 24, 25, pl. 2, figs 1–4 (*non* pl. 1, figs 1–9).

Schackoina multispinata (Cushman and Wickenden). Graham, 1962, p. 108, pl. 20, figs 33a–c.

Schackoina multispinata bicornis Reichel. Neagu, 1966, p. 366, 367, pl. 1, figs 5–8; pl. 2, figs 3–22; Neagu, 1968, text-fig. 2 (53); Neagu, 1970b, p. 63, pl. 15, figs 7 (?), 8 (?)–26; pl. 41, fig. 53; pl. 43, fig. 11.

Description: Test small, planispiral involute but may be a flat trochospiral in the first whorl, biumbilicate; equatorial periphery deeply lobate, stellate; axial periphery pointed to truncate. Chambers inflated, gradually increasing in size, radially elongate with the height exceeding the subequal width and thickness; each of the first 3 in the last whorl with a single tubulospine in the plane of coiling; the ultimate chamber has 2 tubulospines which form an angle ranging from 65°–98°, a plane passing through the tubulospines will be perpendicular to the plane of coiling; 4 chambers in the final whorl, with their radial axes mutually perpendicular. Aperture a low, interiomarginal, umbilical–equatorial arch bordered by a slightly flaring, moderately wide lip. Umbilicus narrow, shallow. Sutures depressed, straight.

Remarks: This four-chambered species was distinguished by Reichel (1948) on the basis of its paired tubulospines on the ultimate chamber and single ones on the previous chambers. The range of intraspecific variation is unknown.

Schackoina bicornis presumably developed from *S. cenomana* (Schacko).

It is possible that this species represents merely one stage in the ontogenetic series of *Schackoina cenomana* (Schacko).

Distribution: This is shown in Fig. 55.

Range: Cenomanian.

Schackoina cenomana (Schacko), Plate 16, figs 1, 2

Siderolina cenomana Schacko, 1897, p. 166, pl. 4, figs 3–5; Egger, 1900, p. 174, pl. 21, fig. 42; Franke, 1928, p. 193, pl. 18, figs 11a–c.

Hantkenina cenomana (Schacko). Cushman and Wickenden, 1930, p. 40, pl. 6, figs 1–3.

Hantkenina (*Schackoina*) *cenomana* (Schacko). Thalmann 1932, p. 288.

Schackoina cenomana (Schacko). Keller, 1939, pl. 1, fig. 1; Noth, 1951, p. 74, pl. 5, figs 9, 10; Bermúdez, 1952, p. 108, pl. 20, figs 1, 2; Subbotina, 1953, p. 133, 134, pl. 1, figs 2, 3 (*non* 1); Montanaro-Gallitelli, 1955, p. 143, tab. 1 (part); Bolli, Loeblich, and Tappan, 1957, p. 26, pl. 2, figs 1a–2; Bykova *et al.*, 1959, p. 300, text-fig. 676; *non* Maslakova, 1959, p. 107, 108, pl. 10, fig. 2; Loeblich and Tappan, 1961, p. 270, 271, pl. 1, figs 4a–c, 5 (*non* 2a, b, 6a–c, 7); Ayala-Castañares, 1962, p. 20, 21, pl. 2, figs 2a–c, (?) 3a–c; pl. 7, figs 3a, b; pl. 8, figs 1a–c; Samuel, 1962, p. 181, pl. 8, figs 2a–3b; Loeblich and Tappan, 1964, p. C658, fig. 526 (8a–c, *non* 9); Takayanagi, 1965, p. 202, 203, pl. 21, figs 21a–c; Saavedra, 1965, p. 322, text-fig. 5; Viterbo, 1965, pl. 10, fig. 2 (5); *non* Salaj and Samuel, 1966, p. 165, 166, pl. 7, figs 8a–c; Marianos and Zingula, 1966, p. 334, pl. 37, figs 2a–c; *non* Pessagno, 1967, p. 279, pl. 48, fig. 6; Scheibnerová, 1969, p. 57, tab. 7, figs 5a–7b; Porthault, 1969, p. 529 pl. 1, fig. 2; *non* Eicher, 1969, text-fig. 3 (part); Douglas, 1969, p. 162, 163, pl. 6, figs 5a, b; *non* Eicher and Worstell, 1970b, p. 298, 300, pl. 9, figs 1, 2, 4.

Schackoina gandolfii Reichel, 1948, p. 397–400, pl. 8, fig. 1, text-figs 3a–g, 6 (3), 7 (3), 8a, 10 (1, 3, 4, 11); Bermúdez, 1952, p. 108, pl. 20, fig. 5; Arouze and de Klasz, 1954, p. 99, text-fig. 1C; Montanaro-Gallitelli, 1955, p. 143, tab. 1 (part); Zanzucchi, 1955, p. 357, pl. 28, figs c, 1–4; Bolli, 1959, p. 263, pl. 20, figs 12, 14–16 (*non* 13, 17, 18); Salaj and Samuel, 1966, p. 166, pl. 7, figs 6a–c; Kalantari, 1969, p. 172, pl. 19, figs 6a, b, text-fig. 25 (16).

Schackoina sp. cf. *S. gandolfii* Reichel. *non* Küpper, 1956, p. 44, pl. 8, figs 4a–c.

Schackoina cenomana cenomana (Schacko). *non* Masella, 1960, p. 24, 25, pl. 1, figs 1–9; pl. 2, figs 1–4; pl. 10, figs 1–4; Neagu, 1966, p. 365, 366, pl. 1, figs 1, 9–12, 14, 15 (*non* 2, 13, 16, 17); Barbieri, 1966, text-fig. 2 (2–6); *non* Bandy, 1967, p. 15, text-fig. 6 (part); Neagu, 1968, text-fig. 2 (51); *non* Neagu, 1970a, fig. 1 (4); Neagu, 1970b, p. 63, pl. 14, figs 17–21, 23 (*non* 15, 16, 22); pl. 41, fig. 51; Gawor-Biedowa, 1972, p. 64, 65, pl. 6, fig. 1.

Schackoina cenomana gandolfii Reichel. (?) Masella, 1960, p. 25, 26, pl. 2, figs 5–12; pl. 3, figs 1–13; pl. 10, figs 6, 7; Luterbacher and Premoli Silva, 1962, p. 270, pl. 22B, figs 1, 2; Boccaletti and Pirini, 1965, p. 55, 56, pl. 1, figs 4–9; text-fig. b; Neagu, 1966, p. 366, pl. 1, figs 3, 4 (?), 18–21; pl. 2, figs 1, 2 (?); Barbieri, 1966, text-fig. 2 (1); Neagu, 1968, text-fig. 2 (52).

Schackoina spp. Boccaletti and Sagri, 1965, p. 474, pl. 3, fig. 2 (part).

Schackoina sp. Boccaletti and Pirini, 1965, p. 58, 59, pl. 1, figs 10, 11; text-fig. c.

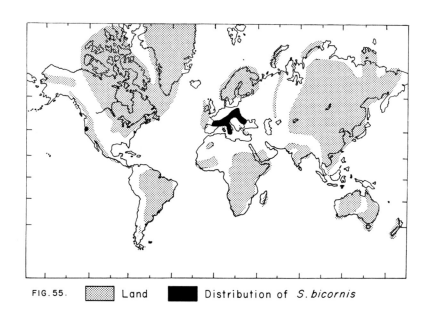

FIG. 55. ▨ Land ■ Distribution of *S. bicornis*

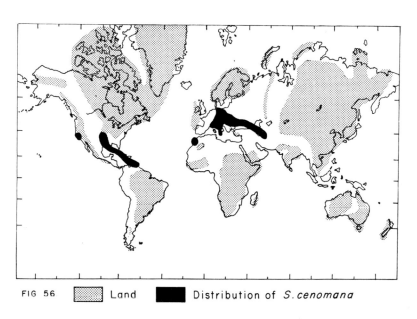

FIG 56 ▨ Land ■ Distribution of *S. cenomana*

Description: Test small, planispiral involute but may possess an early flat trochospire, biumbilicate; equatorial periphery deeply lobate, stellate; axial periphery pointed. Chambers inflated, radially elongate with the height exceeding the other two dimensions; gradually increasing in size with 4 chambers in the last whorl, each with a single tubulospine in the plane of coiling except the first which may lack a tubulospine; radial axes mutually perpendicular. Aperture a low, interiomarginal, umbilical–equatorial arch bordered by a narrow lip. Umbilicus narrow, shallow. Sutures depressed, straight.

Remarks: This species is characterised by the last whorl having four chambers set at 90° intervals, each bearing a single tubulospine. We have no knowledge of the intraspecific variation of this species. The phylogeny of this species is unknown.

Schackoina gandolfii Reichel, 1948, is believed to be conspecific with *S. cenomana*. It has the same number of chambers in the final whorl, each chamber has a single tubulospine, and each is set at an angle of 90°.

Those specimens attributed to *Schackoina cenomana gandolfii* Reichel by Masella (1960) are only tentatively placed in synonymy because some of her specimens in which the tubulospines are broken have bases with rather large diameters, which may indicate *Leupoldina*-like bulbous chamber extensions rather than the narrower tubulospines of *Schackoina*.

Distribution: This is shown in Fig. 56.

Range: Albian–lower Turonian.

Schackoina masellae Luterbacher and Premoli Silva, Plate 16, figs 3, 5

Schackoina gandolfii Reichel. Bolli, 1959, p. 263, pl. 20, figs 13, 18 (*non* 12, 14–17; Todd and Low, 1964, p. 406, 407, pl. 1, figs 5, 6, 8.

Schackoina cenomana pentagonalis Reichel. Masella, 1960, p. 26, 27, pl. 4, figs 5a, b, 9a–14b; pl. 5, figs 1a–5; Barbieri, 1966, text-fig. 2 (12).

Schackoina cenomana masellae Luterbacher and Premoli Silva (nom. nov.), 1962, p. 270.

Schackoina cenomana cenomana (Schacko). Neagu, 1966, p. 365, 366, pl. 1, figs 2, 16, 17 (*non* 1, 9–15); Neagu, 1970b, p. 63, pl. 14, figs 15, 16, 22 (*non* 17–21, 23).

Description: Test small, planispiral, partially involute, biumbilicate; equatorial periphery deeply lobate, stellate; axial periphery pointed. Chambers numbering 5 in the last whorl, gradually increasing in size; each of the first 1–3 chambers in the last whorl have a single, incompletely developed, nipple-like tubulospine; the single tubulospine is fully developed in the ultimate and penultimate chambers; all tubulospines lie in the plane of coiling; the height of the chambers increases more rapidly than the other dimensions. Aperture a low, interiomarginal, umbilical–equatorial arch. Umbilicus relatively wide, shallow. Sutures depressed, straight to slightly curved.

Remarks: This is the only five-chambered species with single tubulospines on each chamber. There is no information concerning variation within this species.

Presumably *Schackoina masellae* evolved from *S. cenomana* (Schacko) with the addition of one chamber per whorl.

In a restudy of *Schackoina pentagonalis* Reichel, 1948, Luterbacher and Premoli Silva (1962) demonstrated that this species possesses bulbous chamber extensions rather than tubulospines. The specimens referred to this species by Masella (1960) were considered as true *Schackoina*, and Luterbacher and Premoli Silva proposed the name *S. masellae* for these forms.

Distribution: This is shown in Fig. 57.

Range: Mid-Albian–lower Turonian.

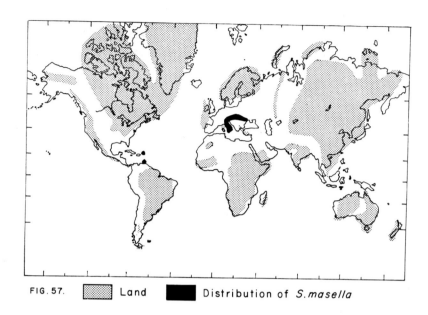

FIG. 57. [ ] Land [■] Distribution of *S. masella*

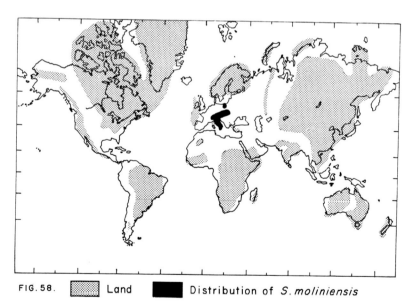

FIG. 58. [ ] Land [■] Distribution of *S. moliniensis*

Schackoina moliniensis (Reichel), Plate 17, fig. 1

Schackoina moliniensis Reichel, 1948, p. 402, 404, text-figs 5a–d, 6 (5), 7 (5), 8c, 10 (13); Bermúdez, 1952, p. 108, pl. 20, fig. 4; Zanzucchi, 1955, p. 373, pl. 28, figs a, 8–10; Montanaro Gallitelli, 1955b, tab. 1 (part); Samuel, 1962, p. 183, pl. 8, figs 7, 8; Pokorný, 1963, p. 381, fig. 427; Boccaletti and Pirini, 1965, p. 54, 55, pl. 1, figs 1 (?), 2 (*non* 3; text-fig. a); Gawor-Biedowa, 1972, p. 66, pl. 6, fig. 3.

Schackoina cenomana cenomana (Schacko). Masella, 1960, p. 24, 25, pl. 1, figs 1–9 (*non* pl. 2, figs 1–4).

Schackoina multispinata (Cushman and Wickenden). Salaj and Samuel, 1966, p. 166, 167, pl. 7, figs 7a, b.

Description: Test small, planispiral involute, may be initially trochospiral, biumbilicate; equatorial periphery deeply lobate, stellate; axial periphery pointed in the first 2 chambers of the last whorl, truncate-pointed in the last two. Chambers higher than wide or thick in the first 2 or 4 in the final whorl, each with a single tubulospine in the plane of coiling; in the last 2 chambers the thickness is greatest, two tubulospines per chamber oriented in a plane perpendicular to that of the coil; positioned in approximately 90° intervals. Aperture a low, interiomarginal, umbilical–equatorial arch. Umbilicus narrow, shallow. Sutures depressed, straight.

Remarks: *Schackoina moliniensis* is a distinctive four-chambered form with the last two chambers bearing paired tubulospines. The range of morphological variation within the species is unknown.

This species may have developed from *Schackoina bicornis* Reichel with the addition of an extra tubulospine on the penultimate chamber.

The stratigraphical range given by Reichel (1948) for *Schackoina moliniensis* may not be an accurate total range. If the range is found to be similar to *S. cenomana* or *S. bicornis*, then *S. moliniensis* is probably a gerontic member of one of these species.

Distribution: This is shown in Fig. 58.

Range: lower Cenomanian to lower Turonian.

Schackoina multispinata (Cushman and Wickenden), Plate 16, figs 4, 6;
Plate 17, fig. 2

Hantkenina multispinata, 1930, Cushman and Wickenden, 1930, p. 40–43, pl. 6, figs 4–6 (*non* 4a–c); Cushman, 1931c, p. 88, pl. 11, fig. 11 (*non* 10).

Schackoina multispinata (Cushman and Wickenden). Cushman, 1933a, pl. 35, fig. 9 (*non* 10); Cushman, 1946b, p. 148, pl. 61, figs 11, 12; Cushman, 1948b, p. 328, Key pl. 35, fig. 9 (*non* 10); Bermúdez, 1952, p. 108, pl. 20, fig. 3; Subbotina, 1953, p. 134, 135, pl. 1, fig. 5a–c (*non* 4a–c); Frizzell, 1954, p. 128, pl. 20, figs 18 (*non* 19); Aurouze and de Klasz, 1954, p. 100, text-fig. 1F, G (*non* D, E); Montanaro-Gallitelli, 1955b, tab. 1 (part); *non* Jones, 1956, text-fig. 9; Bolli, 1959, p. 264, pl. 20, figs 19a, b; Loeblich and Tappan, 1961, p. 271, 272, pl. 1, figs 8a–10b; *non* Ayala-Castañares, 1962, p. 19, 20, pl. 2, figs 1a–d, pl. 7, figs 2a–b; *non* Graham, 1962, p. 108, pl. 20, figs 33a–c; Luterbacher and Premoli Silva, 1962, p. 270, 271, pl. 22C, figs 1–3; Graham and Church, 1963, p. 61, pl. 7, fig. 9; *non* Salaj and Samuel, 1966, p. 166, 167, pl. 7, figs 7a, b; Burckle, Saito and Ewing, 1967, text-fig. 2 (5a, b, 6); Sliter, 1968, p. 100, pl. 15, figs 7a, b; Weaver *et al.*, 1969, pl. 2, fig. 4; Porthault, 1969, p. 529, pl. 1, fig. 3; Eicher, 1969, text-fig. 3 (part); Neagu, 1970a, fig. 1 (6); Eicher and Worstell, 1970b, p. 300, pl. 9, figs 5, 8a, b.

Hantkenina trituberculata Morrow, 1934, p. 195, pl. 29, figs 26–28 (*non* 24).

Schackoina trituberculata (Morrow). Loetterle, 1937, p. 47, 48, pl. 7, figs 7a, b;

Montanaro Gallitelli, 1955b, p. 142, pl. 1, fig. 11; tab. 1 (part); Montanaro Gallitelli, 1958, p. 149, pl. 4, fig. 6; Ansary and Tewfik, 1968, p. 46, pl. 5, figs 9a, b.

Schackoina cenomana multispinata (Cushman and Wickenden). Masella, 1960, p. 28, 29, pl. 8, figs 16a–21; pl. 9, figs 1a–17; pl. 10, figs 13–15; Boccaletti and Pirini, 1965, p. 58, pl. 21 (2), fig. 20; text-fig. g; Bandy, 1967, p. 15, text-fig. 6 (part).

Schackoina cenomana cf. *multispinata* (Cushman and Wickenden). Boccaletti and Pirini, 1965, p. 57, 58, pl. 21 (2); figs 8–19; text-fig. f.

Schackoina cushmani Barr, 1961, p. 565, pl. 69, fig. 3, text-fig. 5a–f.

Description: Test small, planispiral involute possibly with an initial trochospiral stage, biumbilicate; equatorial periphery deeply lobate, spinose; axial periphery broadly rounded to truncate, spinose. Chambers inflated, 3 in the last whorl; earliest chambers without tubulospines, quickly followed by chambers including the first one in the final whorl bearing a single tubulospine in the plane of coiling, the ultimate and penultimate chambers with 2 or more tubulospines each; the single tubulospines from the preceding whorl may penetrate the chambers in the final whorl, frequently broken off leaving a slight circular depression in the later chambers; early chambers subspherical, later with the thickness being the greatest dimension. Aperture a low, interiomarginal, umbilical–equatorial arch bordered by a relatively wide, flaring lip. Umbilicus narrow, shallow. Sutures depressed, straight. Surface covered by numerous, minute pores with an approximate diameter of 250 μm, randomly hispid.

Remarks: This species differs from *Schackoina tappanae* Montanaro Gallitelli by having multiple tubulospines on a chamber. *Schackoina multispinata* has been observed with as few as three to as many as seven tubulospines on the ultimate chamber. Some of these may be spines on earlier chambers which penetrate the ultimate chamber. Typically three to four spines are present.

Schackoina tappanae Montanaro Gallitelli is most probably ancestral to *S. multispinata*.

The possibility that *Schackoina multispinata* is a gerontic stage form of *S. tappanae* Montanaro Gallitelli, cannot be overlooked. Immature specimens of *S. multispinata* are indistinguishable from the latter. Currently, however, *S. tappanae* has been recorded from slightly older strata.

Topotypes of *Hantkenina trituberculata* Morrow, 1934, have been examined and are illustrated herein. The number and distribution of the tubulospines on the ultimate chamber is highly variable. It seems reasonable to consider this species, as a junior synonym of *S. multispinata*.

Distribution: This is shown in Fig. 59.

Range: Upper Cenomanian–Maastrichtian (*Trinitella scotti* Interval).

Schackoina primitiva Tappan

Schackoina primitiva Tappan, 1940, p. 123, pl. 18, figs 14a–c; Frizzell, 1954, p. 128, pl. 20, figs 20a–c; Montanaro Gallitelli, 1955b, p. 144, table 1 (part); Loeblich and Tappan, 1961, p. 272, 273, pl. 1, figs 1a–c.

Description: "Test tiny, biumbilicate, slightly trochoid; chambers inflated at their centre, about three chambers to a whorl, increasing rapidly in size, each with a fairly thick hollow spine, arising from the centre of the chamber on the periphery, about equal in length to one-half the diameter of the test; sutures distinct, slightly depressed, periphery slightly lobulate; wall calcareous, surface smooth; aperture a

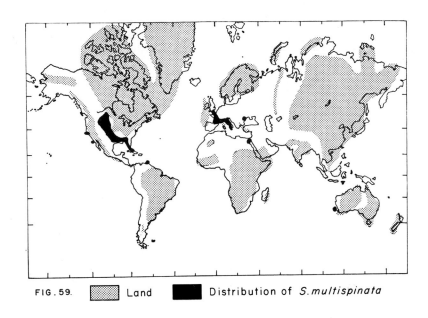

FIG. 59.   ▨ Land   ■ Distribution of *S. multispinata*

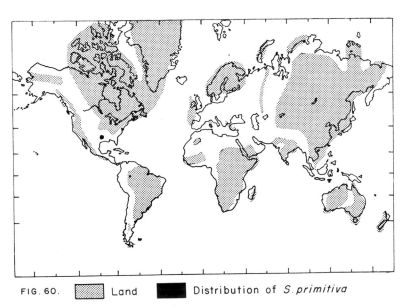

FIG. 60.   ▨ Land   ■ Distribution of *S. primitiva*

small arched slit at the base of the last chamber on the periphery. Greatest diameter of the holotype, not including spines, 0.265 mm; least diameter, 0.17 mm; width through centre, 0.06 mm; width of last chamber, 0.10 mm; length of spine, 0.10 mm''. (Type description.)

Remarks: This species is the only *Schackoina* with appressed chambers and a nearly entire margin. The range of morphological variation within this species is unknown. The phylogeny of this species is unknown.

Only two specimens of *Schackoina tappanae* have ever been recorded in the literature. Both of these were by Tappan (1940). In her remarks, she noted their extreme rarity. The authors own attempts to find additional specimens from Grayson Bluff were fruitless. The holotype was examined and found to be large, atypically so for a *Schackoina*. The shape of the test is also unlike any other *Schackoina*. It is possible that this taxon may in fact not be a *Schackoina*. Unfortunately, not enough is known about this species to analyse it properly, and is for now retained as a *Schackoina*.

Distribution: This is shown in Fig. 60.

Range: Lower Cenomanian.

Schackoina sellaeforma Masters, Plate 17, figs 3, 4

Schackoina sellaeforma Masters, 1976, p. 327, 328, pl. 1, figs. 13–16.

Description: Test small, planispiral, biumbilicate; laterally compressed; equatorial periphery deeply lobate, spinose; axial periphery narrowly rounded, spinose. Chambers initially subspherical, later becoming slightly elongate in a radial direction with a single tubulospine, finally developing a rectangular side view with a tubulospine in both the aboral and adoral positions forming a saddle-shaped chamber, all tubulospines in the plane of coiling; 3 slightly inflated, rapidly enlarging chambers in the final whorl. Aperture a low, interiomarginal, umbilical–equatorial arch with a bordering, wide, flaring lip; relict apertures visible along the umbilical margins. Umbilicus narrow, shallow. Sutures depressed, straight. Surface with abundant, minute pores with an approximate diameter of 250 μm; locally hispid.

Remarks: *Schackoina sellaeforma* has the unique aboral–adoral polarity of its tubulospines. The only significant variation thus far observed is the ontogenetic progression from zero to one to two tubulospines per chamber.

The phylogeny of this species is unknown.

This species has not yet been recognised outside of Alabama, U.S.A. It may prove to be an important stratigraphic marker if sufficiently widespread.

Distribution: This is shown in Fig. 61.

Range: Campanian (FOS *Pseudotextularia carseyae*—FOS *Globotruncana ventricosa* Interval).

Schackoina tappanae Montanaro Gallitelli, Plate 17, figs 5, 6

Hantkenina multispinata Cushman and Wickenden, 1930, p. 40–43, pl. 6, figs 4a–c (*non* 5a–6); Cushman, 1931c, p. 88, pl. 11, fig. 10 (*non* 11).

Schackoina multispinata (Cushman and Wickenden). Cushman, 1933a, p. 35, fig. 10 (*non* 9); Aurouze and de Klasz, 1954, p. 100, text-fig. 1E (*non* D, F G); Jones, 1956, text-fig. 9.

Hantkenina trituberculata Morrow, 1934, p. 195, pl. 29, fig. 24 (*non* 26–28).

Schackoina cf. *trituberculata* (Morrow). Montanaro Gallitelli, 1954, pl. 2 (part).

Schackoina tappanae Montanaro Gallitelli, 1955b, p. 142, 143, pl. 1, figs 1a–10c;

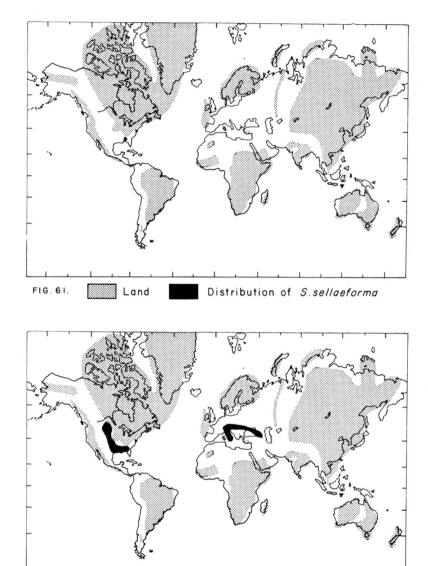

FIG. 61. ▨ Land ■ Distribution of *S. sellaeforma*

FIG. 62. ▨ Land ■ Distribution of *S. tappanae*

tab. 1 (part); Montanaro Gallitelli, 1958, p. 149, pl. 4, figs 5a–c; Cati, 1964, p. 255, pl. 41, figs 10a, b.

Schackoina cenomana (Schacko). Maslakova, 1959, p. 107, 108, pl. 10, fig. 2; Loeblich and Tappan, 1961, p. 270, 271, pl. 1, figs 2a, b, 6a–c, 7 (*non* figs 3–5); Loeblich and Tappan, 1964, p. C658, fig. 526 (9, *non* 8a–c,); Salaj and Samuel, 1966, p. 165, 166, pl. 7, figs 8a–c; Pessagno, 1967, p. 279, pl. 48, fig. 6; Eicher, 1969, text-fig. 3 (part); Eicher and Worstell, 1970b, p. 298, 300, pl. 9, figs 1, 2, 4.

Schackoina cenomana tappanae Montanaro Gallitelli. Masella, 1960, p. 28, pl. 7, figs 1a–10b; 15a, b(?); pl. 8, figs 6–15; pl. 10, figs 8–10 (*non* 11, 12); Boccaletti and Pirini, 1965, p. 57, pl. 1, figs 19, 20; pl. 2, figs 1–7; text-fig. e.

Schackoina cf. *cenomana tappanae* Montanaro Gallitelli. Barbieri, 1966, text-fig. 2 (7–11).

Schackoina cenomana cenomana (Schacko). Bandy, 1967, p. 15, text-fig. 6 (part); Neagu, 1970a, fig. 1 (4).

Schackoina cenomana gandolfii Reichel. Neagu, 1970b, p. 63, pl. 15, figs 1–3 (*non* 4–6).

Description: Test small, planispiral involute possibly with an early trochospiral stage, biumbilicate; equatorial periphery deeply lobate, spinose; axial periphery rounded, spinose. Chambers inflated but thickness generally slightly less than the other two dimensions; 3 in the final whorl; all or only the final 2 may bear single tubulospines in the plane of coiling, the first chamber of the final whorl may be subspherical or with slight radial elongation if not tubulospinose. Aperture a low, interiomarginal, umbilical–equatorial arch bordered by a moderately wide, flaring lip. Umbilicus narrow, shallow. Sutures depressed, straight. Surface with numerous minute pores having diameter of approximately 250 μm; randomly hispid.

Remarks: *Schackoina tappanae* differs from *S. multispinata* (Cushman and Wickenden) by having no more than one tubulospine per chamber. Montanaro Gallitelli (1955) recorded some variation in the tightness of the coiling of this species. The phylogeny of this species is unknown.

Two of the paratypes of *Schackoina tappanae* on deposit at the U.S. National Museum are *S. multispinata* (Cushman and Wickenden) by definition. This again raises the possibility that *S. tappanae* is merely an early ontogenetic stage of *S. multispinata*.

Distribution: This is shown in Fig. 62.

Range: Upper Cenomanian–Santonian (FOS *Globotruncana concavata*—FOS *Globotruncana elevata* Interval).

Schackoina utriculus Masella

Schackoina cenomana subsp. *utriculus* Masella, 1960, p. 27, pl. 5, figs 6a–8b.

Schackoina cenomana subsp. *trinacriae* Masella, 1960, p. 27, 28, pl. 5, figs 9a–10b.

Description: Test small, planispiral, partially involute, biumbilicate; equatorial periphery deeply lobate, stellate; axial periphery pointed. Chambers gradually increasing in size, radially elongate, 5 in the final whorl; each with a single tubulospine in the plane of coiling except for the final 1–2 chambers which have paired tubulospines, a plane passing through these is perpendicular to the plane of coiling. Aperture a low, interiomarginal, umbilical–equatorial arch bordered by a lip. Umbilicus relatively wide, shallow. Sutures depressed, straight.

Remarks: *Schackoina utriculus* differs from *S. masellae* Luterbacher and Premoli Silva by having paired tubulospines on the final one or two chambers.

Schackoina masellae most closely resembles this species and may be ancestral to it. It is suspected that *Schackoina utriculus* is only an ontogenetic stage of *S. masellae*. As yet *S. utriculus* has not been reported outside its type area, so the two species are retained.

Schackoina cenomana trinacriae Masella, 1960, has one additional tubulospine in the final chamber than does *S. utriculus*, and apparently identical stratigraphic ranges. Hence, they are believed to be intraspecific variants, with *S. utriculus* having priority.

Distribution: This is shown in Fig. 63.

Range: Cenomanian.

Family GLOBIGERINIDAE Carpenter, Parker and Jones, 1862

Definition: Test a flat to high trochospire; equatorial periphery may be subcircular to stellate; primary aperture umbilical to extraumbilical–umbilical, accessory apertures may occur along sutural boundaries of gerontic chambers; apertures bordered by imperforate lips or portici; axial periphery porous.

Generic Key

    I. Stellate in peripheral view . . . . *Clavihedbergella*
    II. Subcircular in peripheral view
A. Without accessory apertures . . . . *Globigerina*
B. With accessory apertures . . . . *Guembelitria*

Remarks: No attempt has been made to include in the above definition and key any post-Mesozoic Globigerinidae. In fact application of this key to the younger genera will result in incorrect assignments.

Range: Bathonian–Maastrichtian (*Trinitella scotti* Interval).

Genus CLAVIHEDBERGELLA Banner and Blow, 1959

Praeglobotruncana (Clavihedbergella) Banner and Blow, 1959, p. 8.

Clavihedbergella Banner and Blow. Loeblich and Tappan, 1961a, p. 278.

Type species: *Hastigerinella subcretacea* Tappan, 1943.

Description: Test a low trochospire, superficially appearing planispiral in peripheral view. Shallow, typically wide umbilicus. All chambers spherical except last 2–4 which are radially elongate to occasionally clavate in the plane of coiling. Periphery rounded, with an imperforate band on the spherical chambers which becomes reduced in those chambers of the last whorl, and

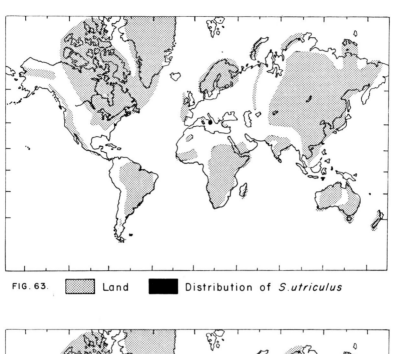

FIG. 63.   Land   Distribution of *S. utriculus*

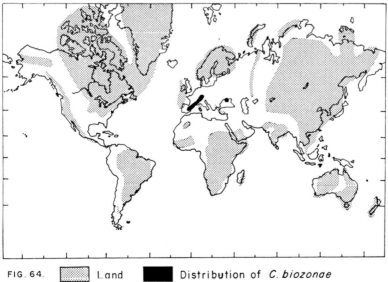

FIG. 64.   Land   Distribution of *C. biozonae*

is absent from the elongate chambers. Aperture a moderately high interio-marginal arch, extending from the umbilicus toward the periphery; bordered by an imperforate lip. Sutures depressed, radial to slightly tangential.

Remarks: *Clavihedbergella* differs from *Globigerina* by possessing radially elongate chambers in the final whorl of mature specimens and an imperforate peripheral band in immature specimens. Mature specimens of *Clavihedbergella* and *Ticinella* can be easily distinguished, but the imperforate band of the former is the only distinguishing structure on immature specimens.

Clavihedbergella is a polyphyletic group arising at different times from various species of *Globigerina* by the development of an imperforate peripheral band and gradual elongation of the chambers in the final whorl.

Range: Aptian–Coniacian. The author has recognised this genus in Maastrichtian samples. As these forms have not been thoroughly studied and speciated, the range of this genus is not extended here.

Clavihedbergella bizonae (Chevalier)

Hastigerinella bizonae Chevalier, 1961, p. 34, 36, pl. 1, figs 24a–28.
Clavihedbergella bizonae (Chevalier). Gorbachik, 1971, pl. 30, figs 4a–c; Kuhry, 1971, p. 231, 232, pl. 3, figs 2A–C.
Hedbergella (*Clavihedbergella*) *bizonae* (Chevalier). Risch, 1971, p. 49, pl. 4, figs 9, 10.

Description: Test a low trochospire, with a wide, shallow umbilicus; serrate in plane view. Chambers of the last whorl slightly more radially elongate than wide, rapidly tapering to a blunt point; earlier chambers spherical, 5–7 in the final whorl. Aperture a low umbilical–extraumbilical arch, bordered by a narrow lip. Sutures radial, depressed.

Remarks: *Clavihedbergella bizonae* is easily distinguished from all other species of this genus by its pointed chambers. Chevalier (1961) reported variation in the number and pointed shape of the chambers, which tend to reduce the "starry" aspect of the test.

The progenitor of *C. bizonae* has not been determined. However, the specimens with less pointed chambers bear a striking resemblance to *Globigerina planispira* Tappan. *Clavihedbergella bizonae* is not known to have given rise to any other species.

The radially pointed chambers never develop true tubulospines. By definition tubulospines maintain a uniform diameter over most of their length and are abruptly reduced in diameter as compared to the chamber.

Distribution: This is shown in Fig. 64.

Range: Aptian.

Clavihedbergella moremani (Cushman), Plate 18, figs 1–4

Hastigerinella moremani Cushman, 1931c, p. 86, pl. 11, figs 1a–c (*non* 2, 3); Cushman, 1946b, p. 147, 148, pl. 61, figs 1a–c (*non* 2, 3); Frizzell, 1954, p. 127, pl. 20, figs 11a, b (*non* 12); Jones, 1956, text-figs 8a, b.
Clavihedbergella moremani (Cushman). Loeblich and Tappan, 1961a, p. 279, pl. 5, figs 12–16; Pessagno, 1967, p. 285, pl. 53, fig. 5; pl. 55, figs 1, 2; Eicher, 1969, text-fig. 3 (part); Eicher and Worstell, 1970b, p. 304, 306, pl. 10, fig. 5.

Description: Test a low trochospire, imparting a pseudoplanispiral appearance in edge view; with a wide, shallow umbilicus; periphery rounded. Early chambers globose, rapidly increasing in size, with imperforate peripheral band; later chambers abruptly becoming elongate–clavate, totally perforate 4–6 chambers per whorl in 1¾–2 whorls. Umbilical–extraumbilical aperture with a narrow imperforate lip. Finely perforate wall with only small, scattered pustules. Sutures depressed, radial.

Remarks: The chambers of *Clavihedbergella moremani* reach maximum elongation for the genus, and no other species of *Clavihedbergella* has clavate chambers. The number of radially elongate clavate chambers can vary between one and six, but typically represent the last two or three chambers. The length of the first elongate–clavate chamber ranges from three to five times the length of the previous chamber. These last chambers are typically straight and cylindrical but may be curved or show a constriction immediately before the bulbous tip.

Clavihedbergella moremani originated from *C. simplex* (Morrow) during the mid-Cenomanian by the development of markedly more elongate and clavate chambers. This is the end member of the lineage.

The elongate chambers are structurally weaker than the globose chamber, particularly at the position of constriction near the tip. As a result these chambers are often incomplete because of breakage.

The holotype and paratypes of *Clavihedbergella moremani* were examined. The paratypes represented by Cushman's figs 2 and 3 are proportionally unlike the holotype. The tests are larger, and the non-clavate chambers are much thicker and shorter than those of the holotype. The paratypes are treated under *C. simplex* (Morrow).

Distribution: This is shown in Fig. 65.

Range: Mid-Cenomanian–Turonian.

Clavihedbergella simplex (Morrow), Plate 19, figs 1–3

Hastigerinella moremani Cushman, 1931, p. 86, pl. 11, figs 2 (?), 3 (*non* 1a–c); Cushman, 1946b, p. 147, 148, pl. 61, figs 2 (?), 3 (*non* 1a–c).

Hastigerinella simplex Morrow, 1934, p. 198, 199, pl. 30, figs 6a, b; *non* Loetterle, 1937, p. 46, pl. 7, figs 5a, b; Cushman, 1946b, p. 148, pl. 61, figs 10a, b; Frizzell, 1954, p. 127, 128, pl. 20, figs 13a, b.

Clavihedbergella simplex (Morrow). Loeblich and Tappan, 1961a, p. 279, 280, pl. 3, figs 11a–c (?), 12a–14c; Ayala-Castañares, 1962, p. 25, 26, pl. 4, figs 1a–c (*non* 2a–3c; pl. 5, figs 1a–c; pl. 9, figs 2a, b (?), 3a, b; pl. 10, figs 1a, b, pl. 12, figs 4a, b); Samuel, 1962, p. 181, pl. 8, figs 1a–c; *non* Wezel, 1965, p. 34, text-fig. 14a, b; *non* Eicher, 1966, p. 27, 28, pl. 6, figs 1a–3c; *non* Salaj and Samuel, 1966, p. 173, 174, pl. 10, figs 4a–c; *non* Pessagno, 1967, p. 285, 286, pl. 52, figs 1, 2; *non* Douglas, 1969, p. 163, 164, pl. 5, figs 4–5c; Eicher, 1969, text-fig. 3 (part); Eicher and Worstell, 1970b, p. 306, pl. 10, figs 4, 6a–7.

Hedbergella (*Clavihedbergella*) sp. aff. *simplex* (Morrow). Moullade, 1966, p. 96, 97, pl. 8, fig. 18 (*non* 19, 20; Moullade, 1969, p. 462, 463, pl. 1, fig. 6 (*non* 4, 5)).

Clavihedbergella subdigitata simplex (Morrow). *non* Bandy, 1967, p. 14, text-figs 6 (part).

Clavihedbergella subcretacea (Tappan). Michael, 1973, p. 212, pl. 3, fig. 9.

Description: Test small, a low trochospire with a pseudoplanispiral edge view; a wide shallow umbilicus; rounded periphery. All chambers globose except the last

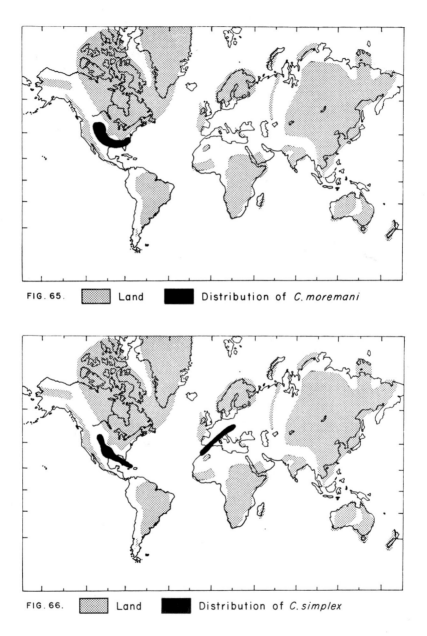

FIG. 65.   Land   Distribution of *C. moremani*

FIG. 66.   Land   Distribution of *C. simplex*

2–3 which are elongate radially in the plane of coiling; globose chambers with an imperforate peripheral band; 4–5 per whorl in 1½–2½ whorls. Relatively broad imperforate lip bordering an umbilical–extraumbilical, low aperture. Few small, scattered pustules on a finely perforate surface. Sutures depressed radial.

Remarks: *Clavihedbergella simplex* lacks the clavate chambers typical for *C. moremani* (Cushman), although an immature specimen of the latter cannot be distinguished from *C. simplex*. It differs from *C. subcretacea* (Tappan) in the angle of chamber placement. Immature specimens of Tappan's species are also indistinguishable from *C. simplex*. A maximum of three radially elongate chambers have been observed, but often only the ultimate chamber is of that type. The length of these elongate chambers is variable, although seldom exceeding twice their diameter. The last three to four chambers are usually, but not always, placed at 90° intervals.

Globigerina simplicissima (Magné and Sigal) gave rise to *Clavihedbergella simplex* through chamber elongation, and complete gradational sequences can be found between these two species.

An apertural lip is present on the holotype, but is not evident on Morrow's type figure. Also, four and one half chambers are visible ventrally.

Distribution: This is shown in Fig. 66.

Range: Cenomanian–Coniacian.

Clavihedbergella subcretacea (Tappan), Plate 19, figs 4–6

Hastigerinella subcretacea Tappan, 1943, p. 513, 514, pl. 83, figs 4a–c; Frizzell, 1954, p. 128, pl. 20, figs 14a–c.

Hastigerinella aff. *subcretacea* Tappan. Bolli, 1959, p. 271, 272, pl. 23, figs 10–13.

Globigerina roblesae Eternod Olvera, 1959, p. 149, pl. 4, figs 4a, b.

Hedbergella amabilis Loeblich and Tappan, 1961, p. 274, pl. 3, figs 8a–c, 10a, b (*non* 1a–7b, 9).

Clavihedbergella subcretacea (Tappan). Saavedra, 1965, p. 324, text-fig. 9; *non* Salaj and Samuel, 1966, p. 174, 175, tab. 33 (11); Kalantari, 1969, text-fig. 25 (18); Caron, 1971, p. 149, 150, text-figs 6a–c; Gorbachik, 1971, pl. 30, figs 3a–c; *non* Michael, 1973, p. 212, pl. 3, fig. 9.

Clavihedbergella simplex (Morrow). Wezel, 1965, p. 34, text-fig. 14a, b; Eicher, 1966, p. 27, 28, pl. 6, figs 2 (?), 3a–c (*non* 1a–c); (?) Salaj and Samuel, 1966, p. 173, 174, pl. 10, figs 4a–c; Douglas, 1969, p. 163, 164, pl. 5, figs 4–5c.

Clavihedbergella subdigitata (Carman). Eicher, 1969, text-fig. 3 (part); Eicher and Worstell, 1970b, p. 306, pl. 10, figs 3a, b, 8a, b; Caron, 1971, p. 150, text-figs 9a–c.

Clavihedbergella simplicissima Magné and Sigal. Gawor-Biedowa, 1972, p. 72, 73, pl. 7, figs 3a–c.

Description: Test of moderate to large size, low trochospire; a wide shallow umbilicus; rounded periphery. Early chambers globose, gradually increasing in size, with an imperforate peripheral band; last 2–3 chambers gradually elongating radially changing to tangentially; 6–6½ chambers per whorl in 2–2½ whorls. Aperture an umbilical–extraumbilical, high arch, with an imperforate lip. Wall finely perforate, with scattered pustules. Sutures depressed, radial tending to become tangential.

Remarks: This species is typically larger in overall size and number of chambers per whorl than other species of *Clavihedbergella*. Moreover, the number of digitate chambers is usually more for *C. subcretacea* than any other in the genus. The number of elongate chambers varies between one and three in the adult, but is typically two. The last one or two chambers are most often placed such that their long axis is nearly tangential to the previous whorl. Some specimens may retain the radially arranged chambers more typical of the genus. Chamber length may range up to three times their diameter.

The globose stage of *Clavihedbergella subcretacea* closely resembles *Globigerina planispira* Tappan, from which it probably evolved.

The holotype of *Clavihedbergella subcretacea* exhibits relict apertures in the umbilicus which were not illustrated in the type figure.

The specimens illustrated by Douglas (1969) as *Clavihedbergella simplex* (Morrow) are referred to *C. subcretacea* because of the greater number of elongate chambers per whorl. His fig. 5 shows evidence of two additional elongate chambers which have been broken off.

Distribution: This is shown in Fig. 67.

Range: Albian–Turonian.

Genus GLOBIGERINA d'Orbigny, 1826

Globigerina d'Orbigny, 1826, p. 277.

Hedbergella Brönnimann and Brown, 1958, p. 16.

Planogyrina Zakharova-Atabekyan, 1961, p. 50.

(?) *Hedbergella* (*Asterohedbergella*) Hamaoui, 1964, p. 133.

Loeblichella Pessagno, 1967, p. 288.

Globigerina (*Globuligerina*) Bignot and Guyader, 1971, p. 80.

Favusella Michael, 1973, p. 212.

Type species: *Globigerina bulloides* d'Orbigny, 1826.

Emended description: Test a low to high trochospiral coil; equatorial periphery lobate, axial periphery rounded, keeless. Chambers spherical to oval. Umbilicus narrow to wide, open, often exhibiting relict apertures. Aperture low to high interiomarginal, umbilical to extraumbilical—umbilical arch bordered by an imperforate lip or flap. Sutures depressed, radial to slightly curved, distinct. Wall with small to large-diameter pores of varying density; may be ornamented with pustules, polygonal ridges, and short to long spines among other surface features.

Remarks: Immature specimens of *Ticinella* may be incorrectly placed with the *Globigerina* because their distinctive accessory apertures develop only in the later ontogenetic stages. Likewise, the absence of the characteristic tegillum on immature specimens of *Rugoglobigerina* causes difficulty in recognition. Other than these two cases, *Globigerina* is easily distinguished by its open umbilicus, lobate equatorial periphery and lack of a keel.

Brönnimann and Brown (1958) erected the new genus *Hedbergella* which has as its diagnostic feature "the extension of the last few chambers into the umbilicus," and as its type species—*Anomalina lorneiana* var. *trocoidea* Gandolfi. The Breggia River material of Gandolfi was examined with the SEM (Masters, in press) and found to be preserved largely as internal molds.

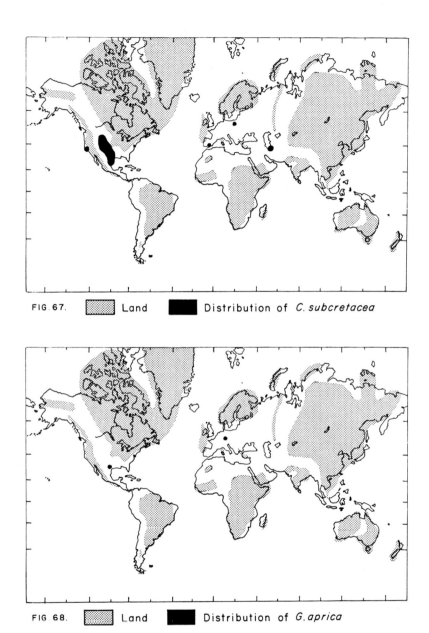

FIG. 67. ▨ Land ◼ Distribution of *C. subcretacea*

FIG 68. ▨ Land ◼ Distribution of *G. aprica*

The "chamber extension" may, then, be an artifact of preservation. On the other hand this feature may simply be the result of the close packing and low convexity of the chambers of this species. At most this can be considered only a species-level characteristic.

Loeblich and Tappan (1964) make no mention of this "chamber extension" in *Hedbergella* as cited by Brönnimann and Brown. Instead, they distinguish this genus from *Globigerina* by the presence of an "interiomarginal, extraumbilical-umbilical" aperture. At the same time, they describe *Globigerina* as possessing an "interiomarginal, umbilical" aperture "with tendency in some species to extend to slightly extraumbilical position . . .".

Many living species of *Globigerina* characteristically have an extraumbilical–umbilical aperture—*G. quinqueloba* Natland, *G. humilis* (Brady) and *G. eggeri* Rhumbler for example. Moreover, an extraumbilical–umbilical aperture may be found in the normal range of intraspecific variation in several species, such as *G. pachyderma* (Ehrenberg). On the other hand, the aperture in some species of *Hedbergella* is found to be umbilical—e.g. *H. hoterivica* (Subbotina). Thus, when critically examined, the criteria established for the distinction between *Globigerina* and *Hedbergella* fails.

There does appear to be a difference in wall structure between Cretaceous and Cainozoic Globigerinacea. Pessagno (1967, 1969b) and Pessagno and Miyano (1968) reported that all Cretaceous species possess a microgranular hyaline wall structure. Whereas, the Cainozoic species may possess a radial hyaline structure. This alone would justify separation of the Mesozoic and the Cainozoic globigerinids. However, Pessagno (1969a) added that the epipelagic individuals of the Cainozoic possess a microgranular wall, while mesopelagic forms either have a microgranular wall covered by a radial crust or an entirely radial wall structure.

The formation of a radial hyaline wall is, it would appear, at least partially environmentally controlled. If so, one must conclude that only since the Mesozoic have planktonic foraminifers occupied the mesopelagic habitat, or that the necessary controls were not operable at that time whatever the reason. This modification of the wall structure is potentially a non-genetic adaptation and cannot be regarded as a valid criterion for the taxonomic separation of *Globigerina* and *Hedbergella*.

The retention of *Hedbergella* seems an unnecessary attempt to restrict the long ranging and structurally primitive *Globigerina* to the post-Cretaceous. *Hedbergella* is a junior synonym of *Globigerina*.

In 1961 Zakharova-Atabekyan proposed the new genus *Planogyrina* with *Globigerina gaultina* Morozova, 1948, as the type species. In this study, *Hedbergella* (= *Globigerina*) was contrasted with *Planogyrina*, which was said to differ by possession of a flat trochospire, distinct lamellar appendages

[= apertural flaps] and a smoother and finer test wall. The height of the spire is an intraspecific or at best only a species-level characteristic. All species of *Globigerina* possess apertural flaps or lips which vary in their degree of development. The smoothness of the surface can be environmentally controlled. None of these features warrant the separation of *G. gaultina* (= *G. planispira* Tappan) into a new species.

Hamaoui (1964) established the new subgenus *Hedbergella* (*Asterohedbergella*), with *H.* (*Asterohedbergella*) *asterospinosa* Hamaoui as the type species. This separation was based on the presence of peripheral tubulospines, and on the attribution of supraspecific value to chamber shape by most authors. This is certainly true in some cases, e.g. *Globigerinelloides* and *Hastigerinoides* or *Globigerina* and *Clavihedbergella*. However, it has depended upon which group is under consideration. Similar structures in different groups cannot be regarded as equivalent in taxonomic importance, and many factors must be considered. Hamaoui observed, *Globotruncana calcarata* Cushman, even though it possesses tubulospines, has never been subgenerically separated from *Globotruncana*. This was subsequently done by El-Naggar (1971a), although not accepted here. Because the value of chamber shape does vary with the taxon, and because *H. asterospinosa* is not known to give rise to a separate lineage, the subgenus *Asterohedbergella* appears unnecessary. Moreover, Hamaoui's type figures are not sufficient to verify the planktonic nature of the species. So it is only tentatively treated as a synonym of *Globigerina*.

Loeblichella Pessagno, 1967, was separated from *Hedbergella* (= *Globigerina*) because it possesses dorsal, sutural supplementary apertures and an extraumbilical to spiroumbilical primary aperture. An examination of the holotype and paratypes of *Praeglobotruncana hessi hessi* Pessagno, the type species of *Loeblichella*, disclosed no supplementary apertures of any type. The development of a spiroumbilical aperture is recorded for the gerontic stage of most species of *Globigerina*. The exact subgeneric status of Pessagno's species, however, is unclear because the primary types are probably internal molds and are poorly preserved.

Bignot and Guyader (1971) proposed the new subgenus *Globuligerina*, with *Globigerina oxfordiana* Grigelis as the type species, basing this determination upon the presence of a not quite umbilical "virguline" aperture and accompanying lip. They also suggest the inclusion of *Globigerina balakhmatovae* Morozova, *G. helvetojurassica* Haeusler, *G. hoterivica* Subbotina and *G. spuriensis* Bars and Ohm. As noted in the generic description above, the aperture is found to range from umbilical to extraumbilical–umbilical in position. The use of the term "virguline" aperture is misleading. This type of aperture is loop-shaped with a tooth plate. No such structure has been identified in *G. oxfordiana* or any of the other early *Globigerina*. Indeed the aperture does appear to be asymmetrical, but the degree of asymmetry is variable. There is a variable amount of asymmetry in the apertures of all Mesozoic planktonics, and this is thought to be an intraspecific trait. Thus, the subgenus *Globuligerina* is judged to be a junior synonym of *Globigerina*.

Michael (1973) proposed the genus *Favusella* with *Globigerina washitensis* Carsey as the type species. The whole concept of this new genus is based upon "its striking pattern of ornamentation . . .". No other modern foraminiferal taxonomist has accepted ornamentation as a supraspecific characteristic. Michael's distinction of *Favusella* from *Globigerina* (presumably Recent representatives) on the basis of the number of pores (numerous vs. one respectively) within the cancellate areas is groundless as two different structures are being compared. High magnification electron-micrographs show that the cancellate structures of *G. washitensis* are surface features (ornamentation) which retain a uniform thickness as new material is added. Each of these polygons encloses a variable area of the chamber surface and, consequently, a variable number of small diameter pores. In contrast, the cancellate structure in Recent *Globigerina* is the chamber surface, which gradually becomes reduced in area as new laminae are deposited. This results in isolated funnel-shaped, wider diameter pores. Separation into two genera is not justified.

Globigerina lobata Terquem, 1883, *G. oolithica* Terquem, 1883, and *G. liasina* Terquem and Berthelin, 1875, are too inadequately described and poorly illustrated for recognition. Each should be considered as a nomen oblitum.

The holotypes of *Hedbergella murphyi* and *H. quadrata* of Marianos and Zingula (1966) are too poorly preserved to be able to adequately determine their taxonomic status. The specimen they refer to as *Globigerina* sp. A is a typical Palaeogene species of perhaps *G. triloculinoides* Plummer or *G. inaequispira* Subbotina affinities.

Range: Bathonian–Recent.

Globigerina aprica (Loeblich and Tappan), Plate 20, figs 1–3
Ticinella aprica Loeblich and Tappan, 1961, p. 292, pl. 4, figs 14–16; Hermes, 1969, p. 42, pl. 4, figs 74–76.
Whiteinella aprica (Loeblich and Tappan). Eicher, 1969, text-fig. 3 (part); (?) Eicher and Worstell, 1970b, p. 314, 316, pl. 11, figs 7a–c, pl. 12, figs 1a–c.
Description: "Test free, low trochospiral coil of two and one-half whorls, broadly umbilicate, periphery rounded, peripheral outline lobulate, five to six globular chambers in the final whorl, rarely seven, chambers increasing gradually in size as added; sutures distinct, strongly constricted, radial, straight; wall calcareous, finely perforate, surface coarsely hispid, but spines less prominent on final chambers; . . . aperture an interiomarginal, umbilical–extraumbilical arch, oriented somewhat

toward the plane of coiling . . .", bordered by a wide apertural flap. (Modified type description.)

Remarks: *Globigerina aprica* can be distinguished from *G. delrioensis* Carsey by its more gradually enlarging chambers. The size of the chambers of the last whorl increases only slightly. It also averages one chamber more in the final whorl. *Globigerina aprica* differs from the smaller *G. planispira* by averaging fewer chambers in the final whorl. Insufficient studies have been made of *Globigerina aprica* to determine the extent of intraspecific variation. The size of the apertural flaps is known to vary, though this may be a secondary phenomenon due to preservation. *Globigerina aprica* evolved from *G. delrioensis* in the late Cenomanian.

Loeblich and Tappan (1961) misinterpreted the apertural flaps of *Globigerina aprica*. The flaps extend from each chamber into the umbilicus, and overlap but are not cemented to the previous flaps. As a result openings exist beneath and around the edges of the flaps. Loeblich and Tappan believed these to be accessory apertures and placed their species into *Ticinella*. These irregular openings are unlike the regular accessory ticinellid apertures in that they do not open directly into the chamber but open merely beneath the flap.

It is suspected that *Globigerina aprica* is a junior synonym of *G. loetterli* Nauss, but because topotypic material was not available during this study, this has not been demonstrated.

One of the specimens attributed to *Whiteinella aprica* (Loeblich and Tappan) by Eicher and Worstell (1970b, pl. 12, fig. 1) possesses a tegillum. The authors note the presence of umbilical cover plates on these specimens which approach the configuration of true tegilla, but do not differentiate between these terms. This is confusing because the terms have heretofore been treated as synonyms. One reason Eicher and Worstell did not place these specimens in *Rugoglobigerina* was that they were not as highly ornamented as other *Rugoglobigerina*. Masters (in press) has demonstrated that the stratigraphically early *Rugoglobigerina* have only traces of costellae, which are often incorrectly assumed to be the distinguishing feature of this genus. The specimens illustrated by Eicher and Worstell are treated tentatively as *Globigerina*, but if a true tegillum is found on these specimens, they must be referred to *Rugoglobigerina*.

Distribution: This is shown in Fig. 68.

Range: Mid-Cenomanian–lower Turonian.

Globigerina (?) *asterospinosa* Hamaoui

Hedbergella (*Asterohedbergella*) *asterospina* Hamaoui, 1964, p. 135–142, pl. 1, figs 1–22; pl. 2, figs 1–7; Arkin and Hamaoui, 1967, pl. 1, figs 8–10; Saint-Marc, 1970, p. 92, pl. 2, figs 1–6.

Description: "Test free, minute, concavo-convex to unequally biconvex, trochospirally coiled, about 10 to 15 (12 on the average) chambers visible in $1\frac{1}{2}$ to $2\frac{1}{2}$ coils on the evolute spiral side and 4 to 6 chambers visible on the last coil of the involute umbilical side. The direction of coiling is random. Chambers rather rapidly increasing in size as added, rarely subspherical, mostly hemispherical to angular conical in axial profile, generally more or less radially elongate. At least each adult chamber drawn out at the periphery into a single conical or flask-shaped, tapering extension which may produce a true tubulospine. The chamber breadth is generally greater than its height, but nearly equals the chamber thickness. Peripheral outline lobulate, petaloid to star-shaped; axial periphery subangular to acute. The usually flattened spiral sides of the chambers are often situated in slightly different planes, producing an imbricate appearance. The pointed extensions of the chambers are usually tubulo-spinate, lying generally backward from the radial mid-line of the chamber and rarely its prolongation. The tubulospines may be oriented in the plane of coiling, but are mostly at an angle to it, being even in alignment with the lateral walls. The axial profile is highly variable: the chambers of the earlier whorls may or may not be elevated over the later whorls on the spiral side, while the umbilical walls of the chambers may be either broadly rounded or angled. Thus, a pseudoumbilicus may or may not be present. A deep, usually narrow, true umbilicus is present. The sutures are slightly to strongly depressed and gently curved on both sides. Wall structure calcitic, radiate, finely perforate, lamellar, very hispid on the surface. Bilamellid structure assumed, but actually not observed. Aperture a low arch, interiomarginal, extraumbilical–umbilical, bordered by a narrow lip." (Type description.)

Remarks: *Globigerina asterospinosa* differs from other *Globigerina* by the possession of tubulospines. Hamaoui reported variations in the degree of elevation of the early coils relative to later ones, in the chamber shape, and the size, position and orientation of the tubulospines.

The phylogeny of this species is unknown.

Several aspects of this species are disturbing. Hamaoui reported the wall structure as being radial, a type unknown in other Cretaceous planktonics, and a bilamellid structure has been assumed though not observed. All recorded planktonics but one in the same stratigraphic interval as *Globigerina asterospinosa* are predominantly dextrally coiled; *G. asterospinosa* is randomly coiled, and bears a strong resemblance to some benthic forms, e.g. *Pararotalia*. The taxon has not been firmly established as a planktonic, and can be only questionably assigned to the genus *Globigerina*.

Distribution: This is shown in Fig. 69.

Range: Middle–Upper Cenomanian.

Globigerina bornholmensis (Douglas and Rankin)

Hedbergella bornholmensis Douglas and Rankin, 1969, p. 193, text-figs 6A–I; (?) Hanzlíková, 1972, p. 100, pl. 25, figs 5a–7b.

Whiteinella baltica Douglas and Rankin, 1969, p. 197, text-figs 9A–I; (?) Hanzlíková, 1972, p. 100, pl. 25, figs 8a–10b.

Description: "Test free, low trochospiral, equatorial periphery lobulate, axial periphery rounded. Chambers subglobular to subspherical, three and one half to four chambers in the final whorl; initial chambers increasing gradually in size, final three or four increasing very rapidly, last formed chamber much larger than penultimate, giving test a trilobate appearance. Spiral and umbilical sutures radial to slightly curved, depressed, distinct. Wall calcareous, perforate, surface covered with small spines. Umbilicus deep and narrow in juvenile stage, becoming shallower in adult. Primary aperture is a high interiomarginal–umbilical arch with a narrow bordering lip." (Type description.)

Remarks: *Globigerina bornholmensis* is separated from *G. delrioensis* Carsey by the more rapid increase in the size of and number of chambers in the last whorl. Apart from the variation in the number of chambers in the last whorl (3.5–4), this species has not been sufficiently reported and studied to determine the total extent of variation.

Although not confirmed, *Globigerina bornholmensis* is believed to have developed from the *G. delrioensis* Carsey stock.

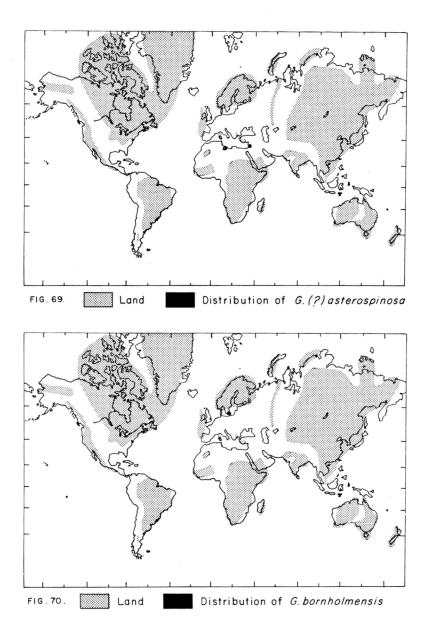

FIG. 69.　☐ Land　■ Distribution of *G. (?) asterospinosa*

FIG. 70.　☐ Land　■ Distribution of *G. bornholmensis*

In the discussion of their new species, *Hedbergella bornholmensis* Douglas and Rankin (1969) observed its similarity to and contrasted it with another of their new species, *Whiteinella baltica*. They remarked that it differs from the latter by, "(1) the trilobate shape of the test, (2) the rapid rate of whorl expansion, and (3) the more extra-umbilical position of the primary aperture". None of these differences appear uniformly in the types of either species, and some are shared by both. The two species appear completely gradational and representative of a single natural taxon.

This species has yet to be recognised outside of the type area. More detailed studies of this taxon may reveal that it belongs to the natural population of *Globigerina delrioensis* Carsey.

Distribution: This is shown in Fig. 70.

Range: Upper Turonian–lower Santonian.

Globigerina delrioensis Carsey, Plate 20, figs 4, 5

Globigerina sp. Moreman, 1925, pl. 18, fig. 9.

Globigerina cretacea var. *del rioensis* Carsey, 1926, p. 43, 44.

Globigerina Hecht, 1938, pl. 23, figs 64–71.

Globigerina infracretacea Glaessner, 1937, p. 28, text-fig. 1; *non* Colom, 1947, pl. 21, figs 1–10 [incorrectly cited as *G. subcretacea* in plates]; *non* Myatlyuk, 1949, p. 216, pl. 5, figs 5a–c; Noth, 1951, p. 73, 74, pl. 7, figs 5a, b; Subbotina, 1953, p. 58, 59, pl. 1, figs 5a–10c; Hofker, 1956a, p. 218–220, pl. 33, figs 4–8 (*non* 1–3); Said and Barakat, 1957a, p. 45, pl. 1, figs 27a–c; Said and Barakat, 1957b, p. 78, 79, pl. 1, figs 29a, b; Parra, 1959, p. 149, pl. 3, figs 6a, b; (?) Cita and Pasquare, 1959, p. 413, 414, text-figs 4 (19, 25–30, 31–36); Maslakova, 1959, p. 105, pl. 10, figs 3a–c; Tollman, 1960, p. 191, pl. 20, fig. 4; *non* Moullade, 1960, p. 136, pl. 2, figs 18–20; Romanova, 1960, in Glazunova *et al.*, p. 117, pl. 21, figs 6a, b, v; *non* Visser and Hermes, 1962, encl. 17, figs 45a–c; Kaptarenko-Chernousova, *et al.* 1963, p. 103, Tab. 13, figs 7a, b; *non* Alekseeva, 1963, p. 44, pl. 8, figs 7a–c; Ugauzzoni and Radrizzani, 1967, p. 1226, pl. 92, figs 7a–8b; (?) Costea and Comsa, 1969, p. 107, 108, pl. 1, figs 3, 4, 7, 11 (?); Hermes, 1969, p. 45, pl. 5, figs 101, 102.

Globigerina cretacea d'Orbigny. Tappan, 1940, p. 121, 122, pl. 19, figs 11a–c; Tappan, 1943, p. 512, pl. 82, figs 16a–c (*non* 17); Crespin, 1953, p. 35, pl. 6, figs 15a–c; Subbotina, 1953, p. 61, 62, pl. 1, figs 13a–15c; Applin, 1955, p. 196, 197, pl. 48, figs 23, 24; Bolin, 1956, p. 292, 293, pl. 39, figs 5, 17 (*non* 4a, b, 6a–8b; 13a, b, 17); Belford, 1960, p. 90, 91, pl. 25, figs 1–3; Rasheed, 1963, p. 237, pl. 4, figs 1–4.

Globigerina portsdownensis Williams-Mitchell, 1948, p. 96, 97, pl. 8, figs 4a–c; *non* Maslakova, 1959, p. 105, 106, pl. 10, figs 6a–c; Bandy, 1967, p. 8, text-fig. 3 (6); Hermes, 1969, p. 49, pl. 7, figs 166–168.

Globigerina gautierensis Brönnimann, 1952a, p. 11–14, pl. 1, figs 1–3, text-figs 2a–c, g–m (*non* d–f); Hermes, 1969, p. 49, 50, pl. 5, figs 121–126.

Rotundina ordinaria Subbotina, 1953, p. 181–183, pl. 3, figs 9a–c (*non* 3a–8c); pl. 4, figs 9a–c (*non* 1a–8c).

Globigerina delrioensis Carsey. Frizzell, 1954, p. 127, pl. 20, figs 1a c; Hermes, 1969, p. 47, pl. 5, figs 118–120.

Praeglobotruncana infracretacea (Glaessner). Bolli, 1959, p. 266, pl. 21, fig. 10 (*non* 9a–c).

Praeglobotruncana crassa Bolli, 1959, p. 265, pl. 21, figs 1a–c, 2; Hermes, 1969, p. 50, 51, pl. 6, figs 153–155.

Praeglobotruncana cf. *gautierensis* (Brönnimann). Bolli, 1959, p. 266, pl. 21, figs 7a–8; Hofker, 1962, text-figs 1a–c; Hermes, 1969, p. 49, 50, pl. 6, figs 143–146.

Praeglobotruncana gautierensis (Brönnimann). Jones, 1960, p. 102, pl. 15, figs 1a–9c; Hermes, 1969, p. 49, 50, pl. 5, figs 127–132.

Globigerina infracretacea subsp. *trochoidea* Moullade, 1960, p. 136, pl. 2, figs 21, 23–25.

Hedbergella delrioensis (Carsey). Loeblich and Tappan, 1961, p. 275, pl. 2, figs 11a–13c; *non* Takayanagi, and Iwamoto 1962, p. 190, 191, pl. 28, figs 10a–12c; Todd and Low, 1964, p. 402, pl. 1, figs 2a–c; Eicher, 1965, p. 904, 905, pl. 106, figs 2a, b (6a, b?); *non* Takayanagi, 1965, p. 204, 205, pl. 21, figs 4a–c; Butt, 1966, p. 173, 174, pl. 2, figs 2a–4c, (?) 6a–7b (*non* 1a–c, 5a–c, 8a–c); Eicher, 1966, p. 27, pl. 5, figs 12a–c (*non* 13a–c); Bartenstein, Bettenstaedt and Bolli, 1966, p. 164, 165, pl. 4, figs 362–365, 368–370 (*non* 360, 361, 366, 367); Salaj and Samuel, 1966, p. 167, pl. 8, figs 5a–c; Arkin and Hamaoui, 1967, pl. 1, figs 22–24; Pessagno, 1967, p. 282, 283, pl. 48, figs 5; Bandy, 1967, p. 10, text-fig. 4 (1); Eicher, 1967, p. 186, pl. 19, figs 6a–c; Wall, 1967, p. 105–107, pl. 3, figs 1–12; pl. 13, figs 13–21; Barr, 1968, pl. 2, figs 5a–c; Bate and Bayliss, 1969, pl. 3, figs 7a–c; Eicher, 1969, text-fig. 3 (part); Hermes, 1969, p. 47, pl. 6, figs 133–142; Kalantari, 1969, p, 199, pl. 22, figs 17a–c; pl. 23, figs 1a–c; Neagu, 1969, p. 139, 140, pl. 13, fig. 14; pl. 14, figs 7–9, 13–15; Eicher and Worstell, 1970b, p. 302, pl. 9, figs 10–11b; Ncagu, 1970a, fig. 1 (1); Neagu, 1970b, p. 64, pl. 20, figs 1–6, 16–18; pl. 21, figs 7–9, 16–17, (*non* pl. 22, figs 4–6); Risch, 1971, p. 48, pl. 4, figs 14–16; El-Naggar, 1971a, pl. 6, figs a–c; Michael, 1973, p. 210, pl. 2, figs 1–3.

Hedbergella amabilis Loeblich and Tappan, 1961, p. 274, pl. 3, fig. 2 (*non* 1, 3–10); Barr, 1968, pl. 2, figs 6a–7c; Douglas, 1969, p. 165, pl. 4, figs 8a–c; Weaver *et al.*, 1969, pl. 3, figs 14a–c; Neagu, 1970b, p. 64, pl. 21, figs 7–9, 13–15.

Hedbergella cf. *amabilis* Loeblich and Tappan. (?) Douglas and Rankin, 1969, p. 196, 197, text-figs 8A–C.

Hedbergella portsdownensis (Williams-Mitchell). Loeblich and Tappan, 1961, p. 277, pl. 5, figs 3a–c; Luterbacher, 1963, in Renz *et al.*, p. 1084, pl. 9, figs 5a–c (cited as *H. delrioensis* on plate); *non* Marianos and Zingula, 1966, p. 337, pl. 38, figs 8a–c; *non* Douglas and Rankin, 1969, p. 194–196, text-figs 7A–F; Gorbachik, 1971, pl. 28, figs 7a–c.

Praeglobotruncana stephani var. 3 Malapris and Rat, 1961, p. 89, 90, pl. 2, figs 3a–c.

Praeglobotruncana ? *gigantea* Lehmann, 1963, p. 140, pl. 2, figs 4a–5c; text-figs 2g (?), 3e, (?) f.

Globigerina ("*Praeglobotruncana*") cf. *gautierensis* Brönnimann. Hofker, 1963, p. 281, pl. 1, figs 1a–d.

Hedbergella trocoida (Gandolfi). Taylor, 1964, p. 593, 594, pl. 84, figs 4–7.

Hedbergella infracretacea (Glaessner). Saavedra, 1965, p. 323, 324, text-fig. 7a; Glaessner, 1966, p. 179–181, pl. 1, figs 1a–3c; Salaj and Samuel, 1966, p. 169, pl. 8, figs 8a–c; Fuchs and Stradner, 1967, p. 331, 332, pl. 17, figs 13a–c; Hermes, 1969, p. 45, pl. 5, figs 103–105; Gorbachik, 1971, pl. 28, figs 5a–c; *non* Kuhry, 1971, p. 230, pl. 2, figs 3A–C; Risch, 1971, p. 46, 47, pl. 4, figs 1–3; *non* Gawor-Biedowa, 1972, p. 69, 70, pl. 6, figs 8a–c.

Hedbergella aptiana Bartenstein, 1965, p. 347, 348, text-figs 3–6; Hermes, 1969, p. 51, pl. 8, figs 192, 193.

Hedbergella beegumensis Marianos and Zingula, 1966, p. 335, 336, pl. 37, figs 7a–c.

Hedbergella loetterlei (Nauss). Kent, 1967, p. 1448, pl. 183, figs 15a–c (*non* 14a–c).
Hedbergella planispira (Tappan). Pessagno, 1967, p. 283, 284, pl. 51, fig. 1; pl. 53, figs 1–4.
Praeglobotruncana lehmanni Porthault, 1969 (new name for *P.* ? *gigantea* Lehmann), p. 538, 539, pl. 2, figs 6a–c.
Archaeoglobigerina bosquensis Pessagno. Douglas and Rankin, 1969, p. 199, 200, text-figs 10A–C, 11A–C.
Hedbergella sp. cf. *H. planispira* (Tappan). Hermes, 1969, p. 45, 46, pl. 1, figs 25–27.
Hedbergella sp. aff. *H. infracretacea* Glaessner. Hermes, 1969, p. 46, 47, pl. 1, figs 19–24; Risch, 1971, p. 47, pl. 4, figs 11–13.
Hedbergella (*Hedbergella*) *sigali* Moullade. Moullade, 1969, p. 462, pl. 1, figs 2, 3.
Hedbergella aprica (Loeblich and Tappan). Neagu, 1969, p. 140, pl. 15, figs 3–5, 8–14.
Hedbergella (*Hedbergella*) *delrioensis* (Carsey). Moullade, 1966, p. 94, 95, pl. 8, fig. 17; Moullade, 1969, p. 462, pl. 1, fig. 1.
Archaeoglobigerina cf. *blowi* Pessagno. North and Caldwell, 1970, p. 50, 51, pl. 4, figs 5a–c.
Whiteinella archaeocretacea Pessagno. Porthault, 1970, in Donze *et al.*, p. 66, pl. 9, figs 10–12.
Hedbergella sp. Belford and Scheibnerová, 1971, pl. 1, figs 15–17.
Hedbergella aptica (Agalarova). Gorbachik, 1971, pl. 28, figs 1a–c.
Clavihedbergella globulifera Krechmar and Gorbachik, 1971, in Gorbachik, p. 136, pl. 30, figs 1a–c.
Clavihedbergella tuschepsensis (Antonova). Gorbachik, 1971, pl. 30, figs 9a–c.
Hedbergella sp. cf. *H. delrioensis* (Carsey). Kuhry, 1971, p. 229, pl. 1, figs 7A–8.
Hedbergella sigali Moullade. Kuhry, 1971, p. 231, pl. 2, figs 6A–C, 8 (?).
Hedbergella planispira (Tappan). El-Naggar, 1971a, pl. 6, fig. e (*non* d, f); Hanzlíková, 1972, p. 101, pl. 25, fig. 15; pl. 26, figs 1a–2.
Hedbergella monmouthensis (Olsson). Hanzlíková, 1972, p. 101, pl. 26, figs 3a–4c, 6, a, b.
Hedbergella. Berger and Rad, 1972, pl. 6, fig. 1.
Clavihedbergella simplicissima Magné and Sigal. Neagu, 1972, p. 215, pl. 6, figs 38, 39.

Description: Test trochospiral with five to rarely six subspherical chambers per whorl, periphery rounded. Chambers rapidly inflated in a very low spire, subspherical. Aperture extraumbilical–umbilical with minor variation. Apertural lip imperforate, extending from near the umbilicus to near the periphery. Apertural face imperforate. Remainder of test perforated by numerous, small, randomly located pores. Surface pustulose, most strongly developed on earlier chambers of last whorl. Pustule growth ultimately covers most pores. Sutures radial, depressed.

Remarks: *Globigerina delrioensis* is recognised by having more rapidly inflated, larger and fewer chambers in the final whorl than has *G. planispira* Tappan, *G. loetterlei* Nauss or *G. aprica* (Loeblich and Tappan). The chamber shape of *G. delrioensis* differs from *G. trocoidea* (Gandolfi), and it is lower spired than *G. paradubia* Sigal. *Globigerina delrioensis* is a moderately variable species. Topotype populations exhibit individuals with six chambers in the final whorl, although five is the usual number. Chamber inflation rate is not always constant. The first chamber of the last whorl may be abruptly larger than the preceding chamber, but followed by subequal to slowly enlarging chambers. The ultimate chamber of gerontic individuals frequently is not in the normal spire of coiling, but is offset toward the umbilicus. This phenomenon has resulted in many workers misinterpreting these forms as being high spired. The shape and position of the aperture is altered by the presence of these atypical final chambers.

G. hoterivica Subbotina is most likely the direct ancestor of *Globigerina delrioensis*. Presumably three changes occurred, (1) an increase in test size, (2) development of a fifth chamber in the final whorl, and (3) migration of the umbilical aperture to an extraumbilical–umbilical position. Intermediate forms have not yet been recognised.

Globigerina portsdownensis Williams-Mitchell (1948) is a junior synonym of *G. delrioensis*. Although some workers have distinguished *G. portsdownensis* as being higher spired, only its ultimate chamber imparts this appearance because it occupies a more umbilical position. This type of umbilical "displacement" is common in gerontic specimens of *G. delrioensis* as has been previously noted.

The primary types of *Globigerina gautierensis* Brönnimann, 1952a, were examined. They are poorly preserved and lack original shell material. There is a tendency in the paratypes for the aperture of the ultimate chamber to extend to the periphery, bestowing a pseudoplanispiral aspect. This trait has been recognised in many other species and genera, and is judged to be a gerontic variant. Brönnimann's species is conspecific to *G. delrioensis*.

Bolli (1959), using apertural position rather than presence or absence of a keel as a primary characteristic, proposed the new species *Praeglobotruncana crassa*. Examination of the holotype of this keeless form has shown that it is identical to *G. delrioensis*.

Globigerina infracretacea subsp. *trochoidea* Moullade, 1960, seems to be another representative of the "displaced" ultimate chamber phenomenon typical of *G. delrioensis*, and is treated as a junior synonym of the latter.

Neither the type figures nor the thin-sections of *Praeglobotruncana ? gigantea* Lehmann, 1963, show evidence of a keel. Specimens examined from one of Lehmann's localities are indistinguishable from *Globigerina delrioensis*.

Hedbergella aptiana Bartenstein, 1965, fits easily into the morphologic concept of *Globigerina delrioensis*; its geological range corresponds with the lower portion of the range of *G. delrioensis*.

Bandy (1967) regarded *Hedbergella brittonensis* Loeblich and Tappan (= *Globigerina paradubia* Sigal) as a junior synonym of *Globigerina portsdownensis* Williams-Mitchell. *Globigerina paradubia* has more chambers per whorl which form a true moderately high spire. The confusion resulted again because of the misinterpretation of the growth pattern caused by the ultimate chamber as noted above.

Masters (1976) has selected a neotype for *Globigerina delrioensis*.

Distribution: This is shown in Fig. 71.

Range: Barremian–Campanian (LOS *Globotruncana concavata*–FOS *Pseudotextularia carseyae* Interval).

Globigerina (?) *graysonensis* Tappan, Plate 21, fig. 1

Globigerina graysonensis Tappan, 1940, p. 122, pl. 19, figs 15a–17; Tappan, 1943, p. 513, pl. 82, figs 15a–c; Loeblich and Tappan, 1950, p. 14, pl. 2, figs 22a–c; Bolli, 1959, p. 270, pl. 23, figs 1a–2b; Bandy, 1967, p. 8, text-fig. 3 (2); *non* Gawor-Biedowa, 1972, p. 87–89, pl. 5, figs 12A–C.

Globigerina portsdownensis Williams-Mitchell. Tzankov *et al.*, 1964, pl. 1, fig. 7.

Hedbergella graysonensis (Tappan). *non* Eicher, 1967, p. 186, pl. 19, figs 8a–9c; Kuhry, 1971, p. 230, pl. 2, figs 2A–C.

Gubkinella graysonensis (Tappan). Michael, 1973, p. 207, 208, pl. 1, figs 1–3.

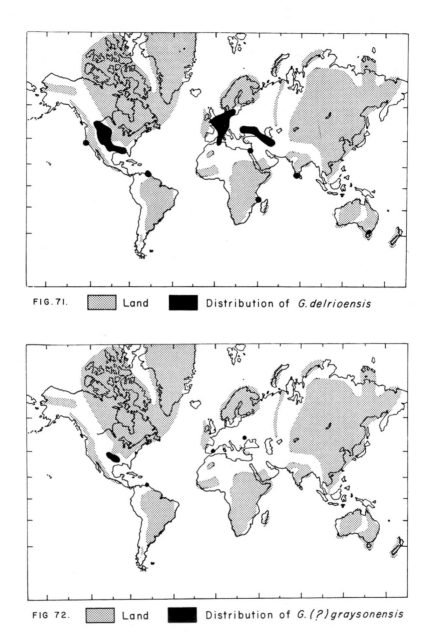

FIG. 71. ▨ Land ■ Distribution of *G. delrioensis*

FIG 72. ▨ Land ■ Distribution of *G. (?) graysonensis*

Description: Minute, high, trochospirally coiled, compact, subspherical to pyriform test. Chambers of the final whorl are $\frac{1}{4}$–$\frac{1}{3}$ spherical segments, such that their widest dimension is at the middle of the chamber and their greatest length parallel to the axis of coiling; closely overlapping the previous whorl; enclosed volume small; $3\frac{1}{2}$ chambers in final whorl comprise 90 % of the test; early chambers small, indistinct. Aperture a narrow, elongate, interiomarginal, extraumbilical–umbilical slit bordered by a narrow lip perpendicular to a vertical apertural face. Umbilicus narrow, shallow, indistinct. Sutures radial, slightly depressed; spiral sutures less depressed. Wall with minute pores, smooth surface.

Remarks: The minute size, compactness of the test, chamber shape and narrow, elongate, interiomarginal, extraumbilical–umbilical aperture make this species easily recognisable. The characteristics of Globigerina (?) graysonensis are rather constant, and only a slight change in spire height has been observed.

Subsequent to its description in 1940, Globigerina (?) graysonensis has been recorded in the literature nine times, and always as a planktonic. A number of characteristics of this taxon lead one to doubt this opinion. The chamber shape and compactness of the test is at least atypical for Mesozoic globigerines. The slit-like aperture is unlike that of any other planktonic. The pore diameter is smaller than that of any other Mesozoic planktonic except those of Schackoina.

This species does not seem to have evolved from any known planktonic. It does occur in fair numbers in some samples but never approaching that of associated planktonics.

Resolution of the taxonomic status of this species must await a more detailed analysis.

Distribution: This is shown in Fig. 72.

Range: Upper Barremian–lowermost Cenomanian.

Globigerina (?) helvetojurassica Haeusler, Plate 21, figs 2–4

Globigerina Helveto-jurassica Haeusler, 1881a, p. 36, pl. 2, figs 44, 44a.

Globigerina bulloides var. Helveto-jurassica Haeusler. Haeusler, 1890, p. 118, pl. 15, fig. 46.

Globigerina ? cf. helveto-jurassica Haeusler. Seibold and Seibold, 1959, p. 64, text-figs 1a–e.

"Globigerina" helvetojurassica Haeusler. Oesterle, 1968, p. 774–778, text-figs 50–52a.

Description: "The test consists of 10–14 globular chambers which are arranged in $2\frac{1}{2}$–3 (?) revolutions. The first 5–6 chambers remain very small, the following become rapidly larger and inflated globular. On the umbilical side only the four last, spherically round chambers are visible, which are divided by means of deep-cut, straight sutures. The semicircular aperture is umbilical to weakly extraumbilical." (Translation of Oesterle, 1968.)

Remarks: Globigerina (?) helvetojurassica is lower spired than G. jurassica Hofman. This species varies in three ways, (1) height of the aperture, (2) height of the spire and (3) degree of inflation of the last four chambers. The variation in spire height is minimal.

The phylogeny of this form is unknown.

According to Oesterle (1969), Haeusler did not select a holotype for Globigerina helvetojurassica, and the specimens in his collection are silicified. Oesterle examined topotype specimens from Eisengraben, Switzerland and found that they were also silicified. He discriminated between the "Globigerines" and the associated Trochammina by apertural characteristics and by the mode of silicification. The "Globigerine" tests are formed entirely of secondary crystals five to ten microns in length. Whereas, the *Trochammina* tests are composed of unequal crystals of thirty to forty microns. Oesterle also found that crystals from the silicified tests of an Epistomine were indistinguishable from those of the "Globigerines". On the basis of this single comparison, he concluded that these "Globigerine" forms were not originally agglutinated.

There is a danger in Oesterle's interpretation. The tests of the topotypes are not simply replaced, but the wall, if any remains, is continuous with the internal filling, and cannot be distinguished as a separate unit. This would suggest that after burial the empty "Globigerine" tests were dissolved with the subsequent precipitation of the minute quartz crystals within the cavity. If this occurred, the size and shape of the crystals would have no relationship to the original test composition and wall structure, but would be controlled by the microenvironment within and around the cavity.

Thus, the taxonomic status of Haeusler's species remains unclear. If it should be proved a *Globigerina*, which seems unlikely, then it will probably be the senior synonym of *G. hoterivica* Subbotina. In the meantime, it is treated as a separate species, questionably placed in *Globigerina*.

Seibold and Seibold (1959) reported the presence of *Globigerina* cf. *helvetojurassica* from the Dogger in south Germany. These have three to four chambers per whorl, and also strongly resemble a *Trochammina*.

A few workers have reported *Globigerina helvetojurassica* from thin-sections. Farinacci and Radoičić (1964) and Radoičić (1966) noted their presence from Bajocian strata of Yugoslavia. Luterbacher (1972) reported specimens from samples in the northwestern Atlantic. Few planktonics can be accurately identified from thin-sections. This is particularly true of primitive species like *G. helvetojurassica*. Often these forms have been recrystallised, which further complicates their recognition even as a planktonic.

Distribution: This is shown in Fig. 73.

Range: Dogger–Malm border—Oxfordian.

Globigerina hoterivica Subbotina, Plate 22, figs 1–3

Globigerina hoterivica Subbotina, 1953, p. 57, 58, pl. 1, figs 1a–4c.

Globigerina ex gr. *bulloides* d'Orbigny. (?) Balakhmatova, 1953, p. 87, 88, text-figs 1a, b, v.

Globigerina oxfordiana Grigelis, 1958, p. 110, 111, text-figs a–c; Premoli Silva, 1966, p. 222, 223, text-figs 2a–c.

Globigerina kugleri Bolli, 1959, p. 270, 271, pl. 23, figs 3a–5; Marianos and Zingula, 1966, p. 335, pl. 37, figs 4a–c; Bandy, 1967, p. 8, text-fig. 3 (3).

Globigerina terquemi Iovčeva and Trifonova, 1961, p. 344–347, pl. II, figs 9–14.

Globigerina (*Eoglobigerina*) *balakhmatovae* Morozova, 1961, in Morozova and Moskalenko, p. 23, 24, pl. 1, figs 1–9, 11, 12.

Hedbergella hoterivica (Subbotina). Salaj and Samuel, 1966, p. 168, 169, pl. 8, figs 7a–c; Gawor-Biedowa, 1972, p. 68, 69, pl. 5, figs 9a–11c.

(?) *Globigerina spuriensis* Bars and Ohm, 1968, p. 582, 583, text-figs 1a–3.

Globigerina bathoniana Pazdrowa, 1969, p. 45–52, text-figs 2a–c; pl. II, figs 6a, b; pl. III, figs 1–3 (?); pl. IV, figs 1–3 (?), (*non* text-figs 1a–c (holotype), 3a–9c; pl. II, figs 1a–5, 7, 9a–c).

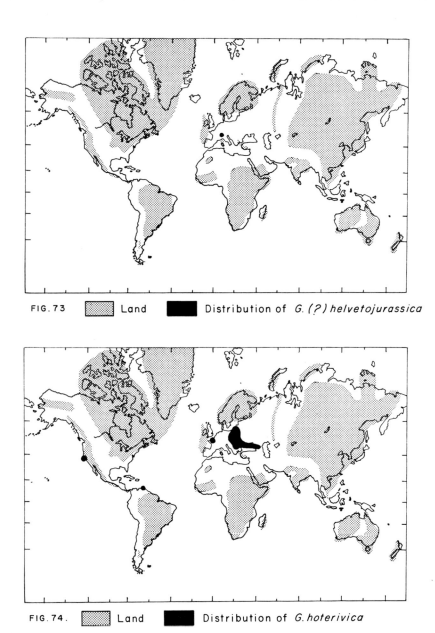

FIG. 73 ▢ Land ■ Distribution of *G. (?) helvetojurassica*

FIG. 74. ▢ Land ■ Distribution of *G. hoterivica*

Globigerina hauterivica Subbotina. Costea and Comsa, 1969, p. 108, pl. 1, figs 1, 2, 5, 6, 9, 10, 13 (?).

Hedbergella kugleri (Bolli). Gorbachik, 1971, pl. 28, figs 8a–c.

Description: Test small, quadrate, a moderate trochospiral coil, lobate periphery, $2\frac{1}{2}$ whorls of 4 chambers each. Chambers subspherical, gradually increasing in size, with the last 4 comprising most of the test. Aperture a moderate to high arch, umbilical, with a narrow lip. Sutures radial, depressed. Finely perforate.

Remarks: This small species is easily recognised by its quadrate shape and umbilical aperture. It differs from *Globigerina jurassica* Hofman by its much lower spire height. As Subbotina (1953) observed, the size of the last four chambers is the most unstable feature of this species. The ultimate chamber may comprise half of the last whorl with the other three chambers being subequal. All four chambers of the last whorl may be subequal. More frequently the size increase is gradual. A second variable is the height of the apertural arch.

Globigerina hoterivica evolved from *G. jurassica* Hofman by the reduction of the spire height.

Globigerina oxfordiana Grigelis, 1958, is said to differ from *G. hoterivica* by having fewer chambers and whorls, and by being smaller. It possesses two whorls with four chambers per whorl. *Globigerina hoterivica* has four chambers per whorl in two and one half whorls. The greatest diameter reported for *G. oxfordiana* ranges between 0.15 mm and 0.21 mm, with a height of 0.13 mm. The holotype of *G. hoterivica* has a diameter of 0.15 mm and a height of 0.10 mm. All these supposed differences are easily encompassed within the range of intraspecific variability. Furthermore, they both possess an umbilical aperture. *Globigerina hoterivica* is regarded as the senior synonym.

Hofker (1969) suggested that the species described by Bignot and Guyader (1966) as *Globigerina oxfordiana* and forms assigned by Premoli Silva (1966) to the same species are most likely species of the benthonic genus *Trochammina*. He claimed that the "pores" attributed to these specimens are merely calcite particles in a less transparent cement. As further evidence he stated that the environments of deposition were near shore, atypical of the Globigerine habitat. However, in the discussion which followed the presentation of Hofker's paper, a worker, who had examined specimens of *G. oxfordiana* from Saratov, U.S.S.R., claimed that the pores were very visible, and that the specimens of Premoli Silva were bilamellar, which would preclude them being a *Trochammina*.

The specimens which Bolli (1959) named *Globigerina kugleri* are identical to *G. hoterivica*.

Globigerina terquemi Iovčeva and Trifonova, 1961, also possesses an umbilical aperture and has the same range of diameters—0.12 mm to 0.20 mm. Although a glauconitic mold, it is believed to be identical to *G. hoterivica*.

Morozova (in Morozova and Moskalenko, 1961) erected the new taxon *Globigerina* (*Eoglobigerina*) *balakhmatovae* which possesses a small, rounded aperture with a weakly developed lip. The author also placed *G.* ex gr. *bulloides* of Balakhmatova into synonymy with *G. balakhmatovae*. Morozova stated that her new species has from three to five and one half chambers per whorl, although all of those illustrated have no less than four in the last whorl. *Globigerina hoterivica* has an aperture similar to *G. balakhmatovae* and four to five chambers in the final whorl. This, along with a coincidental stratigraphic range, justifies placing Morozova's species into synonymy with *G. hoterivica*. Nevertheless, Bignot and Guyader (1966) hold them as separate species, noting that *G. balakhmatovae* has more compressed chambers and a smaller aperture. The amount of compression seems negligible. The aperture of *G. balakhmatovae* is more comparable to that of Grigelis' original figure of *G. oxfordiana* (= *G. hoterivica*) than to those of the specimens illustrated by Bignot and Guyader (figs 3a–6b) under the latter name. All, however, are within acceptable limits of variation as interpreted here, and are regarded as synonyms of *G. hoterivica*.

The holotype of *Globigerina spuriensis* Bars and Ohm, 1968, if it is a planktonic form, seems to be comparable to *H. hoterivica* in chamber arrangement and spire height. However, three other characteristics, the pore diameter, the primary aperture and the "sekundären Kammer über der Nabel-region," are contrary to those of all other recorded Jurassic and early Cretaceous planktonic foraminifers. On these species, the pore-diameters where described are small, not large as in *G. spuriensis*. The primary aperture is not visible in any of the ventral views of this species. Figure 2b of Bars and Ohm shows what appears to be a sutural aperture, a feature which is considered to appear for the first time in the late Albian. An aperture, if indeed this is an aperture, in this position is not regarded as primary. Finally, no other early planktonic foraminifer is known to possess a secondary umbilical chamber. Because topotypic material was not available during this study, doubt remains as to whether *G. spuriensis* is a planktonic, and so can only questionably be treated as a synonym of *G. hoterivica*.

Pazdrowa (1969) erected the new species *Globigerina bathoniana*, in which she placed both high and low spired forms. For the time being each is assumed to be a separate species. The low trochospiral form is synonymous with *G. hoterivica*.

Distribution: This is shown in Fig. 74.

Range: Mid-Bathonian—Mid-Aptian.

Globigerina jurassica Hofman

Globigerina jurassica Hofman, 1958, p. 125, 126, text-figs a–c.

Globigerina conica Iovčeva and Trifonova, 1961, p. 343–347, pl. II, figs 1–8.

Globigerina bathoniana Pazdrowa, 1969, p. 45–52, text figs 1a–c (holotype), 3a–9c; pl. II, figs 1a–5, 7, 9a–c (*non* text-figs 2a–c; pl. II, figs 6a, b; pl. III, figs 1–3 (?); pl. IV, figs 1–3 (?)).

Description: "Tests white, calcareous, trochospiral, usually with diameter nearly equal to height, consisting of 7–13, most often 9 chambers, spherical, rapidly increasing in size, coiled helicoidally in 2–3 whorls. The first whorl around the proloculus consists of 4 chambers, arranged often planispirally [pseudoplanispirally as interpreted here], the next whorls are more trochospiral. The last whorl consists of 3–4 chambers. The last chambers increase their dimension less rapidly. Chambers in the first whorl are less distinctly separated by shallow sutures. During the formation of a younger chamber the wall is formed not only on it, but also on the whole older part of the test. Consequently the external walls of older chambers are thicker and consisting of several layers. Only the last chamber has a single-layer wall. This does not contradict the bilamellar character of the structure of the walls, i.e. the primary layer consists of two lamellae, best visible in septa or in the wall of the last chamber.

"Pores are small and densely distributed, perpendicular to the surface of the wall. Aperture relatively large semicircular, umbilical, enclosed on the peripheral side by a narrow lip nearly perpendicular to the test wall. There is no lip on the umbilical margin. The apertures of the older chambers maintain their shape, lip and position forming the foramen. The last chamber is often broken, the traces of the break being so small that they are visible only when appropriately illuminated. The ratio of sinistrally coiled to dextrally coiled tests is approximately 1:1. The surface of the test is rough, covered by a dense irregular network of tubercles and ridges between which the pores are situated . . .

". . . The diameter of the proloculus is 0.01–0.02 mm, the exceptional limiting values being 0.02 and 0.005 mm. Generally specimens with a large number of chambers have a smaller proloculus, but occasionally a large proloculus is present in specimens consisting even of 12 chambers. . . ." (Type description of *Globigerina bathoniana* Pazdrowa.)

Remarks: *Globigerina hoterivica* Subbotina, the only other species with which *G. jurassica* is associated, possesses a much lower spire height. Pazdrowa (1969) reported diversity in spire height, diameter, total number of chambers and shape of the aperture.

The phylogeny of this species is unknown.

Globigerina conica Iovčeva and Trifonova, 1961, although described from a glauconitic mold, does have a typical globigerine chamber shape and arrangement. The spire height and test diameter of *G. conica* falls within the range given for *G. jurassica*. The number of whorls, chambers per whorl and umbilical aperture is shared. Iovčeva and Trifonova's species is regarded as a synonym of *G. jurassica*.

Morozova (in Morozova and Moskalenko, 1961) proposed the new subgenus *Globigerina (Conoglobigerina)* with their new species *G. (Conoglobigerina) dagestanica* as the type species. Loeblich and Tappan (1964) treated *Conoglobigerina* as a junior synonym of *Gubkinella* Suleimanov, 1955, which they placed in the family Heterohelicidae. The two may well be synonymous, but possibly neither are planktonic. In fact *Gubkinella* has been transferred to the family Discorbidae (Rauzer-Chernousova and Fursenko, 1959). The aperture of *G. dagestanica*, which is a very narrow elongate slit at the umbilical margin of the final chamber, is atypical for a planktonic. The nature of the wall structure is not known. There seems to be an absence of an apertural lip, although this may be attributable to preservation or the artist's rendition. For the present, the subgenus *Conoglobigerina* is not accepted as a planktonic. If this should later be proven an invalid assumption, then *G. (Conoglobigerina) dagestanica* would become a junior synonym of *G. jurassica*.

Pazdrowa (1969) distinguished her new species *Globigerina bathoniana* from *G. jurassica* on the bases that the latter possesses irregularly arranged chambers and a smaller, lower aperture. An "aberrant" final chamber frequently alters the "normal" appearance of the aperture of many other planktonic species. Because the holotype of *G. jurassica* was not examined in this study, the exact nature of the final chamber is not known. But in her very thorough treatment of *G. bathoniana*, Pazdrowa observed specimens of that species which had a low aperture. The irregularly arranged chambers cited by her are not evident from the type figures of *G. jurassica*. The size ranges of these two species are essentially identical. They are considered conspecific.

Distribution: This is shown in Fig. 75.

Range: Bathonian—Tithonian.

Globigerina loetterlei Nauss,

Globigerina loetterli Nauss, 1947, p. 336, 337, pl. 49, figs 11a–c. pl. 22, fig. 4; pl. 23, fig. 1.

Globigerina loetterlei Nauss. Tappan, 1951, p. 4, 5, pl. 1, figs 19a–c.

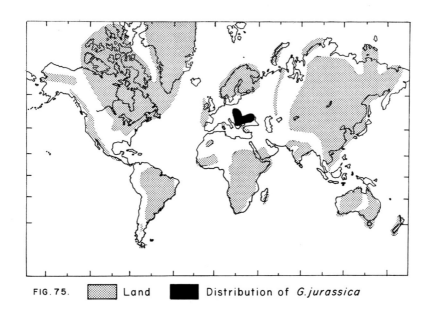

FIG. 75. ▨ Land █ Distribution of *G.jurassica*

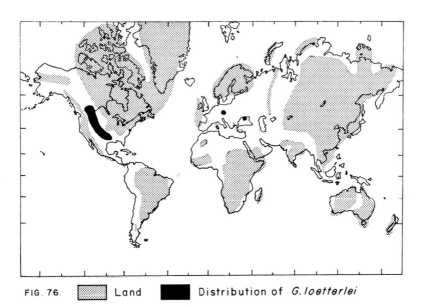

FIG. 76. ▨ Land █ Distribution of *G.loetterlei*

Hedbergella loetterlei (Nauss). Tappan, 1962, p. 196, 197, pl. 55, figs 3–5; Kent, 1967, p. 1448, pl. 183, figs 14a–c (*non* 15a–c); Wall, 1967, p. 107–108, pl. 3, figs 13–15, (?) 16–21; Douglas, 1969, p. 167, pl. 5, figs 3a–c.

Hedbergella globigerinelloides (Subbotina). Salaj and Samuel, 1966, p. 168, pl. 8, figs 3a–c.

Hedbergella planispira (Tappan). Gorbachik, 1971, pl. 28, figs 2a–c.

Archaeoglobigerina bosquensis Pessagno. Hanzlíková, 1972, p. 100, pl. 25, figs 11a–13c.

Hedbergella infracretacea (Glaessner). Gawor-Biedowa, 1972, p. 69, 70, pl. 6, figs 8a–c.

Description: Test a low trochospire, equatorial periphery lobate. Chambers subspherical, 6–7 in final whorl; increasing gradually in size, those in last whorl may be subequal; number of chambers per whorl may progress from an initial 3–5 in the second to 6 or 7 in the final revolution. Sutures radial to slightly curved, depressed. Umbilicus wide, shallow to moderate depth. Aperture interiomarginal, extraumbilical–umbilical with a lip. Wall finely perforate.

Remarks: *Globigerina loetterlei* differs from *G. delrioensis* Carsey in its greater number of less rapidly expanding chambers in the final whorl. *Globigerina paradubia* Sigal is much higher spired. The number of chambers in the final whorl is fairly stable at six, but an occasional specimen with seven can be found. These chambers are typically subequal, though this may vary.

This species is thought to have arisen from *Globigerina delrioensis* Carsey on the basis of intermediate forms. However, forms which appear gradational with *G. paradubia* Sigal are also recorded. So the phylogeny of this species is still not understood.

The holotype of *Globigerina loetterlei* was examined. It was found to be poorly preserved, probably as a calcite cast. No original shell material remains. The ultimate chamber is slightly reduced in size as compared with the penultimate. It is slightly compressed axially, but not keeled. The compression may be real or an artifact of preservation. All other chambers are rounded peripherally.

Distribution: This is shown in Fig. 76.

Range: Turonian—Lower Coniacian.

Globigerina monmouthensis (Olsson)

Globorotalia monmouthensis Olsson, 1960, p. 47, pl. 9, figs 22–24.

Praeglobotruncana (*Hedbergella*) *monmouthensis* (Olsson). Berggren, 1962a, p. 37–41, pl. 8, figs 1a–3c; text-figs 5 (1a–5c).

Hedbergella monmouthensis (Olsson). Olsson, 1964, p. 161, pl. 1, figs 3a–c; Douglas, 1969, p. 167, 168, pl. 9, figs 4a–c; (?) Hanzlíková, 1969, p. 45, pl. 12, figs 4a–5c; Olsson, 1970, pl. 93, figs 1–4; Bang, 1971, pl. 6, figs 1–4; *non* Hanzlíková, 1972, p. 101, pl. 26, figs 3a–4c, 6a–b, *non* Govindan, 1972, p. 173, pl. 2, figs 11–13.

Hedbergella planispira monmouthensis (Olsson). Bandy, 1967, p. 11, 12, text-fig. 4 (5).

Description: "Test small to medium size (seldom exceeds 0.35 mm in maximum diameter), very low—trochospiral, peripheral outline subcircular, peripheral margin distinctly lobulate, axial periphery rounded, umbilicus small, moderately shallow, open (no umbilical tegillum); early chambers strongly depressed, those in last whorl (usually 5–5½) increasing rapidly in size and becoming inflated and globular, 14–16 commonly in 2½–3 whorls on spiral side; sutures on umbilical side depressed and slightly curved, those on spiral side oblique to nearly straight in early chambers, later becoming radial and slightly curved; wall finely perforate, surface covered with dense, hispid ornament; primary aperture a low umbilical–extraumbilical arch (which may be obscured by rounded adumbilical margin of last chamber); relict apertures (earlier foramina) distinctly visible beneath thin, delicate, tapering umbilical portici." (After Berggren, 1962a.)

Remarks: At present no other species of *Globigerina* is considered to occur with *G. monmouthensis*. Douglas (1969) reported that this species is accompanied by *G. crassa* (Bolli) [= *G. delrioensis* Carsey], though this may be a variant of *G. monmouthensis*. The surface of the test is variable and may be smooth or covered with dense, short spines. This may be related to preservation.

Olsson (1964) envisioned his new species *Hedbergella holmdelensis* (= *Globigerina planispira* Tappan) as ancestral to *G. monmouthensis*. Given the stratigraphic distribution as presented in this study, *G. planispira* appears to be the only option. As correctly assessed by Olsson, *G. monmouthensis* represents the last evolving Mesozoic globigerinid.

Govindan (1972) reported specimens of *Hedbergella monmouthensis* which possess a tegillum. He remarked that he did not consider them as being *Rugoglobigerina* because they lacked costellae. Costellae have been shown (Masters, in press) to be of no importance at the generic level, but tegilla are definitely so. Govindan's specimens cannot be referred to *Globigerina*.

Distribution: This is shown in Fig. 77.

Range: Maastrichtian (*Trinitella scotti* Interval).

Globigerina paradubia Sigal, Plate 23, figs 2–4; Plate 24, fig. 1

Pulvinulina sp. Moreman, 1925, pl. 18, fig. 11.

Globigerina cretacea d'Orbigny. Loetterle, 1937, p. 44, pl. 7, figs 2a–c (*non* 1a–c); Bolin, 1956, p. 292, 293, pl. 39, figs 4a, b, 6a–8b (*non* 5, 13a, b, 17).

Globigerina elevata d'Orbigny. Keller, 1946, p. 97, 98, pl. 2, figs 11, 12.

Globigerina paradubia Sigal, 1952, p. 28, text-fig. 28; Hermes, 1969, p. 50, pl. 8, figs 187–189.

Globigerina kelleri Subbotina, 1953, p. 63, 64, pl. 1, figs 16a, b; Hermes, 1969, p. 50, pl. 7, figs 185, 186.

Rotundina ordinaria Subbotina, 1953, p. 181–183, pl. 3, figs 3a–8c (*non* 9a–c); pl. 4, figs 1a–c, 4a–8c (*non* 2a–3c, 9a–c).

Rugoglobigerina ordinaria (Subbotina). Bykova *et al.*, 1959, p. 303, text-figs 693A–C.

Hedbergella brittonensis Loeblich and Tappan, 1961a, p. 274, 275, pl. 4, figs 1–8; Ayala-Castañares, 1962, p. 23, 24, pl. 3, figs 2a–c; pl. 8, figs (?) 3a, b; *non* Fuchs, 1967b, p. 331, pl. 18, figs 1a–c; *non* Pessagno, 1967, p. 282, pl. 52, figs 9–12; Hermes, 1969, p. 51, pl. 7, figs 169–171; Kalantari, 1969, p. 198, 199, pl. 12, figs 21a–c; pl. 23, figs 2a–c; Gawor-Biedowa, 1972, p. 67, 68, pl. 7, figs 1a–2c.

Hedbergella portsdownensis (Williams-Mitchell). Loeblich and Tappan, 1961, p. 277, pl. 5, figs 3a–c; Douglas and Rankin, 1969, p. 194–196, text-figs 7A–F; Eicher, 1969, text-fig. 3 (part); Porthault, 1969, p. 530, 531, pl. 1, figs 5a–c; Eicher and Worstell, 1970b, p. 304, pl. 10, figs 1a–2b.

Praeglobotruncana ? *ordinaria* (Subbotina). Malapris and Rat, 1961, p. 90, 91, pl. 2, figs 4a–c; text-figs 5a, b.

Hedbergella trocoidea (Gandolfi). Ayala-Castañares, 1962, p. 24, 25, pl. 3, figs 3a–c; pl. 9, figs (?) 1a, b.

Rugoglobigerina ? *ordinaria* (Subbotina). *non* Corminboeuf, 1962, p. 496, 497, pl. 2, figs 4a–c.

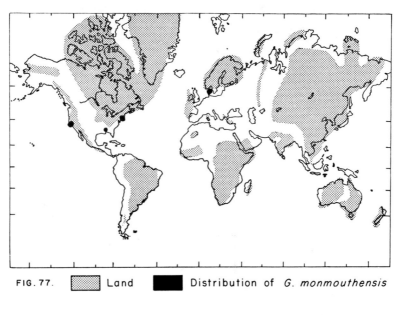

FIG. 77. ▨ Land ■ Distribution of *G. monmouthensis*

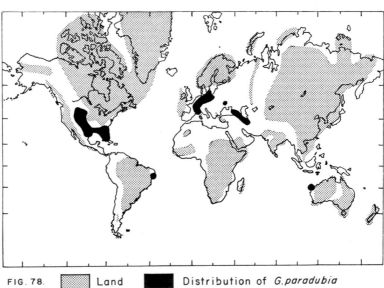

FIG. 78. ▨ Land ■ Distribution of *G. paradubia*

Praeglobotruncana planispira (Tappan). Petri, 1962, p. 121, 122, pl. 16, figs 7a–c (*non* 8a–c).

Hedbergella delrioensis (Carsey). Butt, 1966, p. 173, 174, pl. 2, figs 1a–c, 5a–c (*non* 2a–4c, 6a–8c); Eicher, 1966, p. 27, pl. 5, figs 13a–c (*non* 12a–c).

Rugoglobigerina rugosa ordinaria (Subbotina). Bandy, 1967, p. 21, 22, text-fig. 10 (3).

Archaeoglobigerina bosquensis Pessagno, 1967, p. 317, 318, pl. 70, figs 3–8; pl. 94, figs 4–5; *non* Hanzlíková, 1972, p. 100, pl. 25, figs 11a–13c.

Hedbergella paradubia (Sigal). Porthault, 1970, in Donze *et al.*, p. 64, pl. 9, figs 13–15; Belford and Scheibnerová, 1971, p. 334, pl. 3, figs 12–16.

Description: Test large, moderate to high trochospire, equatorial periphery lobate, axial periphery rounded. Chambers subspherical to spherical, 15–16 in $2\frac{1}{2}$–3 whorls, 5–$6\frac{1}{2}$ per whorl; typically subequal in last whorl, otherwise gradually increasing in size. Apertures a low to moderate extraumbilical–umbilical arch bordered by a narrow lip. Umbilicus deep, relatively narrow. Sutures deeply depressed, radial to slightly curved. Surface finely porous, pustulose.

Remarks: Only three high-spired *Globigerina* are found in the Mesozoic: *G. jurassica* Hofman, *G. washitensis* Carsey and *G. paradubia*. Each is distinctive, and their ranges do not overlap. The lower-spired *G. paradubia* with six chambers per whorl approach the generally smaller *G. loetterlei* Nauss, while those with five chambers are similar to *G. delrioensis* Carsey. The latter can be distinguished by their more rapid chamber inflation. The number of chambers per whorl varies, though the most significant variation occurs in chamber height. The angle formed by the spiral edges of the test when viewed peripherally ranges from 112° (a low spire) to 62° (a high spire).

Globigerina paradubia is believed to have developed from *G. delrioensis* Carsey by gradually becoming higher spired, reducing the rate of chamber inflation and increasing the number of chambers per whorl.

Although some authors (Lehmann, 1962; Caron, 1966; etc.) have placed *Globigerina paradubia* Sigal into the genus *Praeglobotruncana*, neither the outline drawing nor description of the holotype suggest the presence of a keel. Sigal (personal communication, 1972) informed me that *G. paradubia* has neither a keel nor supplementary sutural apertures, and that it fits best into the genus *Hedbergella* [= *Globigerina*].

Subbotina (1953) proposed the new species *Rotundina ordinaria*, which was said to have on occasion, traces of two keels. However, the type figure does not exhibit a keeled periphery. The range of variation of the dorsal convexity of *Globigerina kelleri* Subbotina, 1953, encompasses that of *R. ordinaria*, both of which are considered junior synonyms of *G. paradubia*. The keeled specimens mentioned by Subbotina are not included in this synonymy.

Hedbergella brittonensis Loeblich and Tappan, 1961, is considered a junior synonym of *Globigerina paradubia* because there are no mutually exclusive characteristics.

The description and remarks given by Pessagno (1967) for his new species *Archaeoglobigerina bosquensis* could be substituted for those of *Globigerina paradubia*. The holotype of Pessagno's species was examined and found to be slightly less high-spired than that of *G. paradubia*, but within acceptable limits of intraspecific variation.

Distribution: This is shown in Fig. 78.

Range: Mid-Cenomanian—Santonian.

Globigerina planispira Tappan, Plate 24, figs 2, 3, 5
Globigerina planispira Tappan, 1940, p. 122, pl. 19, figs 12a–c; Tappan, 1943, p. 513, pl. 83, fig. 3; Lozo, 1944, p. 562, pl. 3, fig. 5; Crespin, 1953, p. 35, pl. 6, fig. 16; Frizzell, 1954, p. 127, pl. 20, figs 2a–c; Rasheed, 1963, p. 238, 239, pl. 4, figs 8–10; Hermes, 1969, p. 45, 46, pl. 5, figs 98–100.
Globigerina gaultina Morozova, 1948, p. 41, pl. 2, figs 16–18.
Globigerina globigerinelloides Subbotina, 1949, p. 32, 33, pl. 2, figs 11–15; Subbotina, 1953, p. 59–61, pl. 1, figs 11a–12c; Maslakova, 1959, p. 105, pl. 10, figs 1a–c; Kaptarenko-Chernousova *et al.*, 1963, p. 103, Tab. 13. figs 8a, b; Alekseeva, 1963, p. 44, 45, pl. 8, figs 8a–c; Hermes, 1969, p. 49, pl. 7, figs 182–184.
Globigerina globigerinelliformis Subbotina. Dzhafarov and Agalarova, 1949, p. 70, 71, pl. 4, figs 3a–c.
Globigerina undrizewi Dzhafarov and Agalarova, 1949, p. 71, 72, pl. 4, figs 4a–c.
Globigerina infracretacea Glaessner. Myatlyuk, 1949, p. 216, pl. 5, figs 5a–c; Moullade, 1960, p. 136, pl. 2, figs 18–20; Visser and Hermes, 1962, encl. 17; figs 45a–c; Alekseeva, 1963, p. 44, pl. 8, figs 7a–c.
Ticinella gaultina (Morozova). Subbotina, 1953, p. 167–170, pl. 1, figs 1a–4c; Bykova *et al.*, 1959, p. 301, text-figs 682A, B; Samuel, 1962, p. 183, 185, pl. 8, figs 9a–c; Hermes, 1969, p. 42, pl. 4, figs 71–73.
Globorotalia ? *youngi* Fox, 1954, p. 119, pl. 26, figs 15–18.
Globorotalia youngi Fox, Hermes, 1969, p. 50, pl. 8, figs 194–197.
Globigerina hoelzli Hagn and Zeil, 1954, p. 50, 51, pl. 2, figs 8a–c.
Praeglobotruncana planispira (Tappan). Bolli, Loeblich and Tappan, 1957, p. 39, 40, pl. 9, figs 3a–c; Bolli, 1959, p. 267, pl. 22, figs 3a–4; Petri, 1962, p. 121, 122, pl. 16, figs 8a–c (*non* 7a–c).
Praeglobotruncana infracretacea (Glaessner). Bolli, 1959, p. 266, pl. 21, figs 9a–c (*non* 10).
Praeglobotruncana modesta Bolli, 1959, p. 267, pl. 22, figs 2a–c; Hermes, 1969, p. 50, 51, pl. 6, figs 150–152.
Planomalina escheri (Kaufmann). Bolli, 1959, p. 260, 261, pl. 20, figs 7a–8.
Praeglobotruncana hansbolli Trujillo, 1960, p. 339, pl. 49, figs 7a–c; Graham, 1962, p. 105, pl. 20, figs 23a–c.
Hedbergella planispira (Tappan). Loeblich and Tappan, 1961, p. 276, 277, pl. 5, figs 4–11c; Olsson, 1964, pl. 1, figs 5a–c (*non* 4a–c); Eicher, 1965, p. 905, pl. 106, figs 1a–c; *non* Neagu, 1965, p. 36, pl. 10, figs 1a–4c; Takayanagi, 1965, p. 205, 206, pl. 21, figs 6a–7c; pl. 22, figs 1a–c; Marianos and Zingula, 1966, p. 335, pl. 37, figs 6a–c; Salaj and Samuel, 1966, p. 169, 170, pl. 8, figs 10a–c; Fuchs, 1967b, p. 332, pl. 18, figs 2a–c; Bandy, 1967, p. 10, text-fig. 4 (3); *non* Pessagno, 1967, p. 283, 284, pl. 51, fig. 1; pl. 53, figs 1–4; Eicher, 1967, p. 186, pl. 19, figs 3a–c; Barr, 1968, p. 314, pl. 37, figs 4a–5c; Douglas, 1969, p. 168, pl. 5, figs 1a–c; Eicher, 1969, text-fig. 3 (part); Hermes, 1969, p. 45, 46, pl. 5, figs 106–114; Porthault, 1969, p. 530, pl. 1, figs 4a–c; Weaver *et al.*, 1969, pl. 4, figs 1a–c; Eicher and Worstell, 1970b, p. 302, 303, pl. 9, figs 12–13c; Neagu, 1970b, p. 63, 64, pl. 19, figs 20–22; Neagu, 1970a, fig. 1 (2); *non* Gorbachik, 1971, pl. 28, figs 2a–c; Kuhry, 1971, p. 230, pl. 2, figs 4A–C; ? Risch, 1971, p. 47, 48, pl. 4, figs 4–6; El-Naggar, 1971a, pl. 6, fig. d, (?) f (*non* e); Gawor-Biedowa, 1972, p. 70, 71, pl. 5, figs 8a–c; *non* Hanzlíková, 1972, p. 101, pl. 25, fig. 15; pl. 26, figs 1a–2; Govindan, 1972, p. 172, 173, pl. 2, figs 17–19.
Hedbergella (*Hedbergella*) *planispira* (Tappan). Moullade, 1966, p. 93, pl. 8, figs 4, 5; Moullade, 1969, p. 464, pl. 1, figs 13, 14.

Hedbergella (Hedbergella) sp. aff. *planispira* Tappan. Moullade, 1966, p. 94, pl. 8, figs 1–3; Moullade, 1969, p. 464, pl. 1, figs 15–17.

Globotruncana (Ticinella) gaultina Morozova. Majzon, 1961, p. 763, pl. 6, fig. 10.

Planogyrina gaultina (Morozova). Zakharova-Afabekyan, 1961, p. 50, 51.

Globotruncana (Rugoglobigerina) hoelzli (Hagn and Zeil). *non* Hanzlíková, 1963, pl. 2, figs 2a–c.

Rugoglobigerina hoelzli (Hagn). *non* Kaptarenko-Chernousova, 1963, p. 106, 107, pl. 15, figs 5a, b, v.

Hedbergella holmdelensis Olsson, 1964, p. 160, 161, pl. 1, figs 1a–2c; Sliter, 1968, p. 100, 101, pl. 15, figs 8a–c (*non* 6a–c), *non* Govindan, 1972, p. 173, pl. 2, figs 14–16.

Hedbergelles Caron, 1966, pl. 6, figs 6a–7c.

Hedbergella gaultina (Morozova). Salaj and Samuel, 1966, p. 167, 168, pl. 8, figs 9a–c.

Hedbergella globigerinelloides (Subbotina). *non* Salaj and Samuel, 1966, p. 168, pl. 8, figs 3a–c.

Hedbergella planispira holmdelensis Olsson. Bandy, 1967, p. 10, 11, text-fig. 4 (4).

Hedbergella brittonensis Loeblich and Tappan. Fuchs, 1967b, p. 331, pl. 18, figs 1a–c.

Hedbergella hansbolli (Trujillo). Ansary and Tewfik, 1968, p. 47, pl. 4, figs 2a–3c.

Globigerinelloides escheri escheri (Kaufmann). Neagu, 1970b, p. 62, pl. 25, figs 19–21 (*non* 18; pl. 28, figs 12–14).

Globigerinelloides asper (Ehrenberg). ? Porthault, 1970, in Donze *et al.*, p. 63, pl. 9, fig. 8 (*non* 9).

Hedbergella hoelzli (Hagn and Zeil). Belford and Scheibnerová, 1971, p. 334, pl. 4, figs 1–3 (?), 4–8.

Hedbergella globigerinelloides (Subbotina). Gorbachik, 1971, pl. 28, figs 4a–c.

Hedbergella infracretacea (Glaessner). Kuhry, 1971, p. 230, pl. 2, figs 3A–C.

Hedbergella sp. aff. *H. planispira* (Tappan). Kuhry, 1971, p. 231, pl. 2, figs 5A–C.

(?) *Hedbergella implicata* Michael, 1973, p. 210, pl. 2, figs 4–6.

Hedbergella intermedia Michael, 1973, p. 210, 211, pl. 2, figs 7–9.

Hedbergella ? *punctata* Michael, 1973, p. 212, pl. 3, figs 1–3; pl. 7, figs 1, 2.

Hedbergella pseudotrocoidea Michael, 1973, p. 211, 212, pl. 3, figs 4–8.

Description: Test small, very low trochospiral coil appearing pseudoplanispiral in peripheral view, equatorial periphery lobate, axial periphery rounded. Chambers spherical, 12–18 chambers (typically 16) in $2\frac{1}{2}$ whorls with 6–7 per whorl, only slight increase in size as added, many of last whorl subequal. Umbilicus shallow, wide for test size. Aperture a low interiomarginal, extraumbilical–umbilical arch bordered by a wide, imperforate apertural flap on well preserved specimens; relict apertures indicated by presence of older flaps in umbilicus. Wall finely porous, pustulose. Sutures radial, moderately depressed.

Remarks: *Globigerina planispira* is easily distinguished by its low spire and six to seven gradually enlarging chambers in the final whorl. Most characteristics for this species are stable making it one of the easiest to recognise. The number of chambers in the last whorl is the only variable feature.

Globigerina delrioensis Carsey is the most likely ancestor of *G. planispira*. Small specimens of Carsey's species closely approximate *G. planispira* differing only in the number of chambers per whorl (five and one half vs. six to seven) and their rate of inflation. Change the rate of inflation and the number of chambers per unit space must also change.

Morozova (1948) proposed the name *Globigerina gaultina* for a small, non-keeled, low trochospiral form from the Albian of the southwest Caucasus. Although neither the holotype nor topotypes were examined during this study, the age, description and figures leave no doubt that it is conspecific with *G. planispira*. Subbotina (1953) transferred *G. gaultina* to the genus *Ticinella*. She stated that although no supplementary apertures were seen, there were small depressions which may represent these apertures. However, her illustrations show simple, complete apertural lips typical of *G. planispira*.

Globigerina undrizewi Dzafarov and Agalarova, 1949, was distinguished as having a very low spire and chambers in the last whorl which did not increase in size. The description of this Albian species does not differ from *G. planispira*, and is treated as a synonym of that species.

Examination of the holotype of *Globorotalia* ? *youngi* Fox, 1954, showed it to be identical to *Globigerina planispira*.

Topotypes of *Globigerina hoelzli* Hagn and Zeil, 1954, were found to be identical to *G. planispira*. Hanzlíková (1963) illustrated a specimen attributed to *Rugoglobigerina hoelzli* which does appear to possess a tegillum. It has axially compressed but unkeeled chambers. Hagn and Zeil's species has a rounded axial periphery and no cover plate.

The holotype and reported stratigraphic range of *Praeglobotruncana modesta* Bolli, 1959, show no variation from *G. planispira*.

The holotype and paratypes of *Praeglobotruncana hansbolli* Trujillo, 1960, were examined and found to fall easily within the variation displayed by the topotypic suite of *G. planispira*.

There seems to be no morphologic basis for separating *Hedbergella holmdelensis* Olsson, 1964, from *Globigerina planispira*. Olsson stated that it differs from the latter by having fewer chambers in the final whorl which increase in size more rapidly and which are elongate in the direction of coiling. Interestingly, the form attributed by Olsson to *H.* (*Globigerina*) *planispira* are identical in chamber number, size and shape to his *H. holmsdelensis*. Both are identical to the intraspecific variants found in topotypic suites of *G. planispira*.

Michael (1973) proposed the names *Hedbergella implicata*, *H. intermedia* and *H.* (?) *punctata* for specimens having a "unique" pore pattern, i.e. large pores for the test size. As discussed earlier in this chapter, pore diameter is either environmentally or diagenetically controlled. The latter almost certainly has been partially responsible here. Each of the holotypes is poorly preserved. No surface features remain other than the pores, which probably have been enlarged by secondary solution. These species, with the exception of *H. implicata*, are clearly conspecific with *Globigerina planispira*. *Hedbergella implicata* was further characterised by its overlapping chambers, particularly the ultimate chamber. This type of gerontic-stage development is seen repeatedly in other taxa. In *H. intermedia* the final chamber is reduced in size, imparting a circular outline to the test, while in *H.* (?) *punctata* the final chamber is "normal". Michael described *H. implicata* as having slightly curved sutures. Because only one specimen was illustrated and intraspecific variation was not discussed, it is not certain how stable the feature is. Nevertheless, it is suspected to be variable, in which case it would be synonymous with *G. planispira*.

Hedbergella pseudotrochoidea Michael, 1973, appears to be an internal mold because no surface features or pores are preserved. Michael claims that this species is closely related to *H. trocoidea* (Gandolfi) but makes no attempt to document his statement. This species does not even bear a superficial resemblance to Gandolfi's species, which Michael paradoxically makes clear in his remarks. Working with poorly preserved specimens is difficult at best, but there is little doubt that *H. pseudotrocoidea* is a *Globigerina planispira* with an inflated final chamber.

Distribution: This is shown in Fig. 79.

Range: Lower Aptian—Maastrichtian (LOS *Globotruncana fornicata*–FOS *Trinitella scotti* Interval).

Globigerina simplicissima (Magné and Sigal), Plate 24, figs 4, 6

Globigerina cretacea d'Orbigny. Morrow, 1934, p. 198, pl. 30, fig. 7 (*non* 8, 10a, b); Bolli, 1959, p. 270, pl. 22, figs 8a–9.

Hastigerinella simplicissima Magné and Sigal, 1954, in Cheylan *et al.*, p. 487, 488, pl. 14, figs 11a–c; *non* Kavary and Frizzell, 1963, p. 49, 50, pl. 9, figs 16, 17; Hermes, 1969, p. 50, pl. 8, figs 203–205.

Hedbergella amabilis Loeblich and Tappan, 1961, p. 274, pl. 3, figs 1, 3 (?)–7, 9 (*non* 2, 8a–c, 10a, b); Borsetti, 1962, p. 29, 30, pl. 2, fig. 4; text-figs 15–18 (?); Luterbacher, 1963, in Renz *et al.*, p. 1084, pl. 9, figs 4a–c, 6a–c; Wezel, 1965, p. 33, 34, text-figs 13a–c; Barbieri, 1966, text-fig. 2 (14); Lehmann, 1966b, p. 157, pl. 1, figs 3, 4; Arkin and Hamaoui, 1967, pl. 1, figs 19–21; *non* Pessagno, 1967, p. 281, 282, pl. 52, figs 6–8; *non* Barr, 1968, pl. 2, figs 6a–7c; Bate and Bayliss, 1969, pl. 3, figs 3a–c; *non* Douglas, 1969, p. 165, pl. 4, figs 8a–c; Eicher, 1969, text-fig. 3 (part); Hermes, 1969, p. 51, pl. 7, figs 172–174; (?) Eicher and Worstell, 1970b, p. 300, 302, pl. 9, figs 6, 7, 9a–c.

Clavihedbergella simplex (Morrow). Ayala-Castañares, 1962, p. 25, 26, pl. 4, figs 2a–3c (*non* 1a–c); pl. 5, figs 1a–c; pl. 9, figs 2a, b, (?) 3a, b; pl. 10, figs 1a, b; (*non* pl. 12, figs 4a, b); Todd and Low, 1964, p. 403, 404, pl. 1, figs 1a–c.

Clavihedbergella simplicissima (Magné and Sigal). (?) Samuel, 1962, p. 180, pl. 7, figs 4a–5; (?) Caron, 1966, p. 71, pl. 6, figs 5a–c; *non* Neagu, 1968, text-fig. 2 (54); Neagu, 1969, p. 140, pl. 13, figs 1–6; Neagu, 1970a, fig. 1 (5); ? Neagu, 1970b, p. 64, pl. 19, figs 17–19; pl. 20, figs 7–15; *non* Gawor-Biedowa, 1972, p. 72, 73, pl. 7, figs 3a–c; *non* Neagu, 1972, p. 215, pl. 6, figs 38–39.

Hedbergella (*Clavihedbergella*) sp. aff. *simplex* (Morrow). Moullade, 1966, p. 96, 97, pl. 8, figs 19, 20 (*non* 18); Moullade, 1969, p. 462, 463, pl. 1, figs 4, 5 (*non* 6).

Hedbergella (*Hedbergella*) *sigali* Moullade, 1966, p. 87, 88, pl. 7, figs 20–25.

Hedbergella sigali Moullade. Hermes, 1969, p. 51, 52, pl. 7, figs 180, 181.

Clavihedbergella amabilis (Loeblich and Tappan). (?) Salaj and Samuel, 1966, p. 173, pl. 10, figs 3a–c.

Hedbergella flandrini Porthault, 1970, in Donze *et al.*, p. 64, 65, pl. 10, figs 1a, 1b (misprinted as 10), 2a, b, 3.

Hedbergella simplicissima (Magné and Sigal). Caron, 1971, p. 148, 149, text-figs 3a–c.

Description: Test of medium size, low trochospiral coil of approximately 11 chambers in $2\frac{1}{2}$ whorls with 5, occasionally 6 chambers per whorl; equatorial periphery deeply lobate, axial periphery rounded. Chambers spherical with the last one or two becoming oval to slightly elongate; well separated; moderately increasing in size. Aperture a low interiomarginal, extraumbilical–umbilical arch bordered by an imperforate flap in well-preserved specimens. Umbilicus of moderate diameter and depth, partially covered by successive apertural flaps. Sutures deeply depressed, radial. Surface finely perforate, pustulose.

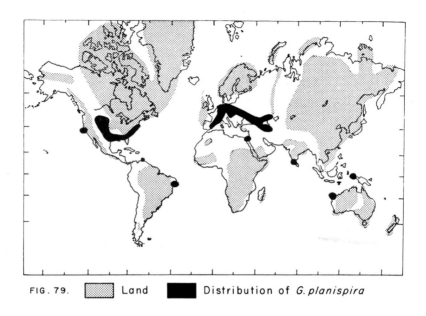

FIG. 79. ☐ Land ■ Distribution of *G. planispira*

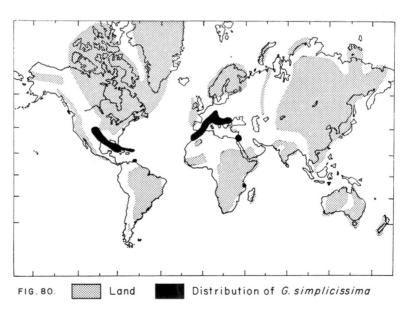

FIG. 80. ☐ Land ■ Distribution of *G. simplicissima*

Remarks: *Globigerina simplicissima* can be distinguished from *G. delrioensis* Carsey by its less rapid chamber inflation, greater separation of its chambers and slight elongation of its last one or two chambers. Apart from a slight variation in the degree of elongation of the ultimate chamber the morphological characteristics of this species are rather stable.

This species appears at about the same time as does *Globigerina delrioensis* together with forms which are intermediate to both. It is possible that they represent a single natural population, but because *G. delrioensis* is known to occur in younger strata, both species are retained as valid. Which preceded the other cannot yet be determined.

The taxon *simplicissima* is assigned to *Globigerina* rather than *Clavihedbergella*. The latter is defined as differing from *Globigerina* by the possession of radially elongate chambers in the ephebic or gerontic ontogenetic stages. The test of *G. simplicissima* has nearly spherical chambers with only the slightest elongation in the ultimate and occasionally the penultimate chambers. The distinction between the two genera is a matter of degree, which is believed insufficient in this case for placement into *Clavihedbergella*.

The specimen assigned to *Clavihedbergella simplicissima* by Caron (1966) has chambers more axially compressed than is typical for this species; this designation is therefore questionably accepted.

Donze *et al.* (1970) described as new the species *Hedbergella flandrini* from the lower Coniacian of Puget-Theniers, France. This species is morphologically identical to and occupies the same stratigraphic interval of *Globigerina simplicissima*.

Eicher and Worstell (1970b) have identified as *Hedbergella amabilis* Loeblich and Tappan a trochospiral specimen with a poreless margin. This assignment is tentatively accepted until the holotype or topotypes of *Globigerina simplicissima* can be examined for the presence of a poreless margin. These authors state that this species grades into *Clavihedbergella simplex* (Morrow), a conclusion which I have confirmed. If a poreless margin should be found in the taxon *simplicissima*, it should be transferred to *Clavihedbergella*. Regardless of its ultimate generic status, *simplicissima* is an important link between *Globigerina* and *Clavihedbergella*. In their discussion of *H. amabilis*, Eicher and Worstell suggest that the poreless margin and the early chambers may be imperforate because of secondary filling. Examination by scanning electron microscopy of neanic individuals, with as few as two and three chambers, of most genera of Mesozoic planktonic foraminifers has shown that the imperforate initial condition of the chambers is primary. In forms which develop a keel, the poreless margin becomes apparent when the pore density increases sufficiently to delineate these areas. This is not to preclude the occurrence of secondary pore filling, but this should be structurally and crystallographically distinguishable in sectional views.

Distribution: This is shown in Fig. 80.

Range: Barremian—lower Santonian.

Globigerina trocoidea (Gandolfi), Plate 25, figs 1–3
Anomalina lorneiana (d'Orbigny). Gandolfi, 1942, p. 98, 99, pl. 4, figs 1, 19; pl. 8, fig. 2; pl. 13, figs 1a, b, 4a, b.
Anomalina lorneiana var. *trocoidea* Gandolfi, 1942, p. 99, pl. 2, figs 1a–c; pl. 4, figs 2, 3; pl. 13, figs 2a, b, 5a, b; Noth, 1951, p. 80, pl. 4, figs 27a, b, 28a, b; Maslakova, 1963a, text-figs 1e, k, l; Hermes, 1969, p. 48, pl. 8, figs 200–202.

Globigerina infracretacea Glaessner. Colom, 1947, pl. 21, figs 1–10.
Globigerina lacera Ehrenberg. (?) Colom, 1947, pl. 21, figs 11, 12.
Globigerina almadenensis Cushman and Todd, 1948, p. 95, 96, pl. 16, figs 18, 19;
Hermes, 1969, p. 49, pl. 8, figs 190, 191.
Pseudovalvulinaria trocoidea (Gandolfi). Hagn, 1953, text-figs 1 (part), 2 (part).
Hedbergina seminolensis (Harlton). Brönnimann and Brown, 1956, p. 529, 530,
pl. 20, figs 4–6; Maslakova, 1963a, text-figs 1v, g, d.
Hedbergella trocoidea (Gandolfi). Brönnimann and Brown, 1958, p. 16, 17, text-
figs 1a–c; *non* Klaus, 1959, p. 792, pl. 1, figs 1a–c; Loeblich and Tappan, 1961, p. 277,
278, pl. 5, figs 1a–c (?), 2a–c; *non* Ayala-Castañares, 1962, p. 24, 25, pl. 3, figs 3a–c;
pl. 9, figs 1a, b; (?) Dallan, 1962, pl. 2, figs 1–3; Graham, 1962, p. 105, pl. 19, figs
18A, B; *non* Takayanagi and Iwamoto, 1962, p. 191, pl. 28, figs 3a–6c; *non* Taylor,
1964, p. 593, 594, pl. 84, figs 4–7; Todd and Low, 1964, p. 403, pl. 2, figs 1a–2c;
Saavedra, 1965, p. 323, 324, text-fig. 7b; (?) Lehmann, 1966b, p. 157, pl. 1, figs 1, 2;
Salaj and Samuel, 1966, p. 172, pl. 8, figs 1a–c; tab. 33 (5); Bandy, 1967, p. 10,
text-fig. 4 (2); Caron and Luterbacher, 1969, p. 23, pl. 7, figs 1a–2c; (?) Eicher,
1969, text-fig. 3 (part); Hermes, 1969, p. 48, 49, pl. 5, figs 115–117; *non* Gorbachik,
1971, pl. 28, figs 3a–c; Risch, 1971, p. 47, pl. 4, figs 20–22.
Hedbergella (*Hedbergella*) *trocoidea* (Gandolfi). Moullade, 1966, p. 90–93,
pl. 7, fig. 26; (?) Moullade, 1969, p. 464, 465, pl. 1, fig. 18.
Praeglobotruncana rohri Bolli, 1959, p. 267, 268, pl. 22, figs 5a–7; Hermes, 1969,
p. 51, pl. 6, figs 147–149.
Praeglobotruncana gautierensis (Brönnimann). Bolli, 1959, p. 265, 266, pl. 21,
figs 3a–6.
Praeglobotruncana seminolensis (Harlton). Maslakova, 1963a, text-figs li, zh, z.

Description: Test a low to moderate trochospire, equatorial periphery moderately
lobate, axial periphery rounded. Chambers inflated, closely packed imparting a
V-shaped outline ventrally; the radial length appears greater but is seldom more
than the maximum width as measured ventrally; 16–22 in 2½ whorls with 7–9 in the
final whorl; subequal to gradual increase in size. Aperture a low extraumbilical–
umbilical arch bordered by a narrow lip. Umbilicus narrow, moderately deep.
Sutures radial to slightly curved, depressed. Surface finely porous, coarsely pustulose.

Remarks: The lower-spired individuals of *Globigerina paradubia* Sigal are the
only forms which may be confused with *G. trocoidea*. Sigal's species generally has
fewer, less tightly packed, more subspherical chambers in the final whorl. Caron
and Luterbacher (1969), in their examination of the original types and of topotypes,
found that the spire height and the intensity of ornamentation varies within this
species.

The origin of *Globigerina trocoidea* is not known, but of the earlier occurring
species, *G. delrioensis* is the most likely candidate.

Brönnimann and Brown (1958) selected the specimen of *Anomalina lorneiana* var.
trocoidea illustrated by Gandolfi on plate 2, fig. 1 as the lectotype for that species
and the type species for their new genus *Hedbergella*. Caron and Luterbacher (1969)
discovered that this specimen was lost, and selected as a replacement lectotype the
specimen illustrated on plate 4, fig. 2 of Gandolfi.

Although the holotype of *Globigerina almadenensis* Cushman and Todd, 1948, is
very poorly preserved, it is believed to be conspecific with the earlier *G. trocoidea*.

Bolli (1959) remarked that his new species *Praeglobotruncana rohri* resembled

Ticinella roberti (Gandolfi), but that it lacks the supplementary apertures of the latter. *Globigerina trocoidea,* also resembles *T. roberti* other than lacking these apertures. Caron and Luterbacher (1969) took notice of the similarity of *P. rohri* and *G. trocoidea,* but retained each as distinct species because the last chamber of Bolli's species does not protrude into the umbilicus. However, by their own redescription of *G. trocoidea,* this protrusion is slight. Nor is it always present. Furthermore, they found intermediate forms in the topotype material. Because the stratigraphic range of *G. trocoidea* encompasses that of *P. rohri* and because of the lack of any distinguishing features, they are considered synonymous.

Distribution: This is shown in Fig. 81.

Range: Mid-Aptian—lower Cenomanian.

> *Globigerina washitensis* Carsey, Plate 25, fig. 4; plate 26, figs 1–3
> *Globigerina* sp. Carpenter, 1925, pl. 17, fig. 5.
> *Globigerina washitensis* Carsey, 1926, p. 44, pl. 7, fig. 10; pl. 8, fig. 2; Plummer, 1931, p. 193, 194, pl. 13, figs 12a, b; Tappan, 1940, p. 122, 123, pl. 19, figs 13a–c; Tappan, 1943, p. 513, pl. 83, figs 1a–c, 2; Lozo, 1944, p. 563, pl. 3, fig. 4; Loeblich and Tappan, 1949, p. 265, pl. 51, figs 4a, b; (?) Colom, 1952, p. 12, 13, pl. 1, figs 21–29; Frizzell, 1954, p. 127, pl. 20, figs 9a–c; Bolin, 1956, p. 293, 294, pl. 39, figs 2a–3c, text-fig. 5 (11a, b); Bolli, 1959, p. 271, pl. 23, figs 6a–7b; Takayanagi, 1960, p. 138, 139, pl. 10, figs 10a–c; Hofker, 1962, p. 82, text-figs 2a–c; Petri, 1962, p. 119, 120, pl. 16, figs 5a–6c; Premoli Silva, 1966, p. 223, 224, text-figs 3a–5c; Bandy, 1967, p. 8, text-figs 3 (4); Hermes, 1969, p. 48, pl. 7, figs 175, 176.
> *Hedbergella hiltermanni* Loeblich and Tappan, 1961a, p. 275, 276, pl. 4, figs 12, 13; Hermes, 1969, p. 51, pl. 7, figs 177–179.
> *Hedbergella (Hedbergella) hiltermanni* Loeblich and Tappan. Arkin and Hamaoui, 1967, pl. 1, figs 14–16.
> *Hedbergella washitensis* (Carsey). Loeblich and Tappan, 1961, p. 278, pl. 4, figs 9–11; (?) Ayala-Castañares, 1962, p. 22, pl. 3, figs 1a–c; pl. 8, figs 2a, b; pl. 12, fig. 1; Postuma, 1962, 7 figs; Takayanagi and Iwamoto, 1962, p. 192, 193, pl. 28, figs 13a–14c; Viterbo, 1965, pl. 9, fig. 3 (4); Salaj and Samuel, 1966, p. 172, 173, pl. 8, figs 6a–c; Pessagno, 1967, p. 284, 285, pl. 49, fig. 1; Barr, 1968, pl. 3, figs 4a–c; Magné and Malmoustier, 1969, p. 38, pl. 2, figs 1 (?, part), 2, 3; Khan, 1970, p. 33, pl. 2, fig. 15; Gorbachik, 1971, pl. 28, figs 6a–c; Risch, 1971, p. 48, 49, pl. 4, figs 17–19.
> *Hedbergella (Catapsydrax) spectrum washitense* (Carsey). Sigal, 1966b, p. 27, 28, pl. 4, figs 7a, b, 9a–10b; pl. 5, figs 1a–10b.
> *Hedbergella (Hedbergella) spectrum washitense* (Carsey). Sigal, 1966b, p. 27, 28, pl. 4, figs 6a, b, 8a, b.
> *Hedbergella (Hedbergella) washitensis* (Carsey). Arkin and Hamaoui, 1967, pl. 1, figs 11–13.
> *Globigerina washitensis hiltermanni* Loeblich and Tappan. Bandy, p. 8, text-fig. 3 (5).
> *Hedbergella* cf. *hiltermanni* Loeblich and Tappan. Bate and Bayliss, 1969, pl. 3, figs 5a–c.
> *Favusella hiltermanni* (Loeblich and Tappan). Michael, 1973, p. 213, 214, pl. 6, fig. 8.
> *Favusella nitida* Michael, 1973, p. 214, pl. 3, figs 10–12.
> *Favusella orbiculata* Michael, 1973, p. 214, pl. 4, figs 1–3.
> *Favusella pessagnoi* Michael, 1973, p. 214, 215, pl. 4, figs 4–6.
> *Favusella quadrata* Michael, 1973, p. 215, pl. 4, figs 7–9.
> *Favusella scitula* Michael, 1973, p. 215, pl. 4, figs 10–12.

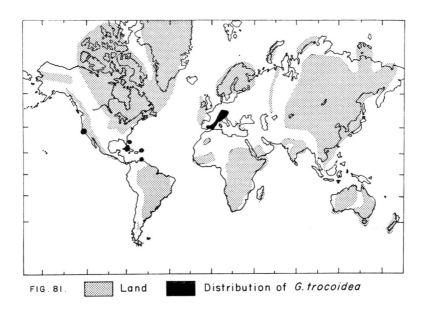

FIG. 81.   Land   Distribution of *G. trocoidea*

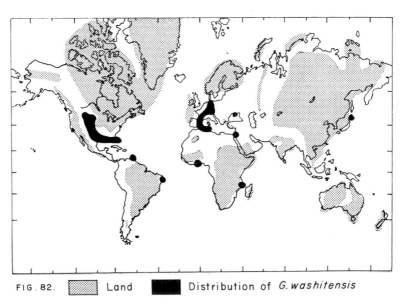

FIG. 82.   Land   Distribution of *G. washitensis*

Favusella washitensis (Carsey). Michael, 1973, p. 215, 216, pl. 5, figs 1–3.
Favusella wenoensis Michael, 1973, p. 216, pl. 5, figs 4–9; pl. 7, figs 3, 4.

Description: Test a low to high trochospiral coil, equatorial periphery lobate, axial periphery rounded. Chambers subspherical to spherical; 11–15 in 2–3 whorls with 3–6 chambers, most commonly 4–5, in the final whorl; moderately to rapidly increasing in size, although chambers in last whorl may be subequal, or last chamber may comprise $\frac{1}{3}$–$\frac{1}{2}$ of test as viewed ventrally. Aperture a low to moderate, interiomarginal, umbilical to extraumbilical–umbilical arch bordered by a narrow lip. Sutures radial to slightly curved, depressed; obscured in early chambers on dorsal side. Surface covered by a reticulated system of fine to coarse ridges dividing the surface into irregular polygons resulting in the characteristic honeycomb pattern; polygons vary greatly in size, some being 10 times the diameter of others, enclosing an equally variable number of very fine pores; polygon walls of uniform thickness, perpendicular to test surface.

Remarks: No other species has the distinctive honeycomb ornamentation. The following morphological characteristics are highly variable: the number of chambers in the final whorl which ranges from three to six, the rate of chamber expansion and the coarseness of the reticulate ornamentation.

Michael (1973) proposed the genus *Favusella* with *Globigerina washitensis* as the type species and simultaneously named several new species within the new genus. These were distinguished from the type species on the bases of the number of chambers per whorl, the rate of their inflation, the spire height and the coarseness of the polygonal ornamentation. All of the species proposed by Michael (i.e. *F. nitida, F. orbiculata, F. pessagnoi, F. quadrata, F. scitula* and *F. wenoensis*) as well as *Hedbergella hiltermanni* Loeblich and Tappan, 1961a, are junior synonyms of *G. washitensis*. Forms similar to any of the above are found associated with "typical" *G. washitensis* over its entire stratigraphic range. This was first noted by Plummer (1931), who, interestingly, was quoted by Michael. Moreover, some of the specimens illustrated by Michael (pl. 5, figs 4–9) show indisputable evidence of post-burial deformation. The resulting flattening and associated features were used by him as species-level characteristics. No evidence produced to date suggests that the many forms are more than intraspecific variants of the single species *G. washitensis*.

Distribution: This is shown in Fig. 82.

Range: Mid-Albian—mid-Cenomanian.

Genus GUEMBELITRIA Cushman, 1933b

Guembelitria Cushman, 1933b, p. 37.
Guembelitriella Tappan, 1940, p. 115.

Type species: *Guembelitria cretacea* Cushman, 1933b.

Description: Test small, initially triserial, or a high trochospire with 3 chambers per whorl, 5–7 whorls, developing supplementary chambers in gerontic stage; distinctly lobate periphery. Chambers inflated, subspherical; frequently not in alignment with corresponding chamber of previous whorl resulting in a slight twisting of the test; proloculus and initial whorl imperforate; gradually to rapidly increasing chamber size, last whorl may comprise half the test. Primary aperture a low to high, interiomarginal arch at base of chamber, with an imperforate, bordering lip; secondary apertures occur at sutural boundaries of gerontic primary chambers or later formed secondary chambers. Surface finely and irregularly perforate with each pore located on a mound, 2–4 such pore mounds may be in close proximity forming an uneven ridge. Sutures deeply depressed.

Remarks: *Guembelitria* is separated from all other Globigerinidae by being the only triserial form which posesses pore mounds.

Guembelitriella has been conclusively shown to be no more than the gerontic stage growth of *Guembelitria* (Masters, in press). This stage of development is found associated with both *Guembelitria* species.

Pokorný (1963) noted without explanation that although it is customary to include *Guembelitria* in the Heterohelicidae, it is not justified. He transferred *Guembelitria* to the Globigerinidae. Brown (1969), concurring with Pokorný, separated *Guembelitria* from the Heterohelicidae on the bases of (1) absence of striae characteristic of that family, and (2) lack of a biserial development in any ontogenetic stage.

One of the reasons *Guembelitria* has been associated with the heterohelicids seems to be that in the minds of most workers "triserial" somehow is more closely related to "biserial" than to "trochospiral". *Guembelitria* should be considered a trochospiral taxon which constantly has three chambers per whorl in the pre-gerontic stages.

High spiring alone cannot be supported as having generic significance. It may be an intraspecific variation (see Intraspecific Variability) or at most of specific rank. No attempt has been made to separate the high-spired *Globotruncana contusa* (Cushman) or *Globigerina brittonensis* (Loeblich and Tappan) from their respective genera on the basis of spire height. Nor in other instances has the presence of three chambers per whorl been given generic level rank, e.g. the mid-Tertiary *Globigerina tripartita* Koch.

Guembelitria is herein placed in the subfamily Globigerininae as a valid generic taxon distinct from the genus *Globigerina* primarily because of the possession of pore mounds.

Olsson (1970) suggested that *Guembelitria* gave rise to *Chiloguembelina*, a lower Tertiary genus regarded by some workers as planktonic, through the development of an asymmetrical aperture and a terminal biserial stage. The supposed "intermediate" forms illustrated by Olsson are merely immature *Chiloguembelina*, and do not possess the *Guembelitria* pore mound structure as claimed. Furthermore, Reiss (1963b) placed *Chiloguembelina* in the newly created Family Chiloguembelinidae belonging to the Superfamily Buliminidea. This he did because *Chiloguembelina*, unlike the Heterohelicidae, has a single-layered wall structure and an asymmetrical aperture with a plate-like flap. Reiss' interpretation is followed here.

Olsson (1970) also proffered the idea that *Guembelitria* and *Globoconusa* (probably a benthonic) are closely related. He stated that the similarity in their wall structure is evident, noting that in *Globoconusa* "there is migration of pores away from spines, but the former close association can be observed". The illustrated specimens purported to demonstrate this close relationship are typical for species of *Globigerina*, and completely unrelated to *Guembelitria*. In all species where pores are randomly located, any ornamentation such as pustules or short spines do occasionally infringe upon the pores, and may even partially surround them. This is clearly the case with Olsson's "*Globoconusa*" and "intermediate" forms with a pore-spine association frequency statistically insignificant. Each pore in *Guembelitria* is always centrally positioned in the pore mound, which is, moreover, not a spine.

The intent on Olsson's part would seem to be an attempt to seek the progenitors of the earliest Palaeocene planktonics from among the last occurring Cretaceous planktonics. The Palaeocene forms may well have had their beginnings here, but the evidence is as yet inconclusive.

Based upon an unpublished dissertation (Davids, Rutgers, 1966), Olsson (1970) advanced the belief that *Guembelitria* had, "At most, a partially planktonic habit . . ". Davids found this genus to be most abundant on the shallow shelf and absent from the continental margin deeper deposits. This dissertation was not seen, but *Guembelitria* is reported in the literature as accompanying nearly every planktonic species over its geographic range. More importantly, however, is the fact that certain living planktonics occur typically and abundantly over the shallow shelf. *Guembelitria*, based upon its frequency and associated species, is assumed to have had a normal planktonic existence.

Specimens younger than Cretaceous that have been attributed to *Guembelitria* do not belong to this genus. The holotype of *Guembelitria* (?) *vivans* Cushman, which has a depressed apertural face with the small aperture of uncertain shape situated at the base of the depression, does not belong to that genus.

Range: Cenomanian—Maastrichtian (*Trinitella scotti* Interval).

Guembelitria cenomana (Keller), Plate 27, figs 1, 3
(?) *Gaudryina* (?) sp. Egger, 1910, p. 99, pl. 5, fig. 22.
Guembelina cenomana Keller, 1935, p. 547, 548, pl. 2, figs 13, 14.
Guembelitria cenomana (Keller). Keller, 1939, pl. 1, figs 3a, b; Maslakova, 1959, p. 117, pl. 15, fig. 6; Lipnik, 1961, p. 42, pl. 2, figs 2a, b; Gawor-Biedowa, 1972, p. 60, 61, pl. 5, fig. 4.
Guembelitria harrisi Tappan, 1940, p. 115, pl. 19, figs 2a, b; Tappan, 1943, p. 507, pl. 81, figs 13–14b; Loeblich and Tappan, 1950, p. 13, pl. 2, figs 15a–18; Frizzell, 1954, p. 110, pl. 15, figs 46a, b; Pessagno, 1967, p. 258, pl. 48, figs 12, 13; Eicher, 1969, text-fig. 3 (part); Michael, 1973, p. 207, pl. 6, fig. 7.
Guembelitriella graysonensis Tappan, 1940, p. 116, pl. 19, figs 3a–c; Frizzell, 1954, p. 110, pl. 15, figs 47a, b; Saavedra, 1965, p. 343, text-fig. 86.

Guembelitria cf. *G. cretacea* Cushman. Arkin and Hamaoui, 1967, pl. 1, figs 25, 26.
Description: As in the genus except with a low apertural arch.
Remarks: *Guembelitria cenomana* has an apertural arch of perhaps half the height of *G. cretacea* Cushman. Variation within the species is confined to the rate of increase in chamber size and the number of supplementary chambers.

The phylogeny of this species is unknown.

Although Keller (1935) described *Guembelina cenomana* as having a "large, gaping aperture situated at the base of the inner margin of the last chamber," the type figure shows one which is much lower than the high arch of *Guembelitria cretacea* Cushman. It is believed that its aperture is comparable to that found on the holotype of *Guembelitria harrisi* Tappan, which then becomes a junior synonym of Keller's species.

Guembelitriella graysonensis Tappan represents the gerontic stage of *Guembelitria cenomana*.

Distribution: This is shown in Fig. 83.

Range: Cenomanian.

Guembelitria cretacea Cushman, Plate 27, fig. 2

Guembelitria cretacea Cushman, 1933b, p. 37, 38, pl. 4, figs 12a, b; Cushman, 1933a, pl. 26, figs 9a, b; Jennings, 1936, p. 28, pl. 3, figs 12a, b; Cushman, 1938a, p. 19, pl. 3, figs 14a, b; Cushman, 1946b, p. 103, pl. 44, figs 14a, b; Cushman, 1948b, p. 256, pl. 21, figs 3a, b, Key pl. 26, figs 9a, b; Harris and Jobe, 1951, p. 38, pl. 7, figs 12a, b; *non* Stelck and Wall, 1954, p. 23, pl. 2, figs 23, 24; Frizzell, 1954, p. 110, pl. 15, figs 45a, b; Jones, 1956, text-fig. 11; Montanaro Gallitelli, 1957, p. 136, 137, pl. 31, figs 1a, b; Montanaro Gallitelli, 1958, p. 143, pl. 3, fig. 2 (1, 3?); Olsson, 1960, p. 27, 28, pl. 4, fig. 8; Hillebrandt, 1962, p. 70, pl. 5, fig. 9; Skinner, 1962, p. 38, pl. 5, fig. 13; Pokorný, 1963, p. 374, fig. 414; Loeblich and Tappan, 1964, p. C652, text-fig 523: 1a, b; Said and Sabry, 1964, p. 390, pl. 3, fig. 32; Saavedra, 1965, p. 343, text-fig. 85; Bandy, 1967, p. 23, text-fig. 12 (6); Pessagno, 1967, p. 258, pl. 87, figs 1–3; *non* Sliter, 1968, p. 94, pl. 13, fig. 16; Ansary and Tewfik, 1968, p. 38, 39, pl. 3, figs 13a, b; Bertels, 1970, p. 28, 29, pl. 1, figs 1–3; text-fig. 7.

Guembelitria columbiana Howe, 1939, p. 62, pl. 8, figs 12, 13; Cushman and Todd, 1945, p. 16, pl. 4, fig. 3.

Guembelitria triseriata (Terquem). Drooger, 1952, p. 95, pl. 15, figs 15, 16.

Guembelitria mauriciana Cole. Hofker, 1960e, text-fig. 37.

Guembelitriella (?) Bertels, 1970, p. 29, 30, pl. 1, figs 4, 5.

Guembelitria (?) sp. Bang, 1971, pl. 4, figs 4a, b; (*non* pl. 5, fig. 6).

Description: As in the genus except with a high, rounded apertural arch.

Remarks: *Guembelitria cretacea* possesses an apertural arch of approximately twice the height of that in *G. cenomana* (Keller). The rate at which the chamber size increases and the number of supplementary chambers are the characteristic variables within this taxon.

This species may have descended from *Guembelitria cenomana* (Keller).

The Cenomanian specimens illustrated by Eicher and Worstell (1970b) as *Guembelitria harrisi* Tappan [= *G. cenomana* (Keller)] do not differ from *G. cretacea* except by occurring in strata older than that of any previously reported *G. cretacea*. These authors remarked that their specimens had a higher-arched aperture than is typical for *G. harrisi*. Two explanations are possible, (1) that *Guembelitria cenomana* is the immediate ancestor of *G. cretacea*, or (2) that the two forms represent a single, long-ranging species in which the apertural height increases through time. In the

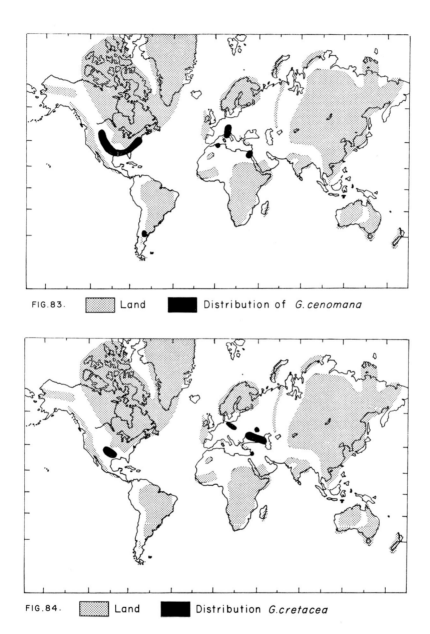

FIG.83.   Land   Distribution of *G.cenomana*

FIG.84.   Land   Distribution *G.cretacea*

absence of a detailed analysis, the true status of both forms cannot be resolved. However, it is suspected that the latter proposal is correct and, if this is the case, the more commonly reported *G. cretacea* will become the junior synonym of *G. cenomana*.

Olsson (1970) described *Guembelitria cretacea* as being covered by low blunt spines which are penetrated by pores. Scanning electron microscopic examination of both fossil and Recent planktonic foraminifers reveals that in no instance are true spines and pores intimately associated in the same structure. Spines are always situated between adjacent pores. For this reason the term "pore mound" is preferable.

The holotype of *Guembelitria colombiana* Howe, 1939, was examined and found to be identical to *G. cretacea*. Howe's Eocene specimen could be the result of contamination, but is more likely due to reworking which has been frequently reported in the Eocene and Miocene strata of the Gulf Coast of the United States.

The specimens assigned to *Guembelitria triseriata* (Terquem) by Drooger (1952) are also suspected to be reworked *G. cretacea*.

Distribution: This is shown in Fig. 84.

Range · Cenomanian—Maastrichtian (*Trinitella scotti* Interval).

Family ROTALIPORIDAE Sigal, 1958.

Definition: Test trochospirally coiled, with or without a single peripheral keel; aperture interiomarginal, umbilical–extraumbilical or umbilical–equatorial in position; with or without accessory apertures formed by the flap-like protrusions of the apertural lips into the umbilicus.

Subfamily ROTALIPORINAE Sigal, 1958.

Definition: Test trochospiral, with a single peripheral keel; aperture interiomarginal, umbilical–extraumbilical with prominent lips extending into the umbilicus which may form accessory apertures along their sutural–umbilical margins.

Generic Key

   I. Without accessory apertures . . . . *Praeglobotruncana*
   II. With accessory apertures . . . . *Rotalipora*

Remarks: Loeblich and Tappan (1964) divided the Rotaliporidae into the subfamilies Hedbergellinae and Rotaliporinae. The Hedbergellinae included the genera *Hedbergella* (= *Globigerina*) and *Clavihedbergella*, which are transferred herein to Globigerinidae, and *Praeglobotruncana*. Under Rotaliporinae were placed the genera *Rotalipora* and *Ticinella*. This arrangement which is similar to those of most other classifications, groups together keeled and non-keeled forms. I consider the keel to be a primary morphologic structure in classification, and one which is more easily determined than the presence of accessory apertures used by Loeblich and Tappan to distinguish these subfamilies (see also Remarks under Subfamily Ticinellinae fam. nov.). Therefore, *Praeglobotruncana* has been placed with *Rotalipora* under the Subfamily Rotaliporinae.

Range: Upper Albian—Turonian.

Genus PRAEGLOBOTRUNCANA Bermúdez, 1952
Praeglobotruncana Bermúdez, 1952, p. 52.
Rotundina Subbotina, 1953, p. 164.
Type species: *Globorotalia delrioensis* Plummer, 1931.

Description: Test trochospiral, biconvex to more commonly a low to high dorsoconvexity; equatorial periphery lobate, but early portion of final whorl may be entire; axial periphery acute. Chambers inflated, sloping convexly from keel to the umbilicus ventrally and to the previous whorl dorsally; with a single, imperforate, peripheral band covered by randomly positioned pustules. Aperture a low, interiomarginal, umbilical–extraumbilical arch bordered by an imperforate lip or flap. Wall finely and densely perforate.

Remarks: *Praeglobotruncana* differs from other members of the Rotaliporidae by lacking accessory apertures.

Reiss (1963b) regarded *Globotruncanella* as a junior synonym of *Praeglobotruncana*. This was followed by El-Naggar (1971a), who also considered *Helvetoglobotruncana* and *Abathomphalus* as synonyms of *Praeglobotruncana*. None of these three genera is structurally similar to *Praeglobotruncana*. Each possesses a tegillum and *Abathomphalus* has a double keel, which is absent in *Praeglobotruncana*. *Globotruncanella* and *Helvetoglobotruncana* each possess a single keel, but the latter is clearly not phyletically related to *Praeglobotruncana*. *Globotruncana citae* Bolli (= *G. havanensis* Voorwijk), the type species of *Globotruncanella*, may have evolved from a *Praeglobotruncana*, but it is equally possible that it arose from a *Globigerina* [see also Remarks under *Globotruncana havanensis*]. Nevertheless, the presence of a true tegillum precludes the treatment of *Globotruncanella* as a synonym of *Praeglobotruncana*.

Globotruncana stephani Gandolfi, the type species of *Rotundina*, shares all its supraspecific characteristics with *Praeglobotruncana delrioensis*. All subsequent workers, except some Russians, have followed Reiss (1963b) in this regard.

In studying the apertures of several species of *Praeglobotruncana*, Caron (1965) observed that:

(1) The apertural flap more or less covers the umbilicus. Each chamber supports a flap, attached above the aperture, and after a supple fold, spreads out over the umbilicus. Successive flaps are not always visible.

(2) When asymmetric, the folds are directed abaperturally.

(3) Small lips form a border along the umbilical periphery.

Caron concluded that the apertures alone are not a sufficient criterion for the separation of *Praeglobotruncana* from *Globotruncana*, but that in addition the former most often has globulose chambers, particularly ventrally. Moreover, she observed that the chambers never overlap one another. The ventral sutures are depressed, radial and straight.

The flexure of the apertural flap, as envisioned by Caron, is not universally present even within a given species. The flap of the *Praeglobotruncana* is nearly as variable in its form as is the tegillum of *Globotruncana*. The difference between the two structures is that the apertural extensions of *Globotruncana* become fused into a complex umbilical cover plate.

Scheibnerová (1960) introduced two new species of *Praeglobotruncana*: *P. oraviensis* and *P. oraviensis trigona*. They appear to be irregular growth forms, but the type figures and descriptions are inadequate to determine whether either represents a new species.

Scheibnerová (1962) described her new species *Praeglobotruncana hagni* as possessing two closely spaced keels. Scanning electron micrographs were given for this species by Belford and Scheibnerová (1971). A careful examination of these reveals not a double keel, but a wide single keel (for comparison see pl. 28, fig. 3). The type figures were stylised making accurate appraisal difficult. Thus, it is not clear whether the holotype does, in fact, have a double keel.

Praeglobotruncana biconvexa biconvexa and *P. biconvexa gigantea* double-keeled forms of Samuel and Salaj (1962) are not accepted as *Praeglobotruncana*, but because of the lack of type material and other essential data, they are not given further consideration here.

Caron (1966) introduced the name *Praeglobotruncana algeriana* for a double-keeled form. She stressed its importance as a transitional stage between the single-keeled *Praeglobotruncana* and the *Globotruncana*. Unfortunately, the critical presence or absence of a tegillum is not discussed by Caron. Because all double-keeled forms have thus far been shown to possess a tegillum, it is suspected that one will be found on better preserved specimens of the taxon *algeriana*. Tegilla are often broken away leaving only a narrow basal remnant which can be mistaken for apertural lips. The taxon *algeriana* is transferred to the genus *Globotruncana*.

Range: Upper Albian—mid-Turonian.

Praeglobotruncana delrioensis (Plummer), Plate 27, figs 4, 5; plate 28, fig. 1

Globorotalia delrioensis Plummer, 1931, p. 199, 200, pl. 13, figs 2a–c; Loeblich and Tappan, 1946, p. 257, text-fig. 4b; Frizzell, 1954, p. 129, pl. 20, figs 27a–c.

Globorotalia marginaculeata Loeblich and Tappan, 1946, p. 257, pl. 37, figs 19–21; pl. 37, figs 19a–21; text-fig. 4A; Frizzell, 1954, p. 129, 130, pl. 20, figs 29a–c.

Globotruncana stephani Gandolfi. Cita, 1948b, p. 159, 160, pl. 4, figs 6a–c; Mornod, 1950, p. 587, 588, text-figs 10 (1a–2c) (*non* 3a–c; pl. 15, figs 9a–c (? figs c–r, 10–17)); (?) Hagn and Zeil, 1954, p. 33, 34, pl. 2, figs 7a–c; pl. 5, figs 7, 8.

Globotruncana appenninica Renz. Jacob and Sastry, 1950, p. 267, text-figs 1a–c.

Praeglobotruncana delrioensis (Plummer). *non* Bermúdez, 1952; p. 52, 53, pl. 7, figs 1a–c; Brönnimann and Brown, 1956, p. 531, 532, pl. 21, figs 8–10, pl. 24, figs 16,

17; Bolli, Loeblich and Tappan, 1957, p. 39, 40, pl. 9, figs 1a–c; Klaus, 1959, p. 793, 794, pl. 6, figs 1a–c; *non* Vinogradov, 1960b, p. 38, 40, pl. 3, figs 13a–c; Loeblich and Tappan, 1961, p. 280, 282, 284, pl. 6, figs 9a–10c, 11 (?), 12a–c (?); Borsetti, 1962, p. 31, 32, pl. 2, fig. 6; text-figs 8–11, 36–39 (?), 40; Lehmann, 1963, p. 140, pl. 2, figs 1a–2c; text-figs 2a, 3a; Pokorný, 1963, p. 386, fig. 432, Cita-Sironi, 1963, text-figs, 16; *non* Küpper, 1964, p. 615, pl. 2, fig 4a–c; Todd and Low, 1964, p. 404, pl. 2, figs 4a–c; Lehmann, 1965, text-fig. 1a; Lehmann, 1966b, p. 157, 158, pl. 1, figs 5, 6; Caron, 1966, p. 72, 73, pl. 2, figs 2a–c; Marianos and Zingula, 1966, p. 337, 338, pl. 37, figs 11a–c; Salaj and Samuel, 1966, p. 188, pl. 15, figs 3a–c, tab. 35 (3); *non* Pessagno, 1967, p. 286, 287, pl. 52, figs 3–5; pl. 100, fig 7; Prosnyakova, 1967, p. 5, 6, pl. figs 3a, b, v; Bate and Bayliss, 1969, pl. 2, figs 4a–c; Pessagno, 1969a, pl. 12, figs A–C; Neagu, 1969, p. 141, pl. 16, figs 4–6; pl. 18, figs 1–3, 7, 8; pl. 21, figs 6–8 (*non* 3–5, pl. 22, figs 1–3); *non* Scheibnerová, 1969, p. 60–62, tab. 8, figs 3a–c. Eicher, 1969, text-fig. 3 (part); *non* Gorbachik, 1971, pl. 29, figs 1a–c; *non* Neagu, 1972, p. 215, 216, pl. 6, figs 40–42; Michael, 1973, p. 216, 217, pl. 5, figs 10–12.

Rotundina stephani (Gandolfi). Subbotina, 1953, p. 179–181, pl. 2, figs 5a–c (*non* 6a–7c); pl. 3, figs 1a–2c; *non* Salaj and Samuel, 1966, p. 195, tab. 33 (8).

Praeglobotruncana cf. *P. delrioensis* (Plummer). Brönnimann and Brown, 1956, p. 511, text-figs 9, 9a–c; 11, 11a, b; 13a, b, d; 15c–f.

Praeglobotruncana (*Praeglobotruncana*) cf. *stephani* (Gandolfi). Banner and Blow, 1959, p. 17, pl. 3, figs 4a–c (*non* text-fig. 1a).

Praeglobotruncana stephani stephani (Gandolfi). Klaus, 1959, p. 794, 795, pl. 6, figs 2a–c; Neagu, 1969, pl. 18, figs 4–6 (*non* 9, 10; pl. 16, figs 1–3, 7–12; pl. 21, figs 9, 10; pl. 23, fig. 3); Neagu, 1970b, p. 64, pl. 25, figs 7–15.

Praeglobotruncana stephani (Gandolfi). Bykova *et al.*, 1959, p. 302, text-figs 687A–C; Klaus, 1961, text-figs 5a–c; Loeblich and Tappan, 1961, p. 284–290, pl. 6, figs 2a–c (*non* 1a–c, 3a–8c); Postuma, 1962, 7, figs. Dabagyan, 1963, p. 115–119, pl. 1, figs 5a–c; Săndulescu, pl. IIIA, figs 1a–c; Eicher, 1966, p. 28, pl. 6, figs 4a–c; Marks, 1967, p. 273, 274, pl. 2. figs 4–6 (*non* 7–12; pl. 3, figs 1–6); Neagu, 1968, text-fig. 2 (49); Gawor-Biedowa, 1972, p. 76–78, pl. 8, figs 1a–c.

Praeglobotruncana stephani var. 2 Malapris and Rat, 1961, p. 89, pl. 2, figs 2a–c.

Globotruncana (*Rotalipora*) *delrioensis* (Plummer). Majzon, 1961, p. 762, pl. 6, fig. 5.

Praeglobotruncana delrioensis forma *globosa* Borsetti, 1962, p. 32, pl. 2, fig. 7; text-figs 12, 13 (?), 14.

Praeglobotruncana marginaculeata (Loeblich and Tappan). Klaus 1960, p. 301, 302; text-figs 1b–d; Caron, 1966, p. 73, pl. 2, figs 2a–c; *non* Salaj and Samuel, 1966, p. 192, pl. 15, figs 4a–c; tab. 35 (4); Neagu, 1970b, p. 65, pl. 27, figs 7–15.

Praeglobotruncana delrioensis delrioensis (Plummer). Bandy, 1967, p. 16, 17, text-fig. 8 (1).

Praeglobotruncana barbui Neagu, 1969, p. 143, pl. 18, figs 11–15; pl. 19, figs 1–3, 7–9 (*non* 4–6, 10–12); pl. 20, figs 1–3.

Praeglobotruncana aff. *delrioensis* (Plummer). Neagu, 1969, pl. 14, figs 10–12.

Praeglobotruncana prahovae Neagu, 1969, p. 144, pl. 21, fig. 11–13; pl. 22, figs 4–6, 9–11 (*non* pl. 23, figs 4–10; pl. 24, figs 1–9).

Praeglobotruncana. Berger and Rad, 1972, pl. 3, fig. 2 (part).

Praeglobotruncana stephani gibba Klaus. Neagu, 1969, p. 141, 142, pl. 20, figs 7–12; pl. 21, figs 1, 2.

Description: Test a low trochospire, biconvex with the dorsoconvexity generally greater; equatorial periphery of early chambers in last whorl may be entire or more frequently moderately lobate throughout; axial periphery acute. Chambers moderately inflated with the greatest thickness occurring approximately $\frac{1}{3}$ of the height from the axis of coiling; width and height equidimensional and approximately 50% greater than thickness; sloping with a gentle convexity from the periphery; 5, occasionally 6 in the final whorl, 11 to 17 chambers in 3 whorls; with an imperforate peripheral band on which is formed coarse pustules that may merge between chambers to reduce the lobate outline. Aperture a low, interiomarginal, umbilical–extraumbilical arch. The apertural flap is a narrow lip-like structure in the extraumbilical position but flares broadly into the umbilicus; shape is highly variable. Wall densely perforate, with random small pustules. Umbilicus moderately wide, shallow. Sutures depressed, straight, radial, may be slightly raised and curved dorsally. Apical angle ranges from 118° to 136°.

Remarks: *Praeglobotruncana delrioensis* has the lowest spire height of any of its congeners. Several minor variations can combine to produce a specimen unlike the holotype. Among these are the coarseness of the pustules on the keel, apical angle or spire height and the number of chambers per whorl.

This species probably evolved from the *Globigerina* stock. From a morphological approach, the most likely ancestor is *G. delrioensis* Carsey. Axial compression and development of a keel could result in *Praeglobotruncana delrioensis*, but this is as yet supposition.

Much confusion has surrounded three *Praeglobotruncana* species—*P. delrioensis* (Plummer), *P. stephani* (Gandolfi) and *P. turbinata* (Reichel). Brönnimann and Brown (1956) treated *P. stephani* as a synonym of *P. delrioensis* but retained *P. turbinata* for high-spired forms. Brown (1962, personal communication) revised his opinion and regarded each as valid taxa.

Loeblich and Tappan (1961a) distinguished between *P. delrioensis* and *P. stephani* but treated *P. turbinata* as a junior synonym of the latter. They noted the association of *P. stephani* and *P. turbinata*, and stated that the greater number of chambers in the final whorl and higher spire "are only functions of the age of the specimen . . ." Pessagno (1967) agreed stating that they "should be regarded as adult or gerontic individuals . . .". He also repeated Loeblich's and Tappan's (1961a) comment that since they have the same geologic and geographic range, it is unnecessary to retain both names.

Although identical geological and geographical ranges are highly suggestive that *P. stephani* and *P. turbinata* are conspecific, this alone does not constitute proof. Geographic distribution is controlled by the environment. Both taxa are morphologically similar, and would be expected to inhabit similar or identical environments. This can be observed in living species where the morphologically more simple forms are found closer to shore and the more complex farther out. Identical geologic ranges again are only suggestive of conspecificity. Many valid taxa have identical geologic ranges.

a. Measurement of apical angle.

The height of the spire is the most obvious morphologic criteria for separating these forms. A numerical value can be assigned to the spire height by measuring the apical angle (Diagram *a*). Application of this procedure to both illustrations and specimens of *P. delrioensis*, *P. stephani* and *P. turbinata* results in three groupings. The first group with the largest apical angle, and therefore, the lowest spire height, ranges from 118°–136°. This group includes the holotype (136°) and topotypes of *P. delrioensis*. Group two has an apical angle range from 92°–104°. This group interestingly includes the holotypes of *P. stephani* (94°) and of *P. turbinata* (104°). The last group, *P.* sp., ranges from 36°–75°, and includes those high-spired forms generally regarded as *P. "turbinata"*.

Loeblich and Tappan (1961a) and Pessagno (1967) considered the high-spiring an ontogenetic process, i.e. more chambers and whorls with maturity. An analysis of the number of chambers and whorls for the three species in question (Table 9) shows little variation. The high-spired forms assigned to *P. turbinata* have one to three more chambers and one half to one more whorl than *P. stephani*. This clearly shows that the height of the spire has no bearing upon either the number of chambers or whorls. Therefore, the spire height is not an ontogenetic trait.

TABLE 9

	Total chambers	Chambers in final whorl	No. of whorls	Apical angle
P. delrioensis	11–17	5–6	3	118°–136°
P. stephani	12–20	5–7	$2\frac{1}{2}$–3	92°–104°
P. sp.	21–23	6–7	$3\frac{1}{2}$–4	36°–75°

However, the possibility remains that the variation in the apical angle may be an environmental or a genetic intraspecific trait. Until such time as this can be demonstrated, all three are considered as valid taxa.

Pustules on the keel are irregularly positioned, but may occasionally be aligned parallel with the imperforate band for a short distance. Two to three pustules may occur across the width of the band. When this happens, that segment may appear in light microscopy as double keel. It is not, although it may be the forerunner of such.

Neagu (1969) described as new *Praeglobotruncana barbui* from the Upper Cenomanian of Romania. Although the holotype was not examined during this study, the type description and figure differ in no way from that of *P. delrioensis*.

Distribution: This is shown in Fig. 85.

Range: Upper Albian—Cenomanian.

Praeglobotruncana prahovae Neagu

Praeglobotruncana prahovae Neagu, 1969, p. 144, pl. 23, figs 4–8; (*non* 9–10; pl. 21, figs 11–13; pl. 22, figs 4–6, 9–11; pl. 24, figs 1–9).

Description: "Test free, trochospiral with chambers arranged in 2 to 3 whorls, with a strongly convex to conical–convex spiral side; sutures of chambers are

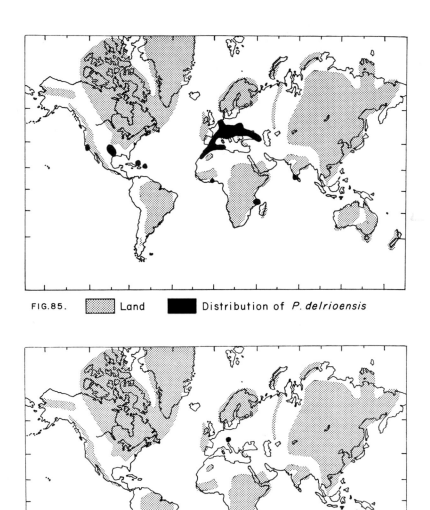

FIG.85. ☐ Land   ■ Distribution of *P. delrioensis*

FIG.86. ☐ Land   ■ Distribution of *P. prahovae*

arcuate, deepened and faintly carinate (particularly on older whorls); surfaces of chambers are convex to markedly convex; the last whorl is composed on 6 to 8 large triangular to rounded chambers, slightly flattened towards the periphery, with radiate and deep sutures the surface being smooth or covered with fine pustules, which are, however, lacking in the 2–3 last chambers. The umbilicus is wide or very wide, deep, and frequently covered with the lamellar extensions of the anterior apertures. Periphery is markedly lobate and provided with a faint keel, which may be absent on the last 2–3 chambers. Aperture consists of a low interiomarginal bow and is protected by a lip that continues umbilically by a lamellar extension." (Type description.)

Remarks: The high degree of chamber inflation is the distinctive characteristic of this species. The range of variability in this recently described species has not yet been established, and its phylogeny is unknown.

Because the significance and persistence of chamber inflation is not understood, *Praeglobotruncana prahovae* may turn out to be an intraspecific variant of *P. stephani* (Gandolfi).

Distribution: This is shown in Fig. 86.

Range: Lower Turonian.

Praeglobotruncana stephani (Gandolfi), Plate 28, figs 2–4

Globotruncana stephani Gandolfi, 1942, p. 130–133, pl. 3, figs 4, 5; pl. 4, figs 36, 37, 41–44; pl. 6, fig. 4 (part); pl. 9, figs 5, 8; pl. 14, fig. 2; Bolli, 1945, p. 224, pl. 9, fig. 2, text-fig. 1 (3, 4); *non* Colom, 1947, p. 97, pl. 25, figs 13, 14, 18, 21; *non* Colom, 1948, p. 27, pl. 4, figs 4, 8–10, 12, 13; *non* Cita, 1948b, p. 159–160, pl. 4, figs 6a–c; Mornod, 1950, p. 587, 588, pl. 15, figs 9a–c (? figs c–r, 10–17), text-fig. 10 (3a–c) (*non* 1a–2c); Carbonnier, 1952, p. 116, 117, pl. 6, figs 2a–c; *non* Hagn and Zeil, 1954, p. 33, 34, pl. 2, figs 7a–c, pl. 5, figs 7, 8; Ayala-Castañares, 1954, p. 411, 412, pl. 11, figs 2a–c; Hiltermann, 1956, text-fig. 2 (part); Książkiewicz, 1956, p. 269, 270, pl. 30, figs 12, 14; text-figs 36 (1–6, 8, 9), (*non* 7); (?) Książkiewicz, 1958, text-fig. 1 (15a–c); Hiltermann and Koch, 1962, p. 330, pl. 47, figs 3a–c; tab. 19 (part).

Globotruncana apenninica var. β Gandolfi, 1942, p. 118, text-fig. 41 (2a, b); Hiltermann, 1956, text-fig. 2 (part).

Globorotalia californica Cushman and Todd, 1948, p. 96, 97, pl. 16, figs 22, 23.

Globotruncana (Globotruncana) stephani Gandolfi. Reichel, 1950, p. 608, 609, pl. 16, fig. 6; pl. 17, fig. 6.

Globotruncana stephani var. *turbinata* Reichel, 1950, p. 609, 610; Mornod, 1950, p. 588, 589, pl. 15, figs 18a–20; text-figs 11 (1e–3c); Ayala-Castañares, 1954, p. 412, 413, pl. 11, figs 3a–c.

Globotruncana (Globotruncana) stephani var. *turbinata* Reichel. Church, 1952, p. 70, text-fig. 1 (lower).

Globigerina aumalensis Sigal, 1952, p. 28, text-fig. 29.

Rotundina stephani (Gandolfi). Subbotina, 1953, p. 179–181, pl. 2, figs 6a–7c (*non* 5a–c; pl. 3, figs 1a–2c); Maslakova, 1959, p. 112, pl. 11, figs 1a–c.

Globotruncana stephani turbinata Reichel. Hagn and Zeil, 1954, p. 34, pl. 2, figs 2a–c; pl. 5, figs 3, 4; (?) Książkiewicz, 1958, text-fig. 1 (14a–c).

Globotruncana (Rotundina) californica (Cushman and Todd). *non* Küpper, 1955, p. 116, 117, pl. 18, figs 7a–c.

Globotruncana (Rotundina) aumalensis (Sigal). Küpper, 1955, p. 116, pl. 18, figs 5a–c.

Globotruncana (Rotundina) stephani stephani (Gandolfi). Küpper, 1955, p. 116, pl. 18, figs 6a–c.

Globotruncana (Praeglobotruncana) renzi subsp. *primitiva* Küpper, 1956, p. 43, pl. 8, figs 2a–c.

Praeglobotruncana cf. *P. delrioensis* (Plummer). Brönnimann and Brown, 1956, p. 514, text-fig. 16d.

Praeglobotruncana stephani (Gandolfi). Bolli, Loeblich and Tappan, 1957, p. 39, 40, pl. 9, figs 2a–c; *non* Bykova *et al.*, 1959, p. 302, text-figs 687A–C; Klaus, 1960, p. 302–304, text-fig. 1e; *non* Klaus, 1961, text-figs 5a–c; Loeblich and Tappan, 1961, p. 284–290; pl. 6, figs 1a–c, 3a–c, 5a–c, 7a–8c (*non* 2a–c, 4a, b, 6); *non* Malapris and Rat, 1961, p. 88, 89, text-fig. 7a; *non* Postuma, 1962, 7 figs; Graham, 1962, p. 105, pl. 20, figs 24Aa–Fc, Ha–c (*non* 24Ga–c); Luterbacher, 1963, in Renz, Luterbacher and Schneider, p. 1086, pl. 9, figs 1a–c, ? 3a–c; *non* Dabagyan, 1963, p. 115–119, pl. 1, figs 5a–c; Takayanagi, 1965, p. 207, 208, pl. 22, figs 3a–c; Saavedra, 1965, p. 324, text-fig. 8; *non* Săndulescu, 1966, pl. IIIa, figs 1a–c; Douglas and Sliter, 1966, p. 107, pl. 4, figs 1a–c; Marianos and Zingula, 1966, p. 337, pl. 37, figs 10a–c; Butt, 1966, p. 176, pl. 3, figs 5a–c; *non* Eicher, 1966, p. 28, pl. 6, figs 4a–c; Prosnyakova, 1967, p. 3, 4, pl. figs 1a, b, v; Marks, 1967, p. 273, 274, pl. 2, figs 7–12 (*non* 4–6); pl. 3, figs 1–3, (?) 4–6; *non* Pessagno, 1967, p. 287, pl. 50, figs 9–11; *non* Neagu, 1968, text-fig. 2 (49); Barr, 1968, pl. 3, figs 5a–c; Radoičić, 1968 pl. 1, fig. 6; pl. 3, fig. 6; Moorkens, 1969, p. 446, pl. 1, figs 5a–c; Eicher, 1969, text-fig. 3 (part); Caron and Luterbacher, 1969, p. 26, pl. 8, figs 7a–c; Douglas, 1969, p. 173, pl. 2, figs 1a–c; *non* Gawor-Biedowa, 1972, p. 76–78, pl. 8, figs 1a–c.

Globotruncana stephani var. *stephani* (Gandolfi). Witwicka, 1958, p. 209, 210, pl. 14, figs 26a–c.

Praeglobotruncana (Praeglobotruncana) stephani (Gandolfi). Banner and Blow, 1959, p. 3, text-fig. 1a (*non* pl. 3, figs 4a–c).

Praeglobotruncana stephani stephani (Gandolfi). *non* Klaus, 1959, p. 794, 795, pl. 6, figs 2a–c; Borsetti, 1962, p. 32, 33, pl. 2, figs 12, 13; (*non* text-figs 45, 46, 76–78, 116?, 117?); Caron, 1966, p. 73, pl. 2, figs 3a–c; Neagu, 1969, p. 141, pl. 16, figs 1–3, 7–12; pl. 18, figs 9, 10 (*non* 4–6); pl. 21, figs 9, 10; pl. 23, fig. 3 (?); Neagu, 1970a, fig. 1 (18); *non* Neagu, 1970b, p. 64, pl. 25, figs 7–15.

Praeglobotruncana stephani turbinata (Reichel). Klaus, 1959, p. 795, pl. 6, figs 3a–c; Borsetti, 1962, p. 33, pl. 2, figs 14–16; (*non* text-figs 102–107); Samuel, 1962, p. 187, pl. 9, figs 2a–c; Broglio Loriga and Mantovani, 1965, pl. III, figs 22, 23; Neagu, 1968, text-fig. 2 (50).

Globotruncana kupperi Thalmann, 1959, p. 130.

Praeglobotruncana delrioensis (Plummer). Vinogradov, 1960b, p. 38, 40, pl. 3, figs 13a–c; Neagu, 1969, p. 141, pl. 21, figs 3–5 (*non* 6–8; pl. 16, figs 4–6; pl. 18, figs 1–3, 7, 8; pl. 22, figs 1–3); Scheibnerová, 1969, p. 60–62, tab. 8, figs 3a–c; Neagu, 1972, p. 215, 216, pl. 6, figs 40–42.

Praeglobotruncana stephani var. *turbinata* Reichel. Vinogradov, 1960b, p. 40, pl. 3, figs 14a–c.

Praeglobotruncana stephani var. *gibba* Klaus, 1960, p. 304–307, text-fig. 1f.

Praeglobotruncana stephani gibba Klaus. Caron, 1966, p. 73, pl. 2, figs 4a–c; Neagu, 1970b, p. 65, pl. 27, figs 1–6.

Praeglobotruncana stephani var. 1 Malapris and Rat, 1961, p. 89, pl. 2, figs 1a–c; text-fig. 4a.

non Praeglobotruncana stephani var. 2 Malapris and Rat, 1961, p. 89, pl. 2, figs 2a–c; text-fig. 4b.

non Praeglobotruncana stephani var. 3 Malapris and Rat, 1961, p. 89, 90, pl. 2, figs 3a–c.

Globotruncana (*Thalmanninella*) *stephani* Gandolfi. Majzon, 1961, p. 762, pl. 5, fig. 6.
Praeglobotruncana turbinata Reichel. Postuma, 1962, 7 figs.
Rotalipora appenninica β (Gandolfi). *non* Borsetti, 1962, p. 35, 36, pl. 1, figs 5, 6;
pl. 2, figs 9, 10; text-figs 3–6, 30–32, 50–56.
"*Praeglobotruncana*" *oraviensis* Scheibnerová. Samuel, 1962, p. 187, 189, pl. 9,
figs 3a–4.
Praeglobotruncana oraviensis oraviensis Scheibnerová. Salaj and Samuel, 1966,
p. 192, 193, pl. 14, figs 4a–c; tab. 35 (11).
Praeglobotruncana oraviensis trigona Scheibnerová. Salaj and Samuel, 1966,
p. 193, pl. 15, figs 1a–c; tab. 35 (12).
Praeglobotruncana gibba Klaus. Salaj and Samuel, 1966, p. 188, 189, pl. 15,
figs 2a–c; tab. 35 (5).
Praeglobotruncana delrioensis stephani (Gandolfi). Bandy, 1967, p. 17, text-fig.
8 (6).
Praeglobotruncana oraviensis Scheibnerová. Sturm, 1969, pl. 10, figs 4a–c.
Praeglobotruncana marginaculeata (Loeblich and Tappan). Neagu, 1969, p. 142,
pl. 16, figs 13–15; pl. 17, figs 1–7.
Praeglobotruncana barbui Neagu, 1969, p. 143, pl. 19, figs 4–6, 10–12 (*non* 1–3,
7–9; pl. 18, figs 11–15; pl. 20, figs 1–3).
Praeglobotruncana prahovae Neagu, 1969, p. 144, pl. 23, figs 9, 10; pl. 24, figs 1 9
(*non* pl. 21, figs 11–13; pl. 22, figs 4–6, 9–11; pl. 23, figs 4–8).
Praeglobotruncana aumalensis (Sigal). Porthault, 1969, p. 537, pl. 2, figs 5a–c.
Praeglobotruncana delrioensis turbinata (Reichel). Scheibnerová, 1969, text-fig. 11.

Description: Test unequally biconvex with a moderately strong dorsal convexity;
equatorial periphery lobate; axial periphery acute. Chambers gradually increase in
size, 12–20 in $2\frac{1}{2}$–3 whorls with 5–6, rarely 7, in the final whorl; dorsal walls nearly
flat to slightly convex, ventral walls more inflated; width and height equidimensional, greater than thickness; triangular-shaped as seen in edge view; with a single
pustulose imperforate peripheral band. Aperture a low, interiomarginal, umbilical–
extraumbilical arch bordered by an umbilically flaring flap. Umbilicus narrow and
moderately deep. Sutures depressed, straight, radial ventrally; raised, curved dorsally.
Surface densely perforate, pustulose. Apical angle ranges between 92° and 104°.

Remarks: This species differs from *Praeglobotruncana prahovae* Neagu by having
less inflated chambers. It is distinguished from all other species by its apical angle.
Minor variations occur in the total number of chambers, the apical angle and the
coarseness of the keel.

Eicher and Worstell (1970b) suspected that *Praeglobotruncana stephani* arose from
P. delrioensis (Plummer). The similarity of these species supports this assumption.

For an analysis of this species, see Remarks under *Praeglobotruncana delrioensis*
(Plummer).

The holotype of *Globorotalia californica* Cushman and Todd, 1948, was examined
and found to be a typical *Praeglobotruncana stephani*.

Caron and Luterbacher (1969) reported that the holotype of *Globotruncana
stephani* var. *turbinata* Reichel is missing. Based upon measurements of the apical
angle made from the type figure (see Remarks under *Praeglobotruncana delrioensis*),
it is conspecific with *P. stephani*. Thus, it is unnecessary to designate a neotype.

Although placed in synonymy by other authors, *Globigerina aumalensis* Sigal is
distinguished by Porthault (1969) on the bases of the possession of a weakly developed keel and the lack of a dorsal sutural rim. The presence or absence of a keel is of
primary taxonomic importance. The strength of the keel, which is related to the

age of the individual and its environment, is not. The sutural rim noted by Port-hault is merely the keel which remains uncovered by later chambers. Its exposure is controlled by the position with which subsequent chambers are added. The apical angle of *G. aumalensis* is approximately 92°, which is at the lower limit of *P. stephani*. The two are judged synonymous.

Thalmann (1959) proposed the new name *Globotruncana kupperi* as a substitute for *G.* (*Praeglobotruncana*) *renzi primitiva* Küpper, 1956, a junior homonym of *G. primitiva* Dalbiez. Loeblich and Tappan (1961) examined Küpper's specimen, find-ing it to be a crushed *P. stephani*.

Praeglobotruncana stephani var. *gibba* Klaus, 1960, is easily within the range of variability of *P. stephani*.

Distribution: This is shown in Fig. 87.

Range: Lower Mid-Cenomanian—Lower Turonian.

Praeglobotruncana sp., Plate 29, figs 1–3

Globotruncana stephani var. *turbinata* Reichel. Mornod, 1950, p. 609, 610, pl. 15, figs 18a–j, 19, 20; Ayala-Castañares, 1954, p. 412, 413, pl. 11, figs 3a–c; Prosnyakova, 1967, p. 4, 5, pl. figs 2a–c.

Praeglobotruncana cf. *P. delrioensis* (Plummer). Brönnimann and Brown, 1956, p. 514, text-fig. 16c, e.

Globotruncana (*Praeglobotruncana*) *stephani turbinata* (Reichel). Küpper, 1956, p. 43, pl. 8, figs 1a–c.

Globotruncana (*Globotruncana?*) *stephani stephani* Gandolfi. Gandolfi, 1957, p. 62, pl. 9, figs 3a–c.

Globotruncana (*Globotruncana?*) *stephani turbinata* (Reichel). Gandolfi, 1957, p. 62, 64, pl. 9, figs 4a, b.

Globotruncana (*Marginotruncana*) *turbinata* Reichel. Majzon, 1961, p. 763, pl. 7, fig. 6.

Praeglobotruncana stephani (Gandolfi). Loeblich and Tappan, 1961, p. 284–290, pl. 6, figs 4a, b, 6 (*non* 1a–3c, 5a–c, 7a–8c); Graham, 1962, p. 105, pl. 20, figs 24Ga–c (*non* 24Aa–Fc, Ha–c).

Globotruncana stephani turbinata Reichel. Küpper, 1964, p. 632, 633, pl. 2, figs 7a–c.

Praeglobotruncana stephani turbinata (Reichel). Săndulescu, 1966, pl. IIIa, figs 2a–c.

Praeglobotruncana stephani gibba Klaus. Neagu, 1969, p. 141, 142, pl. 20, figs 7–12; pl. 21, figs (?) 1, 2; Neagu, 1970a, fig. 1 (19); Belford and Scheibnerová, 1971, p. 332, pl. 2, figs 1–8.

Praeglobotruncana delrioensis (Plummer). Neagu, 1969, p. 141, pl. 22, figs 1–3 (*non* pl. 16, figs 4–6; pl. 18, figs 1–3, 7, 8; pl. 21, figs 3–8).

Description: Test strongly convex dorsally, slightly convex ventrally; equatorial periphery slightly lobate; axial periphery acute. Chambers with little or no inflation dorsally; with slight inflation ventrally, curving with gentle convexity from the periphery to the umbilicus; 21–23 chambers in $3\frac{1}{2}$–4 whorls with 6–7 in the final whorl; gradually increasing in size; height, width and thickness nearly equidimen-sional; with an imperforate peripheral band covered by randomly positioned pustules. Aperture a low, interiomarginal, umbilical–extraumbilical arch, bordered by a flap. Umbilicus narrow, deep. Sutures generally raised, curved dorsally; depressed, straight, radial ventrally. Apical angle 36°–75°. Surface densely perforate, pustulose.

Remarks: This species of *Praeglobotruncana* has the greatest dorsoconvexity, and is distinguished primarily on this factor. This species varies as do the others in chamber number, keel coarseness and apical angle. The range of variation exhibited

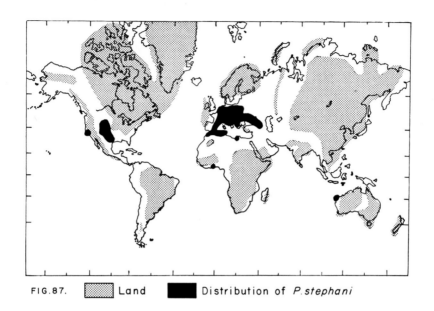

FIG.87. ☐ Land ■ Distribution of *P.stephani*

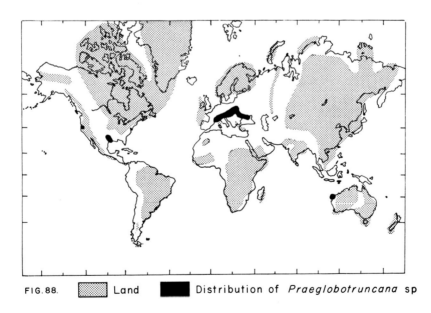

FIG.88. ☐ Land ■ Distribution of *Praeglobotruncana* sp

by the apical angle is two to three times greater than the range shown by other species.

Praeglobotruncana stephani (Gandolfi) is believed to be ancestral to this species because it is the next most high-spired form.

This form has usually been assigned to *Praeglobotruncana turbinata* (Gandolfi) by most workers. The latter is considered here as a synonym of *P. stephani* (Gandolfi) because the apical angle of the holotype is within the clustering of angles typical of *P. stephani* (see Remarks under *P. delrioensis*). This leaves the high-spired forms without a name. As previously mentioned, the possibility exists that spire height may be a reflection of an alternation of generations or an environmental response. If so, those forms with low apical angles can most probably be assigned to *P. stephani*. Until more conclusive evidence is available, it is deemed unwise to propose a new name, a potential synonym, for this high-spired form.

Distribution: This is shown in Fig. 88.

Range: Mid-Cenomanian.

Genus ROTALIPORA Brotzen, 1942

Rotalipora Brotzen, 1942, p. 32.

Thalmanninella Sigal, 1948, p. 101.

Globotruncana (*Rotalipora*) Brotzen. Reichel, 1950, p. 600.

Globotruncana (*Thalmanninella*) Sigal. Reichel, 1950, p. 600.

Rotalipora (*Rotalipora*) Brotzen. Klaus, 1959, p. 812.

Rotalipora (*Thalmanninella*) Sigal. Klaus, 1959, p. 804.

Anaticinella Eicher, 1973, p. 185.

Type species: *Rotalipora turonica* Brotzen, 1942 [= *Rotalipora cushmani* (Morrow), 1934].

Description: Test a flat to high trochospiral coil, more typically biconvex; equatorial periphery entire to deeply lobate; axial periphery acute. Chambers may be inflated both dorsally and ventrally or only ventrally; with a single, pustulose, imperforate, narrow to moderately wide peripheral keel. Primary aperture an interiomarginal, umbilical–extraumbilical arch bordered by a narrow to wide imperforate lip. Accessory aperture located at varying positions along the sutural margin of post-neanic chambers; formed by the arched extension of the chamber and its flap-like projection into the umbilicus or by the flap-like projection alone; bordered by a narrow, imperforate lip. Wall finely and densely perforate, pustulose. Sutures depressed to flush.

Remarks: *Rotalipora* is distinguished from *Ticinella* by its peripheral keel, and from *Praeglobotruncana* by its accessory apertures.

Controversy over the taxonomic value of the position of the accessory apertures has continued undiminished since Sigal introduced the genus *Thalmanninella* in 1948. He characterised this genus as having flush sutures and, particularly, as exhibiting accessory apertures in an intraumbilical position as opposed to the more sutural position on *Rotalipora*. The position of the accessory aperture is judged to be a specific trait controlled by the shape of the chamber and the site of the attachment of the flap-like extension.

The chamber and flap of *Rotalipora cushmani* (Morrow) form a structure, exclusive of secondary calcification, which curves continuously into the umbilicus. The highly inflated chambers accentuate the depressed sutures and leave ample space for accessory apertures along the base of the sutural margins of the chambers. These apertures appear to be positioned more laterally to the chamber than in other species because of the continuity of the chamber wall and the flap. Close examination reveals, however, that the apertures of *R. cushmani* lie at approximately the same level above the umbilical floor as those of any other species.

On the other hand, *Rotalipora greenhornensis* (Morrow) has closely packed chambers with essentially no inflation and flush sutures. This structural combination does not provide the space necessary for the formation of accessory apertures along the sutural margins. Moreover, the umbilical face of each chamber drops abruptly from the carinated umbilical shoulder. The flap is attached near the base of the umbilical face, from whence it makes an asymmetric fold, forming the accessory apertures.

The accessory apertures of both *Rotalipora cushmani* and *R. greenhornensis* are a direct result of the presence of a flap-like extension into the umbilicus and are controlled only by the chamber shape. *Thalmanninella* is considered a junior synonym of *Rotalipora*.

Eicher (1973) proposed the new genus *Anaticinella* with *Globorotalia ? multiloculata* Morrow as the type species. *Anaticinella* was based upon a phyletic interpretation, one not shared here, i.e. a palingenetic origin for this "dominantly non-keeled" morphotype from *Rotalipora greenhornensis* (Morrow). *Globorotalia ? multiloculata* does possess properties common to both *Ticinella* and *Rotalipora*. Although the composite nature of this species makes the decision of generic assignment even more subjective the creation of a new genus does not seem necessary (see Remarks under *Rotalipora multiloculata*). *Anaticinella* is synonymous with the earlier *Rotalipora*.

Range: Mid-Albian—Turonian.

Rotalipora appenninica (Renz), Plate 30, figs 1–3

Globotruncana appenninica O. Renz, 1936, p. 20, 135, pl. 6, figs 1–11; pl. 7, fig. 1; pl. 8, fig. 4; text-figs 2, 7a; O. Renz, 1937, p. 501, text-figs 1a–c; Majzon, 1943, pl. 1, fig. 6; (?) Bolli, 1945, p. 223, 224, pl. 9, fig. 1; text-figs 1, 2; (?) Colom, 1947, p. 97, pl. 18, figs 4, 10, 12; pl. 21, figs 16, 17; pl. 25, figs 1, 3–7, 9–11; (?) Colom, 1948, p. 27, pl. 3, figs 1–10, 13, 14, 16, 19 (*non* 11, 12, 15, 18); Książkiewicz, 1956, p. 268, 269, pl. 30, fig. 11 (*non* 10, 13; text-fig. 35); Barnaba, 1959, text-fig. 17 (part); (?) Raffi and Forti, 1959, pl. 1, fig. 1 (part).

Globotruncana appenninica var. α Gandolfi, 1942, p. 117, text-figs 40a–c; Hiltermann, 1956, text-fig. 2 (part).

Globorotalia almadenensis Cushman and Todd, 1948, p. 98, pl. 16, figs 24a, b.

(?) *Globorotalia decorata* Cushman and Todd, 1948, p. 97, 98, pl. 16, fig. 21.

Rotalipora cushmani var. *evoluta* Sigal, 1948, p. 100, pl. 1, figs 3a–c; pl. 2, figs 2a, b; Ayala-Castañares, 1954, p. 419, 420, pl. 16, figs 3a–c.

Rotalipora appenninica appenninica (Renz). Cita, 1948b, p. 143, 144, pl. 3, figs 1a–c; *non* Said and Barakat, 1957b, p. 79, pl. 1, figs 28a–c; Luterbacher and Premoli Silva, 1962, p. 266, 267, pl. 19, figs 1a–2c; pl. 20, figs 1a–4c; pl. 21, figs 1a–4c; Borsetti, 1962, p. 34, pl. 3, fig. 2 (*non* 1); text-figs 1, 2, 47 (*non* 27–29); Lehmann, 1966b, p. 159, 160, pl. 2, figs 1, 3; text-figs 1q–x; Bandy, 1967, p. 15, text-fig. 7 (4); Neagu, 1969, p. 144, 145, pl. 25, figs 1–6; Sturm, 1969, pl. 9, figs 3a–c; Neagu, 1970a, fig. 1 (12); Neagu, 1970b, p. 65, pl. 21, figs 10–12; pl. 24, figs 7–13; pl. 25, figs 1–6.

Globotruncana (*Rotalipora*) *apenninica* (Renz). Mornod, 1950, p. 578–582, pl. 15, figs 1a–L; text-figs 3 (1a–3c), 4 (IIIa–IVc), 5 (1a–c); Reichel, 1950, p. 604–607, pl. 16, fig. 4; pl. 17, fig. 4; text-figs 3a–c, 4a, b.

Globotruncana (*Rotalipora*) *appenninica* (Renz). Majzon, 1961, p. 762, pl. 1, fig. 5.

Globotruncana (*Rotalipora*) *apenninica* var. *typica* (Gandolfi); Mornod, 1950, p. 582, 583, text-fig. 9 (2a–c); Carbonnier, 1952, p. 119, pl. 7, figs 3a, b (*non* c–e); Church, 1952, p. 69, 70, text-figs 1 (top), 2.

Rotalipora apenninica (Renz). Sigal, 1952, p. 24, text-fig. 23; Książkiewicz, 1958, text-fig. 1 (1a–c); (?) Maslakova, 1959, p. 108, pl. 11, figs 3a–c; *non* Dabagyan, 1963, p. 104–107, pl. 1, figs 1a–c; Caron and Luterbacher, 1969, p. 26, pl. 8, figs 8a–c; Luterbacher, 1972, p. 586, pl. 5, figs 1, 2.

Praeglobotruncana appenninica (Renz). (?) Bermúdez, 1952, p. 52, 53, pl. 27, figs 4a–c.

Rotalipora appenninica (Renz). Subbotina, 1953, p. 171–175, pl. 1, figs 8a–c (*non* 5a–7c); pl. 2, figs 1a–2c; Ayala-Castañares, 1954, p. 416–418, pl. 13, figs 1a–c; Hagn and Zeil, 1954, p. 22, 23, pl. 1, figs 1a–c; pl. 4, figs (?) 11, 12; pl. 5, fig. (?) 1; Bykova *et al.*, 1959, p. 301, text-figs 673A–C; Vinogradov, 1960b, p. 34, pl. 1, figs 1a–4c (*non* 5a–c); Loeblich and Tappan, 1961, p. 296, 297, pl. 7, figs 11a–12c; Reutter and Serpagli, 1961, pl. 9, fig. 2; Postuma, 1962, 7 figs; Cita-Sironi, 1963, text-fig. 9b; *non* Pirini and Radrizzani, 1963, pl. 10, fig. 15 (1); *non* Kaptarenko-Chernousova, 1963, p. 104, pl. 15, figs a, b, v; *non* Alekseeva, 1963, p. 45, pl. 9, figs 1a–c; *non* Todd and Low, 1964, p. 405, pl. 2, figs 6a–c; Boccaletti and Sagri, 1965, p. 473, pl. 1, figs 2a–3b; pl. 6, fig. 2; Săndulescu, 1966, pl. 1a, figs 1a–c; Marianos and Zingula, 1966, p. 338, pl. 38, figs 1a–c; (?) Pessagno, 1967, p. 289–292, pl. 50, figs 1, 2, 4–6; pl. 98, fig. 13; pl. 101, figs 1, 2 (*non* pl. 51, figs 10–12); (?) Neagu (1968), text-fig. 2 (46a); Barr, 1968, pl. 3, figs 1a–c; Bate and Bayliss, 1969, pl. 3, figs 8a–c; Scheibnerová, 1969, p. 64, 65, tab. 10, 4a–c; Sigal, 1969, p. 633, 634, pl. 2, figs 9a–10b; *non* Risch, 1971, p. 53, pl. 5, figs 19–21; Gawor-Biedowa, 1972, p. 78, 79, pl. 9, figs 1a–2c, (?) 3a–c.

Globotruncana (*Rotalipora*) *globotruncanoides* Sigal. Küpper, 1955, p. 113, 114, pl. 18, figs 1a–c; Majzon, 1961, p. 763, pl. 3, fig. 1.

Globotruncana (*Rotalipora*) *apenninica globotruncanoides* (Sigal). Gandolfi, 1957, p. 60, 62, pl. 9, figs 2a–c.

Globotruncana (*Rotalipora*) *apenninica apenninica* (Renz). Küpper, 1955, p. 114, 115, pl. 18, figs 2a–c; Gandolfi, 1957, p. 60, 62, pl. 9, figs 1a–c.

Globotruncana (*Rotalipora*) *evoluta* Sigal. Küpper, 1955, p. 115, pl. 18, figs 3a–c.

Rotalipora apenninica alpha (Gandolfi). Sigal, 1956b, p. 211, text-fig. 3.

Rotalipora micheli Sacal and Debourle, 1957, p. 58, pl. 25, figs 4, 5, 12.

Rotalipora cf. *appenninica* (Renz). Bolli, Loeblich and Tappan, 1957, p. 41, pl. 9, figs 5a–c.

Globotruncana (*Rotalipora*) *apenninica balernaensis* Gandolfi, 1957, p. 60, pl. 8, figs 3a–c; ? Gandolfi, 1957, p. 60, pl. 8, figs 4a, b.

Globotruncana (*Rotalipora*) *apenninica evoluta* (Sigal). Gandolfi, 1957, p. 60, pl. 8, figs 5a–c.

Rotalipora globotruncanoides Sigal. Książkiewicz, 1958, text-fig. 1 (5a–c); Lehmann, 1963, p. 141, pl. 1, figs 3a–c; text-figs 2b (?), 3b; Neagu, 1969, p. 147, pl. 30, figs 6, 7 (?); pl. 31, figs 1–5.

Rotalipora evoluta Sigal. Książkiewicz, 1958, text-fig. 1 (8a–c); Loeblich and Tappan, 1961, p. 298, 299, pl. 7, figs 1a–4c; *non* Ayala-Castañares, 1962, p. 26–28, pl. 5, figs 2a–c; pl. 10, figs (?) 2a–c; pl. 3, figs (?) 1a–e; Graham, 1962, p. 105, pl. 20, figs 27Aa–Dc; Todd and Low, 1964, p. 406, pl. 2, figs 3a–c; Săndulescu, 1966, pl. 1a, figs 2a–c; Pessagno, 1967, p. 294–295, pl. 49, figs 12–14; pl. 53, figs 6–8; pl. 98, fig. 12; Bandy, 1967, text-fig. 7 (6); Michael, 1973, p. 217, pl. 6, figs 4–6.

Rotalipora (*Thalmanninella*) *appenninica balernaensis* (Gandolfi). Klaus, 1959, p. 808, pl. 3, figs 2a–c.

Rotalipora (*Thalmanninella*) *evoluta* (Sigal). Klaus, 1959, p. 810, pl. 4, figs 3a–c.

Rotalipora (*Thalmanninella*) *appenninica appenninica* (Renz). Klaus, 1959, p. 808–810, pl. 3, figs 3a–c; Caron, 1966, p. 72, pl. 1, figs 4a–c, b′; Săndulescu, 1969, p. 194, 195, pl. 38, figs 1a–2c.

Rotalipora balernaensis Gandolfi. Loeblich and Tappan, 1961, p. 297, pl. 8, figs 11a–c; Sigal, 1969, p. 634, pl. 2, figs 2, 4–8.

Thalmanninella brotzeni Sigal. Maslakova, 1961, pl. 3, figs 1a–c; pl. 4, fig. 1; Salaj and Samuel, 1966, p. 178, 179, pl. 11, figs 4a–c.

Rotalipora appenninica evoluta Sigal. Luterbacher and Premoli Silva, 1962, pl. 20, figs 5a–c; Neagu, 1970b, p. 65, pl. 22, figs 1–3.

Thalmanninella appenninica (Renz). Maslakova, 1961, pl. 3, fig. 2a–c; pl. 4, figs 2, 3; Salaj and Samuel, 1966, p. 177, pl. 11, figs 8a–c; Prosnyakova, 1968, p. 20, 21, pl. figs 2a–3c.

Rotalipora appenninica primitiva Borsetti, 1962, p. 37–40, pl. 1, figs 2a–c.

Rotalipora appenninica balernaensis (Gandolfi). Borsetti, 1962, p. 34, 35, pl. 1, figs 1a–c; (*non* pl. 2, fig. 8, text-fig. 33); Bandy, 1967, p. 15, text-fig. 7 (3).

Rotalipora appenninica evoluta Sigal. Luterbacher, 1963, in Renz *et al.*, p. 1088, pl. 7, figs 3a–c; Neagu, 1969, p. 145, pl. 37, figs 6–8.

Rotalipora cf. *micheli* Sacal and Debourle. Luterbacher, 1963, in Renz *et al.*, p. 1088, pl. 7, figs 3a–c.

Rotalipora greenhornensis (Morrow). Graham, 1962, p. 108, pl. 20, figs 28B–Cc (*non* 28Aa–c, Da–Ec).

Rotalipora cf. *appenninica* (Renz). Cita-Sironi, 1963, text-fig. 2b.

Rotalipora (*Thalmanninella*) *appenninica evoluta* (Sigal). Caron, 1966, p. 72, pl. 1, figs 3a–c; Săndulescu, 1969, p. 195, 196, pl. 38, figs 4a–5c.

Thalmanninella balernaensis (Gandolfi). Salaj and Samuel, 1966, p. 178, pl. 12, figs 1a–c.

Thalmanninella evoluta (Sigal). Salaj and Samuel, 1966, p. 179, 180, pl. 11, figs 3a–c; pl. 12, figs 2a–c.

Rotalipora (*Thalmanninella*) *appenninica* (Renz). Caron, 1967, p. 74, pl. 2, figs 1a–c.

Rotalipora (*Thalmanninella*) *appenninica gandolfi* Luterbacher and Premoli Silva. Săndulescu, 1969, p. 195, pl. 38, figs 3a–c.

Rotalipora (*Thalmanninella*) sp. 1 cf. *appenninica appenninica* (Renz). Săndulescu, 1969, p. 195, pl. 41, figs 1a–2b.

Rotalipora montsalvensis Mornod. Sturm, 1969, pl. 9, figs 5a–c.

Rotalipora ticinensis (Gandolfi). Scheibnerová, 1969, p. 63, 64, tab. 10, figs 3a–c (*non* 5a–c); tab. 11, figs 1a–c, 3a–4c.

Rotalipora appenninica gandolfii Luterbacher and Premoli Silva. Neagu, 1970b, p. 65, pl. 23, figs 10–12.

Rotalipora. Berger and Rad, 1972, pl. 3, figs 2 (part), 5 (part).

Description: "Test trochospiral, biconvex and unicarinate, dorsal face always flatter than the ventral face, contour of the test lobate and umbilicus very marked. On the dorsal face, all the chambers (nearly always numbering 11 to 13) in $2\frac{1}{2}$ whorls are visible. In vertical section: chambers rhomboid, dorsally flattened. The growth of the chambers is slow with the exception of the ultimate which, when it is not senile, is always clearly larger than the preceding. The ultimate and the penultimate chamber do not have umbilical protuberances. They dip in a gentle slope into the umbilicus; this in contrast with the preceding which project into the margin of the umbilicus. On the dorsal face, the sutural carina is curved; on the ventral face, the sutures are depressed and radial, but between the last chambers, they can be lightly curved.

"On the septal face nearly vertical from the ultimate chamber, the principal aperture extends from the umbilicus to the vicinity of the periphery. It possesses a lip nearly always well developed. In the first half of the last whorl, the accessory apertures are always situated in the umbilicus. They remain in this position up to the ultimate chamber in the individuals whose coiling is tight, whereas they are displaced towards the periphery in the 2 to 3 last chambers in the more evolute individuals, where the chambers are arranged in a looser manner. The accessory apertures possess lips in their turn. The posterior side of these apertures can be formed by the peristome of the principal aperture of the preceding chamber.

"The 'ornamental' elements—keels and pustules—increasing in thickness with growth are much more accentuated in the first chambers than in the last. This shows especially in the septal carinae of the dorsal face. When they are only slightly accentuated in the last chambers overlapping one another, they thicken in the first chambers of the last whorl, becoming beaded and ending by fusing in large bands.

"The corresponding process observed in the sculpture of the adumbilical protuberances and for the marginal carina which, when it is complete, is composed of two ranges of spiny pustules." (Luterbacher and Premoli Silva, 1962.)

Remarks: The dorsal chamber walls of *Rotalipora appenninica* tend to be flat unlike *R. cushmani* (Morrow). The ventral sutures are generally depressed as opposed to the flush sutures of *R. greenhornensis* (Morrow). It lacks the dorsal flatness and the ventral convexity of *R. deeckei* (Franke), and has fewer chambers per whorl than does *R. ticinensis* (Gandolfi) (see Remarks under *R. ticinensis* for additional comparison). The degree of dorsoconvexity ranges from very low, with an apical angle of 157°, to a moderately high 113°. The accessory apertural position becomes more adumbilically located during ontogeny, but rarely beyond one half the chamber height. There is a tendency in some individuals to become less tightly coiled, and the width of the apertural lip is highly variable.

Caron (1967), on the basis of statistical and stratigraphic analyses, supported the descent of *Rotalipora appenninica* from *R. ticinensis subticinensis* (Gandolfi) [= *R. ticinensis* of this paper]. Data accumulated from other workers also suggest a slightly earlier appearance in the stratigraphic record for *R. ticinensis*.

As reported by other authors (Luterbacher and Premoli Silva, 1962; Caron and Luterbacher, 1969), Marie (1948) selected figure 2 (p. 14) of Renz (1936) to represent the lectotype of *Globotruncana* (= *Rotalipora*) *appenninica*. Apparently unaware of these studies, Pessagno (1967) designated fig. 4, plate 8 of Renz as the lectotype. Pessagno's selection is invalid by priority (I.C.Z.N., Art. 74a).

Caron and Luterbacher (1969) determined that the correct spelling of the species name is with one "p"—*appenninica*, taken from the Latin "apenninicus" not the Italian "appenninico". They referred to Article 32 (I.C.Z.N.), which states that the original spelling is correct unless it fails to satisfy the provisions of Articles 26 to 30, or there is clear evidence of an inadvertent error, or that multiple original spellings exist. None of these apply to the case in point. The name "*appenninica*" is a Latinised non-classical word, and is correct as originally spelled.

The holotypes of *Globorotalia almadenensis* and *G. decorata* of Cushman and Todd (1948) are very poorly preserved with a sugary surface texture. The two paratypes of *G. almadenensis* have accessory apertures, and there is the suggestion of such structures on the holotype. No accessory apertures are preserved on *G. decorata*, but it otherwise resembles *Rotalipora appenninica*. Both are regarded as synonymous with *R. appenninica*.

Lehmann (1966d) distinguished *Rotalipora globotruncanoides* Sigal, 1948, from *R. appenninica* by its more rapid increase of chamber size in the last whorl. The type figure of Sigal's species is identical to the topotypes of *R. appenninica* illustrated by Luterbacher and Premoli Silva (1962). The difference noted by Lehmann is within the range of intraspecific variation of *R. appenninica*. *Rotalipora evoluta* Sigal, 1948, also falls within this distribution and has essentially the same stratigraphic range. Both of Sigal's species are believed synonymous with *R. appenninica*.

Luterbacher and Premoli Silva (1962) treated *Globotruncana* (*Rotalipora*) *apenninica balernaensis* Gandolfi, 1957, as a junior synonym of *Rotalipora appenninica* (Renz), a decision with which this study concurs.

In view of Luterbacher and Premoli Silva's (1962) re-examination of *Rotalipora appenninica*, the forms described by Borsetti (1962) as *R. appenninica primitiva* must also be regarded as a synonym of the former.

Distribution: This is shown in Fig. 89.

Range: Late Albian—Early Cenomanian.

Rotalipora cushmani (Morrow), Plate 30, fig. 4; plate 31, figs 1–4

Globotruncana arca (Cushman). Moreman, 1927, p. 100, pl. 16, figs 16, 17.

Globorotalia cushmani Morrow, 1934, p. 199, pl. 31, figs 2a, b, 4a, b; Cushman, 1946b, p. 152, pl. 62, fig. 9; Kikoïne, 1947, p. 289, 3 text-figs; Gauger, 1953, in Peterson *et al.*, p. 83, pl. 10, figs 1–3; Frizzell, 1954, p. 129, pl. 20, figs 28a–c; Applin, 1955, p. 196, pl. 48, figs 25, 26.

Globorotalia planoconvexa (Seguenza). Keller, 1939, pl. 1, figs 2a–c.

Rotalipora turonica Brotzen, 1942, p. 32, text-figs 10, 11 (4); Sigal, 1948, p. 96, pl. 1, figs 1a–c; Williams-Mitchell, 1948, p. 47, pl. 8, figs 7a–c; Cushman, 1948b, p. 330, Key pl. 54, figs 3a–c; Bermúdez, 1952, p. 100, pl. 18, figs 6a–c; Hagn and Zeil, 1954, p. 27, 28, pl. 1, figs 5a–c; pl. 4, figs 3, 4; Ayala-Castañares, 1954, p. 422, 423, pl. 14, figs 2a–c; Bolli, Loeblich and Tappan, 1957, p. 41, pl. 9, figs 6a–c; Książkiewicz, 1958, text-fig. 1 (9a–c); Maslakova, 1959, p. 109, pl. 11, figs 5a–c; Sigal and Dardenne, 1962, p. 222, pl. 13, fig. 9; Hiltermann and Koch, 1962, p. 329, pl. 49, figs 2a–3; tab. 19, (part); Pokorný, 1963, p. 384, fig. 431; Cita-Sironi, 1963, text-fig. 2a; Săndulescu, 1966, pl. Ia, figs 4a–c.

Globotruncana alpina Bolli, 1945, p. 224, 225, pl. 9, figs 3, 4, text-fig. 1 (5–7);

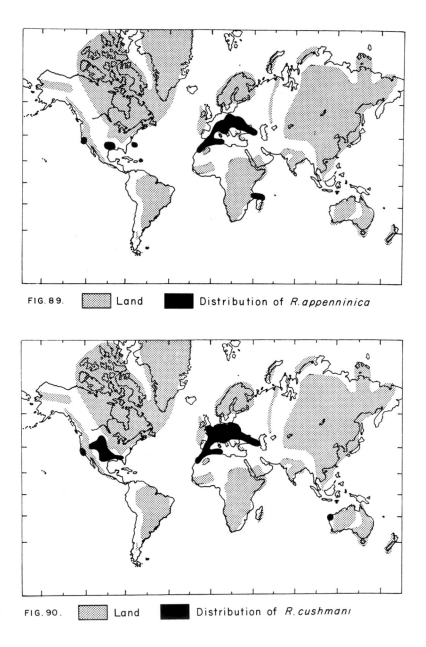

FIG. 89.   [Land]   [Distribution]   Distribution of *R. appenninica*

FIG. 90.   [Land]   [Distribution]   Distribution of *R. cushmani*

Colom, 1947, p. 98, pl. 25, figs 8, 12, 16; Colom, 1948, p. 137, 139, pl. 4, figs 6, 7, 11.

Rotalipora cushmani (Morrow). Sigal, 1948, p. 96, 97, 100, pl. 1, figs 2a–c (*non* pl. 2, figs 1a, b); Hagn and Zeil, 1954, p. 29, 30, pl. 1, figs 3a–c; pl. 4, figs 8–10; Ayala-Castañares, 1954, p. 418, 419, pl. 16, figs 2a–c; Brönnimann and Brown, 1956, p. 537, 538, pl. 20, figs 10–12; Sacal and Debourle, 1957, p. 58, pl. 25, figs 6–8, 13, 16, 17; Książkiewicz, 1958, text-fig. 1 (6a–7c); Vinogradov, 1960b, p. 36, pl. 2, figs 9a–10c; Maslakova, 1961, pl. 3, figs 4a–c; pl. 4, figs 6–9; Loeblich and Tappan, 1961, p. 297, 298, pl. 8, figs 1a–10c; Borsetti, 1962, p. 40, 41, pl. 3, figs 7, 8; text-figs 57–62; Lehmann, 1963, p. 142, pl. 1, figs 5a–c; text-fig. 3c; Postuma, 1962, 7 figs; Graham, 1962, p. 105, pl. 20, figs 26a–c; (?) Dabagayan, 1963, p. 110–113, pl. 1, figs 3a–c; Luterbacher, 1963, in Renz *et al.*, p. 1088, pl. 7, figs 1a–c; Lehmann, 1965, text-fig. 1c; Broglio Lorigo and Mantovani, 1965, pl. 114, fig. 3; Saavedra, 1965, p. 324, text-fig. 10; Marianos and Zingula, 1966, p. 338, 339, pl. 38, figs 2a–c; Săndulescu, 1966, pl. IIa, figs 1a–c; *non* Prosnyakova, 1967, p. 6–8, pl. figs 4a, b, v; Marks, 1967, p. 272, 273, pl. 1, figs 1–12; pl. 2, figs 1–3; Pessagno, 1967, p. 292, 293, pl. 51, figs 6–9; pl. 101, figs 5–7; Bandy, 1967, p. 15, text-fig. 7 (8); Neagu, 1968, text-fig. 2 (46); Moorkens, 1969, p. 444, 445, pl. 1, figs 1a–2c; Porthault, 1969, p. 534, pl. 1, figs 11a–c; Scheibnerová, 1969, p. 66, pl. 11, figs 2a–c, 5a–c; Sturm, 1969, pl. 10, figs 1a–c; Eicher, 1969, text-fig. 3 (part); Douglas, 1969, p. 173, 174, pl. 1, figs 1a–c, 2a–c (?); (?) Bate and Bayliss, 1969, pl. 2, figs 2a–c; *non* Broglio Lorigo and Mantovani, 1970, p. 226, chart (52); Eicher and Worstell, 1970b, p. 310, 311, pl. 12, figs 3a–4b; pl. 13, figs 1a, b; Sidó, 1971, p. 46, pl. 1, fig. 5 (part).

Rosalinella cushmani (Morrow). Marie, 1948, p. 39.

Globotruncana (Rotalipora) turonica (Brotzen). Reichel, 1950, p. 607, 608, pl. 16, fig. 5; pl. 17, fig. 5; Carbonnier, 1952, p. 117, 118, pl. 6, figs 3a–c; Majzon, 1961, p. 762, pl. 2, fig. 12.

Globotruncana (Rotalipora) apenninica Renz. Mornod, 1950, p. 578–582, text-fig. 4 (IVa–c), [*non* text-fig. 3 (1a–3c), 4 (IIIa–c), 5 (Ia–c)].

Globotruncana (Rotalipora) montsalvensis Mornod, 1950, p. 584–586, text-figs 4 (Ia–c), 7 (1a–2c); Colom, 1952, p. 19, 20, pl. 3, figs 12, 13; Majzon, 1961, p. 763, pl. 2, fig. 11.

Globotruncana (Rotalipora) montsalvensis var. *minor* Mornod, 1950, p. 586, text-figs 4 (IIa–c), 8 (1a–4).

Globotruncana (Rotalipora) alpina Bolli. Allemann *et al.*, 1951, p. 160, text-fig. 2; Majzon, 1961, p. 763, pl. 4, fig. 7.

Globotruncana (Rotalipora) cushmani (Morrow). Colom, 1952, p. 19, pl. 3, figs 14–17; Majzon, 1961, p. 762, pl. 4, fig. 6; Sidó, 1961, pl. 3, figs 13, 14.

Globotruncana (Rotalipora) turonica var. *expansa* Carbonnier, 1952, p. 118, pl. 6, figs 4a–c.

Globotruncana (Rotalipora) apenninica (Renz) var. *typica* (Gandolfi). Carbonnier, 1952, p. 119, pl. 7, figs 3a, b.

Rotalipora appenninica (Renz). Subbotina, 1953, p. 171–175, pl. 1, figs 5a–6c (*non* 7a–8c; pl. 2, figs 1a–2c); Dabagyan, 1963, p. 104–107, pl. 1, figs 1a–c; Kaptarenko-Chernousova, 1963, p. 104, pl. 15, figs a, b, v.

Rotalipora montsalvensis Mornod. Hagn and Zeil, 1954, p. 29, pl. 1, figs 4a–c, pl. 5, fig. 2; Ayala-Castañares, 1954, p. 420, 421, pl. 13, figs 2a–c; Książkiewicz, 1958, text-fig. 1 (11a–c); Vinogradov, 1960b, p. 34, 35, pl. 2, figs 6a–8c; Dabagyan, 1963, p. 108, 109, pl. 1, figs 2a–c; Luterbacher, 1963, in Renz *et al.*, p. 1089, pl. 8, figs 1a–c; Küpper, 1964, p. 613, pl. 2, figs 1a–c; Săndulescu, 1966, pl. IIa, figs 2a–c; Porthault, 1969, p. 534, pl. 1, figs 9a–c; *non* Sturm, 1969, pl. 9, figs 5a–c.

Rotalipora turonica Brotzen ssp. *thomei* Hagn and Zeil, 1954, p. 28, pl. 1, figs 6a–c; pl. 4, figs 5, 6.

Rotalipora montsalvensis var. *minor* (Mornod). Ayala-Castañares, 1954, p. 421, pl. 13, figs 3a–c.

Rotalipora cf. *R. appenninica* (Renz). Bonet, 1956, pl. 1, fig. 3 (part).

Globotruncana cf. *turonica* Brotzen. Książkiewicz, 1956, p. 271, pl. 32, figs 15, 18 (*non* 16, 17); text-figs 39, 39a; Alexandrowicz, 1956, p. 53, text-fig. 4.

Globotruncana (*Rotalipora*) cf. *turonica* (Brotzen). Edgell, 1957, p. 109, 110, pl. 1, figs 16–18.

Rotalipora turonica expansa. Carbonnier. Książkiewicz, 1958, text-fig. 1 (10a–c); Küpper, 1964, p. 614, pl. 2, figs 3a–c.

Rotalipora (*Rotalipora*) *turonica* var. *expansa* Carbonnier. Klaus, 1959, p. 816, 817, pl. 5, figs 4a–c.

Rotalipora (*Rotalipora*) *turonica* var. *thomei* Hagn and Zeil. Klaus, 1959, p. 817, pl. 5, figs 5a–c.

Rotalipora (*Rotalipora*) cf. *montsalvensis* var. *minor* Mornod. Klaus, 1959, p. 812, 813, pl. 4, figs 5a–c.

Rotalipora (*Rotalipora*) *montsalvensis* Mornod. Klaus, 1959, p. 812, 813, pl. 5, figs 1a–c.

Rotalipora (*Rotalipora*) *cushmani* (Morrow). Klaus, 1959, p. 814, 815 pl. 5, figs 2a–c.

Rotalipora (*Rotalipora*) *turonica* Brotzen. Klaus, 1959, p. 815, 816, pl. 5, figs 3a–c; Klaus, 1961, text-figs 4a–c.

Rotalipora gr. *cushmani* (Morrow) *turonica* Brotzen. Malapris and Rat, 1961, p. 87, 88, pl. 1, figs 1a–3b.

Rotalipora (*Rotalipora*) cf. *cushmani* (Morrow). Klaus, 1961, text-figs 3a–c.

Rotalipora sp. del gruppo *turonica* Brotzen *cushmani* (Morrow). Sturani, 1962, p. 87, text-fig. 15 (A5–8).

Rotalipora montsalvensis minor Mornod. Luterbacher, 1963, in Renz *et al.*, p. 1089, pl. 7, figs 2a–c.

Rotalipora cf. *turonica* Brotzen. Luterbacher, 1963, in Renz *et al.*, p. 1088, 1089, pl. 7, figs 4a–c.

Rotalipora turonica turonica Brotzen. Küpper, 1964, p. 613, 614, pl. 2, figs 2a–c; Sturm, 1969, pl. 10, figs 2a–c.

Rotalipora cushmani expansa (Carbonnier). Salaj and Samuel, 1966, p. 183, 184, pl. 12, figs 7a–c; Neagu, 1969, p. 146, pl. 28, figs 7–9; pl. 29, figs 3–5 (*non* 1, 2).

Rotalipora cushmani montsalvensis (Mornod). Salaj and Samuel, 1966, p. 184, pl. 13, figs 5a–c; Neagu, 1969, p. 147, pl. 29, figs 6–9; pl. 30, figs 1, 2.

Rotalipora cushmani cushmani (Morrow). Salaj and Samuel, 1966, p. 184, 185, pl. 13, figs 2a–c, 4a–c; Neagu, 1969, p. 145, 146, pl. 25, figs 7–9, pl. 26, figs 1–9; Neagu, 1970a, fig. 1 (13); Neagu, 1970b, p. 65, 66, pl. 22, figs 10–12; pl. 41, fig. 46; (*non* pl. 24, figs 1–6); Gawor Biedowa, 1972, p. 79–81, pl. 10, figs 1a–2c.

Rotalipora cushmani minor (Mornod). Salaj and Samuel, 1966, p. 185, pl. 13, figs 6a–c; Neagu, 1969, p. 147, pl. 30, figs 3–5; pl. 31, figs 6–8.

Rotalipora cushmani thomei Hagn and Zeil. Borsetti, 1962, p. 41, pl. 4, figs 1, 3; text-figs 63–67, 68 (?); Salaj and Samuel, 1966, p. 185, pl. 12, figs 6a–c; Neagu, 1969, p. 146, pl. 28, figs 1–6; Sturm, 1969, pl. 10, figs 3a–c; Gawor-Biedowa, 1972, p. 81, 82, pl. 10, figs 3a 4c.

Rotalipora cushmani turonica (Brotzen). Salaj and Samuel, 1966, p. 185, 186, pl. 13, figs 1a–c; pl. 14, figs 1a–c; Neagu, 1969, p. 146, pl. 27, figs 1–6; Neagu, 1970a, fig. 1 (14); Neagu, 1970b, p. 66, pl. 23, figs 1–3, 7–9 (*non* 4–6).

Rotalipora sp. 1 Salaj and Samuel, 1966, pl. 12, figs 5a–c.

Rotalipora sp. 2 Salaj and Samuel, 1966, pl. 13, figs 3a–c, 7a–c.

Rotalipora sp. 3, Salaj and Samuel, 1966, pl. 14, figs 2a–c.

Description: Test trochospiral; typically biconvex but ranges to a high dorso-convexity; equatorial periphery moderately to deeply lobate; axial periphery acute with a moderately wide, single, pustulose keel. Chambers highly inflated both dorsally and ventrally; numbering 16–18 in 3 whorls with 4–6 in the final whorl; shaped as a sector of a circle ventrally, petaloid dorsally; gradually to rapidly increasing in size. Primary aperture a low to moderately high, interiomarginal, umbilical–extraumbilical arch with a narrow bordering lip which expands as a continuation of the chamber into the umbilicus forming a small, imperforate platform or tegillum-like structure; each covered in turn by succeeding chambers. A single accessory aperture located along the sutural margin of all post-neanic chambers; slightly elevated above the umbilicus. Sutures deeply depressed, straight to sigmoid ventrally; slightly depressed and curved dorsally. Surface densely and finely perforate; secondary calcification in the form of large pustules may be found on the ventral vertex of each chamber, or may form as a ridge along the umbilical extension of the chamber bifurcating to extend as arcs over the primary and accessory apertures. Umbilicus deep, narrow to moderately wide.

Remarks: No other species of *Rotalipora* has as great a chamber inflation or lobate periphery as does *R. cushmani*. The sector-shaped ventral chambers are equally distinctive. A number of features vary within this species. The accessory apertures may be located from a nearly umbilical position to midway along the sutural margin on the same or different individuals. They may be at essentially the same level as the umbilicus or more generally slightly elevated. The narrow lip of either the primary or accessory apertures may or may not be partially rolled back on itself. The degree that the chamber and flap extend into the umbilicus varies considerably. A flat, platform-like structure may project into the umbilicus. The extensions fuse to previous ones, partially or occasionally completely covering the umbilicus. The dorsal apical angle ranges from 81° to 132°. The forms with low apical angles, i.e. higher dorso-convexity, have fewer chambers per whorl and greater dorsal than ventral chamber inflation.

The phylogeny of *R. cushmani* is unknown.

Loeblich and Tappan (1961) pointed out the incorrect assumption by Reichel (1950) that *Rotalipora cushmani* has a flatter spire than *R. turonica* Brotzen, 1942. The two are conspecific.

Although described from thin-section, *Globotruncana alpina* Bolli, 1945, has the characteristic chamber inflation and single keel of *Rotalipora cushmani*, and is judged to be a junior synonym of that species.

Pessagno (1967) concluded that *Globotruncana benacensis* Cita, 1948b, was synonymous with the earlier *Rotalipora cushmani*. This is unlikely because it lacks the inflated chambers typical of that species. Cita's taxon is thought to belong to the *Rotalipora*, but to which species is uncertain.

Porthault (1969) distinguished *Rotalipora montsalvensis* Mornod, 1950, from *R. cushmani* by its less inflated and more radially elongate chambers, the much less lobate outline and because *R. cushmani* possesses the pinched chambers at the keel which are nonexistent or poorly developed in *R. montsalvensis*. All these characteristics fall within the range of intraspecific variation exhibited by the topotypes of *R. cushmani*. *Rotalipora montsalvensis* var. *minor* Mornod, 1950, is also regarded as a junior synonym of *R. cushmani*.

In 1952 Carbonnier established the new variety *Globotruncana* (*Rotalipora*) *turonica* var. *expansa*. He stated that it differed from *R. cushmani* by having a more conical section, absence of sutural rims on the dorsal face, more globulose chambers and granulations on the chambers. All of these features can be observed in a topotypic suite of *R. cushmani*.

Rotalipora turonica ssp. *thomei* Hagn and Zeil, 1954, represents a high-spired form of *R. cushmani*. Although in the minority, such forms are generally present in a given suite of *R. cushmani*.

Distribution: This is shown in Fig. 90.

Range: Cenomanian—Turonian.

Rotalipora deeckei (Franke), Plate 32, figs 1–3

Rotalia deeckei Franke, 1925, p. 90, 91, pl. 8, figs 7a–c; Dalbiez, 1957, p. 187, 188, text-figs 1–5c.

Globotruncana apenninica var. *γ* Gandolfi, 1942, p. 116–123, pl. 6, fig. 6 (part); text-figs 41 (1a, b), 42 (1), 44 (3, 4); Hiltermann, 1956, text-fig. 2 (part).

Globotruncana deeckei (Franke). Majzon, 1943, pl. 1, fig. 15.

Globotruncana apenninica Renz. (?) Colom, 1948, p. 27, pl. 3, figs 11, 12, 15, 18 (*non* 1–10, 13, 14, 16, 19).

Globotruncana reicheli Mornod, 1950, p. 583, 584, pl. 15, figs 2a–p, 3–8; text-figs 5 (IVa–c), 6 (1–6).

Globotruncana (*Rotalipora*) *reicheli* Mornod. Carbonnier, 1952, p. 119, 120, pl. 7, figs 4a, b; Majzon, 1961, p. 763, pl. 2, fig. 8.

Globotruncana (*Rotalipora*) *apenninica* var. *typica* (Gandolfi). Carbonnier, 1952, p. 119, pl. 7, figs 3c–e (*non* a, b); *non* Church, 1952, p. 69, 70, text-figs 1 (top), 2.

Rotalipora reicheli (Mornod). Subbotina, 1953, p. 176–178, pl. 2, figs 4a–c (? 3a–c); Hagn and Zeil, 1954, p. 25, 26, pl. 1, figs 2a–c; pl. 4, figs 1, 2; pl. 7, fig. 11; Ayala-Castañares, 1954, p. 421, 422, pl. 14, figs 1a–c; Książkiewicz, 1958, text-fig. 1 (2a–3c); *non* Maslakova, 1959, p. 108, pl. 11, figs 4a–c; Bykova *et al.*, 1959, p. 301, text-figs 684A–C; Vinogradov, 1960b, p. 38, pl. 3, figs 11a–12c; Loeblich and Tappan, 1961, p. 301, pl. 8, figs 8a–c; Borsetti, 1962, p. 44, pl. 4, figs 7, 8; text-figs 94–97 (?); Postuma, 1962, 7 figs, *non* Kaptarenko-Chernousova, 1963, p. 104, 105, pl. 15, figs 3a, b, v; Săndulescu, 1966, pl. IIa, figs 3a–c; Bandy, 1967, p. 15, text-fig. 7 (7); Neagu, 1968, text-fig. 2 (48); Porthault, 1969, p. 533, pl. 1, figs 10a–c; Neagu, 1969, p. 149, pl. 34, figs 4–6; pl. 35, figs 4–6; pl. 36, figs 1–3; Caron and Luterbacher, 1969, p. 27, pl. 9, figs 10a–c; Neagu, 1970a, fig. 1 (11); Neagu, 1970b, p. 66, pl. 22, figs 7–9; pl. 41, fig. 48, (?) Gawor-Biedowa, 1972, p. 84, 85, pl. 11, figs 1a–c.

Rotalipora appenninica var. *typica* Gandolfi. Ayala-Castañares, 1954, p. 418, pl. 12, figs 3a–c.

Globotruncana apenninica typica Gandolfi. Hiltermann, 1956, text-fig. 2 (part).

Rotalipora (*Thalmanninella*) *deeckei* (Franke). Klaus, 1959, p. 806, text-figs 7 (2a–c); (?) Klaus, 1961, text-figs 1a–c; Săndulescu, 1969, p. 197, 198, pl. 41, figs 3a, b (*non* pl. 42, figs 1a–3b).

Rotalipora (*Thalmanninella*) *reicheli* (Mornod). Klaus, 1959, p. 806–808, pl. 4, figs 2a–c; text-figs 7 (3a–c); Klaus, 1961, text-figs 2a–c; Săndulescu, 1969, p. 198, pl. 43, figs 1a–3c.

Thalmanninella deeckei (Franke). Maslakova, 1961, pl. 3, figs 3a–c; pl. 4, figs 4, 5; Dabagyan, 1963, p. 113–115, pl. 1, figs 4a–c; Salaj and Samuel, 1966, p. 179, pl. 12, figs 4a–c; Prosnyakova, 1968, p. 21, 22, pl. figs 4a–5.

Rotalipora appenninica typica (Gandolfi). *non* Malapris and Rat, 1961, p. 87, pl. 3, figs 4a–c.

Rotalipora appenninica marchigiana Borsetti, 1962, p. 36, 37 (*non* pl. 3, figs 4, 6; text-figs 34, 35, 48, 49).

Rotalipora appenninica gandolfii Luterbacher and Premoli Silva, 1962, p. 267, pl. 19, figs 3a–c; Luterbacher, 1963, in Renz *et al.*, p. 1088, pl. 8, figs 4a–c (*non* 2a–c); Lehmann, 1966b, p. 160, pl. 2, fig. 2; Sturm, 1969, pl. 9, figs 4a–c; *non* Neagu, 1970b, p. 65, pl. 23, figs 10–12.

Thalmanninella brotzeni Sigal. Hiltermann and Koch, 1962, p. 329, 330, pl. 49, figs 5a–c, tab. 19 (part).

Rotalipora deeckei (Franke). (?) Sǎndulescu, 1966, pl. IIa, figs 4a–c; Scheibnerová, 1969, p. 65, tab. 12, figs 1a–3c; (?) Neagu, 1969, p. 148, pl. 36, figs 4–6, pl. 37, figs 1–5; Gawor-Biedowa, 1972, p. 82, 83, pl. 11, figs 2a–3c.

Thalmanninella globotruncanoides (Sigal). Salaj and Samuel, 1966, p. 180, pl. 11, figs 6a–c.

Thalmanninella reicheli (Mornod). Salaj and Samuel, 1966, p. 181, 182, pl. 11, figs 7a–c.

Rotalipora (Thalmanninella) appenninica gandolfii Luterbacher and Premoli Silva. *non* Caron, 1966, p. 72, pl. 1, figs 5a–c; *non* Sǎndulescu, 1969, p. 195, pl. 38, figs 3a–c.

Rotalipora gandolfii Luterbacher and Premoli Silva. Caron and Luterbacher, 1969, p. 26, 27, pl. 9, figs 9a–c.

Globotruncana marianosi Douglas, 1969, p. 341, pl. 39, figs 1a–c; Douglas, 1970, p. 21, pl. 3, figs 1a–c.

Rotalipora micheli (Sacal and Debourle). Neagu, 1969, p. 148, pl. 32, figs 7–9; pl. 33, figs 1–3.

Rotalipora (Thalmanninella) micheli (Sacal and Debourle). Sǎndulescu, 1969, p. 198, 199, pl. 45, figs 1a–3c.

Rotalipora tehamaensis Marianos and Zingula. Neagu, 1972, p. 216, pl. 8, figs 39–41.

Description: Test trochospiral; dorsal side flat to very low spired, final 1–4 chambers may slope concavely toward the previous whorl; ventral side strongly convex; equatorial periphery slightly to moderately lobate; axial periphery acute but approaching 90°, with a single, imperforate pustulose keel. Chambers flat, crescent-shaped to semicircular dorsally, rising steeply on the ventral side to the vertex then sloping abruptly into the umbilicus, appressed; numbering 14–16 in 2½–3 whorls with 6–8 in the final whorl; gradually increasing in size, but final 1–2 chambers may be smaller. Primary aperture a low, interiomarginal, umbilical–extra-umbilical arch bordered by a narrow lip. Accessory apertures result from a fold in the flap-like extension near the base of the umbilical face of the post-neanic chambers; each in line with a suture. Sutures raised, curved dorsally; flush to depressed, straight to slightly curved ventrally, may thicken to form a discontinuous carina on the umbilical shoulder. Surface densely and finely perforate. Umbilicus narrow to moderately wide and moderately deep.

Remarks: *Rotalipora deeckei* is the only member of this genus to possess a flat to concave dorsal side. The degree of dorsal concavity is the most variable feature of this species. It may exhibit a flat spiral side or one with initially a minimal convexity changing to a concave slope in the final whorl.

Forms which appear to be intermediate between *Rotalipora appenninica* and *R. deeckei* have been observed, suggesting a phyletic link between the two.

Dalbiez (1957) illustrated one of Franke's original types of *Rotalia deeckei* which has helped clarify its taxonomic status. Because of the umbilically positioned apertures, Dalbiez assigned this species to *Thalmanninella*, now considered a synonym of *Rotalipora*. He recognised a strong affinity between *R. deeckei* and *R. appenninica* var. *gamma* (Gandolfi).

Borsetti (1962) introduced the new name *Rotalipora marchigiana* for *R. appenninica* var. *γ* (Gandolfi). Upon examination of topotypes of *R. reicheli* (Mornod) and of *R. appenninica* var. *γ*, Caron and Luterbacher (1969) included Gandolfi's species as an intraspecific variant of *R. reicheli*.

Porthault (1969) observed the striking resemblance of *Rotalipora deeckei* to *R. reicheli* (Mornod). In fact there are no features which distinguish the two. Thus, *R. deeckei* has priority over *R. reicheli* and *R. marchigiana* Borsetti [= *R. appenninica* var. *γ* (Gandolfi)].

In 1962 Luterbacher and Premoli Silva proposed the new name *Rotalipora gandolfii* for *R. appenninica* var. *typica* (Gandolfi). Strangely they did not compare it to *R. deeckei*, which it most strongly resembles. Examination of the type figure and numerous topotypes of *R. gandolfii* revealed the dorsal flattening characteristic of *R. deeckei* to which it is synonymous.

Examination of the primary types of *Globotruncana marianosi* Douglas, 1969, disclosed the presence of accessory apertures in the paratype. Considering the otherwise identical morphology of the holotype, it is assumed that the holotype also bears these apertures though they are obscured because of preservation. Douglas' species, then, is conspecific with *Rotalipora deeckei*.

Distribution: This is shown in Fig. 91.

Range: Cenomanian—Mid-Turonian.

> *Rotalipora greenhornensis* (Morrow), Plate 31, figs 5, 6

Globorotalia greenhornensis Morrow, 1934, p. 199, 200, pl. 31, figs 1a–c.

Planulina greenhornensis (Morrow). Cushman, 1940, p. 37, pl. 7, figs 1a–c; Cushman, 1946b, p. 159, pl. 65, figs 3a–c.

Rotalipora globotruncanoides Sigal, 1948, p. 100, 101, pl. 1, figs 4a–c; pl. 2, figs 3a–5; Sigal, 1952, p. 26, text-fig. 24.

Thalmanninella brotzeni Sigal, 1948, p. 102, pl. 1, figs 5a–c; pl. 2, figs 6a–7; Sigal, 1952, p. 26, text-fig. 25; Bermudez, 1952, p. 53, pl. 7, figs 2a, b; Ayala-Castañares, 1954, p. 423, 424, pl. 15, figs 1a–c; Sigal 1956b, p. 212, text-figs 2a, b; *non* Maslakova, 1961, pl. 3, figs 1a–c; pl. 4, fig. 1; *non* Hiltermann and Koch, 1962, p. 329, 330, pl. 49, figs 5a–c; tab. 19 (part); *non* Salaj and Samuel, 1966, p. 178, 179, pl. 11, figs 4a–c; *non* Prosnyakova, 1968, p. 23, 24, pl. fig. 6a–c.

Globotruncana (Thalmanninella) brotzeni Sigal. Mornod, 1950, p. 586, 587, text-figs 9 (1a–c); Colom, 1952, p. 17, pl. 2, figs 12–18; pl. 14, figs 1–7; Majzon, 1961, p. 763, pl. 2, fig. 9.

Globotruncana (Thalmanninella) sp. Küpper, 1955, p. 115, pl. 18, figs 4a–c.

Thalmanninella greenhornensis (Morrow). Brönnimann and Brown, 1956, p. 535, 536, pl. 20, figs 7–9; (?) Salaj and Samuel, 1966, p. 180, 181, text-fig. 15.

Rotalipora brotzeni (Sigal). Bolli *et al.*, 1957, p. 41, pl. 9, figs 7a–c; Lehmann, 1963, p. 142, pl. 1, figs 1a–2c; text-fig. 2c (?); Săndulescu, 1966, pl. 1a, figs 3a–c; Neagu, 1969, p. 148, pl. 32, figs 4–6 (*non* 1–3); pl. 33, figs 5–7; pl. 34, figs 1–3 (*non* pl. 31, figs 9, 10); Neagu, 1970a, fig. 1 (9).

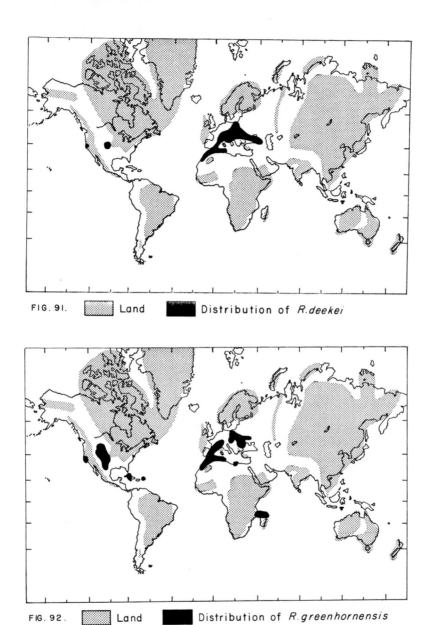

FIG. 91.　▨ Land　█ Distribution of *R.deekei*

FIG. 92.　▨ Land　█ Distribution of *R.greenhornensis*

Rotalipora globotruncanoides Sigal. *non* Książkiewicz, 1958, text-fig. 1 (5a–c); Lehmann, 1962, p. 141, pl. 1, figs 3a–c, text-fig. 3b; Neagu, 1970b, p. 66, pl. 26, figs 4–6 (?), 7–9, (*non* 1–3, 10, 11).

Rotalipora (*Thalmanninella*) *greenhornensis* (Morrow). Klaus, 1959, p. 805, pl. 2, figs 3a–c.

Rotalipora (*Thalmanninella*) *brotzeni* Sigal. Klaus, 1959, p. 805, pl. 3, figs 1a–c; Săndulescu, 1969, p. 196, pl. 39, figs 1a–4c.

Rotalipora (*Thalmanninella*) cf. *brotzeni* Sigal. Caron, 1966, p. 71, pl. 1, figs 1a–2c.

Rotalipora (*Thalmanninella*) *globotruncanoides* (Sigal). Klaus, 1959, p. 805, 806, pl. 4, figs 1a–c; text-fig. 7 (1a–c); Săndulescu, 1969, p. 196, 197, pl. 40, figs 1a–4c.

Rotalipora apenninica typica (Gandolfi). Malapris and Rat, 1961, p. 87, pl. 3, figs 4a–c.

Globotruncana (*Thalmanninella*) *greenhornensis* (Morrow). Majzon, 1961, p. 762, pl. 5, fig. 13 (listed as 12).

Rotalipora greenhornensis (Morrow). Loeblich and Tappan, 1961, p. 299, 301, pl. 7, figs 7a–10c (*non* 5, 6a–c); Ayala-Castañares, 1962, p. 28–30, pl. 5, figs 3a–c; pl. 10, figs (?) 3a, b; pl. 13, figs (?) 2a, b; Postuma, 1962, 7 figs; Graham, 1962, p. 108, pl. 20, figs 28Aa–c, Da–c (*non* B–Cc, Ea–c); Lehmann, 1963, p. 143, pl. 1, figs 6a–c; text-figs 2f (?), 3d; Lehmann, 1965, text-fig. 1d; Saavedra, 1965, p. 324, 325, text-fig. 11; Bandy, 1967, p. 15, text-fig. 7 (5); Pessagno, 1967, p. 295–297, pl. 50, fig. 3 (?); pl. 51, figs 13–21; pl. 101, figs (?) 3, 4; *non* Neagu, 1968, text-fig. 2 (47); Barr, 1968, pl. 3, figs 3a–c; Bate and Bayliss, 1969, pl. 2, figs 3a–c; Douglas, 1969, p. 174, pl. 1, figs 3a–c; Eicher, 1969, text-fig. 3 (part); *non* Porthault, 1969, p. 533, pl. 1, figs 7a–c; Eicher and Worstell, 1970b, p. 312, pl. 12, figs 2a–c; pl. 13, figs 3a, b; Sidó, 1971, p. 46, pl. 1, fig. 5 (part); Gawor-Biedowa, 1972, p. 83, 84, pl. 9, figs 4a–5c; Eicher, 1973, p. 186, pl. 1, figs 1a–4b (*non* 6a, b); pl. 2, fig. 1; text-fig. 2.

Rotalipora evoluta Sigal. Ayala-Castañares, 1962, p. 26–28, pl. 5, figs 2a–c; pl. 10 figs (?) 2a–c; pl. 13, figs (?) 1a–e.

Rotalipora reicheli Mornod. Kaptarenko-Chernousova, 1963, p. 104, 105, pl. 15, figs 3a, b, v.

Rotalipora appenninica (Renz). Todd and Low, 1964, p. 405, pl. 2, figs 6a–c; Pessagno, 1967, p. 289–292, pl. 51, figs 10–12; (*non* pl. 50, figs 1, 2, 4–6; pl. 98, fig. 13; pl. 101, figs 1, 2).

Rotalipora tehamaensis Marianos and Zingula, 1966, p. 339, pl. 38, figs 4a–c; *non* Neagu, 1972, p. 216, pl. 8, figs 39–41.

Thalmanninella globotruncanoides (Sigal). *non* Salaj and Samuel, 1966, p. 180, pl. 11, figs 6a–c.

Praeglobotruncana bronnimanni Pessagno, 1967, p. 286, pl. 49, figs 6–11.

Description: Test trochospiral, generally biconvex but dorsal spire may range from low to moderately high; equatorial periphery of first few chambers of final whorl may be entire, otherwise gently lobate, but may be deeply lobate in the gerontic stage; axial periphery acute with a single, imperforate, pustulose keel. Chambers appressed, triangular in cross-section; sloping dorsally with minimal convexity, crescent-shaped, often elongate; ventral slope gentle to umbilical shoulder then drops abruptly with the umbilical face at 90° or less to the umbilical floor, wider at keel than umbilicus, final few chambers may have a bluntly rounded umbilical shoulder; 15–19 in 3 whorls with 6–9 in the final whorl, gradually increasing in size. Primary aperture a low to moderately high, interiomarginal, umbilical–extraumbilical arch bordered by a narrow lip which extends into the umbilicus forming a flat to gently convex, imperforate platform. Accessory apertures typically aligned with the sutures and just beneath the umbilical shoulder but may be slightly above or well below the shoulder; formed by an asymmetrical fold in the imperforate flap-like extension over the aperture which fuses with the umbilical platform; the loop-like fold may occasionally be double. Sutures of dorsal side curved, limbate, raised; ventral side like dorsal side except in the gerontic stage where sutures may be depressed; carina on the umbilical shoulder formed as an extension of the suture, which itself is merely the ventral segment of a bifurcated keel. Umbilicus narrow to moderately wide. Surface densely and finely perforate with some secondary calcification along the umbilical shoulder and centre of the ventral chambers.

Remarks: *Rotalipora greenhornensis* often has characteristically elongate dorsal chambers. It lacks the flat to concave side typical of *R. deeckei* (Franke), the species which it most nearly resembles. Eicher and Worstell (1970b) documented considerable variation in the dorsoconvexity. There is also variation in the width of the umbilical platform and in the position of the accessory apertures.

The phylogeny of this species is unknown.

The holotype of *Globorotalia greenhornensis* Morrow was found to possess an accessory aperture along the umbilical shoulder of the ultimate chamber. Other apertures are covered by debris filling the umbilicus. The umbilicus is twice the width shown in the type figure.

Douglas (1970) stated that *Rotalipora tehamaensis* Marianos and Zingula, 1966, falls within the range of variability of *R. greenhornensis*. No clear accessory apertures are visible on the holotype of *R. tehamaensis*, but this may be due to preservation. Otherwise, this species does appear identical to *R. greenhornensis*.

The specimen designated as the new species *Praeglobotruncana bronnimanni* by Pessagno (1967) is merely a neanic individual of *Rotalipora greenhornensis*. An examination of a topotypic suite of *R. greenhornensis* reveals that the accessory apertures are not present in the immature stage.

Distribution: This is shown in Fig. 92.

Range: Mid to Upper Cenomanian.

Rotalipora multiloculata (Morrow), Plate 34, figs 2–5

Globorotalia ? *multiloculata* Morrow, 1934, p. 200, pl. 31, figs 3a, b, 5a, b; Cushman, 1946b, p. 153, pl. 62, figs 10a–11c; (?) Gauger, 1953, in Peterson *et al.*, p. 83, 84, pl. 10, fig. 4; *non* Mallory, 1959, p. 256, pl. 34, figs 11a–c.

Globotruncana ticinensis var. α Gandolfi, 1942, p. 114, pl. 2, figs 4a–c; pl. 4, figs 8, 9, 21, 22; pl. 8, fig. 3; pl. 11, fig. 5 (part); pl. 13, figs 9, 10.

Globotruncana (*Thalmanninella*) *ticinensis* var. *alpha* (Gandolfi). Reichel, 1950, p. 603, 604, pl. 16, figs 2, 3; pl. 17, figs 2, 3.

Globotruncana (*Ticinella*) *roberti* (Gandolfi). Noth, 1951, p. 75, pl. 4, figs 29a, b.

Rugoglobigerina multiloculata (Morrow). Bermudez, 1952, pl. 22, figs 1a–c.

Ticinella roberti (Gandolfi). (?) Sigal, 1952, p. 23, text-fig. 19.

Thalmanninella ticinensis var. α (Gandolfi). Ayala-Castañares, 1954, p. 425, pl. 15, figs 2a–c.

Thalmanninella multiloculata (Morrow). Brönnimann and Brown, 1956, p. 534, 535, pl. 20, figs 1–3.

Rotalipora (*Thalmanninella*) *multiloculata* (Morrow). *non* Klaus, 1959, p. 804, 805, pl. 2, figs 2a–c; *non* Caron, 1967, p. 73, 74, pl. 2, figs 3a–c.

Ticinella multiloculata (Morrow). Loeblich and Tappan, 1961, p. 292, 294, pl. 6, figs 13a–c; *non* Salaj and Samuel, 1966, p. 175, 176, pl. 11, figs 1a–c; Eicher, 1969, text-fig. 3 (part); Hermes, 1969, p. 41, 42, pl. 3, figs 61–67; pl. 4, figs 68–70; Eicher and Worstell, 1970b, p. 312, 313, pl. 13, figs 2a–c, 4a–c.

   Globotruncana (*Thalmanninella*) *multiloculata* (Morrow). Majzon, 1961, p. 762, pl. 5, fig. 12 (listed as 13).

   Ticinella roberti multiloculata (Morrow). Bandy, 1967, p. 15, text-fig. 7 (2).

   Rotalipora greenhornensis (Morrow). Eicher, 1973, p. 186, pl. 1, figs 6a, b (*non* 1a–4b; pl. 2, fig. 1).

   Anaticinella multiloculata (Morrow). Eicher, 1973, p. 185, 186, pl. 1, figs 5a, b; pl. 2, figs 2a–7b.

   Description: Test trochospiral, biconvex with a low to moderately high dorsal side; equatorial periphery slightly lobate in the early chambers of the last whorl, later becoming moderately lobate; axial periphery broadly rounded, may exhibit a narrow, imperforate pustulose keel on first half of last whorl. Chambers inflated, closely packed in early portion of last whorl, less so later; circular to oval in cross-section with the thickness exceeding the other dimensions in the late ephebic and gerontic stages; numbering 17–19 in 3 whorls with 5–9 in the final whorl; gradually increasing in size but may be subequal in the final whorl; ventral surface of early chambers slopes gently into the umbilicus, later chambers curve sharply into the umbilicus. Primary aperture a low to moderately high, interiomarginal umbilical–extraumbilical arch bordered by a narrow lip which continues as a wide flap from the base of the umbilical face out into the umbilicus forming a flat to slightly convex imperforate platform. Accessory apertures result from an asymmetrical fold in the projected flap and may approach a 90° angle to the platform; the resulting loop-like apertures are aligned with the sutures at or just below the chamber surface. Dorsal and ventral sutures straight to slightly curved, moderately depressed becoming deeply depressed in last 3–6 chambers. Umbilicus moderately to very wide. Surface perforated by abundant, randomly distributed, small diameter pores; apertural face with less density.

   Remarks: *Rotalipora multiloculata* is the only *Rotalipora* with a broadly rounded axial periphery. The most significant variable feature of this species is the peripheral keel. Eicher (1973) reported that approximately one third of the specimens he examined possessed a keel in the early chambers of the final whorl.

   Eicher (1973) proposed that *Rotalipora greenhornensis* (Morrow) was ancestral to *R. multiloculata*. He based this opinion on certain morphologic similarities (see Remarks) and the absence of members of the genus *Ticinella* in strata younger than Albian. *Ticinella roberti* (Gandolfi) has been recorded from the lower Cenomanian, however, and may be ancestral.

   Although Brönnimann and Brown (1956) redrew the holotype of *Rotalipora multiloculata* illustrating a weakly developed keel, Loeblich and Tappan (1961) regarded it as non-keeled and placed it in the genus *Ticinella*. A narrow keel is present in the holotype, which precludes its placement in *Ticinella*.

   The specimen illustrated by Caron (1967): *Rotalipora* (*Thalmanninella*) *multiloculata* has a very strong, well-developed keel, whereas this taxon has only a narrow, poorly-developed keel. It also lacks sutural and spiral dorsal carinae of Caron's specimen.

   The genus *Anaticinella* was proposed by Eicher (1973) with *Globorotalia* ? *multiloculata* Morrow as the type species, and an attempt to demonstrate a phyletic connection with *Rotalipora greenhornensis* (Morrow) was made. Among the features he cited as shared between the two species are the loop-like supplementary (here termed accessory) apertures, the occasional doubling of these apertures, a prominent umbilical platform, late chambers that protrude umbilically and similarity in coiling direction. Each of these are discussed below.

In his remarks of the accessory apertures, Eicher stated that the doubling of these apertures is found in other species of *Rotalipora* and "is not indicative of close affinity" for the species in question. He characterised the apertures of *Ticinella* as being more "subdued," and those of *G. ? multiloculata* as more like those of *R. greenhornensis*. Caron (1967, 1971), among others, has illustrated several species of *Ticinella* and *Rotalipora* in which the structure of the accessory aperture is identical, even to the bordering lip.

The umbilical platform is a flap-like continuation of the apertural lip that is analogous to the portici and the tegilla of other genera. Each of these structures is variable, and the degree of variation increases with complexity. The platform, developed to a greater or lesser degree, is present on several taxa assigned to *Ticinella* and *Rotalipora*. It cannot be considered indicative of a phyletic relationship between this taxon and *R. greenhornensis*.

The development of umbilically protruding chambers is more likely homeomorphic rather than denoting a tie between two species. Such a chamber form is commonly found in gerontic species of *Rotalipora*, *Praeglobotruncana* and *Globotruncana*, and so is believed not to have taxonomic significance. It is most likely in response to cytological and/or environmental changes.

Eicher stated that the dextral coiling directions for *Globorotalia ? multiloculata* and *Rotalipora greenhornensis* (90.7% and 88.6% respectively) "may be taken as evidence of genetic affinity". The summary of planktonic coiling directions given by Bolli (1971) show that nearly all post-Albian species are 90–100% dextrally coiled. Moreover, as presented earlier in this paper, coiling direction can be related directly to the method of reproduction in some species. Thus, the similarity in coiling direction cannot be taken to signify a phyletic relationship.

The author agrees with the observation by Eicher that "*Anaticinella multiloculata* and *Rotalipora greenhornensis* appear superficially so different . . . that a close genetic affinity does not immediately suggest itself". But if not *R. greenhornensis*, what then was the progenitor of *Globorotalia ? multiloculata*, and to what genus does it belong? The similarity of this species to those of *Ticinella* has been noted by many authors, but this genus has often been interpreted as being restricted to pre-Cenomanian strata. However, if the stratigraphic diagnoses in the various papers used in this study are correct, then *T. roberti* (Gandolfi) ranges into the lower Cenomanian to approximately the level of the lowermost occurrence of *G. ? multiloculata*. *Ticinella roberti* is like *G. ? multiloculata* in every way except the possession of a keel. Eicher found that approximately $\frac{1}{3}$ of his species of *G. ? multiloculata* possess a keel in the last whorl. A few thin-sections of individuals which lack a keel in the last whorl show the preence of this structure in the earlier growth stages.

It is suggested that *Ticinella roberti* may have given rise to *Globorotalia ? multiloculata* by the development of a whole keel. Furthermore, although this keel is not always present in the last whorl, it appears to be present in the earlier stages. It is conceivable that some specimens referred to *T. roberti* are *R. multiloculata* which lack the keel in the final whorl. The loss of the keel in the last few chambers is comparable to the reduction from a double keel to a single keel, and occasionally its total loss in some *Globotruncana*.

Distribution: This is shown in Fig. 93.

Range: Mid to upper Cenomanian.

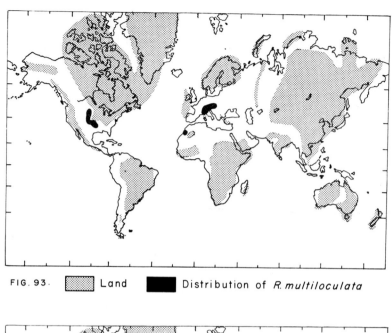

FIG. 93 . ░░░ Land ███ Distribution of *R. multiloculata*

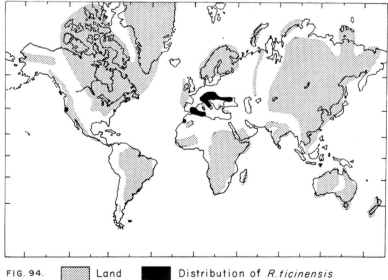

FIG. 94. ░░░ Land ███ Distribution of *R. ticinensis*

Rotalipora ticinensis (Gandolfi), Plate 32, fig. 4; plate 33, figs 1–4; plate 34, fig. 1
Globotruncana ticinensis Gandolfi, 1942, p. 113–115, pl. 2, figs 3a–c; pl. 3, figs 10, 11.
Globotruncana ticinensis var. α Gandolfi, 1942, p. 114, pl. 3, figs 8, 9; Hiltermann, 1956, text-fig. 2 (part).
Globotruncana ticinensis ticinensis Gandolfi. Cita, 1948b, p. 161, 162, pl. 4, figs 8a–c.
Globotruncana (Thalmanninella) ticinensis Gandolfi. Reichel, 1950, p. 603, 604, pl. 16, figs 2, 3; pl. 17, figs 2, 3; Majzon, 1961, p. 762, pl. 2, fig. 7.
Thalmanninella ticinensis (Gandolfi). Sigal, 1952, p. 23, fig-text. 20; Ayala-Castañares, 1954, p. 424, 425, pl. 15, figs 3a–c; Pokorný, 1963, p. 384, fig. 430; Salaj and Samuel, 1966, p. 182, 183, pl. 11, figs 5a–c; tab. 35 (2); Prosnyakova, 1968, p. 19, pl. fig. 1a–c.
Thalmanninella ticinensis var. α (Gandolfi). Ayala-Castañares, 1954, p. 425, pl. 15, figs 2a–c.
Globotruncana ticinensis typica Gandolfi. Hiltermann, 1956, text-fig. 2 (part).
Rotalipora ticinensis ticinensis (Gandolfi). (?) Bolli, 1957, p. 56, pl. 12, figs 1a–c; Marianos and Zingula, 1966, p. 339, pl. 38, figs 3a–c; Risch, 1971, p. 52, pl. 5, figs 10–12.
Globotruncana (Thalmanninella) ticinensis ticinensis Gandolfi. Gandolfi, 1957, p. 59, 60, pl. 8, figs 2a–c.
Globotruncana (Thalmanninella) ticinensis subticinensis Gandolfi, 1957, p. 59, pl. 8, figs 1a–c.
Rotalipora (Thalmanninella) ticinensis subticinensis (Gandolfi). *non* Klaus, 1959, p. 803, 804, pl. 1, figs 4a–c; Caron, 1967, p. 70, pl. 1, figs 1a–c; Caron, 1967, pl. 1, figs 2a–c (passing to *Rotalipora (Thalmanninella) apenninica*); Caron, 1967, pl. 1, figs 4a–c (passing to *Rotalipora (Thalmanninella) ticinensis ticinensis*).
Rotalipora (Thalmanninella) ticinensis ticinensis (Gandolfi). Klaus, 1959, p. 804, pl. 2, figs 1a–c; Caron, 1967, p. 70–73, pl. 1, figs 5a–c; text-figs 21, 22; (*non* pl. 1, figs 6a–c).
Thalmanninella ticinensis subticinensis Gandolfi. *non* Malapris and Rat, 1961, p. 87, pl. 3, figs 3a–c; *non* Sturm, 1969, pl. 9, figs 2a–c.
Rotalipora subticinensis (Gandolfi). Postuma, 1962, 7 figs; Lehmann, 1966b; p. 158, pl. 2, fig. 5; text-fig. 1b.
Rotalipora ticinensis (Gandolfi). Postuma, 1962, 7 figs; Lehmann, 1966b, p. 158, pl. 2, fig. 6; text-fig. 1e; Caron and Luterbacher, 1969, p. 25, pl. 8, figs 6a–c; *non* Scheibnerová, 1969, p. 63, 64, tab. 10, figs 3a–c; *non* Kuhry, 1971, p. 232, 233, pl. 3, figs 3A–C.
Rotalipora ticinensis subticinensis (Gandolfi). Borsetti, 1962, p. 44, 45, pl. 1, figs 3a–c; Hermes, 1969, p. 44, 45, pl. 1, figs 7–9; text-fig. 1; (?) Risch, 1971, p. 52, 53, pl. 4, figs 23–25.
(?) *Rotalipora klausi* Lehmann, 1966b, p. 159, pl. 2, fig. 4; text-figs 1a, j–n.
Thalmanninella ticinensis ticinensis (Gandolfi). Sturm, 1969, pl. 9, figs 1a–c.
Globotruncana ticinensis var. *alpha* Gandolfi. Hermes, 1969, p. 44, 45, pl. figs 92–94.
Description: Test trochospiral, generally biconvex but dorsal side may have a very low to a high spire; equatorial periphery slightly lobate in earlier chambers of the last whorl, later moderately so; axial periphery sharply acute in all but last 1–2 chambers which may approach 90°, with a wide, imperforate, pustulose, single

keel. Chambers nearly flat to gently convex dorsally with a semicircular shape; gently convex in early chambers becoming moderately so in the last few, trapezoidal to nearly triangular in outline ventrally; 18–21 in 3 whorls with 6–10 in the final whorl; gradually increasing in size, may be subequal in last whorl; early chambers slope gently into the umbilicus, final few slope more steeply. Primary aperture a low to moderately high, interiomarginal, umbilical–extraumbilical arch bordered by a narrow, imperforate lip which continues out into the umbilicus as a flap that fuses to the previous projections. Accessory apertures formed by an asymmetric fold on the distal side of the flap, aligned with the sutures but at the margin of the umbilicus. Dorsal sutures curved, limbate, raised except in last 1–6 chambers in which they may be depressed and narrow; ventral sutures in early chambers may be flush later becoming depressed, straight to slightly curved. Umbilicus narrow to moderately wide. Surface densely perforate except for the imperforate apertural face; secondary calcification may occur on the chamber vertex bordering the umbilicus, early chambers may be pustulose.

Remarks: *Rotalipora ticinensis* most closely resembles *R. appenninica* (Renz). So much so that they may be merely intraspecific variants. Some topotypic individuals seem identical. This species may have a few more chambers and chambers per whorl than are generally reported for *R. appenninica*. Its final few chambers tend to be more broadly acute and more inflated than *R. appenninica*. However, these features may not be taxonomically distinctive. Two features of this species are particularly variable—the dorsoconvexity and the apertural flap. The apical angle of different specimens ranges from 104° to 155°. Basically the flap adjoins the chamber along its umbilical face from which it projects as a hyperbolic paraboloid. The distal edge and particularly the proximal edge may be convoluted. The greater the length the flatter it becomes until it is a gently convex structure, that may or may not form an umbilical platform.

According to Sigal (1966a) *Rotalipora ticinensis* evolved from *Ticinella praeticinensis* Sigal. This was agreed to by Caron (1967), and seems to be a reasonable conclusion.

Topotypes of *Rotalipora klausi* Lehmann, 1966b, were examined and compared to those of *R. ticinensis*. Only two slight differences were found. The ventral side of the chambers has slightly greater curvature in the plane of coiling in *R. ticinensis* than in *R. klausi*. The apertural face of the latter is always perforated by several large diameter pores. This has not been observed in *R. ticinensis*. The two are tentatively regarded as synonymous.

Caron and Luterbacher (1969) refigured the holotype of *Rotalipora ticinensis*, which lacks the accessory apertures, presumably due to poor preservation. They also reported the loss of the lectotype of *R. subticinensis* (Gandolfi).

Rotalipora subticinensis (Gandolfi) has generally been regarded as a taxon, species or subspecies, distinct from but closely related to *R. ticinensis*. This separation has been based primarily upon spire height, with the type figure of *R. subticinensis* having an apical angle of approximately 104° and that of *R. ticinensis* being approximately 148°, or 98° for the refigured holotype.

As noted in the section Intraspecific Variability, spire height has been documented as a genetic variant in which the schizont is larger and lower spired than the gamont. *Rotalipora subticinensis* is lower spired and usually has a few more chambers per whorl than *R. ticinensis*. They differ in no other way, as even their total stratigraphic ranges are identical. Thus, it is concluded that they do represent genetic variants of a single species. *Rotalipora ticinensis* has priority.

Distribution: This is shown in Fig. 94.
Range: Upper Albian.

Subfamily TICINELLINAE Masters, fam. nov.

Definition: Test trochospiral throughout or later becoming planispiral, periphery non-keeled; primary aperture interiomarginal, umbilical–extra-umbilical or umbilical–equatorial; with a prominent apertural lip which extends into the umbilicus forming accessory apertures along the distal edge; if planispiral the lip projects over the periphery to the other side, successive lips mark the position of relict apertures.

Generic Key

I. Trochospiral with accessory apertures only . . . . . *Ticinella*

II. Planispiral with accessory and relict apertures . . . *Biticinella*

Remarks: The decision to place *Ticinella* and *Biticinella* into a new subfamily separate from *Rotalipora* was based upon two premises. First, the keel is believed to be second only to mode of coiling in taxonomic importance in the planktonics. Second, the accessory apertures, present on all three genera, are highly variable structures. One finds that the extended lip, which forms the accessory apertures, grades on the one hand with the portici of the Globigerinidae and on the other hand with the tegilla of the Globotruncanidae. Furthermore, these apertural extensions are very delicate structures which are easily broken away, and being in the umbilicus, are often covered by sediment. Whereas, the keel either is or is not present, a fact easily determined because of the location of the structure.

Range: Upper Aptian—lower Cenomanian.

Genus BITICINELLA Sigal, 1956

Biticinella Sigal, 1956, p. 35.
Ticinella (*Biticinella*) Sigal. Sigal, 1966a, p. 189.

Type species: *Anomalina breggiensis* Gandolfi, 1942.

Description: Test a flat trochospiral in early whorls becoming planispiral in last whorl; ephebic and gerontic stages biumbilicate, involute with tendency to become evolute; equatorial periphery lobate, axial periphery rounded, non-keeled. Chambers inflated, appressed, greatest width parallel to axis of coiling. Primary aperture a low interiomarginal, umbilical–equatorial arch; often asymmetric, flaring toward one side; successive apertures remain as relict apertures only on one side. Accessory apertures are formed on the other side by extensions of the apertural lip into the umbilicus. Sutures depressed, radial. Surface perforate.

Remarks: *Biticinella* possesses true accessory apertures in one umbilicus and relict apertures in the other; whereas *Globigerinelloides* has relict apertures in each umbilicus.

Loeblich and Tappan (1961a, 1964) treat *Biticinella* and *Globigerinelloides* as synonymous because they both lack a keel. This decision was most likely reached without examination of type material, for as noted above their apertural characteristics differ.

Luterbacher and Premoli Silva (1962) distinguish *Biticinella* from *Globigerinelloides* by the fact that the former becomes planispiral only in the last whorl, that it is never truly biumbilicate and finally, that the "dorsal" accessory apertures correspond to those of *Globigerinelloides* while those on the "ventral" side are like those of *Ticinella*. Their second reason is related directly to their first, neither of which can serve as a distinguishing criterion because the species of *Globigerinelloides* also arose from trochospiral forms with which they are gradational. However, the presence of accessory ticinellid-like apertures is a significant enough departure from *Globigerinelloides* to warrant a separate genus, a conclusion also reached by Hermes (1969).

Luterbacher and Premoli Silva (1962) frequently observed transitional forms between *Biticinella* and *Ticinella*. This was corroborated by Sigal (1966a) who concluded that *Ticinella* was ancestral to *Biticinella*. This conclusion is supported by this study.

Sigal (1966a) considered *Biticinella* as a subgenus (morphogenus of Sigal) of *Ticinella* because of the gradational nature of the two, with the former differing primarily by the development of relict apertures on the "dorsal" side. So far the occurrence of both relict and accessory apertures has been frequently shown in only one taxon, *B. breggiensis* (Gandolfi), although Luterbacher reported (Renz *et al.*, 1963) the same occurrence for rare individuals of his *T. primula*. This development is not believed to be ontogenetic, although the possibility cannot yet be conclusively discarded, because individuals with both relict and accessory apertures are thus far described from only a short segment of the total stratigraphic range of their ancestors. Furthermore, because such an apertural combination has been recognised in but two of the *Ticinella* species, the more logical approach is to consider this development as palingenetic. *Biticinella* is here regarded as a genus separate from *Ticinella*, the two being differentiated solely by the presence or absence of the relict and accessory aperture combination.

It is to a degree a matter of personal preference as to how to treat closely related or gradational species, in this case genus or subgenus. Many genera and species have been shown to be gradational and yet have been assigned equal rank with their ancestors, for example *Ticinella praeticinensis* Sigal, which Sigal (1966a) proposed as the ancestral form of *Rotalipora ticinensis* (Gandolfi).

Range: Upper Albian.

Biticinella breggiensis (Gandolfi), Plate 35, figs 1–3

Anomalina breggiensis Gandolfi, 1942, p. 102, 103, pl. 3, figs 6a–c; pl. 5, fig. 3 (?); pl. 9, fig. 1, pl. 13, figs 7a–8b; text-figs 34 (1–4); Hermes, 1969, p. 45, pl. 5, figs 95–97.

Biticinella breggiensis (Gandolfi). Sigal, 1956a, p. 35, 36, text-figs 1a–c; Luterbacher and Premoli Silva, 1962, p. 272–274, pl. 23, figs 2a–4c; Lehmann, 1966b, p. 156, 157, pl. 1, fig. 8; Salaj and Samuel, 1966, p. 164, pl. 6, figs 6a–c; tab. 33 (3); Caron and Luterbacher, 1969, p. 25, pl. 7, figs 4a–c; Hermes, 1969, p. 45, pl. 1, figs 4–6.

Biticinella ? *breggiensis* (Gandolfi). Klaus, 1959, p. 830, 831, pl. 8, figs 6a, b; *non* Takayanagi and Iwamoto, 1962, p. 188–190, pl. 28, figs 7a–9c.

Globigerinelloides breggiensis (Gandolfi). Postuma, 1962, 8 figs; Moullade, 1966, p. 126–128, pl. 9, figs 12–14; Kuhry, 1971, p. 228, pl. 1, figs 6A–C.

Ticinella (*Ticinella*) spectrum *breggiense* (Gandolfi). Sigal, 1966a, p. 192, 193, pl. 1, figs 8a–10b; [replaced by *T.* (*T.*) *breggiensis* (Gandolfi), p. 214].

Ticinella (*Biticinella*) spectrum *breggiense* (Gandolfi). Sigal, 1966a, p. 193–195, pl. 1, figs 1a–7b; pl. 2, figs 2a, b; [replaced by *T.* (*B.*) *subbreggiensis* Sigal, p. 214].

Ticinella (*Biticinella*) cf. *breggiensis* (Gandolfi). Sigal, 1966a, pl. 2, figs 1a, b.

Ticinella (*Biticinella*) *breggiensis* (Gandolfi). Risch, 1971, p. 52, pl. 4, figs 7, 8.

Description: Test initially a flat trochospire becoming planispiral in the last whorl; biumbilicate in ephebic and gerontic stages; involute tending to become evolute in the final 1–2 chambers; equatorial periphery lobate, axial periphery rounded, perforate. Chambers inflated, appressed, greatest width parallel to coiling axis; 9–11 in final whorl; gradually, but irregularly increasing in size, many subequal. Primary aperture a low, interiomarginal, asymmetric, equatorial arch, bordered by an apertural lip which extends into each umbilicus; relict apertures remain visible in one umbilicus; true accessory apertures occur along the margin of the other umbilicus, forming on the abapertural side of the chamber extension. Sutures depressed nearly radial. Surface with large diameter pores.

Remarks: *Biticinella* is monotypic. Some specimens have a tendency to become evolute in the gerontic stage. The degree of asymmetry in the primary aperture is variable. The most fluctuating feature is the number of chambers which form the planispiral portion of the coil.

Specimens gradational between *Biticinella breggiensis* and the ancestral *Ticinella roberti* (Gandolfi) are common.

In 1956, Sigal ceased to regard *Biticinella breggiensis* as synonymous with *Globigerinelloides bentonensis* pending further study. It is established here that they indeed are two different taxa.

Luterbacher and Premoli Silva (1962) state that large pores are characteristic of *Biticinella breggiensis*. This is certainly possible. However, pore diameter is also known to be environmentally as well as diagenetically controlled. More detailed assemblage analyses are needed to determine the cause in this case.

Sigal (1966a) envisioned two morphotypes of the species *breggiensis*. One he placed in the subgenus *Ticinella*, which includes the holotype; the other he placed in the subgenus *Biticinella*. In an addendum, he proposed the new species *Ticinella* (*Biticinella*) *subbreggiensis* for the latter morphotype. The sole distinction was that "dorsal" apertures were visible on *subbreggiensis*. As Sigal noted, the poor preservation of the holotype of *Anomalina breggiensis* Gandolfi precludes recognition of "dorsal" apertures. He illustrated a homeomorphoholotype which supposedly lacks these apertures. But from the photographs of his specimen (pl. 1, figs 10a, b), relict

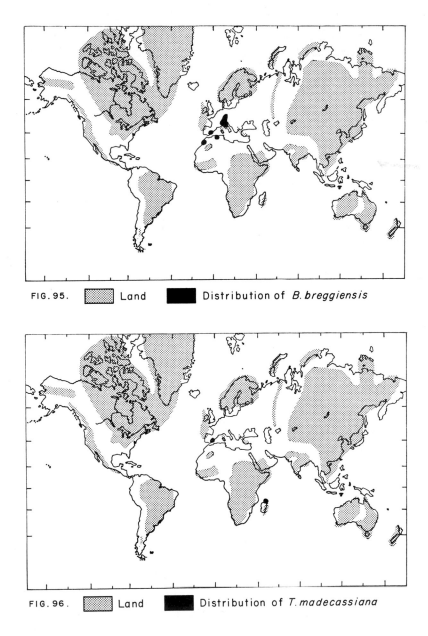

FIG. 95.   ▨ Land   ■ Distribution of *B. breggiensis*

FIG. 96.   ▨ Land   ■ Distribution of *T. madecassiana*

apertures are visible on this "dorsal" side. Topotypes of *A. breggiensis* possess apertures on the two sides. Thus, the creation of a new species for those specimens with apertures on both sides is unwarranted.

Distribution: This is shown in Fig. 95.

Range: Upper Albian.

Genus TICINELLA Reichel, 1950

Globotruncana (Ticinella) Reichel, 1950, p. 600.

Ticinella Reichel. Bermúdez, 1952, p. 116.

Rotalipora (Ticinella) Reichel. Klaus, 1959, p. 803.

Ticinella (Ticinella) Reichel. Sigal, 1966a, p. 189.

Claviticinella El-Naggar, 1971a, p. 436.

Type species: *Anomalina roberti* Gandolfi, 1942.

Description: Test trochospiral with a dorsoconvexity ranging from high to nearly flat, imparting a pseudoplanispiral appearance; equatorial periphery lobate; axial periphery rounded, perforate. Chambers inflated, subspherical. Primary aperture a low, interiomarginal, umbilical–extraumbilical arch bordered by a narrow, imperforate lip which merges into a flap-like extension; this extension projects from the lower, umbilical margin of the adult chambers. Accessory aperture situated on the distal edge of the flap near the margin of the chamber and in alignment with the suture. Sutures depressed, straight to curved dorsally and ventrally. Umbilicus shallow.

Remarks: *Ticinella* differs from *Praeglobotruncana* and *Rotalipora* by having a perforate peripheral margin rather than a keel. The dorsal side lacks the relict apertures present on *Biticinella*.

The most comprehensive study of the genus *Ticinella* was given by Sigal (1966a). In this excellent work, he applied the concept of the "spectra" to *Ticinella* and *Biticinella* and their respective species. By so doing, Sigal was in fact recognising intraspecific variation. As he remarked in an addendum to his paper (p. 214), the terms "spectrogenus," "morphogenus," etc., are not recognised as valid designations under the Code, and he offered appropriate substitutions. But his insight into the morphologic and taxonomic transitions of this group is without parity.

Sigal (1966a) treated *Biticinella* as a subgenus of *Ticinella*, a position not held here (see Remarks under *Biticinella*). This was based upon the transitional sequence between *Ticinella roberti* (Gandolfi) and *Biticinella breggiensis* (Gandolfi). Here one meets again the problem of delimiting species and genera when dealing with gradational individuals. The approach here is to restrict to *Ticinella* those forms which possess only accessory apertures on the ventral side and to *Biticinella* those with both "ventral" accessory apertures and relict "dorsal" apertures.

Also envisioned by Sigal was the phylogenetic development of *Ticinella* from a *Hedbergella* (= *Globigerina*) that in turn gave rise to *Rotalipora*.

El-Naggar (1971a) judged the presence of radially elongate chambers as taxonomically more important than accessory apertures. In so doing, he established the new genus *Claviticinella* with *Ticinella raynaudi* var. *digitalis* Sigal as the type species. It is difficult to determine which morphologic feature to rank as more important than another. The choice is complicated by the varying importance of certain homologous features from genus to genus. The author does not necessarily disagree with El-Naggar's decision, but as noted in the remarks under *Ticinella raynaudi* Sigal, the variety *digitalis* has not been determined to be a distinct taxon as opposed to an intraspecific variant. Moreover, it was invalidly named [I.C.Z.N., Art. 45e (ii)]. For these reasons, *Claviticinella* is deemed a junior synonym of *Ticinella*.

Range: Upper Aptian–lower Cenomanian.

Ticinella madecassiana Sigal

Ticinella madecassiana Sigal, 1966a, p. 197, 198, pl. 3, figs 7a–10b; Hermes, 1969, p. 44, pl. 3, figs 48–50.

Description:

Test of small size, finely and densely perforated wall; the number of chambers in the last whorl is 6–7; about 2 whorls; they regularly and rapidly increase the size of the last whorl; the coiling is subsymmetric, the last whorl overhangs in an important way the rest of the spire, whose trochoidicity is weak; the periphery is largely rounded, the contour of the test is strongly lobed.

The spiral face shows sutures rather rectilinear, oblique, more and more depressed towards the last chambers; the surface of the chambers is strongly inflated.

The umbilical face shows a deep and rather narrow umbilicus; the surface of the chambers is inflated, the sutures radial rectilinear, and strongly depressed; the lamellar expansions join in an unimportant plate, in the form of a funnel.

The terminal face is strongly inflated, wider than high.

The ornamentation is little pronounced, one observes it uniquely on the whorl of the internal spire.

The principal aperture is high and largely climbing towards the peripheral margin; the lip which surmounts it is little developed.

The supplementary aperture is small, practically in the intraumbilical position.

The coiling is apparently dextrally dominant (66%) (contrarily to that which would be implied from the figured forms).

(Translation of the type description.)

Remarks: In this species according to Sigal (1966a) . . . "the last whorl overhangs in an important way the rest of the spire, whose trochoidicity is weak . . .", the variations "are unimportant (number of chambers in the final whorl: 6–7, umbilicus more or less narrow), except in that which concerns the relative position of the internal spire and the last whorl on the dorsal face on the one hand—leading to some nearly symmetric tests . . .—on the other hand an occasional perceptible elongation of the chambers in the direction of the spire."

As mentioned above, Sigal (1966a) stated that the distinguishing feature of this species was its weak trochoidicity in which the chambers of the last whorl tend to be more dorsally inflated and, thus, extending over those of the earlier whorls. He also remarked that this characteristic leads to some analogy with *Ticinella primula* Luterbacher. In fact it is questionable that the two are distinguishable. Unfortunately topotypes of *T. madecassiana* could not be obtained during this study, so it is tentatively retained as valid.

Distribution: This is shown in Fig. 96.

Range: Upper Albian.

Ticinella praeticinensis Sigal, Plate 35, fig. 4; plate 36, figs 1, 2

Ticinella praeticinensis Sigal, 1966a, p. 195, 196, pl. 2, figs 3a–8b; pl. 3, figs 1a–6b; Moullade, 1966, p. 101, 102, pl. 8, figs 21–23; Caron, 1967, p. 68, 69, text-fig. 18; Moullade, 1969, pl. 1, figs 24–26; Hermes, 1969, p. 39, 43, 44, pl. 3, figs 51–55; Risch, 1971, p. 50, pl. 5, figs 16–18.

Rotalipora praebalernaensis Sigal, 1969, p. 635–637, pl. 1, figs 1–8.

Description:

Test of low height, finely and densely perforate; the trochospiral coil is very little raised, giving the test a nearly symmetric and very flattened profile; one observes $2\frac{1}{2}$–3 whorls of regularly and very slowly growing chambers, particularly on the end; one counts 7–8 in the last whorl; the last half-whorl (delimited by a line departing from the base of the last chamber and passing through the proloculus) comprises no more than 4 chambers; the periphery is reinforced by an accumulation of papillae or pustules, from which the importance decreases little by little, the last 4 to 5 chambers show finally only a scarcely angular, perforated margin; the exterior contour is clearly lobed.

The spiral face shows a final whorl in general flattened by relation to the internal whorls to the level of the last chambers; the surface of the chambers is strongly inflated, the sutures are subradial and strongly depressed; the spiral suture is strongly depressed.

The umbilical face shows chambers rather strongly inflated, separated by depressed, radial and rectilinear sutures, the form is typically triangular, they slope slowly towards the umbilicus. The umbilicus is rather deep, large. The lamellar expansions fuse in a helicoid ramp which largely covers the umbilicus.

The terminal face is strongly inflated, nearly isodiametric, small.

The ornamentation is uniquely formed by the pustules and papillae, densely distributed on the spiral face of the internal whorls and the 3 to 4 first chambers of the last whorl; they are slightly developed on the umbilical face; at the periphery they are densely assembled and frequently very strong, simulating a carinal band, the wall remains perforated between them.

The principal aperture is lengthened, rather high, reaching nearly the peripheral margin; it is surmounted by a wide lip.

The supplementary apertures are very small, in the intraumbilical position, except the last 1–2 which migrate slightly into the suture; they are surmounted by a thin lip.

Coiling: approximately $\frac{2}{3}$ sinistral, $\frac{1}{3}$ dextral.

Size: the large diameter varies habitually between 0.35 and 0.40 mm.

(Translated from the type description.)

Remarks: In comparing *Ticinella praeticinensis* to other members of the genus, Sigal (1966a) remarked that only one species, *T. roberti bejaouaensis* Sigal (a junior synonym of *T. roberti*), may be similar. *Ticinella praeticinensis* is distinguished by fewer chambers in the last half-whorl and the development of the pseudo-keel. Sigal reported that the profile of the spiral face ranges from flat to strongly trochoid, the degree of ornamentation varies on the spiral face of the inner whorls and finally that the number of chambers in the last whorl rarely drops below seven.

Sigal (1966a) suggested that *Ticinella praeticinensis* evolved from a *Hedbergella* rather than another *Ticinella*.

Ticinella praeticinensis is characterised by a pseudocarinate periphery. In 1969, Sigal proposed the name *Rotalipora praebalernaensis* for a form which strongly resembles *T. praeticinensis*, but which possesses a keel formed by an "assemblage

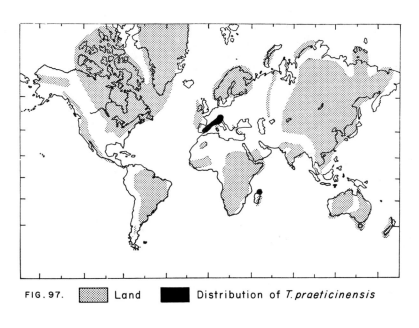

FIG. 97. ▨ Land ■ Distribution of *T. praeticinensis*

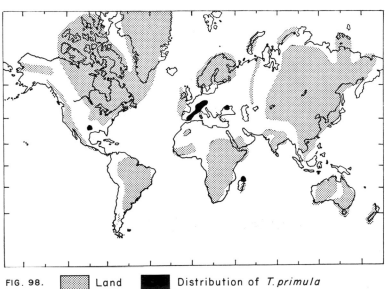

FIG. 98. ▨ Land ■ Distribution of *T. primula*

of protuberances" in the early chambers of the last whorl, later becoming a narrow, less densely perforate band in the last two or three chambers. From his description, the "keel" seems to be formed of aligned pustules; in other words, surface ornamentation. Thin-section examination would be necessary to confirm this. However, on the basis of available information, *R. praebalernaensis* is believed to be conspecific with *T. praeticinensis*.

Distribution: This is shown in Fig. 97.

Range: Upper Albian.

Ticinella primula Luterbacher, Plate 36, figs 3, 4; plate 37, fig. 1, 3

Ticinella primula Luterbacher, 1963, in Renz *et al.*, p. 1085, 1086, text-fig. 4; Sigal, 1966a, p. 198, 199, pl. 3, figs 11a–13b, (?) 14a, b; pl. 4, figs 1a–9b; (?) Mantovani Uguzzoni and Pirini Radrizzani, 1967, p. 1228, pl. 92, figs 14a–16c; pl. 93, figs 5a–c; Hermes, 1969, p. 43, pl. 4, figs 77–79; Risch, 1971, p. 51, pl. 6, figs 1–3; Caron, 1971, p. 146, 147, 152, 153; text-figs 1a–c, 10a–13c; Michael, 1973, p. 217, pl. 6, figs 1–3.

Ticinella roberti Gandolfi. Gorbachik, 1971, pl. 29, figs 2a–c.

Description:

Test of average size, rather loose in coiling, composed of $2\frac{1}{2}$–(3) whorls, the last numbering about 6.5 chambers; these last increase the size regularly and rather rapidly; in lateral view the trochoidicity is weak, the trochospiral coiling is very slightly asymmetric, in its entirety the test tends to be subsymmetric; the peripheral margin is broadly rounded; the exterior contour of the test is rather strongly lobed.

The spiral face is flat to slightly concave, the inner whorls were in a slight depression; the chambers are globulose; the sutures radial, rectilinear or slightly curved, as also the spiral line are strongly depressed; the surface of the chambers (except for the last) is a little rugose.

The umbilical face shows an umbilicus of large size, shallow (and covered by the above-umbilical plate), the chambers are inflated, the sutures radial and strongly depressed.

The terminal face is inflated, wider than high, and it shows an important principal aperture.

The ornamentation is very weakly expressed, the surface of the chambers, up to the first 2–3 of the last whorl, is covered by small rugosities uniformly distributed.

The principal aperture is interiomarginal and extraumbilical, it continues broadly to the base of the terminal face, bordered by a clear lip.

The supplementary aperture is located in the axis of the suture, a little intrasutural in position between the last 2–3 chambers, otherwise in the intraumbilical position; it is surmounted by a rather important lip which terminates in a short tunnel.

The coiling, estimated from a limited number of individuals, seems predominantly dextral (75%).

Size: the holotype is small (greatest diameter 0.26, thickness 0.13 mm), but nine recognoises individuals of larger size (up to 0.45).

(Translated from Sigal, 1966a.)

Remarks: *Ticinella primula* generally has a flatter dorsoconvexity and fewer chambers in the final whorl than does *T. roberti* (Gandolfi). It lacks the pseudo-carinal periphery of *T. praeticinensis* Sigal. Among the variations observed for this species by Sigal (1966a) are size; chambers numbering from six to eight in the final whorl, with a tendency towards pseudoplanispiral coiling as the number increases; and very rarely (two individuals) the presence of a single aperture along the spiral suture between the last two chambers.

The phylogeny of this species is unknown. It might be pointed out that Sigal (1966a) noted a certain similarity between the accessory apertures of *Ticinella primula* and *Rotalipora cushmani* (Morrow) while recognising the stratigraphic hiatus which separates them.

As recorded above, rare specimens with a dorsal aperture have been found. This may be comparable to the *Ticinella roberti–Biticinella breggiensis* morphologic sequence as Sigal (1966a) has suggested as being possible. Because of the rarity of these forms in this case, Sigal declined to formally propose their separation into two subgenera until the number of specimens warranted it. However, he did suggest that *Ticinella* (*Ticinella*) *primula* be applied to those possessing only umbilical accessory apertures and *Ticinella* (*Biticinella*) *subprimula* Sigal be used for those forms with apertures on both sides.

Unfortunately, confusion is apt to occur because Sigal did publish a new species name and select a holotype for it. Nevertheless, it must be emphasised that Sigal treated his remarks on the above designations in the future tense, that these will be proposed "if further discoveries confirm it". The Code is not clear on matters such as this, but the author interprets this as though the names have never appeared in print. Moreover, should Sigal decide later that the numbers are sufficient to warrant the distinctions which he suggested as possible, then the new species would take as its date of publication not 1966 but the later one. Also the holotype would then have to be redesignated and redescribed to be valid.

Distribution: This is shown in Fig. 98.

Range: Lower to mid-Albian.

Ticinella raynaudi Sigal

Ticinella raynaudi var. *raynaudi* Sigal, 1966a, p. 201, 202, pl. 5, figs 10a, b; pl. 6, figs 1a–5.

Ticinella raynaudi var. *aperta* Sigal, 1966a, p. 202, 203, pl. 6, figs 11a–13b; Hermes, 1969, p. 44, pl. 4, figs 86–88; Caron, 1971, text-figs 20a–c.

Ticinella raynaudi var. *digitalis* Sigal, 1966a, p. 202, pl. 6, figs 6a–8b; Hermes, 1969, p. 44, pl. 4, figs 83–85; Caron, 1971, p. 154, 156, text-figs 8a–c, 21a–c.

Ticinella raynaudi digitalis Sigal. (?) Hermes, 1969, p. 40, pl. 1, figs 10–12; Risch, 1971, p. 51, pl. 6, figs 7–9.

Ticinella raynaudi raynaudi Sigal. Hermes, 1969, p. 44, pl. 4, figs 80–82; Risch, 1971, p. 51, pl. 6, figs 4–6; Caron, 1971, text-figs 19a–c.

Ticinella raynaudi aperta Sigal. Risch, 1971, p. 51, pl. 6, figs 10–12.

Clavihedbergella aff. *simplex* (Morrow). (?) Caron, 1971, p. 148, 149, text-figs 5a–c.

Description:

Test of small size, wall finely and densely perforated; the number of chambers in the last whorl is from 6–9 . . .; on about two and a half whorls, they regularly increase in size; the trochospiral coiling is a little asymmetric, the test is relatively thin, with spiral and umbilical faces flattened, the trochoidicity is weak, the periphery is regularly rounded; the contour of the test is strongly lobed.

The spiral face shows a last whorl occasionally a little flattened by relation to the inner whorls at the level of the last chambers; the surface of the chambers is globulose, the sutures radial, a little curved, as also the spiral line are strongly depressed.

The umbilical face shows an umbilicus of large size, very open and shallow; the surface of the chambers is inflated; the sutures radial, rectilinear and strongly depressed; the chambers are prolonged by the lamellar extensions which join in an umbilical plate in the form of a funnel.

The terminal face is inflated . . ..

The ornamentation is always present, but relatively little developed, under the form of small rounded pustules, more prominent in approaching the periphery, where they group themselves in more prominent and more irregular protuberances; it is present on the inner whorls and on 2–3 chambers of the last whorl.

The principal aperture is interiomarginal, high, large and goes up more than half of the distance to the umbilical–spiral face. It is protected by a thin prominent lip; their entirity, in the umbilicus, forms by joining a plate in the form of a funnel.

The supplementary aperture is partially in the intrasutural position, in general low and wide, bordered by a slight rim.

The coiling grants an important place (at least 50%) to the sinistral forms.

(Translated from the type description.)

Remarks: This species is generally lower spired than *Ticinella roberti*. Sigal (1966a) reports the number of chambers ranging from 6–9 for the different varieties. Also there is a tendency for the chambers in the last whorl of some individuals to become more elongate.

The phylogeny of this species is unknown.

It is suspected that *Ticinella raynaudi* is conspecific with *T. primula* Luterbacher as there is nothing in the description of Sigal's (1966a) species which would distinguish it from the latter. Because no topotypes were available for direct comparison, *T. raynaudi* is tentatively retained.

Sigal (1966a) separated *Ticinella raynaudi* var. *aperta* from *T. raynaudi* s. str. by its more rapid growth of chambers in the last whorl. The photographs of the primary types show this difference to be insignificant, even at the subspecific level. It may or may not have been Sigal's intent, but according to the Code (I.C.Z.N., Art. 45c–e), a post-1960 useage of the term "variety" relegates that taxon to an infrasubspecific rank. This is equally true of *T. raynaudi* var. *digitalis* Sigal. The final chambers of this form are more radially elongate than those of any other Ticinellid. Unfortunately only one specimen was illustrated by Sigal. It is not known whether this is an intraspecific form or whether it becomes a separate lineage. If the latter, then this form must be validly named. It should be pointed out that Caron illustrated (1971, text-figs 21a–c) a specimen of this variety which is midway in degree of chamber elongation between the type figures of *T. raynaudi* s. str. and the variety *digitalis*. This along with an identical stratigraphic range suggests that the variety is merely an intraspecific variant, and is so treated here.

Hermes (1969) did not recognise any accessory apertures on the specimens he attributed to *Ticinella raynaudi digitalis* Sigal. He stated that "Sigal's variety is the only taxon of this size in the late Albian which shows radially elongated chambers, and it would be too much of a coincidence if in the same stratigraphic interval two species occurred which are almost exactly alike, except that one is a *Ticinella*, the other a *Hedbergella*". To the first point, this specimen could potentially be assigned to *Clavihedbergella subcretacea* (Tappan), which is found throughout the Albian. As to the "coincidence "mentioned by Hermes, this sort of occurrence should be the rule rather than the exception, if we are to accept evolution as a principle. For a new taxon to arise from a continuing linage, initially both would be expected to occur together. Thus, Hermes' assignment must be questioned.

The specimen referred by Caron (1971) to (?) *Clavihedbergella* aff. *simplex* (Morrow) shows only slightly more elongation to the ultimate chamber than does the holotype of *Ticinella raynaudi* var. *digitalis* Sigal, and otherwise seems to be identical. Without the presence of the accessory apertures, this specimen would be identical to *C. subcretacea* (Tappan).

Distribution: This is shown in Fig. 99.

Range: Middle to upper Albian.

Ticinella roberti (Gandolfi), Plate 36, figs 5, 6; plate 37, fig. 2

Anomalina roberti Gandolfi, 1942, p. 100, 101, pl. 2, figs 2a–c; pl. 4, figs 4–7, 20 (?); pl. 5, fig. 1 (?); pl. 13, figs 3, 6; text-fig. 1 (?), 2 (?).

Globotruncana (*Ticinella*) *roberti* (Gandolfi). Reichel, 1950, p. 600–603, pl. 16, fig. 1; pl. 17, fig. 1; text-figs 1a–c, 2a–c.

Ticinella roberti (Gandolfi). Sigal, 1952, p. 23, text-fig. 19; Ayala-Castañares,

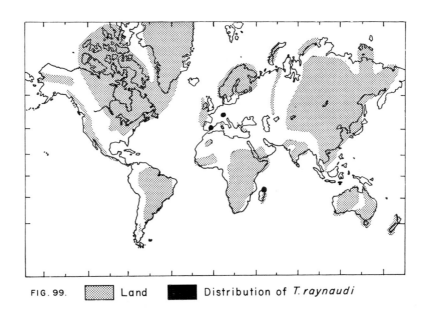

FIG. 99.   [Land]   [Distribution]   Distribution of *T. raynaudi*

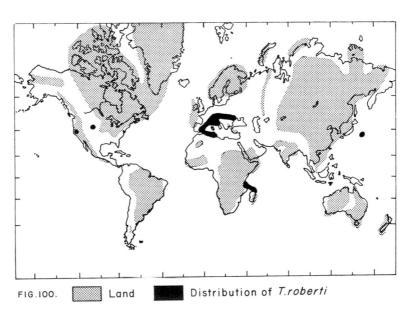

FIG. 100.   [Land]   [Distribution]   Distribution of *T. roberti*

1954, p. 426, 427, pl. 16, figs 1a–c; Banner and Blow, 1959, p. 8, pl. 3, fig. 3; Loeblich and Tappan, 1961, p. 294, 296, pl. 6, figs 14a–c; Postuma, 1962, 7 figs; Pokorný, 1963, p. 384, fig. 429; Savaedra, 1965, p. 325, text-fig. 12; Ewing *et al.*, 1966, text-fig. 5 (2a–c); Marianos and Zingula, 1966, p. 337, pl. 37, figs 9a–c; Moullade, 1966, p. 99, 100, pl. 9, figs 9–11; Sigal, 1966a, p. 203–207, pl. 4, figs 10a–12b; pl. 5, figs 1a–4b; Mantovani Uguzzoni and Pirini Radrizzani, 1967, p. 1229, pl. 93, figs 6a–8c; Hermes, 1969, p. 37–40, pl. 1, figs 1–3; pl. 2, figs 28–47; pl. 4, figs 89–91; (?) Scheibnerová, 1969, p. 63, tab. 10, figs 1a–2c; Moullade, 1969, pl. 1, figs 21–23; Caron and Luterbacher, 1969, p. 25, pl. 7, figs 3a–c; Neagu, 1970a, fig. 1 (7); Douglas, 1971, pl. 2, fig. 3; pl. 3, figs 9–11; Risch, 1971, p. 49, 50, pl. 5, figs 13–15; Kuhry, 1971, p. 233, 234, pl. 3, figs 4A–C; *non* Gorbachik, 1971, pl. 29, figs 2a–c; Caron, 1971, p. 153, 154, text-figs 14a–16c.

  Rotalipora roberti (Gandolfi). Bolli *et al.*, 1957, p. 41, pl. 10, figs 1a–c.
  Rotalipora (*Ticinella*) *roberti* (Gandolfi). Klaus, 1959, p. 803, pl. 1, figs 2a–3c.
  Globotruncana roberti Gandolfi. Majzon, 1961, p. 762, pl. 4, fig. 17.
  Ticinella bejaouaensis Sigal. Moullade, 1966, p. 103, pl. 9, figs 4, 5; Hermes, 1969, p. 42, 43, pl. 3, figs 59, 60; Moullade, 1969, pl. 1, figs 27, 28; Risch, 1971, p. 50, 51, pl. 5, figs 1–9.
  Ticinella roberti var. *bejaouensis* Sigal, 1966a, p. 207, 208, pl. 5, figs 5a–9b; Hermes, 1969, p. 42, 43, pl. 3, figs 56–58.
  Hedbergella roberti (Gandolfi). Salaj and Samuel, 1966, p. 170, 171, tab. 33 (6).
  Ticinella roberti roberti (Gandolfi). Bandy, 1967, p. 15, text-fig. 7 (1).
  Description: Test trochospiral, moderately biconvex but ranges to a high dorso-convexity; equatorial periphery moderately lobate; axial periphery broadly rounded. Chambers subspherical, highly inflated dorsally and ventrally; increasing gradually in size; 16–19 chambers in 3 whorls with 7–11 in the final whorl; width of ultimate chamber may exceed its other dimensions. Primary aperture is a low, interiomarginal, umbilical–extraumbilical arch bordered by a narrow, imperforate lip which flares into the umbilicus forming a flap-like extension that may fuse with earlier extensions. The accessory aperture forms along the distal edge of the flap immediately adjacent to the chamber wall and aligned with the suture; forming a small, but relatively high arch usually at a level slightly above the primary aperture. Umbilicus relatively wide, shallow to moderately deep, with the apertural flaps essentially covering the entire width. Sutures depressed, straight to slightly curved both dorsally and ventrally, spiral suture depressed. Surface perforated by numerous, small to moderate diameter pores; secondary calcification occurs on the earlier chambers creating an irregularly roughened surface, at the same time decreasing the pore density while increasing their diameter.
  Remarks: The chambers of *Ticinella roberti* are more closely packed and more slowly increasing in size than those of *T. primula* Luterbacher, which also tends to have a flatter spire. It lacks the concentrated peripheral pustules characteristic of *T. praeticinensis* Sigal. The most significant fluctuation in morphology occurs in the dorsoconvexity. Sigal (1966a) listed other variations including the umbilical width, which is controlled by the change in the dorsoconvexity and the number of chambers in the last whorl. He also remarked on "the 'precarinal' rugosity of the inner whorls . . . ". Scanning electron micrographs have established that this "rugosity" covers the entire chamber surface and bears no relationship to any "pre-carinal" development.
  Ticinella roberti may have evolved from *Globigerina trocoidea* (Gandolfi), from which it differs primarily by the posesssion of accessory apertures.

Sigal (1966a) recorded four individuals which possess "supplementary apertures" between the last three chambers on the dorsal side and opening along the spiral suture. Unfortunately, it is difficult to see these apertures on the specimen illustrated by Sigal. The author has seen such "spiral apertures" in light microscopy which turn out to be only deep depressions when viewed by electron microscopy. Those noted by Sigal may be true accessory apertures, and if so, must be treated accordingly.

Ticinella roberti var. *bejaouaensis* Sigal, 1966a, was separated from *T. roberti* s. str. on the presence of slightly more chambers in the last whorl (averaging nine), a more open umbilicus and a slower increase in chamber size. Each of these differences are present in a topotypic suite of *T. roberti* and are believed to be of an intraspecific nature. Sigal proposed *bejaouaensis* as a variety of *T. roberti*. This must be treated at the infrasubspecific rank according to the I.C.Z.N., Art. 45c–e.

Caron and Luterbacher (1969) refigured the holotype of *Anomalina roberti* Gandolfi, but the accessory apertures were not visible because of sediment in the umbilicus.

Distribution: This is shown in Fig. 100.

Range: Upper Aptian—lower Cenomanian.

Family GLOBOTRUNCANIDAE Brotzen, 1942

Definition: Trochospiral test; with or without an imperforate peripheral band or keel, which may be double or single; aperture interiomarginal, umbilical to umbilical–extraumbilical; umbilicus always covered by a tegillum; chambers may be radially elongate.

Generic Key

I. With a peripheral band or keel.
A. Peripheral band or keel throughout .   .   .   . *Globotruncana*
B. Peripheral band or keel on last few chambers only .   .   .   . *Trinitella*
   II. Without a peripheral band or keel.
A. Subspherical chambers .   .   .   . *Rugoglobigerina*
B. Radially elongate chambers   .   .   .   . *Plummerita*

Remarks: The tegillar apparatus is a highly variable structure even at the species level. The infralaminal apertures and the intralaminal apertures (Bolli *et al.*, 1957) may or may not be present. These result from incomplete fusing of one apertural flap to those pre-existing. They are termed tegillar apertures or openings herein, and have no taxonomic status.

Range: Turonian–Maastrichtian (*Trinitella scotti* Interval).

Genus GLOBOTRUNCANA Cushman, 1927

Globotruncana Cushman, 1927c, p. 81.
Rosalinella Marie, 1941, p. 237, 256, 258.
Marginotruncana Hofker, 1956c, p. 319.

Rugotruncana Brönnimann and Brown, 1956, p. 546.
Bucherina Brönnimann and Brown, 1956, p. 557.
Globotruncanella Reiss, 1957b, p. 135.
Globotruncanita Reiss, 1957b, p. 136.
Helvetoglobotruncana Reiss, 1957b, p. 137.
Abathomphalus Bolli, Loeblich and Tappan, 1957, p. 43.
Whiteinella Pessagno, 1967 (part), p. 298.
Archaeoglobigerina Pessagno, 1967 (part), p. 315, 316.
Praeoglobotruncana (*Dicarinella*) Porthault, 1970, p. 70.
Plummerita (*Radotruncana*) El-Naggar, 1971a, p. 434.
Type species: *Pulvinulina arca* Cushman, 1926.

Description: Test a trochospiral coil; biconvex, spiroconvex, umbilico-convex or concavo-convex; equatorial periphery lobate to circular in outline; axial periphery rounded with an imperforate peripheral band, acute with a single keel or truncate with a double keel, keels beaded often spinose. Chambers spherical, oval, hemispherical, angular rhomboid or angular truncate. Aperture a low to high interiomarginal, umbilical to extraumbilical–umbilical arch; well developed, irregularly shaped, imperforate apertural flaps extend over umbilicus fusing to previous extensions to form a tegillum; commonly with irregular tegillar apertures; apertural face imperforate, continuous with peripheral band of keel. Sutures radial, curved abaperturally or sigmoid, depressed to raised and beaded, narrow to wide. Surface with narrow to moderate diameter pores, variously ornamented with pustules, spines and costellae.

Remarks: The presence of both the tegillum and the keel set this genus apart. *Rotalipora*, although keeled, lacks a tegillum and has accessory sutural apertures. *Rugoglobigerina* has a tegillum but no keel.

The keel of the *Globotruncana* has been variously referred to as an imperforate peripheral band, a poreless margin and a single or double keel. The first two are synonymous and describe the type of keel characterised by a narrow, peripherally located imperforate band particularly typical of the more primitive species with spherical or oval chambers, e.g. some *G. cretacea* d'Orbigny. The other two keel types also have an imperforate peripheral band, but to this is added one or two carinae. In the case of the single keel, a single carina is situated over the band, which is noticeable when the carina has not developed—as on the ultimate chamber. The imperforate band is bordered by two carinae in the double-keeled variety. Either the dorsal or ventral carina may disappear during ontogeny, but the remaining carina can be distinguished from a true single keel by observing its position relative to the band.

The status of *Globotruncana spinea* Kikoïne, 1947, is not clear. It is suspected, however, that the holotype may well represent an already valid species. The characteristic feature of *G. spinea* is the "spines" which project from the dorsal keel of the first half of the final whorl. In all probability, these "spines" are merely the sutural remnants of chambers which were broken away, and not really spines at all. Such a broken specimen was attributed to *G. spinea* by Edgell (1957). A re-examination of Kikoïne's species is in order.

Two features were used to distinguish *Bucherina* Brönnimann and Brown, 1956, from *Globotruncana*: the lack of a tegilla and the final two chambers which have been displaced ventrally out of the "normal" spiral of coiling. Topotypes of *Bucherina sandidgei* Brönnimann and Brown, the type species of that genus, have been examined, and it is believed that the absence of a tegilla is due to preservation. The "abnormal" chamber displacement is frequently encountered in gerontic individuals of other species and genera. The reason for this is not understood. However, all degrees of displacement can be found in a given assemblage, as is true for *B. sandidgei*.

Bolli *et al.* (1957) selected *Globotruncana mayaroensis* Bolli as the type species for *Abathomphalus*. This new genus was stated as differing from *Globotruncana* by being almost nonumbilicate, by having from the last chamber a single tegillar extension with only inframinal accessory apertures, and by possessing an extraumbilical aperture. The size and shape of an umbilicus is dependent upon the number, size and shape of the chambers and upon the degree of spiroconvexity of the test. These are highly variable features throughout the Globotruncanidae. The ventral side of the chambers in *G. mayaroensis* is flat to very slightly convex, so that this surface slopes gently into the umbilicus. This creates an illusion of a wide, shallow umbilicus, a feature unacceptable as a supra-specific criterion.

Tegilla are morphologically so diverse, even intraspecifically, that any attempt to use their variations taxonomically would be totally unsatisfactory. Some individuals are known to have a single tegillar extension from only the final chamber, while others of the same species have extensions from many chambers; e.g. *Globotruncana fornicata* Plummer, *G. gansseri* Bolli and *G. marginata* (Reuss) as illustrated by Masters (in press). Contrary to the statements by Bolli *et al.* (1957), even specimens of *G. mayaroensis* have been illustrated (Brönnimann and Brown, 1956) with both inframinal and intralaminal tegillar apertures. Nor is the position of the primary aperture in *G. mayaroensis* restricted to an extraumbilical position. Subbotina (1953) figured a specimen, in which the tegillum was missing, which clearly showed an extraumbilical–umbilical apertural position.

Dupeuble (1969) remarked that the differences between *Globotruncana* and *Abathomphalus* rest essentially on the large umbilicus and the extraumbilical aperture, which is a direct effect of the umbilical size. He maintained that these differences are not sufficient to justify retaining the genus *Abathomphalus*. This is in agreement with the conclusions reached here.

Porthault (in Donze *et al.*, 1970) established the new subgenus *Praeglobotruncana* (*Dicarinella*) with *Globotruncana indica* Jacob and Sastry as the type species. A translation of the original French follows:

Definition:

> Test calcareous perforate, bilamellar, in a trochospiral coil. Primary aperture interiomarginal, umbilical–extraumbilical. Peripheral imperforate carinal band comprised of two visible carinae at least on certain chambers of the last whorl. On the dorsal face, the sutures are more or less oblique, depressed or in relief and bordered by a rim; on the ventral face they are always radial and depressed.
> Some lamellated expansions issue from the adumbilical portion of the chambers ('portici') forming a ring around the umbilical cavity, which they cover more or less completely making between them some infralaminal 'accessory apertures'. There does not exist supplementary sutural apertures.
> Remarks: The subgenus *Praeglobotruncana* (*Dicarinella*) nov. subgen. is distinguished from *Praeglobotruncana* s. str. by the existence of two carinae, at least on certain chambers of the last whorl and by the more important development of 'portici'. The primary aperture presents on the other hand an extraumbilical extension more limited than in *Praeglobotruncana* s. str.
> With regard to *Marginotruncana*, *Praeglobotruncana* (*Dicarinella*) nov. subgen. is differentiated by the form of sutures which are radial and depressed on the ventral face.

Praeglobotruncana s. str. is confined to those forms which have a single peripheral keel and lack a porticus or a tegillum. A porticus is defined as a distinctly asymmetrical apertural lip, while a tegillum is a chamber extension comparable to a highly developed apertural flap extending across the umbilicus, thus covering the primary aperture (Loeblich and Tappan, 1964). True *Globotruncana* possess tegilla. Yet scanning electron micrographs have shown that these structures on *Globotruncana* are highly variable, interspecifically and intraspecifically. Tegilla, being thin, are often broken leaving only a narrow remnant still attached to the chamber. These remnant structures were misinterpreted by Porthault as portici.

Porthault distinguishes *Dicarinella* from *Marginotruncana* (= *Globotruncana*) by the presence of radially depressed ventral sutures. This feature is taxonomically significant only at the species level. Thus, *Praeglobotruncana* (*Dicarinella*) is considered to be a junior synonym of *Globotruncana*.

El-Naggar (1971a) placed forms with "Later chambers radially elongate, clavate or tubulospinate" into the genus *Plummerita*. He proposed the new subgenus *P.* (*Radotruncana*), with *Globotruncana calcarata* Cushman as the type species. *Globotruncana calcarata* is morphologically and phylogenetically *totally* unrelated to *Plummerita*. Indeed, the erection of a subgenus for this species based solely on the possession of peripheral tubulospines is unnecessary and unjustifiable. These spines are only characteristic at the species-level.

The remaining synonyms of *Globotruncana* have been treated previously (Masters, in press).

In 1954 Nakkady and Osman proposed several new species and varieties which were accompanied by brief descriptions and inadequate illustrations. Without access to the types, the following cannot be recognised: *Globotruncana torensis*, *G. pseudofornicata*, *G. sudrensis*, *G. sudrensis* var. *parallela*, *G. ansarii*, *G. quadrata* and *F. quadrata* var. *plata*.

When the types of Gandolfi (1955) and a topotypic suite from Colombia are examined, it is readily apparent that there is a very high percentage of environmentally induced, atypical growth forms. Gandolfi's failure to recognise these intraspecific morphovariants lead him to propose many unwarranted new species. Sudden increase or decrease in chamber size is a common phenomenon in his material. Another problem was discovered during the examination of his specimens. Many of the type figures and descriptions of these species and subspecies are misleading. In fact a few of the holotypes are difficult to reconcile with the type figures. Keels are illustrated which are not present. Edge views and occasionally side views are incorrectly presented. Evaluation of Gandolfi's taxa must be made directly from the holotypes.

El-Naggar (1966) proposed the new taxa *Globotruncana fareedi*, *G. orientalis* and *G. sharawnaensis*, which are indistinguishable except for the nature of their respective keels. The first species is single-keeled throughout. *Globotruncana orientalis* has a double keel only in the first one or two chambers of the final whorl with the remainder being single-keeled. The last species, *G. sharawnaensis*, has a double keel only in the ultimate chamber. An important observation made by El-Naggar concerning *G. orientalis* and *G. sharawnaensis* is that individuals of both species are known to be single-keeled throughout. He failed to mention how these morphotypes can be distinguished from one another or from *G. fareedi*. All three may form an integrating population of a single species. A detailed examination of topotypic suites in thin-section and in whole mounts is required before the status of these three species can be determined.

Salaj and Maamouri (1971) proposed the new subspecies *Globotruncana calcarata globulosa*. It would appear that the authors have added to the type figure tubulospinose chamber extensions which are not actually present (or preserved?) on the holotype. The chambers do seem to be somewhat radially elongated. In the type description, they noted the presence of traces of two pustulose keels. Yet, in the comparison, they stated that their subspecies differs from *G. calcarata* s. str. by the absence of a keel. These contradictory statements dictate the need for a more careful morphologic analysis before it can be taxonomically validated. The type figure is reminiscent of both *Rugoglobigerina reicheli* Brönnimann and *Plummerita hantkeninoides* Brönnimann, if the possible double keel is disregarded. However, the known stratigraphic ranges of Brönnimann's species are younger than that of *G. calcarata globulosa*.

Range: Turonian—Maastrichtian.

Globotruncana aegyptiaca Nakkady, Plate 37, fig. 4
Globotruncana nov. sp. van Wessem, 1943, p. 48, pl. 2, figs 1, 2.
Globotruncana aegyptiaca Nakkady, 1950, p. 690, pl. 90, figs 20–22; Nakkady and Osman, 1954, p. 75, 76, pl. 20, figs 20a–c; Lehmann, 1966, p. 312, pl. 1, fig. 6; Pessagno, 1967, p. 319–321, pl. 79, figs 2–4 (*non* pl. 83, figs 8–10).
Globotruncana gagnebini Tilev, 1952, p. 50–56, text-figs 14a–e, pl. 3, figs 2a–5d (*non* text-figs 15a–17d); Majzon, 1961, p. 763, pl. 3, fig. 10; Herm, 1962, p. 79, 80, pl. 7, fig. 5; Postuma, 1962, 7 figs; *non* Borsetti, 1964, p. 674, 675, pl. 79, fig. 15, text-fig. 8; El-Naggar, 1966, p. 111–113, pl. 2, figs 1a–4d; pl. 3, figs 3a–d (*non* 1a–d, 6); Longoria, 1970, p. 51–54, pl. 5, figs 1–3; pl. 9, figs (?) 5, 6; pl. 15, figs (?), 5, 6; *non* Youssef and Abdel-Aziz, 1971, pl. 1, figs 6a, b; *non* Govindan, 1972, p. 179, pl. 4, figs 7–9; pl. 5, figs 1–6.
Globotruncana pennyi subpennyi Gandolfi, 1955a, p. 73, pl. 7, figs 7a–c.
Globotruncana (*Globotruncana*) *wiedenmayeri wiedenmayeri* Gandolfi, 1955a, p. 71, pl. 7, figs 4a–c; *non* Salaj and Samuel, 1966, p. 221, pl. 24, figs 3a–c.
Globotruncana (*Globotruncana*) *wiedenmayeri magdalaensis* Gandolfi, 1955a, p. 72, pl. 7, figs 3a–c.
Rugotruncana sp. 1, Seiglie, 1958, p. 73, pl. 5, figs 1a–2c.
Globotruncana ventricosa ventricosa White. Ashworth, 1959, p. 498, text-figs 2a–c.
Globotruncana (*Rugotruncana*) *gansseri dicarinata* Pessagno, 1960, p. 103, pl. 2, figs 9–11; pl. 3, figs 1–3; pl. 5, fig. 2.
Globotruncana gansseri dicarinata Pessagno. Pessagno, 1962, pl. 4, figs 12, 15, 16 (? figs 9, 10).
Globotruncana zargos Kavary, 1963, p. 50, 51, pl. 9, figs 23–25.
Globotruncana gansseri (Bolli). Brönnimann and Rigassi, 1963, pl. 16, figs 1a–c.
Globotruncana bahijae El-Naggar, 1966, p. 86, 87, pl. 6, figs 2a–d.
Rugotruncana subpennyi (Gandolfi). Pessagno, 1967, p. 370, 371, pl. 76, figs 12–14.
Globotruncana aegyptiaca aegyptiaca Nakkady. Ansary and Tewfik, 1968, p. 49, pl. 6, figs 8–c; *non* Youssef and Abdel-Aziz, 1971, pl. 1, figs 1a, b.
Globotruncana sp. cf. *G. gansseri* Bolli. Bertels, 1970, p. 38, 39, pl. 5, figs 4a–c.
Description: Test plano-convex with the dorsal side flat to very slightly convex, ventral side strongly convex; equatorial periphery lobate, axial periphery dorsally angled, ventrally rounded. Chambers 11–13 in $2\frac{1}{2}$ whorls with 4–6 (usually 5) chambers which rapidly increase in size in the final whorl; neanic and occasionally early ephebic chambers subspherical rapidly becoming dorsally flattened and ventrally elongated in final whorl; as viewed on edge the peripheral wall of the final 2–3 chambers is approximately at a 90° angle to the dorsal side; narrow, beaded, double keel at dorsal angle of chambers, ventral carina borders imperforate apertural face sweeping abaperturally to form a carina on the umbilical shoulder of the last 1–3 chambers. Aperture a low interiomarginal, umbilical to slightly extraumbilical–umbilical arch. Tegillum covers a wide, deep umbilicus. Sutures on

ventral side radial, slightly depressed; dorsally curved, formed by dorsal carina of keel. Surface finely and densely perforate, pustules scattered over entire surface.

Remarks: This species is most likely to be confused with *Globotruncana ventricosa* White. It differs from White's species by possessing fewer chambers in the final whorl, and the peripheral wall of last few chambers forms an angle of 90° with the dorsal side as opposed to a distinctly acute angle for *G. ventricosa*. The characteristics of *Globotruncana aegyptiaca* are rather stable. Only minor fluctuation in the number of chambers in the final whorl has been observed.

Globotruncana duwi Nakkady is the most likely ancestor of *G. aegyptiaca*.

El-Naggar (1966), in a detailed discussion of *Globotruncana aegyptiaca*, noted that it differs from *G. gagnebini* Tilev, 1952. The latter he declared "has a less lobulate, more tightly coiled, distinctly elongate test; chambers which increase very rapidly in size, a much larger or smaller last chamber and a greater number of chambers in the last whorl". There is no difference in the "lobulate" appearance of the two species. El-Naggar does not clarify what he means by an elongate test. But the ratio of the small diameter to the large diameter given by Tilev for *G. gagnebini* ranges from 0.75 to 0.90, while that calculated for *G. aegyptiaca* is 0.83. El-Naggar reported four to four and one half or rarely five chambers in the last whorl of *G. aegyptiaca* and four to six, most commonly five, in *G. gagnebini*. The number of chambers is not a distinction. The remaining differences offered by El-Naggar are all environmentally controlled. The taxa are conspecific.

The holotypes of *Globotruncana wiedenmayeri* s. str. and the subspecies *magdaienaensis* of Gandolfi (1955) were examined and are considered to be synonyms of *G. aegyptiaca*. The former synonym possesses the maximum number of chambers known for *G. aegyptiaca*.

The holotype of *Globotruncana zagros* Kavary, 1963, was judged to be synonymous with *G. aegyptiaca*.

Globotruncana bahijae El-Naggar, 1966, is also a junior synonym of Nakkady's species. Its slight differences are attributable to environmental controls, and it is found at the same stratigraphic levels as *G. aegyptiaca*.

Distribution: This is shown in Fig. 101.

Range: Maastrichtian (FOS *Globotruncana aegyptiaca*—LOS *Rugoglobigerina hexacamerata* Interval—*Trinitella scotti* Interval).

Globotruncana arca (Cushman), Plate 38, figs 1, 2, 4

Pulvinulina arca Cushman, 1926b, p. 23, pl. 3, figs 1a–c.

Globotruncana arca (Cushman). Cushman, 1927e, p. 169, pl. 28, figs 15a–c; *non* Moreman, 1927, p. 100, pl. 16, figs 16, 17; Cushman, 1928b, p. 311, pl. 48, figs 11a–c; pl. 49, figs 1a–c; Cushman and Church, 1929, p. 518, pl. 41, figs 1–3; Cushman, 1931d, p. 59, pl. 11, figs 6a–c; *non* Plummer, 1931, p. 195–198, pl. 13, figs 7a–9c, 11a–c; Sandidge, 1932d, p. 285, pl. 44, figs 6–8; Cushman, 1933a, pl. 35, figs 17a–c, *non* Cushman, 1936, p. 419, pl. 1, figs 14a–c; *non* Jennings, 1936, p. 37, pl. 4, figs 14a, b; Glaessner, 1937, p. 36, 37, pl. 1, figs 10a–c; Majzon, 1943, pl. 1, fig. 12; (?) Hermes, 1945, p. 36, 37, pl. 5, figs 8a–c; Cushman, 1946b, p. 150, pl. 62, figs 4a–c (*non* 5a–c); Cushman, 1948b, p. 329, 330, pl. 27, figs 11a–c, Key pl. 35, figs 17a–c; Cita, 1948b, p. 145, 147, pl. 3, figs 2a–c; Bandy, 1951, p. 509, pl. 75, figs 1a–c; Noth, 1951, p. 77, 78, pl. 8, figs 15a–c; Bermúdez, 1952, p. 53, 54, pl. 7, figs 3a–c; Tilev, 1952, p. 57–62, text-figs 18a–19d; Subbotina, 1953, p. 209–213, pl. 9, figs 1a–5c, pl. 10, figs 1a–5c; Hagn, 1953, p. 97, 98, pl. 8, figs 11a–c; text-

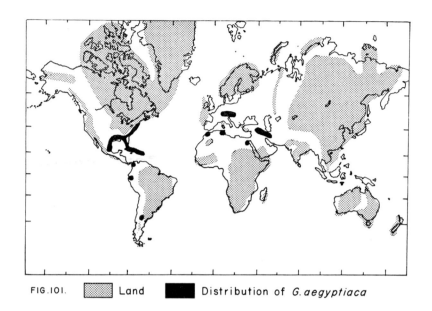

FIG.101. ▨ Land ■ Distribution of *G. aegyptiaca*

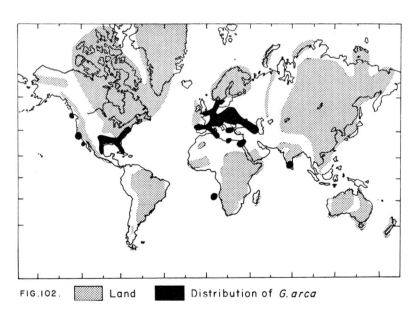

FIG.102. ▨ Land ■ Distribution of *G. arca*

figs 20, 21; *non* Frizzell, 1954, p. 128, pl. 20, figs 22a–c; *non* Veljković-Zajec, 1954, p. 253, pl. 3, fig. 4; Brönnimann and Brown, 1955, p. 539, 540, pl. 23, figs 10–12; *non* Dalbiez, 1955, text-figs 5a–c; Książkiewicz, 1956, p. 281, 282, text-figs 58, 59 (?); (*non* pl. 30, figs 4–6); Bolli *et al.*, 1957, p. 44–46, pl. 11, figs 6–11c; Edgell, 1957, p. 110, 111, pl. 3, figs 4–5, (*non* pl. 1, figs 10–12, pl. 3, figs 13–15; McGugan, 1957, pl. 34, figs 15a–c; Sacal and Debourle, 1957, p. 59, pl. 27, fig. 8 (*non* 6, 10, 12); Bieda, 1958, p. 60, 61, text-figs 24a–c; Martirosyan, 1958, p. 12–14, pl. 3, figs 3a–c (non figs 1a–2c; pl. 2, figs 2a–c); Seiglie, 1958, p. 66, pl. 1, figs 1a–c; *non* Witurcka, 1958, p. 224, 225, pl. 18, figs 35a–c; Rauzer-Chernousova and Fursenko, 1959, p. 302, text-figs 685a–c; Klaus, 1959, p. 824, 825, pl. 7, figs 5a–c; Maslakova, 1959, p. 109, pl. 13, figs 3a–c; Vinogradov, 1960a, pl. 5, figs 27a–c; Barr, 1961, p. 567, pl. 69, figs 8a–c; Majzon, 1961, p. 762, pl. 1, fig. 12; Postuma, 1962, 7 figs; Cita-Sironi, 1963, p. 126, text-fig. 15; van Hinte, 1963, p. 80, 81, pl. 6, figs 2a–c, 4a–c; Maslakova, 1963b, pl. 6, fig. 3; Pokorný, 1963, p. 386, figs 433a–c; Loeblich and Tappan, 1964, p. C662, C663, figs 1a–2; Martin, 1964, p. 79, pl. 9, figs 4a–c; Olsson, 1964, p. 162, 163, pl. 4, figs 1a–3c; van Hinte, 1965a, p. 20, pl. 2, figs 3a–c, pl. 3, figs 3a–c; Takayanagi, 1965, p. 209, 211, pl. 22, figs 6a–c; pl. 23, figs 1a–2c; Christodoulou and Marangoudakis, 1966, p. 304, 305, pl. 1, figs 5a–c, 7a–c; Douglas and Sliter, 1966, p. 107, 108, pl. 2, figs 6a–7c; El-Naggar, 1966, p. 83–86, pl. 1, figs 1, a; Wille-Janoschek, 1966, p. 96–98, pl. 5, figs 1a, 4–6 (*non* 3a–c); Salaj and Samuel, 1966, p. 199, 200, pl. 21, figs 3a–c; Săndulescu, 1966, pl. VIa, figs 2a–c; Pessagno, 1967, p. 321–323, pl. 79, figs 5 (?), 6–8, pl. 90, figs 6–8; Bandy, 1967, p. 20, text-fig. 9 (5); Christodoulou and Marangoudakis, 1967, p. 311, 312, pl. 1, figs 5a–c, 7a–c; Barr, 1968, p. 315, pl. 39, figs 3a–c; Sliter, 1968, p. 101, 102, pl. 15, figs 9a–10c; Ansary and Tewfik, 1968, p. 53, pl. 7, figs 5a–c; *non* Rao *et al.*, 1968, p. 19–21, pl. 1, figs 1–3; Hanzlíková, 1969, p. 47, 48, pl. 10, figs 1a–3c; Douglas, 1969, p. 176, pl. 9, figs 1a–3c (*non* pl. 10, figs 4a–7c); Dupeuble, 1969, p. 155, 156, pl. 2, figs 4a–d (*non* 5); pl. 3, figs 7a–d, 9a–d; Funnell *et al.*, 1969, p. 28, pl. 2, figs 11–13; pl. 3, figs 1–3; text-figs 11a–c; *non* Kalantari, 1969, p. 200, pl. 25, figs 9a–c; Moorkens, 1969, p. 453, 454, pl. 3, figs 4a–c; Weaver *et al.*, 1969, pl. 6, figs 5a–c; Scheibnerová, 1969, p. 75, 76, tab. 17, figs 4a–c; Longoria, 1970, p. 25–28, pl. 3, figs 1–3; pl. 8, figs (?) 1–3; pl. 17, figs (?) 4–6; Neagu, 1970b, p. 68, pl. 16, figs 13–17; pl. 17, figs 10–12 (*non* 13, 14); Sturm, 1970, p. 111, pl. 6, figs 3a–c; Todd, 1970, p. 153, pl. 6, figs 3a–4c (*non* 2a–c); *non* Al-Shaibani, 1971, p. 106, pl. 1, figs 1a–c; Berggren, 1971, pl. 1, figs 8–14; El-Naggar, 1971a, pl. 5, figs g–1; Hanzlíková, 1972, p. 102, pl. 26, figs 11a–13c; Caron, 1972, p. 551, pl. 2, figs 2a–c; (?) Govindan, 1972, p. 175, pl. 4, figs 1–6.

Globotruncana convexa Sandidge, 1932d, p. 285, 286, pl. 44, figs 9–11; Majzon, 1943, pl. 1, fig. 13; Keller, 1946, p. 99, 101, pl. 3, fig. 8; Majzon, 1961, p. 762, pl. 1, fig. 13; Borsetti, 1964, p. 671, pl. 79, fig. 12; text-fig. 3.

Globotruncana linnaeana (d'Orbigny). (?) Voorwijk, 1937, p. 195, pl. 1, figs 23, 27, 28.

Globotruncana cretacea Cushman, 1938b, p. 67, 68, pl. 11, figs 6a–c; *non* Cita, 1948b, p. 152, 153, pl. 3, figs 7a–c; Bermudcz, 1952, p. 53, 54, pl. 7, figs 6a–c; Tilev, 1952, p. 62–67, text-figs 21a–d (?), (*non* 20a–d); Ayala-Castañares, 1954, p. 390, 391, pl. 4, figs 2a–c; Frizzell, 1954, p. 128, 129, pl. 20, figs 25a–c.

Globotruncana linnei (d'Orbigny). Colom, 1948, pl. 2, figs 1–3.

Globotruncana leupoldi Bolli. Olsson, 1960, p. 50, pl. 11, figs 1–3; El-Naggar, 1966, p. 121, 122, pl. 1, figs 4a–c; text-figs 9 (27), 10 (part).

Globotruncana mariei Banner and Blow, 1960, p. 8 (*nom.* [rom.] nov. pro. *G. cretacea* Cushman); Olsson, 1964, p. 167, 168, pl. 4, figs 8a–c (*non* 9a–10c); *non*

Caron, 1966, p. 85, 87, pl. 6, figs 1a–c; (?) Rao *et al.*, 1968, p. 28, pl. 1, figs 4–6; *non* Hanzlíková, 1969, p. 49, pl. 11, figs 2a–c; Govindan, 1972, p. 177, 178, pl. 5, figs 7–12.

Globotruncana (*Globotruncana*) *mariei* Banner and Blow. Berggren, 1962a, p. 54–56, pl. 9, figs 5a–c.

Globotruncana (*Globotruncana*) *arca* (Cushman). Berggren, 1962a, p. 49–51, pl. 9, figs 1a–2c; van Hinte, 1965b, p. 83, pl. 1, figs 3a–c.

Globotruncana cf. *arca* (Cushman). McGugan, 1964, p. 948, pl. 152, figs 8a–c; Caron, 1966, p. 83, pl. 5, figs 5a–c (?), 6a–c.

Globotruncana ex. gr. *arca* (Cushman). *non* Küpper, 1964, p. 616, 617, pl. 3, figs 6a–c.

Globotruncana angusticarinata Gandolfi. Takayanagi, 1965, p. 208, 209, pl. 22, fig. 5 (*non* 4a–c).

Globotruncana coronata Bolli. Takayanagi, 1965, p. 212, 213, pl. 24, figs 1a–c (*non* 2a–c); Douglas and Sliter, 1966, p. 109, pl. 4, figs 5a–c (*non* 4a–c).

Globotruncana linneiana (d'Orbigny). Takayanagi, 1965, p. 217, 218, pl. 25, figs 6a–c; pl. 26, figs 1a–2c.

Globotruncana (*Globotruncana*) *rosetta* (Carsey). van Hinte, 1965b, p. 85, pl. 3, figs 1a–c.

Globotruncana cf. *convexa* Sandidge. El-Naggar, 1966, p. 97, 98, pl. 1, figs 5a–c; text-figs 9 (12), 10 (part); (?) Youssef and Abdel-Aziz, 1971, pl. 1, figs 2a, b.

Globotruncana stephensoni Pessagno, 1967, p. 354–356, pl. 69, figs 1–7, Pessagno, 1969b, p. 506, pl. 4, figs 1, 4; pl. 5, fig. 1.

Abathomphalus mayaroensis (Bolli). Ansary and Tewfik, 1968, p. 48–9, pl. 6, figs 1a–c.

Globotruncana rosetta (Carsey). Rao *et al.*, 1968, p. 25, pl. 4, figs 1–3.

Globotruncana rugosa (Marie). Neagu, 1970b, p. 69, pl. 17, figs 15–17.

(?) *Globotruncana pessagnoi* Longoria, 1973, p. 97–100 pl. 1, figs 1–9; text-fig. 1.

Description: Test a moderate trochospiral coil; unequally biconvex with the dorsal side being more convex; equatorial periphery slightly lobate, axial periphery truncate with widely spaced, beaded, double keel. Chambers of first whorl sub-spherical, later chambers flat dorsally, slightly inflated ventrally with umbilical shoulder being maximum; subequal to gradually increasing in size; ventrally imbricate and elongate in axis of coiling, petaloid dorsally; 15–17 in 2½–3 whorls with 6–7 chambers in final whorl. Aperture a low to high interiomarginal, umbilical arch in the steep slope beneath the umbilical shoulder. Umbilicus moderately deep, wide, covered by a tegillum. Sutures formed by the respective dorsal and ventral keels; dorsally curved, thickened and raised, ventrally curved, continuous with umbilical shoulder carina. Surface densely penetrated by small to medium diameter pores, typically nonpustulose.

Remarks: The widely spaced double keel, the spiroconvexity and the imbricate ventral chambers characterise *Globotruncana arca*. A similar species, *G. lapparenti* Brotzen has a nearly flat trochospiral coil. The degree of spiroconvexity varies slightly as do the number of chambers in the final whorl, which is due to the rate of size increase. The keel may narrow to become a single keel in the final one or two chambers.

Globotruncana arca evolved from *G. lapparenti* by an increase in the degree of spiroconvexity.

Examination of the holotype revealed that the ventral view of the type figure is inaccurately illustrated in that the imbricate chambers and curved sutures were not shown.

Brönnimann and Brown (1956) envisioned *Globotruncana cretacea* Cushman as an incipient *G. rosetta* (Carsey). This view was followed by Pessagno (1967). Olsson (1964) noted the similarity of Cushman's species to *G. arca* but distinguished it, as did Berggren (1962a), as being less convex dorsally. Olsson further commented that when the sizes of the two are considered, there is little difference in the spacing of their keels. Berggren added that *G. cretacea* Cushman has fewer chambers (five) in the last whorl than does *G. arca*. The neanic whorls of adult *G. arca* typically have 5 chambers per whorl. Thus, *G. cretacea* Cushman is interpreted as merely an immature *G. arca*.

Globotruncana cretacea Cushman, 1938a, a junior homonym of *G. cretacea* (d'Orbigny), 1840, was given the new name *G. mariei* by Banner and Blow (1960) in honour of Pierre Marie. *Rosalinella globigerinoides* Marie, 1941, now considered a *Globotruncana* and a junior homonym of *G. globigerinoides* Brotzen, 1936, was given the new name *G. mariai* by Gandolfi, also in honour of Pierre Marie. According to Appendix D, III, of the I.C.Z.N., the genitive singular "-i" is to be added to the *entire* name if it is that of a man. It would appear that Gandolfi's spelling, by changing the "e" to an "a" before adding the "i", was an inadvertent error. Thus, *G. mariei* Banner and Blow becomes a junior homonym of *G. mariei* Gandolfi after the latter is corrected. Nevertheless, a new name is not being proposed for the Banner and Blow species because each is regarded as a junior synonym of another taxon.

The holotype and two paratypes of *Globotruncana pessagnoi* Longoria, 1973, are immature specimens. The ventral carina of the double keel is not well developed, the tegillum is absent rather than broken, and the ventral sutures are straight. These are characteristics of juvenile *Globotruncana*. The ventral carina sweeps up and back along the umbilical shoulder of the ultimate chamber in all three specimens, but is particularly well developed in the second paratype (fig. 9). This is an incipient imbricate structure, which along with the dorsoconvexity and the nature of the keel is strong evidence in support of *G. arca* as the adult form.

Distribution: This is shown in Fig. 102.

Range: Campanian (*Globotruncana calcarata* Interval)—Maastrichtian (*Trinitella scotti* Interval).

Globotruncana calcarata Cushman, Plate 38, fig. 3; plate 39, fig. 1

Globotruncana calcarata Cushman, 1927b, p. 115, 116, pl. 23, figs 10a, b; White, 1928b, p. 285, pl. 38, figs 6a–c; Cushman, 1933a, pl. 35, figs 14a–c; Majzon, 1943, pl. 1, fig. 11; Cushman, 1946b, p. 151, 152, pl. 62, figs 8a–c; Cushman, 1948b, p. 329, 330, Key pl. 35, figs 14a–c; Cushman, 1948a, p. 266, pl. 26, figs 3a, b; Bartenstein, 1948, p. 244–246, text-fig. 1; Wicher, 1949, p. 62, pl. 5, fig. 17; Noth, 1951, p. 78, 79, pl. 8, figs 14a–c; Bermúdez, 1952, p. 53, 54, pl. 7, figs 8a–c; Sigal, 1952, p. 40, text-fig. 43; Reiss, 1952, p. 270–272, text-figs 1a–c; Hamilton, 1953, p. 232, pl. 29, figs 4, 5; Frizzell, 1954, p. 128, pl. 20, figs 23a–c; Sacal and Debourle, 1957, p. 60, pl. 28, fig. 4; Danilova, 1958, p. 225, pl. 25, figs 1, 2; Eternod Olvera, 1959, p. 100, 101, pl. 7, figs 11, 13, 14; Thalmann and Ayala-Castañares, 1959, pl. 4, fig. 10; Majzon, 1961, p. 762, pl. 1, fig. 11; Herm, 1962, p. 67, 68, pl. 6, fig. 3; Postuma, 1962, 7 figs; Cita-Sironi, 1963, p. 123, text-fig. 14b; van Hinte, p. 74, 75, pl. 4, figs 2a–c, pl. 5, fig. 3; *non* Pirini and Radrizzani, 1963, pl. 42, fig. 59; Olsson, 1964, p. 163, pl. 5, figs 1a–c; Cati, 1964, p. 255, 256, pl. 42, figs 2a–c; Săndulescu, 1966, pl. VIIIa, figs 3a–c; Salaj and Samuel, 1966, p. 200. 201, pl. 24, fig. 4a–c; Pessagno, 1967, p. 326–328, pl. 64, figs 18–20; pl. 72, figs 5, 6; Bandy, 1967, p. 21, text-fig. 9 (13); Liszkowa, 1967, pl. 4, fig. 1; Pessagno, 1969a, p. 609, pl. 9, fig. B; Bate and Bayliss, 1969, pl. 3, figs 2a–c; Sturm, 1969, pl. 12, figs 3a–c; Cita, 1970, pl. 4, fig. 2; Longoria, 1970, p. 32–34, pl. 6, figs 10–12; pl. 11, figs (?) 3, 4; pl. 18, figs (?) 1, 2; Cita and Gartner, 1971, pl. 1, figs 1, 2.

Rugotruncana calcarata (Cushman). Brönnimann and Brown, 1955, p. 548, 549; pl. 23, figs 1–3; Danilova, 1958, p. 225, pl. 23, fig. 3 (part); Brönnimann and Rigassi, 1963, pl. 17, figs 1a–c.

Globotruncana aff. *calcarata* Cushman. Lucini, 1959, text-fig. 3 (6).

Globotruncana arca Cushman. Küpper, 1964, p. 617, pl. 4, figs 8a–c.

Description: Test nearly planoconvex with the dorsal side flat to barely convex and the ventral side strongly convex; equatorial periphery scalloped between tubulo-spines, axial periphery acute with a single keel. Chambers dorsally flat, ventrally high, steeply sloping up to the carinate umbilical shoulder with vertical slope into the umbilicus; typically both the adapertural and the abapertural periphery project into one side of a tubulospine with the other side formed by the adjacent chamber, the abapertural projection is usually larger and may form a tubulospine by itself on later chambers, tubulospines develop only in final whorl; 15–16 chambers in 3 whorls with 5–7 in the last whorl. Aperture interiomarginal, umbilical, situated low on the vertical umbilical wall. Umbilicus deep, wide, covered by a tegillum. Sutures on ventral side radial to slightly curved, depressed to slightly raised, narrow; dorsal side radial, curved, sigmoid, raised and wide in last whorl, flush to depressed, radial in earlier whorls. Surface densely perforate, some pustules or rugosities.

Remarks: *Globotruncana calcarata* is the most easily recognised taxon of this genus. No other *Globotruncana* possesses tubulospines. The tubulospines are the most variable feature of *Globotruncana calcarata*. Occurring only in the last whorl, they are longer and of greater diameter in the earlier chambers, becoming smaller toward the ultimate chamber. The last chamber may have both an abapertural and an adapertural projection, or may lack the latter or both. These projections of adjacent chambers appear to merge into a single tubulospine in some individuals. In others each projection is a tubulospine. In this case the adapertural one is usually smaller and overlain by the abapertural tubulospine of the succeeding chamber.

This species arose from *Globotruncana elevata* (Brotzen) by developing tubulo-spines.

Distribution: This is shown in Fig. 103.

Range: Campanian (*Globotruncana calcarata* Interval).

Globotruncana concavata (Brotzen), Plate 39, figs 2, 3

Rotalia concavata Brotzen, 1934, p. 66, pl. 3, fig. b; Kuhry, 1970, p. 229–302, text-fig. 6.

Globotruncana asymetrica Sigal, 1952, p. 34, 35, text-fig. 35; Majzon, 1961, p. 763, pl. 3, fig. 7.

Globotruncana ventricosa White. (?) Carbonnier, 1952, p. 116, pl. 6, figs 1a–c; Maslakova, 1959, p. 110, pl. 13, figs 2a–c.

Globotruncana aff. *concavata* (Brotzen). de Klasz, 1953, p. 236, 237, pl. 6, figs 2a–c.

Globotruncana fundiconulosa Subbotina, 1953, p. 230, 231, pl. 14, figs 1a–4c; Majzon, 1961, p. 764, pl. 2, fig. 5.

Globotruncana ventricosa carinata Dalbiez, 1955, p. 168, 169, text-figs 8a–d.

Globotruncana ventricosa ventricosa White. Dalbiez, 1955, p. 168, text-figs 7a–d.

Globotruncana concavata (Brotzen). Bolli, 1957, p. 57, pl. 13, figs 3a–c; Książ-kiewicz, 1958, text-fig. 2 (6a–c); Barr, 1961, p. 569, pl. 71, figs 4a–c; Majzon, 1961, p. 762, pl. 6, fig. 1; Herm, 1962, p. 70, 71, pl. 5, fig. 4; Lehmann, 1963, p. 147, 148, pl. 6, figs 2a–3c, 4b; Postuma, 1962, 7 figs; Lehmann, 1965, text-figs 1m, o, p, q (?), s; Christodoulou and Marangoudakis, 1966, p. 305, text-figs 2a–c, pl. 1, figs 1a–c;

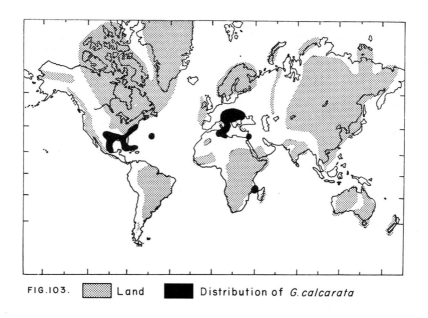

FIG.103.   ▨ Land   ■ Distribution of *G. calcarata*

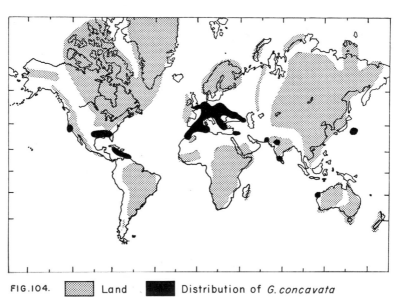

FIG.104.   ▨ Land   ■ Distribution of *G. concavata*

Wille-Janoschek, 1966, p. 99, 100, pl. 7, figs 5a–c; Banerji, 1966, p. 60, 61, pl. 3, figs 1–3; Dekker, 1966, text-fig. 4; Lehmann, 1966c, p. 169, 170, text-fig. 2; Bandy, 1967, p. 20, text-fig. 9 (11); Christodoulou and Marangoudakis, 1967, p. 312, 313, pl. 1, figs 1a–c; text-figs 2a–c; Kock, 1968, p. 65, pl. 59, figs 1–2b; Esker, 1969, p. 213, 214, pl. 1, figs 4–6; *non* Moorkens, 1969, p. 452, pl. 3, figs 3a–c; Kuhry, 1970, p. 299–302, pl. 2, figs 16–18; Longoria, 1970, p. 34–36; pl. 4, figs 4–9; pl. 12, figs 1, 2, 5, 6; Rahhali, 1970, p. 60, text-fig. 3e.

Globotruncana araratica Martirosyan, 1958, p. 14, 15, pl. 4, figs 1a–c (*non* 2a–c).

Globotruncana concavata var. *carinata* (Dalbiez). Edgell, 1962, p. 41–50, pl. 1, figs 1–5.

Globotruncana cf. *ventricosa* White. Lehmann, 1963, p. 150, pl. 6, figs 5a–c.

Globotruncana lamellosa Sigal. Lehmann, 1963, p. 149, pl. 9, figs 1a–2c; text-fig. 2s (?).

Globotruncana carinata Dalbiez. Postuma, 1962, 7 figs; Scheibnerová, 1968, p. 45–50, pl. 2, figs 1a–4c, 6a–7c (*non* 5a–c); Kuhry, 1970, p. 302, 303, pl. 2, figs 10–12, 19–21.

Praeglobotruncana concavata (Brotzen). Sturani, 1962, p. 86, text-fig. 15 (E2).

Globotruncana concavata concavata (Brotzen). Küpper, 1964, p. 617, 618, pl. 4, figs 5a–c; Broglio Lorigo and Mantovani, 1965, pl. 114, fig. 8; *non* Caron, 1966, p. 85, pl. 6, figs 2a–c; Broglio Lorigo and Mantovani, 1970, p. 230, pl. 7, figs 7–9, 11, 14, 18 (*non* 10, 12, 16, 22, 23); chart (71); Latif, 1970, p. 34, pl. 3, figs 1–3.

Globotruncana cf. *concavata primitiva* Dalbiez. Küpper, 1964, p. 619, pl. 4, figs 3a–c.

Globotruncana concavata carinata Dalbiez. Küpper, 1964, p. 618, 619, pl. 4, figs 4a–c; Broglio Lorigo and Mantovani, 1965, pl. 112, fig. 8; pl. 114, fig. 8; Gelati and Dario Passeri, 1967, pl. 69, fig. 2; Sturm, 1969, pl. 11, figs 4a–c; Scheibnerová, 1969, p. 72–74, tab. 14, figs 4a–c; tab. 15, figs 1a–6c; Broglio Lorigo and Mantovani, 1970, p. 230, pl. 7, figs 17, 19 (?) 20; chart (72).

Globotruncana ventricosa concavata (Brotzen). Salaj and Samuel, 1966, p. 218, 219, text-fig. 17.

Globotruncana vridhachalensis Banerji, 1966, p. 67, 68, pl. 5, figs 1, 2, 5.

Praeglobotruncana concavata carinata (Dalbiez). Săndulescu, 1966, pl. IVa, figs 3a–c.

Marginotruncana concavata (Brotzen). Pessagno, 1967, p. 304, 305, pl. 58, figs 1–9.

Globotruncana cf. *concavata carinata*. Dalbiez. Latif, 1970, p. 34, pl. 3, figs 1–3.

Praeglobotruncana (*Dicarinella*) *concavata* (Brotzen). Porthault, 1970, in Donze *et al.*, 1970, p. 73, pl. 10, figs 7, 8; pl. 13, fig. 25.

Globotruncana ex gr. *carinata* Dalbiez. Hanzlíková, 1972, p. 103, pl. 27, figs 6a–c.

Description: Test concavo–convex, dorsal side slightly to moderately concave, ventrally strongly convex; equatorial periphery lobate, axial periphery dorsally angular. Chambers of last whorl hemispherical with the flat dorsal side sloping toward earlier chambers, which are subspherical to spherical and are raised above the surface of inner edge of the chambers of the last whorl; 9–18 chambers in 2–3 whorls with 4–6, most commonly 6, chambers in the last whorl; a narrow, beaded, double keel situated on the dorsal edge of the chambers, becoming more dorsally positioned in the last 2–4 chambers reaching as much as a 60° angle out of the axis of coiling; ventral keel frequently curves back along ventral crest of most of the chambers in the final whorl to form a carina. Aperture a low to moderate, interiomarginal, umbilical to extraumbilical–umbilical arch. Umbilicus wide, moderately deep, covered by a tegillum. Sutures depressed, radial to curved ventrally; depressed, radial to curved in early whorls of dorsal side, later curved, flush to raised. Surface finely and densely perforate, pustulose.

Remarks: *Globotruncana concavata* is distinguished from the younger *G. aegyptiaca* Nakkady by its concave dorsal side. It differs from *G. ventricosa* White by possessing ventrally convex hemispherical chambers and a narrower double keel as well as its dorsal convexity. There is a wide diversity in the number of chambers in the final whorl and in their rate of inflation. In some individuals, the chambers rapidly increase in size, but in others are subequal. The ventral carina on the crest of the chambers may be found on none or all of those in the final whorl with any intermediate number.

This species is believed to have evolved from *Globotruncana primitiva* Dalbiez by becoming dorsally concave and developing more ventrally hemispherical chambers.

The presence or absence of a carina on the chambers has been used to distinguish between *Globotruncana carinata* Dalbiez and *G. concavata* respectively. This structure is unquestionably an ontogenetic development. The carina is absent on immature individuals with fewer whorls and a slightly concave dorsal side. The two morphotypes are found associated together as a normal population. There is no stratigraphic difference.

In a re-examination of the type material of *Rotalia concavata* Brotzen, Kuhry (1970) selected a lectotype for that species. The lectotype is certainly within the range of variability of this species, but represents an early ephebic individual. It has a carina developed on the last two chambers. As Kuhry observed, the presence of a carina cannot then be a distinguishing criterion. He proposed that both *Globotruncana concavata* and *G. carinata* Dalbiez be retained because of stratigraphic differences, and that the two be separated by the presence of the carina on more than two chambers for *G. carinata* and two or less for *G. concavata*. This is highly artificial and meaningless and completely ignores ontogenetic change.

The specimen illustrated as *Globotruncana ventricosa* White by Carbonnier (1952) resembles *G. concavata*. However, it is reportedly from Upper Cenomanian strata, too old for *G. concavata*, and is questionably placed here.

Figures 1a–c of *Globotruncana araratica* Martirosyan, 1958, is strongly reminiscent of a neanic *G. concavata*, and is treated as a junior synonym of that species. Figures 2a–c of Martirosyan's species may represent a *G. primitiva* Dalbiez. But this is not certain because Dalbiez's species was figured only in edge view.

Banerji (1966) recorded the new taxon *Globotruncana vridhachalensis* from the Santonian and lower Campanian of South India. He distinguished this species from *G. concavata* by it "having a slightly more convex dorsal side and a relatively less convex ventral side, keels are more widely separated and developed," and the sutures which "are more strongly raised and thick . . .". His illustration does not show keels that are more widely spaced. The sutures are no different than those of *G. concavata*, and the degree of convexity is well within the range of intraspecific variability as shown by Lehmann (1966) and others. Furthermore, its stratigraphic range is identical to that of *G. concavata*, and it is judged a junior synonym of that species.

Distribution: This is shown in Fig. 104.

Range: Santonian—Campanian (FOS *Globotruncana elevata*—LOS *Globotruncana concavata* Interval).

Globotruncana conica White

Globotruncana conica White, 1928b, p. 285, pl. 38, figs 7a–c; Majzon, 1943, pl. 1, fig. 16 (*non* pl. 2, figs 4a–c); Cushman, 1946b, p. 151, pl. 61, figs 20a–c; (?) Cita, 1948b, p. 149, 150, pl. 3, figs 5a–c; Bolli, 1951, p. 196, pl. 34, figs 13–15; Tilev, 1952, p. 67–71, text-figs 22a–d; *non* Obradović, 1953, p. 74, pl. 1, fig. 1; *non* Subbotina, 1953, p. 215–218, pl. 11, figs 1a–2c; Said and Kenaway, 1956, p. 150, pl. 5, figs 16a–c; Eternod Olvera, 1959, p. 103, 104, pl. 8, figs 1–3; Majzon, 1961, p. 762, pl. 1, fig. 15; *non* Malapris and Rat, 1961, p. 92, pl. 3, figs 2a–c; Postuma, 1962, 7 figs; El-Naggar, 1966, p. 87–90, pl. 12, figs 2a–d; text-figs 9 (7), 10 (part); Pessagno, 1967, p. 328–330, pl. 65, figs 8–10, pl. 82, figs 1–5; pl. 93, figs 12, 13; Ansary and Tewfik, 1968, p. 51, pl. 6, figs 5a–c; *non* Dupeuble, 1969, p. 15, pl. 2, figs 6a–d; (?) Funnell *et al.*, 1969, p. 29, pl. 3, figs 4–6; text-figs 12a–c; Longoria, 1970, p. 40–43, pl. 1, figs 10–12; pl. 9, figs (?) 7, 9, 10; pl. 12, figs (?) 3, 4; pl. 16, fig. (?) 3; *non* Cita and Gartner, 1971, pl. 5, figs 1a–c.

Globotruncana esnehensis Nakkady. Said and Kenawy, 1956, p. 150, pl. 5, figs 21a–c.

Globotruncana stuarti (de Lapparent). Said and Kenawy, 1956, p. 151, pl. 5, figs 22a–c.

Globotruncana stuarti conica White. Bandy, 1967, p. 21, text-fig. 9 (15).

Description: Test unequally biconvex to nearly planoconvex with the dorsal side strongly convex and the ventral side nearly flat to slightly convex; equatorial periphery nearly entire resulting in a circular outline, axial periphery acute. Chambers angular, dorsal walls flat, ventral walls flat to slightly inflated; gradually increasing in size; some or all those in last whorl of the dorsal side tend to become rectangular with the long dimension in the axis of coiling, 22–27 chambers in $3\frac{1}{2}$–4 whorls with 6–8 chambers in final whorl; single beaded keel bifurcates to form dorsal and ventral sutures, sweeps up and back to form carina on the umbilical shoulder. Aperture a low, interiomarginal umbilical arch. Umbilicus wide, deep, covered by a tegillum. Sutures on ventral side narrow, slightly curved, depressed to slightly raised; on dorsal side raised, wide, curved; spiral suture scalloped becoming a smooth curve in final whorl. Surface finely perforate, smooth to lightly pustulose.

Remarks: The only species which equals or exceeds the spiroconvexity of *Globotruncana conica* is *G. contusa*, which can be distinguished by its double keel and frequent fluting of its chambers. The total number of chambers is the main variable character. Some variation in the number of rectangular chambers in the last whorl is also observed.

Globotruncana elevata (Brotzen) is the most probable ancestor of *G. conica*, but no completely gradational forms have been recorded. An increase in the spiroconvexity with simultaneous decrease in the convexity of the ventral side of *G. elevata* would produce *G. conica*.

Distribution: This is shown in Fig. 105.

Range: Campanian (FOS *Globotruncana ventricosa*—FOS *Pseudoguembelina costata* Interval)—Maastrichtian (*Trinitella scotti* Interval).

Globotruncana contusa (Cushman), Plate 40, figs 1–4

Rosalina linnei d'Orbigny, de Lapparent, 1918, mutation caliciforme, p. 5, text-fig. 2j, pl. 1, fig. 2; Colom, 1931, p. 32, 33, text-fig. 1 (5); Colom, 1931, mutation caliciforme, p. 32, 33, text-fig. 1 (1).

Pulvinulina arca var. *contusa* Cushman, 1926b, p. 23.

Globotruncana conica var. *plicata* White, 1928b, p. 285, 286; Cushman, 1946b, p. 151, pl. 61, figs 21a–c; Cushman and Renz, 1947, p. 50, pl. 12, fig. 13; Sellier de Civrieux, 1952, p. 282, pl. 9, figs 10a–c.

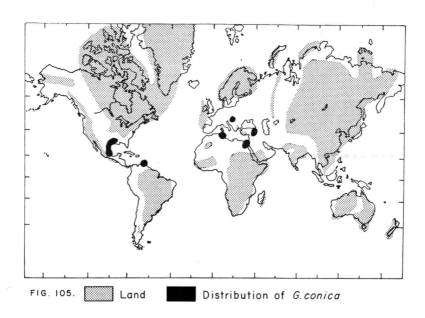

FIG. 105.   ▨ Land   ■ Distribution of *G. conica*

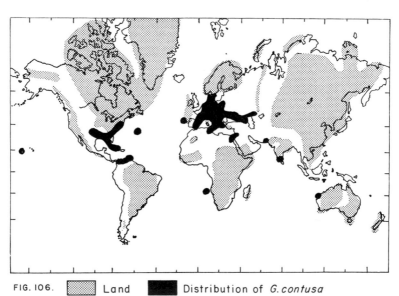

FIG. 106.   ▨ Land   ■ Distribution of *G. contusa*

Globotruncana lapparenti Brotzen, 1936, p. 175, 176, text-fig. 2 (*non* 1a–h; 2a–h, k–n; 3a, b; 5d).

Globotruncana arca var. *contusa* (Cushman). Morozova, 1939, p. 80, pl. 1, figs 1–3; Cushman, 1946b, p. 150, 151, pl. 62, figs 6a–c; Napoli Alliata, 1948, p. 21, 22, text-figs 2a–c.

Globotruncana linnei caliciformis. Vogler, 1941, p. 288, 289, pl. 24, fig. 23.

Globotruncana linnei caliciformis (de Lapparent). Colom, 1948, p. 28, pl. 2, fig. 7; pl. 8, fig. 7.

Globotruncana arca contusa (Cushman). van Wessem, 1943, p. 47–8, pl. 1, figs 54–56.

Globotruncana conica White. Majzon, 1943, p. 77, pl. 2, figs 4a–c (*non* pl. 1, 16); Keller, 1946, p. 102, 103, pl. 3, figs 3, 4; Subbotina, 1953, p. 215–8, pl. 11, figs 1a–2c.

Globotruncana contusa (Cushman). Cita, 1948b, p. 150, 151, pl. 3, figs 6a–c; Bolli, 1951, p. 196, pl. 34, figs 7–9; text-fig. 1f; Noth, 1951, p. 79, pl. 8, figs 17a–c; Hamilton, 1953, p. 232, pl. 29, figs 14–16; Subbotina, 1953, p. 218–222, pl. 11, figs 3a–c; pl. 12, figs 1a–2c; Nakkady and Osman, 1954, p. 78, 79, text-figs a–c; Troelsen, 1955, p. 80, 81, text-figs 2a–g; Cita, 1956, p. 440, section 5, 1493.50–1492.70, fig. a; section 6, 375, fig. a, section 7, 364, fig. s; Wicher, 1956, pl. 12, figs 5, 6; Edgell, 1957, p. 111, 112, pl, 2 figs 10–12; pl. 4, figs 1–3; Sacal and Debourle, 1957, p. 59, pl. 28, figs 11–13; Bieda, 1958, p. 63–65, text-figs 26a–c; Hofker, 1958, pl. 8, figs 44a–c; Seiglie, 1958, p. 66, 67, pl. 2, figs 1a–c; Haque, 1959, p. 19, pl. 1, figs 8a–c; Eternod Olvera, 1959, p. 105, 106, pl. 5, figs 6–8; Rauzer-Chernousova and Fursenko, 1959, p. 302, text-figs 686a–c; Maslakova, 1959, p. 110, pl. 13, figs 4a–c; Montanaro Gallitelli, 1959, p. 187, 188, pl. 2, fig. 4, text-fig. 10; Olsson, 1960, p. 50, pl. 10, figs 25, 26; Christodoulou, 1960, p. 11, text-fig. 14 (part); Danilova, 1960, p. 144, pl. 2, fig. 4; Vinogradov, 1960a, pl. 4, figs 23a–24c; pl. 5, figs 25a–c; Majzon, 1961, p. 762, pl. 2, fig. 13; Reutter and Serpagli, 1961, pl. 1, fig. 1 (part); Malaroda, 1962, text-fig. 4 (9); Postuma, 1962, 7 figs; Cita-Sironi, 1963, p. 119, text-fig. 5c; p. 120, text-fig. 8e; Maslakova, 1963, Tab. 6, fig. 4; Brönnimann and Rigassi, 1963, pl. 16, figs 2a–c; Hofker, 1963e, text-figs 22a–c; Pirini and Radrizzani, 1963, pl. 12, fig. 17 (L); pl. 43, fig. 61; Pokorný, 1963, p. 386, fig. 434; Olsson, 1964, p. 163, 164, pl. 2, figs 5a–c; pl. 3, figs 6a–c; 9a–c, Lehmann, 1966, p. 311, pl. 1, fig. 2; Wille-Janoschek, 1966, p. 100–102, pl. 8, figs 1–3; Săndulescu, 1966, pl. VIIa, figs 2a–c; Pessagno, 1967, p. 330–333, pl. 75, figs 18–20; pl. 77, figs 1–9; pl. 78, figs 6–11; pl. 92, figs 10–12; Christodoulou, 1967, pl. 1, figs 1 (part), 2 (part); *non* Pieri, 1967, p. 1280, pl. 95, fig. 4; Mabesoone *et al.*, 1968, p. 175, fig. 5, 7a–c; Neagu, 1968, text-fig. 3, fig. 59; Rao *et al.*, 1968, p. 21, 22, pl. 3, figs 1–3; Dupeuble, 1969, p. 154, 155, pl. 1, figs 2a–d; Funnell *et al.*, 1969, p. 29, 30, pl. 3, figs 7–9, text-figs 13a–c; Hanzlíková, 1969, p. 48, pl. 10, figs 4a–6c; Scheibnerová, 1969, p. 78, tab. 18, figs 2a–3c; Cita, 1970, pl. 4, figs 4, 5; Longoria, 1970, p. 36–40, pl. 4, figs 1–3; pl. 10, figs 1–3; pl. 16, figs 1, 2; Neagu, 1970a, fig. 1 (26); Neagu, 1970b, p. 69, pl. 18, figs 7–9; pl. 42, fig. 69; Berggren, 1971, pl. 2, figs 1–10; Cita and Gartner, 1971, pl. 2, figs. 1, 2; Hanzlíková, 1972, p. 104, pl. 27, figs 4a–5c; pl. 28, figs 2a–c.

Globotruncana caliciformis (de Lapparent). Subbotina, 1953, p. 213–215, pl. 10, figs 6a–c; Nakkady and Osman, 1954, p. 77, 78, pl. 20, figs 22a–c; Said and Kenawi, 1956, p. 150, pl. 5, figs 18a–c; Eternod Olvera, 1959, p. 101, 102, pl. 4, fig. 15; pl. 5, figs 1, 2; Majzon, 1961, p. 761, pl. 7, fig. 9; Cita-Sironi, 1963, p. 119, text-fig. 5b; Pirini and Radrizzani, 1963, pl. 13, fig. 19 (1); Broglio Lorigo and Mantovani, 1965, pl. 112, fig. 19; *non* Ansary and Tewfik, 1968, p. 52, pl. 7, figs 3a–c.

Globotruncana contusa contusa (Cushman). Gandolfi, 1955a, p. 53, 54, pl. 4, figs 3a–c; Herm, 1962, p. 72, 73, pl. 1, fig. 4; pl. 9, figs 1–5; Borsetti, 1962, p. 50, 52, pl. 7, figs 5, 6; text-figs 235, 245; El-Naggar, 1966, p. 90–93, pl. 7, figs 2a–3c; Salaj and Samuel, 1966, p. 202, pl. 25, figs 1a–c; Bandy, 1967, p. 20, text-fig. 9 (8).

Globotruncana contusa patelliformis Gandolfi, 1955a, p. 54, 55, pl. 4, figs 2a–c; El-Naggar, 1966, p. 93–95, pl. 8, figs 1a–c.

Marginotruncana contusa (Cushman). Hofker, 1956d, p. 53, text-fig. 9.

Globotruncana cf. *contusa* (Cushman). Edgell, 1957, pl. 3, figs 7–9; Bukowy and Geroch, 1957, pl. 29, figs 5a–c; Seiglie, 1958, pl. 2, figs 2a, b; Marrocu *et al.*, 1959, pl. 99, pl. 100.

Globotruncana cf. *caliciformis* (de Lapparent). Bukowy and Geroch, 1957, pl. 29, figs 6a, b; Montanaro-Gallitelli, 1959, p. 187, pl. 2, fig. 2, text-fig. 9.

Globotruncana plicata White. Sacal and Debourle, 1957, p. 61, pl. 28, figs 9, 10.

Globotruncana cf. *falsostuarti* Sigal. Radoičić, 1958, p. 125, pl. 11, fig. 10.

Globotruncana calcarata Cushman. Marrocu *et al.*, 1959, pl. 101.

Globotruncana contusa cf. *patelliformis* Gandolfi. Corminboeuf, 1961, p. 112, pl. 1, figs 1a–c.

Globotruncana cfr. *caliciformis* (de Lapparent). Reuter and Serpagli, 1961, pl. 12, fig. 2.

Globotruncana contusa subsp. *galeoidis* Herm, 1962, p. 74, 75, pl. 1, fig. 3; pl. 9, figs 6–14; Salaj and Samuel, 1966, p. 202, 203, pl. 25, figs 4a–c.

Globotruncana (*Globotruncana*) *contusa* (Cushman). Berggren, 1962a, p. 51–54, pl. 9, figs 3a–c (?), (*non* 4a, b).

Globotruncana caliciformis caliciformis (de Lapparent). Borsetti, 1962, p. 48, pl. 7, figs 3, 4; text-figs 213, 214 (?), 218 (?), 219, 220 (?), 221 (?), 233, 244; El-Naggar and Haynes, 1967, p. 5–8, pl. 1, figs 2a–3c.

Globotruncana plicata caliciformis Vogler. van Hinte, 1963, p. 64, 65, pl. 3, figs 2a–c.

Globotruncana contusa witwickae El-Naggar, 1966, p. 95–97, pl. 7, figs 1a–c.

Globotruncana caliciformis witwickae El-Naggar. El-Naggar and Haynes, 1967, p. 10, 11, pl. 1, figs 4a–c.

Globotruncana caliciformis galeoidis Herm. El-Naggar and Haynes, 1967, p. 8, pl. 2, figs 2a–5c; Ansary and Tewfik, 1968, p. 52, pl. 7, figs 1a–c.

Globotruncana caliciformis patelliformis Gandolfi. El-Naggar and Haynes, 1967, p. 8–10, pl. 1, figs 1a–c.

Globotruncana contusa (Cushman) n. subsp. (?) Kock, 1968, p. 652, 653, pl. 59, figs 5a–c.

Globotruncana walfischensis Todd, 1970, p. 153, pl. 5, fig. 8.

Description: Test large, planoconvex with the ventral side essentially flat to slightly concave, the dorsal side strongly convex; equatorial periphery lobate to polygonal, axial periphery acute. Chambers ventrally flat with wall sloping toward umbilicus; dorsally flat, curved and elongate in direction of coiling, develops a slight to deep fold in later chambers, those of first whorl may be subspherical; gradually increasing in size, those of last whorl often subequal; 14–18 chambers in $3\frac{1}{2}$–4 whorls with 4–7 in final whorl; with narrow, beaded, double keel on the acute periphery. Aperture a low, interiomarginal, umbilical arch. Umbilicus moderately wide, very deep, covered by a tegillum. Sutures raised, curved, beaded; carina on umbilical shoulder continuous with ventral keel. Surface finely and densely perforate, smooth.

Remarks: This species differs from *Globotruncana conica* White by having a double keel and chamber folding. Separation from *G. fornicata* Plummer is arbitrarily based upon the increased degree of spiroconvexity and folding. *Globotruncana contusa* exhibits three important variations. The chambers of the first whorl may or may not be subspherical. The spire height varies over a considerable range. The degree of chamber folding or fluting has wide variation. Any one or a combination of these features can impart a distinctly different appearance to the test, and have led to the establishment of new taxa. All forms can be found in a normal population. A complete gradational sequence exists between *Globotruncana contusa* and the older *G. fornicata* Plummer.

Berggren (1962a) stated that it was not possible to determine whether the thin-section of *Globotruncana linnei caliciformis* Vogler, 1941, was conspecific to either *G. contusa* (Cushman) or *G. conica* White. It can be determined. An axial section of *G. conica* has a lower, smoothly curved dorsal surface. Whereas, *G. contusa* is higher and more irregular in its spiroconvexity. The latter is caused by the folding of the chambers which does not occur in *G. conica*. Upon examination of the type figure of *G. linnei caliciformis*, an outline typical for *G. contusa* can be seen. Furthermore, El-Naggar and Haynes (1967) demonstrated by use of both thin-sections and free individuals that *G. linnei caliciformis* is conspecific with *G. contusa*. Unfortunately, these authors incorrectly believed that the *Rosalina linnei* "mutatión caliciforme" of de Lapparent (1918) has priority, when in fact it was *not* a validly proposed taxon until so done by Vogler, a fact continued to be ignored by many workers.

El-Naggar (1966) proposed the new subspecies *Globotruncana contusa witwickae* commenting that it was morphologically slightly different and stratigraphically older than *G. contusa contusa*. *Globotruncana contusa* and *G. fornicata* form a completely gradational bioseries wherever a continuous sequence of Maastrichtian strata is found, and El-Naggar's taxon is in this series, as he suggested. Attempts to statistically quantify the end members and particularly the intermediate forms, have met with little success. At some point an arbitrary cutoff must be chosen. As interpreted here, El-Naggar's subspecies is synonymous with *G. contusa*.

Globotruncana walfischensis Todd, 1970, was distinguished from *G. contusa* by its early globular, inflated chambers. As established above this characteristic is merely an intraspecific variant of *G. contusa*.

Distribution: This is shown in Fig. 106.

Range: Maastrichtian (LOS *Globotruncana fornicata*—FOS *Trinitella scotti* Interval to *Trinitella scotti* Interval).

Globotruncana coronata Bolli, Plate 41, figs 1, 2
Rosalina linnei d'Orbigny. de Lapparent, 1918, p. 4, text-figs 1a, b.
Globotruncana linnei (d'Orbigny). Glaessner, 1937, p. 38, pl. 1, figs 11a–c; Gandolfi, 1942, p. 125–130, text-figs 46–2a–c (*non* 46–1a–c, pl. 3, figs 2a–3c, pl. 4, fig. 18); Bermúdez, 1952, p. 53, 54, pl. 7, figs 7a–c.
Globotruncana lapparenti subsp. *coronata*. Bolli, 1945, p. 233, text-figs 1:21, 22, pl. 9, fig. 15 (*non* 14); Colom, 1947, p. 101, 102, pl. 24, figs 20, 21; Colom, 1948, p. 27, pl. 6, figs 14, 15, 18, 20 (?); Cita, 1948b, p. 156, 157, pl. 4, figs 3a–c; Mornod, 1950, p. 591, 592, text-figs 13:1a–d; Hagn and Zeil, 1954, p. 43, 44, pl. 3, figs 4a, b; pl. 7, figs 1–3; Książkiewicz, 1956, p. 273, 274, pl. 31, figs 17, 18; text-figs 43, 44 (1–4); Książkiewicz, 1958, text-fig. 2 (3a–c); *non* Witurcka, 1958, p. 217, 218, pl. 16, figs 30a–c; Tollmann, 1960, p. 194, pl. 21, fig. 2; Sturani, 1962, p. 89, text-fig.

15 (B1 [?], 2–5), [non text-fig. 15 (C3, F20, 21)]; Küpper, 1964, p. 625, 626, pl. 3, figs 2a–c; Caron, 1966, p. 80, pl. 4, figs 1a–c; Christodoulou and Marangoudakis, 1967, p. 314, 315, text-figs 3a–c; Radoičić, 1968, pl. 2, figs 6, 8; pl. 3, fig. 8; pl. 4, fig. 1; pl. 5, fig. 1 (non 3); pl. 10, figs 5, 6, 8; Kalantari, 1969, p. 205, 206, pl. 24, figs 3a–c; Neagu, 1970b, p. 67, pl. 17, figs 1–3.

Globotruncana lapparenti tricarinata (Quereau). Cita, 1948b, p. 157, 158, pl. 4, figs 4a–c; Książkiewicz, 1956, p. 274, 275, text-fig. 47 (lower fig.); [non pl. 31, figs 10–12; text-figs 47 (upper), 48 (1–4)]; Neagu, 1970b, p. 67, pl. 16, figs 4–6.

Globotruncana coronata Bolli. Sigal, 1952, p. 34, text-fig. 36; Subbotina, 1953, p. 201–203, pl. 8, figs 1a–c; Majzon, 1961, p. 763, pl. 2, fig. 1; Herm, 1962, p. 76, pl. 6, fig. 5; Lehmann, 1963, p. 150, pl. 8, figs 2a–3c; Postuma, 1962, 7 figs; van Hinte, 1963, p. 81–83, pl. 7; figs 1a–c; *non* Takayanagi, 1965, p. 212, 213, pl. 24, figs 1a–2c; Douglas and Sliter, 1966, p. 109, pl. 4, figs 4a–c (non 5a–c); pl. 5, figs 7a–c (non 8a–c); Salaj and Samuel, 1966, p. 203, 204, pl. 19, figs 5a–c; Barr, 1968, pl. 3, figs 7a–c); Douglas, 1969, p. 177, 178, pl. 3, figs 6a–c, 8a–c (non 5a–c, 7a–c; Douglas and Rankin, 1969, p. 202, 203, text-figs 13A–C; Moorkens, 1969, p. 449, pl. 2, figs 5a–c; Rahhali, 1970, p. 62, text-fig. 4g (non e–f); Hanzlíková, 1972, p. 104, pl. 28, figs 1a–c.

Globotruncana lapparenti Brotzen. Sigal, 1952, p. 35, text-fig. 38; Nagappa, 1959, pl. 6, figs 10a–c; Wille-Janoschek, 1966, p. 108–110, pl. 1, figs 3a–c.

Globotruncana (*Globotruncana*) *lapparenti coronata*. Bolli. Papp and Küpper, 1953, p. 36, 37, pl. 1, figs 2a–c.

Globotruncana lapparenti bulloides Vogler. Hagn and Zeil, 1954, p. 4–6, pl. 2, figs 5a–c.

Globotruncana lapparenti lapparenti Brotzen. Hagn and Zeil, 1954, p. 39–43, pl. 3, figs 3a–c; pl. 6, figs 5, 8; Książkiewicz, 1958, text-fig. 2 (1a–c); Küpper, 1964, p. 624, pl. 3, figs 1a–c.

Globotruncana (*Globotruncana*) *tricarinata* subsp. *desioi* Gandolfi, 1955a, p. 27.

Globotruncana tricarinata (Quereau). Gandolfi, 1955a, text-fig. 6 (1a, b).

Globotruncana linneiana coronata Bolli. Barr, 1961, p. 572, 573, pl. 70, figs 1a–c; *non* Banerji, 1966, p. 63, pl. 4, figs 1–3; Bandy, 1967, p. 19, text-figs 8 (10); *non* Bate and Bayliss, 1969, pl. 3, figs 6a–c.

Globotruncana linneiana (d'Orbigny). Graham and Clark, 1961, p. 113, pl. 5, figs 11a–c; Douglas and Sliter, 1966, p. 112, pl. 4, figs 8a–c (non 6a–7c), pl. 5, figs 9a–c (non 4a–c, 6a–c); Sliter, 1968, p. 104, pl. 17, figs 3a–4c; Hanzlíková, 1972, p. 107, 108, pl. 29, figs 6a–7c.

Globotruncana cf. *coronata* Bolli. Lehmann, 1963, p. 150, pl. 4, figs 3a–c (non pl. 5, figs 3a–c).

Globotruncana linneiana linneiana (d'Orbigny). van Hinte, 1963, p. 75–78, pl. 5, figs 2a–c (non 1a–c); Salaj and Samuel, 1966, p. 209, pl. 19, figs 1a–c.

Globotruncana tricarinata desioi Gandolfi. Salaj and Samuel, 1966, p. 217, 218, pl. 19, figs 3a–c.

Globotruncana tricarinata tricarinata (Quereau). Salaj and Samuel, 1966, p. 218, pl. 19, figs 3a–c.

Marginotruncana coronata (Bolli). Pessagno, 1967, p. 305, 306, pl. 65, figs 11–13; (?) Esker, 1969, p. 214, pl. 1, figs 9–11; Porthault, 1970, in Donze *et al.*, p. 78, 79, pl. 11, figs 1–3; pl. 13, fig. 20; *non* Douglas, 1971, pl. 2, figs 1, 2.

Globotruncana pseudolinneiana Pessagno. Douglas, 1969, p. 185, pl. 3, figs 2a–c (non 3a–4c).

Globotruncana angusticarinata Gandolfi. Hanzlíková, 1972, p. 103, pl. 27, figs 2a–3c.

Description: Test gently biconvex with nearly equal curvature on each side; equatorial periphery slightly lobate, axial periphery truncate. Chambers flat, with mild slope toward periphery; petaloid dorsally, imbricate ventrally, greatest thickness at umbilical shoulder; 16–18 chambers in $2\frac{1}{2}$–3 whorls with 5–8 subequal to gradually expanding chambers in final whorl; narrow, beaded, double keel, ventral keel curved back to form carina on umbilical shoulder, dorsal keel forms dorsal sutures; carina usually discontinuous from chamber to chamber. Aperture a low to moderate, interiomarginal, umbilical to extraumbilical–umbilical arch. Umbilicus wide, shallow, covered by a tegillum. Sutures raised. Surface finely and densely perforate, smooth to pustulose.

Remarks: This species has more closely spaced double keels than G. lapparenti Brotzen. Another similar species, G. renzi Gandolfi, differs by having more rapidly expanding chambers with more inflation both ventrally and dorsally. The truncate periphery may become gradually narrower relative to the chamber size through the ontogenetic development. The last one or two chambers may lack the dorsal and ventral carina of the keel, and with the remaining imperforate peripheral band occupying the entire width of the narrow margin, it may appear as a single keel.

Globotruncana coronata is morphologically closer to G. lapparenti than any other species, and so is believed to have evolved from it.

Gandolfi (1955a) in a footnote proposed the new subspecies Globotruncana tricarinata desioi for the form illustrated by Cita (1948b, pl. 4, fig. 4) under the name G. tricarinata (Quereau). This subspecies is invalid because it fails to satisfy Article 13a (i–iii) of the I.C.Z.N. The morphotype is a G. coronata.

Distribution: This is shown in Fig. 107.

Range: Turonian—Campanian (FOS Globotruncana elevata—LOS Globotruncana concavata Interval).

Globotruncana cretacea (d'Orbigny), Plate 41, figs 3, 4; plate 42, fig. 1
Globigerina cretacea d'Orbigny, 1840, p. 34, pl. 3, figs 12–14; (?) Reuss, 1845, p. 36, pl. 8, figs 55a, b; (?) Brown, 1853, p. 241, pl. IX, figs 8a, b; non Tutkovskii, 1887, p. 353–355, pl. 4, figs 4a, b; pl. 5, figs 4c, d; pl. 7, fig. e; non Haeusler, 1890, p. 119, pl. 15, fig. 47; (?) Beissel, 1891, p. 71, pl. 13, figs 43–47; non Woodward and Thomas, 1893, p. 41, pl. D, figs 18, 19; (?) Chapman, 1896, p. 588, 589, pl. 13, figs 5a–6; (?) Franke, 1925, p. 93, pl. 8, figs 15a–c; non Berry and Kelly, 1930, p. 11, pl. 3, figs 7–9; (?) Cushman, 1934, p. 63, pl. 8, figs 13a–c; (?) Dain, 1934, p. 42, pl. 4, figs 47a–c; non Morrow, 1934, p. 198, pl. 30, figs 7, 8, 10a, b; non Tappan, 1943, p. 512, pl. 82, figs 16a–17; (?) Montanaro Gallitelli, 1947, p. 193, 194, text-fig. 1 (17); text-fig. 2 (10); non Brönnimann, 1952a, p. 14–16, text-figs 3a–m; non Veljković-Zajec, 1955, p. 328, 329, pl. 2, fig. 1; non Bolin, 1956, p. 292, 293, pl. 39, figs 4–8, 13, 17; (?) Smitter, 1957, p. 199, text-figs 19a, b; non Bolli, 1959, p. 270, pl. 22, figs 8a–9; Hofker, 1961c, text-fig. 67; (?) Hofker, 1962, p. 93, text-fig. 15a–d; Cati, 1964, p. 250, 251, pl. 41, figs 1a–c.
Globigerina cretacea var. saratogaensis Applin, 1925, in Applin et al., p. 98, pl. 3, figs 8a–c.
Globigerina White, 1928b, pl. 38, fig. 4.
Globotruncana globigerinoides Brotzen, 1936, p. 177, pl. 12, figs 3a–c; pl. 13, fig. 3; Noth, 1951, p. 76, pl. 5, figs 4a, b; Hagn, 1953, p. 94, 95, pl. 8, figs 9a–c; text-figs 14 (?), 15 (?); Liszka, 1955, p. 185, 186, pl. 13, figs 13a–c; Książkiewicz, 1956, p. 280, pl. 30, figs 2, 3; text-figs 56, 57; Edgell, 1957, p. 112, 113, pl. 2, figs

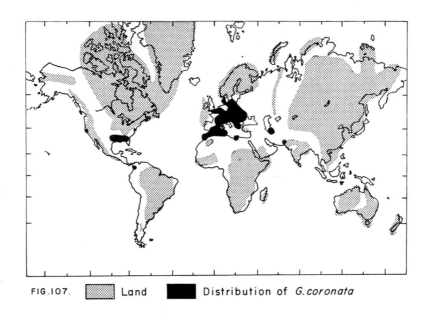

FIG.107. ☐ Land ■ Distribution of *G. coronata*

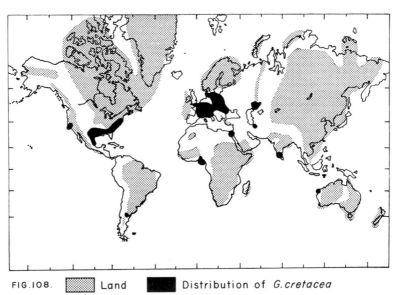

FIG.108. ☐ Land ■ Distribution of *G. cretacea*

13–15; Sacal and Debourle, 1957, p. 59, pl. 26, figs 1, 11; Bieda, 1958, p. 65, 66, text-figs 27a–c; *non* Witwicka, 1958, p. 221, 222, pl. 18, figs 33a–c; Eternod Olvera, 1959, p. 110, 111, pl. 5, fig. 12; pl. 6, figs 1, a; *non* Maslakova, 1959, p. 110, pl. 12, figs 3a–c; Belford, 1960, p. 99, pl. 29, figs 12–14; *non* Jurkiewicz, 1961, pl. 24, figs 7a–c; Herm, 1962, p. 80, 81, pl. 5, fig. 6; *non* Küpper, 1964, p. 622, 623, pl. 3, figs 7a–c; Banerji, 1966, p. 61, 62, pl. 5, figs 3, 4, 7; Săndulescu, 1966, pl. VIa, figs 1a–c; *non* Ansary and Tewfik, 1968, p. 55, pl. 8, figs 2a–c; Kalantari, 1969, p. 202, 203, pl. 25, figs 4a–c.

Globotruncana marginata (Reuss). Kalinin, 1936, p. 52, pl. 6, figs 94–96; Jírová, 1956, Tab. 2, obr. 3a–c; Belford, 1960, p. 100, pl. 30, figs 1–7.

Rosalinella globigerinoides (Brotzen). Schijfsma, 1946, p. 96, 97, pl. 7, figs 9a–c.

Globotruncana saratogaensis (Applin). Brönnimann and Brown, 1953, p. 544, 545, pl. 21, figs 1–3; Majzon, 1961, p. 761, pl. 5, fig. 4; Salaj and Samuel, 1966, p. 214, pl. 9, figs 1a–c.

Globigerina saratogaensis Applin. Frizzell, 1954, p. 127, pl. 20, figs 7a–c.

Marginotruncana pura Hofker. Hofker, 1956c, text-fig. 7.

Globotruncana cretacea (d'Orbigny). Banner and Blow, 1960, p. 8–10, pl. 7, figs 1a–c; Barr, 1961, p. 567–568, pl. 69, fig. 9a–c; van Hinte, 1963, p. 85, 86, pl. 6, figs 3a–c; Borsetti, 1964, p. 672, pl. 79, fig. 11 (*non* 10); text-fig. 5c (*non* 5a, b, d, e); van Hinte, 1965a, p. 21, pl. 3, figs 1a–c; *non* Takayanagi, 1965, p. 213, 214, pl. 24, figs 3a–c; *non* Rasheed and Govindan, 1968, p. 82, pl. 7, figs 4–6; Douglas and Rankin, 1969, p. 200–202, text-figs 12A–C; *non* Rahhali, 1970, p. 65, text-figs 5e–i; Hanzlíková, 1972, p. 105, pl. 28, figs 3a–5c.

Globotruncana cf. *pura* Hofker. Reyment, 1960, p. 76, pl. 15, figs 5a, b.

Rugoglobigerina kingi Trujillo, 1960, p. 339, 340, pl. 49, figs 5a–c; Graham, 1962, p. 108, pl. 20, figs 29a–c.

Marginotruncana globigerinoides (Brotzen). Hofker, 1961b, p. 124, text-figs 2a–c.

Globotruncana (Marginotruncana) globigerinoides Brotzen. Hofker, 1961c, text-figs 65, 66.

Globotruncana (Ticinella) globigerinoides Brotzen. Majzon, 1961, p. 762, pl. 6, fig. 9.

Globotruncana pura Hofker. Majzon, 1961, p. 764, pl. 7, fig. 7; Takayanagi, 1965, p. 221, pl. 27, figs 1a–c.

(?) *Rotalipora bicarinata* Samuel and Salaj, 1962, p. 318, pl. 10, figs 6a–c; Salaj and Samuel, 1966, p. 183, pl. 12, figs 3a–c.

Globotruncana mariai Gandolfi. Martin, 1964, p. 82, pl. 9, figs 7a–c.

Rugoglobigerina kingi Trujillo. *non* Marianos and Zingula, 1966, p. 339, 340, pl. 38, figs 6a–c; *non* Scheibnerová, 1969, p. 80, tab. 19, figs 5a–c.

Rugoglobigerina loetterli subloetterli (Gandolfi). Salaj and Samuel, 1966, p. 195, 196, pl. 9, figs 2a–c.

Archaeoglobigerina blowi Pessagno, 1967, p. 316, pl. 59, figs 5–7 (*non* 1–4, 8–10).

Archaeoglobigerina cretacea (d'Orbigny). Pessagno, 1967, p. 317, 318, pl. 70, figs 3–7; Bertels, 1970, p. 36–38, pl. 1a–4c.

Rugoglobigerina cretacea (d'Orbigny). Bandy, 1967, p. 21, text-fig. 10 (1).

Globotruncana lapparenti bulloides (Vogler). Ansary and Tewfik, 1968, p. 54, pl. 7, figs 7a–c.

Hedbergella kingi (Trujillo). *non* Douglas, 1969, p. 166, 167, pl. 4, figs 6a–7c; *non* Belford and Scheibnerová, 1971, p. 334, pl. 4, figs 10–15.

Globigerinelloides cretacea (d'Orbigny). *non* Neagu, 1970b, p. 63, pl. 28, figs 7–11.

Rotundina cretacea (d'Orbigny). Bellier, 1971, p. 87, 88, pl. 1, figs 1a–2c, (?) 3a–c.

Rugoglobigerina (*Archaeoglobigerina*) *blowi* (Pessagno). El-Naggar, 1971a, pl. 4, figs a–c.

Rugoglobigerina (*Archaeoglobigerina*) *cretacea* (d'Orbigny). El-Naggar, 1971a, pl. 4, figs (?) g, h, (*non* d–f, i–l).

Description: Test small, a low trochospire, equatorial periphery distinctly lobate, axial periphery rounded to slightly truncate. Chambers spherical to subspherical, tending toward elongation in axis of coiling; 13–14 chambers in 2½ whorls of 5 chambers each, rarely 6 in final whorl; gradually increasing in size; usually well separated; with a wide, imperforate peripheral band, often bordered by low, thin, double carinae; ventral carina may be abaperturally curved back along the umbilical shoulder. Aperture a low, interiomarginal, umbilical to extraumbilical–umbilical arch. Umbilicus moderately wide, deep, covered by a tegillum. Sutures radial to very slightly curved, well depressed, narrow. Surface finely perforate, pustulose.

Remarks: *Globotruncana marginata* (Reuss) is the most similar species to *G. cretacea*, but differs in having a more strongly developed double keel and less spherical chambers which tend to be more closely packed. The keel is the most unstable characteristic of *Globotruncana cretacea*. It may possess only an imperforate peripheral band throughout, or on all but the last two or three chambers which are completely perforate. A weakly developed double keel may be found on other individuals, or found only on the earlier chambers of the last whorl followed by loss of the carinae on the later chambers.

This species differs from its ancestor *Globigerina delrioensis* Carsey only in having a keel and a tegillum.

Banner and Blow (1969) selected a lectotype from among the specimens in d'Orbigny's original syntypic series, and were the first to discover that the taxon *cretacea* was keeled and a *Globotruncana*. Contrary to Bandy (1967), this species cannot be placed in the genus *Rugoglobigerina* because it is keeled.

Although several species of *Globotruncana* first appear at the base of the Turonian, *G. cretacea* is certainly one of the most primitive and most closely related to the *Globigerina*-stock.

Woodward (1894) recognised the presence of *Globigerina cretacea* in New Jersey, but did not illustrate the species. In synonymy with this species, he placed the following of Ehrenberg's species: *Globigerina foveolata* (pars), *G. libani*, *Planulina pachyderma*, *Rotalia pertusa*, *R. aspera*, *R. globulosa*, *R. densa*, *R. quaterna*, *R. rosa*, *R. pachyomphala*, *R. tracheotetras*, *R. perforata*, *R. protacmaea*, *R. laxa*, and *R. centralis*. There is no indication that Woodward personally examined these types, and Ehrenberg's figures are so poorly illustrated as to prevent recognition. The taxonomic status of these species cannot be ascertained and should be considered nomen oblitum.

Examination of the holotype of *Rugoglobigerina kingi* Trujillo, 1960, revealed the presence of an imperforate peripheral band on the chambers of the last whorl. The costellae characteristically present on *Rugoglobigerina* are not found on Trujillo's species, instead the surface is pustulose. Although no remnant of a tegillum could be detected, *R. kingi* is regarded as a junior synonym of *Globotruncana cretacea*.

The presence of sutural accessory apertures on *Rotalipora bicarinata* Samuel and Salaj, 1962, must be questioned. The umbilicus of the type figure is filled. The authors have illustrated some sort of projections through this adhering sediment in the approximate position of the assumed accessory apertures. It is more likely that these projections are remnants of a tegillum. Often the thin tegillum is broken away leaving only the narrow, arched basal portion of each apertural flap extension. This, with the double keel, is sufficient evidence to remove this species from the genus *Rotalipora* and re-assign it to *Globotruncana*, tentatively with *G. cretacea*.

Distribution: This is shown in Fig. 108.

Range: Turonian–Maastrichtian (FOS *Globotruncana aegyptiaca*—FOS *Rugoglobigerina hexacamerata* Interval).

Globotruncana duwi Nakkady, Plate 43, figs 1–5

Globigerina rosetta Carsey. Plummer, 1926, p. 36, pl. 2, figs 9a–c.

Globotruncana rosetta (Carsey). Majzon, 1943, pl. 1, figs 14a–c.

Globotruncana aegyptiaca var. *duwi* Nakkady, 1950, p. 690, pl. 90, figs 17–19; (?) Nakkady and Osman, 1954, p. 76, pl. 20, figs 21a–c.

Rugotruncana skewesae Brönnimann and Brown, 1956, p. 550, 551, pl. 23, figs 4–6.

Globotruncana (*Rugotruncana*) *skewesae* Brönnimann and Brown. Majzon, 1961, p. 764, pl. 5, fig. 9.

Globotruncana aegyptiaca duwi Nakkady. El-Naggar, 1966, p. 80, 81, pl. 3, figs 5a–c; text-figs 9 (3), 10 (part).

Globotruncana duwi Nakkady. *non* Bandy, 1967, p. 20, text-fig. 9 (10).

Description: Test small, planoconvex with a nearly flat dorsal side and a strongly convex ventral side; equatorial periphery lobate, axial periphery slightly truncate. Dorsal chambers of the last whorl flat, earlier ones slightly inflated; ventral ones convexly curved up to the umbilical shoulder, then steeply down into the umbilicus; 15–16 chambers in about 3 whorls with 5, rarely 6, chambers per whorl; with a narrow double keel becoming wider in later chambers, situated on dorsal edge of periphery, ventral carina may become reduced and disappear in last 1–2 chambers; first 2–3 chambers in last whorl typically exhibit well-developed short, thick spines perpendicular to the periphery as part of the carinae. Aperture a high, interiomarginal, umbilical arch. Umbilicus moderately deep and wide for the test size, covered by a tegillum, bordered by a carinate shoulder. Sutures dorsally curved, flush to raised; ventrally radial, depressed. Dorsal surface smooth to slightly pustulose, ventral surface hispid to pustulose; finely perforate.

Remarks: *Globotruncana aegyptiaca* Nakkady is the only species with which *G. duwi* may be confused. *Globotruncana duwi* is, however, smaller and has a relatively wider keel for its size than does *G. aegyptiaca*. Within this species the peripheral spines are developed to a greater or lesser degree as is the ventral carina of the keel.

This species may have evolved from *G. ventricosa* White with the development of more hemispherical-shaped chambers from the more steeply angled type of the latter.

It is possible that *Globotruncana duwi* and *G. aegyptiaca* Nakkady are intraspecific variants of one population. A more detailed study is necessary before this is determined. If this is proven, *G. aegyptiaca* becomes the senior synonym, and *G. ventricosa* the most likely ancestor for it.

The type (Nakkady, 1950) and subsequent (Nakkady and Osman, 1954) illustrations of *Globotruncana aegyptiaca duwi* impart very little information. Those given by El-Naggar (1966), are excellent and there is little doubt that the specimen figured is conspecific with *Rugotruncana* (= *Globotruncana*) *skewesae* (Brönnimann and Brown). He noted its appearance at the base of the *Globotruncana gansseri* Zone (= LOS *G. lapparenti*–FOS *G. aegyptiaca* Interval), which corresponds to the first appearance of *G. skewesae*. However, Nakkady and Osman (1954) recorded the presence of *G. duwi* in Campanian strata. A restudy of these older specimens is

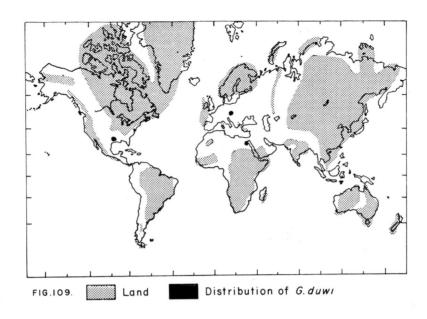

FIG.109.   Land   ▇ Distribution of *G.duwi*

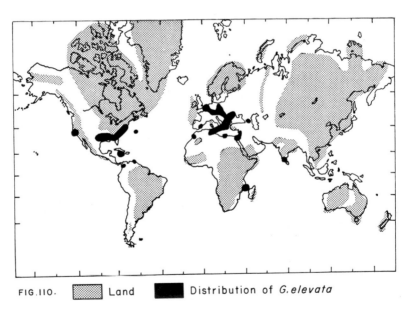

FIG.110.   Land   ▇ Distribution of *G.elevata*

necessary to determine if they have been correctly identified and whether the age determination is accurate.

Distribution: This is shown in Fig. 109.

Range: Maastrichtian (LOS *Globotruncana lapparenti*—FOS *Globotruncana aegyptiaca* Interval to *Trinitella scotti* Interval).

Globotruncana elevata (Brotzen), Plate 42, figs 2–4

Rotalia elevata Brotzen, 1934, p. 66, pl. 3, fig. C; Kuhry, 1970, p. 292–299, fig. 1.

Globotruncana andori de Klasz, 1953a, p. 233–235, pl. 6, figs 1a–c; Bolli, 1957, p. 59, pl. 14, figs 6a–c; Majzon, 1961, p. 764, pl. 6, fig. 2; Herm, 1962, p. 64, 65, pl. 5, fig. 1; Kuhry, 1970, p. 292–299, text-figs 4e, f.

Globotruncana rosetta (Carsey). Subbotina, 1953, p. 226–230, pl. 13, figs 6a–c (*non* 5a–c).

Globotruncana stuarti (de Lapparent). Subbotina, 1953, p. 231–234, pl. 15, figs 3a–5c; Vinogradov, 1960a, pl. 6, figs 30a–c (*non* 29a–c); Wille-Janoschek, 1966, p. 115, 116, pl. 7, figs 4a–c.

Globotruncana elevata elevata (Brotzen). Dalbiez, 1955, p. 169, text-figs 9a–c; Knipscheer, 1956, p. 4, 5, pl. 4, figs 1a–3b, 5a–c, text-fig. 1; van Hinte, 1963, p. 70, 71, pl. 5, figs 4a–c; Graham and Church, 1963, p. 63, pl. 7, figs 13a–c; Christodoulou and Marangoudakis, 1966, p. 306, pl. 1, figs 10a–11c; Săndulescu, 1966, pl. VIIa, figs 3a–c; Christodoulou and Marangoudakis, 1967, p. 315, pl. 1, figs 10a–c, 11a–c; text-figs 4 (1a–6c, 7a–c (?), 8a–10c); (?) Sturm, 1969, pl. 12, figs 1a–c; Kuhry, 1970, p. 292–299, text-figs 4a–b.

Globotruncana rosetta insignis Gandolfi, 1955a, p. 67, 68, pl. 6, figs 2a–c; Salaj and Samuel, 1966, p. 211, 212, pl. 24, figs 5a–c.

Globotruncana stuarti parva Gandolfi, 1955a, p. 65, pl. 5, figs 7a–c; El-Naggar, 1966, p. 131–133, pl. 9, figs 2a–d.

Globotruncana stuarti stuarti (de Lapparent). Gandolfi, 1955a, p. 64, 65, pl. 5, figs 6a–c.

Globotruncana ventricosa stuartiformis Dalbiez, 1955, p. 169, text-figs 10a–c.

Globotruncana elevata elevata (Brotzen) × *falsostuarti* Sigal. Knipscheer, 1956, pl. 4, figs 6a, b, 8a, b, 11a, b.

Globotruncana elevata stuartiformis Dalbiez. Knipscheer, 1956, p. 52, pl. 4, figs 9a–10b, 12a, b, 14a, b; Herm, 1962, p. 77, pl. 8, fig. 2; van Hinte, 1963, p. 68–70, pl. 1, figs 3a–c; Küpper, 1964, p. 620, 621, pl. 4, figs 9a–c; Săndulescu, 1966, pl. VIIa, figs 4a–c; *non* Salaj and Samuel, 1966, p. 204, 205, pl. 22, figs 4a–c; Kuhry, 1970, p. 292–299, text-figs 4c, d.

Globotruncana stuartiformis Dalbiez. *non* Hofker, 1956e, text-fig. 1; Postuma, 1962, 7 figs; *non* Lehmann, 1963, p. 151, pl. 10, figs 2a–c; Wille-Janoschek, 1966, p. 103–105, pl. 7, figs 6–8; *non* Douglas and Sliter, 1966, p. 114, 115, pl. 3, figs 3a–c; Pessagno, 1967, p. 357–359, pl. 92, figs 1 3; Barr, 1968a, p. 318, pl. 40, figs 1a–2c; Sliter, 1968, p. 106, 107, pl. 18, figs 5a–c (6a–c ?); Douglas, 1969, p. 186, 188, pl. 8, figs 6a–c (*non* 7a–8c); Longoria, 1970, p. 76–79, pl. 6, figs 4–6; pl. 11, figs 5, 6; pl. 18, fig. (?) 3; pl. 19, figs (?) 1, 3; Cita and Gartner, 1971, pl. 2, fig. 4; pl. 5, figs 5a–c; *non* Hanzlíková, 1972, p. 111, pl. 31, figs 3a–c; *non* Govindan, 1972, p. 177, pl. 6, figs 1–6.

Marginotruncana elevata (Brotzen). Hofker, 1956b, fig. 70.

Globotruncana subspinosa Pessagno, 1960, p. 101, 102, pl. 1, figs 1–9; Caron, 1972, p. 555, pl. 1, figs 3–4c.

Globotruncana elevata (Brozten). Majzon, 1961, pl. 7, fig. 11; Postuma, 1962,

7 figs; Douglas and Sliter, 1966, p. 110, pl. 3, figs 2a–c; Salaj and Samuel, 1966, p. 204, pl. 22, figs 3a–c; Wille-Janoschek, 1966, p. 102, 103, pl. 7, figs 1–3; Pessagno, 1967, p. 336–338, pl. 78, figs 12–14; pl. 80, figs 1–3; pl. 81, figs 9–14; Sliter, 1968, p. 102, 103, pl. 16, figs 3a–c; Douglas, 1969, p. 179, pl. 1, figs 6a–d; Esker, 1969, pl. 2, figs 16–18; Scheibnerová, 1969, p. 76, 77, tab. 16, figs 5a–c; tab. 17, figs 1a–2c; text-fig. 25; Kuhry, 1970, p. 292–299, pl. 1, figs 1–9; text-figs 4g, h; Longoria, 1970, p. 43–47, pl. 6, figs 7–9; pl. 11, fig. 7; pl. 18, fig. (?) 5; pl. 19, fig. (?) 4; Cita and Gartner, 1971, pl. 1, fig. 7; *non* Caron, 1972, p. 554, pl. 1, fig. 1 (?); text-fig. 3a.

   Globotruncana stuarti elevata (Brotzen). Pessagno, 1962, p. 362, pl. 1, fig. 10; pl. 2, figs 10, 11; Olsson, 1964, p. 169, pl. 5, figs 7a–c; Ansary and Tewfik, 1968, p. 50, pl. 6, figs 2a–c; *non* Neagu, 1970b, p. 68, pl. 18, figs 10–14.

   Globotruncana stuarti subspinosa Pessagno. Pessagno, 1962, p. 362, pl. 2, figs 7–9; Bate and Bayliss, 1969, pl. 3, figs 1a–c.

   Globotruncana cf. *andori* de Klasz. Lehmann, 1962, p. 151, pl. 10, figs 1a–c.

   Globotruncana elevata subspinosa Pessagno. van Hinte, 1963, p. 71, pl. 3, figs 3a–c; pl. 4, figs 1a–c.

   Globotruncana stuarti stuartiformis Dalbiez. Pessagno, 1962, p. 362, pl. 2, figs 4–6; Olsson, 1964, p. 170, pl. 5, figs 6a–c, 8a–c; El-Naggar, 1966, p. 136–139, pl. 9, figs 3a–d; Ansary and Tewfik, 1968, p. 50, pl. 6, figs 7a–c; *non* Rao *et al.*, 1968, p. 27, 28, pl. 2, figs 1–3; *non* Funnell *et al.*, 1969, p. 34, 35, pl. 4, figs 10–12; text-figs 18a–c; Neagu, 1970b, p. 68, 69, pl. 18, figs 4–6.

   Globotruncana gr. *elevata* (Brotzen). Christodoulou and Marangoudakis, 1966, p. 307, text-figs 1a–4c, 6a–c, 8a–10c (*non* 5a–c, 7a–c); pl. 1, figs 8a–c, 13a–c (*non* 4a–c, 9a–c).

   Description: Test large, unequally biconvex with the ventral side strongly convex and the dorsal side slightly; earlier whorls with greater spiroconvexity grading into the last whorl which is flat to concave, which results in a higher central cone; equatorial periphery slightly lobate to nearly circular, axial periphery acute on dorsal edge of chambers. Chambers flat dorsally, with ventral side a steep (60°–90°), straight to convex slope up to the umbilical shoulder, then reverses sharply into the umbilicus; slightly to strongly imbricate ventrally and elongate in direction of coiling; dorsally the shape may be entirely triangular, semicircular or a combination, large specimens often have crescentic chambers elongate in the direction of coiling, often exhibiting an undulating surface; 18–23 chambers in 3 whorls with 5–7 chambers in the last whorl. A single, wide, beaded, keel on the acute periphery which bifurcates to form the sutures. Dorsal sutures, curved to angular, raised, wide; ventral sutures slightly raised to flush in earlier chambers of final whorl to depressed in the last few, curved abaperturally to form a carina on the umbilical shoulder. Umbilicus wide, moderately deep, covered with a tegillum. Aperture a low to moderately high, interiomarginal, umbilical to extraumbilical–umbilical arch. Surface densely perforated by fine to moderate diameter pores, dorsal side smooth, ventral side lightly pustulose near umbilical shoulder.

   Remarks: The range of *Globotruncana elevata* overlaps that of only one single-keeled species which is likely to be confused with it—*G. stuarti* (de Lapparent). The latter has a more evenly arched dorsal side and biconvex test, and lacks the undulatory dorsal surface of *G. elevata*. Within *G. elevata* the undulating dorsal surface may or may not be present. The convexo–concave dorsal side varies in degree of the final concavity. Perhaps the most significant variation is in the shape of the chambers as viewed dorsally. This latter factor has been used to split this taxon into two species. The shape ranges from triangular to crescentic.

Globotruncana elevata arose from *G. rosetta* (Carsey) with loss of ventral keel.
Kuhry (1970) re-examined the type material of *Rotalia elevata*, and selected a lectotype which agrees well with the accepted concept of the species. He discussed the status of *Globotruncana stuartiformis* Dalbiez, and concluded that it is synonymous with *G. elevata*. They have been shown to intergrade (Pessagno, 1967) and their stratigraphic ranges are identical (Masters, in press). Kuhry illustrated a topotype of *G. elevata* (pl. 1, figs 7–9) which is identical with the type figure of *G. stuartiformis*.

Distribution: This is shown in Fig. 110.

Range: Campanian (FOS *Globotruncana elevata*—FOS *Globotruncana concavata* Interval)—Maastrichtian (*Trinitella scotti* Interval).

Globotruncana ellisi Brönnimann and Brown, Plate 44, figs 1–3
Rugotruncana ellisi Brönnimann and Brown, 1956, p. 547, pl. 22, figs 7–9.
Globotruncana (Rugotruncana) ellisi Brönnimann and Brown. Majzon, 1961, p. 764, pl. 5, fig. 2.
Globotruncana subcircumnodifer Gandolfi. (?) Takayanagi, 1965, p. 224, 225, pl. 28, figs 1a–4c.

Description: "The low trochospirally coiled test is covered with an incipient development of costellae. All chambers are inflated and globigerine-like, but later ones show a slight degree of compression. The last whorl is scalloped and composed of five or six chambers. Some or all later chambers exhibit a very weak double-keeled, imperforate, peripheral band. The two keels are very faint and may be missing from a few of the last chambers. The costellae, which are barely discernible, give the surface a roughened appearance. The principal aperture is rounded interiomarginal, and opens into a rather large umbilicus. Long apertural flaps extend into the umbilicus, and in later chambers form a protruded, imperforate, umbilical cover-plate with accessory apertures." (Type description.)

Remarks: *Globotruncana ellisi* differs from *G. cretacea* (d'Orbigny) by having somewhat more closely packed chambers, particularly as viewed dorsally. *Globotruncana subcircumnodifer* Gandolfi has more axially compressed, less inflated chambers. Within *G. ellisi* the rate of chamber size increase varies slightly. The costellae may or may not be present.

This species most likely developed from *Globotruncana cretacea* (d'Orbigny) by its chambers becoming more tightly packed.

A comparison between the holotype and the type figure revealed that both the keel, particularly on the first chamber of the last whorl, and the costellae are too coarsely drawn. The costellae were given too much importance by Brönnimann and Brown. Brown later reversed his position by stating that the "costellae aren't significant enough to establish a separate genus," (personal communication, 1962). It is believed that this surface ornamentation is not necessarily even significant at the species level in the *Globotruncana*.

The possibility exists that *Globotruncana ellisi* and *G. subcircumnodifer* Gandolfi will be found to be synonymous after more intense study.

Distribution: This is shown in Fig. 111.

Range: Campanian (*Globotruncana calcarata* Interval)—Maastrichtian (*Trinitella scotti* Interval).

Globotruncana esnehensis Nakkady
Globotruncana arca var. *esnehensis* Nakkady, 1950, p. 690, pl. 90, figs 23–26.

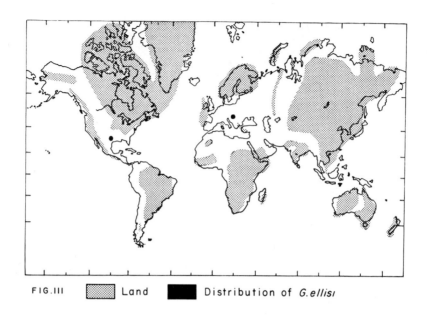

FIG.III [Land] Land [■] Distribution of *G. ellisi*

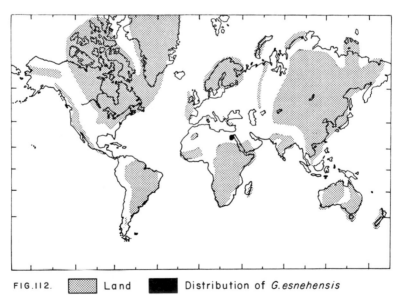

FIG.II2. [Land] Land [■] Distribution of *G. esnehensis*

Globotruncana esnehensis Nakkady. Nakkady and Osman, 1954, p. 79, 80, pl. 19, figs 3a–c; *non* Said and Kenawy, p. 150, pl. 5, figs 2a–c; El-Naggar, 1966, p. 98–100, pl. 12, figs 1a–d; text-figs 9 (13), 10 (part).

Remarks: In 1954 Nakkady and Osman raised the rank of the taxon *esnehensis* to the species level. In so doing, they incorrectly believed that they were creating a new name which should bear both of their names. Any change in the rank of a taxon does not affect the authorship (I.C.Z.N. Art. 50b). Therefore, this taxon is correctly attributed only to Nakkady, the original author.

Berggren (1962a), following Hofker (1956f), judged *Globotruncana esnehensis* to be a junior synonym of *G. intermedia* Bolli. El-Naggar (1966) took issue with this position, stating that the two species were "too remote to be related . . .". The validity of this statement without an explanation escapes the reader. However, El-Naggar did establish that *G. esnehensis* possesses a single keel throughout. Bolli (1951) reported that *G. intermedia* has a single and sometimes a double keel.

The type figures and description of *Globotruncana esnehensis* are totally inadequate for recognition. El-Naggar's (1966) figures are much improved, but the specimens he assigned to this species may not be *G. esnehensis*. All of the chambers of his specimens are highly inflated, subspherically shaped both dorsally and ventrally. This is a characteristic not shown, although it may exist, on the type figure of Nakkady's species.

One of the above interpretations is obviously incorrect. Until a careful redescription of the holotype and an examination of a topotypic suite is made, the status of *Globotruncana esnehensis* remains uncertain. The synonymies listed here will have to be reinterpreted in the light of such an examination. Because type material was not available during this study, any description and differentiation would be premature and potentially misleading.

Distribution: This is shown in Fig. 112.

Potential range: Maastrichtian (LOS *Globotruncana lapparenti*–FOS *Globotruncana aegyptiaca* Interval to *Trinitella scotti* Interval).

Globotruncana falsocalcarata Kerdany and Abdelsalam

Globotruncana falsocalcarata Kerdany and Abdelsalam, 1969, p. 261–264, pl. 1, figs 1–4; pl. 2, figs 1a–2c.

Description: "Test trochospiral, planoconvex, single keeled; up to sixteen or seventeen chambers arranged in two to two and [one] half whorls; five to six chambers in the last whorl (mostly five): early chambers inflated and globigeriniform, later with a hollow peripheral spine. This spine becomes well developed in the early chambers of the second whorl and disappears in the later chambers. It exhibits progressive shifting toward the anterior end of the chambers, which become triangular in shape. Flattening of the dorsal side increases gradually; the dorsal side of the later chambers becomes sometimes slightly concave. Rugosities developed on the early chambers on both the dorsal and ventral sides, particularly on those chambers which exhibit a peripheral spine and decrease progressively toward the adult chambers, not oriented in a meridional pattern. Sutures in the early part depressed, later becoming raised and sometimes beaded; on the dorsal side straight and oblique, ventrally often curved. Umbilicus wide, deep, usually surrounded by umbilical shoulders which are sometimes beaded; apparently covered by a spiral system of tegillae with accessory apertures. Primary aperture interiomarginal umbilical. Wall calcareous, perforate, with imperforate marginal keel." (Type description.)

Remarks: The authors suggested that *G. falsocalcarata* descended from a *Plummerita*. This is doubtful. The tubulospines of *G. falsocalcarata* are found on the angular periphery of the chamber. As in *G. calcarata* Cushman, the keel is an integral part of the tubulospine resulting in the structure being almost entirely imperforate. On the other hand, the chambers in *Plummerita* are elongate, spine-shaped structures which are thought to be perforate throughout, exclusive of the surface ornamentation. Moreover, *Plummerita* lacks a keel. The ancestor of *G. falsocalcarata* is not yet recognised.

Kerdany and Abdelsalam distinguished their new species from *Globotruncana calcarata* Cushman on the basis of the adapertural position of the tubulospines. They may also be found in this position on Cushman's species. However, in the first few tubulospinate chambers of *G. falsocalcarata*, the spines appear centrally positioned on the periphery, which is unknown in *G. calcarata*. The latter may also lack tubulospines in the last one or two chambers. The most significant difference between these two species is their ages. The fauna associated with *G. falsocalcarata* is latest Maastrichtian. If this species is not a Campanian contaminant, then it is a near homeomorph to *G. calcarata*, and a valid species.

Distribution: This is shown in Fig. 113.

Range: Maastrichtian (*Trinitella scotti* Interval).

Globotruncana falsostuarti Sigal

Globotruncana falsostuarti Sigal, 1952, p. 43, text-fig. 46; *non* Knipscheer, 1956, p. 54, pl. 4, figs 13a–c, 16a–17c; text-fig. 4; Majzon, 1961, p. 763, pl. 3, fig. 4; Postuma, 1962, 7 figs; *non* Salaj and Samuel, 1966, p. 205, pl. 23, figs 2a–c; Sändulescu, 1966, pl. VIIIa, figs 1a–c; Dupeuble, 1969, p. 156, 157, pl. 3, figs 10a–d; Funnell *et al.*, 1969, p. 30, 31, pl. 3, figs 10–12; text-figs 14a–c; Bate and Bayliss, 1969, pl. 4, figs 7a–c.

Description: Test a large, low, biconvex trochospire with the ventral side slightly more convex than the dorsal; equatorial periphery moderately lobate, axial periphery truncate to angular. Chambers flat dorsally with a steeply sloping, flat to gently convex ventral surface; petaloid dorsally, imbricate, slightly elongate in axis of coiling ventrally; 19–27 chambers in $3-3\frac{1}{2}$ whorls with 7–8 chambers in the last whorl; periphery with a double keel in earlier chambers of last whorl becoming a single keel in the last 2–7 chambers with the loss of the ventral carina; ventral and dorsal carina may adjoin at their midpoint of those chambers with a double keel; subequal to gradually increasing in size. Sutures wide, raised, curved dorsally; ventrally depressed to raised, curved forming a carina on the umbilical shoulder. Umbilicus wide, moderately deep, covered with a tegillum. Aperture a low, interiomarginal, umbilical arch. Wall finely perforate, smooth.

Remarks: Postuma (1962) as a result of conversations with Sigal considers the adjoined keels to be the most characteristic feature of *Globotruncana falsostuarti*, although this may not be exhibited in those specimens which lack the ventral carina. The large number of chambers in the test and in the final whorl is also distinctive. The main variations in this species are in the number of chambers with a single keel and with depressed ventral sutures in the last whorl. The appressed chambers cause the sutures to be depressed and the umbilical shoulder carina to appear to be discontinuous with that of the keel.

The predecessor of this species has not yet been determined. *Globotruncana arca* (Cushman) is a possible candidate. By increasing the number of chambers per whorl

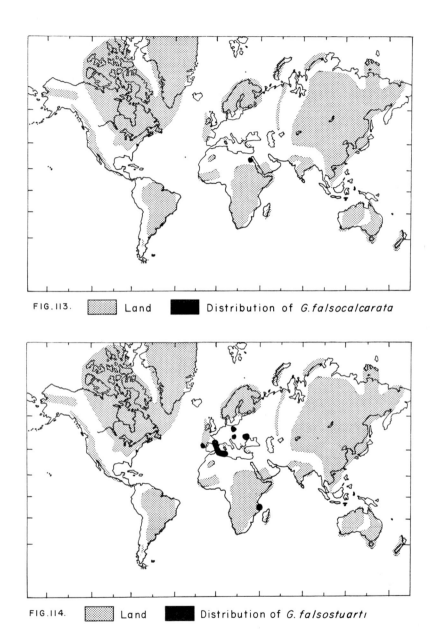

FIG.113. ▨ Land ■ Distribution of *G.falsocalcarata*

FIG.114. ▨ Land ■ Distribution of *G. falsostuarti*

and altering the shape of the keel, a *G. falsostuarti* could be produced. This species is not related to *G. stuarti* (de Lapparent).

Distribution: This is shown in Fig. 114.

Range: Maastrichtian (LOS *Globotruncana calcarata*—LOS *Globotruncana lapparenti* Interval to *Trinitella scotti* Interval).

Globotruncana fornicata Plummer, Plate 44, figs 4–6

Pulvinulina sp. Moreman, 1925, pl. 18, fig. 6.

Globotruncana fornicata Plummer, 1931, p. 198, 199, pl. 13, figs 4a–6; Sandidge, 1932d, p. 285, pl. 44, figs 12, 13; *non* Jennings, 1936, p. 37, pl. 4, fig. 13; Cushman and Hedberg, 1941, p. 99, pl. 23, figs 18a–c; Majzon, 1943, pl. 1, fig. 7; Cushman, 1944a, p. 15, pl. 3, fig. 11; Cushman, 1946b, p. 149, pl. 61, figs 19a–c; Cushman, 1948a, p. 265, pl. 26, fig. 1; Cita, 1948b, p. 153, 154, pl. 3, figs 8a–c; Drooger, 1951, p. 71, figs 9a–c; Bermúdez, 1952, p. 53, 54, pl. 7, figs 4a–c; Sellier de Civrieux, 1952, p. 281, pl. 9, fig. 4; Sigal, 1952, p. 35, text-fig. 39; Tilev, 1952, p. 75–79, text-figs 23a–d, pl. 3, fig. 7; Subbotina, 1953, p. 206–209, pl. 8, figs 3a–5c; Hagn, 1953, p. 98, pl. 8, figs 8a–c; (?) text-figs 22, 23; Frizzell, 1954, p. 129, pl. 20, figs 26a–c; Brönnimann and Brown, 1956, p. 542–544, pl. 21, figs 7, 14, 15; Sacal and Debourle, 1957, p. 61, pl. 28, fig. 6 (*non* 8, 14, 15); Książkiewicz, 1958, text-fig. 2 (8a–c); (?) Witwicka, 1958, p. 220, 221, pl. 17, figs 32a–c; Bieda, 1958, p. 61–63, text-figs 25a–c; *non* Martirosyan, 1958, p. 10, 11, pl. 2, figs 1a–c; Seiglie, 1958, p. 67–69, pl. 2, figs 3a, b (?), 4a–5b; Eternod Olvera, 1959, p. 106–108, pl. 5, figs 9–11; Lucini, 1959, text-fig. 3 (1); Klaus, 1959, p. 825, pl. 8, figs 4a–c; Takayanagi, 1960, p. 135, pl. 10, figs 3a–c; (?) Pessagno, 1960, pl. 1, fig. 6; Barr, 1961, p. 570, 571, pl. 69, figs 6a–c; Graham and Clark, 1961, p. 112, pl. 5, figs 10a–c; Majzon, 1961, p. 762, pl. 1, fig. 8; Herm, 1962, p. 78, 79, pl. 7, fig. 2; Lehmann, 1963, p. 148, pl. 7, figs 1a–4c; Pessagno, 1962, p. 362, pl. 4, figs 4, 5, 11; Postuma, 1962, 7 figs; Kaptarenko-Chernousova *et al.*, 1963, p. 105, tab. 21, figs 2a–c; Brönnimann and Rigassi, 1963, pl. 17, figs 3a–c; Graham and Church, 1963, p. 63, 64, pl. 7, figs 14a–c; Olsson, 1964, p. 164, 165, pl. 2, figs 3a–4c; pl. 3, figs 7a–8c; Cati, 1964, p. 256, 257, pl. 42, figs 3a–c; Küpper, 1964, p. 621, 622, pl. 4, figs 7a–c; van Hinte, 1965a, p. 21–23, pl. 1, figs 1a–c; pl. 2, figs 1a–2c; Lehmann, 1965, text-figs 1n, r, t; Takayanagi, 1965, p. 214, 215, pl. 24, figs 4a–c; Caron, 1966, p. 80, pl. 4, figs 6a–c; Douglas and Sliter, 1966, p. 110, 111, pl. 2, figs 2a–c (*non* 1a–c, 3a–4c); Wille-Janoschek, 1966, p. 105, 106, pl. 4, figs 1–9; Săndulescu, 1966, pl. VIa, figs 3a–c; pl. VIIIa, figs 2a–c; Pessagno, 1967, p. 338, 341, pl. 63, figs 1–9; pl. 80, figs 7–9; Bandy, 1967, p. 20, text-fig. 9 (6); Barr, 1968, p. 315, pl. 39, figs 1a–2c; Sliter, 1968, p. 103, pl. 16, figs 6a–c (*non* 5a–c, 7a–c); (?) Rasheed and Govindan, 1968, p. 82, pl. 7, figs 7–9; text-figs 11–13; Esker, 1969, pl. 3, figs 6, 7; *non* Douglas, 1969, p. 179, pl. 7, figs 6a–c; Kalantari, 1969, p. 202, pl. 25, figs 2a–c (?), 8a–c; Pessagno, 1969a, pl. 11, fig. A; Bate and Bayliss, 1969, pl. 4, figs 5a–c; Scheibnerová, 1969, p. 70, 71, tab. 18, figs 4a–c; Sturm, 1970, p. 110, 111, pl. 6, figs 4a–c; Latif, 1970, p. 33, pl. 2, figs 1–3; Neagu, 1970b, p. 68, pl. 17, figs 7–9; (*non* pl. 42, fig. 66); Rahhali, 1970, p. 65, text-figs 5j; 6a–f; Porthault, 1970, in Donze *et al.*, p. 82, 83, pl. 11, figs 14–16; Cita, 1970, pl. 4, fig. 3; Longoria, 1970, p. 47–51, pl. 1, figs 7–9; pl. 8, figs (?) 5, 6; pl. 17, figs (?) 1–3; Cita and Gartner, 1971, pl. 1. figs 4, 5; Hanzlíková, 1972, p. 106, pl. 28, figs 6a–7c; Govindan, 1972, p. 175, 176, pl. 5, figs 13–15.

Globotruncana (*Globotruncana*) *fornicata* Plummer. Papp and Küpper, 1953, p. 39, pl. 2, figs 1a–c; van Hinte, 1965b, p. 83, pl. 1, figs 6a–c.

Globotruncana caliciformis (de Lapparent). Hamilton, 1953, p. 232, pl. 29, figs 6–8; Săndulescu, 1966, pl. VIa, figs 4a–c.

Globotruncana fornicata fornicata Plummer. Gandolfi, 1955a, p. 40, 41, pl. 2, figs 2a–c; van Hinte, 1963, p. 61–63, pl. 1a–2c; El-Naggar, 1966, p. 105–108, pl. 13, figs 5a–6; Salaj and Samuel, 1966, p. 205, 206, pl. 21, fig. 6a–c; Ansary and Tewfik, 1968, p. 56, pl. 8, figs 6a–c; Ansary and Tewfik, 1969, pl. 1, figs 4a, b; Youssef and Abdel-Aziz, 1971, pl. 1, figs 5a, b.

Globotruncana contusa scutilla Gandolfi, 1955a, p. 54, pl. 4, figs 1a–c; Küpper, 1964, p. 619, 620, pl. 3, figs 4a–c; Salaj and Samuel, 1966, p. 203, pl. 25, figs 2a–c; Săndulescu, 1966, pl. VIIa, figs 1a–c; El-Naggar and Haynes, 1967, p. 10, pl. 2, figs 1a–c.

Globotruncana fornicata manaurensis Gandolfi, 1955a, p. 41, 42, pl. 2, figs 1a–c; El-Naggar, 1966, p. 109, 110, pl. 13, figs 2a–c.

Globotruncana (*Globotruncana*) *fornicata* subsp. *plummerae* Gandolfi, 1955a, p. 42, pl. 2, figs 3a–4c.

Globotruncana (*Globotruncana*) *fornicata* subsp. *cesarensis* Gandolfi, 1955a, p. 45, pl. 2, figs 10a–c.

Globotruncana (*Globotruncana*) *thalmanni thalmanni* Gandolfi, 1955a, p. 60, 61, pl. 4, figs 4a–c (*non* 5a–c).

Globotruncana fornicata motai Petri, 1962, p. 126, pl. 19, figs 2a–c.

Globotruncana (*Globotruncana*) *caliciformis* Vogler. van Hinte, 1965b, p. 83, pl. 3, figs 2a–c.

transitional *Globotruncana fornicata fornicata* Plummer to *Globotruncana fornicata globulocamerata* El-Naggar. El-Naggar, 1966, pl. 14, figs 1a–c.

transitional *Globotruncana fornicata menaurensis* Gandolfi to *Globotruncana tricarinata tricarinata* (Quereau). El-Naggar, 1966, pl. 14, figs 7a–c.

Globotruncana thalmanni thalmanni Gandolfi. *non* Salaj and Samuel, 1966, p. 216, 217, pl. 21, figs 4a–c.

Globotruncana sp. ex gr. *G. fornicata* Plummer. Moorkens, 1969, p. 452, 453, pl. 2, figs 4a–c (*non* pl. 3, figs 9a–c).

Marginotruncana sinuosa Porthault, 1970, in Donze *et al.*, p. 81, 82, pl. 11, figs 11–13.

Globotruncana fornicata Plummer *caliciformis* (de Lapparent). Latif, 1970, p. 33, pl. 2, figs 4–6.

Globotruncana fornicata ackermanni Gandolfi. (?) Youssef and Abdel-Aziz, 1971, pl. 1, figs 3a, b.

Globotruncana fornicata cesarensis Gandolfi. (?) Youssef and Abdel-Aziz, 1971, pl. 1, figs 4a, b.

Description: Test a low to moderate trochospiral coil; equatorial periphery slightly lobate, axial periphery truncate. Chambers of first whorl subspherical, later dorsal chambers flattened with an undulating surface developing in last whorl, crescentic and very elongate in direction of coiling; ventral side of chambers slightly arched to undulating, imbricate; 12–16 chambers in 3 whorls with 4–6, usually 5, chambers in the final whorl; periphery with a narrow, beaded double keel which becomes wider adaperturally on each chamber, and the keel of each successive chamber beginning on the wide portion of the previous giving the impression of the older bifurcating around the younger. Aperture a low, interiomarginal, umbilical to extraumbilical–umbilical arch. Umbilicus wide, moderately deep, covered by a tegillum. Sutures dorsally curved, wide, flush to raised; ventrally wide, raised, sharply curved back to form a carina on the umbilical shoulder. Surface finely perforate, smooth to lightly pustulose.

Remarks: *Globotruncana fornicata* is distinguished from *G. contusa* (Cushman) more or less arbitrarily on the basis of the spire height and deep infolding of the dorsal chambers. It differs from all other members of *Globotruncana* by its elongate, crescentic, gently folded dorsal chambers. The height of the spire varies considerably from rather low to moderately high, grading into the high spire characteristic of *Globotruncana contusa* (Cushman). The undulating dorsal surface is present in varying degrees of intensity. Masters (in press) observed forms with keel widths double those most commonly found.

Globotruncana fornicata evolved from *G. renzi* Gandolfi by developing more elongate, crescent-shaped chambers which are slightly folded.

The holotypes of *Globotruncana fornicata* subsp. *plummerae* and *G. fornicata* subsp. *cesarensis* of Gandolfi (1955a) possess undulating dorsal chambers, and are considered as immature specimens of *G. fornicata* s. str. *Globotruncana thalmanni* Gandolfi, 1955, does have a reduced ventral keel through the last whorl, and is conspecific with Plummer's species.

Porthault (in Donze *et al.*, 1970) referred forms with an extraumbilical–umbilical apertural position to the new species *Marginotruncana sinuosa*. These forms were stated to be otherwise identical to *Globotruncana fornicata*. This exemplifies the danger of strict reliance upon apertural position for taxonomic distinction because topotypes of *G. fornicata* exhibit apertures which range from umbilical to extraumbilical in position. Porthault's species is a junior synonym of *G. fornicata*.

Distribution: This is shown in Fig. 115.

Range: Coniacian—Maastrichtian (FOS *Rugoglobigerina hexacamerata*—LOS *Globotruncana fornicata* Interval).

Globotruncana gansseri Bolli, Plate 45, figs 1–3

Globotruncana gansseri Bolli, 1951, p. 196, 197, pl. 35, figs 1–3; Hamilton, 1953, p. 232, pl. 29, figs 18–20; Nakkady and Osman, 1954, p. 80, pl. 20, figs 24a–c; Said and Kenawy, 1956, p. 150, pl. 5, figs 17a–c; Ansary and Fakhr, 1958, p. 135, 136, pl. 2, figs 17a–c; Eternod Olvera, 1959, p. 108–110, pl. 8, figs 4–6; Said and Kerdany, 1961, p. 331, pl. 2, figs 16a–c; Majzon, 1961, p. 763, pl. 4, fig. 1; Herm, 1962, p. 80, pl. 8, fig. 4; Skinner, 1962, p. 44, pl. 6, figs 21, 22; Postuma, 1962, 7 figs; Ansary *et al.*, 1962, pl. 4 (part); van Hinte, 1963, p. 72, 73, pl. 3, figs 4a–c; Olsson, 1964, p. 165, 166, pl. 3, figs 2a–5c; *non* Borsetti, 1964, p. 675, 676, pl. 79, figs 13, 14; text-figs 10a, b; Pessagno, 1967, p. 341–343, pl. 92, figs 13–18; Bandy, 1967, p. 21, text-fig. 9 (16); Rao *et al.*, 1968, p. 22, 23, pl. 3, figs 4–6; Koch, 1968, p. 653, pl. 59, figs 6–76; Funnell *et al.*, 1969, p. 31, 32, pl. 4, figs 1–3; text-figs 15a–c; Hanzlíková, 1969, p. 48, pl. 11, figs 1a–c; Longoria, 1970, p. 54–57, pl. 5, figs 4–6; pl. 10, figs 6, 7; pl. 15, fig. (?) 4; Al-Shaibani, 1971, p. 106, 107, pl. 1, figs 3a–c; Berggren, 1971, pl. 1, figs 15–17; El-Naggar, 1971a, pl. 5, figs d–f; (?) Hanzlíková, 1972, p. 107, pl. 28, figs 8a–c; Govindan, 1972, p. 176, pl. 4, figs 13–15; pl. 5, figs 20–22.

Globotruncana lugeoni Tilev, 1952, p. 41–46, text-figs 10a–12e, pl. 1, figs 5, 6; Edgell, 1957, p. 113, 114, pl. 2, figs 7–9; Majzon, 1963, p. 763, pl. 3, fig. 9; El-Naggar, 1966, p. 122, 123, pl. 6, figs 1a–d.

Globotruncana lugeoni var. *angulata* Tilev, 1952, p. 46–50, text-figs 13a–d, pl. 3, figs 1a–d.

Globotruncana gabeliatensis Nakkady and Osman, 1954, p. 83, text-figs a–c.

Globotruncana gansseri gansseri (Bolli). Gandolfi, 1955a, p. 69, 70, pl. 6, figs 5a–6c, 8a–c, text-figs 11, 12; El-Naggar, 1966, p. 117–119, pl. 5, figs 1a–d; *non* Salaj

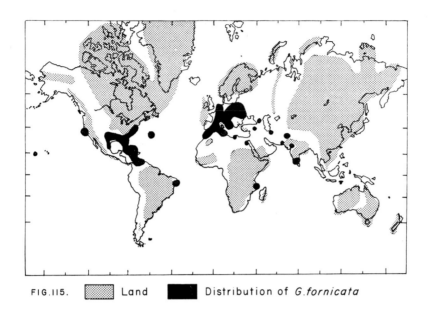

FIG.115. ▨ Land ▮ Distribution of *G.fornicata*

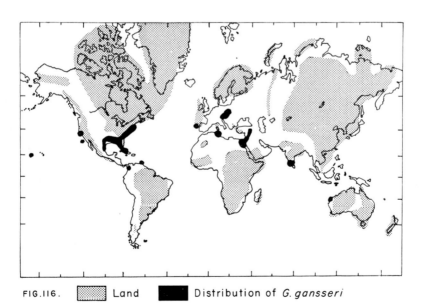

FIG.116. ▨ Land ▮ Distribution of *G. gansseri*

and Samuel, 1966, p. 206, 207, pl. 22, figs 1a–2c; Ansary and Tewfik, 1968, p. 51, pl. 6, figs 3a–c; Ansary and Tewfik, 1969, pl. 1, figs 3a–b; Youssef and Abdel-Aziz, 1971, pl. 1, figs 7a–c.

Globotruncana gansseri subgansseri Gandolfi, 1955a, p. 70, pl. 6, figs 7a–c; El-Naggar, 1966, p. 119, 120, pl. 5, figs 3a–d.

Globotruncana rosetta pettersi Gandolfi, 1955a, p. 68, pl. 6, figs 3a–4c, text-figs 11–1; *non* Salaj and Samuel, 1966, p. 212, pl. 21, figs 5a–c.

Rugotruncana gansseri (Bolli). Brönnimann and Brown, 1955, p. 549, 550, pl. 23, figs 7–9; Seiglie, 1958, p. 71, 72, pl. 4, figs 2a–c.

Globotruncana (*Rugoglobigerina*) *rotundata* subsp. *subrotundata* Gandolfi, 1955a, p. 70, 71, pl. 7, figs 1a–c.

Bucherina sandidgei Brönnimann and Brown, 1956, p. 557, 558, pl. 21, fig. 19; pl. 22, figs 19–21; text-fig. 24.

Globotruncana monmouthensis Olsson, 1960, p. 50, 51, pl. 10, figs 22–24.

Globotruncana (*Bucherina*) *sandidgei* Brönnimann and Brown. Majzon, 1961, p. 764, pl. 5, fig. 7.

Globotruncana gansseri dicarinata (Pessagno). Olsson, 1964, p. 166, pl. 3, figs 1a–c.

Globotruncana arabica El-Naggar, 1966, p. 81–83, pl. 6, figs 3a–d.

Globotruncana gansseri gandolfi El-Naggar, 1966, p. 115, 116, pl. 5, figs 2a–d.

Rugoglobigerina rotundata subrotundata (Gandolfi). *non* Salaj and Samuel, 1966, p. 198, pl. 9, figs 5a–c.

Globotruncana youssefi El-Naggar, 1966, p. 144, 145, pl. 6, figs 4a–d.

Rugoglobigerina rotundata Brönnimann. Funnell *et al.*, 1969, p. 37, pl. 5, figs 7–9; text-figs 21a–c.

Rugoglobigerina scotti (Brönnimann). Funnell *et al.*, 1969, p. 37, 38, pl. 5, figs 10–12; text-figs 22a–c.

Globotruncana helvetica Bolli. Weaver *et al.*, 1969, pl. 4, figs 4a–c.

Description: Test planoconvex with the dorsal side flat and the ventral side very convex; equatorial periphery slightly to moderately lobate, axial periphery dorsally acute forming approximately a 90° angle. Chambers strongly hemispherical; flat to faintly convex on the dorsal side with the ventral wall perpendicular to the dorsal, evenly curved to form a high dome; gradually increasing in size; 12–15 chambers in $2\frac{1}{2}$–3 whorls of 5 chambers each; with a single, beaded keel on the acute dorsal periphery. Aperture a high, interiomarginal, umbilical to extraumbilical–umbilical arch. Umbilicus wide, deep, covered by a tegillum; shoulder typically not carinate. Sutures radial, depressed on ventral side, depressed to slightly raised, curved dorsally. Surface finely perforate, densely pustulose ventrally, smooth dorsally.

Remarks: This single-keeled, hemispherical chambered species is readily distinguished from other species of this age. It is nearly homeomorphic with the much older *Globotruncana helvetica* Bolli. An occasional specimen with 4 or 6 chambers in the final whorl is found, otherwise there is little variation within this species.

The progenitor of *Globotruncana gansseri* is not known.

Contrary to Nakkady and Osman (1954), their new species *Globotruncana gabeliatensis* is not distinguishable from juvenile specimens of *G. gansseri*.

The holotype of *Globotruncana rotundata* subsp. *subrotundata* Gandolfi, 1955, is a gerontic specimen of *G. gansseri*. It possesses a single keel throughout and has a reduced ultimate chamber.

Globotruncana arabica and *G. youssefi* of El-Naggar (1966) are but slight intra-specific variants of *G. gansseri*, and occupy the same stratigraphic range.

Distribution: This is shown in Fig. 116.

Range: Maastrichtian (LOS *Globotruncana lapparenti*—FOS *Globotruncana aegyptiaca* Interval to *Trinitella scotti* Interval).

Globotruncana havanensis Voorwijk, Plate 45, figs 4–6
Globotruncana havanensis Voorwijk, 1937, p. 195, pl. 1, figs 25, 26, 29; Cormin-boeuf, 1961, p. 112, 113, pl. 1, figs 2a–c; Majzon, 1961, p. 762, pl. 7, fig. 8; Loeblich and Tappan, 1964, p. C662, 663, figs 3a–c; (?) Salaj and Samuel, 1966, p. 207, pl. 24, figs 1a–2c; *non* Douglas and Sliter, 1966, p. 111, pl. 1, figs 9a–10c; Pessagno, 1967, p. 373, pl. 84, figs 1–3; *non* Sliter, 1968, p. 103, 104, pl. 17, figs 1a–2c; Dupeuble, 1969, p. 154, pl. 3, figs 8a–c; Funnell *et al.*, 1969, p. 32, 33, pl. 4, figs 4–6; text-figs 16a–c; Longoria, 1970, p. 57, 58, pl. 3, figs 7–10; pl. 10, figs 4, 5; pl. 14, figs 1–5; Neagu, 1970b, p. 69, pl. 18, figs 18–25; pl. 19, figs 1–5; Hanzlíková, 1972, p. 105, 106, pl. 29, figs 2a–5.
Globorotalia nov. sp. van Wessem, 1943, p. 48, pl. 2, figs 3, 4.
Globorotalia pschadae Keller, 1946, p. 99, pl. 2, figs 4–6; Subbotina, 1953, p. 235, 236, pl. 16, figs 1a–6c.
Globotruncana citae Bolli, 1951, p. 197, pl. 35, figs 4–6; Edgell, 1957, p. 111, pl. 1, figs 13–15; *non* Ansary and Fakhr, 1958, p. 134, 135, pl. 2, figs 15a–c; Hofker, 1960e, text-figs 20a–c; Majzon, 1961, p. 763, pl. 4, fig. 3; Săndulescu, 1966, pl. IXa, figs 2a–c.
Globotruncana (Globotruncana) citae Bolli. Papp and Küpper, 1953, p. 38, pl. 1, figs 4a–c.
Rugotruncana havanensis (Voorwijk). Brönnimann and Brown, 1955, p. 552, pl. 22, figs 4–6.
Globotruncana (Rugoglobigerina) petaloidea subsp. *subpetaloidea* Gandolfi, 1955a, p. 52, 53, pl. 3, figs 12a–c; text-figs 8a–c.
Globotruncana (Rugoglobigerina) petaloidea petaloidea Gandolfi, 1955a, p. 52, pl. 3, figs 13a–c.
Globorotalia membranacea (Ehrenberg). Buckowy and Geroch, 1957, pl. 28, figs 11a–c; Maslakova, 1959, p. 111, 112, pl. 12, figs 2a–c.
Praeglobotruncana coarctata Bolli, 1957, p. 55, pl. 12, figs 2a–3c.
Globotruncana inornata Bolli, 1957, p. 57, 58, pl. 13, figs 5a–6c; Majzon, 1961, p. 764, pl. 6, fig. 3; Youssef and Abdel-Aziz, 1971, pl. 1, figs 8a, b.
Globotruncanella havanensis (Voorwijk). Seiglie, 1958, p. 73, 74, pl. 4, figs 3a–5c; Eternod Olvera, 1959, p. 94–96, pl. 4, figs 12–14; Thalmann and Ayala-Castañares, 1959, pl. 2, figs 5 (part), 6, 7, 10, 11; van Hinte, 1963, p. 94–96, pl. 10, figs 3a–c; pl. 11, figs 4a–5c; pl. 12, figs 1a–c; Brönnimann and Rigassi, 1963, pl. 17, figs 2a–c; *non* Douglas, 1969, p. 190, 192, pl. 10, figs 3a–c; Berggren, 1971, pl. 3, figs 5–8.
Rugoglobigerina jerseyensis Olsson, 1960, p. 49, pl. 10, figs 19–21.
Globotruncana (Praeglobotruncana) coarctata Bolli. Majzon, 1961, p. 764, pl. 6, fig. 7.
Praeglobotruncana (Praeglobotruncana) havanensis (Voorwijk). Berggren, 1962a, p. 25–30, pl. 7, figs 1a–c.
Praeglobotruncana citae (Bolli). *non* Borsetti, 1962, p. 31, pl. 6, fig. 8; text-figs 179, 180; Postuma, 1962, 7 figs; Cita and Gartner, 1971, pl. 4, figs 3a–c.
Globotruncana petaloidea Gandolfi. Douglas and Sliter, 1966, p. 113, 114, pl. 1, figs 11a–c.
Globotruncanella citae (Bolli). Maslakova, 1963b, pl. 7, fig. 2.
Globotruncana (Marginotruncana) intermedia Bolli. Hofker, 1963, p. 282, pl. 2, figs 6a–d.

Globotruncana (*Marginotruncana*) *citae* Bolli. Hofker, 1963, p. 282, pl. 2, figs 7a–d.

Praeglobotruncana petaloidea (Gandolfi). Olsson, 1964, p. 162, pl. 1, figs 7a–c (*non* 6a–c; pl. 2, figs 1a–c); Ansary and Tewfik, 1968, p. 48, pl. 5, figs 7a, b.

Globotruncana (*G.*) sp. cf. *G.* (*G.*) *havanensis* (Voorwijk). Coleman, 1966, p. 445, pl. 8, figs 12–14.

Rugoglobigerina petaloidea subpetaloidea (Gandolfi). Salaj and Samuel, 1966, p. 196, 197, pl. 10, figs 2a–c.

Globotruncana cf. *G. inornata* Bolli. Banerji, 1966, p. 62, 63, pl. 3, figs 7–9.

Praeglobotruncana inornata (Bolli). *non* Salaj and Samuel, 1966, p. 190, pl. 15, figs 5a–c; Bandy, 1967, p. 17, text-fig. 8 (3); Eicher, 1969, text-fig. 3 (part); Scheibnerová, 1969, p. 57, 58, tab. 7, figs 8a–13c; text-fig. 10; Eicher and Worstell, 1970b, p. 310, pl. 11, figs 1a–c.

Loeblichella coarctata (Bolli). Pessagno, 1967, p. 288, 289, pl. 62, figs 1–3, 6–8 (*non* pl. 48, figs 14, 16, 20; pl. 61, figs 4, 5; pl. 76, figs 7–9); Bertels, 1970, p. 35, 36, pl. 3, figs 3 (?), 4a–c.

Praeglobotruncana havanensis petaloidea (Gandolfi). Bandy, 1967, p. 18, text-fig. 10 (12).

Praeglobotruncana havanensis havanensis (Voorwijk). Bandy, 1967, p. 18, text-fig. 10 (11).

Praeglobotruncana havanensis subpetaloidea (Gandolfi). Bandy, 1967, p. 18, 19, text-fig. 10 (13).

Praeglobotruncana havanensis (Voorwijk). Barr, 1968a, p. 314, pl. 37, figs 1a–2c; Hanzlíková, 1969, p. 46, pl. 9, figs 1a–3c; Neagu, 1970a, fig. 1 (22).

Globotruncanella petaloidea (Gandolfi). Douglas, 1969, p. 192–194, pl. 7, figs 2a–c.

Rugoglobigerina petaloidea (Gandolfi). Scheibnerová, 1969, p. 79, text-fig. 26.

Rugoglobigerina petaloidea petaloidea (Gandolfi). Al-Shaibani, 1971, p. 107, pl. 1, figs 4a–c.

Description: Test small, biconvex with the dorsal side occasionally somewhat more convex; equatorial periphery distinctly lobate, axial periphery rounded on earlier chambers becoming acute in the last whorl. Chambers of earlier whorls spherical to subspherical; in final whorl, inflated, oval in side view with the long axis in the direction of coiling; dorso-ventrally compressed final whorl with a narrow to relatively wide imperforate peripheral band over which may be developed, to a greater or lesser extent, a single, discontinuous carina composed of short spines or of pustules; 12–14 chambers in $2\frac{1}{2}$–3 whorls of 4–5 chambers each; rapidly increasing in size. Aperture a low to moderate, interiomarginal extraumbilical–umbilical arch. Umbilicus relatively wide, shallow, covered by a tegillum. Sutures depressed, radial on ventral side, radial to curved dorsally. Surface finely perforate, with fine spines and pustules.

Remarks: The single-keeled, axially compressed but inflated chambers of this species distinguishes it from all other of its congeners. The spire height ranges from low to moderate, rarely higher. The degree of equatorial lobation may also vary enough that the extremes, without the intermediate forms, might be considered as distinct species. The imperforate peripheral band exists throughout, but may be partially or completely covered in the final whorl by a single carina.

There are two potential candidates for the ancestor of *Globotruncana havanensis*. The first, *Praeglobotruncana delrioensis* (Plummer), early in its lineage is very similar to *G. havanensis*, with the exception of the tegillum on the latter. However, during its later history *P. delrioensis* has the tendency to develop more coarsely ornamented and appressed chambers with a coarser keel.

Globigerina delrioensis (Carsey), the second candidate, is equally similar in side view to *Globotruncana havanensis*. It has been documented that *G. delrioensis* gave rise to the *Globotruncana cretacea* (d'Orbigny) lineage through the development of an imperforate peripheral band and a tegillum. It is just as likely that *G. havanensis* developed with the addition of axial compression.

Although intermediate forms between *Globotruncana havanensis* and either of the two above-mentioned species have not yet been recorded. The author believes *Globigerina delrioensis* is the probable ancestor.

Berggren (1962a) was the first to argue for the placement of *Globotruncana havanensis* in the genus *Praeglobotruncana* on the bases that (1) it lacks a tegillum; (2) the primary aperture is umbilical–extraumbilical; (3) it lacks raised, beaded or pustulose ventral sutures; and (4) it has a thinner shell structure and finer pores than do the *Globotruncana*.

Although not all specimens of this species exhibit a tegillum, many do (e.g. Douglas, 1969). The position of the aperture in unquestioned *Globotruncana* has been observed in this study to vary from umbilical to extraumbilical–umbilical. This is true of other genera as well (see Remarks under *Globigerina*), and is dependent in part upon such factors as chamber shape and size and the degree of spiroconvexity of the test.

To suggest that the *Globotruncana* must have raised, beaded or pustulose ventral sutures, would be to displace many bona fide members of this genus; e.g. *G. schneegansi* Sigal, *G. cretacea* (d'Orbigny), *G. gansseri* Bolli, *G. linneiana* (d'Orbigny), *G. marginata* (Reuss), *G. aegyptiaca* Nakkady, *G. ventricosa* White, plus others. Even in those species having this suture type, it is most typically developed on progressively older portions of the test. This is directly attributed to and accentuated by the successive addition of laminae. The thickness of the wall and the pore density have been shown by studies of living foraminifers to be strongly influenced by the environment, and cannot then be regarded as generic characters.

In 1946 Keller introduced the species *Globorotalia pschadae* from Senonian strata of Sochi, U.S.S.R. Although somewhat larger and more dorsally convex than the typical *Globotruncana havanensis*, it is believed to be within the limits of variation.

It was determined through the examination of the holotype that *Globotruncana citae* Bolli, 1951, is identical to the earlier *G. havanensis*.

The peripheral view of the holotype of *Globotruncana petaloidea petaloidea* Gandolfi is poorly drawn. The penultimate chamber was not sketched in. The specimen is a calcite cast, but there is a suggestion of a single keel. All views of the holotype of *G. petaloidea subpetaloidea* Gandolfi are completely misleading. All chambers of the holotype expand at a uniform rate; are more inflated, particularly the ultimate and penultimate ones; and have a single keel throughout. Both of these subspecies are conspecific with *G. havanensis*.

The holotypes of Bolli's (1957) *Globotruncana inornata* and *Praeglobotruncana coarctata* were examined. The keel of *G. inornata* extends throughout and is pustulose on the earlier chambers of the last whorl which are more inflated. His second species also is keeled throughout, but its chambers are more compressed dorso-ventrally. Both species appear to fall within the range of variation observed for *G. havanensis*.

Berggren (1962a) placed *Rugoglobigerina jerseyensis* Olsson into synonymy with *Globotruncana petaloidea* Gandolfi, and transferred them to *Praeglogotruncana*

(*Hedbergella*). Olsson did not mention a keel in his description or comparison of
G. citae (= *G. havanensis*). Berggren remarked that his Scandinavian specimens,
like *R. jerseyensis*, have no trace of a keel, but a perforate pseudocarina. However,
there is an imperforate peripheral band in all of the chambers of the final whorl of the
holotype of *R. jerseyensis*, and a pustulose carina on all but the ultimate chamber.
Olsson's species is synonymous with *G. havanensis*.

Distribution: This is shown in Fig. 117.

Range: Turonian—Maastrichtian (*Trinitella scotti* Interval).

Globotruncana helvetica Bolli, Plate 46, figs 1, 2

Globigerina Hecht, 1938, pl. 23, figs 57a–59b, 63.

Globotruncana helvetica Bolli, 1945, p. 226, 227, pl. 9, figs 6–8; text-figs 1 (9–12);
Colom, 1947, p. 98, 99, pl. 24, figs 1, 2, 4; Colom, 1948, p. 27, pl. 4, figs 1, 3, 5;
Cita, 1948b, p. 154, 155, pl. 4, figs 1a–c; Sigal, 1952, p. 31, text-fig. 32; Hagn and
Zeil, 1954, p. 30, 31, pl. 3, figs 1a–c; pl. 5, figs 5, 6; Ayala-Castañares, 1954, p. 397,
pl. 6, figs 3a–c; Schijfsma, 1955, text-fig. 2 (part); Alexandrowicz, 1956, p. 53, 54,
text-figs 5A, B; Książkiewicz, 1956, p. 270, 271, pl. 30, fig. 17; text-figs 38 (1–7);
Bolli, 1957, p. 56, pl. 13, figs 1a–c; Książkiewicz, 1958, text-fig. 1 (13a–c); Trujillo,
1960, p. 341, 342, pl. 50, figs 2a–c; Majzon, 1961, p. 763, pl. 3, fig 5; Borsetti, 1962,
p. 53, 54, pl. 4, figs 5 (?), 6; text-figs 79, 80, 82, 83 (?), 84, 85 (*non* 81); Samuel,
1962, p. 189, 191, pl. 10, figs 2a–3c; Sigal and Dardenne, 1962, p. 221, pl. 13, fig. 5;
Postuma, 1962, 7 figs; Graham, 1962, p. 103, pl. 19, figs 8a–c; Lehmann, 1963, p. 143,
pl. 3, figs 1a–c; text-figs, 2h, 3g, h; Küpper, 1964, p. 623, 624, pl. 2, figs 9a–c; (?)
Saavedra, 1965, p. 325, 326, text-fig. 13; Broglio Lorigo and Mantovani, 1965,
pl. 111, fig. 21 (*non* 18–20); Lehmann, 1965, text–figs 1g, h; Marianos and Zingula,
1966, p. 340, pl. 39, figs 2a–c; *non* Weaver *et al.*, 1969, pl. 4, figs 4a–c; Kalantari,
1969, p. 203, pl. 23, figs 3a–4c; Magné *et al.*, 1969, pl. 8, fig. 2 (part); Bate and Bayliss,
1969, pl. 3, figs 4a–c; Broglio Lorigo and Mantovani, 1970, p. 229, pl. 6, figs 12 (?),
13; chart (63).

Globotruncana helvetica Bolli var. Schijfsma, 1955, text-fig. 2 (part).

Praeglobotruncana helvetica (Bolli). Scheibnerová, 1958, p. 189, pl. 4, figs 1–3,
5, 8 (*non* 4, 6 ?); text-figs 1a–2c; Săndulescu, 1966, pl. IVa, figs 4a–c; *non* Douglas
and Sliter, 1966, p. 105, 106, pl. 5, figs 1a–c; Salaj and Samuel, 1966, p. 189, figs
1a–c; tab. 35 (9); Burckle *et al.*, 1967, text-fig. 2 (2a–c); Bandy, 1967, p. 17, text-fig
8 (2); Radoičić, 1968, pl. 1, figs 3, 7, 9; pl. 2, figs 1, 2; Scheibnerová, 1969, p. 58, 59,
tab. 8, figs 2a–c; Moorkens, 1969, p. 447, 448, pl. 1, figs 3a–c; pl. 2, figs 6a–c; pl. 3, fig.
11a–c; Sturm, 1969, pl. 11, figs 1a–c; Neagu, 1970a, fig. 1 (20); Belford and Scheibnerová, 1971, p. 334, pl. 2, figs 9–12; Gawor-Biedowa, 1972, p. 73, 74, pl. 8, figs 4a–c.

Globotruncana cf. *helvetica* Bolli. Dufaure, 1959, pl. 3, figs 28 (part), 29 (?).

Rugoglobigerina praehelvetica Trujillo, 1960, p. 340, 341, pl. 49, figs 6a–c; Graham,
1962, p. 108, pl. 20, figs 31a–c.

Globotruncana carpathica Scheibnerová, 1963, p. 140, 141, text-figs 2a–c; Scheibnerová, 1969, p. 68, tab. 12, figs 4a–c.

Globotruncana helvetica posthelvetica Hanzlíková, 1963, p. 325–327, pl. 1, figs
1a–4c.

Praeglobotruncana (?) *helvetica* (Bolli). (?) Caron, 1966, p. 74, pl. 3, figs 2a–c.

Praeglobotruncana praehelvetica (Trujillo). *non* Marianos and Zingula, 1966, p.
338, pl. 38, figs 9a–c.

Praeglobotruncana (?) *praehelvetica* (Trujillo). *non* Caron, 1966, p. 74, pl. 3,
figs 3a–c.

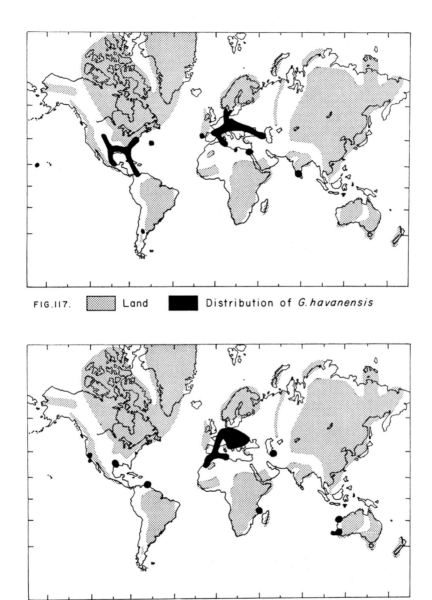

FIG.117. ▨ Land ■ Distribution of *G.havanensis*

FIG.118. ▨ Land ■ Distribution of *G.helvetica*

Hedbergella (*Hedbergella*) spectrum *helveticum* (Bolli) mut. prae *Helvetoglobotruncana* Sigal, 1966b, p. 28–30, pl. 6, figs 7a–8b.

Hedbergella (*Helvetoglobotruncana*) spectrum *helveticum* (Bolli). Sigal, 1966b, p. 28–30, pl. 6, figs 9a–10b.

Hedbergella (*Catapsydrax*) spectrum *helveticum* (Bolli). Sigal, 1966b, p. 28–30, pl. 6, figs 11a–12b, pl. 7, figs 1a–2b.

Hedbergella (? *Trinitella*) spectrum *helveticum* (Bolli). Sigal, 1966b, p. 28–30, pl. 7, figs 3a, b, 5a, b.

Globotruncana cachensis Douglas, 1966, in Douglas and Sliter, p. 108, pl. 5, figs 5a–c; Douglas, 1970, p. 25, pl. 4, figs 1a–4c.

Rugoglobigerina aff. *praehelvetica* Trujillo. Salaj and Samuel, 1966, p. 197, pl. 10, figs 1a–c.

Marginotruncana helvetica (Bolli). Pessagno, 1967, p. 306, pl. 53, figs 9–13; pl. 54, figs 1–3; pl. 99, fig. 4; pl. 100, fig. 4.

Description: Test plano-convex with the dorsal side nearly flat to slightly concave and the ventral side being strongly convex; equatorial periphery lobate, axial periphery abruptly angled on dorsal edge. Chambers hemispherical; on the dorsal side, chambers of the first whorl are subspherical, slightly raised above the later flat chambers, last few occasionally sloping toward earlier whorls; ventrally the chamber walls form a 90°–110° angle with the dorsal side, highly arched with the vertex at midpoint of chamber; 16–19 chambers in 3 whorls with 5–8 chambers in the final whorl; gradually expanding to subequal; with a single, beaded keel on the dorsal edge. Aperture a low, interiomarginal, umbilical arch. Umbilicus wide, moderately deep, covered by a tegillum. Sutures on dorsal side raised, curved; ventrally depressed, radial. Surface pustulose, particularly on ventral side; finely perforate.

Remarks: No other species of *Globotruncana* has a single, dorsally positioned keel on hemispherical chambers except *G. gansseri* Bolli, which is stratigraphically much younger. The number of chambers/whorl has a fairly wide range. Occasionally the last 1 or 2 chambers will be positioned more ventrally out of the axis of coiling.

The phylogeny of this species is unknown.

Bolli (1945) described *Globotruncana helvetica* in thin-section from strata of eastern Switzerland. It is one of the few species which can be identified in this manner.

According to Scheibnerová (1963) her *Globotruncana carpathica* differs from *G. helvetica* in three ways: (1) it has 6–8 chambers in the last whorl, (2) it has a younger stratigraphic range and (3) *G. helvetica* is considered a true *Praeglobotruncana*. The laws of physics dictate that as the diameter of the spire increases, chamber size must also increase if the same number per whorl are retained and the plane of spire is flat or nearly so. *Globotruncana helvetica* and *G. carpathica* have a nearly flat trochospire, and the chambers of the last whorl of each are subequal. Because foraminifers continue to grow until reproduction occurs, a delay would simply result in an increase in test size. Given these limitations, the end result would be more chambers per whorl. The total range of *G. helvetica* encompasses that of *G. carpathica*. The mere presence of a single keel does not predicate placement of the taxon *helvetica* into *Praeglobotruncana*. It possesses a tegillum, a characteristic of *Globotruncana* not *Praeglobotruncana*. Scheibnerová's species is a synonym of *G. helvetica*.

Hanzlíková (1963) proposed the name *Globotruncana helvetica posthelvetica* for forms said to be phylogenetically younger than *G. helvetica* Bolli by possessing such characteristics as (1) sporadically occurring macrospheric individuals, (2) the predominance of evolutely coiled tests and (3) increase in size. It is not clear what bearing the sporadic occurrence of macrospheric individuals has on phylogeny. However, the slight size difference can easily be attributed to environmental influences.

She noted the presence of evolutely coiled forms. An accepted definition of the word "evolute" is: "tending to uncoil; chambers nonembracing" (Loeblich and Tappan, 1964). This definition is not applicable to the holotype nor three paratypes of Scheibnerová's species as illustrated. The holotype does have a somewhat aberrant growth form with the last chamber being out of the normal axis of coiling. This growth habit often occurs in planktonic foraminifers, and should not be regarded as unique here. The two are synonymous.

The holotype of *Globotruncana cachensis* Douglas, 1966, was examined and determined to be a junior synonym of *G. helvetica*. There is no ventral keel as shown in the type figure. Douglas noted that juvenile forms were similar to *G. helvetica*. The supposed distinguishing spiroconvexity of *G. cachensis* is well within the range of variability of Bolli's species. The last two chambers of *G. cachensis* lack a keel, but this characteristic is common in gerontic specimens of *G. helvetica* and other species of *Globotruncana*.

Douglas (1970) determined that *Rugoglobigerina praehelvetica* Trujillo and *Globotruncana helvetica* are ontogenetic end members of a single species. The holotype and paratypes of *R. praehelvetica* were examined during this study. On the holotype, the last three chambers are more inflated dorsally and slope toward the previous whorl imparting an overall concave appearance to the dorsal side of the test. At the break-in-slope, which is situated dorsally on each chamber of the final whorl, is located an imperforate ridge. It is possible that this ridge may be a secondary surficial deposit rather than a true keel, and only thin-sections would clarify this. The umbilical chamber extension shown in the type figure is exaggerated, although a tegillar extension, like those illustrated by Postuma (1962), is present, and traces of these can be seen on earlier chambers. The author concurs with Douglas that these two species belong to the same ontogenetic series, with Trujillo's species as the junior synonym.

Distribution: This is shown in Fig. 118.

Range: Turonian—lower Coniacian.

Globotruncana imbricata Mornod

Globotruncana imbricata Mornod, 1950, p. 589, 590, pl. 15, figs (?) 21–34; text-figs 5 (IIa–IIId); ? Hagn and Zeil, 1954, p. 34, 35, pl. 2, figs 6a–c; pl. 5, figs 9, 10; Ayala-Castañares, 1954, p. 398, 399, pl. 7, figs 1a–c; (?) Książkiewicz, 1958, text-fig. 1 (16a–c); Majzon, 1961, p. 763, pl. 2, fig. 2; (?) Samuel, 1962, p. 191, pl. 11, figs 1a–c; Postuma, 1962, 7 figs; Cita-Sironi, 1963, text-fig. 10b; Hanzlíková, 1963, pl. 2, figs 3a–c; Takayanagi, 1965, p. 215, 216, pl. 24, figs 5a–c; pl. 25, figs 1a–2c; *non* Douglas and Sliter, 1966, p. 111, 112, pl. 4, figs 2a–c; Douglas, 1969, p. 180, 181, pl. 2, figs 4a–c, 7a–c (*non* 5a–6c).

Globotruncana indica Jacob and Sastry, 1950, p. 267, text-figs 2a–c.

Globotruncana (Globotruncana) imbricata Mornod. Gandolfi, 1957, p. 64, pl. 9, figs 5a–c.

Rotundina imbricata (Mornod). Maslakova, 1959, p. 112, pl. 11, figs 2a–c.

Praeglobotruncana imbricata (Mornod). Caron, 1966, p. 76, pl. 6, figs 4a–c; Săndulescu, 1966, pl. IVa, figs 1a–c; *non* Salaj and Samuel, 1966, p. 190, pl. 16, figs 2a–c; tab. 35 (6); Scheibnerová, 1969, p. 59, 60, tab. 8, figs 4a–c; tab. 9, figs 1a–5c, 7a–c (*non* 6a–c); Gawor Biedowa, 1972, p. 74, 75, pl. 8, figs 2a–3c; Hanzlíková, 1972, p. 102, pl. 26, figs 7a–10c.

Hedbergella sp. 1 Douglas and Sliter, 1966, p. 105, pl. 5, figs 10a–c.

Globotruncana sp. C Marianos and Zingula, 1966, p. 341, pl. 39, figs 4a–c.
Marginotruncana imbricata (Mornod). *non* Pessagno, 1967, p. 306, 307, pl. 57, figs 3–5.
Marginotruncana indica (Jacob and Sastry). *non* Pessagno, 1967, p. 307, pl. 55, figs 3, 8–10; pl. 57, figs 6–9; pl. 98, fig. 2.
Praeglobotruncana loeblichae Douglas, 1969, p. 170, 171, pl. 5, figs 6a–7c; Douglas, 1970, p. 20, 21, pl. 3, figs 3a–4c.
Praeglobotruncana (Dicarinella indica (Jacob and Sastry). *non* Porthault, 1970, in Donze *et al.*, p. 70, 71, pl. 10, figs 4a–5.

Description: Test convexo–concave with the dorsal side being moderately convex and the ventral side tending to become concave; equatorial periphery lobate, axial periphery truncate. Chambers on dorsal side inflated, slightly elongate in the spiral axis; on ventral side very slightly inflated sloping toward the umbilicus, appressed in the shape of a truncate triangle with the apex missing; 12–15 chambers in 3 whorls with 5–6 chambers in the final whorl; gradually increasing in size; with a beaded, double keel occupying most of the periphery, becoming slightly wider adaperturally. Aperture a low to moderate, interiomarginal, umbilical arch. Umbilicus wide, shallow, covered by a tegillum. Sutures raised, wide, curved on dorsal side and radial, depressed on the ventral. Surface finely perforate, slightly pustulose.

Remarks: This species differs from all others of this stratigraphic interval by possessing truncated–triangular umbilically sloping ventral chambers. The often-mentioned imbricate shape of the dorsal chambers is no more prominent than in the majority of other *Globotruncana*, and is not judged as a distinctive criterion for *G. imbricata*. The dorsal side is typically convex but may range down to a very low trochospire. The keel varies in width, but usually occupies the entire periphery of the chamber.

The phylogeny of this species is unknown.

Jacob and Sastry (December, 1950) proposed the new species *Globotruncana indica* from the Uttattur Stage of South India. The Cenomanian age of this species was based solely upon the presence of *Rotalipora appenninica* (Renz). The *R. appenninica* is a misidentified *Praeglobotruncana delrioensis* (Plummer). *Globotruncana indica* is regarded as a junior synonym of *G. imbricata*, the latter having a six-month priority.

The holotype of *Praeglobotruncana loeblichae* Douglas, 1969, was examined and is believed to be conspecific with *Globotruncana imbricata*. The dorsal side has an atypically low convexity, although it is potentially within the range of *G. imbricata*. The lack of a tegillum may be due to preservation or immaturity. The specimens Douglas attributed to *G. imbricata* also do not display a tegillum.

Distribution: This is shown in Fig. 119.

Range: Turonian—lower Santonian.

Globotruncana intermedia Bolli

Globotruncana intermedia Bolli, 1951, p. 197, 198, pl. 35, figs 7–9; Ayala-Castañares, 1954, p. 399, pl. 7, figs 2a–c; Majzon, 1961, p. 763, pl. 4, fig. 2; *non* Herm, 1962, p. 81, 82, pl. 2, fig. 6; (?) Bandy, 1967, p. 19, 20, text-fig. 9 (2).
Rugotruncana intermedia (Bolli). Brönnimann and Brown, 1956, p. 553, pl. 22, figs 13–15.
Praeglobotruncana (Praeglobotruncana) intermedia (Bolli). Berggren, 1962a, p. 31, 32, pl. 7, figs 2a–c.

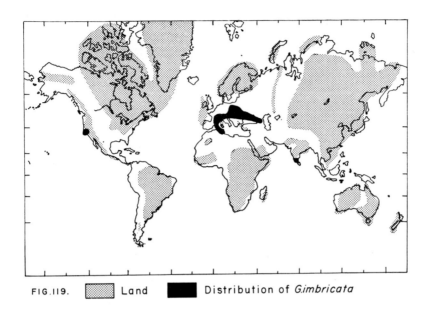

FIG.119.   [Land]   [Distribution of] *G.imbricata*

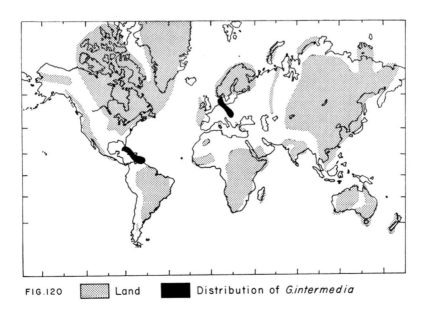

FIG.120   [Land]   [Distribution of] *G.intermedia*

Globotruncana (*Marginotruncana*) *intermedia* Bolli. *non* Hofker, 1963, p. 282, pl. 2, figs 1a–d.
Praeglobotruncana intermedia (Bolli). Hanzlíková, 1969, p. 46, 47, pl. 9, figs 5a–6c (*non* 4a–c).
Abathomphalus intermedia (Bolli). *non* Neagu, 1970b, p. 70, pl. 19, figs 9–11, *non* Govindan, 1972, p. 180, pl. 6, figs 10–11; Hanzlíková, 1972, p. 113, pl. 32, figs 1a–3c.

Description: Test a low trochospire, slightly convex dorsally, nearly flat to slightly concave ventrally; equatorial periphery lobate, axial periphery truncate to acute. Chambers in first whorl of dorsal side subspherical, later inflated but elongate in axis of coiling; ventrally elongate as later dorsal side, slightly inflated, sloping toward umbilicus; gradually increasing in size to subequal in the final whorl; 11–15 chambers in 2½–3 whorls with 4 to most commonly 5 chambers in the last whorl; with a single keel in early whorls, later developing a narrow double keel. Umbilicus wide, shallow, with a tegillum. Aperture a low interiomarginal, umbilical arch. Sutures depressed radial on ventral side; curved, raised, beaded dorsally. Surface finely perforate, may possess short, radial costellae on ventral surface, pustules on dorsal surface.

Remarks: *Globotruncana intermedia* differs from *G. mayaroensis* Bolli by having more dorsally inflated chambers, greater spiroconvexity and a much narrower double keel. Bolli (1951) observed that the ventral keel may not be developed on all chambers of the final whorl.

Berggren (1962a) agreed with Bolli's belief that *Globotruncana intermedia* descended from *G. havanensis* Voorwijk. The author also concurs with this view.

Bolli (1951) named this species *intermedia* upon the assumption that it is intermediate between *Globotruncana havanensis* Voorwijk and *G. mayaroensis* Bolli. This has since been confirmed. He observed specimens which were gradational between these three species. This will require an arbitrary separation of these species which will tend to make their use as zonal fossils less accurate and less desirable.

Distribution: This is shown in Fig. 12.

Range: Maastrichtian (FOS *Rugoglobigerina hexacamerata*—LOS *Globotruncana fornicata* Interval to *Trinitella scotti* Interval).

Globotruncana lamellosa Sigal

Globotruncana lamellosa Sigal, 1952, p. 42, 43, text-fig. 45; Majzon, 1961, p. 763, pl. 3, fig. 6; Lehmann, 1963, p. 149, pl. 9, figs 1a–2c; Postuma, 1962, 7 figs; *non* Rahhali, 1970, p. 63–65, text-figs 5a–d.

Description: Test nearly plano-convex, with the dorsal side very slightly convex and the ventral side strongly convex; equatorial periphery lobate, axial periphery truncate to acute. Chambers flat dorsally; sloping steeply up to the umbilical shoulder ventrally; 14–18 chambers in 3 whorls with 6, rarely 5, chambers in the last whorl; gradually increasing in size to subequal in final whorl; a narrow, beaded, double keel on dorsal margin, the ventral carina may not be developed on final 1–2 chambers, with a carina developed on the umbilical shoulder. Aperture a moderate to high interiomarginal, umbilical arch. Umbilicus moderately deep, very wide, covered by a tegillum. Sutures on dorsal side curved, raised, wide; ventrally radial to curved or sigmoid, depressed. Surface generally smooth.

Remarks: *Globotruncana lamellosa* is separated from *G. ventricosa* White by its narrower keel. It differs from *G. aegyptiaca* Nakkady in lacking hemispherically shaped chambers. The morphological characteristics of this species are fairly stable.

Globotruncana ventricosa White is the probable ancestor of *G. lamellosa*. A careful examination of topotypic suites of both *Globotruncana lamellosa* and *G. ventricosa* White may show them to be intraspecific variants, in which case White's species would have priority.

Distribution: This is shown in Fig. 121.

Range: Maastrichtian (LOS *Globotruncana calcarata*—LOS *Globotruncana lapparenti* Interval to FOS *Rugoglobigerina hexacamerata*—LOS *Globotruncana fornicata* Interval).

Globotruncana lapparenti Brotzen, Plate 46, fig. 4

Globotruncana lapparenti Brotzen, 1936, p. 175, 176, text-figs 2m (*non* 1a–c, 2a–c, e, f, h–j); Reichel, 1950, p. 613, pl. 16, fig. 9; pl. 17, fig. 9; Subbotina, 1953, p. 198–201, pl. 6, figs 5a–6c; pl. 7, figs 1a–5c; Martirosyan, 1958, p. 8, 9, pl. 1, figs 1a–2c; Eternod Olvera, 1959, p. 111–113, pl. 6, figs 3–5; *non* Nagappa, 1959, pl. 6, figs 10a–c; Maslakova, 1959, p. 109, pl. 12, figs 4a–c; Belford, 1960, p. 96, 97, pl. 27, figs 7–11; Postuma, 1962, 7 figs; Visser and Hermes, 1962, encl. 17, figs 47a–c; Saavedra, 1965, p. 325, 326, text-fig. 14; Wille-Janoschek, 1966, p. 108–110, pl. 1, figs 1, 2, 4–13 (*non* 3); *non* Pessagno, 1967, p. 344–346, pl. 71, figs 6–13; Porthault, 1970, in Donze *et al.*, p. 84, pl. 11, figs 17–19; Longoria, 1970, p. 59–62, pl. 2, figs 1–3; pl. 7, figs (?) 4–6; pl. 13, figs 1–3.

Globotruncana linnei (d'Orbigny). Gandolfi, 1942, p. 125–130, text-figs 46—1a–c (*non* 46—2a–c), pl. 3, figs 2a–3c; pl. 4, fig. 18.

Globotruncana lapparenti lapparenti Brotzen. (?) Colom, 1947, p. 100, 101, pl. 24, figs 3, 5–9; Cita, 1948b, p. 155, 156, pl. 4, figs 2a–c; Hagn, 1953, p. 96, 97, pl. 8, figs 12a–c; (?) text-figs 16, 17; *non* Hagn and Zeil, 1954, p. 39–42, pl. 3, figs 3a–c; pl. 6, figs 5, 8; Książkiewicz, 1956, p. 274, pl. 31, figs 7–9; text-figs 45, 46 (1–4); *non* Książkiewicz, 1958, text-fig. 2 (1a–c); *non* Haque, 1959, p. 20, 21, pl. 1, figs 1a–c, 6a–c; Klaus, 1959, p. 822, pl. 8, figs 2a–c; *non* Tollmann, 1960, p. 192, 193, pl. 20, figs 10, 11; Vinogradov, 1960a, pl. 5, figs 28a–c; *non* Leischner, 1961, p. 30, pl. 8, fig. 10; *non* Malapris and Rat, 1961, p. 92, pl. 3, figs 1a–c; Herm, 1962, p. 82–84, pl. 6, fig. 2; Hiltermann and Koch, 1962, p. 332, pl. 49, figs 8a–9; tab. 19 (part); *non* Ansary *et al.*, 1962, pl. 4 (part); *non* Küpper, 1964, p. 624, pl. 3, figs 1a–c; Caron, 1966, p. 80, pl. 5, figs 4a–c; Christododoulou and Marangoudakis, 1966, p. 305, pl. 1, figs 2a–c, 6a–c; Marianos and Zingula, 1966, p. 340, pl. 39, figs 3a–c; Săndulescu, 1966, pl. Va, figs 1a–c; Caron, 1966, p. 80, pl. 5, figs 4a–c; Christododoulou and Marangoudakis, 1967, p. 313, 314, pl. 1, figs 2a–c, 6a–c; Ansary and Tewfik, 1968, p. 54, pl. 7, figs 8a–c; *non* Rao *et al.*, 1968, p. 23, 24, pl. 2, figs 7–9; Sturm, 1970, p. 109, 110, pl. 6, figs 1a–c; Latif, 1970, p. 31, 32, pl. 1, figs 1–6; Neagu, 1970b, p. 66, 67, pl. 16, figs 7–9; pl. 41, fig. 57.

Globotruncana lapparenti tricarinata (Quereau). (?) Colom, 1947, p. 101, pl. 24, figs 10–15; (?) Kochansky-Devide, 1951, p. 114, 116, pl. 1, figs 1–11; Hagn, 1953, p. 97, pl. 8, figs 13a–c; (?) text-figs 18, 19; Książkiewicz, 1956, p. 274, 275, pl. 31, figs 10–12, text-figs 47 (upper, *non* lower), 48 (1–4); Książkiewicz, 1958, text-fig. 2 (2a–c); Bieda, 1958, p. 55–58, text-figs 22a–c; Eternod Olvera, 1959, p. 114, 115, pl. 6, figs 9–11; Klaus, 1959, p. 823, pl. 8, figs 3a–c; Tollmann, 1960, p. 193, 194, pl. 21, fig. 1; Belford, 1960, p. 97, pl. 28, figs 1–6; Hiltermann and Koch, 1962, p. 332, pl. 50, figs 8a–c; tab. 19 (part); Küpper, 1964, p. 626, 627, pl. 3, figs 3a–c; Caron, 1966, p. 83, pl. 5, figs 1a–c; Wille-Janoschek, 1966, p. 110, 111, pl. 2, figs 1, 2 (*non* 3–11); Săndulescu, 1966, pl. Va, figs 2a–c; Caron, 1966, p. 83, pl. 5, figs 1a–c.

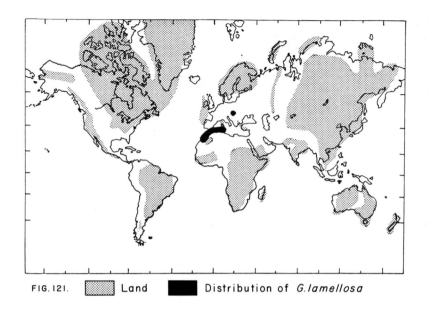

FIG. 121.   [Land]   [Distribution of *G. lamellosa*]

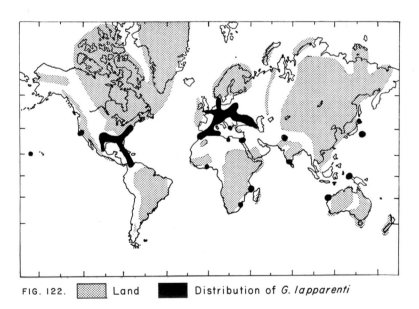

FIG. 122.   [Land]   [Distribution of *G. lapparenti*]

Globotruncana canaliculata (Reuss). Asano, 1950, p. 162, pl. 1, figs 3a–c (*non* 1a–c); Smitter, 1957, p. 201, text-figs 21a–b.

Globotruncana (*Globotruncana*) *lapparenti tricarinata* (Quereau). Papp and Küpper, 1953, p. 37, 38, pl. 1, figs 3a–c.

Globotruncana lapparenti coronata Bolli. Hagn and Zeil, 1954, p. 43, 44, pl. 3, figs 4a, b; Săndulescu, 1966, pl. Va, figs 3a–c; Latif, 1970, p. 35, pl. 3, figs 7–9.

Globotruncana lapparenti bulloides Vogler. Hagn and Zeil, 1954, p. 45, 46, pl. 2, figs 5a–c; pl. 7, fig. 4 (?).

Globotruncana lapparenti longilocula Gandolfi, 1955a, p. 17, 18, pl. 1, figs 1a–c.

Globotruncana tricarinata colombiana Gandolfi, 1955a, p. 20–22, pl. 1, figs 3a–4c; text-fig. 6 (4a, b); Küpper, 1964, p. 627, pl. 3, figs 5a–c; Salaj and Samuel, 1966, p. 217, pl. 20, figs 1a–c.

Globotruncana sp. (ex. gr. *arca* Cushman). Hiltermann and Koch, 1955, p. 373, pl. 27, figs 2a–c.

Globotruncana marginata (Reuss). Jírová, 1956, Tab. 1, obr. 3a–c.

Globotruncana lapparenti subsp. *lapparenti* Brotzen. Edgell, 1957, p. 113, pl. 1, figs 7–9.

Globotruncana majzoni Sacal and Debourle, 1957, p. 62, pl. 26, figs 6, 21, 22.

Globotruncana lapparenti var. *lapparenti* Brotzen. Witwicka, 1958, p. 212–215, pl. 15, figs 28a–c.

Globotruncana lapparenti var. *tricarinata* (Quereau). Witwicka, 1958, p. 215–217, pl. 16, figs 29a–c.

Globotruncana lapparenti var. *coronata* Bolli. Witwicka, 1958, p. 217, 218, pl. 16, figs 30a–c.

Globotruncana erevanica Martirosyan, 1958, p. 9, 10, pl. 1, figs 3a–c.

Globotruncana riojae Eternod Olvera, 1959, p. 116, 117, pl. 7, figs 4, 5.

Globotruncana linneiana (d'Orbigny). Klaus, 1959, p. 823, 824, pl. 8, figs 1a–c; Trujillo, 1960, p. 342, 343, pl. 49, figs 8a–c; Lehmann, 1962, p. 149, 150, pl. 8, figs 4a–5c; Kaptarenko-Chernousova *et al.*, 1963, p. 105, 106, Tab. 17, figs 3a–c; Brönnimann and Rigassi, 1963, pl. 17, figs 5a–c; Martin, 1964, p. 81, pl. 10, figs 3a–c; Cati, 1964, p. 257, pl. 42, figs 4a–c; Caron, 1966, p. 83, pl. 5, figs 3a–c; Douglas and Sliter, 1966, p. 112, pl. 5, figs 4a–c; 6a–c (*non* 9a–c); Rao *et al.*, 1968, p. 24, 25, pl. 4, figs 10–12; Douglas, 1969, p. 181, 182, pl. 3, figs 1a–c; Esker, 1969, pl. 2, figs 1–3; Scheibnerová, 1969, p. 69, 70, tab. 16, figs 3a–c; text-figs 13–17; Rahhali, 1970, p. 60, text-figs 3a–d; Longoria, 1970, p. 65–69, pl. 2, figs 4–6; pl. 7, figs (?) 9–10; pl. 13, figs (?) 5, 6; El-Naggar, 1971a, pl. 7, figs g, j, m.

Globotruncana linneiana linneiana (d'Orbigny). Barr, 1961, p. 571, 572, pl. 69, figs 7a–c; van Hinte, 1963, p. 75–78, pl. 5, figs 1a–c (*non* 2a–c); Salaj and Samuel, 1966, p. 209, pl. 19, figs 1a–c; Bandy, 1967, p. 19, text-fig. 8 (11).

Globotruncana linneiana tricarinata (Quereau). Barr, 1961, p. 573, 574, pl. 70, figs 2a–c; Graham and Clark, 1961, p. 112, pl. 5, figs 8a–c; van Hinte, 1963, p. 79, 80, pl. 6, figs 1a–c; Banerji, 1966, p. 63, 64, pl. 4, figs 7–9; Bandy, 1967, p. 19, text-fig. 8 (12), 9 (4).

Globotruncana (*Globotruncana*) *tricarinata* (Quereau). Berggren, 1962a, p. 64–66, pl. 10, figs 3a–c.

Globotruncana tricarinata (Quereau). Herm, 1962, p. 93, 94, pl. 6, fig. 4; Takayanagi, 1965, p. 225, 226, pl. 27, figs 4a–c; van Hinte, 1965a, p. 23, pl. 3, figs 2a–c; Douglas and Sliter, 1966, p. 115, pl. 4, figs 9a–10c; Barr, 1968a, p. 319, pl. 37, figs 8a–c; pl. 38, figs 1a–c; Rasheed and Govindan, 1968, p. 83, pl. 7, figs 13–15.

Globotruncana (Globotruncana) linneiana linneiana (d'Orbigny). van Hinte, 1965b, p. 84, pl. 2, figs 4a–c.

Globotruncana (Globotruncana) linneiana tricarinata (Quereau). van Hinte, 1965b, p. 84, pl. 2, figs 1a–c.

Globotruncana ventricosa White. Banerji, 1966, p. 66, 67, pl. 5, figs 6, 8, 9; Sliter, 1968, p. 107, pl. 18, figs 8a–c.

Globotruncana tricarinata tricarinata (Quereau). Salaj and Samuel, 1966, p. 218, pl. 19, figs 3a–c.

Marginotruncana pseudolinneiana Pessagno, 1967, p. 310, pl. 65, figs 24–27; pl. 76, figs 1–3; Porthault, 1970, in Donze *et al.*, p. 79, pl. 11, figs 9a–10; pl. 13, (?) figs 19, 22.

Globotruncana coronata Bolli. Douglas 1969, p. 177, 178, pl. 3, figs 5a–c (*non* 6a–8c); Rahhali, 1970, p. 72, text-figs e, f (*non* g).

Globotruncana pseudolinneiana Pessagno. Douglas, 1969, p. 185, pl. 3, figs 3a–4c (*non* 2a–c); Hanzlíková, 1972, p. 109, pl. 29, figs 8a–9c.

Globotruncana orientalis El-Naggar. Esker, 1969, pl. 1, figs 1–3.

Globotruncana spinea Kikoïne. Scheibnerová, 1969, p. 75, tab. 16, figs 1a–2c (*non* 4a–c).

Globotruncana linneiana coronata Bolli. Bate and Bayliss, 1969, pl. 3, figs 6a–c.

Description: Test a very low trochospire with both dorsal and ventral sides nearly flat to very slightly convex; equatorial periphery lobate, axial periphery truncate. Chambers flat to slightly inflated, rectangular in axial section; imbricate ventrally, petaloid to crescent-shaped dorsally; 15–17 chambers in 3–3½ whorls with 5–7 chambers in the final whorl; with a widely spaced, beaded double keel typically occupying the entire peripheral margin, the ventral carina of which recurves to form a carina on the umbilical shoulder. Umbilicus shallow, wide with a tegillum. Aperture a low interiomarginal, umbilical to extraumbilical–umbilical arch. Sutures on either side curved, raised, wide. Surface finely perforate, smooth to pustulose.

Remarks: This species is separated from *Globotruncana linneiana* (d'Orbigny) on the basis of its distinctive imbricate ventral chamber pattern. In some specimens the test diameter is so wide that when the tegillum is broken out the previous whorls are visible in the wide umbilicus. The keel width may vary slightly, but is typically wide.

Specimens of *Globotruncana linneiana* (d'Orbigny) are found to grade into *G. lapparenti*. Although they appear in the stratigraphic record at approximately the same time, *G. linneiana* is believed to be structurally more primitive and, therefore ancestral.

Brotzen (1936) stated that the form-group that de Lapparent (1918) described as *Rosalina linnei* d'Orbigny must be considered as a new species. For these forms, he proposed the name *Globotruncana lapparenti* without selecting a holotype. Brotzen did present a brief description in table form of his new species, and compared it to *G. linneiana* (d'Orbigny), *G. canaliculata* (Reuss) and *G. ventricosa* White.

Pessagno (1967) selected text-figure 2n of de Lapparent (1918) as representative of the lectotype for *Globotruncana lapparenti*. This specimen and the paralectotype, figure 2m, are thin-sections. Article 73c (i) of the I.C.Z.N. states that "Syntypes may include specimens . . . not seen by the author [in this case, Brotzen] but which were the bases of previously published descriptions or figures upon which he founded his taxon in whole or in part". There is no evidence to support Pessagno's contention that Brotzen based his new species upon de Lapparent's thin-sections. In fact, the morphologic features briefly described by Brotzen had to have been from whole, free specimens. Such features as the narrow, attenuated (= imbricate) chamber shape can be seen only in a three-dimensional specimen. For these reasons, Pessagno's designation of a lectotype is rejected.

Pessagno (1967) remarked that his selection of a lectotype for *Globotruncana lapparenti* should stabilise the taxonomic status of that species and "leave little doubt concerning the identification of this species in thin-section". Stabilisation would not result even if his designation were valid, because this species cannot with certainty be identified from thin-section.

A lectotype should be selected from among de Lapparent's free specimens, but only after careful examination of these specimens. Any one of several appears suitable.

As observed above, it is seldom possible from thin-section of a species to visualise a three-dimensional image sufficient for recognition. *Pulvinulina tricarinata* Quereau, 1893, is a case in point. It was described from thin-section and distinguished on the basis of its "third keel". This structure is merely the extension of the ventral carina of a double keel to the umbilical shoulder—a feature common to the majority of *Globotruncana* including *G. lapparenti*. Thus, it cannot be determined whether or not *P. tricarinata* is the senior synonym of *G. lapparenti*, even though their thin-sections appear similar. However, most of the specimens subsequently attributed to Quereau's species are conspecific with *G. lapparenti*.

The specimen in the slide marked as the holotype of *Globotruncana marginata austinensis* Gandolfi, 1955a, is not the one which was illustrated nor described. The specimen labelled as the holotype has a widely spaced double keel on the five chambers comprising the final whorl. The test has a nearly flat trochospire with slight ventral chamber inflation. The specimen is conspecific with *G. lapparenti*. The fate of the holotype is not known.

Globotruncana erevanica Martirosyan, 1958, is a neanic individual of *G. lapparenti*.

Pessagno (1967) established the new species *Marginotruncana pseudolinneiana* on the bases of its extraumbilical–umbilical aperture, crescent-shaped dorsal chambers, planiform test and lack of a strongly beaded keel. Each of these features is within the known range of intraspecific variability for *Globotruncana lapparenti*. Pessagno regarded the absence of ultragranular hyaline material in the keel as a distinctive feature. The significance of the presence or absence of ultragranular deposits is not understood, nor has it been fully investigated. However, the topotype material from Edelbachgraben, Austria, appears to have a high percentage of recrystallised specimens, which could account for the absence of ultragranular material. *Marginotruncana pseudolinneiana* is a junior synonym of *G. lapparenti*.

Distribution: This is shown in Fig. 122.

Range: Turonian—Campanian (*Globotruncana calcarata* Interval).

   Globotruncana linneiana (d'Orbigny), Plate 46, figs 3, 5, 6
Rosalina linneiana d'Orbigny, 1839, p. 106, pl. 5, figs 10–12.
Rosalina canaliculata Reuss, 1854, p. 70, pl. 26, figs 4a–b.
(?) *Discorbina bi-concava* Parker and Jones, 1862, in Carpenter *et al.*, p. 201, figs 10a–c, 32q.
Globigerina linnaeana (d'Orbigny). Jones, 1895, p. 285, pl. 7, figs 23a–c.
Rosalina linnei d'Orbigny. *non* Parejas, 1926, p. 58, 59, text-fig. 1; *non* Marin *et al.*, 1934, p. 657, text-figs 3 (3a–c).
Globotruncana canaliculata (Reuss). White, 1928b, p. 282, 284, pl. 38, figs 3a–c; Dampel, 1934, p. 26, 27, pl. 3, figs 7a–c; Majzon, 1943, p. 75, 76, pl. 1, fig. 2; pl. 2,

figs 2a–c; Cushman, 1946b, p. 149, pl. 61, figs 17a–c (*non* 18a–c); Asano, 1950a, p. 162, pl. 1, figs 1a–c (*non* 3a–c); (?) Asano, 1950b, p. 22, pl. 3, fig. 16a, b; Hagn, 1953, p. 95, 96, pl. 8, figs 14a–c; (?) text-figs 14, 15; Frizzell, 1954, p. 128, pl. 20, figs 21a–c; Majzon, 1961, p. 761, pl. 1, fig. 2.

Globotruncana linneiana (d'Orbigny) Cushman, 1931b, p. 90; Keller, 1939, pl. 2, figs 6a–c; Subbotina, 1953, p. 195–198, figs 7a–9c, pl. 6, figs 1a–4c; Bönnimann and Brown, 1955, p. 540–542, pl. 20, figs 13–17; pl. 21, figs 16–18; Seiglie, 1958, p. 69, pl. 1, figs 2a–c; *non* Klaus, 1959, p. 823, 824, pl. 8, figs 1a–c; Graham and Clark, 1961, p. 113, pl. 5, figs 11a–c; Majzon, 1961, p. 761, pl. 1, fig. 1; *non* Lehmann, 1963, p. 149, 150, pl. 8, figs 4a–5c; Cita-Sironi, 1963, p. 113, text-figs 1a–c; *non* Kaptarenko-Chernousova *et al.*, 1963, p. 105, 106, pl. 17, figs 3a–c; *non* Olsson, 1964, p. 166, 167, pl. 2, figs 6a–8c; *non* Cati, 1964, p. 257, pl. 42, figs 4a–c; *non* Takayanagi, 1965, p. 217, 218, pl. 25, figs 6a–c; pl. 26, figs 1a–2c; *non* Caron, 1966, p. 83, pl. 5, figs 3a–c; Pessagno, 1967, p. 346–349, pl. 72, figs 1–4, 7–9; *non* Ansary and Tewfik, 1968, p. 53, pl. 7, figs 2a–c; Rasheed and Govindan, 1968, p. 82, pl. 7, figs 10–12; *non* Rao *et al.*, 1968, p. 24, 25, pl. 4, figs 10–12; *non* Scheibnerová, 1969, p. 69, 70, tab. 16, figs 3a–c; text-figs 13–17; *non* Kalantari, 1969, p. 208, pl. 23, figs 11a–c; pl. 24, figs 2a–c; Moorkens, 1969, p. 450, pl. 3, figs 5a–c; *non* Esker, 1969, pl. 2, figs 1–3; *non* Rahhali, 1970, p. 60, text-figs 3a–d; *non* Longoria, 1970, p. 65–69, pl. 2, figs 4–6; pl. 7, figs (?) 9, 10; pl. 13, figs (?) 5, 6; (?) Cita and Gartner, 1971, pl. 1, fig. 6; *non* El-Naggar, 1971a, pl. 7, figs g, j–m; *non* Hanzlíková, 1972, p. 107, 108, pl. 29, figs 6a–7c.

Globotruncana linnaeana (d'Orbigny). *non* Voorwijk, 1937, p. 195, pl. 1, figs 23, 27, 28; Morozova, 1939, p. 79, pl. 1, figs 4–6; Majzon, 1943, p. 74, 75, pl. 1, fig. 1.

Globotruncana linnei (d'Orbigny). *non* Glaessner, 1937, p. 38, pl. 1, figs 11a–c.

Globigerina (*Globotruncana*) *marginata linnaeana* Olbertz, 1942, p. 135, pl. 5, figs 4a–c.

Globotruncana lapparenti Brotzen. Majzon, 1943, pl. 1, fig. 3; Majzon, 1961, p. 762, pl. 1, fig. 6.

Globotruncana lapparenti tricarinata (Quereau). Noth, 1951, p. 77, pl. 8, figs 16a–c.

Globotruncana cf. *linnei* (d'Orbigny). Hofker, 1956a, p. 190, 192, pl. 29, figs 1–3.

Globotruncana lapparenti lapparenti Brotzen. Tollmann, 1960, p. 192, 193, pl. 20, figs 10, 11.

Globotruncana (*Marginotruncana*) *linneiformis* Hofker. Majzon, 1961, p. 764, pl. 7, fig. 1.

Globotruncana aspera Hofker. Hofker, 1961c, text-fig. 69.

Globotruncana (*Globotruncana*) *tricarinata* (Quereau). Berggren, 1962a, p. 64–67, pl. 10, figs 3a–c.

Globotruncana (*Globotruncana*) *linneiana linneiana* (d'Orbigny). *non* van Hinte, 1965b, p. 84, pl. 2, figs 4a–c.

Globotruncana linneiana linneiana (d'Orbigny). *non* Salaj and Samuel, 1966, p. 209, pl. 19, figs 1a–c; *non* Bandy, 1967, p. 19, text-fig. 8 (11).

Marginotruncana canaliculata (Reuss). Pessagno, 1967, p. 302–304, pl. 74, figs 5–8.

Globotruncana pseudolinneiana Pessagno. Douglas and Sliter, 1969, p. 207, 208; text-figs 16A–C; 17A–C; Weaver *et al.*, 1969, pl. 3, figs 9a–c.

Praeglobotruncana (*Dicarinella*) *canaliculata* (Reuss). Porthault, 1970, in Donze *et al.*, p. 72, pl. 10, figs 9–11; pl. 13, fig. 24 (?).

Description: Test a low trochospire with both dorsal and ventral sides nearly flat, ventral side may be slightly convex; equatorial periphery lobate, axial periphery truncate. Chambers flat to very slightly inflated both dorsally and ventrally; rectangular in axial section; gradually increasing in size to subequal in final whorl; 14–16 chambers in $2\frac{1}{2}$ whorls with 5–7, typically 6, chambers in the final wherl; with a wide, beaded, double keel occupying the entire periphery. Aperture a low, interiomarginal, umbilical arch. Umbilicus shallow, wide, covered by a tegillum. Sutures curved, raised, wide on dorsal side; radial, depressed to flush or slightly raised ventrally. A carinate umbilical shoulder may develop on later chambers. Surface finely perforate, pustulose.

Remarks: *Globotruncana linneiana* has radial ventral sutures, whereas *G. lapparenti* Brotzen has strongly curved sutures resulting in imbricate ventral chambers. The carinate umbilical shoulder is a late ontogenetic development and is thus absent from less mature individuals.

This species evolved from *Globotruncana cretacea* (d'Orbigny) into which gradational species are found.

Brönnimann and Brown (1956) considered the various spellings of the specific name, and concluded that the correct form is *linneiana*. Inasmuch as the holotype is lost, these authors selected a neotype which is in accordance wth d'Orbigny's original specimen.

Masters (in press) confirmed the synonomy of *Rosalina canaliculata* Reuss with the older *Globotruncana linneiana*.

Although neither the holotype nor a topotypic suite of *Globigerina (Globotruncana) marginata linnaeana* Olbertz, 1942, was examined, there is little doubt that it is a synonym of *Globotruncana linneiana*.

Distribution: This is shown in Fig. 123.

Range: Turonian—Campanian (*Globotruncana calcarata* Interval).

Globotruncana lobata de Klasz

Globotruncana lobata de Klasz, 1955, p. 43, 44, pl. 7, figs 2a–c; Majzon, 1961, p. 764, pl. 3, fig. 12.

Description: "A double-keeled *Globotruncana* with the following peculiarities: Dorsal side almost completely flattened, with only a slight central elevation, umbilical side strongly convex. The periphery of the test is lobate. On the dorsal side there are in general three whorls visible, the chambers of the innermost whorl appearing indistinct. The sutures of the dorsal side are strongly curved opposite to the direction of coiling. The number of chambers in the last-formed whorl is 7 to 9; on the umbilical side they are very slightly overlapping, their sutures curved. The sutures of the central whorls of the dorsal side and on 'some specimens those of the first 3 or 4 chambers of the last formed whorl on the umbilical side are beaded. These last-mentioned 3 or 4 chambers on most specimens show small, irregular rugosities; on some specimens the 3 or 4 last-formed chambers of the test tend to drop below the normal plane of the dorsal side, in such a manner, that their upper side is lower than that of the other chambers. Because of the high number of chambers and the relatively high test the species appears robust." (Type description.)

Remarks: The large size of the double-keeled test with its appressed chambers distinguish this species from its congeners. The range of intraspecific variation is unknown.

De Klasz (1955) presumed that *Globotruncana lobata* evolved from *G. lapparenti tricarinata* (Quereau), but did not list his reasons. Quereau's species is a most improbable progenitor for *G. lobata*. It more likely developed from *G. ventricosa*

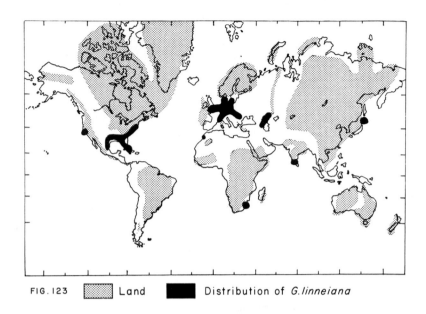

FIG. 123   [▒] Land   [■] Distribution of *G.linneiana*

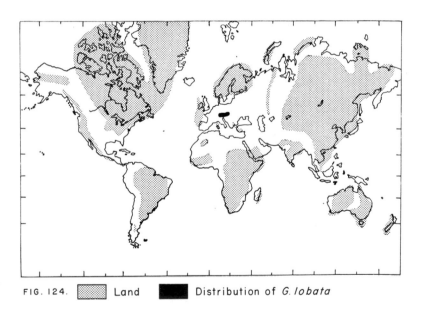

FIG. 124.   [▒] Land   [■] Distribution of *G.lobata*

White, with which it shares several characteristics, but this has not been confirmed. The holotype and topotypes of *Globotruncana lobata* are poorly preserved. With only two occurrences described in the literature, very little is known concerning this taxon.

Distribution: This is shown in Fig. 124.

Range: Campanian (*Globotruncana calcarata* Interval).

Globotruncana lunaris Masters, Plate 47, figs 1, 2

Globotruncana lunaris Masters, 1976, p. 328, 399, pl. 2, figs 4–6.

Description: Test a low trochospire, slightly biconvex; equatorial periphery gently lobate; axial periphery truncate with a well-developed, moderately wide, beaded double keel. Chambers of dorsal side unarched, elongate and crescent-shaped with a wide, imperforate, lunate area adjacent to and sloping sharply toward the inner margin of the previous chamber; ventral chambers inflated with the wall sloping steeply to the carinated umbilical shoulder, then dropping abruptly into the umbilicus; 8–10 chambers in 2–2½ whorls with 4 in the last whorl, rapidly increasing in size. Aperture a low, interiomarginal, umbilical arch. Umbilicus wide, moderately deep, covered by a tegillum. Dorsal sutures curved, raised near periphery but become flush with the wall near lunate areas into which they merge, here although flush the sutures lie in a depression; ventral sutures slightly curved, depressed but curve back to form the carinae on the umbilical shoulder. Wall densely perforate, pustulose particularly on the imperforate lunate areas.

Remarks: Only *Globotruncana fornicata* Plummer has elongate, crescentic dorsal chambers similar to this species. *Globotruncana lunaris* is lower-spired and lacks the characteristic dorsal arching of the chambers, and has the imperforate lunate areas which are absent on *G. fornicata*. No variation has been observed within this species.

The phylogeny of *G. lunaris* is unknown.

The lunate areas are interpreted as secondary calcification which occurred very quickly after chamber formation because even the ultimate chamber of various individuals show traces of this structure.

Distribution: This is shown in Fig. 125.

Range: Campanian (FOS *Pseudotextularia carseyae*—FOS *Globotruncana ventricosa* Interval).

Globotruncana marginata (Reuss), Plate 47, figs 3, 4

Rosalina marginata Reuss, 1845, p. 36, pl. 8, figs 54a, b, 74a, b; pl. 13, figs 68a, b; Reuss, 1846, in Geinitz, p. 676, 677, pl. 24, figs 57a, b.

Globigerina marginata (Reuss). Heron-Allen and Earland, 1910, p. 424, pl. 9, figs 1–3; Franke, 1925, p. 93, pl. 8, figs 16a–c; *non* Liebus, 1927, p. 374, 375, pl. 14, figs 1a–c; Eichenberg, 1933, p. 23, pl. 2, figs 21a, b; Loetterle, 1937, p. 44, 45, pl. 7, figs 3a–c.

Globotruncana ventricosa White. Brotzen, 1936, p. 137, pl. 13, figs 4a–c; Majzon, 1943, pl. 1, fig. 5; Subbotina, 1953, p. 222–226, pl. 13, fig. 1a–4c; Majzon, 1961, p. 762, pl. 1, fig. 7.

Globotruncana marginata (Reuss). Morozova, 1939, p. 79, 80, pl. 1, figs 7–9; Cushman, 1946b, p. 150, pl. 62, figs 1a–2c; *non* Cushman, 1948a, p. 265, pl. 26, figs 2a, b; Peterson *et al.*, 1953, p. 82, pl. 9, figs 27–29; Hagn, 1953, p. 93, 94, pl. 8, figs 10a–c, text-figs 10, 11; Frizzell, 1954, p. 129, pl. 20, figs 24a–c; Hagn and Zeil, 1954, p. 46, 47, pl. 2, figs 4a–c; pl. 7, figs (?) 5, 6; Jírová, 1956, p. 241, pl. 1, figs 1a–c; pl. 3, figs 1a–c (*non* pl. 1, figs 2a–3c; pl. 2, figs 1a–3c); Książkiewicz, 1956, p. 278–280, pl. 32, figs (?) 1–6, (*non* 7–12); text-figs 52 (1–3) 53, 54 (1, 2, 4–7), [*non* 54 (3, 8), 55]; Edgell, 1957, p. 114, pl. 2, figs 4–6; Sacal and Debourle, 1957,

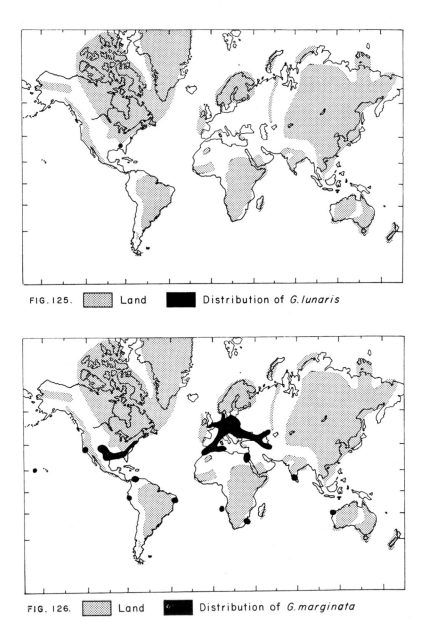

FIG. 125. ░░░ Land ███ Distribution of *G. lunaris*

FIG. 126. ░░░ Land ███ Distribution of *G. marginata*

pl. 59, pl. 26, figs 3, 5, 12, 13; Smitter, 1957, p. 201, text-figs 22a, b; Witwicka, 1958, p. 222, 223, pl. 17, figs 34a–c; Bieda, 1958, p. 58, 59, text-figs 23a–c; *non* Seiglie, 1958, p. 69, 70, pl. 3, figs 1a–3; Barr, 1961, p. 574, 575, pl. 70, figs 3a–c; Jurkiewicz, 1961, pl. 24, figs 5a–c; Herm, 1962, p. 85, 86, pl. 5, fig. 5; Lehmann, 1963, p. 151, pl. 10, figs 3a–c; *non* Hiltermann and Koch, 1962, p. 330, 331, pl. 47, figs 2a–c; tab. 19 (part); Kaptarenko-Chernousova *et al.*, 1963, p. 106, pl. 17, figs 2a–c; van Hinte, 1963, p. 83–85, pl. 7, figs 2a–c; pl. 8, figs 3a–4c; Küpper, 1964, p. 628, pl. 4, figs 2a–c; van Hinte, 1965a, p. 23, pl. 1, figs 2a–c; *non* Takayanagi, 1965, p. 218; 219, pl. 26, figs 3a–4c; Banerji, 1966, p. 64, 65, pl. 5, figs 10–12; Douglas and Sliter, 1966, p. 112, pl. 2, figs 9a–c; *non* Pessagno, 1967, p. 307–310, pl. 54, figs 7–12, 16–18; pl. 56, figs 10–12; Kent 1967, p. 1449, pl. 184, figs 1a–c; *non* Sliter, 1968, p. 104, 105, pl. 77, figs 5a–6c; Douglas, 1969, p. 182, 183, pl. 8, figs 4a–5c; Kalantari, 1969, p. 208, 209, pl. 23, figs 8a–c; Moorkens, 1969, p. 450, pl. 3, figs 1a–c; Douglas and Rankin, 1969, p. 203–207, text-figs 14A–E, 15A–F; Rahhali, 1970, p. 61, 62, text-fig. 4d (*non* a–c); Neagu, 1970b, p. 67, 68, pl. 16, figs 10–12; pl. 41, fig. 55; Govindan, 1972, p. 179, 180, pl. 4, figs 10–12; Hanzlíková, 1972, p. 108, pl. 30, figs 3a–4c.

Globotruncana linnei bulloides Vogler, 1941, p. 287, pl. 23, figs 33–35 (*non* 32, 36–39).

Rosalinella globigerinoides form. typica Marie, 1941, p. 239, 240, pl. 36, figs 338a–c.

Rosalinella globigerinoides var. *sublaevigata* Marie, 1941, p. 239, 240, pl. 36, figs 339a–c.

Rosalinella marginata (Reuss). Schijfsma, 1946, p. 97, 98, pl. 7, figs 10a–c.

Rotundina marginata (Reuss). Subbotina, 1953, p. 183–187, pl. V, figs 1a–c, 5a–c (*non* 2a–4c, 6a, b).

Globotruncana pooleyi Nakkady and Osman, 1954, p. 82, pl. 19, figs 4a–c.

Globotruncana lapparenti bulloides Vogler. *non* Hagn and Zeil, 1954, p. 45, 46, pl. 2, figs 5a–c; pl. 7, fig. 4; Weiss, 1955, p. 307, pl. 1, figs 3–5; Książkiewicz, 1956, p. 275, 276, text-fig. 49 (*non* pl. 31, figs 13–15); Ansary and Fakhr, 1958, p. 136, 137, pl. 2, figs 18a–c; Belford, 1960, p. 97, 98, pl. 28, figs 7–11; Herm, 1962, p. 84, pl. 6, fig. 6; Ansary *et al.*, 1962, pl. 4 (part); Küpper, 1964, p. 625, pl. 3, figs 8a–c; Sǎndulescu, 1966, pl. Va, figs 4a–c; *non* Ansary and Tewfik, 1968, p. 54, pl. 7, figs 7a–c; *non* Rao *et al.*, 1968, p. 23, pl. 1, figs 7–9; Sturm, 1970, p. 110, pl. 6, figs 2a–c; Neagu, 1970b, p. 67, pl. 17, figs 4–6.

Globotruncana (*Globotruncana*) sp. aff. *tricarinata* (Quereau). Gandolfi, 1955b, p. 257, 258, text-figs 6–9.

Globotruncana (*Rugoglobigerina*) *loetterli subloetterli* Gandolfi, 1955a, p. 36–38, pl. 1, figs 14a–c.

Marginotruncana paraventricosa Hofker. Hofker, 1956b, fig. 14; Hofker, 1961b, p. 125, text-figs 3a, b.

Globotruncana wilsoni Bolli, 1957, p. 58, 59, pl. 14, figs 4a–c; Majzon, 1961, p. 764, pl. 6, fig. 8.

Globigerina subcretacea Lomnicki. Le Roy and Schielz, 1958, p. 2453, text-fig. 8 (17, 18).

Globotruncana lapparenti var. *bulloides* Vogler. Witwicka, 1958, p. 218, 219, pl. 17, figs 31a–c.

Globotruncana culverensis Barr 1961, p. 569, 570, pl. 71, figs 1a–c; Kalantari, 1969, p. 201, 202, pl. 24, figs 1a–c.

Globotruncana (*Marginotruncana*) *paraventricosa* Hofker. Majzon, 1961, p. 764, pl. 7, fig. 2.

Globotruncana globigerinoides Brotzen. Maslakova, 1959, p. 110, pl. 12, figs 3a–c; Jurkiewicz, 1961, pl. 24, figs 7a–c; Hofker, 1962, text-figs 11a–d; Küpper, 1964, p. 622, 623, pl. 3, figs 7a–c; Ansary and Tewfik, 1968, p. 55, pl. 8, figs 2a–c.

Globotruncana (Marginotruncana) ventricosa White. Hofker, 1961c, text-fig. 70.

Globotruncana lapparenti lapparenti Ansary *et al.*, 1962, pl. 4 (part) Brotzen.

Globotruncana linneiformis Hofker. Hofker, 1962, p. 115, text-figs 47a–d.

Globotruncana campbelli Petri, 1962, p. 126, 127, pl. 19, figs 3a–c.

Globotruncana bulloides Vogler. Postuma, 1962, 7 figs; Pessagno, 1967, p. 324–326, pl. 64, figs 16–17; pl. 67, figs 1–3; pl. 75, figs 4–8; pl. 97, figs 14, 15; Porthault, 1970, in Donze *et al.*, p. 83, pl. 11, figs 20–22; *non* Longoria, 1970, p. 29–32, pl. 3, figs 4–6, pl. 7, figs 1–3; pl. 19, figs 5, 6.

Globotruncana paraventricosa (Hofker). Hiltermann and Koch, 1962, p. 331, 332, pl. 49, figs 6a–7; tab. 19 (part).

Globotruncana sp. aff. *G. marginata* (Reuss). Graham and Church, 1963, p. 64, pl. 7, figs 15a–16c.

Globotruncana linneiana (d'Orbigny). Olsson, 1964, p. 166, 167, pl. 2, figs 7a–c (*non* 6a–c, 8a–c); Douglas and Sliter, 1966, p. 112, pl. 4, figs 6a–7c (*non* 8a–c); Kalantari, 1969, p. 208, pl. 23, figs 11a–c; pl. 24, figs 2a–c (?).

Globotruncana (Globotruncana) marginata (Reuss). van Hinte, 1965b, p. 84, pl. 2, figs 2a–c.

Globotruncana cf. *marginata* (Reuss). Lehmann, 1966, p. 312, pl. 1, fig. 4.

Globotruncana coronata (Bolli). Douglas and Sliter, 1966, p. 109, pl. 5, figs 8a–c (*non* 7a–c); Douglas, 1969, p. 177, 178, pl. 3, figs 7a–c (*non* 5a–6c, 8a–c).

Globotruncana linneiana marginata (Reuss). Salaj and Samuel, 1966, p. 210, pl. 20, figs 3a–c, 5a–c.

Rugoglobigerina loetterli subleotterli (Gandolfi). *non* Salaj and Samuel, 1966, p. 195, 196, pl. 9, figs 2a–c.

Globotruncana hilli Pessagno, 1967, p. 343, 344, pl. 64, figs 9–14, 21–23; pl. 94, fig. 1; pl. 97, fig. 7.

Globotruncana linneiana bulloides Vogler. Bandy, 1967, p. 19, text-fig. 8 (13).

Globotruncana marginata marginata, (Reuss). Bandy, 1967, p. 19, text-fig. 9 (1).

Praeglobotruncana wilsoni (Bolli). Bandy, 1967, p. 17, text-fig. 8 (5).

Globotruncana lapparenti subsp. *bulloides* Vogler. Kalantari, 1969, p. 205, pl. 25, figs 3a–c, 10a–c.

Globotruncana lapparenti subsp. *lapparenti* (Brotzen). Kalantari, 1969, p. 206, pl. 25, figs 1a–c.

Globotruncana arca (Cushman). Douglas, 1969, p. 176, pl. 10, figs 5a–c, 7a–c (*non* 4a–c, 6a–c); Todd, 1970, p. 153, pl. 6, figs 2a–c (*non* 3a–4c).

Globotruncana cretacea (d'Orbigny). Rasheed and Govindan, 1968, p. 82, pl. 7, figs 4–6; Rahhali, 1970, p. 65, text-figs 5e–i.

Globotruncana (?) *cretacea* (d'Orbigny). Porthault, 1969, p. 439, 440, pl. 2, figs 10a–c.

Marginotruncana marginata (Reuss). Porthault, 1970, in Donze *et al.*, p. 74, 75, pl. 10, figs 18–20; pl. 13, figs 21–23.

Globotruncana tricarinata (Quereau). Govindan, 1972, p. 179, pl. 4, figs 16–18.

Description: Test a low trochospire, slightly to moderately biconvex; equatorial periphery distinctly lobate, axial periphery truncate. Chambers moderately inflated, more ventrally than dorsally in some specimens; subspherical to oval in side view with the long dimension in the axis of coiling; gradually increasing in size to subequal in the final whorl; 15–18 chambers in 3 whorls with 5–8 chambers in the final whorl;

with a moderately wide to narrow, beaded double keel occupying most of the periphery when widely spaced decreasing to $\frac{1}{4}$ when narrow; may be aligned from chamber to chamber or in echelon arrangement. Aperture a low, interiomarginal, umbilical to extraumbilical–umbilical arch. Umbilicus shallow, wide, covered by a tegillum. Sutures on ventral side radial, generally depressed but may be flush or slightly raised in early chambers of last whorl; on dorsal side curved, raised although may be in a depression because of chamber inflation. Surface densely and finely perforate; pustulose ventrally and dorsally, pustules may cluster on the vertex of the dorsal side to form an oval-shaped imperforate area.

Remarks: *Globotruncana marginata* is distinguishable from *G. linneiana* (d'Orbigny) by being generally more biconvex and having more highly inflated chambers. It differs from *G. cretacea* (d'Orbigny) by having less spherically shaped chambers and a more prominently developed keel. This species is highly variable. The characters of keel width and position, number of chambers per whorl and degree of chamber inflation may combine to produce specimens which have different appearances. However, any or all intergrading combinations may be found in a given sample.

Globotruncana marginata is gradational with *G. cretacea* (d'Orbigny), which is regarded as its ancestor.

Jírová (1956) selected a neotype from Lužice in northern Bohemia for *Globotruncana marginata*. The author illustrated several specimens which show the wide morphological variation of this species.

The two taxa *Rosalinella globigerinoides* form. typica and *R. globigerinoides* var. *sublaevigata* of Marie (1941) fall within the concept of *Globotruncana marginata* as presented here.

The type description and figure of *Globotruncana pooleyi* Nakkady and Osman, 1954, are very poor. However, this species is believed to be conspecific with *G. marginata* for in comparing this species with *G. globigerinelloides* Brotzen [= *G. cretacea* (d'Orbigny)], these authors stated that *G. pooleyi* possesses a more distinct double keel and less inflated chambers.

The type figure of *Globotruncana (Rugoglobigerina) loetterli subloetterli* Gandolfi, 1955a, is not an accurate representation of the holotype. The first chamber in the final whorl of this poorly preserved specimen is larger than the succeeding one. Otherwise the chambers are subequal. It is an intraspecific variant of *G. marginata*.

Although the holotype of *Globotruncana wilsoni* Bolli, 1957, has only four chambers in its last whorl, it is in other respects like *G. marginata* to which it is referred as a junior synonym.

Globotruncana hilli Pessagno, 1967, is treated as synonymous with *G. marginata*. The keel width of the latter has been shown to be variable, and that of Pessagno's species falls within the range of *G. marginata*.

Distribution: This is shown in Fig. 126.

Range: Turonian—Maastrichtian (FOS *Rugoglobigerina hexacamerata*—LOS *Globotruncana fornicata* Interval).

Globotruncana mayaroensis Bolli, Plate 48, figs 1–5

Globotruncana mayaroensis Bolli, 1951, p. 25, pl. 35, figs 10–12; Subbotina, 1953, p. 203–206, pl. 8, figs 2a–c; Ayala-Castañares, 1954, p. 407, 408, pl. 10, figs 1a–c; *non* Gandolfi, 1955a, p. 18–20, pl. 1, figs 2a–c, text-figs 4 (10a, b); Wicher, 1956, p. 104, 105, pl. 13, figs 7, 8; Said and Kenawy, 1956, p. 151, pl. 5, figs 23a–c; Witwicka, 1958, p. 225, 226, pl. 18, figs 36a–c; *non* Ansary and Fakhr, 1958, p. 137, 138, pl. 2, figs 20a–c; Maslakova, 1959, p. 111, pl. 12, figs 5a–c; Vinogradov, 1960a, pl. 5, figs 26a–c; Majzon, 1961, p. 763, pl. 2, fig. 4; Herm, 1962, p. 86, 87, pl. 6, fig. 1; Postuma, 1962, 7 figs; Bandy, 1967, p. 20, text-fig. 9 (3); Scheibnerová, 1969,

p. 78, 79, tab. 19, figs 1a–4c; Sturm, 1969, pl. 12, figs 5a–c; Depeuble, 1969, p. 154, pl. 1, figs 3a–d; Al-Shaibani, 1971, p. 106, pl. 1, figs 2a–c; Berggren, 1971, pl. 3, figs 1–4, 11–14.

Rugotruncana mayaroensis (Bolli). Brönnimann and Brown, 1956, p. 553, 554, pl. 22, figs 10–12; Seiglie, 1958, p. 72, 73, pl. 7, figs 1a–c.

Abathomphalus mayaroensis (Bolli). Bolli *et al.*, 1957, p. 43, pl. 11, figs 1a–c; Saavedra, 1965, p. 326, text-fig. 16; Săndulescu, 1966, pl. IXa, figs 3a–c; Salaj and Samuel, 1966, p. 221, pl. 23, figs 4a–c; Pessagno, 1967, p. 372, pl. 92, figs 4–9; pl. 95, fig. 5; *non* Ansary and Tewfik, 1968, p. 48, 49, pl. 6, figs 1a–c; Funnell *et al.*, 1969, p. 26, 27, pl. 2, figs 5–7; text-figs 9a–c; Hanzlíková, 1969, p. 47, pl. 9, figs 7a–8c; Ansary and Tewfik, 1969, pl. 1, figs 2a, b; Neagu, 1970a, fig. 1 (27); Neagu, 1970b, p. 69, 70, pl. 19, figs 6–8; Cita, 1970, pl. 4, fig. 1; Cita and Gartner, 1971, pl. 1, fig. 3; pl. 4, figs 1a–c; Govindan, 1972, p. 180, pl. 6, figs 7–9; Hanzlíková, 1972, p. 114, pl. 32, figs 4a–6c.

Globotruncana (*Globotruncana*) *planata* Edgell, 1957, p. 115, pl. 4, figs 7–9.

Praeglobotruncana (*Praeglobotruncana*) *mayaroensis* (Bolli). Berggren, 1962a, p. 32–36, pl. 8, figs 3a–c.

Globotruncana (*Marginotruncana*) *mayaroensis* (Bolli). (?) Hofker, 1963, p. 281, 282, pl. 1, figs 4a–g.

Praeglobotruncana intermedia (Bolli). Hanzlíková, 1969, p. 46, 47, pl. 9, figs 4a–c (*non* 5a–6c).

Description: Test a low trochospire with the dorsal side flat to slightly convex and the ventral side slightly concave; equatorial periphery lobate, axial periphery broadly truncate. Chambers on dorsal side nearly flat to slightly inflated, elongate in the direction of coiling; ventrally chambers slope from the keel toward the umbilicus, often arched along a radial median line; 13–17 chambers in 3 whorls with 4–6, commonly 5, chambers in the final whorl; chamber size increases gradually to rapidly; with a wide but variably spaced double keel, the greatest width occurring at midpoint of the chamber corresponding to the ventral arch of the chamber; on both the dorsal and ventral carinae are short, radial costellae; imperforate peripheral band tends to slope toward ventral side such that both carinae may be viewed ventrally. Aperture a low, interiomarginal, umbilical to extraumbilical–umbilical arch. Umbilicus shallow, width difficult to determine because of inward slope of chambers, covered by a tegillum. Sutures on dorsal side curved, wide, often raised above chamber level; ventrally radiate, flush to depressed. Surface finely perforate; pustules or short, radial costellae on either side.

Remarks: The costellae keel, chamber shape and overall test shape permit easy recognition of this species. The characteristics of this distinctive *Globotruncana* are rather constant. Ventral chamber arching and associated expansion of the keel width are ontogenetic developments.

The *Globotruncana havanensis-G. intermedia-G. mayaroensis* lineage has been well established by several workers (Bolli, 1951; Berggren, 1962a; Pessagno, 1967).

Bolli *et al.* (1957) introduced the genus *Abathomphalus* using *Globotruncana mayaroensis* as the type species. *Abathomphalus* is clearly a junior synonym of *Globotruncana* (see Remarks under the genus *Globotruncana*). The authors misinterpreted the taxonomic value of the narrow umbilicus and the single tegillar extension from the final chamber which was supposed to have only infralaminal accessory apertures. The first is of specific level rank, and the latter is untrue. Specimens of *G. mayaroensis* have been illustrated with tegillar extensions from all chambers in the last whorl, and which bear both infralaminal and intralaminal accessory apertures (Brönnimann and Brown, 1956; Postuma, 1962; Corminboeuf, 1962).

Berggren (1962a) placed the taxon *mayaroensis* into the genus *Praeglobotruncana* because it possesses "a low-trochospiral test and radial, depressed sutures on the umbilical side, imperforate, asymmetric portici along the umbilical margin of the chambers and an extraumbilical aperture". He also noted the strong evidence to support a *havanensis–intermedia–mayaroensis* phyletic development. Because the first two taxa are considered to be *Praeglobotruncana* by him, this was used as evidence for the inclusion of the taxon *mayaroensis* in that genus. The test shape and the nature of the sutures are also characteristic of many true *Globotruncana*, not just *Praeglobotruncana* (see Remarks under *G. havanensis*). Berggren's latter reasoning, based upon the phyletic lineage, is circular. Even if the taxon *mayaroensis* had developed from a *Praeglobotruncana*, this would not preclude its placement in the *Globotruncana* providing it fits morphologically, which it does. Berggren restricted the *Globotruncana* to those forms which possess a bifurcating keel with an umbilical extension of the ventral carina. He used the lack of this as additional evidence for excluding *mayaroensis* from *Globotruncana*. This type of keel structure is not a definitive characteristic of the *Globotruncana*. It may or may not occur. Interestingly, he stated that through palingenesis this species became homeomorphic to the *Globotruncana* by the development of a double keel, sutural carina and the umbilical tegillum. In which case by his own reasoning, *mayaroensis* is a *Globotruncana*. With fossilised organisms, the morphology must differ between genera if they are to be separated.

Globotruncana planata Edgell, 1957, based upon the type figures and description, is believed to be conspecific with *G. mayaroensis*.

Distribution: This is shown in Fig. 127.

Range: Maastrichtian (*Trinitella scotti* Interval).

Globotruncana nothi (Brönnimann and Brown), Plate 49, figs 1–3
Rugotruncana nothi Brönnimann and Brown, 1956, p. 551, pl. 22, figs 16–18
Globotruncana (*Rugotruncana*) *nothi* Brönnimann and Brown. Majzon, 1961, p. 764, pl. 5, fig. 8.
Globotruncana nothi (Brönnimann and Brown). Olsson, 1964, p. 168, pl. 4, figs 9a–c; *non* Pessagno, 1967, p. 350, pl. 67, figs 4–9; pl. 68, figs 6–8; Hanzlíková, 1972, p. 109, pl. 29, figs 10a–c.

Description: Test biconvex with the dorsal side more convex than the ventral, which is slightly convex to nearly flat; equatorial periphery lobate, axial periphery truncate. Chambers nearly flat to moderately inflated dorsally, slight to moderate inflation ventrally; 13–16 chambers in 3 whorls with 5–7 chambers in the final whorl; gradually to rapidly expanding in size; with a narrowly to moderately spaced, beaded, double keel. Aperture a low, interiomarginal, umbilical to extraumbilical–umbilical arch. Umbilicus moderately wide and shallow, covered by a tegillum. Sutures radial, depressed ventrally; radial, depressed in early whorls dorsally, becoming curved and raised in last whorl. Surface of both sides and keel typically covered with short, rapidly tapering spines and with pustules.

Remarks: The moderately high spire with well-developed spinose ornamentation of this species distinguish it from all other *Globotruncana*. The degree of spiroconvexity varies somewhat. The divergent nature of the keels which was thought to be diagnostic by Brönnimann and Brown (1956) is not a stable characteristic.

Globotruncana nothi is thought to have descended from *G. arca* (Cushman), but this has not been confirmed.

Distribution: This is shown in Fig. 128.

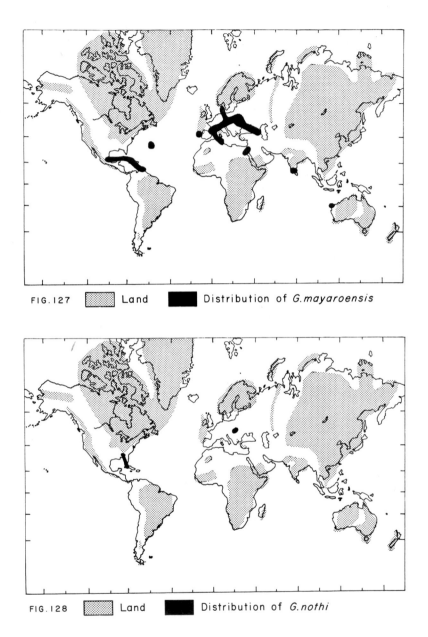

FIG.127　[ ] Land　　[■] Distribution of *G.mayaroensis*

FIG.128　[ ] Land　　[■] Distribution of *G.nothi*

Range: Campanian (*Globotruncana calcarata* Interval)—Maastrichtian (*Trinitella scotti* Interval).

Globotruncana obliqua Herm, Plate 49, figs 4–6
Globotruncana (Marginotruncana) pauperata Hofker. Majzon, 1961, p. 764, pl. 7, fig. 3.
Globotruncana sp. Herm, 1962, p. 85, pl. 4, fig. 6; Küpper, 1964, p. 633, pl. 4, figs 1a–c.
Globotruncana linneiana (d'Orbigny). van Hinte, 1965a, p. 23, pl. 1, figs 3a–c.
Globotruncana linneiana obliqua Herm, 1965, p. 336–338, pl. 8, figs 1–4, text-fig. 14.
Globotruncana loeblichi Pessagno, 1967, p. 349, 350, pl. 73, figs 1–4.
Globotruncana obliqua Herm. Barr, 1968a, p. 316–318, pl. 39, figs 5a–6d; Sturm, 1970, p. 111, pl. 7, figs 1a–c.

Description: Test a flat trochospire with slight biconvexity; equatorial periphery lobate, axial periphery broadly truncate. Chambers of last whorl positioned askew to the direction of coiling; flat to generally inflated dorsally, flat to slightly arched ventrally; 12–15 chambers in 2½–3 whorls with 5–6 chambers in the final whorl; gradually to rapidly increasing in size; ventral chambers imbricate, dorsal petaloid; with a very wide, beaded double keel. The flat spire along with the askewness and dorsal inflation of the chambers may combine to create a dorsal pseudoumbilicus. Aperture a low, interiomarginal, umbilical to extraumbilical arch. Umbilicus wide, shallow, covered by a tegillum. Sutures on ventral side raised and curved back to form the carinate umbilical shoulders; dorsally early chambers have depressed, radial sutures, later chambers have curved sutures which are raised but become depressed toward the previous whorls. Surface with fine to moderately coarse pores, pustulose, and may develop secondarily imperforate areas on the dorsal vertex of the chambers.

Remarks: The distinctive positioning of the chambers in this species, i.e. askew to the direction of coiling, enables easy discrimination from the otherwise similar *Globotruncana lapparenti* Brotzen. The degree of the askewness of the chambers varies markedly. It may be slight to extreme. In the latter situation, this could be misinterpreted as aberrantism.

Globotruncana obliqua grades into *G. lapparenti* Brotzen.
Distribution: This is shown in Fig. 129.
Range: Campanian (FOS *Globotruncana obliqua*—FOS *Globotruncana calcarata* Interval to *Globotruncana calcarata* Interval).

Globotruncana primitiva Dalbiez

Globotruncana (Globotruncana) ventricosa primitiva Dalbiez, 1955, p. 168, text-fig. 6.
Globotruncana primitiva Dalbiez. Lehmann, 1963, p. 145, pl. 4, figs 1a–c; text-fig. 3L; Postuma, 1962, 7 figs.
Praeglobotruncana concavata primitiva Dalbiez Săndulescu, 1966, pl. IVa, figs 2a–c.
Globotruncana linneiana coronata Bolli. Banerji, 1966, p. 63, pl. 4, figs 1–3.
Globotruncana ventricosa primitiva Dalbiez. Salaj and Samuel, 1966, p. 220, pl. 20, figs 4a–c.
Globotruncana concavata primitiva Dalbiez. Scheibnerová, 1968, p. 46, pl. 1, figs 1a–5c; *non* Sturm, 1969, pl. 11, figs 5a–c; Scheibnerová, 1969, p. 71, 72, tab. 14, figs (?) 2a–3c, 6a–c (*non* 1a–c, 5a–c).

Description: Test a very low trochospire with the dorsal side nearly flat to slightly

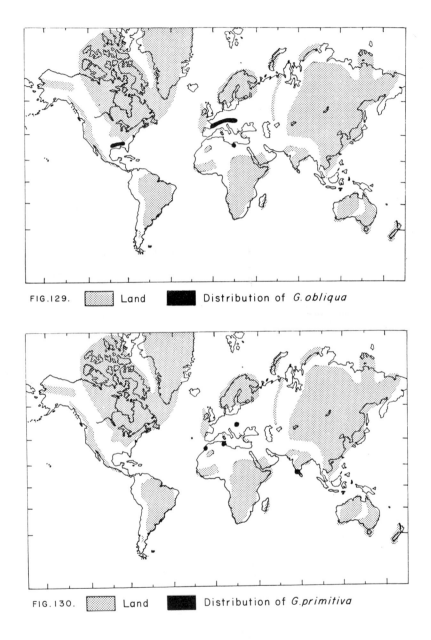

FIG.129. ⬚ Land   ⬛ Distribution of *G. obliqua*

FIG.130. ⬚ Land   ⬛ Distribution of *G. primitiva*

convex, ventral side equally convex in all but the last 2–3 chambers which are much more convex; equatorial periphery lobate, axial periphery truncate to acute. Chambers mildly inflated both dorsally and ventrally except for the final 2–3 which are flat on the dorsal side; 14–16 chambers in 3 whorls with 5–6 chambers in the final whorl; gradually increasing in size; with a narrow, beaded double keel dorsally particularly on later chambers. Aperture a low, interiomarginal, umbilical to extraumbilical–umbilical arch. Umbilicus shallow, wide, covered with a tegillum. Sutures on ventral side gently curved, depressed; dorsally curved, raised. Surface finely perforate, pustulose.

Remarks: *Globotruncana primitiva* differs from *G. concavata* (Brotzen) in being less ventrally convex and in lacking the dorsal concavity. It is distinguished from *G. marginata* (Reuss) by having a more dorsally located double keel. It differs from *G. imbricata* Mornod by being less spiroconvex, having a dorsally positioned keel and having greater ventral convexity. Ventral convexity varies as does the degree of dorsal flatness of the test.

The most likely ancestor of *Globotruncana primitiva* is *G. marginata* (Reuss). Only the development in the later chambers of dorsal flattening and a dorsally positioned keel distinguishes *G. primitiva* from *G. marginata*.

Remarks: Dalbiez (1955) presented only a single edge-view of his new species *Globotruncana primitiva*. The best illustration of this species is by Postuma (1962).

Distribution: This is shown in Fig. 130.

Range: Coniacian—lower Santonian.

Globotruncana pustulifera Masters, Plate 50, figs 1, 3

Globotruncana pustulifera Masters, 1976, p. 329, pl. 2, figs 7–9.

Description: Test trochospiral, unequally biconvex with the dorsal side being moderately high-spired; equatorial periphery gently lobate; axial periphery acutely truncate with a narrow, beaded, double keel. Chambers inflated, 14–16 in 3 whorls with 5–7 in the final whorl, gradually increasing in size; chambers of the final whorl tend to be rhomboidal dorsally, trapezoidal ventrally. Aperture a low, interiomarginal, umbilical to extraumbilical–umbilical arch. Umbilicus wide, deep. Sutures depressed, dorsally curved, ventrally straight to slightly curved. Surface with small diameter pores, abundant pustules.

Remarks: *Globotruncana pustulifera* differs from *G. conica* White by being lower-spired with more inflated chambers and depressed sutures. The main morphological features of this species appear to be stable.

The phylogeny of this species is unknown.

Distribution: This is shown in Fig. 131.

Range: Campanian (*Globotruncana calcarata* Interval)—Maastrichtian (LOS *Globotruncana lapparenti*—FOS *Globotruncana aegyptiaca* Interval).

Globotruncana renzi Gandolfi, Plate 50, fig. 2; plate 51, figs 1, 3

Globotruncana renzi Gandolfi, 1942, p. 124, 125, pl. 3, figs 1a–c; pl. 4, 16 (*non* 15; text-figs 45a–c; (?) Colom, 1947, p. 98, pl. 24, figs 18, 19; pl. 25, figs 15, 17, 19, 20, 22, 23; (?) Colom, 1948, p. 27, pl. 4, figs 14–22 (*non* pl. 2, fig. 8); *non* Hagn and Zeil, 1954, p. 37–39, pl. 3, figs 2a–c; pl. 6, figs 3, 4; Hiltermann, 1956, text-fig. 2 (part); *non* Bolli, 1957, p. 58, pl. 14, figs 3a–c; *non* Książkiewicz, 1958, text-fig. 2 (5a–c); *non* Trujillo, 1960, p. 343, pl. 50, figs 3a–4c; Majzon, 1961, p. 762, pl. 2, fig. 3; Graham, 1962, p. 105, pl. 19, figs 12Aa–Bc; Postuma, 1962, 7 figs; Küpper, 1964, p. 628–630, pl. 2, figs 10a–c; Caron, 1966, p. 77–79, pl. 4, figs 4a–c; text-figs 5a–c (*non* 4a–c);

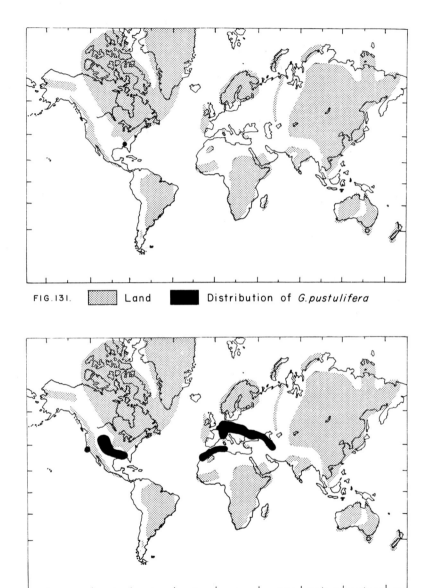

FIG. 131. [Land] Land   [black]   Distribution of *G. pustulifera*

FIG. 132. [Land] Land   [black]   Distribution of *G. renzi*

non Salaj and Samuel, 1966, p. 211, pl. 18, figs 1a–c; *non* Pessagno, 1967, p. 310–313, pl. 55, figs 4–7; pl. 65, figs 20–23; pl. 98, figs 3, 4; Scheibnerová, 1969, p. 66, 67, tab. 13, figs 1a–4c; text-fig. 12—upper (*non* lower); Caron and Luterbacher, 1969, p. 27–29, pl. 9, figs 12a–c; Kalantari, 1969, p. 209, pl. 25, figs 7a–c; Neagu, 1970b, p. 66, pl. 16, figs 1–3; *non* Gawor-Biedowa, 1972, p. 86, 87, pl. 11, figs 4a–c; Hanzlíková, 1972, p. 109, 110, pl. 30, figs 1a–2c.

Globotruncana linnei var. *angusticarinata* Gandolfi, 1942, p. 127, text-fig. 46–3a–c, pl. 18, fig. 17; *non* Kalantari, 1969, p. 207, pl. 23, figs 7a–c.

Globotruncana (*Globotruncana*) sp. aff. *renzi* Thalmann-Gandolfi. Reichel, 1950, p. 612, 613, pl. 16, fig. 8; pl. 17, fig. 8.

Globotruncana angusticarinata Gandolfi. Sigal, 1952, p. 34, text-fig. 37; *non* Sacal and Debourle, 1957, p. 60, pl. 27, figs 2, 4, 23, 24; Majzon, 1961, p. 762, pl. 3, fig. 3; Lehmann, 1963, p. 145, 146, pl. 4, figs 5a–c; Postuma, 1962, 7 figs; Takayanagi, 1965, p. 208, 209, pl. 22, figs 4a–c (*non* 5); Caron, 1966, p. 79, 80, pl. 4, figs 5a–c; Salaj and Samuel, 1966, p. 199, pl. 19, figs 4a–c; *non* Pessagno, 1967, p. 300, 301, pl. 65, figs 17–19; *non* Hanzlíková, 1972, p. 103, pl. 27, figs 2a–3c.

Globotruncana lapparenti angusticarinata (Gandolfi). Książkiewicz, 1956, p. 272, 273, pl. 31, figs (?) 1–3, 6; text-figs 41, 42 (?); Ziegler, 1957, p. 78, pl. 1, figs 11a–c; Tollmann, 1960, p. 192, pl. 20, figs 8, 9; Neagu, 1970b, p. 67, pl. 15, figs 27–29.

Globotruncana coldreriensis Gandolfi, 1957, p. 64, pl. 9, figs 7a–c; Lehmann, 1963, p. 144, pl. 3, figs 3a–c; text-fig. 3j; *non* Salaj and Samuel, 1966, p. 201, 202, pl. 18, figs 2a–c.

Globotruncana fungicamerata Martirosyan, 1958, p. 11, 12, pl. 2, figs 3a–c; Caron, 1966, p. 83, pl. 5, figs 2a–c.

Globotruncana renzi angusticarinata Gandolfi. van Hinte, 1963, p. 67, 68, pl. 2, figs 3a–c; Bandy, 1967, p. 19, text-fig. 8 (9).

Globotruncana tarafayaensis Lehmann, 1963, p. 146, 147, pl. 5, figs 4a–c; text-fig. 2i.

Praeglobotruncana renzi (Gandolfi). Douglas and Sliter, 1966, p. 106, pl. 4, figs 3a–c; (?) Eicher 1966, p. 28, 29, pl. 6, figs 9a–c; Douglas, 1969, p. 172, 173, pl. 2, figs 8a–c.

Globotruncana mariei Banner and Blow. Caron, 1966, p. 85, 87, pl. 6, figs 1a–c.

Globotruncana renzi renzi Gandolfi. Bandy, 1967, p. 19, text-fig. 8 (8).

Marginotruncana renzi angusticarinata (Gandolfi). (?) Esker, 1969, p. 214, 216, pl. 3, figs 10–12.

Marginotruncana renzi renzi (Gandolfi). *non* Esker, 1969, p. 216, pl. 1, figs 12–14.

Globotruncana concavata (Brotzen). Moorkens, 1969, p. 452, pl. 3, figs 3a–c.

Praeglobotruncana difformis (Gandolfi). Eicher and Worstell, 1970b, p. 308, 309, pl. 11, figs 4a–6c.

Marginotruncana angusticarinata (Gandolfi). (?) Porthault, 1970, in Donze *et al.*, p. 76, 77, pl. 10, figs 15–17 (*non* pl. 13, fig. 18).

Marginotruncana tarfayaensis (Lehmann). (?) Porthault, 1970, in Donze *et al.*, p. 80, 81, pl. 11, figs 6–8.

Marginotruncana renzi (Gandolfi). (?) Porthault, 1970, in Donze *et al.*, p. 75, 76, pl. 10, figs 13a–14.

Description: Test biconvex with a moderately strong dorsal convexity and a lesser ventral convexity; equatorial periphery slightly lobate; axial periphery a truncated acute margin. Chambers nearly flat to slightly inflated on dorsal side, ventrally sloping with or without minimal convexity to the umbilical shoulder; elongate in the direction of coiling, crescent-shaped dorsally, imbricate ventrally with slight arching; 14–17 chambers in 3 whorls with 5–7 chambers in the last whorl; gradually expanding to subequal in size; with a narrow, beaded, double keel. Aperture a low, interiomarginal, umbilical to extraumbilical–umbilical arch. Umbilicus wide, moderately deep, covered by a tegillum. Sutures on dorsal side curved, raised; ventrally raised to flush or depressed in later chambers, curved abaperturally to form a carinate umbilical shoulder. Surface densely and finely perforate; smooth to pustulose both ventrally and dorsally, with the latter side developing secondary imperforate areas on the chamber vertex.

Remarks: *Globotruncana renzi* has a narrower keel and a higher spire than *G. coronata* Bolli. The spire height varies from moderately low to moderately high. Otherwise, the features of this species are rather stable. The supposed single keel on the last one or two chambers results from the progressively decreasing and ultimate lack of the two carinae bordering the imperforate peripheral band. This is attributed to a weaker laminal buildup on the younger chambers and not to a true ontogenetic loss of the double keel.

This species is thought to have evolved from *G. coronata* Bolli by developing the above-mentioned features.

Pessagno (1967) reviewed the taxonomic history of *Globotruncana renzi*. He noted that although Thalmann's (1942) use of the name predated that of Gandolfi, it was a nomen nudem, which Thalmann did not validly propose until 1945. In the meantime, Gandolfi (1942) did properly establish the taxon *G. renzi*, making Thalmann's *renzi* the junior homonym. The situation became more confused when Gandolfi (1957) proposed the name *G. coldreriensis* as a substitute for his *G. renzi*, because he incorrectly thought Thalmann's usage had priority.

The fact that Gandolfi did not select a holotype from among those specimens he illustrated in 1942 prompted Pessagno (1967) to select the specimen represented by text-fig. 45a–c of Gandolfi as the lectotype of *G. renzi*. Caron and Luterbacher (1969) concluded that this specimen is too badly preserved to serve as the type specimen and, thereby, rejected it, a move which can be done only with the approval of the International Commission on Zoological Nomenclature. Those authors found that the specimen illustrated on plate 3, figures 1a–c had been labelled by Gandolfi as the holotype, although this in itself does not constitute a valid designation. But because of these facts, Caron and Luterbacher regarded this specimen (pl. 3, figs 1a–c, Gandolfi, 1942) as the lectotype.

Pessagno (1967) did not accept the interpretation of Ellis and Messina (Catalogue of Foraminifera) that the holotype of *Globotruncana coldreriensis* Gandolfi is automatically the lectotype of *G. renzi* Gandolfi. Article 74a (i) of the I.C.Z.N. states that "The first published designation of a lectotype fixes the status of the specimen . . .". This Gandolfi (1957) did when he designated the holotype for *G. coldreriensis* because this was merely a new name for his *G. renzi*, which he regarded as distinct from Thalmann's. Furthermore, because Gandolfi's *G. renzi* is the senior homonym and a validly designated new species, *G. coldreriensis*, as an unnecessary replacement name, becomes a junior *objective* synonym of *G. renzi* Gandolfi. Type-fixation of *G. coldreriensis* must also, then, apply to *G. renzi* Gandolfi, paralleling the generic-level name replacement (I.C.Z.N., Art. 67i). Thus, figs 1a–c on plate 3 of Gandolfi (1942) is the type specimen for his *G. renzi* by subsequent designation, making the selections of both Pessagno and of Caron and Luterbacher invalid. This is particularly fortunate in Pessagno's case because his designated lectotype does not coincide with the majority interpretation of this species.

Globotruncana angusticarinata Gandolfi, 1942, is a junior synonym of *G. renzi*. The only difference between the type figures of the two is that the former has one additional chamber in the final whorl. Caron and Luterbacher (1969) report that the holotype of *G. angusticarinata* is lost, and that the remaining specimens in Gandolfi's collection have fewer chambers in the last whorl. They remarked that specimens intermediate between the two are frequent. In a brief description of *G. angusticarinata*, they commented that its keel [carina] bifurcates and is distinctly separated at the contact with the succeeding chamber. This feature is common to all species in which the carina pass from the periphery to the dorsal and ventral sides as raised sutures. Those specimens attributed to *G. angisticarinata* are merely ontogenetically older individuals of *G. renzi*.

In text-fig. 46 (3a–c) of Gandolfi (1942), *Globotruncana angusticarinata* was spelled *G. angusticarenata*. This is an apparent typographical error for all other citations are spelled with an "i" instead of an "e".

There seem to be no morphologic or stratigraphic differences to distinguish *Globotruncana fungicamerata* Martirosyan from *G. renzi*. Thus, it is treated as a junior synonym of the latter.

Lehmann (1963) distinguished his new species *Globotruncana tarfayaensis* from *G. coronata* on the basis that it possesses a closely spaced double keel, a difference which also separates *G. renzi* from the latter. Moreover, the number of chambers and their morphology, and the spire height of *G. tarfayaensis* are identical to those of *G. renzi*, which is considered the senior synonym.

The specimens identified as *Praeglobotruncana difformis* (Gandolfi) by Eicher and Worstell (1970b) are probably *Globotruncana renzi*.

Distribution: This is shown in Fig. 132.

Range: Turonian—Campanian (FOS *Globotruncana obliqua*—FOS *Globotruncana calcarata* Interval).

Globotruncana repanda Bolli

Globotruncana repanda Bolli, 1957, p. 56, pl. 13, figs 2a–c; Majzon, 1961, p. 764, pl. 6, fig. 4.

Praeglobotruncana repanda (Bolli). Bandy, 1967, p. 18, text-fig. 8 (4).

Description: "Shape of test: very low trochospiral, spiral side concave, umbilical side strongly inflated equatorial periphery lobate, early chambers of last whorl with double keel, which may be absent in the ultimate and penultimate chambers. Wall: calcareous, perforate, surface in well-preserved specimens slightly rugose, especially on the umbilical side. Chambers globular to hemispherical; 12–15; arranged in 2–3 whorls, the 4 chambers of the last whorl increase rapidly in size, earlier whorls small by comparison. Sutures: spiral side almost radial, depressed; umbilical side radial, depressed. Umbilicus: deep, wide. Apertures: primary apertures interiomarginal, umbilical; tegilla with accessory apertures not preserved in Trinidad material, but present in specimens of this species from the Gulf Coast. Coiling: the 25 specimens counted all coiled dextrally." (Type description.)

Remarks: This species is the only *Globotruncana* with a dorsally positioned double keel on generally subspherical chambers. The range of variation within this species is unknown.

Globotruncana repanda shows a striking similarity to immature specimens of *G. concavata*, but a phyletic relationship is as yet unverified.

Distribution: This is shown in Fig. 133.

Range: Campanian (FOS *Globotruncana elevata*—LOS *Globotruncana concavata* to FOS *Globotruncana obliqua*—FOS *Globotruncana calcarata*).

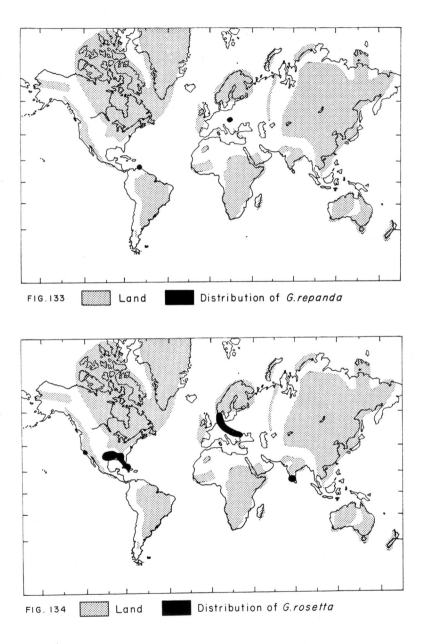

FIG.133 [Land] [Distribution of] *G.repanda*

FIG.134 [Land] [Distribution of] *G.rosetta*

Globotruncana rosetta (Carsey), Plate 51, fig. 2; plate 52, fig. 1
Globigerina rosetta Carsey, 1926, p. 44, 45, pl. 5, figs 3a, b.
Globotruncana arca (Cushman). Plummer, 1931, p. 195–198, pl. 13, figs 9a–c, 11a–c (*non* 7a–8c).
Globotruncana rosetta (Carsey). Majzon, 1943, pl. 1, fig. 14; (?) Cita, 1948b, p. 158, 159, pl. 4, figs 5a–c; (?) Hagn, 1953, p. 98, 99, pl. 8, figs 16a–c; text-figs 24–25; Brönnimann and Brown, 1956, p. 545, 546, pl. 21, figs 11–13; *non* Sacal and Debourle, 1957, p. 60, pl. 28, figs 1, 3; Seiglie, 1958, p. 70, pl. 5, figs 3a–c; Majzon, 1961, p. 762, pl. 1, fig. 14; Barr, 1962, p. 575, 576, pl. 70, figs 4a–c; *non* Lehmann, 1963, p. 147, pl. 8, figs 1a–c; Martin, 1964, p. 83, pl. 10, figs 5a–c; *non* Olsson, 1964, p. 168, 169, pl. 4, figs 5a–c, 7a–c; *non* Săndulescu, 1966, pl. IXa, figs 1a–c; Esker, 1968b, p. 170, 171, text-figs 1–3; *non* Barr, 1968, p. 316, pl. 39, figs 4a–c; *non* Sliter, 1968, p. 105, 106, pl. 18, figs 3a–4c; *non* Rao *et al.*, 1968, p. 25, pl. 4, figs 1–3; *non* Ansary and Tewfik, 1968, p. 55, 56, pl. 8, figs 3a–c; *non* Hanzlíková, 1969, p. 49, pl. 11, figs 3a–5c; (?) Sturm, 1970, p. 112, pl. 7, figs 2a–c; *non* Rahhali, 1970, p. 60, 61, text-figs 3f, g; *non* Longoria, 1970, p. 69–72, pl. 5, figs 7–9; pl. 9, fig. 8; pl. 16, figs 4, 5; *non* Hanzlíková, 1972, p. 110, pl. 30, figs 5a–7c.
(?) *Globotruncana flexuosa* van der Sluis, 1950, p. 21, pl. 1, figs 7a–8c; (?) Ayala-Castañares, 1954, p. 391, 392, pl. 4, figs 3a–c; (?) Majzon, 1961, p. 763, pl. 4, fig. 4; *non* Herm, 1962, p. 77, 78, pl. 7, fig. 7.
Globotruncana rosetta rosetta (Carsey). *non* Gandolfi, 1955a, p. 66, 67, pl. 6, figs 1a–c, text-figs 10a–c; van Hinte, 1963, p. 89–91, pl. 9, figs 3a–c (*non* 1a–2c; pl. 10, figs 1a–c); (?) Salaj and Samuel, 1966, p. 212, 213, text-fig. 16; Bandy, 1967, p. 20, text-fig. 9 (12).
Globotruncana (Globotruncana) rosetta (Carsey). Berggren, 1962a, p. 56–60, pl. 10, figs 1a–c; *non* van Hinte, 1965b, p. 85, pl. 3, figs 1a–c.
Globotruncana rosetta falsostuarti Sigal. van Hinte, 1963, p. 91, 92, pl. 10, figs 2a–c.
Globotruncana putalensis Takayanagi, 1965, p. 221–223, pl. 27, figs 2a–c.
Globotruncana thalmanni flexuosa (van der Sluis). *non* Salaj and Samuel, 1966, p. 216, pl. 22, figs 5a–c.
Globotruncana stuarti stuarti (de Lapparent). Neagu, 1970b, p. 68, pl. 18, figs 1–3.
Description: Test biconvex with the ventral side being strongly convex and the dorsal side slightly convex; equatorial periphery mildly lobate, axial periphery acute. Chambers flat dorsally; with a high angle slope to the umbilical shoulder ventrally, then drops sharply into umbilicus; 14–16 chambers in 3 whorls with 5–6 chambers in final whorl; gradually increasing in size; with a beaded, double keel in the early chambers of last whorl, ventral carina not developed in final 2–3 chambers; crescent-shaped on dorsal side, imbricate ventrally and slightly elongate in direction of coiling. Aperture a low, interiomarginal, umbilical arch. Umbilicus moderately deep, wide, covered by a tegillum. Sutures on dorsal side curved, raised; ventrally curved to form carinae on umbilical shoulder, raised to flush. Surface finely perforate pustulose ventrally.
Remarks: *Globotruncana stuarti* (de Lapparent) differs from *G. rosetta* in being single-keeled throughout. *Globotruncana ventricosa* White has a well-developed, more widely spaced double keel throughout, and is less convex dorsally than *G. rosetta*. The main variable characteristic of this species is the extent of the double keel in the last whorl.
G. coronata is a likely candidate for the ancestor of *Globotruncana rosetta*.

Although no intermediate forms have yet been recognised, progressive diminution of the ventral carina and increase of the ventral convexity could have occurred.

Van der Sluis (1950) studied a Cretaceous fauna which he dated as Maastrichtian on the basis of a single benthonic species, *Palmula reticula* (Reuss). He listed but did not illustrate species of planktonics which are typical of the Maastrichtian, including an illustrated specimen of *Gublerina*. From this fauna, he proposed a new species— *Globotruncana flexulosa*. This species bears a striking resemblance to *G. sigali* Reichel, but it appears much too late to be that species. Although *G. flexulosa* is primarily single-keeled, van der Sluis did report the occasional presence of a weakly developed double keel, which is suggestive of *G. rosetta*. The type figure shows *G. flexulosa* to be less convex ventrally than *G. rosetta*, but this may be due to the quality of the illustration. *Globotruncana flexulosa* is tentatively referred to *G. rosetta*.

Douglas (1969) examined the holotype of *Globotruncana putahensis* Takayanagi and found it to possess only a single keel, contrary to the type description. He decided that it was a junior synonym of *G. stuartiformis* Dalbiez. However, in 1970, Douglas reversed his position claiming that *G. putahensis* was a valid new taxon because he found otherwise identical forms, which possess two keels, from the type formation of this species. The author examined the holotype of *Globotruncana putahensis* Takayanagi, and found it to be identical to *G. rosetta*.

The type figures of *G. putahensis* show a well-developed double keel in all but the final two chambers of the last whorl. This is in error. The ventral keel is weakly developed and present only in the first chamber of the last whorl. The remaining chambers have only the dorsal keel and the imperforate peripheral band. Several of the chambers exhibit a pustulose ventral surface. The pustules are distributed randomly over the surface, but never on the imperforate band. The pustules, when adjacent to the imperforate band, are discrete, never merging into a keel, but could be easily misinterpreted as such.

Distribution: This is shown in Fig. 134.

Range: Campanian (FOS *Globotruncana elevata*—FOS *Globotruncana concavata* Interval)—Maastrichtian (*Trinitella scotti* Interval).

Globotruncana schneegansi Sigal

Globotruncana schneegansi Sigal, 1952, p. 33, text-fig. 34; (?) Hagn and Zeil, 1954, p. 37, pl. 5, fig. 12; *non* Bolli, 1957, p. 58, pl. 14, figs 1a–c; Trujillo, 1960, p. 343, 344, pl. 49, figs 9a–c; Majzon, 1961, p. 763, pl. 3, fig. 2; *non* Lehmann, 1963, p. 144, 145, pl. 4, figs 4–c; text-fig. 3k; Postuma, 1962, 7 figs; Graham, 1962, p. 105, pl. 19, figs 14a–c; Lehmann, 1965, text-fig. 1k; Carson, 1966, p. 83, 85, pl. 3, figs 1a–c; (?) Salaj and Samuel, 1966, p. 215, pl. 17, figs 4a–c; Bandy, 1967, p. 19, text-fig. 8 (7); Kalantari, 1969, p. 209, 210, p. 23, figs 9a–c; Moorkens, 1969, p. 448, pl. 2, figs 1a–d; *non* Sturm, 1969, pl. 11, figs 3a–c.

Praeglobotruncana schneegansi (Sigal). Klaus, 1959, p. 796, 797, pl. 6, figs 5a–c; Sǎndulescu, 1966, pl. IIIa, figs 3a–c.

Globotruncana aff. *schneegansi* Sigal. Hanzlíková, 1963, pl. 2, figs 4a–c.

Description: Test equally biconvex or with the ventral side being more convex; equatorial periphery lobate, axial periphery sharply acute. Chambers flat dorsally, sloping in a low to moderate angle to the umbilical shoulder ventrally; semicircular on dorsal side, early chambers of last whorl on ventral side are equidimensional in width and length but become slightly and gradually elongate in direction of coiling; 17–20 chambers in 3–3½ whorls with 5, usually 6 final chambers; with a single, beaded keel, bifurcating to form sutures. Aperture a low, interiomarginal umbilical to extraumbilical–umbilical arch. Umbilicus moderately deep, wide, with a tegillum. Sutures of dorsal side raised, curved; ventrally depressed to slightly raised, slightly curved, forming carinae on umbilical shoulder. Surface pustulose.

Remarks: This species lacks the thicker keel and distinctly imbricate ventral chambers of *Globotruncana sigali* Reichel, which is the only species with which it might be confused. Minor variations in ventral convexity and in ventral suture elevation are known to occur. Often the carina forming the ventral suture is covered by the succeeding chamber.

Intermediate forms between *Globotruncana schneegansi* and *G. sigali* are known. The ranges of these two species are similar, and they may prove to be intraspecific variants. The parent stock may have been within *Rotalipora*. Several features are shared between *G. schneegansi* and the *Rotalipora*. Particularly significant is the simple tegillum found on some specimens which is only slightly modified from the chamber structure of the *Rotalipora*.

Distribution: This is shown in Fig. 135.

Range: Lower Turonian—lower Santonian.

Globotruncana semsalensis Corminboeuf

Globotruncana (?) *semsalensis* Corminboeuf, 1962, p. 493–496, pl. 2, figs 1a–3c.

Remarks: The type figures of this species are stylised, but that of the holotype and one paratype (figs 2a–c) are strikingly similar to *Praeglobotruncana delrioensis* (Plummer) and the other paratype (figs 3a–c) to *P. stephani* (Gandolfi). However, Corminboeuf's species is recorded from the upper Campanian, which is much too late for the *Praeglobotruncana*. *Globotruncana semsalensis* must be re-examined along with topotypic suites to determine the possibility of contamination or reworking.

Globotruncana sigali Reichel, Plate 52, figs 2, 3; plate 53, fig. 1

Globotruncana (*Globotruncana*) *sigali* Reichel, 1950, p. 610–612, pl. 16, fig. 7; pl. 17, fig. 7; text-figs 5a–c, 6.

Globotruncana sigali Reichel. Sigal, 1952, p. 32, 33, text-fig. 33; (?) Carbonnier, 1952, p. 117, pl. 7, figs 1a–c; Hagn and Zeil, 1954, p. 35, 36, pl. 2, figs 1a–c; pl. 6, fig. 2 (?); Ayala-Castañares, 1954, p. 409, 410, pl. 10, figs 3a–c; Książkiewicz, 1958, text-fig. 1 (12a–c); *non* Klaus, 1959, p. 819, 820, pl. 7, figs 1a–c; Majzon, 1961, p. 763, pl. 2, fig. 10; Postuma, 1962, 7 figs; (?) Lehmann, 1963, p. 143, 144, pl. 3, figs 2a–c; text-fig. 3i; Hanzlíková, 1963, pl. 2, figs 1a–c; Caron, 1966, p. 77, pl. 4, figs 3a–c; Salaj and Samuel, 1966, p. 214, 215, pl. 17, figs 5a–c; Moorkens, 1969, p. 449, pl. 1, figs 4a–c; Douglas, 1969, p. 185, 186, pl. 1, figs 4a–c.

Globotruncana schneegansi Sigal. Lehmann, 1963, p. 144, 145, pl. 4, figs 4a–c; text-fig. 3k; Sturm, 1969, pl. 11, figs 3a–c.

Globotruncana undulata Lehmann, 1963, p. 148, pl. 9, figs 3a–c.

Globotruncana aff. *undulata* Lehmann. Lehmann, 1963, p. 148, 149, pl. 9, figs 4a–c.

Marginotruncana sigali (Reichel). *non* Pessagno, 1967, p. 313, 314, pl. 54, figs 4–6; pl. 56, figs 1–3; pl. 57, figs 1, 2; pl. 98, figs 6, 7; *non* Esker, 1969, p. 216, pl. 3, figs 14–16; Porthault, 1970, in Donze *et al.*, p. 79, 80, pl. 11, figs 4a–5.

Description: Test equally biconvex or with dorsal side slightly more convex than ventral; equatorial periphery gently lobate, axial periphery acute. Chambers of dorsal side flat or with minimal convexity, slope at a low angle up to the umbilical shoulder on ventral side; crescent-shaped or semicircular dorsally, imbricate and slightly elongate in direction of coiling ventrally; 15–18 chambers in 3–3½ whorls with 5–7 chambers in the final whorl; gradually increasing in size to subequal in

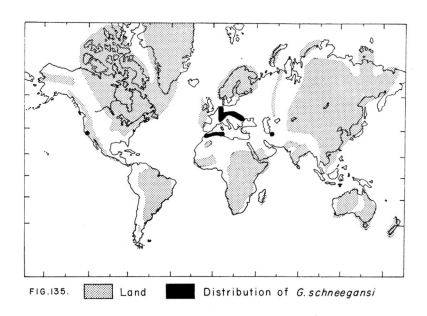

FIG.135. ▨ Land ▮ Distribution of *G. schneegansi*

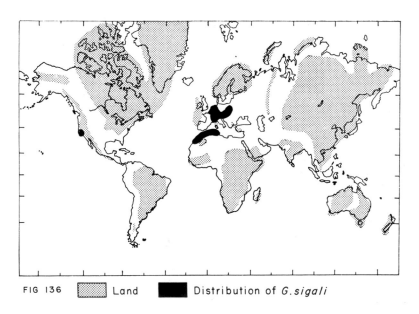

FIG 136 ▨ Land ▮ Distribution of *G. sigali*

final whorl; with a single, beaded keel which bifurcates adapertually to form the sutures. Aperture a low, interiomarginal, umbilical to extraumbilical–umbilical arch. Umbilicus wide, moderately deep, covered with a tegillum. Sutures raised, curved dorsally; raised, curved to sigmoid forming carinated umbilical shoulder ventrally, last 1–2 may be depressed. Surface finely perforate, smooth.

Remarks: *Globotruncana sigali* is distinguished from *G. schneegansi* Sigal by its imbricate ventral chambers. In very large specimens, the dorsal walls of the last few chambers tend to undulate.

For phylogeny see Remarks under *Globotruncana schneegansi* Sigal.

Topotypes of *Globotruncana undulata* Lehmann, 1963, are identical with *G. sigali*. The former merely represents an ontogenetically older and larger individual.

Esker (1969) reported that *Marginotruncana* (= *Globotruncana*) *sigali* has two closely spaced keels in the early portion of the final whorl. Reichel's species is clearly single-keeled throughout as demonstrated by his thin-section (text-fig. 6). The specimens of Esker are incorrectly assigned to this species.

Distribution: This is shown in Fig. 136.

Range: Turonian—Santonian.

Globotruncana stuarti (de Lapparent), Plate 53, figs 2, 3; Plate 54, fig. 1

Rosalina stuarti de Lapparent, 1918, p. 11, 12, text-fig. 4, pl. 1, figs 5–7; Colom, 1931, p. 33, text-fig. 2; pl. 1 (part); pl. 2 (part); *non* Murgeanu, 1933, pl. fig. 6; (?) Colom, 1934, pl. 1, fig. 40 (part); (?) Marin *et al.*, 1934, p. 656, text-figs 3 (1a–c).

Globotruncana rosetta (Carsey). White, 1928b, p. 286, pl. 39, figs 1a–c; Noth, 1961, p. 78, pl. 8, figs 13a–c.

Globotruncana (?) *stuarti* (de Lapparent). *non* Arni, 1933, p. 11, text-figs 4e, f.

Globotruncana stuarti (de Lapparent). Glaessner, 1937, p. 39, 40, pl. 1, figs 13a–c; Majzon, 1943, p. 76, 77, pl. 1, fig. 10; pl. 2, figs 3a–c; Keller, 1946, p. 102, pl. 2, fig. 13 (*non* pl. 3, fig. 7); Cita, 1948b, p. 160, 161, pl. 4, figs 7a–c; Reichel, 1950, p. 613, 614, text-fig. 7a, pl. 16, fig. 10; pl. 17, fig. 10; *non* Lipparini, 1950, p. 172, 173, text-fig. 2; Bolli, 1951, p. 196, pl. 34, figs 10–12; Sigal, 1952, p. 40, text-fig. 42; Tilev, 1952, p. 34–41, text-figs 7a–9d, pl. 1, fig. 3; *non* Subbotina, 1953, p. 231–234, pl. 15, figs 3a–5c; Dalbiez, 1955, text-figs 4a–c; Knipscheer, 1956, p. 52, 53, text-figs 2, 3, pl. 4, figs 19a–20c; *non* Said and Kenawy, 1956. p. 151, pl. 5, figs 22a–c; Martirosyan, 1958, p. 15, 16, pl. 4, figs 3a–c; Seiglie, 1958, p. 70, 71, pl. 1, figs 4a–c; *non* Witwicka, 1958, p. 211, 212, pl. 15, figs 27a–c; Nagappa, 1959, pl. 7, figs 1a–c; Eternod Olvera, 1959, p. 120–122, pl. 8, figs 7–9; *non* Haque, 1959, p. 21, pl. 2, figs 4a–c; Thalmann and Ayala-Castañares, 1959, pl. 1, figs 10, 11 (?); pl. 2, figs 1 (?), 2, 8; Maslakova, 1959, p. 110, 111, pl. 14, figs 1a–c; Christodoulou, 1960, p. 11, text-fig. 15 (part); Vinogradov, 1960a, pl. 6, figs 29a–c (*non* 30a–c); Noth, 1961, p. 78, pl. 8, figs 12a–c; Majzon, 1961, p. 761, pl. 1, fig. 10; Reutter and Serpagli, 1961, pl. 14, fig. 2; Visser and Hermes, 1962, encl. 17, figs 46a–c; Postuma, 1962, 7 figs; Pokorný, 1963, p. 386, figs 433d, e; Pirini and Radrizzani, 1963, pl. 41, fig. 52; pl. 42; figs 56, 57 (?); Farinacci and Radoičić, 1964, p. 278, pl. 14, fig. 2 (part); *non* Said and Kenawy, 1965, p. 151, pl. 5, figs 22a–c; Broglio Lorigo and Mantovani, 1965, pl. 112, figs 15, 16; pl. 113, fig. 3; pl. 114, fig. 5; Lehmann, 1966, p. 310, 311, pl. 1, fig. 1; Salaj and Samuel, 1966, p. 215, 216, pl. 23, figs 3a–c; Săndulescu, 1966, pl. VIIa, figs 5a–c; Pessagno, 1967, p. 356, 357, pl. 81, figs 1–6; Mabesoone *et al.*, 1968, p. 175, figs 5: 8a–c; Scheibnerová, 1969, p. 77, tab. 17, figs 3a–c; tab. 18, figs 1a–c; Dupeuble, 1969, p. 155, pl. 1, figs 1a–d; Sturm, 1969, pl. 12, figs 4a–c; Funnell *et al.*, 1969, p. 33, 34, pl. 4, figs 7–9; text-figs 17a–c; Hanzlíková, 1969, p. 50, pl. 11, figs 6a–c; Bate and Bayliss, 1969, pl. 4, figs 6a–c; Sturm, 1970, p. 112, pl. 8, figs 1a–c; Longoria, 1970, p. 72–76, pl. 6, figs 1–3; pl. 11, figs 1, 2; pl. 18, fig. 4; Berggren, 1971, pl. 1, figs 1–7; Hanzlíková, 1972, p. 111, pl. 31, figs 1a–2c; Govindan, 1972, p. 176, 177, pl. 6, figs 13–15.

    Globotruncana arca (Cushman). Cushman, 1946, p. 150, pl. 62, figs 5a–c.

    Globotruncana linnei stuarti (de Lapparent). Colom, 1948, p. 28, pl. 2, figs 13, 15–18.

    Globotruncana (*Globotruncana*) *stuarti* (de Lapparent). Reichel, 1950, p. 613–615, pl. 17, fig. 10; pl. 18, fig. 10; text-figs 7a, b; Papp and Küpper, 1953, p. 39, 40, pl. 2, figs 2a–c; Papp, 1955, p. 331, text-fig. 4 (3); *non* Berggren, 1962a, p. 60–64, pl. 10, figs 2a–c.

    Globotruncana stuarti stuarti (de Lapparent). *non* Gandolfi, 1955a, p. 64, 65, pl. 5, figs 6a–c; Herm, 1962, p. 89–91, pl. 8, fig. 1; Pessagno, 1962, pl. 2, figs 1–3; Borsetti, 1962, pl. 7, figs 1 (?), 2; text-figs 234, 236, 237 (*non* 223); Olsson, 1964, p. 169, 170, pl. 5, figs 9a–c; El-Naggar, 1966, p. 133–135, pl. 8, fig. 4a–d; pl. 9, fig. 1a–d; Bandy, 1967, p. 21, text-fig. 9 (14); *non* Ansary and Tewfik, 1968, p. 49, 50, pl. 7, figs 4a–c; *non* Neagu, 1970b, p. 68, pl. 18, figs 1–3.

    Globotruncana armenica Martirosyan, 1958, p. 12, pl. 2, figs 4a–c.

    Globotruncana sp. Pirini and Radrizzani, 1963, pl. 12, fig. 17(2).

    Globotruncanita stuarti (de Lapparent). Maslakova, 1963b, pl. 7, figs 1, 3 (?).

    Globotruncana stuarti subspinosa Pessagno. El-Naggar, 1966, p. 139, 140, pl. 10, figs 2a–3c.

    transitional *Globotruncana stuarti subspinosa* to *Globotruncana stuarti stuarti* (de Lapparent). El-Naggar, 1966, pl. 10, figs 1a–c.

    Globotruncana stuarti stuartiformis Dalbiez. Funnell *et al.*, 1969, p. 34, 35, pl. 4, figs 10–12, text-figs 18a–c.

    Globotruncana atlantica Caron, 1972, p. 553, pl. 1, figs 5a–c; text-figs 1a–3c.

    Globotruncana stuartiformis Dalbiez. Govindan, 1972, p. 177, pl. 6, figs 1–6.

    Description: Test biconvex with the ventral side typically more strongly so; equatorial periphery gently lobate to nearly entire, axial periphery acute. Chambers on dorsal side flat, sloping steeply to umbilical shoulder ventrally; crescent-shaped to semicircular dorsally, becoming nearly rhomboidal in final whorl; ventrally slightly imbricate and elongate in direction of coiling; gradually increasing in size to subequal in final whorl; 18–23 chambers in 3–3½ whorls with 5–7 in the final whorl; with a single, beaded keel throughout, bifurcating to form dorsal and ventral sutures and the carinated umbilical shoulder. Aperture a moderately high, interiomarginal, umbilical arch. Umbilicus moderately wide, deep, covered by a tegillum. Sutures raised, curved on dorsal side; flush to raised, curved ventrally.

    Surface densely perforate, smooth dorsally, lightly pustulose ventrally.

    Remarks: This species differs from *Globotruncana elevata* (Brotzen) by having a uniformly convex dorsal side without the undulating surface of the latter. It differs from *G. conica* by lacking the high spiroconvexity, the slightly convex ventral side and the rectangular-shaped dorsal chambers of the final whorl. The degree of spiroconvexity is variable, often low, but never flat, to rather high.

    Three species have been considered by various workers as ancestral to *Globotruncana stuarti*: *G. rosetta* (Carsey) through the loss of the ventral keel, *G. conica* White by a decrease in the spiroconvexity and *G. elevata* (Brotzen) from an increase in the spiroconvexity.

    Globotruncana rosetta (Carsey) is double-keeled in all but the last few chambers where the ventral carina becomes reduced and finally disappears. Thin-sections of *G. stuarti* demonstrate that it is single-keeled throughout. No gradational forms are known.

There do appear to be intermediate forms between *Globotruncana conica* White and *G. stuarti*. One such form was illustrated by de Lapparent (1918). Associated with the decrease in spiroconvexity is a reduction in chamber length, which gives rise to their rectangular shape.

Gradational forms with *Globotruncana elevata*, which loses the ontogenetically late developing spiroconvexity, have also been recognised. Thus, *G. stuarti* is probably polyphyletic.

Pessagno (1967) designated the specimen represented by the lower three figures of text-figure 4 of de Lapparent (1918) as the lectotype of *Globotruncana stuarti*.

Globotruncana armenica Martirosyan, 1958, is regarded as a neanic individual of *G. stuarti*.

Berggren (1962a) described *Globotruncana stuarti* as having a single dorsal keel which bifurcates at the distal chamber margin to form ventral sutural carinae. However, the specimen he illustrated shows a double keel, and is not a *G. stuarti*.

Caron (1972) illustrated an unequally biconvex, single-keeled form as the new species *Globotruncana atlantica*. She stated that it differs from *G. stuarti* by having triangular-shaped dorsal chambers and in being more spiroconvex. The chambers are no more "triangular" than those of *G. stuarti*. Although *G. atlantica* has a spiroconvexity greater than the average for *G. stuarti*, it is believed to be within the limits of intraspecific variation.

Distribution: This is shown in Fig. 137.

Range: Maastrichtian (LOS *Globotruncana fornicata*—FOS *Trinitella scotti* Interval to *Trinitella scotti* Interval).

Globotruncana subcircumnodifer Gandolfi

Globotruncana cf. *globigerinoides* Brotzen. Bolli, 1951, p. 198, pl. 35, figs 16–18.

Globotruncana (*Rugoglobigerina*) *circumnodifer subcircumnodifer* Gandolfi, 1955a, p. 44, pl. 2, figs 8a–c.

Globotruncana (*Rugoglobigerina*) *rugosa* subsp. *subrugosa* Gandolfi, 1955a, p. 72, 73, pl. 7, figs 5a–c.

Rugotruncana tilevi Brönnimann and Brown, 1956, p. 547, 548, pl. 22, figs 1–3.

Rugotruncana sp. 2 Seiglie, 1958, p. 73, pl. 7, figs 4a–c.

Globotruncana (*Rugotruncana*) *tilevi* Brönnimann and Brown. Majzon, 1961, p. 764, pl. 5, fig. 3.

Globotruncana (*Rugotruncana*) *subcircumnodifer* Gandolfi. *non* Berggren, 1962a, p. 67–69, pl. 10, figs 4a–c.

Globotruncana subcircumnodifer Gandolfi. Olsson, 1964, p. 170, 171, pl. 6, figs 1a–2c, 4a–5c (*non* 6a–c); *non* Rao et al., 1968, p. 28, pl. 3, figs 7–9; Longoria, 1970, p. 79–82, pl. 2, figs 7–12; pl. 7, figs (?) 7, 8; pl. 13, fig. (?) 4; pl. 14, fig. (?) 6; pl. 15, figs (?) 1–3; Govindan, 1972, p. 178, 179, pl. 5, figs 16–19; Hanzlíková, 1972, p. 111, 112, pl. 30, figs 8a–10c.

Globotruncana subrugosa Gandolfi. (?) Olsson, 1964, p. 171, pl. 6, figs 7a–8c.

Rugotruncana subcircumnodifer (Gandolfi). Pessagno, 1967, p. 369, 370, pl. 62, figs 14–16; pl. 74, figs 1–3.

Rugoglobigerina rugosa subrugosa (Gandolfi). Bandy, 1967, p. 21, text-fig. 10 (2).

Rugotruncana subpennyi (Gandolfi). Bertels, 1970, p. 39–41, pl. 6, figs 1a–c.

Description: Test a small, low trochospire with dorsal side nearly flat; equatorial

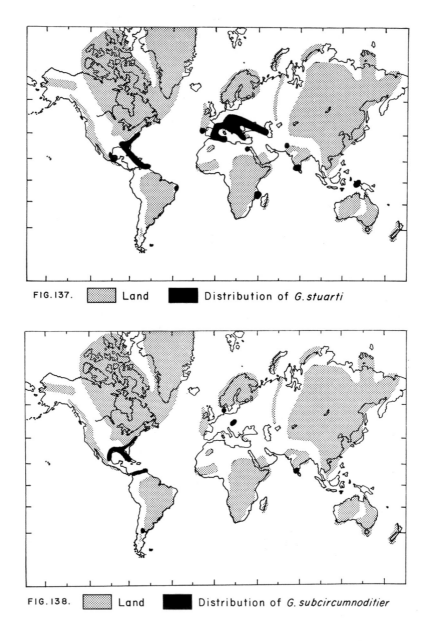

FIG.137. ▨ Land   ■ Distribution of *G. stuarti*

FIG.138. ▨ Land   ■ Distribution of *G. subcircumnoditier*

periphery deeply lobate, axial periphery narrowly truncate. Chambers subspherical becoming slightly elongate in the direction of coiling; well inflated in early whorls, later slightly compressed axially; rapidly increasing in size; 10–12 chambers in $2\frac{1}{2}$–3 whorls with 5–6 chambers in the final whorl; with a narrow, double keel on the periphery. Aperture a low, interiomarginal, umbilical arch. Umbilicus shallow, wide for test size, covered by a tegillum. Sutures depressed, radial to curved dorsally, radial ventrally. Surface finely perforate, pustulose or with fine costellae.

Remarks: *Globotruncana subcircumnodifer* has chambers which are more axially compressed than *G. cretacea* (d'Orbigny) or *G. ellisi* (Brönnimann and Brown). The features of this species appear to be rather stable.

With increased axial compression, *Globotruncana cretacea* (d'Orbigny) probably gave rise to *G. subcircumnodifer*.

During an examination of Gandolfi's (1955) types, Douglas (13 July 1965) reported the holotype of *Globotruncana subcircumnodifer* as missing. The holotype was discovered to be lodged between the cover slip and the slide. Unfortunately, the ultimate chamber had been broken off subsequent to the original description. This damage complicates the interpretation of the taxonomic status of this species, but it is believed to be the senior synonym of *Rugotruncana tilevi* Brönnimann and Brown, 1956.

Comparison of *Globotruncana subcircumnodifer* with *G. rugosa* subsp. *subrugosa* Gandolfi, 1955, revealed no distinguishing differences.

Distribution: This is shown in Fig. 138.

Range: Campanian (upper *Globotruncana calcarata* Interval)—Maastrichtian (*Trinitella scotti* Interval).

Globotruncana trinidadensis Gandolfi, Plate 54, figs 2, 3

Globotruncana caliciformis (de Lapparent). Bolli, 1951, p. 195, 196, pl. 34, figs 4–6; Hamilton, 1953, p. 232, pl. 29, figs 6–8.

Globotruncana caliciformis trinidadensis Gandolfi, 1955a, p. 47, pl. 3, figs 2a–c.

Globotruncana caliciformis sarmientoi Gandolfi, 1955a, p. 47, 48, pl. 3, figs 3a–c.

Globotruncana caliciformis caliciformis (de Lapparent). Gandolfi, 1955a, p. 46, 47, pl. 3, figs 1a–c.

Globotruncana trinidadensis Gandolfi. Pessagno, 1967, p. 359–362, pl. 84, figs 4–12; pl. 90, figs 9–10; text-fig. 57.

Description: Test biconvex with the dorsal side being more convex than the ventral; equatorial periphery lobate, axial periphery acute. Chambers of dorsal side flat, sloping at a low angle from the dorsal carina of the keel to the umbilical shoulder ventrally arched in the direction of coiling; gradually increasing in size, closely appressed; crescent-shaped to semicircular dorsally, elongate in direction of coiling ventrally; 12–15 chambers in $2\frac{1}{2}$–3 whorls with 4–6 chambers in the final whorl; with a closely spaced, beaded, double keel, the ventral carina of which tends to become reduced or remain undeveloped in the final whorl. Aperture a low, interiomarginal umbilical arch. Umbilicus shallow, moderately wide, with a tegillum. Sutures raised, curved dorsally; depressed, radial to curved ventrally. Surface finely perforate, pustulose.

Remarks: This species lacks the undulating dorsal surface characteristic of *Globotruncana fornicata* Plummer. It lacks the well-developed double keel of *G. arca* (Cushman), and instead has a subperipheral double keel, only the dorsal carina of which is peripheral. The ventral carina of the double keel may be absent from all the chambers of the final whorl or only the last chamber.

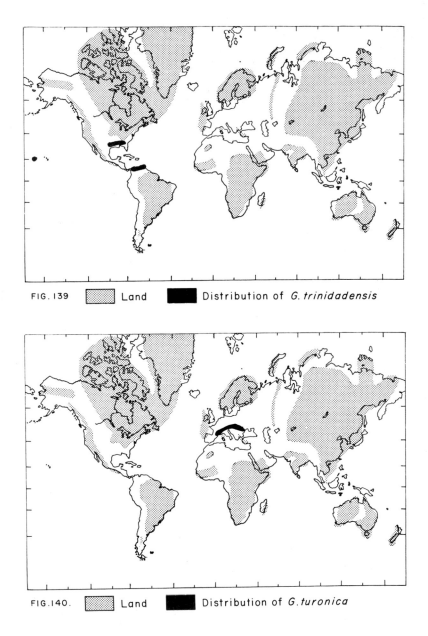

FIG. 139   Land   Distribution of *G. trinidadensis*

FIG. 140.   Land   Distribution of *G. turonica*

Globotruncana arca (Cushman) may be ancestral to *G. trinidadensis* as suggested by Pessagno (1967).

The poorly illustrated holotype was refigured by Pessagno (1967). Even this second type illustration does not show sufficient ventral inflation.

Distribution: This is shown in Fig. 139.

Range: Maastrichtian (LOS *Globotruncana calcarata*—LOS *Globotruncana laparenti* Interval to *Trinitella scotti* Interval).

Globotruncana turonica Samuel and Salaj

Globotruncana (*Globotruncana*) aff. *renzi* Thalmann-Gandolfi. Reichel, 1950, p. 612, 613, pl. 16, fig. 8; pl. 17, fig. 8.

Globotruncana turonica Samuel and Salaj, 1962, p. 317, 318, pl. 10, figs 1a–4.

Praeglobotruncana turonica (Samuel and Salaj). Salaj and Samuel, 1966, p. 194, pl. 18, figs 4a–5c; tab. 35 (13).

Praeglobotruncana algeriana Caron, 1966, p. 74, 75; Porthault, 1969, p. 537, 538, pl. 2, figs 4a–c; Neagu, 1969, p. 142, pl. 22, figs 7, 8; pl. 23, figs 1, 2 (*non* pl. 17, figs 8–15; pl. 20, figs (?) 4–6).

Praeglobotruncana cf. *algeriana* Caron. Caron, 1966, p. 74, 75, pl. 2, figs 5a–c.

Description: Test nearly equally biconvex with the dorsal side occasionally having greater convexity; equatorial periphery lobate, axial periphery narrowly truncate to acute. Chambers flat dorsally, inflated ventrally with a curved slope up to the vertex of the chamber, then down into the umbilicus; gradually increasing in size; crescentic to petaloid dorsally, appressed in the shape of a truncated triangle ventrally; 15–17 chambers in 3 whorls with 6 chambers in the final whorl; with a narrow, beaded double keel in which the ventral carina may not be fully developed in the last few chambers. Aperture a low, interiomarginal, extraumbilical–umbilical arch. Umbilicus narrow, shallow. Sutures raised, curved dorsally; depressed, radial ventrally. Surface finely perforate, pustulose.

Remarks: The presence of a narrower double keel distinguish *G. turonica* from *G. imbricata* Mornod. However, subsequent study may prove it to be an intraspecific variant of Mornod's species. The variation within the species is unknown.

The phylogeny of *G. turonica* is unknown.

Samuel and Salaj (1962) proposed the name *Globotruncana turonica* for a double-keeled form from Turonian beds of the Slovakian Carpathians. These authors later (Salaj and Samuel, 1966) transferred their species to the genus *Praeglobotruncana*. The descriptions and illustrations of this taxon do not make clear whether it possesses portici or a tegillum. Because no undisputed *Praeglobotruncana* is known to possess a double keel, *G. turonica* is retained as originally assigned until further study reveals the nature of the umbilical chamber extensions.

Caron (1966) designated *Globotruncana* (*Globotruncana*) aff. *renzi* Thalmann-Gandolfi of Reichel (1950) as the holotype of her new species *Praeglobotruncana algeriana*. She asserted that this taxon was phylogenetically important because it possesses a double keel, forming a link between *Praeglobotruncana* and *Globotruncana*. The primary difference between these two genera is that the former has a single keel and portici, while the latter has either a single or a double keel and a tegillum. Neither Reichel nor Caron, who followed his original description, discussed the type of chamber extensions, i.e. portici vs. tegillum. *Praeglobotruncana algeriana* is believed to be conspecific with *G. turonica*.

Distribution: This is shown in Fig. 140.
Range: Mid-Turonian.

Globotruncana ventricosa White, Plate 55, figs 1–3
Globotruncana canaliculata var. *ventricosa* White, 1928b, p. 284, pl. 38, figs 5a–c;
non Plummer, 1931, p. 199, pl. 13, figs 10a–b.
Globotruncana ventricosa White. *non* Brotzen, 1936, p. 137, pl. 13, figs 4a–c;
Cushman, 1946b, p. 150, pl. 62, figs 3a–c; *non* Cita, 1948b, p. 162, 163, pl. 4,
figs 9a–c; Mornod, 1950, p. 590, 591, text-figs 12: 1a–2c; *non* Carbonnier, 1952,
p. 116, pl. 6, figs 1a–c; *non* Subbotina, 1953, p. 222–226, pl. 13, figs 1a–4c;
Książkiewicz, 1956, p. 276, 278, text-fig. 50 (1, 2); (*non* pl. 30, fig. 1; text-fig. 51);
non Alexandrowicz, 1956, p. 55, text-figs 7A, B; Sacal and Debourle, 1957, p. 62,
pl. 27, figs 5, 13 (*non* 3, 14); Bolli, 1957, p. 57, pl. 13, figs 4a–c; (?) Seiglie, 1958,
p. 71, pl. 4, figs 1a–c; Książkiewicz, 1958, text-fig. 2 (10a–c); *non* Maslakova, 1959,
p. 110, pl. 13, figs 2a–c; Eternod Olvera, 1959, p. 122, 123, pl. 7, figs 9, 10, 12;
Belford, 1960, p. 98, 99, pl. 29, figs 5–11; *non* Graham and Clark, 1961, p. 112,
pl. 5, figs 9a–c; *non* Samuel, 1961, p. 193, pl. 13, figs 1a–c; *non* Jurkiewicz, 1961,
pl. 24, figs 8a–c; *non* Majzon, 1961, p. 762, pl. 1, fig. 7; Sturani, 1962, p. 91, text-fig. 15
(F3); Borsetti, 1962, p. 70–71, pl. 6, fig. 6; text-figs 162, 163; *non* Samuel, 1962, p.
193, pl. 13, figs 1a–c; Postuma, 1962, 7 figs; Kavary and Frizzell, 1963, p. 56, 57, pl.
11, figs 14–16; van Hinte, 1963, p. 86, 87, pl. 7, figs 3a–c; Takayanagi, 1965, p. 226,
227, pl. 29, figs 1a–d; *non* Banerji, 1966, p. 66, 67, pl. 5, figs 6, 8, 9; Douglas and
Sliter, 1966, p. 115, 116, pl. 3, figs 1a–c; Pessagno, 1967, p. 362–364, pl. 75, figs
21–26; pl. 79, figs 9–14; Bandy, 1967, p. 20, text-fig. 9 (9); Rao *et al.*, 1968, p. 27, pl. 2,
figs 4–6; Ansary and Tewfik, 1968, p. 53, pl. 7, figs 6a–c; Barr, 1968a, p. 319, pl. 40,
figs 3a–c; Sliter, 1968, p. 107, pl. 18, figs 7a–c (*non* 8a–c); Douglas, 1969, p. 188,
pl. 7, figs 1a–3c; Pessagno, 1969a, p. 609, pl. 9, fig. D; (?) Sturm, 1969, pl. 12, figs
2a–c; Longoria, 1970, p. 82–85, pl. 1, figs 1–3; pl. 8, figs 4, 7, 8; pl. 16, fig. 6; Cita
and Gartner, 1971, pl. 4, figs 2a–c; Hanzlíková, 1972, p. 113, pl. 31, figs 5a–7a;
Caron, 1972, p. 555, pl. 2, figs 3a–c.
    Globotruncana canaliculata (Reuss). Cushman, 1946b, p. 149, pl. 61, figs 18a–c
(*non* 17a–c).
    Globotruncana gagnebini Tilev, 1952, text-figs 15a–17d (*non* 14a–e); Bolli, 1957,
p. 59, pl. 14, figs 5a–c; Olsson, 1964, p. 165, pl. 4, figs 4a–c.
    Globotruncana marginata (Reuss). (?) Jírová, 1956, tab. 1, figs 2a–c.
    Globotruncana ventricosa ventricosa White, *non* Ashworth, 1959, p. 498, text-figs
2a–c; Tollmann, 1960, p. 195, 196, pl. 21, fig. 7 (? 8); Salaj and Samuel, 1966, p. 220,
pl. 20, figs 6a–c.
    Globotruncana tricarinata colombiana (?) Gandolfi. Corminboeuf, 1961, p. 116,
pl. 2, figs 1a–c.
    Globotruncana aegyptiaca Nakkady. van Hinte, 1963, p. 87–89, pl. 8, figs 1a–2c.
    Globotruncana concavata (Brotzen). Christodoulou and Marangoudakis, 1966,
p. 305, pl. 1, figs 1a–c; text-figs 2a–c.
    Globotruncana lapparenti tricarinata (Quereau). Wille-Janoschek, 1966, p. 110,
111, pl. 2, figs 3–11 (*non* 1, 2).
    Globotruncana tricarinata (Quereau). (?) Rao *et al.*, 1968, p. 26, 27, pl. 3, figs 10–12.
    Description: Test planoconvex with the dorsal side being flat and the ventral side
strongly convex; equatorial periphery lobate, axial periphery broadly truncate.
Chambers flat dorsally; on ventral side early chambers of last whorl only slightly

above the keel, later with a straight, steep slope up to the umbilical shoulder; gradually increasing in size to subequal in final whorl; crescent-shaped to semi-circular dorsally, slightly imbricate and elongate in final 2–4 chambers; 14–16 chambers in 3 whorls with 5–7 chambers in the final whorl; with a widely spaced, beaded, double keel on the dorsal margin. Aperture a moderately high, interio-marginal, umbilical to extraumbilical–umbilical arch.

Remarks: *Globotruncana ventricosa* can be distinguished from its congeners by its dorsally located, widely spaced double keel. In addition, *G. aegyptiaca* Nakkady, a similar form, has convexly sloping ventral chambers in peripheral view as compared with the straight-sloping chambers of *G. ventricosa*. The number of chambers and their rate of increase in size varies slightly, but other features remain stable.

This species probably developed from *Globotruncana linneiana* (d'Orbigny) through increased ventral convexity.

The taxon described by Porthault (in Donze *et al.*, 1970) as the new species *Marginotruncana paraconcavata* is strikingly similar to *Globotruncana ventricosa*. Should they prove to be synonymous, the range of the latter would be extended into the upper Coniacian. Porthault envisioned his species to be phyletically linked to *G. concavata* (Brotzen).

Distribution: This is shown in Fig. 141.

Range: Campanian (FOS *Globotruncana ventricosa*—FOS *Pseudoguembelina costata* Interval)—Maastrichtian (*Trinitella scotti* Interval).

Genus PLUMMERITA Brönnimann, 1952b

Rugoglobigerina (*Plummerella*) Brönnimann, 1952a, p. 37 [non *Plummerella* de Long, 1942].

Rugoglobigerina (*Plummerita*) Brönnimann, 1952b, p. 146 [pro *R*. (*Plummerella*) Brönnimann].

Plummerita Brönnimann. Brönnimann and Brown, 1956, p. 555, 556.

Plummerita (*Plummerita*) Brönnimann. El-Naggar, 1971a, p. 434.

Type species: *Rugoglobigerina* (*Plummerella*) *hantkeninoides hantkeninoides* Brönnimann.

Description: Test a very low trochospiral coil; equatorial periphery lobate and stellate in varying degrees, axial periphery correspondingly rounded or pointed. Chambers of early whorls subspherical becoming radially elongate and pointed in early portion of final whorl, typically being subspherical in the ultimate and penultimate chambers. Aperture a low, interiomarginal, umbilical-extraumbilical arch. Umbilicus shallow, narrow to moderately wide, possibly covered by a tegillum. Surface ornamented with costellae.

Remarks: *Plummerita* is distinguished from other members of the Globo-truncanidae by its drawn-out, spine-like chambers in the final whorl.

An unquestionable tegillum has yet to be illustrated on *Plummerita*. If it should prove to be a structure not developed by this taxon, it would have to be transferred from the Globotruncanidae to the Globigerinidae.

El-Naggar (1971a) proposed dividing *Plummerita* into two subgenera. He

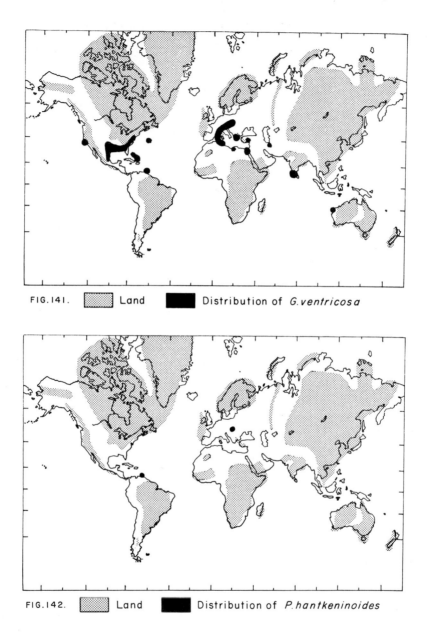

FIG.141. ▨ Land ■ Distribution of *G.ventricosa*

FIG.142. ▨ Land ■ Distribution of *P. hantkeninoides*

resurrected *P.* (*Plummerita*) and established *P.* (*Radotruncana*), with *Globotruncana calcarata* Cushman as the type species. These two taxa are entirely unrelated, and cannot be associated as subgenera.

Range: Maastrichtian (*Trinitella scotti* Interval).

Plummerita hantkeninoides Brönnimann

Rugoglobigerina (*Plummerella*) *hantkeninoides hantkeninoides* Brönnimann, 1952a, p. 37–39, pl. 3, figs 1–3, text-fig. 17; Pokorný, 1963, p. 376, fig. 418.

Rugoglobigerina (*Plummerita*) *hantkeninoides hantkeninoides* Brönnimann. Brönnimann, 1952b, p. 146.

Rugoglobigerina (*Plummerella*) *hantkeninoides inflata* Brönnimann, 1952a, p. 40–42, pl. 3, figs 7–9, text-fig. 19.

Rugoglobigerina (*Plummerita*) *hantkeninoides costata* Brönnimann, 1952a, p. 39, 40, pl. 3, figs 4–6, text-fig. 18.

Plummerita hantkeninoides hantkeninoides Brönnimann. Brönnimann and Brown, 1956, p. 555, 556.

Plummerita hantkeninoides inflata Brönnimann. Brönnimann and Brown, 1956, p. 556, pl. 21, figs 4–6.

Rugoglobigerina hantkeninoides Brönnimann. Bolli *et al.*, 1957, p. 43, 44, pl. 11, figs 5a–c.

Globotruncana (*Plummerita*) *hantkeninoides* (Brönnimann). Majzon, 1961, p. 763, pl. 5, fig. 5.

Plummerita hantkeninoides (Brönnimann). Saavedra, 1965, p. 327, text-fig. 18; Bandy, 1967, p. 22, text-fig. 10 (10).

Description: Test a flat trochospire; equatorial periphery may be partially stellate, partially rounded. Chambers initially subspherical, developing varying degrees of radial elongation in the final whorl; elongate chambers bluntly to sharply tapered, with greatest development in the first 2–3 chambers in the final whorl; the ultimate and penultimate chambers typically exhibit only short, blunt spine-like projections or none at all; 5 chambers in the last whorl; rapidly increasing in size. Aperture a low, interiomarginal, umbilical–extraumbilical arch. Umbilicus shallow, narrow to moderately wide. A tegillum may be present. Surface ornamented with meridionally oriented costellae. Sutures depressed, straight to slightly curved.

Remarks: A monotypic genus within which chamber elongation and inflation are the most variable features.

Plummerita hantkeninoides may have arisen from *Rugoglobigerina rugosa* (Plummer), but this has not been confirmed.

Brönnimann (1952a) distinguished three subspecies of *Plummerita hantkeninoides* based primarily upon the degree of inflation of the ultimate and penultimate chambers. The holotypes were examined and deemed intraspecific variants of a single species. The inflation is regarded as most probably environmentally controlled. Moreover, the stratigraphic ranges of the subspecies coincide.

Distribution: This is shown in Fig. 142.

Range: Maastrichtian (*Trinitella scotti* Interval).

Genus RUGOGLOBIGERINA Brönnimann, 1952a

Rugoglobigerina (*Rugoglobigerina*) Brönnimann, 1952a, p. 16.
Globotruncana (*Rugoglobigerina*) Brönnimann. Gandolfi, 1955a, p. 15.
Kuglerina Brönnimann and Brown, 1956, p. 557.
Type species: *Globigerina rugosa* Plummer, 1927.

Description: Test trochospiral with a flat to high dorsal side; equatorial periphery lobate; axial periphery broadly rounded, unkeeled. Chambers inflated, subspherical to ellipsoid; with meridionally arranged fine to coarse, discontinuous ridges (costellae). Aperture a low, interiomarginal, umbilical-extraumbilical arch covered by a highly variable tegillum. Sutures depressed, radial to slightly curved. Umbilicus small to moderate in diameter, shallow to moderately deep. Pores numerous, of a fine to moderate diameter.

Remarks: Loeblich and Tappan (1964) distinguished *Rugoglobigerina* from *Globotruncana* primarily on the presence of costellae on the former, and completely overlooked the more important absence of a peripheral keel in *Rugoglobigerina*. This genus lacks the acute to keeled final few chambers of *Trinitella* and the stellate equatorial periphery of *Plummerita*.

Gandolfi's (1955) concept of *Rugoglobigerina* being a "mutated," "pelagic" form being repeatedly developed from various species of *Globotruncana* is totally rejected. Reiss (1957b) has given a detailed analysis of Gandolfi's claim. *Rugoglobigerina* is believed to have descended not from *Globotruncana* but *Globigerina* (Masters, in press).

El-Naggar (1971) treated *Archaeoglobigerina* as a subgenus of *Rugoglobigerina*. This is not acceptable because the type species of *Archaeoglobigerina* has a keel and must be placed in the *Globotruncana*. The non-keeled members belong in the *Globigerina*.

Range: Santonian (FOS *Globotruncana concavata*—FOS *Globotruncana elevata* Interval)—Maastrichtian (*Trinitella scotti* Interval).

Rugoglobigerina hexacamerata Brönnimann, Plate 56, figs 2, 3
   Rugoglobigerina reicheli hexacamerata Brönnimann, 1952a, p. 23–25, pl. 2, figs 10–12, text-fig. 8a–m; Cati, 1964, p. 253, pl. 41, figs 7a–c.
   Rugoglobigerina rugosa pennyi Brönnimann, 1952a, p. 34, pl. 4, figs 1–3; text-figs 14a–i; Bandy, 1967, p. 22, text-fig. 10 (4).
   Globotruncana (*Rugoglobigerina*) *hexacamerata* subsp. *subhexacamerata* Gandolfi, 1955a, p. 34, pl. 1, figs 11a–c.
   Rugoglobigerina plana Belford, 1960, p. 95, 96, pl. 27, figs 1–4.
   Rugoglobigerina hexacamerata hexacamerata (Brönnimann). Herm, 1962, p. 57, pl. 4, fig. 5.
   Rugoglobigerina pennyi Brönnimann. Berggren, 1962a, p. 75, pl. 12, figs 1a–3c; (?) Hanzlíková, 1969, p. 51, pl. 13, figs 1a–2c (*non* 3a–c); Govindan, 1972, p. 174, pl. 3, figs 7–9; Hanzlíková, 1972, p. 114, 115, pl. 33, figs 1a–3c.
   Rugoglobigerina rugosa rugosa (Plummer). Brönnimann and Rigassi, 1963, pl. 18, figs 1a–c.

Globotruncana (*Rugoglobigerina*) *rugosa* (Plummer). van Hinte, 1965b, p. 85, pl. 3, figs 4a–c.

Rugoglobigerina pennyi pennyi Brönnimann. *non* Salaj and Samuel, 1966, p. 196, pl. 9, figs 4a–c.

Rugoglobigerina rotundata Brönnimann. Douglas and Sliter, 1966, p. 116, pl. 1, figs 5a–c.

Rugoglobigerina hexacamerata Brönnimann. Pessagno, 1967, p. 364–5, pl. 91, figs 5–7.

Rugoglobigerina rugosa (Plummer). Douglas, 1969, p. 175, 176, pl. 6, figs 2a–c (*non* 1a–c); Scheibnerová, 1969, p. 80, 81, tab. 20, figs 4a–c (*non* 2a–3c; text-figs 27, 28); Hanzlíková, 1969, p. 52, pl. 13, figs 4a–5c.

Rugoglobigerina (*Rugoglobigerina*) *hexacamerata* Brönnimann. El-Naggar, 1971b, p. 485, 486, pl. 9, figs 1–3, 5–12; pl. 10, figs 1a–2d.

Rugoglobigerina (*Rugoglobigerina*) *pennyi* Brönnimann. El-Naggar, 1971b, p. 488, 489, pl. 9, figs 4, 13–16; (*non* pl. 6, fig. 19; pl. 8, figs 1, 2; pl. 11, figs 1–10; pl. 17, fig. 12).

Rugoglobigerina (*Rugoglobigerina*) *badryi* El-Naggar, 1971b, p. 482, 483, pl. 14, figs 2, 6, 13 (*non* 1, 5, 9, 10); pl. 16, figs 2–8, 10, 12 (*non* 1, 9, 11, 13; pl. 17 (?) figs 11, 14).

Description: Test a low to flat trochospire; equatorial periphery moderately lobate; axial periphery broadly rounded. Chambers highly inflated, subspherical except for final 1–2 which may be ellipsoidal with the width being the long dimension; numbering 13–15 in $2\frac{1}{2}$–3 whorls with 5–7, generally 6, chambers in the final whorl; slowly increasing in size, several chambers in the final whorl may be subequal; each with meridionally arranged costellae which are continuous to discontinuous. Aperture a low, interiomarginal, umbilical–extraumbilical arch. Umbilicus shallow, moderately wide, covered by a tegillum. Sutures depressed, generally straight but may be slightly curved both dorsally and ventrally. Small to moderate diameter pores randomly distributed between costellae.

Remarks: *Rugoglobigerina hexacamerata* differs from other species by generally having six slowly enlarging chambers in the final whorl. The final chamber tends to be slightly more ventrally positioned than the other chambers. This is regarded as a gerontic trait which is recognised in many other taxa. The costellae of the earlier chambers are coarser due to normal secondary calcification. They also tend to be more numerous on successive chambers. As with all tegillated taxa, the tegillum varies from individual to individual by shape, degree of fusing and number of tegillar apertures.

Specimens transitional between this species and *Rugoglobigerina rugosa* (Plummer), its presumed ancestor, have been observed.

The holotypes of *Rugoglobigerina hexacamerata* and *R. rugosa pennyi* Brönnimann were compared. The only detectable difference between them was the chambers in the last whorl of the latter are more inflated. This is believed to be an environmentally rather than a genetically controlled feature. *Rugoglobigerina hexacamerata* is the senior synonym.

Contrary to the description and figure given by Gandolfi (1955a) for his new *Globotruncana hexacamerata* subsp. *subhexacamerata,* the holotype does not possess a double keel or an imperforate peripheral band. Its chambers have a perforate periphery and typical rugoglobigerinid costellae. This subspecies is deemed a junior synonym of *Rugoglobigerina hexacamerata.*

El-Naggar (1971b) distinguished his *Rugoglobigerina badiyr* from *R. hexacamerata* on the basis of "its smaller, thinner, less regular, ovoid test; rapidly increasing

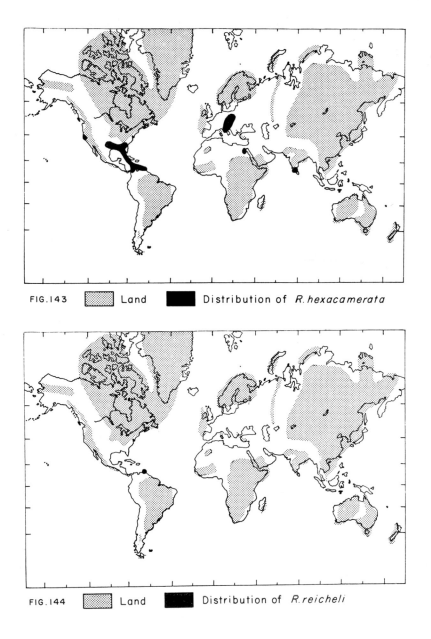

FIG.143 [Land] Land   [black] Distribution of *R. hexacamerata*

FIG.144 [Land] Land   [black] Distribution of *R. reicheli*

chambers in the outer whorl; large last chamber which is somewhat elongated is the direction of coiling; relatively smaller umbilicus, and depressed spiral suture". The size and relative thickness of the test is ontogenetically or environmentally controlled. The rapidity of the increase of chamber size, certainly of the holotype, is not at variance with that of *R. hexacamerata*, both of which are gradual. Other cited differences are of an intraspecific nature. *Rugoglobigerina badryi* is conspecific with *R. hexacamerata*.

Distribution: This is shown in Fig. 143.

Range: Maastrichtian (FOS *Rugoglobigerina hexacamerata*—LOS *Globotruncana fornicata* Interval to *Trinitella scotti* Interval)

Rugoglobigerina (?) *reicheli* Brönnimann

Rugoglobigerina reicheli reicheli Brönnimann, 1952a, p. 18, 19, pl. 3, 10–12; text-figs 4a–m, 5a–i.

Rugoglobigerina (*Rugoglobigerina*) *reicheli* Brönnimann. *non* El-Naggar, 1971b, p. 490, 491, pl. 6, fig. 18; pl. 8, figs 3, 6–9.

Description: "The last volution of the small- to medium-sized low trochoidal test comprises 5 to 6 chambers. Umbilical and spiral side are well defined. About 2 whorls can be counted on the centrally slightly depressed spiral side. No details of the initial portion are discernible due to the coarsely rugose surface. The ultimate chamber can be larger or of the same size or even smaller than the penultimate one and is displaced toward the umbilical side. The first 2 or 3 chambers of the last whorl are of conic shape. The adjoining chambers are peripherally rounded and truncate at the apertural side. The umbilicus is deep, usually filled with matrix. Remains of the delicate covering plate were noted. The straight sutures are depressed, thus producing a lobulate outline. The larger arcuate aperture of the end chamber with a small liplike projection opens into the umbilicus. The apertures of the preceding chambers are not known. The walls appear to be thick, and the surface is coarsely rugose. The rugosities of the inflated last chambers are arranged in meridional rows radiating from a centre on the surface toward the edges of the aperture. The investigated specimens are invariably dextrally coiled." (Type description.)

Remarks: No other *Rugoglobigerina* is known to possess radially elongate chambers. The degree of radial chamber elongation exhibited in the holotype and paratypes varies considerably.

The phylogeny of this species is unknown.

Rugoglobigerina reicheli has not been recognised outside of its type locality nor recorded since it was first named. Thus, a more exact study has not been made. The elongation of the chambers on the primary types does appear to be more than coalesced spines.

It is questionably retained in the genus *Rugoglobigerina*, because of the unavailability of topotype material during this study. It most probably should be transferred to *Plummerita*.

Distribution: This is shown in Fig. 144.

Range: Maastrichtian (*Trinitella scotti* Interval).

Rugoglobigerina rotundata Brönnimann

Rugoglobigerina rugosa rotundata Brönnimann, 1952a, p. 34, 35, pl. 4, figs 7–9; text-figs 15a–e, 16a–c; Seiglie, 1958, p. 75, pl. 7, figs 3a–c; Bandy, 1967, p. 22, text-fig. 10 (5); *non* Al-Shaibani, 1971, p. 107, pl. 2, figs 1a–c.

Rugoglobigerina rotundata Brönnimann. Postuma, 1962, 7 figs; Pessagno, 1967

p. 365, 366, pl. 65, figs 1–3; pl. 68, figs 1–3; *non* Funnell *et al.*, 1969, p. 37, pl. 5, figs 7–9; text-figs 21a–c; (?) Hanzlíková, 1969, p. 51, 52, pl. 11, figs 8a–9c; Cita and Gartner, 1971, pl. 4, figs 4a–c; *non* Hanzlíková, 1972, p. 115, pl. 33, figs 8a–9c.

Rugoglobigerina rotundata rotundata Brönnimann. (?) Salaj and Samuel, 1966, p. 197, 198, pl. 9, figs 3a–c.

Rugoglobigerina (Rugoglobigerina) rotundata Brönnimann. El-Naggar, 1971b, p. 491, 492, pl. 10, figs 3a–d.

Rugoglobigerina pustulata Brönnimann. Hanzlíková, 1972, p. 115, pl. 33, figs 7a–c.

Description: Test large, trochospiral with a flat spire except for the final 1–2 chambers which are more ventrally positioned; equatorial periphery gently lobate; axial periphery very broadly rounded. Chambers of early whorls subspherical, gradually increasing in size; those of the final whorl abruptly larger, highly inflated, subspherical to ellipsoidal where the long dimension may be either the width or, as in the case of the final 1–2 chambers, the thickness; 14–15 in number, $2\frac{1}{2}$–3 whorls, 5–6 in the final whorl; covered by meridionally arranged costellae. Aperture a moderately high, interiomarginal, umbilical–extraumbilical arch. Umbilicus narrow, deep, covered by a tegillum. Sutures depressed, straight to slightly curved both dorsally and ventrally.

Remarks: No other *Rugoglobigerina* has such large inflated chambers, the final two of which are displaced ventrally. The main diversity in this species lies in the degree to which the ultimate and penultimate chambers have moved out of the otherwise flat spire.

Brönnimann (1952) correctly considered *Rugoglobigerina rotundata* to be an offshoot of *R. rugosa* (Plummer).

It is possible that *Rugoglobigerina rotundata* may be an intraspecific variant of *R. rugosa* (Plummer). The displaced chambers have been treated as a gerontic trait in other taxa, and the chamber inflation can be environmentally caused. It is retained as a distinct species because of its more restricted stratigraphic range.

Distribution: This is shown in Fig. 145.

Range: Maastrichtian (LOS *Globotruncana fornicata*—FOS *Trinitella scotti* Interval to *Trinitella scotti* Interval).

Rugoglobigerina rugosa (Plummer), Plate 56, figs 1, 4, 5; plate 57, figs 1, 2

Globigerina rugosa Plummer, 1927, p. 38, 39, pl. 2, figs 10a–d; Frizzell, 1954, p. 127, pl. 20, figs 3–6; (?) Hofker, 1956e, text-fig. 6; Hofker, 1956d, p. 53, text-fig. 1; Hofker, 1960e, text-figs 14a–c, 24a–c; Hofker, 1962, text-figs 3a–c.

Globigerina lacera (Ehrenberg). Cushman, 1931d, p. 58, 59, pl. 11, figs 1–4.

Rugoglobigerina macrocephala macrocephala Brönnimann, 1952a, p. 25–27, pl. 2, figs 1–3, text-figs 9a–s; Corminboeuf, 1961, p. 117, 118, pl. 2, figs 3a–c; Brönnimann and Rigassi, 1963, pl. 18, figs 2a–c; Ansary and Tewfik, 1968, p. 58, 59, pl. 5, figs 3a–c.

Rugoglobigerina macrocephala ornata Brönnimann, 1952a, p. 27, 28, pl. 2, figs 4–6, text-figs 10a–j; Corminboeuf, 1961, p. 118, pl. 2, figs 4a–c; Ansary and Tewfik, 1968, p. 59, pl. 5, figs 4a–c.

Rugoglobigerina reicheli pustulata Brönnimann, 1952a, p. 20–23, pl. 2, figs 7–9, text-figs 6a–m, 7g–j (*non* 7a–f); Olsson, 1960, p. 50, pl. 10, figs 13–15.

Rugoglobigerina rugosa rugosa (Plummer). Hamilton, 1953, p. 227, pl. 30, figs 1–3; Olsson, 1960, p. 50, pl. 10, figs 16–18; Corminboeuf, 1961, p. 119, pl. 2, figs 5a–c; Herm, 1962, p. 60, pl. 3, fig. 2; *non* Brönnimann and Rigassi, 1963, pl. 18, figs 1a–c;

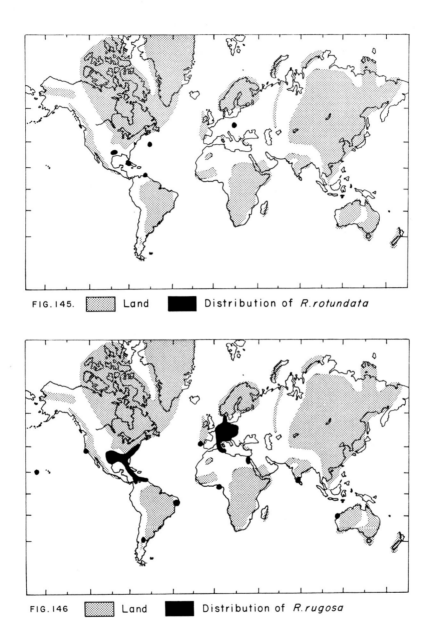

FIG.145. ☐ Land ■ Distribution of *R.rotundata*

FIG.146 ☐ Land ■ Distribution of *R.rugosa*

Said and Sabry, 1964, p. 386, pl. 3, figs 18a–c; *non* Salaj and Samuel, 1966, p. 198, 199, pl. 9, figs 6a–c; Bandy, 1967, p. 22, text-fig. 10 (6); Ansary and Tewfik, 1968, p. 58, pl. 5, figs 1a–c; Neagu, 1970b, p. 70, pl. 28, figs 21–23.

Rugoglobigerina rugosa (Plummer). Weiss, 1955, p. 307, pl. 1, fig. 9; Bolli *et al.*, 1957, p. 43, 44, pl. 11, figs 2a–c; *non* Edgell, 1957, p. 115, 118, pl. 4, figs 10–12; Reyment, 1960, p. 75, pl. 15, figs 1–3b; pl. 16, figs 5a–f; Graham and Clark, 1961, p. 111, pl. 5, figs 1a–c; Berggren, 1962a, p. 71–75, pl. 11, figs 1a–5b; text-fig. 8 (1a–5b); Postuma, 1962, 7 figs; Graham, 1962, p. 108, pl. 20, figs 32a–c; Pessagno, 1962, p. 360, pl. 5, fig. 13; Skinner, 1962, p. 45, pl. 6, fig. 18; Cati, 1964, p. 253, pl. 41, figs 6a–c; Loeblich and Tappan, 1964, p. C663, figs 3a–c; Saavedra, 1965, p. 326, text-fig. 15; Takayanagi, 1965, p. 228, 229, pl. 29, figs 5a–c; Douglas and Sliter, 1966, p. 116, 117, pl. 1, figs 2a–c; Pessagno, 1967, p. 366, 367, pl. 75, figs 2, 3; Rasheed and Govindan, 1968, p. 84, pl. 8, figs 7–9; Mabesoone, 1968, p. 175, fig. 5:5a–c; Sliter, 1968, p. 108, pl. 19, figs 2a–c; Douglas, 1969, p. 175, 176, pl. 6, figs 1a–c; *non* Dupeuble, 1969, p. 157, pl. 4, figs 11a–d; Scheibnerová, 1969, p. 80, 81, tab. 20, figs 2a–3c (*non* 4a–c); text-figs 27, 28; *non* Hanzlíková, 1969, p. 52, pl. 13, figs 4a–5c; (?) Khan, 1970, p. 32, 33, pl. 2, figs 16a, b; Bertels, 1970, p. 41, pl. 6, figs 2a–c; Neagu, 1970a, fig. 1 (17); Govindan, 1972, p. 173, 174, pl. 3, figs 1–3, 13–18; Hanzlíková, 1972, p. 116, pl. 33, figs 5a–6c.

Globotruncana (*Rugoglobigerina*) *macrocephala* subsp. *submacrocephala* Gandolfi, 1955a, p. 46, pl. 2, figs 11a–c.

Globigerina (*Rugoglobigerina*) *rugosa* Plummer. Hofker, 1958, pl. 7, figs 38a–c; Hofker, 1960d, text-fig. 1 (10).

Globigerina (*Rugoglobigerina*) cf. *rugosa* Plummer. Hofker, 1959, text-figs 4a–6c.

Rugoglobigerina bulbosa Belford, 1960, p. 93–95, pl. 26, figs 1–8; Takayanagi, 1965, p. 227, 228, pl. 29, figs 2a–3c.

Rugoglobigerina pilula Belford, 1960, p. 92, 93, pl. 25, figs 7–13; Douglas and Sliter, 1966, p. 116, pl. 1, figs 3a–4c; Douglas, 1969, p. 175, pl. 6, figs 8a–9c; Porthault, 1970, in Donze *et al.*, p. 68, 69, pl. 9, figs 21–23; Hanzlíková, 1972, p. 115, pl. 33, figs 4a–c.

Globotruncana (*Rugoglobigerina*) *rugosa* (Plummer). Majzon, 1961, p. 762, pl. 4, fig. 18; *non* van Hinte, 1965b, p. 85, pl. 3, figs 4a–c.

Rugoglobigerina macrocephala Brönnimann. Berggren, 1962a, p. 76–78, pl. 12, figs 4a–6c; text-figs 9 (1a–5); Petri, 1962, p. 122, 123, pl. 17, figs 3a–c; Olsson, 1964, p. 172, pl. 6, figs 9a–c; Pessagno, 1967, p. 366, pl. 66, figs 1, 2; Mabesoone *et al.*, 1968, p. 175, figs 5a–c; Hanzlíková, 1969, p. 50, pl. 12, figs 6a–10b; Bertels, 1970, p. 42, pl. 6, figs 3a–c; Govindan, 1972, p. 174, pl. 3, figs 4–6, 19–21; Hanzlíková, 1972, p. 114, pl. 33, figs 10a–11c.

Rugoglobigerina pustulata Brönnimann. Berggren, 1962a, p. 78–80, pl. 13, figs 1a–c; [*non* text-figs 10 (1–12)]; Ansary and Tewfik, 1968, p. 59, pl. 5, figs 6a–c; Funnell *et al.*, 1969, p. 36, 37, pl. 5, figs 4–6; text-figs 20a–c; *non* Hanzlíková, 1969, p. 51, pl. 11, figs 7a–c; Govindan, 1972, p. 174, 175, pl. 3, figs 10–12.

Globotruncana rugosa (Plummer). van Hinte, 1963, p. 92–94, figs 1a–3c.

Rugoglobigerina (*Rugoglobigerina*) *rugosa rugosa* (Plummer). Pokorný, 1963, p. 376, fig. 417.

Hedbergella sp. 2. Douglas and Sliter, 1966, p. 105, pl. 1, figs 1a–c.

Rugoglobigerina rotundata Brönnimann. Douglas and Sliter, 1966, p. 116, pl 1, figs 6a–c (*non* 5a–c).

Rugoglobigerina rugosa macrocephala Brönnimann. Bandy, 1967, p. 22, text-fig. 10 (7).

Rugoglobigerina rugosa pustulata Brönnimann. Bandy, 1967, p. 22, text-fig. 10 (8).
Rugoglobigerina rugosa rotundata Brönnimann. Ansary and Tewfik, 1968, p. 58, pl. 5, figs 2a–c; (?) Al-Shaibani, 1971, p. 107, pl. 2, figs 1a–c.
Rugoglobigerina ? *alpina* Porthault, 1969, p. 535, 536, pl. 2, figs 1a–3b.
Rugoglobigerina sp. cf. *R. bulbosa* Belford. Bertels, 1970, p. 42, pl. 7, figs 1a–c.
Rugoglobigerina sp. Bertels, 1970, p. 42, 43, pl. 7, figs 2–3c.
Rugoglobigerina aff. *bulbosa* Belford. Porthault, 1970, in Donze *et al.*, p. 68, pl. 9, figs 18–20.
Rugoglobigerina (*Rugoglobigerina*) *rugosa* (Plummer). El-Naggar, 1971a, pl. 5, figs a–c; El-Naggar, 1971b, p. 492, 493, pl. 1, figs 1–11; pl. 2, figs 1–8, 10, 11; pl. 3, figs 1–16; pl. 4, figs 1–15; pl. 6, fig. 17; pl. 8, figs 4, 5.
Rugoglobigerina (*Rugoglobigerina*) *ornata* Brönnimann. El-Naggar, 1971b, p. 487, 488, pl. 5, figs 1–12; pl. 14, figs 8, 12.
Rugoglobigerina (*Rugoglobigerina*) *macrocephala* Brönnimann. El-Naggar, 1971b, p. 486, 487, pl. 6, figs 1–16; pl. 14, fig. 4.
Rugoglobigerina (*Rugoglobigerina*) *reicheli* Brönnimann. El-Naggar, 1971b, p. 490, 491, pl. 6, fig. 18; pl. 8, figs 3, 6–9.
Rugoglobigerina (*Rugoglobigerina*) *pennyi* Brönnimann. El-Naggar, 1971b, p. 488, 489, pl. 6, fig. 19; pl. 8, figs 1, 2; pl. 17, fig. 12.
Rugoglobigerina (*Rugoglobigerina*) *pustulata* Brönnimann. El-Naggar, 1971b, p. 489, 490, pl. 7, figs 1a–5, 7–9.
Rugoglobigerina (*Rugoglobigerina*) *rotundata* Brönnimann. El-Naggar, 1971b, p. 491, 492, pl. 10, figs 3a, b, (?) c, (?) d.
Rugoglobigerina (*Rugoglobigerina*) *badryi* El-Naggar, 1971b, p. 482, 483, pl. 14, figs 9, 10 (*non* 1, 2, 5, 6, 13); pl. 16, figs 1, 9, 11, 13 (*non* 2–8), 10, 12).
Rugoglobigerina (*Rugoglobigerina*) *browni* El-Naggar, 1971b, p. 484, 485, pl. 17, figs 1–10, 13.
Rugoglobigerina (*Rugoglobigerina*) *arwae* El-Naggar, 1971b, p. 481, 482, pl. 15, figs 2, 5a–9 (*non* 3a, b).

Description: Test a flat trochospire; equatorial periphery deeply lobated; axial periphery broadly rounded. Chambers subspherical but tend to be wider and thicker than high in the ultimate and penultimate chambers; increasing rapidly in size, highly inflated to the point where the earlier whorls may be slightly recessed dorsally; 13–17 chambers in 2½–3 whorls with 4, typically 5 chambers in the last whorl; with fine to coarse meridionally arranged costellae. Aperture a moderate to high, interiomarginal, umbilical–extraumbilical arch bordered by a very narrow imperforate lip. Umbilicus narrow to moderately wide, moderately deep, covered by a highly variable tegillum. Sutures deeply depressed, straight to slightly curved dorsally and ventrally. Fine to moderately coarse pores with a tendency for those of the initial and the final chambers to be finer; proloculus and initial whorl imperforate except for a few pores along the spiral suture.

Remarks: *Rugoglobigerina rugosa* is easily recognised by its four to five rapidly enlarging chambers in the final whorl and its deeply lobate periphery. Not only do the costellae of the earlier chambers of this species become coarser ontogenetically, i.e. as each new chamber is added, but the author has also demonstrated (in press) that they become coarser phylogenetically. This is the only recognisable change in this species through time. The morphoseries is totally gradational and must represent a single species.

Globigerina delrioensis Carsey gave rise to *Rugoglobigerina rugosa*. The specimens of *R. rugosa* at its lowermost range are indistinguishable from *G. delrioensis* if it were not for the presence of the costellae and tegillum.

Examination of the holotype of *Rugoglobigerina reicheli pustulata* Brönnimann, 1952, revealed that it is conspecific with *R. rugosa*.

The holotype of *Globotruncana* (*Rugoglobigerina*) *macrocephala submacrocephala* Gandolfi, 1966, possesses neither a keel nor an imperforate peripheral band. All chambers bear well developed costellae. This subspecies is a junior synonym of *Rugoglobigerina rugosa*.

Porthault (1969) erected the new species *Rugoglobigerina* ? *alpina* which is said to differ from other species by having weakly developed costellae and lacking a tegillum. This latter structure is delicate and often not preserved. As noted above, the costellae of the stratigraphically early members of *R. rugosa* are thin. Porthault's species is a *R. rugosa*.

El-Naggar (1971b) proposed the new taxa *Rugoglobigerina arwae* and *R. browni*, each of which merely represent intraspecific variants of *R. rugosa*. *Rugoglobigerina arwae* supposedly differs from *R. rugosa*. "by its smaller, thinner, more closely coiled test; much smaller early part, and less rapidly increasing chambers in the outer whorl; smaller umbilicus; more curved, depressed sutures; less rugose surface and less well developed meridional costellae". Each of these differences are found in the topotypic suite of *R. rugosa*. He also remarked that the "rugosity on *R. arwae* is in the form of stout spicules with pointed ends . . ." as illustrated on one paratype (pl. 15, figs 3a, b). This paratype is regarded as not being conspecific with the holotype, which has typical costellae. *Rugoglobigerina browni* was stated to differ from *R. rugosa* "by its smaller, thinner, quadrate test, much narrower umbilicus, and less rugose surface". Again the differences are clearly intraspecific.

Distribution: This is shown in Fig. 146.

Range: Santonian (FOS *Globotruncana concavata*—FOS *Globotruncana elevata* Interval)—Maastrichtian (*Trinitella scotti* Interval).

Rugoglobigerina tradinghousensis Pessagno, Plate 58, figs 1, 3
Rugoglobigerina rugosa (Plummer). Olsson, 1964, p. 173, pl. 7, figs 3a–4c.
Rugoglobigerina tradinghousensis Pessagno, 1967, p. 367, 368, pl. 64, figs 1–8.

Description: Test trochospiral with a moderately high dorsoconvexity; equatorial periphery deeply lobate; axial periphery broadly rounded. Chambers inflated, subspherical except in the final few chambers which may have a proportionally reduced height; gradually enlarging, to the final chambers which may be subequal; 15–18 in 3 whorls with 4–5 in the final whorl; covered by discontinuous, meridionally arranged costellae. Aperture a low to moderately high, interiomarginal, umbilical–extraumbilical arch. Umbilicus narrow, deep, covered by a tegillum. Sutures deeply depressed, straight to slightly curved. Proloculus and initial chambers imperforate except for associated spiral pores, others with small diameter pores of moderately low concentration.

Remarks: *Rugoglobigerina tradinghousensis* is the only member of this genus that is truly high-spired. *Rugoglobigerina rotundata* Brönnimann has a flat spire except for the final one or two chambers which are more ventrally positioned. This species has been too infrequently reported to fully determine its morphologic diversity.

Rugoglobigerina tradinghousensis may have evolved directly from the *Globigerina* stock. However, it most likely descended from *R. rugosa* (Plummer), from which it differs in being higher spired.

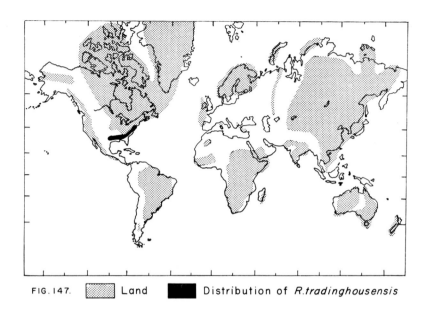

FIG. 147.   Land   Distribution of *R.tradinghousensis*

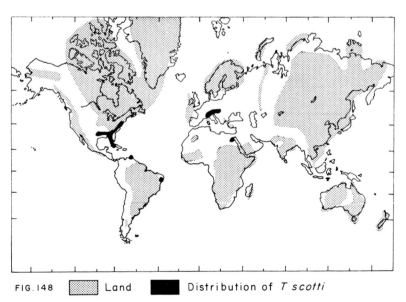

FIG. 148   Land   Distribution of *T scotti*

It is conceivable that this species is a genetic variant of *Rugoglobigerina rugosa* (Plummer). At present it appears to have a shorter stratigraphic range, and so is retained as a valid species.

Distribution: This is shown in Fig. 147.

Range: Upper Santonian (FOS *Globotruncana concavata*—FOS *Globotruncana elevata* Interval)—lower Maastrichtian (LOS *Globotruncana calcarata*—LOS *Globotruncana lapparenti* Interval).

Genus TRINITELLA Brönnimann, 1952a

Trinitella Brönnimann, 1952a, p. 56.

Rugoglobigerina (*Trinitella*) Brönnimann. Banner and Blow, 1959, p. 11.

Globotruncana (*Trinitella*) Brönnimann. Majzon, 1961, p. 763.

Type species: *Trinitella scotti* Brönnimann, 1952a.

Description: Test coiled in a flat trochospire; equatorial periphery lobate; axial periphery rounded to acute. Chambers subspherical, inflated except last few which are dorsally flattened and exhibit a single keel or imperforate peripheral band; excepting the nepionic chambers, all are covered by meridionally arranged costellae. Aperture a low to moderately high, interiomarginal arch bordered by an imperforate flap which extends into the umbilicus where it fuses irregularly to earlier flaps forming a tegillum.

Remarks: *Trinitella* differs from *Rugoglobigerina* by possessing chambers in the last whorl which develop a keel. It is distinguished from *Globotruncana* by the absence of a keel in all but the final few chambers.

El-Naggar (1971a), in his proposed classification of the Globigerinacea, subdivided the subfamily Globotruncaninae into the genera *Rugoglobigerina*, *Globotruncana* and *Plummerita*. He restricted the *Rugoglobigerina* to those forms possessing chambers which are "globigerine inflated throughout". Yet, in the subgenus *R.* (*Rugoglobigerina*), he questionably assigns without comment the genus *Trinitella* as a junior synonym. This contradicts his later statement (1971b, p. 494, 495) that *Trinitella* has a compressed, keeled final one or two chambers. Thus, the reader is left in a quandary.

Since its origin the genus *Trinitella* has variously been treated as a subgenus of *Rugoglobigerina* and of *Globotruncana* or as a synonym of either. It possesses characteristics of both genera; the costellae, the tegillum and in the early stages the subspherical, non-keeled chambers of *Rugoglobigerina*; the tegillum and in the final stages the keeled periphery of *Globotruncana*.

I have retained *Trinitella* at its original generic status for two main reasons. First, to alter the definitions of either *Globotruncana* or particularly *Rugoglobigerina* so as to include *Trinitella* would render them meaningless. Second, the keel, central in all modern classifications, is seen to develop phylogenetically in several separate lineages. This palingenetic structure is awarded the same importance with *Trinitella* as with other taxa.

Range: Maastrichtian (*Trinitella scotti* Interval).

Trinitella scotti Brönnimann, Plate 58, figs 2, 4

Trinitella scotti Brönnimann, 1952a, p. 57, pl. 4, figs 4–6, text-figs 30a–m; Brönnimann and Brown, 1956, p. 555, pl. 23, figs 13–15; Seiglie, 1958, p. 75, pl. 6, figs 1a–3c; Corminboeuf, 1961, p. 119, 120, pl. 2, figs 6a–c; Brönnimann and Rigassi, 1963, pl. 16, figs 3a–c; Loeblich and Tappan, 1964, p. C663, figs 2a–c; Olsson, 1964, p. 173, 174, pl. 6, figs 10a–c; Saavedra, 1965, p. 326, text-fig. 17.

Rugoglobigerina scotti (Brönnimann). Bolli *et al.*, 1957, p. 43, 44, pl. 11, figs 3a–4c; Herm, 1962, p. 62–64, pl. 2, fig. 2; Postuma, 1962, 7 figs; Bandy, 1967, p. 22, text-fig. 10 (9); Pessagno, 1967, p. 367, pl. 74, figs 9–14; pl. 76, figs 4–6; Mabesoone, *et al.*, 1968, p. 175, figs 5: 4a–c; *non* Ansary and Tewfik, 1968, p. 60, pl. 5, figs 5a–c; *non* Funnell *et al.*, 1969, p. 37, 38, pl. 5, figs 10–12; text-figs 22a–c.

Globotruncana (Trinitella) scotti Brönnimann. Majzon, 1961, p. 763, pl. 5, fig. 10.

Rugoglobigerina (Rugoglobigerina) pennyi Brönnimann. El-Naggar, 1971b, p. 488, 489, pl. 11, figs 1–10.

Rugoglobigerina (Rugoglobigerina) scotti Brönnimann. El-Naggar, 1971b, p. 494, 495, pl. 13, figs 3a–c.

Rugoglobigerina (Rugoglobigerina) bronnimanni El-Naggar, 1971b, p. 483, 484, pl. 12, figs 1–7, 9; pl. 13, figs 1–2e; pl. 14, (?) figs 14–16.

Rugoglobigerina (Rugoglobigerina) badryi El-Naggar, 1971b, p. 482, 483, pl. 14, figs 1, 5 (*non* 2, 6, 9, 10, 13; pl. 16, figs 1–13; pl. 17, figs 11, 14).

Description: Test a low biconvex trochospire in which the early whorls are below the level of the final whorl on the dorsal side; equatorial periphery moderately to deeply lobate; axial periphery rounded in early stages, later acute with an imperforate peripheral band or keel. Chambers subspherical, inflated except for the final 1–4 which become progressively wider and more flattened dorsally; increase moderately to rapidly in size; 16–18 chambers in 3 whorls with 5–6 in the final whorl; neanic chambers pustulose, later chambers with meridionally oriented, discontinuous costellae which may extend into short thick spines in the early chambers of the last whorl. Aperture a low to moderately high, interiomarginal arch bordered by an umbilically protruding flap which along with previous flaps form a tegillum. Umbilicus wide, rather shallow, covered by the tegillum. Sutures depressed, straight to slightly curved, spiral suture depressed. Pores numerous, of small to moderate diameter.

Remarks: The genus is monotypic. The number of keeled chambers ranges from one to four, with the higher number occurring on the six-chambered variety. The costellae, as would be expected, are progressively less coarse as the ultimate chamber is approached. Likewise, the pores become less wide and more numerous. As with all tegillated forms, the tegilla are highly varied.

The author has suggested (in press) that *Rugoglobigerina hexacamerata* Brönnimann may be ancestral to *Trinitella scotti*. Except for the keel in *Trinitella* the two taxa are very similar, and most *T. scotti* seem more often to have six chambers in the final whorl as does *R. hexacamerata*. However, five-chambered *T. scotti* are not uncommon. Thus, *R. rugosa* (Plummer) may also be ancestral to *T. scotti* as it is to *R. hexacamerata*.

El-Naggar (1971b) observed "a certain degree of similarity" between his new species *Rugoglobigerina bronnimanni* and *Trinitella scotti*. He distinguished his species by its larger size and absence of the "compressed, keeled later chamber or chambers and the angular periphery" of *T. scotti*. However, not all the compressed chambers in a given population of *T. scotti* have developed a keel periphery. Moreover, the specimens El-Naggar illustrated for his species do have axially compressed chambers in the gerontic stage. Size, of course, can be environmentally or reproductively controlled. The two species are considered as conspecific.

Distribution: This is shown in Fig. 148.

Range: Maastrichtian (*Trinitella scotti* Interval).

Acknowledgements

Funds which partially offset the expense of this project were made available through the generosity of the Hartwick College, Oneonta, New York and College Center of the Finger Lakes, Corning, New York, U.S.A.

Katherine V. W. Palmer graciously placed at my disposal the primary types reposited at the Palaeontological Research Institution, Ithaca, New York, U.S.A. The collections of the U.S. National Museum, Washington, D.C., were made available through the courtesy of Richard Cifelli.

I am indebted to the following people for arranging the loan of holotypes: Fredrick Collier, U.S. National Museum; Raymond L. Ethington, University of Missouri; Allan D. Griesemer, University of Nebraska State Museum; James C. Ingle, Jr., Stanford University; Alvin M. Phillips, Jr., Louisiana State University; LouElla Saul, University of California, Los Angeles.

The success of a project of this type is strongly dependent upon the availability of topotypes and residues for comparative purposes. I gratefully acknowledge the following people who freely assisted me in obtaining this material: Hans M. Bolli, Zurich, Switzerland; Michèle Caron, Fribourg, Switzerland; Maurice Hamaoui, Pau, France; H. G. Kugler, Basle, Switzerland; Roger Lehmann, Begles, France; H. G. Lorenz, Rabat, Morocco; H. P. Luterbacher, Begles, France; Hans Oesterle, Djakarta, Indonesia; Olga Pazdro, Warsaw, Poland; Viera Scheibner, Sydney, N.S.W., Australia; Jacques Sigal, Vincennes, France.

The electron micrographs were taken using an AMR-900 (American Metals Research Corporation). Technical assistance was provided by Thomas Fernalld and James Lyon through the courtesy of James Sorauf, Department of Geology, State University of New York at Binghampton.

I must give especial thanks to my wife Shirley who assisted in all phases of this project except for the interpretive, which is mine alone. Without her dedication, a project of this scope would have taken double the time for completion.

Space does not permit the individual acknowledgement of all who kindly made reprints of their publications available. Through their courtesy the delays encountered in seeking inter-library loans were reduced.

References

Adams, C. G. (1960). A note on the age of the Laig Gorge beds, Eigg. *Geol. Mag.* **97**(4), 322–325. Text-fig. 2.

Albritton, C. C. (1937). Upper Jurassic and Lower Cretaceous foraminifera from the Malone Mountains, Trans-Pecos, Texas. *J. Paleont.* **11**(1), 19–23. Plate 4.

Alekseeva, L. V. (1963). Foraminifera of the Lower Cretaceous of western Turkmen. *Akad. Nauk. SSSR, Inst. Geol. i Razrab. goryuch. Iskop.* 3–56. Plates 1–9, 1 text-fig., 1 tab. (in Russian).

Alexandrowicz, S. (1956). Zespoly globotrunkan w turonie okolic Krakowa. *Acta geol. pol.* **6**(1), 41–63. 12 text-figs., 2 tabs.,

Alexandrowicz, S. (1971). Stratygrafia i mikrofauna górnego cenomanu niecki pólnocnosudeckiej. *Roczn. pol. Tow.* geol. **41**(2), 321–335. 3 text-figs.

Aliyulla, Kh. (1965). The state of the knowledge of the family Heterohelicidae and the way of its subsequent study. *Vop. Mikropaleont.* **9**, 215–228. 1 plate, 4 text-figs. (in Russian).

Al-Shaibani, S. K. (1971). An Upper Maestrichtian foraminiferal fauna from Dörfles, Lower Austria. *Jb. geol. Bundesanst. Wien. Sonderbd.* **17**, 105–119. 3 plates.

Ansary, S. E., Andrawis, S. F. and Fahmy, S. E. (1962). Biostratigraphic studies of the Upper Cretaceous sections in the G.P.C. wells in the Eastern Desert and Sinai. *Fourth Arab Petrol. Congress, Bierut* Paper 24 (B-3), 339–347. 5 plates.

Ansary, S. E. and Fakhr, B. Y. (1958). Maestrichtian foraminifera from Um El Huetet area, west of Safaga. *Egypt. J. Geol.* **2**(2), 105–146. Plates 1, 2, 1 text-fig., 1 tab.

Ansary, S. E. and Tewfik, N. M. (1968). Planktonic foraminifera and some new benthonic species from the surface Upper Cretaceous of Ezz El Orban area, Gulf of Suez. *J. Geol. Un. Arab. Repub.* **10**(1), 37–76. Plates 1–9.

Ansary, S. E. and Tewfik, N. M. (1969) Biostratigraphy and time stratigraphy of subsurface Upper Cretaceous of Ezz El Orban area, Gulf of Suez region, U.A.R. *Proc. Third African Micropaleont. Colloq. Cairo* 95–106. 2 plates, 2 tabs.

Applin, E. R. (1955). A biofacies of Woodbine Age in southeastern Gulf Coast region. *Prof. Pap. U.S. geol. Surv.* **264-I**, 187–197. Plates 48, 49, text-fig. 41, tab. 1.

Applin, E. R., Ellisor, A. C. and Kniker, H. T. (1925). Subsurface stratigraphy of the Gulf Coastal Plain of Texas and Louisiana. *Bull. Am. Ass. Petrol. Geol.* **9**(1), 79–122.

Arkin, Y. and Hamaoui, M. (1967). The Judea Group (Upper Cretaceous) in central and southern Israel. *Bull. geol. Surv. Israel* **42**, 1–17. 17 text-figs., 2 plates.

Arni, P. (1933). Foraminiferen des Senons und Untereocäns im Prätigauflysch. *Beitr. geol. Karte Schweiz*, N.F. **65**, 1–18. 5 plates, 11 text-figs.

Arnold, Z. M. (1955). An unusual feature of miliolid reproduction. *Contr. Cushman Fdn foramin. Res.* **6**(3), 94–96.

Arnold, Z. M. (1964). Biological observations on the foraminifer *Spiroloculina hyalina* Schulze. *Univ. Calif. Publ. Zool.* **72**, 78 p. 7 plates, 14 text-figs. 3 tabs.

Arnold, Z. M. (1968). The uniparental species concept in the foraminifera. *Trans. Amer. Microsc. Soc.* **87**(4), 431–442. 35 text-figs.

Asano, K (1950a). Upper Cretaceous foraminifera from Japan. *Pacif. Sci.* **4**(2), 158–163. Plate 1.

Asano, K. (1950b). Cretaceous foraminifera from Teshio, Hokkaido. *Short Pap. Inst. Geol. Paleont. Tohoku Univ.* **2**, 13–22. Plate 3.

Ashworth, E. T. (1959). Occurrence of *Globotruncana ventricosa* in northwestern Peru. *Micropaleont.* **5**(4), 497–499. Text-figs. 1, 2.

Aurouze, G. and de Klasz, I. (1954). Sur la présence de Schackoines dans le Crétacé superieur de France, de Bavière et de Tunisie. *Bull. Soc. geol. Fr.* ser. 6, **4**(1–3), 97–103. Plate VIa.

Ayala-Castañares, A. (1954). El género *Globotruncana* Cushman, 1927. *Boln. Asoc. mex. Geol. petrol.* **VI**(11–12), 353–440. Lamina 1–16.

Ayala-Castañares, A. (1959). Estudio de Algunos microfosiles planctonicos de las Calizas del Cretacico Superior de la Republica de Haiti. *Paleont. mex.* **4**, 1–41. 12 lamina, 2 tf.

Ayala-Castañares, A. (1962). Morfologia y estructura de algunos foraminiferos planctonicos del Cenomaniano de Cuba. *Boln. Soc. geol. mex.* **25**(1), 1–63. 13 plates, 1 text-fig.

Bailey, J. W. (1841). American Polythalmia from the upper Mississippi, and also from the Cretaceous formation on the upper Missouri. *Am. J. Sci.* **41**(2), 50, 51. 4 text-figs.

Balakhmatova, V. T. (1953). On the Globigerinidae and Globorotaliidae of the middle Jurassic. *In Paleontology and Stratigraphy. Trudy, vses. nauchno-issled. geol. Inst.* 86–89, Gosgeolizdat, Moscow, 4 text-figs. (in Russian).

Bandy, O. L. (1951). Upper Cretaceous foraminifera from the Carlsbad area, San Diego County, California. *J. Paleont.* **25**(4), 488–513. Plates 72–75, 2 text-figs.

Bandy, O. L. (1960). Planktonic foraminiferal criteria for paleoclimatic zonation. *Sci. Rep. Tohoku Univ.*, Ser. 2 *Geol.*, **Spec. Vol.** (Hanzawa Memorial Volume), 1–8. 2 text-figs.

Bandy, O. L. (1967). Cretaceous planktonic foraminiferal zonation. *Micropaleont.* **13**(1), 1–31. 13 text-figs.

Banerji, R. K. (1966). The genus *Globotruncana* and biostratigraphy of the lower Ariyalur stage (Upper Cretaceous) of Vridhachalam, South India. *J. geol. Soc. India* **7**, 51–69. Plates 3–5, 3 text-figs., 1 tab.

Banerji, R. K. (1968). Late Cretaceous foraminiferal biostratigraphy of Pondicherry area, South India. *In Cretaceous-Tertiary formations of South India. Mem. geol. Soc. India* **2**, 30–49. 4 text-figs.

Bang, I. (1971). Planktonic foraminifera of the lowermost Danian. *In* (Farinacci, A., ed.), *Proc. II Plankt. Conf.* **1**, 17–25. 6 plates, 1 text-fig. Tecnoscienza, Roma.

Banner, F .T. and Blow, W. H. (1959). The classification and stratigraphical distribution of the Globigerinaceae: Part 1. *Palaeontology* **2**(1), 1–27. Plates 1–3, 3 text-figs.

Banner, F. T. and Blow, W. H. (1960). Some primary types of species belonging to the superfamily Globigerinaceae. *Contr. Cushman Fdn foramin. Res.* **11**(1), 1–41. Plates 1–8, figs. 1, 2.

Barbieri, F. (1966). Segnalazione dei generi *Gabonella e Grimsdaleinella* (foraminiferi) nel cretacico dell' Appennino settentrionale. *Boll. Soc. geol. ital.* **85**(1), 13–20. 2 plates.

Barker, R. W. (1960). Taxonomic notes on the species figured by H. B. Brady in his report on the Foraminifera dredged by H.M.S. Challenger during the years 1873–1876. *Spec. Publs. Soc. econ. Paleont. Miner., Tulsa* **9**, 1–238. 115 plates.

Barnaba, P. F. (1959). Geologia dei Monti di Gubbio. *Boll. Soc. geol. ital.* **77**(3) (1958), 39–70. 2 plates, 18 text-figs.

Barr. F. T. (1961). Upper Cretaceous planktonic foraminifera from the Isle of Wight, England. *Palaeontology* **4**(4), 552–580. Plates 69–72, text-figs. 1–4.

Barr, F. T. (1966). Upper Cretaceous foraminifera from the Ballydeenlea Chalk, County Kerry, Ireland. *Palaeontology* **9**(3), 492–510. Plates 77–79, 2 text-figs.

Barr, F. T. (1968a). Late Cretaceous planktonic foraminifera from the coastal area east of Susa (Apollonia), northeastern Libya. *J. Paleont.* **42**(2), 308–321. Plates 37–40, 5 text-figs.

Barr, F. T. (1968b). Upper Cretaceous stratigraphy of Jabal Al Akhdar, Northern Cyrenaica. In *Geology and archaeology of Northern Cyrenaica, Libya. Petrol. Explor. Soc. Libya* 131–142. 3 plates, 9 text-figs.

Barr, F. T. (1972). Cretaceous biostratigraphy and planktonic foraminifera of Libya. *Micropaleont.* **18**(1), 1–46. Plates 1–10, 10 text-figs.

Barr, F. T. and Cordey, W. G. (1964). Some Upper Cretaceous foraminifera from the Chapman Collection (1892). *J. Paleont.* **38**(2), 306–310. Plate 49.

Bars, H. and Ohm, U. (1968). Der Dogger des Profils Rocchetta, Prov. Trient., Italien "*Globigerina spuriensis*" n. sp. *Neues Jb. Geol. Palaont. Monatshefte* **10**, 577–590. 4 text-figs.

Bartenstein, H. (1948). *Globotruncana calcarata* Cushman, foraminifere caractéristique du Campanien. *C. r. somm. Seanc. Soc. geol. Fr.* **12**, 244–246. 1 text-fig.

Bartenstein, H. (1965). Taxionomische Revision und Nomenklator zu Franz E. Hecht "Standard-Gliederung der Nordwest-deutschen Unterkreide nach Foraminiferen" (1938). Teil 4: Alb. Mit Beschreibungen von Arten aus verschiedenen Unterkreide-Niveaus. *Senckenberg. leth.* **46**(4/6), 327–366. 7 text-figs., 3 tabs.

Bartenstein, H., Bettenstaedt, F. and Bolli, H. M. (1966). Die Foraminiferen der Underkreide von Trinidad, W.I. Zweiter Teil: Maridale-Formation (Typlokalitat). *Eclog. geol. Helv.* **59**(1), 129–177. 4 plates 1 text-fig.

Bartenstein, H. and Brand, E. (1937). Mikro-paläontologische Untersuchungen zur Stratigraphie des nordwest-deutschen Lias und Doggers. *Senckenberg. leth.* **439**, 224 p. 20 plates, 20 text-figs., 5 tabs.

Bartlett, G. A. (1969). Cretaceous biostratigraphy of the Grand Banks of Newfoundland. *Marit. Sediments* **5**(1), 4–14. Text-figs. 1–4.

Bate, R. H. and Bayliss, D. D. (1969). An outline account of the Cretaceous and Tertiary foraminifera and of the Cretaceous ostracods of Tanzania. *Proc. Third African Micropaleont. Colloq., Cairo* 113–164. 8 plates, 7 text-figs.

Bé, A. W. H. (1965). The influence of depth on shell growth in *Globigerinoides sacculifera* (Brady). *Micropaleont.* **11**(1), 81–97. Plates 1, 2, 12 text-figs.

Bé, A. W. H. (1968). Shell porosity of Recent planktonic foraminifera as a climatic index. *Science* **161** (3844), 881–884. 3 text-figs.

Bé, A. W. H. and Hemleben, C. (1970). Calcification in a living planktonic foraminifer *Globigerinoides sacculifer* (Brady). *Neues Jb. Geol. Palaont. Abh.* **134**(3), 221–234. Plates 25–32, 1 text-fig., 1 tab.

Beissel, I. (1891). Die Foraminiferen der Aachener Kreide. *Abh. preuss. geol. Landesanst.* n. s, **3**, 1–78. 1 tab. Atlas, 16 plates.

Belford, D. J. (1960). Upper Cretaceous foraminifera from the Toolonga Calcilutite and Gingin Chalk, western Australia. *Bull. Bur. Miner. Resour. Geol. Geophys. Aust.* **57**, 1–198. 35 plates, 14 text-figs., 1 chart.

Belford, D. J. and Scheibnerová, V. (1971). Turonian foraminifera from the Carnarvon Basin, western Australia, and their palaeogeographical significance. *Micropaleont.* **17**(3), 331–344. 4 plates, 1 text-fig., 1 tab.

Bellier, J. P. (1971). Les foraminifères planctoniques du Turonien-type. *Rev. Micropaleont.* **14**(2), 85–90. 1 plate, 2 text-figs.

Berger, W. H. and Rad, U. von (1972). Cretaceous and Cenozoic sediments from the Atlantic Ocean. *Initial Repts. Deep Sea Drilling Proj.* **14**, 787–886, 944–954. 48 plates, 59 text-figs., 11 tabs., 5 append.

Berggren, W. A. (1962a). Some planktonic foraminifera from the Maastrichtian and the type Danian stages of southern Scandinavia. *Stockh. Contr. Geol.* **9**(1), 1–106. 14 plates, 14 text-figs.

Berggren, W. A. (1962b). Stratigraphic and taxonomic-phylogenetic studies of Upper Cretaceous and Paleogene planktonic foraminifera. *Stockh. Contr. Geol.*, **9**(2), 107–129, 3 text-figs.

Berggren, W. A. 1971). Paleogene planktonic foraminiferal faunas on Legs I–IV (Atlantic Ocean), JOIDES Deep Sea Drilling Program—a synthesis. *In* (Farinacci, A., ed.), *Proc. II Plankt. Conf.* **1**, 57–77. 5 plates, 3 text-figs., 6 tabs. Tecnoscienza, Roma.

Bergquist, H. R. (1943). Clay County .Geology. *Bull. Miss. St. geol. Surv.* **53**, 11–70. 14 figs.

Bergquist, H. R. (1971). Biogeographical review of Cretaceous foraminifera of the Western Hemisphere. *In* (Yochelson, E. L., ed.), *Cretaceous biogeography. Proc. North American Paleont. Convention, Chicago*, 1969. 1565–1609, 4 text-figs.

Bermúdez, P. J. (1952). Estudio sistematico de los foraminíferos Rotaliformes. *Boln Geol. Minist. Minas Venez.* **2**(4), 1–153, 35 lamina.

Bernoulli, D. and Renz, O. (1970). Jurassic carbonate facies and new ammonite faunas from western Greece. *Eclog. geol. Helv.* **63**(2), 573–607. 6 plates, 6 text-figs., 2 tabs.

Berry, W. and Kelley, L. (1930). The foraminifera of the Ripley Formation on Coon Creek, Tennessee. *Proc. U.S. natn. Mus.* **76**(19), 1–17. 3 plates.

Bertels, A. (1970). Los foraminiferos planctonicos del a cuenca Cretacico-Terciaria en Patagonia Septentrional (Argentina), con consideraciones sobre la estratigrafia de Fortin General Roca (Provincia de Rio Negro). *Ameghiniana* **7**(1), 1–56. Plates 1–9, text-figs. 1–10, 2 tabs.

Bettenstaedt, F. and Wicher, C. A. (1956). Stratigraphic correlation of Upper Cretaceous and Lower Cretaceous in the Tethys and Boreal by aid of microfossils. *Proc. Fourth World Petrol. Congr.* (1955). Sect. I/D, **5**, 493–516. 5 plates, 1 tab.

Bieda, E. (1958). Otwornice przewodnie i wiek kredy piszacej Mielnika. *Biul. Inst. geol.* **121**(3), 17–89. 27 text-figs.

Bignot, G. and Guyader, J. (1966). Découverte de foraminifères planctoniques dans l'Oxfordien du Havre (Seine-Maritime). *Rev. Micropaleont.* **9**(2), 104–110. 1 plate, 1 text-fig.

Bignot, G. and Guyader J. (1971). Observations nouvelles sur *Globigerina oxfordiana* Grigelis. *In* (Farinacci, A., ed.), *Proc. II Plankt. Conf.* **1**, 79–83. 2 plates. Tecnoscienza, Roma.

Boccaletti, M. and Pirini, C. (1965). Ritrovamento di *Schackoina* nel complesso argilloso-calcareo di S. Fiora. *Palaeontogr. ital.* **LIX** (n. ser. XXIX) 1964, 53–61. 2 tav., 7 text-figs.

Boccaletti, M. and Sagri, M. (1965). Strutture caotiche dell'Appennino. 1) Eta', assetto e giacitura del complesso argilloso-calcareo affiorante nella parte occidentale del F°. 129 "S. Fiora." *Boll. Soc. geol. ital.* **83**(4) (1964), 461–510. 5 plates, 30 text-figs.

Bolin, E. J. (1952). Microfossils of the Niobrara Formation of southeastern South

Dakota. *Rep. Invest. S. Dak. geol. nat. Hist. Surv.* **70**, 1–74. 5 plates, 1 text-fig., 1 tab.

Bolin, E. J. (1956). Upper Cretaceous foraminifera, ostracoda, and radiolaria from Minnesota. *J. Paleont.* **30**(2), 278–298. Plates 37–39, 5 text-figs.

Bolli, H. (1945). Zur Stratigraphie der Oberen Kreide in den höheren helvetischen Decken. *Eclog. geol. Helv.* **37**(2), 217–329. Plate 9, text-figs. 1–6.

Bolli, H. (1951). The genus *Globotruncana* in Trinidad, B. W. I. *J. Paleont.* **25**(2), 187–199. Plates 34, 35, 1 text-fig.

Bolli, H. (1957). The genera *Praeglobotruncana, Rotalipora, Globotruncana* and *Abathomphalus* in the Upper Cretaceous of Trinidad, B. W. I. *In* (Loeblich, A. R., Jr. *et al.*), Studies in Foraminifera. *Bull. U.S. natn. Mus.* **215**, 51–60. Plates 12–14, text-fig. 10.

Bolli, H. (1958). The foraminiferal genera *Schackoina* Thalmann, emended and *Leupoldina*, n. gen. in the Cretaceous of Trinidad B. W. I. *Eclog. geol. Helv.* **50**(2) (1957), 271–278. Plates 1, 2, 1 text-fig.

Bolli, H. (1959). Planktonic foraminifera from the Cretaceous of Trinidad, B. W.I. *Bull. Am. Paleont.* **39**(179), 258–277. Plates 20–23, 1 chart.

Bolli, H. (1960). Planktonic foraminifera as index fossils in Trinidad West Indies and their value for worldwise stratigraphic correlation. *Eclog. geol. Helv.* **52**(2) (1959), 627–637. 1 tab.

Bolli, H. (1966). Zonation of Cretaceous to Pliocene marine sediments based on planktonic foraminifera. *Boln Informativo Asoc. Venez. Geol., Min. Petrol.* **9**(1), 1–26. 4 tabs.

Bolli, H. (1971). The direction of coiling in planktonic foraminifera. *In* (Funnell, B. M. and Riedel, W. R., eds), *The Micropaleontology of Oceans.* 639–648. 6 text-figs. Cambridge University Press.

Bolli, H., Loeblich, A. R., Jr. and Tappan, H. (1957). Planktonic foraminiferal families Hantkeninidae, Orbulinidae, Globorotaliidae and Globotruncanidae. *In* (Loeblich, A. R., Jr. *et al.*), Studies in Foraminifera. *Bull. U.S. natn. Mus.* **215**, 3–50. Plates 1–11. text-figs. 1–9.

Bonet, F. (1956). Zonification microfaunistica de las calizas cretacicas del este de Mexico. *Boln. Asoc. mex. Geol. petrol.* **8**(7–8), 389–488. 31 plates, 3 text-figs., 3 tabs.

Borsetti, A. M. (1962). Foraminiferi planctonici di una serie Cretacea dei dintorni di Piobbico (prov. di Pesero). *G. Geol.* ser. 2, **29** (1960–1961), 19–75. 7 plates, 245 text-figs., 2 tabs.

Borsetti, A. M. (1964). Una microfauna a *Globotruncana* del Flysch alloctono di Castel di Casio (Bologna). *G. Geol.* ser. 2, **32**(2), 667–685. Plate 79, text-figs. 1–15.

Borza, K. (1959). Geologicko-pertografické pomery mezozoika Belanských Tatier a masívu Širokej. Die geologisch-petrographischen Verhaltnisse des Mesozoikums des Gebirges Balenske Tatry und des Massive Široka. *Geol. Sb., Bratisl.* **10**(1), 133–170. Plates 8–15, text-figs. 12, 13.

Borza, K. (1969). *Die Mikrofazies und Mikrofossilien des oberjuras und der unterkreide der Klippenzone der Westkarpaten.* Verlag Slowak. Akad. Wiss. Bratislava, 301 pp., 88 plates, 12 text-figs., 4 tabs.

Bradshaw, J. S. (1957). Laboratory studies on the rate of growth of the foraminifer, "*Streblus beccarii* (Linne) var. *tepida* (Cushman)." *J. Paleont.* **31**(6), 1138–1147. 5 text-figs.

Brady, G. S., Robertson, D. and Brady, H. B. (1870). The ostracoda and foraminifera of Tidal Rivers. With an analysis and descriptions of the foraminifera. *Ann. Mag. nat. Hist.* ser. 4, **6**(34), 273–309. Plates 11, 12, 1 tab.

Broglio Loriga, C. and Mantovani, M. G. (1965). Le biofacies del Cretacico della Valle del Vaiont (Belluno). *Riv. ital. Paleont. Stratigr.* **71**(4), 1225–1248. Plates 110–114, 2 text-figs.

Broglio Loriga, C. and Mantovani, M. G. (1970). Microbiostratigrafia della serie affiorante nella massa scivolata dal M. Toc (Vaiont) il 9 ottobre 1963 ed alcune osservazioni su Foraminiferi, Radiolari, Calcisfere e *Nannoconus*. *Studi trent. Sci. nat.* sez. A, **47**(2), 202–286. 7 plates, 5 text-figs., 2 tabs., chart.

Bronn, H. G. (1853–1856). *Lethaea Geognostica oder Abbildung und Beschreibung der für die Gebirgs-Formationen bezeichnendsten Versteinerungen.* Dritter Band. **4,** Caeno-Lethaea: VI. Theil: Mollassen-Periode. E. Schweizerbart'sche Verlagshandlung und Druckerei, Stuttgart, 1130 pp. Atlas (1850–1856), plates I-LXIII.

Brönnimann, P. (1952a). Globigerinidae from the Upper Cretaceous (Cenomanian-Maestrichtian) of Trinidad, B. W. I. *Bull. Am. Paleont.* **34**(140), 1–70. Plates 1–4, text-figs. 1–30.

Brönnimann, P. (1952b). *Plummerita*, new name for *Plummerella* Brönnimann, 1952 (non *Plummerella* DeLong, 1942). *Contr. Cushman Fdn foramin. Res.* **3**(68), 146.

Brönnimann, P. and Brown, N. K., Jr. (1953). Observations on some planktonic Heterohelicidae from the Upper Cretaceous of Cuba. *Contr. Cushman Fdn foramin. Res.* **4**(4), 150–156. 14 text-figs.

Brönnimann, P. and Brown, N. K., Jr. (1954). Synonyms of *Gublerina*. *Contr. Cushman Fdn foramin. Res.* **5**(2), 62.

Brönnimann, P. and Brown, N. K., Jr. (1956). Taxonomy of the Globotruncanidae. *Eclog. geol. Helv.* **48**(2), 503–561. Plates 20–34, 24 text-figs.

Brönnimann, P. and Brown, N. K., Jr. (1958a). *Hedbergella*, a new name for a Cretaceous planktonic foraminiferal genus. *J. Wash. Acad. Sci.* **48**(1), 15–17. 1 text-fig.

Brönnimann, P. and Brown, N. K., Jr. (1958b). Taxonomy of the Globotruncanidae-remarks. *Micropaleont.* **4**(2), 201–203.

Brönnimann, P. and Rigassi, D. (1963). Contribution to the geology and paleontology of the area of the city of La Habana, Cuba, and its surroundings. *Eclog. geol. Helv.* **56**(1), 193–490. 26 plates, 75 text-figs.

Brönnimann, P. and Wernli, R. (1971). Les "Globigérines" du Dogger du Jura méridional (France). *In* (Farinacci, A., ed.), *Proc. II Plankt. Conf.* **1,** 117–128. 5 plates, 1 text-fig. Tecnoscienza, Roma.

Brotzen, F. (1934). Foraminiferen aus dem Senon Palastinas. *Z. dt. PalastVer.* **57,** 28–72. Plates 1–4.

Brotzen, F. (1936). Foraminiferen aus dem Schwedischen, Untersten Senon von Eriksdal in Schonen. *Sver. geol. Unders. Avh.* ser. C, N(0) 396 (Ars. 30(3)), 1–206. 14 plates.

Brotzen, F. (1942). Die Foraminiferengattung *Gavelinella* nou. gen. und die Systematik der Rotaliiformes. *Sver. geol. Unders. Avh.* ser. C, N(0), 451 (Års. 36(8)), 1–60. 1 plate.

Brown, J. (1853). Note on the artesian well at Colchester; and remarks on some of the microscopic fossils from the Colchester Chalk. *Ann. Mag. nat. Hist.* ser. 2, **12**(70), 240–242. Plates 8, 9.

Brown, N. K., Jr. (1969). Heterohelicidae Cushman, 1927, amended, a Cretaceous planktonic foraminiferal family. *In* (Brönnimann, P. and Renz, H. H., eds.) *Proc. I. Plankt. Conf.*, **2**, 21–67. Plates 1–4, 15 text-figs.

Bukowy, S. and Geroch, S. (1957). O wieku zlepeńców egzotykowych w Kruhelu Wielkim. *Roczn. pol. Tow. geol.* **26**(4) (1956), 297–329. Plates 28–31. 5 text-figs.

Burckle, L. H., Saito, T., and Ewing, M. (1967). A Cretaceous (Turonian) core from the Naturaliste Plateau southeast Indian Ocean. *Deep-Sea Res.* **14**, 421–426. 4 text-figs.

Burma, B. H. (1959). Status of the genera *Hedbergina* and *Hedbergella*. *Contr. Cushman Fdn foramin. Res.* **10**(1), 15.

Butler, E. A. (1963) Species concept and foraminifera. *Trans. Gulf Coast Ass. Geol. Soc.* **13**, 121–125.

Butt, A. A. (1966). Foraminifera of the type Turonian. *Micropaleont.* **12**(2), 168–182. Plates 1–4.

Bykova, N. K. and Subbotina, N. N. (1959). Order Heterohelicida [in Russian] *In* (Rauzer-Chernousova, D. M. and Fursenko, A. V., eds), *Osnovy Paleontologii, Prosteishiye* Akad. Nauk SSSR, Moscow, 332–338. Text-figs 832–859.

Calvin, S. (1895). Composition and origin of Iowa chalk. *Rep. Iowa geol. Surv.* 2nd *Ann.* 1893, **3**, 211–236. Plate 19, figs 1–14.

Carbonnier, A. (1952). Sur un gisement de foraminifères d'age Cénomanien supérieur de la région de Taza (Maroc). *Bull. Soc. geol. Fr.* scr. 6, **2**(1–3), 111–122. Plates 5–7.

Carman, K. (1929). Some foraminifera from the Niobrara and Benton Formations of Wyoming. *J. Paleont.* **3**(3), 309–315. Plate 34.

Caron, M. (1965). Les ouvertures chez le genre *Praeglobotruncana* (Foraminifères). *C.r. somm. Seanc. Soc. geol. Fr.* **7**, 217–218.

Caron, M. (1966). Globotruncanidae du Crétacé supérieur du synclinal de la Gruyère (Préalpes Médianes, Suisse). *Rev. Micropaleont.* **9**(2), 68–93. 6 plates, 6 text-figs.

Caron, M. (1967). Étude biométrique et statistique de plusieurs populations de Globotruncanidae. 2-le sous-genre *Rotalipora* (*Thalmanninella*) dans l'Albien supérieur de la Breggia (Tessin). *Eclog. geol. Helv.* **60**(1), 47–79. 2 plates, 22 text-figs., 1 tab.

Caron, M. (1971). Quelques cas d'instabilité des caractères génériques chez les foraminifères planctoniques de l'Albien. *In* (Farinacci, A., ed.), *Proc. II Plankt. Conf.* **1**, 145–157. 21 text-figs. Tecnoscienza, Roma.

Caron, M. (1972). Planktonic foraminifera from the Upper Cretaceous of Site 98, Leg II, D.S.D.P. *Initial Repts. Deep Sea Drilling Proj.* **11**, 551–559. 2 plates, 3 text-figs.

Caron, M. and Luterbacher, H. P. (1969). On some type specimens of Cretaceous planktonic foraminifera. *Contr. Cushman Fdn foramin. Res.* **20**(1), 23–29. Plates 7–9, text-fig. 1.

Carpenter, N. (1925). Micrology of the upper Washita. *In* The geology of Denton County. *Bull. Univ. Tex. Bur. econ. Geol. Technol.* **2544**, 71–74. Plates 16, 17, 23 (1).

Carpenter, W. B., Parker, W. K. and Jones, T. R. (1862). Introduction to the study of foraminifera. *Ray Soc. Publs.* 1–319. 22 plates.

Carsey, D. O. (1926). Foraminifera of the Cretaceous of central Texas. *Bull. Univ. Tex. Bur. econ. Geol. Technol.* **2612**, 1–56. 8 plates.

Cati, F. (1964). Una microfauna campaniana dei Monti Berici (Vicenza). *G. Geol.*, ser. 2, **32**(1), 199–271. Plates 34–42, 1 text-fig., 1 tab.

Chapman, F. (1892). Microzoa from the phosphatic chalk of Taplow. *Q. Jl. geol. Soc. Lond.* **48**, 514–518. Plate XV.

Chapman, F. (1896). The foraminifera of the Gault of Folkestone.-IX *Jl. R. microsc. Soc.* 581–591. Plates 12, 13.

Chapman, F. (1917). XI. Monograph of the foraminifera and ostracoda of the Gingin Chalk. *Bull. geol. Surv. West. Aust.* (*Paleont. Contr. Geol.* ser **6**, no. 11–12), 1–81. Plates 1–14, map.

Chapman, F. (1926). The Cretaceous and Tertiary foraminifera of New Zealand, with an appendix on the ostracoda. *Paleont. Bull.*, *Wellington* **11**, 1–119, 22 plates.

Chevalier, J. (1961). Quelques nouvelles espèces de Foraminifères dans le Crétacé inférieur Méditerranéen. *Rev. Micropaleont.* **4**(1), 30–36. 1 plate.

Cheylan, G., Magné, J., Sigal, J. and Grekoff, N. (1954). Résultats géologiques et micropaléontologiques du sondage d'El Krachem (Hauts plateaux Algérois). Description of quelques espèces mouvelles. *Bull. Soc. geol. Fr.* ser. 6, **3** (1953), fasc. 4–6, 471–492, Plate 14, 1 text-fig., 1 tab.

Christodoulou, G. (1960). Geologische und mikropaläontologische Untersuchungen auf der Insel Karpathos (Dodekanes). *Palaeontographica* **115**, pt. A:1–6, 1–143. Plates 1–16, 22 text-figs.

Christodoulou, G. (1967). Some remarks on the geology of Kythera Island and a micropaleontological analysis of its Neogene formations. *Boll. geol. Soc. Greece* **6**(2), 385–399. 4 plates, 2 text-figs. (in Greek).

Christodoulou, G. and Marangoudakis, N. (1966). A *Globotruncana* fauna from Kiveri-Argos area, Greece. *Eclog. geol. Helv.* **59**(1), 303–308. 1 plate, 4 text-figs.

Christodoulou, G. and Marangoudakis, N. (1967). Study of a rich *Globotruncana* fauna isolated from the Upper Campanian sediments of the Olonos Series near Kiveri Village, Argos area (NE Peloponnesus). *Boll. geol. Soc. Greece* **6**(2), 308–318. Plate 1, 4 text-figs. (in Greek).

Church, C. C. (1952). Cretaceous foraminifera from the Franciscan Calera Limestone of California. *Contr. Cushman Fdn foramin. Res.* **3**(2), 68–70. 2 text-figs.

Cita. M. B. (1948a). Ricerche stratigrafiche e micropaleontologiche sul Cretacico e sull'Eocene di Tignale (Lago di Garda). Parte II: Paleontologia. *Riv. ital. Paleont. Stratigr.* **54**(3), 117–134. Plate 2.

Cita, M. B. (1948b). Ricerche stratigrafiche e micropaleontologiche sul Cretacico e sull'Eocene di Tignale (Lago di Garda) (I) *Riv. ital. Paleont. Stratigr.* **54**(4), 143–168. Plates 3, 4.

Cita, M. B. (1956). The Cretaceous-Eocene boundary in Italy. *Proc. Fourth World Petrol. Congr.* (1955). Sect. I/D, **2**, 427–452. 9 text-figs.

Cita, M. B. (1963). Tendances évolutives des foraminifères planctiques (Globotruncanae) du Crétacé supérieur. *Evolutionary Trends in Foraminifera*, p. 112–138, 19 text-figs., 3 plates. Elsevier Publ. Co., Amsterdam.

Cita, M. B. (1970). Observations sur quelques aspects paleoecologiques de sondages suboceaniques effectués dans l'Atlantique Nord. *Rev. Micropaleont.* **12**(4), 187–201. 4 plates, 2 text-figs.

Cita, M. B. and Gartner, S., Jr. (1971). Deep sea Upper Cretaceous from the western North Atlantic. *In* (Farinacci, A., ed.), *Proc. II Plankt. Conf.* **1**, 287–319. 8 plates, 7 text-figs. Tecnoscienza, Roma.

Cita, M. B. and Pasquare, G. (1959). Osservazioni micropaleontologische sul Cretaceo delle Dolomiti. *Riv. ital. Paleont. Stratigr.* **65**(4), 385–443. Plates 25–28, 6 text-figs., 1 tab.

Cole, W. S. (1938). Stratigraphy and micropaleontology of two deep wells in Florida. *Bull. Fla St. geol. Surv.* **16**, 1–48. 12 plates, 3 text-figs.

Coleman, P. J. (1966). Upper Cretaceous (Senonian) bathyal pelagic sediments with *Globotruncana* from the Solomon Islands. *J. geol. Soc. Aust.* **13**(2), 439–447. Plate 8, 1 text-fig.

Colom, G. (1931). Las margas rojas con Rosalinas del senoniense de Vélez-Blanco (Prov. de Almería). *Butll. Inst. catal. Hist. nat.* **31**(1) (ser. 2, vol. 11), 28–35. Plates 1, 2, 3 text-figs.

Colom, G. (1934). Contribucion al conocimiento de las facies litopalaeontológicas del Cretacio de las Baleares y del S.E. de España. *In* Geologie des pays Catalans. *Assoc. Etude Géol. Méditerranee Occidentale* **3** (1930–1934), pt. 5, no. 2, 1–11. 1 plate, 3 text-figs.

Colom, G. (1935). Estudios litologicos sobre el Jurassico de Mallorca. *Assoc. Etude Géol. Méditerranee Occidentale* **1**–3(4), pt. 4

Colom, G. (1947). Estudios sobre la sedimentación profunda de las Baleares desde el Lias superior al Cenomanense-Turonense. *Inst. "Lucas Mallada" de Invest. Geol. Consejo Superior de Invest. Cient.* Mono. no. 8, 147 p., 28 plates, 22 text-figs.

Colom, G. (1948). Los microforaminiferos fosiles y su utilidad en geologica estratigrafica. *Boln Inst. geol. min. Esp.* **60**, 113–151. Lamina 1–8, 10 text-figs.

Colom, G. (1952). Los caracteres micropaleontológicos de algunas formaciones del Secundario de España. *Boln. Inst. geol. min. Esp.* **64**, 1–88. 14 plates, 3 text-figs.

Colom, G. (1955). Jurassic-Cretaceous pelagic sediments of the western Mediterranean zone and the Atlantic area. *Micropaleont.* **1**(2), 109–124. 5 plates, 4 text-figs.

Colom, G. and Rangheard, Y. (1966). Les couches a protoglobigérines de l'Oxfordien supérieur de l'Ile d'Ibiza et leurs équivalents a Majorque et dans le domaine Subbétique. *Rev. Micropaleont.* **9**(1), 29–36. 2 plates, 2 text-figs.

Corminboeuf, P. (1961). Tests isolés de *Globotruncana mayaroensis* Bolli, *Rugoglobigerina*, *Trinitella* et Heterohelicidae dans le Maestrichtien des Alpettes. *Eclog. geol. Helv.* **54**(1), 107–122. Plates 1, 2, 1 text-fig.

Corminboeuf, P. (1962). Association de *Belemnitella* et de Globotruncanidae dans le Campanien supérieur des Alpettes (Préalpes externes fibourgeoises). *Eclog. geol. Helv.* **54**(2) (1961), 491–498. Plates 1–2, text-fig. 1.

Costea, I. and Comsa, D. (1969). Organismes planktoniques a la limite Jurassique-Crétacé dans la Plate-forme Moesienne (Roumanie). *In* (Brönnimann, P. and Renz, H. H., eds), *Proc. I Plankt. Conf.* **2**, 100–122. Plates 1–4, text-figs. 1–8. E. J. Brill, Leiden.

Crespin, I. (1953). Lower Cretaceous foraminifera from the Great Artesian Basin, Australia. *Contr. Cushman Fdn foramin. Res.* **4**(1), 26–36. Plates 5, 6.

Cushman, J. A. (1925). The genera *Pseudotextularia* and *Guembelina*. *J. Wash. Acad. Sci.* **15**(6), 133, 134.

Cushman, J. A. (1926a). The foraminifera of the Velasco shale of the Tampico Embayment. *Bull. Am. Ass. Petrol. Geol.* **10**(6), 581–612. Plates 15–21.

Cushman, J. A. (1926b). Some foraminifera from the Mendez Shale of eastern Mexico. *Contr. Cushman Lab. foramin. Res.* **2**(1), 16–24. Plates 2, 3.

Cushman, J. A. (1927a). The American Cretaceous foraminifera figured by Ehrenberg. *J. Paleont.* **1**(3), 213–217. Plates 34–36.

Cushman, J. A. (1927b). New and interesting foraminifera from Mexico and Texas. *Contr. Cushman Lab. foramin. Res.* **3**(2), 111–117. Plates 22, 23.

Cushman, J. A. (1927c). An outline of a re-classification of the foraminifera. *Contr. Cushman Lab. foramin. Res.* **3**(1), 1–105. Plates 1–21.

Cushman, J. A. (1927d), Phylogenetic studies of the foraminifera. Part I. *Am. J. Sci.* **13**(76), 315–326. Text-figs. A, B.

Cushman, J. A. (1927e). Some characteristic Mexican fossil foraminifera. *J.Paleont.* **1**(2), 147–172. Plates 23–28.

Cushman, J. A. (1927f). Some new genera of the foraminifera. *Contr. Cushman Lab. foramin. Res.* **2**(2), 77–81.

Cushman, J. A. (1928a). Additional genera of the foraminifera. *Contr. Cushman Lab. foramin. Res.* **4**(1), 1–10. Plate 1.

Cushman, J. A. (1928b). Foraminifera, their classification and economic use. *Spec. Publs. Cushman Lab.* **1**, 1–401. 59 plates, text-figs., charts.

Cushman, J. A. (1931a). Cretaceous foraminifera from Antigua, B. W. I. *Contr. Cushman Lab. foramin. Res.* **7**(2), 33–46. Plates 5, 6.

Cushman, (1931b). The foraminifera of the Atlantic Ocean. Part 8. *Bull. U.S. natn. Mus.* **104**, 1–179. 26 plates.

Cushman, J. A. (1931c). *Hastigerinella* and other interesting foraminifera from the Upper Cretaceous of Texas. *Contr. Cushman Lab. foramin. Res.* **7**(4), 83–90. Plates 11, 12.

Cushman, J. A. (1931d). A preliminary report on the foraminifera of Tennessee. *Bull. Tenn. Div. Geol.* **41**, 1–62. 13 plates, 1 chart.

Cushman, J. A. (1932). The foraminifera of the Annona Chalk. *J. Paleont.* **6**(4), 330–345. Plates 50, 51.

Cushman, J. A. (1933a). An illustrated key to the genera of the foraminifera. *Spec. Publs Cushman Lab.* **5**. 40 plates.

Cushman, J. A. (1933b). Some new foraminiferal genera. *Contr. Cushman Lab. foramin. Res.* **9**(2), 32–38. Plates 3, 4.

Cushman, J. A. (1933c). Foraminifera, their classification and economic use. *Spec. Publs Cushman Lab.* **4**, 1–349. 31 plates, text-figs., charts, maps.

Cushman, J. A. (1934). A recent *Gumbelitria* (?) from the Pacific. *Contr. Cushman Lab. foramin. Res.* **10**(4), 105. Plate 13, figs. 9, 10a, b.

Cushman, J. A. (1936). Geology and paleontology of the Georges Bank canyons; Part IV—Cretaceous and Late Tertiary foraminifera. *Bull. geol. Soc. Am.* **47**(3), 413–440. 5 plates, 1 tab.

Cushman, J. A. (1938a). Cretaceous species of *Guembelina* and related genera. *Contr. Cushman Lab. foramin. Res.* **14**(1), 2–28. 4 plates.

Cushman, J. A. (1938b). Some new species of rotaliform foraminifera from the American Cretaceous. *Contr. Cushman Lab. foramin. Res.* **14**(3), 66–71. Plates 11, 12.

Cushman, J. A. (1939). New American Cretacous foraminifera. *Contr. Cushman Lab. foramin. Res.* **15**(4), 89–93.

Cushman, J. A. (1940). American Upper Cretaceous foraminifera of the family Anomalinidea. *Contr. Cushman Lab. foramin. Res.* **16**(2), 27–40. Plates 5–7.

Cushman, J. A. (1944a). Foraminifera from the Aquia Formation of Virginia. *Contr. Cushman Lab. foramin. Res.* **20**(1), 17–28. Plates 3, 4.

Cushman, J. A. (1944b). Foraminifera of the lower part of the Mooreville chalk of the Selma Group of Mississippi. *Contr. Cushman Lab. foramin. Res.* **20**(4), 83–96. Plates 13–16.

Cushman, J. A. (1944c). The foraminiferal fauna of the type locality of the Pecan Gap Chalk. *Contr. Cushman Lab. foramin. Res.* **20**(1), 1–16. Plates 1, 2.

Cushman, J. A. (1946a). The species of *Globigerina* described between 1839 and 1850. *Contr. Cushman Lab. foramin. Res.* **22**(1), 15–21. Plates 3, 4.

Cushman, J. A. (1946b). Upper Cretaceous foraminifera of the Gulf Coastal Region of the United States and adjacent areas. *Prof. Pap. U.S. geol. Surv.* **206**, 1–241. 66 plates.

Cushman, J. A. (1947). A foraminiferal fauna from the Santa Anita Formation of Venezuela. *Contr. Cushman Lab. foramin. Res.* **23**(1), 1–18. Plates 1–4.

Cushman, 1948a). Foraminifera from the Hammond Well. *In* Cretaceous and Tertiary subsurface geology. *Bull. Md Dep. Geol. Mines* **2**, 213–267. Plates 15–26, figs. 25–27.

Cushman, J. A. (1948b). *Foraminifera, Their Classification and Economic Use*, 4th ed. 605 pp., 55 plates, 31 text-pls., 9 text-figs. Harvard University Press, Cambridge, Mass.

Cushman, J. A. (1949). The foraminiferal fauna of the Upper Cretaceous Arkadelphia marl of Arkansas. *Prof. Pap. U.S. geol. Surv.* **221–A**, 1–10. 4 plates.

Cushman, J. A. and Church, C. C. (1929). Some Upper Cretaceous foraminifera from near Coalinga, California. *Proc. Calif. Acad. Sci.* 4th ser., **18**(16), 497–530. Plates 36–41.

Cushman, J. A. and Deaderick, W. H. (1942). Cretaceous foraminifera from the Brownstown marl of Arkansas. *Contr. Cushman Lab. foramin. Res.* **18**(3), 50–66. Plates 9–15.

Cushman, J. A. and Deaderick, W. H. (1944). Cretaceous foraminifera from the Marlbrook marl of Arkansas. *J. Paleont.* **18**(4), 328–342. Plates 50–53.

Cushman, J. A. and Dusenbury, A. N., Jr. (1934). Eocene foraminifera of the Poway Conglomerate of California. *Contr. Cushman Lab. foramin. Res.* **10**(3), 51–65. Plates 7–9.

Cushman, J. A. and Goudkoff, P. P. (1944). Some foraminifera from the Upper Cretaceous of California. *Contr. Cushman Lab. foramin. Res.* **20**(3), 53–64. Plates 9, 10.

Cushman, J. A. and Hedberg, H. D. (1941). Upper Cretaceous foraminifera from Santander del Norte, Columbia, S.A. *Contr. Cushman Lab. foramin. Res.* **17**(4), 79–100. Plates 21–23.

Cushman, J. A. and Jarvis, P. W. (1932). Upper Cretaceous foraminifera from Trinidad. *Proc. U.S. natn. Mus.* **80**(2914), 1–60. 16 plates.

Cushman, J. A. and Parker, F. L. (1940). New species of *Bulimina. Contr. Cushman Lab. foramin. Res.* **16**(2), 44–48. Plates 7, 8.

Cushman, J. A. and Renz, H. H. (1946). The foraminiferal fauna of the Lizard Springs Formation of Trinidad, British West Indies. *Spec. Publs Cushman Lab.* **18**, 1–48. 8 plates.

Cushman, J. A. and Renz, H. H. (1947). Further notes on the Cretaceous foraminifera of Trinidad. *Contr. Cushman Lab. foramin. Res.* **23**(2), 31–51. Plates 11, 12.

Cushman, J. A. and ten Dam. A. (1948). *Globigerinelloides*, a new genus of the Globigerinidae. *Contr. Cushman Lab. foramin. Res.* **24**(2), 42, 43. Plate 8, figs. 4–6.

Cushman, J. A. and Todd, R. (1943). Foraminifera of the Corsicana Marl. *Contr. Cushman Lab. foramin. Res.* **19**(3), 49–72. Plates 9–12.

Cushman, J. A. and Todd, R. (1945). A foraminiferal fauna from the Lisbon Formation of Alabama. *Contr. Cushman Lab. foramin. Res.* **21**(1), 11–21. Plates 3, 4.

Cushman, J. A. and Todd, R. (1948). A foraminiferal fauna from the New Almaden District California. *Contr. Cushman Lab. foramin. Res.* **24**(4), 90–98. Plate 16.

Cushman, J. A. and Wickenden, R. T. D. (1930). The development of *Hantkenina* in the Cretaceous with a description of a new species. *Contr. Cushman Lab. foramin. Res.* **6**(2), 39–43. Plate 6.

Cuvillier, J. (1961). *Stratigraphic correlations by microfacies in Western Aquitaine*, 3rd ed., 34 pp., 100 plates. E. J. Brill, Leiden.

Dabagyan, N. V. (1963). Some Cenomanian planktonic foraminifera from the Utesovoi zone of the eastern Carpathians. *In Geology and Petroleum Gas Content of the Soviet Carpathians. Trudy ukr. nauchno-issled. geologo-razv. Inst.* **6**, 102–120. 1 plate, 1 text-fig. (In Russian).

Dain, L. G. (1934). Foraminifera of the Upper Jurassic and Cretaceous deposits located in Dzhaksy-bai Temirskovo region. *Trudy neft. geol.-razv. Inst.* ser. A, **43**, 1–62. 5 plates, map.

Dalbiez, F. (1955). The genus *Globotruncana* in Tunisia. *Micropaleont.* **1**(2), 161–171. 10 text-figs., 2 charts.

Dalbiez, F. (1957). The generic position of *Rotalia deeckei* Franke, 1925. *Micropaleont.* **3**(2), 187, 188. Text-figs. 1–5.

Dallan, L. (1962). Contributo alla geologia dell'Appennino Tosco-Emiliano; II-Richerche micropaleontologiche nei flysch dei dintorni di Pievepelago (Appennino Modenese). *Boll. Soc. geol. ital.* **81**(3), 77–115. Plates 1–6, 3 text-figs.

ten Dam, A. and Sigal, J. (1949). Sur une espèce nouvelle du genre de Foraminifères *Biglobigerinella* Lalicker, 1948. *C. r. somm. Seanc. Soc. geol. Fr.* **11**, 234, 235, 3 text-figs.

Dampel, N. I. (1934). Foraminifera from the Upper Cretaceous of the Karaton Field (Emba region). *Trudy neft. geol.-razv. Inst.* ser. A, **50**, 1–35. 4 plates, 2 text-figs., 1 tab. (In Russian).

Dana, J. D. (1863). *Manual of geology: treating of the principles of the science with special reference to American geological history, for the use of colleges, academies, and schools of science*, 798 pp., 984 text-figs., tabs., chart. Theodore Bliss & Co., Philadelphia. Pa.

Dana, J. D. (1874). *New text-book of geology. Designed for schools and Academies*, 3rd ed., 366 p., 409 text-figs., appendix. Ivison, Blakeman, Taylor, & Co., New York.

Dana, J. D. (1875). *The geological story briefly told. An introduction to geology for the general reader and for beginners in the science*, 263 pp., 245 text-figs. Ivison, Blakeman, Taylor, & Co., New York.

Danilova, A. (1958). Mikropaleontološki prikaz zona višeg senona u Boki Kotorskoj. *Geol. glasnik Zav. za geol. istr. Crne Gore, Titograd* **2**, 223–232. Plates 23–27, text-fig. 1.

Danilova, A. (1960). Gornjosenonska mikrofauna krečnjaka Čitluk potoka (Fruška gora). *Vesnik Geol. Zav. za Geol. i Geof. istr.* Ser. A, **18**, 141–147. 2 plates.

Das, R. M. and Chattarjee, B. P. (1963). Upper Cretaceous smaller foraminifera from the gypseous clays of the Uttatteir Group, Trichinopoly District, Madras State. *Bull. geol. min. metall. Soc. India* **28**, 1–20. 1 plate, 1 text-fig.

Datta, A. K. (1969). Recent developments in the concept of foraminiferal biostratigraphic principles and practices *Selected Lectures on Petroleum Exploration*, vol. **1**, Inst. Petrol. Expl., 217–233. 4 tabs.

Davids, R. N. (1965). *A paleoecologic and paleo-biogeographic study of Maestrichtian*

(*Upper Cretaceous*) *planktonic foraminifera*, 241 pp., 37 text-figs, 14 maps. Unpub. Ph.D. dissertation, Rutgers.

Dawson, G. M. (1875). Notes on the occurrence of foraminifera, coccoliths, etc. in the Cretaceous rocks of Manitoba. *Can. Nat.* n.s., **7** (1874), 252–257. 2 text-figs.

Dawson, J. W. (1876). On some new specimens of fossil Protozoa from Canada. *Proc. Am. Ass. 1875* **24**(2), 100–105. 5 text-figs.

Dekker, L. (1966). Report on the discovery of a rudist in Upper Cretaceous pelagic limestones near La Parroquia (Prov. Murcia-S.E. Spain). *Geologie Mijnb.* **45**, 386–390, 6 text-figs.

Dilley, F. C. (1971). Cretaceous foraminiferal biogeography. *In Faunal provinces in space and time*, p. 169–190, 7 text-figs. Seel House Press, Liverpool, England.

Donze, P., Porthault, B., Thomel, G. and Villoutreys, O. de. (1970). Le Sénonien inferieur de Puget-Theniers (Alpes-Maritimes) et sa microfaune. *Geobios* **3**(2), 41–106. Plates 8–13, 4 text-figs.

Douglas, R. G. (1969). Upper Cretaceous planktonic foraminifera in northern California. Part 1—Systematics. *Micropaleont.* **15**(2), 151–209. 11 plates, 4 text-figs.

Douglas, R. G. (1970). Planktonic foraminifera described from the Upper Cretaceous of California. *Contr. Cushman Fdn foramin. Res.* **21**(1), 18–27. Plates 3–5, 1 tab.

Douglas, R. G. (1971). Cretaceous foraminifera from the northwestern Pacific Ocean: Leg 6, Deep Sea Drilling Project. *Initial Repts. Deep Sea Drilling Proj.* **6**, 1027–1053. 3 plates, 1 text-fig., 4 tabs.

Douglas, R. G. (1972). Paleozoogeography of Late Cretaceous planktonic foraminifera in North America. *J. foramin. Res.* **2**(1), 14–34. 14 text-figs., 1 tab., 2 appendices.

Douglas, R. G. and Rankin, C. (1969). Cretaceous planktonic foraminifer from Bornholm and their zoogeographic significance. *Lethaia* **2**, 185–217. 18 text-figs., 2 tabls.

Douglas, R. G. and Sliter, W. V. (1966). Regional distribution of some Cretaceous Rotaliporidae and Globotruncanidae (Foraminiferida) within North America. Tulane Stud. Geol., **4**(3), 89–131. 5 plates, 1 text-fig., 2 tabs.

Drooger, C. W. (1951). Upper Cretaceous foraminifera of the Midden-Curaçao beds near Hato, Curaçao (N.W.I.). *Proc. sect. Sci. K. ned. Akad. wet.* ser. B **54**(1), 66–72. 1 plate, 1 text-fig.

Drooger, C. W. (1952) Foraminifera from Cretaceous-Tertiary-transitional strata of the Hodna Mountains. Algeria. *Contr. Cushman Fdn foramin. Res.* **3**(2), 89–103. Plates 15–16, text-figs. 1–3, tab. 1.

Dufaure, Ph. (1959). Problèmes stratigraphiques dans le Crétacé supérieur des pays de Bigorre et de Comminges. *Rev. Micropaleont.* **2**(2), 99–112. 3 plates, 1 tab.

Dupeuble, P-A. (1969). Foraminifères planktoniques (Globotruncanidae et Heterohelicidae) du Maestrichtien supérieur en Aquitaine occidentale. *In* (Brönnimann, P. and Renz, H. H., eds), *Proc. I. Plankt. Conf.* **2**, 153–162. Plates 1–4. E. J. Brill, Leiden.

Dzhafarov, D. I. and Agalarova, D. A. (1949). *Microfauna of Albian deposits of Azerbaidzhan.* Aznefteizdat (Azerbaidzhan Publishing House for Petroleum Literature) Baku, 104 pp., 4 plates, 6 text-figs., 5 tables, 3 diagrams (in Russian).

Dzhafarov, D. I., Agalarova, D. A. and Khalilov, D. M. (1951). *Investigation into the microfauna of the Cretaceous deposits in Azerbaidzhan.* Aznefteizdat (Azerbaidzhan Publishing House for Petroleum Literature) Baku, 128 pp., 17 plates, 11 text-figs. (in Russian).

Edgell, H. S. (1957). The genus *Globotruncana* in northwest Australia. *Micropaleont.* **3**(2), 101–126. Plates 1–4, text-figs. 1–4, tab. 1.

Edgell, H. S. (1962). A record of *Globotruncana concavata* (Brotzen) in northwest Australia. *Rev. Micropaleont.* **5**(1), 41–50. 1 plate, 2 tabs.

Egger, J. G. (1857). Die Foraminiferen der Miocän-Schichten bei Ortenburg in Nieder-Baynern. *Neues Jahr. Min. Geog., Geol., Petref.-Kunde* 266–311. Plates 5–15.

Egger, J. G. (1899–1902). Foraminiferen und Ostrakoden aus den Kreidemergeln der Oberbayerischen Alpen. *Abh. bayer. Akad. Wiss. Math.-phys. Kl.* **21**, 1–230. Tafeln 1–27.

Egger, J. G. (1908). Mikrofauna der Kreideschichten des westlichen bayer. Waldes und des Gebietes um Regensburg. *Ber. naturw. Ver. Passau* **20** (1905–1907), 1–75. 10 plates.

Egger, J. G. (1909). Foraminiferen der Seewener Kreideschichten. *Sber. bayer. Akad. Wiss. Math.-phys. Kl.* **11**, 1–52. 6 tafeln.

Egger, J. G. (1910). Ostrakoden und Foraminiferen des Eybrunner Kreidemergels in der Umgegend von Regensburg. *Ber. naturw. Ver. Regensburg*, **12** (1907–1909), 86–133. Plates 1–6.

Ehremeeva, A. I. and Belousova, N. A. (1961). Stratigrafija i Fauna Foraminifer Melovykh i Paleogenovykh Otlozhenij Vostochnogo Sklona Urala, Zaural'ja i Severnogo Kazakhstana. *Mater. Geol. polez. Iskop. Urala* **9**, 1–113. Plates 1–38, 7 text-figs.

Ehrenberg, C. G. (1838). Über die Bildung der Kreidefelsen und des Kreidemergels durch unsichtbare Organismen. *Abh. preuss. Akad. Wiss. Berlin. Phys.-math. Kl.* 59–147. Tafeln 1–4, 2 tab.

Ehrenberg, C. G. (1840). Die Bildung der europaischen, libyschen und arabischen Kreidefelsen und des Kreidemergels aus mikroskopischen Organismen. *Abh. preuss. Akad. Wiss. Berlin. Phys-math. Kl.* 1–91. 3 tabs.

Ehrenberg, C. G. (1843). Verbreitung und Einfluss des Mikroskopischen Lebens in Süd- und Nord-Amerika. *Abh. preuss. Akad. Wiss. Berlin Phys.-math. Kl.* (1841), pt. 1, 291–446. Plates 1–4.

Ehrenberg, C. G. (1844). Eine Mittheilung über 2 neue Lager von Gebirgsmassen aus Infusorien als Meeres-Absatz in Nord-Amerika und eine Vergleichung derselben mit den organischen Kreide-Gebilden in Europa und Afrika. *Ber. K. preuss. Akad. Wiss. Berlin* 57–98.

Ehrenberg, C. G. (1854). *Mikrogeologie. Das Erden und Felsen Schaffende wirken des unsichtbar Kleinen selbstandigen Lebens auf der Erde.* Verlag von Leopold Voss, Leipzig, in 2 vols. and atlas. Vol. 1: 374 pp. Vol. 2: 88 pp. Atlas: 40 tafeln.

Ehrenberg, C. G. (1856). Über den Grünsand und seine Erläuterung des Organischen Lebens. *Abh., Jb. K. Akad. Wiss. Berlin. Phys.* (1855), 85–176. Plates 1–7.

Eichenberg, W. (1933). Die Erforschung der Mikroorganismen, unsbesondere der Foraminiferen der norddeutschen Erdolfelder; Teil I—Die Foraminiferen der Unterkreide; Folge 1—Foraminiferen aus dem Albien von Wenden am Mittellandkanal. *Jber. niedersachs. geol. Ver.* **25** (1932–1933), 1–32. Plates 1–8, 5 text-figs.

Eichenberg, W. (1935). Mikrofaunen-Tafeln zur Bestimmung von Unterkreide-Horizonten in Bohnkernen norddeutscher Oelfelder. *Ol Kohle Erdol Teer* **11**(23), 388–398. 14 plates, 1 tab.

Eicher, D. L. (1965). Foraminifera and biostratigraphy of the Graneros shale. *J. Paleont.* **39**(5), 875–909. Plates 103–106, 6 text-figs.

Eicher, D. L. (1966). Foraminifera from the Cretaceous Carlisle Shale of Colorado. *Contr. Cushman Fdn foramin. Res.* **17**(1), 16–31. Plates 4–6, text-figs. 1, 2.

Eicher, D. L. (1967). Foraminifera from Belle Fourche Shale and equivalents, Wyoming and Montana. *J. Paleont.* **41**(1), 167–188. Plates 17–19, 6 text-figs.

Eicher, D. L. (1969). Cenomanian and Turonian planktonic foraminifera from the western interior of the United States. *In* (Brönnimann, P and Renz, H. H., eds), *Proc. I Plankt. Conf.* **2**, 163–174. 5 text-figs. E. J. Brill, Leiden.

Eicher, D. L. (1973). Phylogeny of the Late Cenomanian planktonic foraminifer *Anaticinella multiloculata* (Morrow). *J. foramin. Res.* **2**(4) (1972), 184–190. 2 plates, 4 text-figs.

Eicher, D. L. and Worstell, P. (1970a). *Lunatriella*, a Cretaceous heterohelicid foraminifer from the western interior of the United States. *Micropaleont.* **16**(1), 117–121. 1 plate, 2 text-figs.

Eicher, D. L. and Worstell, P. (1970b). Cenomanian and Turonian foraminifera from the Great Plains, United States. *Micropaleont.* **16**(3), 269–324. Plates 1–13, 12 text-figs.

Elay, H. (1859). *Geology in the garden; or the fossils in the flint pebbles*, 121 pp., 12 plates, 4 cross-sections, 2 maps. Bell & Daldy, London.

El-Naggar, Z. R. (1966). Stratigraphy and planktonic foraminifera of the Upper Cretaceous-Lower Tertiary succession in the Esna-Idfu region, Nile Valley, Egypt, U.A.R. *Bull. Br. Mus. nat. Hist.* Geology, Supp. **2**, 1–291, 23 plates, 18 text-figs.

El-Naggar, Z. R. (1971a). On the classification, evolution and stratigraphical distribution of the Globigerinacea. *In* (Farinacci, A., ed.), *Proc. II Plankt. Conf.* **1**, 421–476. 7 plates, 2 text-figs., 1 tab. Tecnoscienza, Roma.

El-Naggar, Z. R. (1971b). The genus *Rugoglobigerina* in the Maestrichtian Sharawna Shale of Egypt. *In* (Farinacci, A., ed.), *Proc. II Plankt. Conf.* **1**, 477–537. 19 plates, 2 text-figs. Tecnoscienza, Roma.

El-Naggar, Z. R. and Haynes, J. 1967. *Globotruncana caliciformis* in the Maestrichtian Sharawna shale of Egypt. *Contr. Cushman Fdn foramin. Res.* **18**(1), 1–13. Plates 1–4.

Ericson, D. B., Ewing, M. and Wollin, G. (1963). Pliocene-Pleistocene boundary in deep-sea sediments. *Science* **139**(3556), 727–737. 14 text-figs.

Esker, G. C. III. (1968a). A new species of *Pseudoguembelina* from the Upper Cretaceous of Texas. *Contr. Cushman Fdn foramin. Res.* **19**(4), 168, 169. Text-figs. 1–3.

Esker, G. C. III. (1968b). Designation of a lectotype of *Globotruncana rosetta* (Carsey). *Contr. Cushman Fdn foramin. Res.* **19**(4), 170, 171. Text-figs. 1–3.

Esker, G. C. III. (1969). Planktonic foraminifera from St. Ann's Great River Valley, Jamaica. *Micropaleont.* **15**(2), 210–220. 3 plates, 2 text-figs.

Eternod Olvera, Y. (1959). Foraminiferos del Cretácico Superior Tampico-Tuxpan. *Boln Asoc. mex. Geol. petrol.* **11**(3–4), 63–134. Lamina 1–8.

Ewing, M., Saito, T., Ewing, J. I. and Burckle, L. H. (1966). Lower Cretaceous sediments from the northwest Pacific. *Science* **152**, 751–755. 5 text-figs.

Farinacci, A. and Radoičić, R. (1964). Correlazions fra serie giuresi e cretacae dell'Appennino centrale e dell Dinaridi esterne. *Ricerca scient.* Rend. A, **7**(2), ser. 2, 269–300. 15 plates. 4 text-figs.

Farinacci, A. and Sirna, G. (1960). Livelli a Saccocoma nel Malm dell'Umbria e della Sicilia. *Boll. Soc. geol. ital.* **79**(1), 59–70. 6 plates, 2 text-figs.

Flandrin, J., Moullade, M. and Porthault, B. (1962). Microfossiles caractéristiques du Crétacé inférieur Voçontien. *Rev. Micropaleont.* **4**(4), 211–228. 3 plates, 1 text-fig., 1 tab.

Fox, S. K., Jr. (1954). Cretaceous foraminifera from the Greenhorn, Carlile and Cody Formations, South Dakota, Wyoming. *Prof. Pap. U.S. geol. Surv.* **254**–E, 97–124. Plates 24–26, 4 tabs.

Franke, A. (1925). Die Foraminiferen der pommerschen Kreide. *Abh. geol.-palaeont. Inst. Greifswald* **6**, 1–96. 9 plates.

Franke, A. (1928). Die Foraminiferen der Oberen Kreide Nord- und Mitteldeutschlands. *Abh. preuss. geol. Landesanst.* n.f. **3**, 1–207. 18 tafeln, 2 text-figs.

Frerichs, W. E. (1971). Evolution of planktonic foraminifera and paleotemperatures. *J. Paleont.* **45**(6), 963–968. 5 text-figs.

Frerichs, W. E., Heiman, M. E., Borgman, L. E. and Bé, A. W. H. (1972), Latitudinal variations in planktonic foraminiferal test porosity: Part 1. Optical studies. *J. foramin. Res.* **2**(1), 6–13, 9 text-figs., 3 tabs.

Frizzell, D. L. (1943). Upper Cretaceous foraminifera from northwestern Peru. *J. Paleont.* **17**(4), 331–353. Plates 55–57, 2 text-figs.

Frizzell, D. L. (1954). Handbook of Cretaceous foraminifera of Texas. *Rep. Invest. Univ. Tex. Bur. econ. Geol. Technol.* **22**, 1–232. 21 plates, 2 text-figs., 4 tabs.

Fuchs, W. (1967a). Über Ursprung und Phylogenie der Trias- "Globigerinen" und die Bedeutung dieses Formenkreises für das echte Plankton. *Verh. geol. Bundesanst.*, *Wien* **1–2**, 135–176. 8 plates, 2 text-figs.

Fuchs, W. (1967b). Die Foraminiferenfauna eines Kernes der höheren Mittel-Alb der Tiefbohrung Delft 2-Niederlande. *Jb. geol. Bundesanst.*, *Wien* **110**(2), 255–341. 99 tafeln.

Fuchs, W. (1969). Zur Kenntnis des Schalenbaues der zu den Trias- "Globigerinen" zählenden Foraminiferengattung *Praegubkinella*. *Verh. geol. Bundesanst. Wien* **2**, 158–162. 3 plates, 1 text-fig.

Fuchs, W. (1970). Eine alpine, tiefliassische Foraminiferenfauna von Hernstein in Niederösterreich. *Verh. geol. Bundesanst.*, *Wien* **1**, 66–145. 10 plates, 2 text-figs.

Fuchs, W. (1971). Eine alpine Foraminiferenfauna des tieferen Mittle-Barrême aus der Drusbergschichten vom Ranzenberg bei Hohenems in Vorarlberg (Österreich). *Abh. geol. Bundesanst.*, *Wien* **27**, 1–49. 11 plates, 5 text-figs.

Funnell, B. M., Friend, J. K. and Ramsay, A. T. S. (1969). Upper Maestrichtian planktonic foraminifera from Galicia Bank, west of Spain. *Palaeontology* **12**(1), 19–41. Plates 1–5, Text-figs. 1–22.

Galloway, J. J. (1933). *A manual of foraminifera*, James Furman Kemp Memorial Series Publ. no. 1, 483 pp., 42 pls. Principia Press, Inc., Bloomington, Indiana.

Gandolfi, R. (1942). Ricerche Micropaleontologiche e Stratigrafiche. Sulla Scaglia e sul Flysch Cretacici. Dei Dintorni di Balerna (Canton Ticino). *Mem. Riv. ital. Paleont.* **48**(4), 1–160. 14 plates, 49 text-figs.

Gandolfi, R. (1955a). The genus *Globotruncana* in northeastern Columbia. *Bull. Am. Paleont.* **36**(155), 1–118. 10 plates, 12 text-figs.

Gandolfi, R. (1955b). A *Globotruncana* fauna from the Pecan Gap chalk of Texas. *Micropaleont.* **1**(3), 257–259. 9 text-figs.

Gandolfi, R. (1957). Notes on some species of *Globotruncana*. *Contr. Cushman Fdn foramin. Res.* **8**(2), 59–65. Plates 8, 9.

Gauger, D. J. (1953). Microfauna of the Hilliard. *In* Peterson, R. H., Gauger, D. J. and Lankford, R. R., Microfossils of the Upper Cretaceous of northeastern Utah and southwestern Wyoming. *Bull. Utah geol. miner. Surv.* **47** (Contr. Micropaleont. no. 1), 51–90. Plates 4–11.

Gawor-Biedowa, E. (1972). The Albian, Cenomanian and Turonian foraminifers of Poland and their stratigraphic importance. *Acta Paleont. pol.* **17**(1), 1–155. 20 plates, 14 text-figs., 4 tabs.

Geinitz, H. B. (1846). *Gundriss der Versteinerungskunde.* Arnoldische Buchhandlung, Dresden und Leipzig, 815 pp., 26 plates, 1 tab.

Gelati, R. and Dario Passeri, L. (1967). Il Flysch di Bergamo nuova formazione. Cretacica delle Prealpi Lombarde. *Riv. ital. Paleont. Stratigr.* **73**(3), 835–850 Plate 69, 6 text-figs.

Geodakchan, A. A. and Aliyulla, Kh. (1959). Representatives of the genus *Gumbelina* in Upper Cretaceous deposits of Azerbaidzhan. *Uchen. Zap. azerb. gos. Univ.*, **1**, 51–62. 1 plate (in Russian).

Germeraad, J. H. (1946). *Geological, petrographical and palaeontological results of explorations, carried out from September* 1917 *till June* 1919 *in the Island of Ceram by L. Rutten and W. Hotz.* 3rd ser.: Geology, no. 2, Geology of central Seran, 135 pp., 12 plates, 1 map, 5 tabs. J. H. de Bussy, Amsterdam.

Gianotti, A. (1958). Deux facies due Jurassique supérieur en Sicile. *Rev. Micropaleont.* **1**(1), 38–51. 2 plates, 5 text-figs.

Glaessner, M. (1936). Die Foraminiferengattungen *Pseudotextularia* und *Amphimorphina. Problemy Paleont.* Moscow Univ., USSR, **1**, 95–134. 2 tafeln, 3 text-figs.

Glaessner, M. (1937). Planktonforaminiferen aus der Kreide und dem Eozän und ihre stratigraphische Bedeutung. *Etyudy Mikropaleont.* Moscow Univ., **1**(1), 27–52. 2 plates, 6 text-figs.

Glaessner, M. (1948). *Principles of Micropaleontology,* 296 pp., 14 plates, 64 text-figs., 7 tabs. John Wiley & Sons, Inc., New York.

Glaessner, M. (1966). Notes on Foraminifera of the genus *Hedbergella. Eclog. geol. Helv.* **59**(1), 179–184. Plate 1.

Glazunova, A. Ye., Balakhmatova, V. T., Lipman, R. Kh., Romanova, V. I. and Khokhlova, I. A. (1960). Stratigraphy and fauna of the Cretaceous deposits of western Siberian lowlands. *Trudy, vses. nauchno-issled. geol. Inst.*, nov. ser., **29**, 1–347. 52 plates, 13 text-figs., 3 tabs. (in Russian).

Glintzboeckel, C. and Magné, J. (1955). Sur la répartition stratigraphique de *Globigerinelloides algeriana* Cushman et ten Dam, 1948. *Micropaleont.* **1**(2), 153–155. Text-figs. 1–3.

Gohrbandt, K. H. A. (1967). The geologic age of the type locality of *Pseudotextularia elegans* (Rzehak). *Micropaleont.* **13**(1), 68–74. 1 plate.

Gorbachik, T. N. (1964). Variability and microstructure of the wall of the test of *Globigerinelloides algeriana. Paleont. Zh.* **4**, 32–37. Text-figs. 1–6 (in Russian).

Gorbachik, T. N. (1971). On early Cretaceous foraminifera from the Crimea. *Vop. Mikropaleont.* **14**, 125–139. Plates 2–6, 1 tab. (in Russian).

Gorbachik, T. N. and Krechmar, V. (1970). Features of the articulation of chambers by representatives of the genus *Leupoldina* (Foraminifera). *Paleont. Zh.* **3**, 143–146. Plate 13, text-figs. 1a–e (in Russian).

Gorbachik, T. N. and Krechmar, V. (1971). Wall structure of some early Cretaceous planktonic foraminifera. *Vop. Mikropaleont.* **14**, 17–24. Plates 21–30. 1 text-fig. (in Russian).

Govindan, A. (1972). Upper Cretaceous planktonic foraminifera from the Pondicherry area, south India. *Micropaleont.* **18**(2), 160–193. 6 plates, 2 text-figs., 4 tabs.

Graham, J. J. (1962). A review of the planktonic foraminifera from the Upper Cretaceous of California. *Contr. Cushman Fdn foramin. Res.* **13**(3), 100–109. Plates 19, 20, tab. 1.

Graham, J. J. and Church, C. C. (1963). Campanian foraminifera from the Stanford University campus California. *Stanford Univ. Publs, Geological Sciences* **8**(1), 1–107. 8 plates, 2 text-figs.

Graham, J. J. and Clark, D. K. (1961). New evidence for the age of the "G-1 Zone" in the Upper Cretaceous of California. *Contr. Cushman Fdn foramin. Res.* **12**(3), 107–114. Plate 5, text-figs. 1, 2, tab. 1.

Grigelis, A. A. (1958). *Globigerina oxfordiana* sp. n.—Discovery of a globigerine in the Upper Jurassic deposits of Lithuania. *Nauch. Dokl. vyssh. Shk., Geol.-Geogr. Nauki* **3**, 109–111. 1 text-fig. (in Russian).

Grigsby, R. D. (1964). *Hedbergella*, an Oklahoma genus of foraminifera.Translation of N.I. Maslakova's 1963 article: On the classification of the genus *Hedbergella*. *Geol. Notes* **24**(6), 130–136. Text-fig. 2, 1 tab.

Guha, D. K. and Mohan, M. (1965). A note on Upper Cretaceous microfauna from the Middle Andaman Island. *Bull. geol. min. metall. Soc. India* **33**, 1–4. Plate 1, 1 text-fig.

Haake, F. W. (1962). Untersuchungen an der Foraminiferen-Fauna im Wattgebiet zwischen Langeoog und dem Festland. *Meyniana* **12**, 24–64. 6 plates, 4 text-figs., 5 tabs.

Haeusler, R. (1881a). *Untersuchungen die microscopischen Structurverhaltnisse der Aargauer Jurakalke mit besonderer Berücksichtigung ihrer Foraminiferenfauna*, 47 pp., 2 pls. Dissertation, Univ. Zurich, Druck von Fisch, Wild & Co.

Haeussler, R. (1881b). Note sur une zone a Globigérines dans le terrain jurassique de la Suisse. *Annls Soc. r. malacol. Belg.* (3) **1**. Bruxelles, 188–190.

Haeusler, R. (1890). Monographie der Foraminiferen-Fauna der schweizerischen Transversarius-zone. *Abh. schweiz. palaont. Ges.* **17**, 1–134. 15 plates.

Hagn, H. (1953). Die Foraminiferen der Pinswanger Schichten (unteres Obercampan); Ein Beitrag zur Mikropaläontologie der Helvetischen Oberkreide Sudbayerns. *Palaeontographica* **104**(A), fasc. 1–3, 1–119. 8 plates, 27 text-figs.

Hagn, H. and Zeil, W. (1954). Globotruncanen aus dem Ober-Cenoman und Unter-Turon der Bayerischen Alpen. *Eclog. geol. Helv.* **47**(1), 1–60. 7 tafeln, 3 text-figs., 1 tab.

Halkyard, E. (1918). The fossil foraminifera of the Blue Marl of Côte des Basques, Biarritz. *Mem. Proc. Manchr. lit. phil. Soc.* **62**(6) (1917), 1–145. 9 plates.

Hamaoui, M. (1964). On a new subgenus of *Hedbergella* (Foraminiferida). *Israel Jnl. Earth-Sci.* **13**, 133–142. 2 plates.

Hamilton, E. L. (1953). Upper Cretaceous, Tertiary, and Recent planktonic foraminifera from mid-Pacific flat-topped seamounts. *J. Paleont.* **27**(2), 204–237. Plates 29–32, 5 text-figs.

Hanzlíková, E. (1963). *Globotruncana helvetica posthelvetica* n. subsp. from the Carpathian Cretaceous. *Vest. ustred. Ust. geol.* **38**(5), 325–328. Plates 1, 2, 2 text-figs.

Hanzlíková, E. (1969). The foraminifera of the Frydek Formation (Senonian). *Sb. Geol. Věd, Paleont. rada P*, **11**, 7–84. 20 plates, 3 text-figs.

Hanzlíková, E. (1972). Carpathian Upper Cretaceous Foraminiferida of Moravia (Turonian-Maestrichtian). *Roz. ustred. Ust. geol.* **39**, 1–160. 40 plates, 5 text-figs., 1 tab.

Haque, A. F. M. M. (1959). Some Late Cretaceous smaller foraminifera from West Pakistan. *Mem. geol. Surv. Pakist. Palaeont. Pakist.* **2**(3), 1–33. 2 plates, 1 text-fig., 3 tabs.

Harris, R. W. and Jobe, B. I. (1951). *Microfauna of basal Midway outcrops near Hope, Arkansas.* Transcript Press, Norman, Okla., 85 pp., 14 pls., 1 text-fig., tabs. 2–5.

Harting, P. (1851). *Die Macht des Kleinen sichtbar in der Bildung der Rinde unseres Erdballs oder Uebersicht der Gestaltung, der geographischen und geolischen Berbreitung der Polypen, Foraminiferen und tiefelschaligen Bacillarien,* 171 pp., 40 text-figs., 1 pl. Verlag von Wilhelm Engelmann, Leipzig.

Hartwig, G. L. (1866). *The sea and its living wonders. A popular account of the marvels of the deep,* 3rd ed., 518 pp., 12 pls., text-figs., map. Longmans, Green & Co.

Haynes, J. (1956). Certain smaller British Paleocene foraminifera. Part I. Nonionidae, Chilostomellidae, Epistominidae, Discorbidae, Amphistegenidae, Globigerinidae, Globorotaliidae and Guembelinidae. *Contr. Cushman Fdn foramin. Res.* **7**(3), 79–101. Plates 16–18, 2 text-figs.

Hecht, F. E. (1938). Standard-Gliederung der Nordwest-deutschen Unterkreide nach Foraminiferen. *Senckenberg. leth.* **443**, 1–42. 23 plates, 1 text-fig., 4 tabs.

Hecht, A. D. and Savin, S. M. (1972). Phenotypic variation and oxygen isotope ratios in Recent planktonic foraminifera. *J. foramin. Res.* **2**(2), 55–67. 14 text-figs., 3 tabs.

Heer, O. (1865). *Die Urwelt der Schweiz,* 622 pp., 10 pls., 368 text-figs. Druck und Verlag von Friedrich Schulthesz, Zurich.

Herm, D. (1962). Stratigraphische und mikropaläontologische Untersuchungen der Oberkreide im Lattengebirge und Nierential. *Abh. bayer, Akad. Wiss. Math.-phys. Kl.* n.f. **104**, 1–119. 11 tafeln, 9 abb.

Herm, D. (1965). Mikropaläontologisch-stratigraphische Untersuchungen im Kreiderflysch zwischen Deva und Zumaya (Prov. Guipuzcoa Nordspanien). *Z. dt. geol. Ges.* **115**(1), 277–348.

Hermes, J. J. (1945). Geology and paleontology of east Camaguey and west Oriente, Cuba. *Utrecht, Univ., Geogr. Geol. Meded., Physiogr.-Geol. Reeks* ser. 2, **7**, 1–63. 5 plates, 2 text-figs., 8 tabs.

Hermes, J. J. (1966). Lower Cretaceous planktonic foraminifera from the Subbetic of southern Spain. *Geologie Mijnb.* **45**(5), 157–164. 1 plate, 4 text-figs.

Hermes, J. J. (1969). Late Albian foraminifera from the Subbetic of southern Spain. *Geologie Mijnb.* **48**(1), 35–65. 8 plates, 1 text-fig., 2 tabs.

Heron-Allen, E. and Earland, A. (1910). On the Recent and fossil foraminifera of the shore-sands of Selsey Bill, Sussex; Part V—The Cretaceous foraminifera. *Jl R. microsc. Soc.* 401–426. Plates 6–11.

Hillebrandt, A. von (1962). Das Paleozän und seine Foraminiferenfauna im Becken von Reichenhall und Salzburg. *Abh. bayer, Akad. Wiss. Math.-phys. Kl.* n.f., **108**, 1–182. Plates 1–15, 13 text-figs.

Hiltermann, H. (1956). Biostratigraphie der Oberkreide auf Grund von Mikrofossilien. *Palaont. Z.* **30**, 19–32. Abb. 1–6.

Hiltermann, H. and Koch, W. (1955). Biostratigraphie der Grenzschichten Maestricht/Campan in Lüneburg und in der Bohrung Brunhilde. 2 Teil: Foraminiferen. *Geol. Jb.* **70**, 357–383. Plates 27–29, text-figs. 5–7, 2 tabs.

Hiltermann, H. and Koch, W. (1956). Mikropalaontologische Feinhorizontierung von Santon-Profilen durch das Erzlager Legende-Broistedt. *Palaont. Z.* **30**, 33–44. Tafeln 1'–3', abb. 1–6.

Hiltermann, H. and Koch, W. (1962). Oberkreide des nördlichen Mitteleuropa. *In* *Leitfossilien der Mikropalaontologie*, pp. 299–338. plates 42–51. text-fig. 25, tab. 19. Gebruder Borntraeger, Berlin.

Hinte, J. E. van (1963). Zur stratigraphie und Mikropaläontologie der Oberkreide und des Eozäns des Krappfeldes (Kärnten). *Jb. geol. Bundesanst.*, *Wien* **8**, 1–147. Abb. 1–15, Tabellen 1–6, Beilagen 1–4, 2 phototafeln, 22 tafeln.

Hinte, J. E. van (1965a). The type Campanian and its planktonic foraminifera I. *Proc. Sect. Sci. K. ned. Akad. wet.* ser. B, **68**(1), 8–28. 3 plates, 9 text-figs.

Hinte, J. E. van. (1965b). Remarks on the Kainach Gosau (Styria, Austria). *Proc. Sect. Sci. K. ned. Akad. wet.* ser. B, **62**(2), 72–92. 4 plates, 4 text-figs.

Hinte, J. E. van (1969a). A *Globotruncana* zonation of the Senonian Subseries. *In* (Brönnimann, P. and Renz, H. H., eds), *Proc. I Plankt. Conf.* **2**, 257–266, 3 text-figs. E. J. Brill, Leiden.

Hinte, J. E. van (1969b). The nature of biostratigraphic zones. *In* (Brönnimann, P. and Renz, H. H., eds), *Proc. I Plankt. Conf.* **2**, 267–272. 4 text-figs. E. J. Brill, Leiden.

Hinte, J. E. van. (1972). The Cretaceous time scale and planktonic-foraminiferal zones. *Proc. Sect. Sci. K. ned. Akad. wet.* ser. B, **75**(1), 1–8, 1 text-fig.

Hofker, J. Sr. (1954). Chamber arrangement in foraminifera. *Micropaleont.* **7**(1), 30–32.

Hofker, J. Sr. (1956a). Foraminifera Dentata; Foraminifera of Santa Cruz and Thatch-Island, Virginia-Archipelago, West Indies. *Spolia zool. Mus. haun.* **15**, 1–237. Plates 1–35.

Hofker, J. Sr. (1956b). Les foraminifères de la zone de contact Maestrichtien-Campanien dans l'est de la Belgique et le sud des Pays-Bas. *Annls Soc. geol. Belg.* **80** (1956–1957), no 3, B191–B233. 14 plates, 2 text-figs.

Hofker, J. Sr. (1956c). Die Globotruncanen von Nordwest-Deutschland und Holland. *Neues Jb. Geol. Palaont. Abh.* **103**(3), 312–340. 26 text-figs.

Hofker, J. Sr. (1956d). Foraminifera from the Cretaceous of southern Limberg, Netherlands; XIX—Planctonic foraminifer of the chalk tuff of Maestricht and environments. *Natuurh. Maandbl.* **45**(5–6), 51–57. Text-figs. 1–24, 1 chart.

Hofker, J. Sr. (1956e). The structure of *Globorotalia*. *Micropaleont.* **2**(4), 371–373. 7 text-figs.

Hofker, J. Sr. (1956f). Die *Pseudotextularia*-Zone in der Bohrung Maabüll I und ihre Foraminiferen-Fauna. *Palaont. Z.* **30**, 59–79. Tafeln 5'–10', abb. 1.

Hofker, J. Sr. (1957). Foraminiferen der Oberkreide von Nordwestdeutschland und Holland, *Beih. geol. Jb.* **27**, 1–464. 495 abb.

Hofker, J. Sr. (1958). Les foraminifères du Crétacé supérieur de Glons. *Annls Soc. geol. Belg.* **81**(6), B467–B493. Plates 1–8, 3 text-figs.

Hofker, J. Sr. (1959). Foraminifera from the Cretaceous of South-Limberg, Netherlands; XLIII. Globigerines and related forms in the Cretaceous and Lower Paleocene of South Limberg. *Natuurh. Maandbl.* **48**(7–8), 89–95. 9 text-figs.

Hofker, J. Sr. (1960a). Le problème du Dano-Paléocène et le passage Crétacé-Tertiaire. *Rev. Micropaleont.* **3**(2), 119–130. 3 plates, 1 text-fig.

Hofker, (1960b). Planktonic foraminifera in the Danian of Denmark. *Contr. Cushman Fdn foramin. Res.* **11**(3), 73–86. Text-figs. 1–38, 6 tabs.

Hofker, J. Sr. (1960c). The taxonomic status of *Praeglobotruncana*, *Planomalina*, *Globigerinella*, and *Biglobigerinella*. *Micropaleont.* **6**(3), 315–322. 2 plates, 1 text-fig.

Hofker, J. Sr. (1960d). The type localities of the Maestrichtian (Maestrichtian Chalk Tuff) and of the Montian (Tuffeau de Ceply, Calcaire de Mons, Lagunar, and Lacustre Montian). *J. Paleont.* **34**(3), 584–588. 1 text-fig.

Hofker, J. Sr. (1960e). The foraminifera of the lower boundary of the Danish Danian. *Meddr dansk geol. Foren.* **14**(3), 212–242. 47 text-figs.

Hofker, J. Sr. (1961a). The gens *Globigerina cretacea* in northwestern Europe. *Micropaleont.* **7**(1), 95–100. 1 plate.

Hofker, J. Sr. (1961b). Globotruncanidae Brotzen, 1942, as toothplate foraminifera. *Contr. Cushman Fdn foramin. Res.* **12**(4), 123–126. Text-figs. 1–4.

Hofker, J. Sr. (1961c). Les foraminifères du tuffeau arénacé de Folx-les-Caves. *Annls Soc. geol. Belg.* **84** (1960–1961), 549–580. 73 text-figs.

Hofker, J. Sr. (1962a). Correlation of the Tuff Chalk of Maestricht (type Maestrichtian) with the Danske Kalk of Denmark (type Danian), the stratigraphic position of the type Montian, and the planktonic foraminiferal faunal break. *J. Paleont.* **36**(5), 1051–1089. 28 text-figs.

Hofker, J. Sr. (1962b). Studien an planktonischen Foraminiferen. *Neues Jb. Geol. Palaont. Abh.* **114**(1), 81–134. Text-figs. 1–85.

Hofker, J. Sr. (1963). Mise au point concernant les genres *Praeglobotruncana* Bermúdez, 1952, *Abathomphalus* Bolli, Loeblich and Tappan, 1957. *Rugoglobigerina* Brönnimann, 1952, et quelques espèces de *Globorotalia. Rev. Micropaleont.* **5**(4), 280–288. 2 plates.

Hofker, J. Sr. (1969). "Globigérines" du Jurassiques Supérieur. *In* (Brönnimann, P. and Renz, H. H., eds.) *Proc. I Plankt. Conf.* **2**, 287–290. 1 text-fig.; E. J. Brill, Leiden.

Hofker, J. Sr. (1972). Is the direction of coiling in the early stages of an evolution of planktonic foraminifera at random? (50% right and 50% left). *Revta. esp. Micropaleont.* **4**(1), 11–17. 1 text-fig. 2 plates.

Hofman, E. A. (1958). New discovery of Jurassic globigerines. *Nauch. Dokl. vyssh. Shk., Geol.-Geogr. Nauki* **2**, 125–126. 1 text-fig (in Russian).

Howe, H. V. (1939). Louisiana Cook Mountain Eocene Foraminifera. *Bull. geol. Surv. La.* **14**, 1–122. 14 plates, 2 tabs., 1 map.

Huss, F. (1957). Stratigrafia jednostki Węglówki na podstawie mikrofauny. *Acta geol. pol.* **7**(1), 29–69. 11 plates, 7 text-figs., 3 tabs.

The International Commission on Zoological Nomenclature. (1964). *International code of zoological nomenclature adopted by the XV International Congress of Zoology*, 176 pp., appendices. Published by the International Trust for Zoological Nomenclature, London.

Iovčeva, P. and Trifonova, E. (1961). Tithonian *Globigerina* from north-west Bulgaria. *Trud. Geol. Bulg. ser. Paleont.* **3**, 343–351. 3 plates.

Itzhaki, J. (1952). Séries de variabilité di *Pseudotextularia* (Rzehak) d'après la forme du tests et ses tendances évolutives. *C.r. somm. Seanc. Soc. geol. Fr.* **10**, 187–189. 11 text-figs., 1 tab.

Jacob, K. and Sastry, M. V. A. (1950). On the occurrence of *Globotruncana* in Uttattur Stage of the Trichinopoly Cretaceous, south India. *Sci. Cult.* **16**(6), 266–268. 2 text-figs.

Jedlitschka, H. (1935). Beitrag zur Kenntnis der Microfauna der subbeskidischen Schichten. *Mitt. naturw. Ver. Troppau* **40**(27), 31–48. 19 text-figs., 2 tabs.

Jennings, P. H. (1936). A microfauna from the Monmouth and basal Rancocas Groups of New Jersey. *Bull. Am. Paleont.* **23**(78), 3–77. 7 plates, 1 text-fig.

Jírová, D. (1956). The genus *Globotruncana* in upper Turonian and Emscherian of Bohemia. *Univ. carol. Geologica* **2**(3), 239–255. 3 plates.

Jones, D. J. (1956). *Introduction to microfossils*, 406 pp., text-figs. Harper & Brothers, Publishers, New York.

Jones, J. I. (1960). The significance of variability in *Praeglobotruncana gautierensis* Brönnimann, 1952, from the Cretaceous Eagle Ford Group of Texas. *Contr. Cushman Fdn foramin. Res.* **11**(3), 89–103. Plate 15, 7 text-figs., 4 tabs.

Jones, T. R. (1895). A monograph of the foraminifera of the Crag. Part II. *Palaeontogr. Soc.* (Monogr.), i-vii, 75–210. Plates 5–7, 22 text-figs.

Jordan, R. R. (1962). Planktonic foraminifera and the Cretaceous-Tertiary boundary in central Delaware. *Rep. Invest. Delaware geol. Surv.* **5**, 1–13. 2 plates.

Jurkiewicz, H. (1961). Fauna otwornicowa nizszej części warstw czarnorzeckich Centralnej Depresji Karpackkiej. *Acta geol. pol.* **11**(4), 507–524. Plates 23, 24, 1 tab.

Kalantari, A. (1969). Foraminifera from the Middle Jurassic-Cretaceous successions of Koppet-Dagh region (N. E. Iran). *Nat. Iranian Oil Co., Geol. Lab. Publ.* No. 3, 1–298. 26 plates, 28 text-figs.

Kalinin, N. A. (1937). Foraminifera from the Cretaceous of Baktygaryn (Aktiubinsk Province USSR). *Etyudy Mikropaleont.* **1**(2), 7–61. 8 plates, 2 tabs. (in Russian).

Kaptarenko-Chernousova, O. K. (1954). On the subject of the article of V. T. Balakhmatova "On the Globigerinidae and Globorotaliidae of the middle Jurassic." *Geol. Zh.* **14**(4), 88, 89 (in Russian).

Kaptarenko-Chernousova, O. K. (1963). Atlas of characteristic foraminifera of the Jurassic, Cretaceous, and Paleogene of the platform-like part of the Ukraine. *Trudy Inst. geol. Nauk. Kiev. Ser. Strat. i Paleont.* **45**, 60–199. Plates 12–29.

Kaufmann, F. J. (1865). Die Zeit der Kreidebildung. A. Polythalamien des Geewertaltes. *In* Heer, O., *Die Urwelt der Schweiz*, p. 198, text-fig. 110a. Druck und Verlag von Friedrich Schulthesz, Zurich.

Kavary, E. and Frizzell, D. L. (1963). Upper Cretaceous and Lower Cenozoic foraminifera from west-central Iran. *Bull. Univ. Mo. School of Mines and Metal. Tech. Ser.* **102**, 1–89. Plates 1–13, 3 text-figs., 1 tab.

Keller, B. M. (1935). Microfauna of the upper Cretaceous of the Dnieper-Donets Valley and some other adjoining regions. *Byull. mosk. Obshch. Ispyt. Prir.*, n.s., **43** (Sect. Geol., vol. 13:4), 522–558. 3 plates, 3 tabs. (in Russian).

Keller, B. M. (1939). The foraminifera of the Upper Cretaceous deposits of the U.S.S.R. *Trudy neft. geol.-razv. Inst. ser.* A, **116**, 7–30. 2 plates, 4 tabs. (in Russian).

Keller, B. M. (1946). Foraminifera from the Upper Cretaceous deposits of the Sochi region. *Byull. mosk. Obshch. Ispyt. Prir.* n.s. **51** (Sect. Geol., 21:3), 83–108. 3 plates, 2 tabs. (in Russian).

Kent, H. C. (1967) Microfossils from the Niobrara Formation (Cretaceous) and equivalent strata in northern and western Colorado. *J. Paleont.* **41**(6), 1433–1456. Plates 183, 184, 8 text-figs.

Kent, H. C. (1969). Distribution of keeled Globotruncanidae in Coniacian and Santonian strata of the western interior region of North America. *In* (Brönnimann, P. and Renz, H. H., ed.), *Proc. I Plankt. Conf.* **2**, 323–327. 2 text-figs. E. J. Brill, Leiden.

Kerdany, M. T. and Abdelsalam, H. (1969). *Globotruncana falso calcarata* n. sp. from the Quseir area, Eastern Desert, U.A.R. *Proc. Third African Micropaleont. Colloq., Cairo*, 1968, 261–267. 2 plates, 1 text-fig.

Kesling, R. V. (1941). *Zonation of the larger foraminifera of the Upper Cretaceous Selma, Ripley, and Prairie Bluff Formations in Western Alabama*, 19 pp., 20 plates. Unpubl. M.S. thesis, University of Illinois, Urbana, Illinois.

Khan, M. H. (1970). Cretaceous and Tertiary rocks of Ghana with a historical account of oil exploration. *Bull. Ghana Geol. Surv.* **40**, 1–55. 5 plates, 6 figs.

Kikoïne, J. (1947). Mise au point sur la nomenclature de *Globorotalia cushmani* Morrow. *C. r. somm. Seanc. Soc. geol. Fr.* **13/14**, 287–299. 3 text-figs.

Kikoïne, J. (1948). Les Heterohelicidae du Crétacé supérieur Pyreneen. *Bull. Soc. geol. Fr.* ser. 5, **18**(1–3), 15–35. Plates 1, 2, 1 text-fig.

Kiskyras, D. (1941). Über ein Oberkreide-Vorkommen mit *Globotruncana* in Nauplion (Argolis Griechenland). *Zentbl. Miner. Geol. Palaont.* Abt. B, **2**, 33–40. 5 text-figs.

Klasz, I. de (1953a). Einige neue oder wenig bekannte foraminiferen aus der helvetischen Oberkreide der bayerischen Alpen südlich Traunstein (Oberbayern). *Geologica bav.* **17**, 223–239. Plates 4–7.

Klasz, I. de (1953b). On the foraminiferal genus *Gublerina* Kikoïne. *Geologica bav.* **17**, 245–249. Plate 8.

Klasz, I. de (1955). A new *Globotruncana* from the Bavarian Alps and North Africa. *Contr. Cushman Fdn foramin. Res.* **6**(1), 43, 44. Plate 7, fig. 2a–c.

Klasz, I. de, Le Calvez, Y. and Rcfat, D. (1969). Nouveaux foraminifères du bassin sédimentaire du Gabon (Afrique Équatoriale). *Proc. Third African Micropaleont. Colloq.*, *Cairo* 269–283, 2 plates, 1 text-fig.

Klaus, J. (1959). Le "complexe schisteux intermédiaire" dans le synclinal de la Gruyère (Préalpes medianes) Stratigraphie et micropaléontologie avec l'étude speciale des Globotruncanidés de l'Albien, du Cénomanien et du Turonien. *Eclog. geol. Helv.* **52**(2), 753–851. Plates 1–8, 9 text-figs., 2 tabs.

Klaus, J. (1960). Étude biométrique et statistique de quelques espèces de Globotruncanides: 1. Les espèces du genre *Praeglobotruncana* dans le Cénomanien de la Breggia (Tessin, Suisse méridionale). *Eclog. geol. Helv.* **53**(1), 285–308. 3 text-figs., 1 tab.

Klaus, J. (1961). Rotalipores et Thalmanninelles d'un niveau des couches rouges de l'Anticlinal d'Ai. *Eclog. geol. Helv.* **53**(2) (1960), 704–709. 5 text-figs.

Kline, V. H. (1943). Fossils. Midway foraminifera and ostracoda. *Bull. Miss. St. geol. Surv.* **53**, 1–78. 8 plates.

Kneeland, S. (1874). Foraminifera. In *American cyclopaedia: a popular dictionary of general knowledge*, p. 311, 3 text-figs. D. Appleton & Co., New York.

Knipscheer, H. C. G. (1956). Biostratigraphie in der Oberkreide mit Hilfe der Globotruncanen. *Palaont. Z.* **30**, 50–56. Abb. 1–4, Tafel 4'.

Kochansky-Devide, V. (1951). *Globotruncana* from the vicinity of Bor in eastern Serbia. *Geoloski Anali balk. Poluost.* **19**, 113–117. 1 plate.

Kock, W. (1968). Zur Mikropaläontologie und Biostratigraphie der Oberkreide und des Alttertiars von Jordanien. I. Oberkreide. *Geol. Jb.* **85**, 627–668. 4 plates, 4 text-figs., 1 tab.

Kristan-Tollmann, E. (1964). Die Foraminiferen aus den rhätischen Zlambachi-mergelen der Fischerwiese bei Ausse im Salzkammergut. *Jb. geol. Bundesanst., Wien* **10**, 1–189. 39 plates, 5 text-figs.

Książkiewicz, M. (1956). Jura i Kreda Bachowic. *Roczn. pol. Tow. geol.* **24**(2–3) (1954), 119–405. Plates 11–32, 61 text-figs., 4 tabs.

Książkiewicz, M. (1958). On the Turonian in the Pieniny Klippes Belt. *Bull. Acad. pol. Sci. Ser. Sci. chim. geol. geogr.* **6**(8), 537–544. 3 text-figs., 1 tab.

Kuhry, B. (1970). Some observations on the type material of *Globotruncana elevata* (Brotzen) and *Globotruncana concavata* (Brotzen). *Revta. esp. Micropaleont.* **2**(3), 291–304. 2 plates, 7 text-figs.

Kuhry, B. (1971). Lower Cretaceous planktonic foraminifera from the Miravetes, Argos, and Repressa Formations (S.E. Spain). *Revta esp. Micropaleont.* **3**(3), 219–237. 3 plates, 2 text-figs., 1 tab.

Kummel, B. (1970). *History of the earth. An introduction to historical geology*, 2nd ed., 707 pp., text-figs., tabs., appendices. W. H. Freeman & Co., San Francisco, California.

Küpper, I. (1964). Mikropaläontologische Gliederung der Oberkreide des Beckenuntergrundes in den oberösterreichischen Molassebohrungen. *Mitt. geol. Ges. Wien* **56**(2) (1956), 591–651. 4 plates, 3 tabs.

Kupper, K. (1955). Upper Cretaceous foraminifera from the "Franciscan Series," New Almaden district, California. *Contr. Cushman Fdn foramin. Res.* **6**(3), 112–118. Plate 18.

Kupper, K. (1956). Upper Cretaceous pelagic foraminifera from the "Antelope Shale", Glenn and Colusa Counties, California. *Contr. Cushman Fdn foramin. Res.* **7**(2), 40–47. Plate 8, 1 text-fig.

Lalicker, C. G. (1948). A new genus of foraminifera from the Upper Cretaceous. *J. Paleont.* **22**(5), 624. Plate 92.

Lapparent, J. de (1918). *Étude lithologique des Terrains Crétacés de la Région D'Hendage.* Mémoires pour servir à L'Explication de la Carte Géologique Détaillée de la France, 155 pp., 10 plates, 27 text-figs.

Latif, M. A. (1970). Micropaleontology of the Chanali Limestone, Upper Cretaceous, of Hazara, West Pakistan. *Jb. geol. Bundesanst., Wien* **15**, 25–61. Plates 1–8, 2 text-figs.

Le Calvez, J. (1950). Recherches sur les foraminifères. 2. Place de la méiose et sexualité *Archives Zool. Exper. Générale* **87**(4), 211–244. 1 plate, 4 text-figs., 2 tabs.

Lehmann, R. (1963). Étude des Globotruncanidés du Crétacé supérieur de la province de Tarfaya (Maroc Occidental). *Notes Mem. Serv. geol. Maroc.* **21**(156) (1962), 133–181. Plates 1–10, text-figs. 1–3.

Lehmann, R. (1965). Résultats d'une étude des Globotruncanidés du Crétacé supérieur de la province de Tarfaya (Maroc Occidental). *Mem. Bur. Rech. Geol. Min.* no. 32, Colloque Int. Micropaleont. (1963). 113–117. 1 text-fig.

Lehmann, R. (1966a). Description des Globotruncanidés et Hétérohelicidés d'une faune Maestrichtienne du Prérif (Maroc). *Eclog. geol. Helv.* **59**(1), 309–317. 2 plates.

Lehmann, R. (1966b). Les foraminifères pelagiques de Crétacé du Basin Cotier de Tarfaya. 1. Planomalinidae et Globotruncanidae du Sondage de Puerto Cansado (Albien Supérieur, Cénomanien Inférieur). *Notes Mem. Serv. geol. Maroc.* **2**(175), Paleontologie, 153–167. Plates 1, 2, text-fig. 1.

Lehmann, R. (1966c). Les foraminifères pelagique du Crétacé du Basin Cotier de Tarfaya. 2. *Globotruncana concavata* (Brotzen) dans la coupe de la Sebkha Tah. *Notes Mem. Serv. geol. Maroc.* **2**(175), Paleontologie, 169–171. Text-fig. 2.

Lehmann, R. (1966d). Les foraminifères pelagiques du Crétacé du Bassin Cotier de Tarfaya. 3. Discussion taxonomique et répartition stratigraphique de quelques foraminifères pelagiques Crétacés du Bassin de Tarfaya. *Notes Mem. Serv. geol. Maroc.* **2**(175), Paleontologie, 173–175.

Leischner, W. (1961). Zur Kenntnis der Mikrofauna und -flora der Salzburger Kalkalpen. *Neues Jb. Geol. Palaont. Abh.* **112**(1), 1–47. Plates 1–14.

LeRoy, L. W. (1953). Biostratigraphy of the Maqfi section, Egypt. *Mem. geol. Soc. Am.* **54**, 1–73. 14 plates, 4 text-figs.

LeRoy, L. W. and Schieltz, N. C. (1958). Niobrara-Pierre boundary along Front Range. Colorado. *Bull. Am. Ass. Petrol. Geol.* **42**(10), 2444–2464. 12 figs.

Liebus, A. (1927). Neue Beiträge zur Kenntnis der Eozänfauna des Krappfeldes in Kärnten. *Jb. geol. Bundesanst., Wien* **77**, 333–392. Plates 12–14, 4 text-figs.

Lipnik, O. S. (1961). Foraminifera and stratigraphy of the upper Cretaceous of the Dnieper-Donetz depression. *Trudy Inst. geol. Nauk, Kiev Ser. Stratigr. Paleont.* **35**, 5–65. 7 plates, 2 tabs. (in Russian).

Lipparini, T. (1950). "*Globotruncana stuarti*" (de Lapp.) nel livello fosfatico Campaniano-Maestrichtiano della Tripolitania orientale. *Boll. uff. geol. ital.* **70** (1945–46), pt. 1, 171–173. 2 text-figs.

Liszka, S. (1955). Foraminifera of the lower Senonian in the vicinity of Krakow. *Roczn. pol. Tow. geol.* **23** (1953), 165–190. Plates 12, 13. (in Polish).

Liszkowa, J. (1967). A microfauna of the Upper Cretaceous marls in the Sub-Silesian Series of the Wadowice region (western Carpathians). *Biul. Inst. geol.* **211**, 341–353, 5 plates, 1 text-fig.

Loeblich, A. R., Jr. (1951). Coiling in the Heterohelicidae. *Contr. Cushman Fdn foramin. Res.* **2**(35), 106–110. Plate 12, text-figs. 1, 2.

Loeblich, A. R., Jr. and Tappan, H. (1946). New Washita foraminifera. *J. Paleont.* **20**(3), 238–258. Plates 35–37, 4 text-figs.

Loeblich, A. R., Jr. and Tappan, H. (1949). Foraminifera from the Walnut Formation (Lower Cretaceous) of northern Texas and southern Oklahoma. *J. Paleont.* **23**(3), 245–266. Plates 46–51.

Loeblich, A. R., Jr. and Tappan, H. (1950). Foraminifera from the type Kiowa Shale, Lower Cretaceous, of Kansas. *Paleont. Contr. Univ. Kans.* **3**, 1–15. Plates 1, 2.

Loeblich, A. R., Jr. and Tappan, H. (1961a). Cretaceous planktonic foraminifera: Part 1–Cenomanian. *Micropaleont.* **7**(3) 257–304. 8 plates.

Loeblich, A. R., Jr. and Tappan, H. (1961b). The genera *Microaulopora* Kuntz, 1895, and *Guembelina* Kuntz, 1895, and the status of *Guembelina* Egger, 1899. *J. Paleont.* **35**(3), 625–627. 1 text-fig.

Loeblich, A. R., Jr. and Tappan, H. (1962). Type localities of some American Cretaceous foraminiferal genotype species described by Ehrenberg. *J. Paleont.* **36**(2), 352–354.

Loeblich, A. R., Jr. and Tappan, H. (1964). Sarcodina, chiefly "Thecamocbians" and Foraminiferida. *Treatise on Invertebrate Paleontology*, Part C. Protista **2**. geol. Soc. Am. 2 vols., 900 pp. 653 text-figs.

Loetterle, G. J. (1937). The micropaleontology of the Niobrara Formation in Kansas, Nebraska, and South Dakota. *Bull. geol. Surv. Neb.* **12**, 1–73. 11 plates, 3 text-figs.

Longoria, J. F. (1970). Estudio en seccion delgada de algunas especies del genero *Globotruncana* Cushman del Este de Mexico. Inst. mex. Petroleo, Publ. no. 70A1/057, 1–135. 19 plates, 2 text-figs., 19 tabs.

Longoria, J. F. (1973). A new species of *Globotruncana* from the Upper Cretaceous of Texas and Mexico. *Micropaleont.* **19**(1), 97–100. 1 plate, 2 text-figs.

Low, D. (1964). Redescription of *Anomalina eaglefordensis* Moreman. *Contr. Cushman Fdn foramin. Res.* **15**(3), 122, 123. 1 text-figs.

Lucini, P. (1959). Su due microfaune del flysch del versante tirrenico della Basilicata. *Boll. Soc. geol. ital.* **77**(3) (1958), 173–181. 4 text-figs.

Luterbacher, H. (1972). Foraminifera from the Lower Cretaceous and Upper Jurassic of the northwestern Atlantic. *Initial Repts. Deep Sea Drilling Proj.* **11**, 561–593. 8 plates, 6 text-figs., 3 charts.

Luterbacher, H. and Premoli-Silva, I. (1962). Note preliminaire sur une revision du profil de Gubbio. Italie. *Riv. ital. Paleont. Stratigr.* **68**(2), 253–288. Plates 19–23, Text-figs. 1–3.

McClung, C. E. (1898). Microscopic organisms of Upper Cretaceous. *The Univ. geol. Surv. Kans.*, IV. *Paleontology* Part I. Upper Cretaceous, 415–427. Plate LXXXV.

McGowan, J. A. (1971). Oceanic biogeography of the Pacific. *In* (Funnell, B. M. and Riedel, W. R., eds), *The Micropalaeontology of Oceans*, 3–74, 47 text-figs, 6 tables, Cambridge University Press.

McGugan, A. (1957). Upper Cretaceous foraminifera from northern Ireland. *J. Paleont.* **31**(2), 329–348. Plates 31–35, 4 text-figs.

McGugan, A. (1964). Upper Cretaceous zone foraminifera, Vancouver Island, British Columbia, Canada, *J. Paleont.* **38**(5), 933–951. Plates 150–152, 4 text-figs.

McKerrow, W. S. (1956). Fossil species and the rules of nomenclature. *In* The species concept in paleontology. *Publs Syst. Ass.* **2**, 122.

Mabesoone, J. M., Tinoco, I. M. and Coutinho, P. N. (1968). The Mesozoic-Tertiary boundary in northeastern Brazil. *Palaeogeog. Palaeoclimatol. Palaeoecol.* **4**, 161–185. 7 text-figs., 6 tables.

Macfadyen, W. A. (1933). Fossil foraminifera from the Burdwood Bank and their geological significance. *Discovery Rep.* Cambridge, **8**, 1–16. 2 text-figs., 1 tab.

Magné, J. and Malmoustier, G. (1969). Le genre *Colomiella* Bonet dans l'Albien d'Aquitaine et des Pyrenees occidentales. *In* (Brönnimann, P. and Renz, H. H., eds), *Proc. I Plankt. Conf.* **2**, 378–382. 3 plates, 2 text-figs. E. J. Brill, Leiden.

Magné, J., Paquet, J. and Sigal, J. (1969). Crétacé et passage au Tertiaire dans le Prebetique et le Subbetique externe (Cordilleres betiques, zones de Caravaca-Calasparra, Province de Murcie, Espagne). *Annls. Soc. geol. N.* **89**(2), 177–189. Plates 5–9, 4 text-figs., 4 tabs.

Majzon, L. (1943). Adatok egyes Kárpátaliai flis-rétegekhez tekintettel a globotrun-canákra. (Beiträge zur Kenntnis einiger Flysch-Schichten de Karpatenvorlandes mit Rücksicht auf die Globotruncanen). *Magy. allami foldt. Intez. Evk.* **37**(1), 1–169. 2 plates, 6 text-figs., 2 tabs.

Majzon, L. (1961). *Globotruncana*-bearing sediments in Hungary. *Magy. kiralyi foldt. Intez. Evk.* **49**(3), Mater. Confer., Mesozoique, 745–787. 7 plates, 4 tabs., 2 maps (in Russian).

Malapris, M. and Rat, P. (1961). Données sur les Rosalines du Cénomanien et du Turonien de Côte-d'Or. *Rev. Micropaleont.* **4**(2), 85–98. 3 plates, 9 text-figs.

Malaroda, R. (1962). Gli "hard-grounds" al limite tra Cretaceo ed Eocene nei Lessini occidentali. *Memorie Soc. geol. ital.* **3**, 111–135. 9 text-figs.

Mallory, V. S. (1959). *Lower Tertiary Biostratigraphy of the California Coast Ranges. Am. Ass. Petrol. Geol.* 416 pp., 42 plates, 7 text-figs., 19 tabs.

Malumian, N. (1968). Foraminiferos del Cretacico superior y Terciario del subsuelo de la Provincia Santa Cruz, Argentina. *Revta Asoc. paleont. argent.* **5**(6), 191–227, 8 lamina, 1 fig.-texto.

Mantell, G. A. (1854). *The medals of creation; or, first lessons in geology, and the study of organic remains,* 2nd ed. Vol. 1, 446 pp., 6 plates, 139 text-figs. Henry G. Bohn, London.

Mantovani-Uguzzoni, M. P. and Pirini-Radrizzani, C. (1967). I foraminiferi delle Marne a Fucoidi. *Riv. ital. Paleont. Stratigr.* **73**(4), 1181–1256. Plates 85–94.

Marianos, A. W. and Zingula, R. P. (1966). Cretaceous planktonic foraminifers from Dry Creek, Tehama County, California. *J. Paleont.* **40**(2), 328–342. Plates 37–39, 3 text-figs.

Marie, P. (1938). Zones à foraminifères de l'Aturien dans la Mesogée. *C. r. somm. Seanc. Soc. geol. Fr.* 341–343.

Marie, P. (1941). Les Foraminifères de la Craie a *Belemnitella mucronata* du Bassin de Paris. *Mem. Mus. natn. Hist. nat., Paris* n. ser., **12**(1), 1–296. 37 plates, 5 text-figs.

Marie, P. (1948). A propos de *Rosalinella cushmani* (Morrow). *Soc. geol. France, Compte Rendu Somm.* No. 2, p. 39–42.

Marin, A., Pastora, J. L. and Lizaur, J. de (1934). Premières recherches de pétrole sur la cote atlantique de le territoire du Protectorat Espagnol au Maroc. *Bull. Soc. geol. Fr.* ser. 5, **4**(8–9), 649–673. 3 text-figs.

Marks, P. (1967). *Rotalipora* et *Globotruncana* dans la craie de Theligny (Cénomanien; Dept. de la Sarthe). *Proc. Sect. Sci. K. ned. Akad. wet.* ser. B, **70**(3), 264–275. 3 plates, 4 text-figs.

Marrocu, P. M. R., *et al.* (1959). *Microfacies italiane (dal carbonifero al miocene medio).* AGIP Mineraria, S. Donato Milanese, Italy, 35 pp., 145 pls.

Martin, L. (1964). Upper Cretaceous and Lower Tertiary foraminifera from Fresno County, California. *Jb. geol. Bundesanst. Wien* **9**, 1–128, 16 plates, text-figs. A–D, 8 tabs., 3 figs.

Martin, S. E. (1972). Reexamination of the Upper Cretaceous planktonic foraminiferal genera *Planoglobulina* Cushman and *Ventilabrella* Cushman. *J. foramin. Res.* **2**(2), 73–92. 4 plates, 6 text-figs.

Martirosyan, Yu. A. (1958). The *Globotruncana* of the Upper Cretaceous deposits of the southwest part of Armenia SSR. *Izv. Akad. Nauk armyan. SSR* **11**(6), 7–17. Plates 1–4. (in Russian).

Masella, L. (1959). Una nuova specie di *Heterohelix* del Cretaceo della Sicilia. *Riv. Miner. sicil.* **55**, 15–17. 10 figs.

Masella, L. (1960). Le *Schackoina* (Foraminifera, Globigerinacea) del Cretaceo di Patti (Messina). *Riv. Miner. sicil.* **11**(61), 16–30. 9 tav., 41 text-figs.

Maslakova, N. I. (1959). Foraminifera. *In* (Moskvin, M. M., ed.), Atlas of the Upper Cretaceous fauna of the northern Caucasus and Crimea. *Glavnoe Upravlenie Gazovoi Promyshelennosti Pri Sovete Ministrov SSSR,* VNIIGAZ, Gostoptek-hizdat, Moscow, Trudy, 87–129. 15 plates, text-figs. 2–6, tab. 6. (in Russian).

Maslakova, N. I. (1961). Systematics and phylogeny of the genera *Thalmanninella* and *Rotalipora* (foraminifera). *Paleont. Zh.* **1**, 50–55. Plates III–IV. (in Russian).

Maslakova, N. I. (1963a). On the systematics of the genus *Hedbergella.* *Paleont. Zh.* **4**, 112–116. 2 figs., 1 tab. (in Russian).

Maslakova, N. I. (1963b). Structure of the wall of the test of globotruncanids. *Vop. Mikropaleont.* **7**, 138–149. 7 plates, 6 text-figs. (in Russian).

Maslakova, N. I. (1964). Contribution to the systematics and phylogeny of the Globotruncanids. *Vop. Mikropaleont.* **8**, 102–117. 5 text-figs., 2 tabs. (in Russian).

Maslakova, N. I. (1969). Individual development of Globotruncanids. *Vop. Mikropaleont.* **12,** 95–107. 2 text-figs. (in Russian).

Maslakova, N. I. (1970). On the structure and taxonomic significance of the aperture of the Globotruncanid test. *Vop. Mikropaleont.* **13,** 84–87. Plate 22, 1 text-fig. (in Russian).

Maslakova, N. I. (1971). Classification of the globotruncanids. *In* New in the taxonomy of microfauna. *Trudy, vses. nauchno-issled. geol.-razv. neft. Inst.* **291,** 55–62. (in Russian).

Masters, B. A. (1976). Planktic foraminifera from the Upper Cretaceous Selma Group, Alabama. *J. Paleont.* **50**(2), 318–330. 2 plates, 2 text figs.

Mayr, E., Linsley, E. G., and Usinger, R. L. (1953). *Methods and principles of systematic zoology,* 328 pp., 45 text-figs., 15 tabs. McGraw-Hill Book Co., Inc., New York.

Mello, J. F. (1969). Foraminifera and stratigraphy of the upper part of the Pierre Shale and lower part of the Fox Hills Sandstone (Cretaceous), north-central South Dakota. *Prof. Pap. U.S. geol. Surv.* **611,** 1–121. 12 plates, 14 text-figs. 2 tabs.

Mello, J. F. (1971). Foraminifera from the Pierre Shale (Upper Cretaceous) at Red Bird, Wyoming. *Prof. Pap. U.S. geol. Surv.* **393–C,** 54 pp., 7 plates, 4 text-figs., 5 tabs.

Michael, F. Y. (1973). Planktonic foraminifera from the Comanchean Series (Cretaceous) of Texas. *J. foramin. Res.* **2**(4) (1972), 200–220. 7 plates, 7 text-figs.

Montanaro Gallitelli, E. (1943). Per la geologia della argille ofiolitifere appenniniche; Nota II—Foraminiferi dell' argilla scagliosa di Varana. *Mem. Soc. Tosc. Sci. nat. Pisa* **52,** 51–67. 2 plates, 1 text-fig.

Montanaro Gallitelli, E. (1947). Per la geologia delle argille ofiolitifere appenniniche; Nota III—Foraminiferi dell' argilla scagliosa di Castelvecchio (Modena). *Mem. Soc. Tosc. Sci. nat. Pisa* **54,** 175–196. 2 text-figs.

Montanaro Gallitelli, E. (1954). Marne ed argille a *Schackoina* e *Gumbelina* nella formazione a fucoidi ed elmintoidee di Serramazzoni (Modena). *Atti Memorie Accad. Sci. Lett., Modena* ser. V, **12,** 201–210. 2 plates.

Montanaro Gallitelli, E. (1955a). Una revisione della famiglia Heterohelicidae Cushman. *Atti Memorie Accad. Sci. Lett., Modena* ser. V, **13,** 213–223.

Montanaro Gallitelli, E. (1955b). *Schackoina* from the Upper Cretaceous of the northern Apennines, Italy. *Micropaleont.* **1**(2), 141–146. Plate 1, tab. 1.

Montanaro Gallitelli, E. (1956). *Bronnimannella, Tappanina,* and *Trachelinella,* three new foraminiferal genera from the Upper Cretaceous. *Contr. Cushman Fdn foramin. Res.* **7**(2), 35–39. Plate 7.

Montanaro Gallitelli, E. (1957). A revision of the foraminiferal family Heterohelicidae. *In* (Loeblich, A. R., Jr. *et al.*), Studies in Foraminifera. *Bull. U.S. natn. Mus.* **215,** 133–154. Plates 31–34.

Montanaro Gallitelli, E. (1958). Specie nuove e note di Foraminiferi del Cretaceo superiore di Serramazzoni (Modena). *Atti Memorie Acad. Sci. Lett., Modena* ser. V, **16,** 127–152. Plates 1–4.

Montanaro Gallitelli, E. (1959). Globotruncane campaniano-maestrichtiane nella formazione a facies di flysch di Serramazzoni nell'Appennino settentrionale Modenese. *Boll. Soc. geol. ital.* **77**(2) (1958), 171–191. 3 plates, 10 text-figs.

Moorkens, T. L. (1969). Quelques Globotruncanidae et Rotaliporidae du Cénomien, Turonien et Coniacien de la Belgique. *In* (Brönnimann, P. and Renz, H. H., eds), *Proc. I Plankt. Conf.* **2,** 435–459. Plates 1–3, 2 tabs. E. J. Brill, Leiden.

Moreman, W. L. (1925). Micrology of the Woodbine, Eagle Ford, and Austin Chalk. *In The geology of Denton County. Bull. Univ. Tex. Bur. econ. Geol. Technol.* **2544,** 74–78. Plates 18, 19, 26, 27.

Moreman, W. L. (1927). Fossil zones of the Eagle Ford of North Texas. *J. Paleont.* **1**(1), 89–101. Plates 13–16, text-fig. 1.

Mornod, L. (1950). Les Globorotalidés du Crétacé supérieur du Montsalvens (Préalpes fribourgeoises). *Eclog. geol. Helv.* **42**(2) (1949), 573–575. 14 text-figs.

Morozova, V. G. (1939). On the stratigraphy of the Upper Cretaceous and Lower Tertiary deposits in the Emba oil-bearing district according to the foraminiferal fauna. *Byull. mosk. Obshch. Ispyt. Prir.* n.s., **47** (Sect. Geol., vol. 17), no. 4–5, 59–86. 2 plates, 1 text-fig., 2 tabs. (in Russian).

Morozova, V. G. (1948). Foraminifera of the Lower Cretaceous deposits in the region of the Sochi Mountains (southwest Caucasus). *Byull. mosk. Obshch. Ispyt. Prir.* **23**(3), 23–43. Plates 1, 2, 1 tab. (in Russian).

Morozova, V. G. (1957). Foraminiferal superfamily Globigerinidea, superfam. nov. and certain of its representatives. *Dokl. Akad. Nauk. SSSR* **114**(5), 1109–1112. 1 text-fig. (in Russian).

Morozova, V. G. and Moskalenko, T. A. (1961). Planktonic foraminifers from adjacent deposits of Bajocian and Bathonian of Daghestan (northeast Caucasus). *Vop. Mikropaeont.* **5,** 3–30, 2 plates, 9 text-figs., 1 tab. (in Russian).

Morris, R. W. (1971). Upper Cretaceous foraminifera from the upper Mancos Formation, the Mesaverde Group, and the basal Lewis Formation, northwestern Colorado. *Micropaleont.* **17**(3), 257–296. 7 plates, 5 text-figs.

Morrow, A. L. (1934). Foraminifera and ostracoda from the Upper Cretaceous of Kansas. *J. Paleont.* **8**(2), 186–205. Plates 29–31. 1 tab,

Mosna, S. (1963). "Globigerine" in termini calcarei del Cretaceo inferiore basale affioranti nell'area del Trentino centrale. *Stud. trent. Sci. nat.* **40**(2), 167–175. 5 plates.

Moullade, M. (1960). Sur quelques Foraminifères du Crétacé inférieur des Baronnies (Drôme). *Rev. Micropaleont.* **3**(2), 131–142. 2 plates, 2 text-figs.

Moullade, M. (1961). Quelques foraminifères et ostracodes nouveaux du Crétacé inférieur Vocontien. *Rev. Micropaleont.* **3**(4), 213–216. 1 plate.

Moullade, M. (1965). Nouvelles propositions pour l'establissement d'une zonation micropaleontologique de l'Aptien et de l'Albien vocontiens. *C. r. somm. Seanc. Soc. geol. Fr.* **2,** 48–50. Chart.

Moullade, M. (1966). Étude stratigraphique et micropaleontologique de Crétacé micropaleontologique de Crétacé inférieur de la "Fosse vocontienne." *Lyons, Fac. Sci., Lab. Geol., Docums* **15**(1–2), 1–369. 17 plates, 27 text-figs.

Moullade, M. (1969). Sur l'importance des phénomènes de convergence morphologique chez les foraminifères planctoniques du Crétacé inférieur. *In* Brönnimann, P. and Renz, H. H., eds), *Proc. I Plankt. Conf.* **2,** 460–467. 1 plate, 1 text-fig. E. J. Brill, Leiden.

Murgeanu, G. (1933). Sur l'importance des marnes a Rosalines dans la zone de recouvrements de Comarnic. *C. r. Seanc. Inst. geol. Roum.* **19,** 82–88. 1 plate.

Myatlyuk, Ye. V. (1949). Material to monograph study of foraminiferal fauna of the lower Cretaceous deposits of South-Emenskovo oil-bearing region. *In* Microfauna of petroleum deposits, SSSR. *Trudy vses. nauchno-issled. geol.-razv. neft. Inst.* n.s. **34,** 187–222. 5 plates, 1 table (in Russian).

Nagappa, Y. (1959). Foraminiferal biostratigraphy of the Cretaceous-Eocene succession in the India-Pakistan-Burma region. *Micropaleont.* **5**(2), 145–192. Plates 1–11, text-figs. 1–11, tabs. 1–9, charts 1–4.

Nakkady, S. E. (1950). A new foraminiferal fauna from the Esna shales and Upper Cretaceous chalk of Egypt. *J. Paleont.* **24**(6), 675–692. Plates 89 and 90, 4 text-figs.

Nakkady, S. E. and Osman, A. (1954). The genus *Globotruncana* in Egypt; taxonomy and Stratigraphical value. 19*th Int. Geol. Congr.* Sec. XIII, Fasc. XV, 75–95. Plates 19, 20, text-figs.

Napoli Alliata, E. di (1948). *Globotruncana* nell'Eocene della Sicilia centrale. *Riv. ital. Paleont. Stratigr.* **54**(1), 19–28. Text-fig. 2.

Nauss, A. W. (1947). Cretaceous microfossils of the Vermilion area, Alberta. *J. Paleont.* **21**(4), 329–343. Plates 48, 49, text-figs. 1–3, tab. 1.

Neagu, T. (1965). Albian foraminifera of the Rumanian Plain. *Micropaleont.* **11**(1), 1–38. Plates 1–10.

Neagu, T. (1966). *Schackoina* from the Cenomanian of the eastern Carpathians. *Micropaleont.* **12**(3), 365–369. 2 plates.

Neagu, T. (1968). Biostratigraphy of Upper Cretaceous deposits in the southern Eastern Carpathians near Brasov. *Micropaleont.* **14**(2), 225–241. Plates 1, 2.

Neagu, T. (1969). Cenomanian planktonic foraminifera in the southern part of the eastern Carpathians. *Roczn. pol. Tow. geol.* **39**(1–3), 133–155. Plates 13–37, 1 text-fig.

Neagu, T. (1970a). Microbiostratigraphy of the Cenomanian deposits from the southern part of eastern Carpathians (with some evolutionary-phylogenetic considerations regarding the planktonic foraminifera). *Revue roum., ser. Geol.* **14**(2), 171–188. 1 text-fig., 1 tab.

Neagu, T. (1970b). Micropaleontological and stratigraphical study of the Upper Cretaceous deposits between the upper valleys of the Buzau and Riul Negru Rivers (eastern Carpathians). *Mem. Inst. Geol. Bucarest* **12**, 1–109. 44 plates, 5 text-figs.

Neagu, T. (1972). The Eo-Cretaceous foraminiferal fauna from the area between the Ialomitza and Prahova Valleys (Eastern Carpathians). *Revta. esp. Micropaleont.* **4**(2), 181–224. 8 plates.

Netskaya, A. I. (1948). About some foraminifera of the upper Senonian deposits of western Siberia. *In* Microfauna of the petroleum deposits, SSSR. Part I. Eastern Baku and western Siberia. *Trudy. vses. nauchno-issled. geol.-razv. neft. Inst.* nov. ser. **31**, 211–229. 3 plates. (in Russian).

Newell, N. D. (1956). Fossil populations. *In* The species concept in paleontology. *Publs Syst. Ass.* **2**, 63–82. 5 text-figs.

North, B. R. and Caldwell, W. G. E. (1970). Foraminifera from the Late Cretaceous Bearpaw Formation in the South Saskatchewan River Valley. *Rep. Sask. Res. Coun., Geol. Dev.* **9**, 117 pp. 6 plates, 7 text-figs.

Noth, R. (1951). Foraminiferen aus Unter-und Oberkreide des Österreichischen anteils an Flysch. Helvetikum und Vorland vorkommen. *Jb. geol. Bundesanst., Wien* **3**, 1–91. 9 plates, 2 tabs.

Oberhauser, V. R. (1960). Foraminiferen und Mikrofossilien "incertae sedis" der ladinischen und karnischen Stufe der Trias aus den Ostalpen und aus Persien. *Jb. geol. Bundesanst., Wien* **5**, 5–46. Plates 1–6, 5 text-figs.

Obradović, S. N. (1953). Microfauna of the Upper Cretaceous in the vicinity of Belgrade (Kijevo, Resnik, Ripanj and Klenje). *Zborn. Rad. geol. Inst.* **6**, 67–87. 5 plates, 1 text-fig., map. (in Serbian).

Oesterle, H. (1968). Foraminiferen der Typlokalität der Birmenstorfer-Schichten, unterer Malm. (Teilrevision der Arbeiten von J. Kübler and H. Zwingli 1866–1870 und von R. Haeusler 1881–1893). *Eclog. geol. Helv.* **61**(2), 695–792. 53 text-figs.

Oesterle, H. (1969). A propos de *"Globigerina" helveto-jurassica* Haeusler, 1881. *In* (Brönnimann, P. and Renz, H. H., eds), *Proc. I Plankt. Conf.* **2**, 492, E. J. Brill, Leiden.

Olbertz, G. (1942). Untersuchungen zur Mikrostratigraphie der Oberen Kreide Westfälens (Turon-Emscher-Untersenon). *Palaont. Z.* **23**(1–2), 74–156. Plates 4, 5; 1 text-fig., 1 tab.

Olsson, R. K. (1960). Foraminifera of latest Cretaceous and earliest Tertiary age in the New Jersey coastal plain. *J. Paleont.* **34**(1), 1–58. Plates 1–12, 2 text-figs.

Olsson, R. K. (1963). Latest Cretaceous and earliest Tertiary stratigraphy of New Jersey coastal plain. *Bull. Am. Ass. Petrol. Geol.* **47**(4), 643–665. 6 figs., 3 tabs.

Olsson, R. K. (1964). Late Cretaceous planktonic foraminifera from New Jersey and Delaware. *Micropaleont.* **10**(2), 157–188. 7 plates, 3 text-figs.

Olsson, R. K. (1970). Planktonic foraminifera from base of Tertiary, Millers Ferry, Alabama. *J. Paleont.* **44**(4), 598–604. Plates 91–93, 2 text-figs.

Orbigny, A. D. d' (1826). Tableau méthodique de la classe des Céphalopodes. *Ann. Sci. Nat. Paris* ser. 1, **7**, 245–314. Atlas: plates 10–17.

Orbigny, A. D. d' (1839). Foraminifères. *In* (Sagra, Ramon de la) *Histoire physique, politique et naturelle de l'ile de Cuba.* Paris, xlviii + 224 pp., atlas, 12 plates.

Orbigny, A. D. d' (1840). Mémoire sur les Foraminifères de la Craie Blanche du Bassin de Paris. *Mem. Soc. geol. Fr.* **4**(1), 1–51. Plates 1–4.

Orlini, A. (1949). Sulla presenza di *Globotruncana calcarata* Cushman, in Italia. *Riv. ital. Paleont. Stratigr.* **55**(1), 35, 36.

Papp, A. (1955). Die Foraminiferenfauna von Guttaring und Klein St. Paul (Kärnten). IV. Biostratigraphische Ergebnisse in der Oberkreide und Bermerkungen über die Lagerung des Eozäns. *Sber. ost. Akad. Wiss. math.-naturwiss. Kl.* Abt. I, **164**, 317–334. 4 text-figs., 1 tab.

Papp, A. and Kupper, K. (1953). Die Foraminiferenfauna von Guttaring und Klein St. Paul (Kärnten). I. Über Globotruncanen sudlich Pemberger bie Klein St. Paul. *Sber. ost. Akad. Wiss., math.-naturwiss. Kl.* pt. 1, **162**(1–2), 31–48. 2 plates.

Parejas, E. (1926). Sur la présence de *Rosalina linnei* d'Orb. et de *Rosalina stuarti* J. de Lapp. dans le Crétacé supérieur de Piatigorsk (Caucase). *C. r. Seanc. Soc. Phys. Hist. nat. Geneve* **43**, 57–59. 2 text-figs.

Parra, J. O. de la (1959). Foraminiferos de la Formacion de Pena. *Boln. Asoc. mex. Geol. petrol.* **11**(3–4), 135–153. Lamina 1–5.

Pazdrowa, O. (1969). Bathonian *Globigerina* of Poland. *Roczn. pol. Tow. geol.* **39**(1–3), 41–56. Plates 2–4, 16 text-figs.

Perlmutter, N. M. and Todd, R. (1965). Correlation and foraminifera of the Monmouth Group (Upper Cretaceous) Long Island, New York. *Prof. Pap. U.S. geol. Surv.* **483–I**, 1–24. 8 plates, 5 tables.

Perner, J. (1892). Foraminifery Českého Cenomanu. *Palaeontogr. Bohem.* **2**(1) (1891), 1–65, 10 plates, 6 text-figs., 1 tab.

Perner, J. (1897). Foraminifery vrstev bělohoroských. *Palaeontogr. Bohem.* **2**(4) (1897), 1–73. 7 plates, 14 text-figs., 1 tab.

Pessagno, E. A. Jr. (1960a). Stratigraphy and micropaleontology of the Cretaceous and Lower Tertiary of Puerto Rico. *Micropaleont.* **6**(1), 87–110. Plates 1–5, text-figs. 1–2, charts 1–3.

Pessagno, E. A. Jr. (1960b). Thin-sectioning and photographing small foraminifera. *Micropaleont.* **6**(4), 419–423. 2 plates, 3 text-figs.

Pessagno, E. A. Jr. (1962). The Upper Cretaceous stratigraphy and micropaleontology of south-central Puerto Rico. *Micropaleont.* **8**(3), 349–368. Plates 1–6.

Pessagno, E. A. Jr. (1967). Upper Cretaceous planktonic foraminifera from the western Gulf Coastal Plain. *Palaeontogr. am.* **5**(37), 245–445. Plates 48–101, text-figs. A, B, 1–63, 2 tabs.

Pessagno, E. A. Jr. (1969a). Mesozoic planktonic foraminifera and radiolaria. *Initial Repts. Deep Sea Drilling Proj.* **1**, 607–621. Plates 4–12, text-fig. 8.

Pessagno, E. A. Jr. (1969b). Scanning electron microscope analyses of Globigerinacea wall structure. *In* (Brönnimann, P. and Renz, H. H., eds), *Proc. I Plankt. Conf.* **2**, 504–508. 8 plates. E. J. Brill, Leiden.

Pessagno, E. A. Jr. and Brown, W. R. (1969). The microreticulation and sieve plates of *Racemiguembelina fructicosa* (Egger). *Micropaleont.* **15**(1), 116, 117. Plate 1.

Pessagno, E. A. Jr and Miyano, K. (1968). Notes on the wall structure of the Globigerinacea. *Micropaleont.* **14**(1), p. 38–43, pls. 1–7, text-figs. 1, 2.

Peterson, R. H., Gauger, D. J. and Lankford. R. R. (1953). Mirofossils of the Upper Cretaceous of northeastern Utah and southwestern Wyoming. *Bull. Utah geol. miner. Surv.* **47** (Contributions to Micropaleontology no. 1), 1–158. 16 plates, 8 text-figs., 4 tabs.

Petri, S. (1954). Foraminíferos fósseis da Bacia do Marajó. *Bolm Fac. Filos. Cienc.Univ S. Paulo* **176**, (Geol. no. 11), 1–170, 14 plates, 10 text-figs., 3 tabs.

Petri, S. (1962). Foraminíferos Cretaceos de Sergipe. *Bolm Fac. Filos. Cienc. Univ. S. Paulo* **265** (Geol. no. 20), 1–140. Plates 1–21, text-figs. 1–3, tabs. 1–8.

Pflaumann, U. (1971). Porositäten von Plankton-Foraminiferen als Klimaanzeiger? *"Meteor" Forsch.-Ergebnisse* Ser C, **7**, 4–14. 3 plates, 5 text-figs.

Pieri, M. (1967). Caratteristiche sedimentologiche del limite Cretacico-Terziario nella zona di Monterosso Almo (Monti Iblei, Sicilia sud-orientale). *Riv. ital. Paleont. Stratigr.* **73**(4), 1259–1294. Plates 95, 96, 11 text-figs.

Pirini, C. and Radrizzani, S. (1963). Stratigrafia del Foglio 118 "Ancona." *Boll. Serv. geol. Ital.* **83** (1962), 71–110. 45 plates, 5 text-figs.

Plummer, H. J. (1927). Foraminifera of the Midway Formation in Texas. *Bull. Univ. Bur. econ. Geol. Technol.* **2644**, 1–206. 15 plates, 13 text-figs., chart.

Plummer, H. J. (1931). Some Cretaceous foraminifera in Texas. *Bull. Univ. Tex. Bur. econ. Geol. Technol.* **3101**, 109–203. Plates 8–15. text-fig. 12.

Pokorný, V. (1963). *Principles of Zoological Micropaleontology*, Vol. 1, 652 pp., 548 text-figs. Pergamon Press, New York.

Porthault, B. (1969). Foraminifères planctoniques et biostratigraphie du Cénomanien dans le sud-est de la France. *In* (Brönnimann, P. and Renz, H. H., eds), *Proc. I Plankt. Conf.* **2**, 526–546. Plates 1, 2, text-figs. 1, 2. E. J. Brill, Leiden.

Postuma, J. A. (1962). *Manual of Planktonic Foraminifera. Part I. Cretaceous (Albian-Maestrichtian)*, 10 pp. 42 plates, 1 chart. Bataafse Internationale Petroleum Maatschappij N.V., Exploration and Production, The Hague.

Premoli-Silva, I. (1966). La struttura della parete di alcuni Foraminiferi planctonici. *Eclog. geol. Helv.* **59**(1), 219–233. 3 plates, 6 text-figs.

Premoli-Silva, I. and Luterbacher, H. P. (1966). The Cretaceous-Tertiary boundary in the Southern Alps (Italy). *Riv. ital. Paleont. Stratigr.* **72**(4), 1183–1266. Plates 91–99, 28 text-figs.

Prosnyakova, L. V. (1967). Planktonic foraminifera (*Praeglobotruncana* and

Rotalipora) from the Cenomanian of the Crimea Plain. *Paleont. Sb.* **4**(2), 3–9. 1 plate, (in Russian).

Prosnyakova, L. V. (1968). The species of genus *Thalmanninella* Sigal (foraminifera) from the upper Albian and Cenomanian of the Crimean Plain. *Paleont. Sb.* **5**(1), 18–24. 1 plate. (in Russian).

Quereau, E. C. (1893). Die Klippenregion von Iberg (Sihlthal). *Beitr. geol. Karte Schweiz* **33**, 3–153. Plate 5.

Radoičić, R. (1958). Rezultati prvih mikropaleontoloških proučavanja flišnih sedimenata Durmitora. *Geol. glasnik Zav.* za geol. istr. Crne Gore, Titograd, **2**, 119–139. Plates 10–17, 1 text-fig., 1 chart.

Radoičić, R. (1966). Microfacies du Jurassique Dinarides externes de la Yougoslavie. *Geol. Razprave in Porocila, Letnik* **9**, 5–23. 165 plates, 1 text-fig., 11 tabs., 2 charts.

Radoičić, R. (1968). Globotrunkanide u nekim serijama gornje krede zapadne Srbije i Sumadije. *Vesnik Geol. Zav. za Geol. i Geof. istr* Ser. A, **24/25** (1966/1967), 297–329. Plates 1–11, 1 tab.

Raffi, G. and Forti, A. (1959). Micropaleontological and stratigraphical investigations in "Montagna del Morrone" (Abruzzi-Italy). *Rev. Micropaleont.* **2**(1), 8–20. 2 plates, 3 text-figs., 1 tab.

Rahhali, I. (1970). Foraminifères benthoniques et pélagiques du Crétacé supérieur du synclinal d'El-Koubbat (Moyen Atlas Maroc). *Notes Mem. Serv. geol. Maroc.* **30**(225), 51–98. Plates 1–5, 16 text-figs.

Rao, B. R. J., Mamgain, V. D. and Sastry, M. V. A. (1968). *Globotruncana* in Ariyalur Group of Trichinopoly Cretaceous, south India. *In* Cretaceous-Tertiary formations of south India. *Mem. geol. Soc. India* **2**, 18–29. Plates 1–4, tabs. 1, 2.

Rasheed, D. A. (1963). Some calcareous foraminifera belonging to the families Rotaliidae, Globigerinidae, Globorotaliidae and Anomalinidae from the Cullygoody (Dalmiapuram) Limestone, Trichinopoly Cretaceous of South India. Part 3. *J. Madras Univ.* Sect. B, **33**(3), 231–248. Plates 1–4.

Rasheed, D. A. and Govindan, A. (1968). Upper Cretaceous foraminifera from Vridhachalam, south India. *In* Cretaceous-Tertiary formations of south India. *Mem. geol. Soc. India* **2**, 66–84. Plates 1–8, text-figs. 1–16.

Rauzer-Chernousova, D. M. and Fursenko, A. V. (1937). *Guide to the foraminifera of the oil-bearing regions of the USSR, Part I.* ONTI, NKTP, SSR, Glavnaya Redaktsiya Gorno-Toplivnoi Literatury, Leningrad and Moscow, 320 pp., 241 text-figs., 14 tabs. (in Russian).

Rauzer-Chernousova, D. M. and Fursenko, A. V. (1959). *Osnovy Paleontologii* (*Fundamentals of paleontology*). *Part I. Protista.* Izdatel'stvo Akademii Nauk SSSR, Moscow, 368 pp., 13 plates, 894 text-figs. (in Russian).

Reichel, M. (1948). Les Hantkéninidés de la Scaglia et des Couches rouges (Crétacé supérieur). *Eclog. geol. Helv.* **40**(2) (1947), 391–409. Text-figs. 1–11.

Reichel, M. (1950). Observations sur les *Globotruncana* du gisement de la Breggia (Tessin). *Eclog. geol. Helv.* **42**(2), 596–617. Plates 15–17, 7 text-figs.

Reiss, Z. (1952). On the occurrence of *Globotruncana calcarata* Cushman 1927 in the Upper Cretaceous of Israel. *Bull. Res. Coun. Israel* **2**(3), 270–272. Text-fig. 1.

Reiss, Z. (1957a). Notes on foraminifera from Israel. *Bull. Res. Coun. Israel* **6B**, 239–244.

Reiss, Z. (1957b). The Bilamellidea, nov. superfam., and remarks on Cretaceous Globorotaliids. *Contr. Cushman Fdn foramin. Res.* **8**(4), 127–145. 2 plates, 7 text-figs., 1 tab.

Reiss, Z. (1958a). Classification of lamellar foraminifera. *Micropaleont.* **4**(1), 51–70. Plates 1–5, tab. 1.

Reiss, Z. (1958b). Notes on foraminifera from Israel; 8. The systematic position of *Sigalia* and *Bolivinoides*. *Bull. geol. Surv. Israel.* **17**, 5–7. Text-fig. 1.

Reiss, Z. (1963a). Note sur la structure des Foraminifères planctoniques. *Rev. Micropaleont.* **6**(3), 127–129. 1 plate.

Reiss, Z. (1963b). Reclassification of perforate foraminifera. *Bull. geol. Surv. Israel* **35**, 1–111. 8 plates.

Reiss, Z. (1971). Progress and problems of foraminiferal systematics. *In* (Funnell, B. M. and Riedel, W. R., eds), *The Micropaleontology of Oceans*, 633–638. Cambridge University Press.

Renz, H. H. (1962). Stratigraphy and paleontology of the type section of the Santa Anita group, and overlying Merecure group, R10 Querecual, State of Anzoategui, northeastern Venezuela. *Boln Informativo Asoc. Venez. Geol., Min. Petrol.* **5**(4), 89–108. 1 plate, 2 text-figs.

Renz, O. (1936). Stratigraphische und mikropaläontologische Untersuchung der Scaglia (Obere Kreide—Tertiär) im zentralen Apennin. *Eclog. geol. Helv.* **29**(1), 1–149. 15 Taf., 14 text-figs.

Renz, O. (1937). Über Globotruncanen im Cenomanien des Schweizerjura. *Eclog. geol. Helv.* **29**(2) (1936), 500–503. 1 text-fig.

Renz, O., Luterbacher, H. P. and Schneider, A. (1963). Stratigraphisch-paläontologische Untersuchungen im Albien und Cenomanien des Neuenburger Jura. *Eclog. geol. Helv.* **56**(2), 1073–1116. 9 plates, 4 text-figs.

Reuss, A. E. (1845). *Die Versteinerungen der bökmischen Kreideformation*, 58 pp., 13 plates. Abth. 1. E. Schweizerbart'sche Verlagsbuchhandlung und Druckerei, Stuttgart.

Reuss, A. E. (1854). Beitrage zur charateristik der Kreideschichten in der Ostalpen, besonders in Gosauthale und am Wolfgangsee. *Denkschr. Akad. Wiss., Wien. Math.-nat. Kl.* **7**(1), 1–156. 31 plates.

Reuss, A. E. (1860). Die Foraminiferen der Westphälischen Kreideformation. *Sbr. Akad. Wiss., Wien. Math.-nat. Kl.* **40**, 147–238. Plates 1–13.

Reutter, K. J. and Serpagli, E. (1961). Micropaleontologia stratigrafica sulla "Scaglia Rossa" di Val Gordana (Pontremoli-Appennino Settentrionale). *Boll. Soc. geol. ital.* **1**(2), 10–30. Plates 9–14, 2 text-figs., 2 tabs.

Reyment, R. A. (1960). Notes on some Globigerinidae, Globotruncanidae, and Globorotaliidae from the Upper Cretaceous and Lower Tertiary of Western Nigeria. *Rec. geol. Surv. Nigeria* (1597), 68–86. Plates 15–17, 2 text-figs.

Rickwood, F. K. (1955). The geology of the western highlands of New Guinea, *J. geol. Soc. Aust.* **2**, 63–82.

Risch, H. (1971). Stratigraphie der höheren Unterkreide der Bayerischen Kalkalpen mit Hilfe von Mikrofossilien. *Palaeontographica* Abt. A, **138**(1–4), 1–180. 8 plates. 8 text-figs.

Ruggieri, G. (1963). *Globigerinelloides algeriana* nell'Aptiano della Sicilia. *Boll. Soc. paleont. ital.* **2**(2), 75–78. 3 text-figs.

Rzehak, A. (1886). Die Foraminiferenfauna der Neogenformation der Umgebung von Mähr-Ostrau. *Verh. naturf. Ver. Brunn* **24**, 77–126. Plate 1.

Rzehak, A. (1888). Die Foraminiferen des Kieseligen Kalkes von Nieder-Hollabrunn und des Melettamergels der Umgebung von Bruderndorf in Niederösterreich. *Annln. naturh. Mus. Wien* **3**(3), 257–270. Tafel XI.

Rzehak, A. (1891). Die Foraminiferenfauna der alttertertiaren Ablagerungen von Bruderndorf in Nieder-osterreich, mit Berücksichtigung des angeblichen Kreidevorkommens von Leitzerdorf. *Annln. naturh. Mus. Wien* **6**, 1–12.

Rzehak, A. (1895). Ueber einige merkwurdige Foraminiferen aus österreichischen Tertiär. *Annln. naturh. Mus. Wien* **10**, 213–230. Plates 6, 7.

Saavedra, J. L. (1965) La evolución de los Globigerináceos. *Boln. R. Soc. esp. Hist. nat.* (Biol.), **63**, 317–349. 95 text-figs.

Sacal, V. and Debourle, A. (1957). Foraminifères d'Aquitaine. II. Peneroplidae a Victoriellidae. *Mem. Soc. geol. Fr.* n.s., **36**(78), 1–81. Plates 25–28.

Said, R. and Barakat, M. G. (1957a). Lower Cretaceous foraminifera from Khashm el Mistan, northern Sinai, Egypt. *Micropaleont.* **3**(1), 39–47. Plate 1, text-figs. 1, 2, tab. 1.

Said, R. and Barakat, M. G. (1957b). Cenomanian foraminifera from Gebel Asagil, northern Sinai, Egypt. *Egypt. J. Geol.* **1**(1), 65–83. Plate 1, text-figs. 1, 2.

Said, R. and Kenawy, A. (1956). Upper Cretaceous and Lower Tertiary foraminifera from northern Sinai, Egypt. *Micropaleont.* **2**(2), 105–173. Plates 1–7, text-figs. 1–6.

Said, R. and Kerdany, M. T. (1961). The geology and micropaleontology of the Farafra Oasis, Egypt. *Micropaleont.* **7**(3), 317–336. 2 plates, 13 text-figs., 1 chart.

Said, R. and Sabry, H. (1964). Planktonic foraminifera from the type locality of the Esna Shale in Egypt. *Micropaleont.* **10**(3), 375–395. 3 plates, 2 text-figs., 2 tables.

Saint-Marc, P. (1970). Sur quelques foraminifères Cénomaniens et Turoniens du Liban. *Rev. Micropaleont.* **13**(2), 85–94. 2 plates, 2 text-figs.

Salaj, J. (1969). Zones planctiques du Crétacé et du Paléogène de Tunisie. *In* (Brönnimann, P. and Renz, H. H., eds), *Proc. I. Plankt. Conf.* **2**, 588–593. 2 tables. E. J. Brill, Leiden.

Salaj, J. (1970). Quelques remarques sur les problèmes de microbiostratigraphie du Crétacé supérieur et du Paléogène. *Proc. Fourth African Micropaleont. Colloq.* 357–374. 4 text-figs.

Salaj, J. and Maamouri, A. L. (1971). Remarques microbiostratigraphiques sur le Sénonien supérieur de l'anticlinal de l'Oued Bazina (Région de Béja, Tunisie septentrionale). *Notes Serv. geol. Tunis.* **32** (1970), 65–78. 5 text-figs.

Salaj, J. and Samuel, O. (1963). Mikrobiostratigrafia strednej vrchnej kriedy z vychodnej časti bradloveho pasma. *Geol. Pr. Bratisl.* **30**, 93–112. Plates 6–8, text-figs. A-C, 1 tab.

Salaj, J. and Samuel. O. (1966). *Foraminifera der Westkarpaten—Kreide.* Geol. Ustav Dionyza Stura, Bratislava, 291 s., 48 tafeln, 18 text-figs., 37 tab., 6 beilagen.

Samuel, O. (1962). Mikrobiostratigrafické Pomery Kriedových sedimentov vnutorného Bradlového pásma v okolí Beňatíny. *Geol. Pr. Bratisl.* **24** (1961), 153–197. Tabs. 1–13.

Samuel, O. and Salaj, J. (1962). Nové Druhy Foraminifer z Kreidy A Paleogenu Západnych Karpát. *Geol. Pr. Bratisl.* **62**, 313–320. Tab. IX-X.

Sandidge, J. R. (1932a). Significant foraminifera from the Ripley Formation of Alabama. *Am. Midl. Nat.* **13**(4), 190–202. Plate 19.

Sandidge, J. R. (1932b). Fossil foraminifera from the Cretaceous Ripley Formation of Alabama. *Am. Midl. Nat.* **13**(5), 312–317. Plate 29.

Sandidge, J. R. (1932c). Additional foraminifera from the Ripley Formation in Alabama. *Am. Midl. Nat.* **13**(6), 333–376. Plates 31–33.

Sandidge, J. R. (1932d). Foraminifera from the Ripley Formation of western Alabama. *J. Paleont.* **6**(3), 265–287. Plates 41–44.

Săndulescu, J. (1966). Biostratigrafia si faciesurile Cretacicului superior şi Paleo-genului den Ţara Bîrsei (Carpaţii Orientali). *C. r. Seanc. Inst. geol. Roum.* **52**(2) (1964–1965), 241–278. Plates 1–5, 1a–9a, 1 text-fig. (Resumé in French).

Săndulescu, J. (1969). Globotruncanidae zones in the Upper Cretaceous within the Tara Birsei area (Crystalline-Mesozoic zone, eastern Carpathians). *Roczn. pol. Tow. geol.* **39**(1–3), 182–212. Plates 38–45, 1 text-fig., tabs. A–E.

Schacko, G. (1897). Beitrag über Foraminiferen aus der Cenoman-Kreide von Moltzow in Mecklenburg. *Arch. Ver. Freunde Naturg. Mecklenb.* **50** (1896), 161–168. Plate 4.

Scheibnerová, V. (1958). *Globotruncana helvetica* Bolli v kysuckom vývine pieninskej série vnútorného bradlového pásma v Západných Karpatoch. *Geol. Sb., Bratisl.* **9**(2), 188–194. Plate 4, 2 text-figs.

Scheibnerová, V. (1960). Poznámky k rodu *Praeglovotruncana* Bermudez z kysuckých vrstiev bradlového pásma. *Geol. Sb., Bratisl.* **11**(1), 85–90. Text-figs. 4, 5.

Scheibnerová, V. (1962). Stratigrafia strednej a vrchnej kriedy tétydnej oblasti na základe globotrunkaníd. *Geol. Sb., Bratisl.* **13**(2), 197–226. 7 text-figs.

Scheibnerová, V. (1963). Some new foraminifera from the middle Turonian of the Klippen belt of West Carpathians in Slovakia. *Geol. Sb., Bratisl.* **14**(1), 139–143. 3 text-figs.

Scheibnerová, V. (1967). Genera *Tappanina, Eouvigerina, Gublerina* and *Aragonia* in the Cretaceous of the West Carpathian Klippen Belt. *Čas. min. geol.* **12**(3), 261–269, 11 text-figs.

Scheibnerová, V. (1968). *Globotruncana concavata* (Brotzen) de la région de la Téthys. *Rev. Micropaleont.* **11**(1), 45–50, 2 plates, 4 text-figs.

Scheibnerová, V. (1969). Middle and Upper Cretaceous microbiostratigraphy of the Klippen Belt (West Carpathians). *Acta geol. geogr. Univ. Comenianae., Geol.* no. 17, 5–98. Tabs. 1–20, text-figs. 1–28. 3 charts, 1 map.

Scheibnerová, V. (1970). Some notes on the palaeoecology and palaeogeography of the Great Artesian Basin, Australia, during the Cretaceous. *Search* **1**(3), 125, 126.

Scheibnerová, V. (1971a). Foraminifera and their Mesozoic biogeoprovinces. *Records, Geol. Surv. New South Wales* **13**(3), 135–174.

Scheibnerová, V. (1971b). Implications of Deep Sea Drilling in the Atlantic for studies in Austrialia and New Zealand—some new views on Cretaceous and Cainozoic palaeogeography and biostratigraphy. *Search* **2**(7), 251–254. 2 text-figs.

Scheibnerová, V. (1972). Some new views on Cretaceous biostratigraphy, based on the concept of foraminiferal biogeoprovinces. *Records* **14**(1), 85–87. 1 tab.

Schijfsma, E. (1946). *The Foraminifera from the Hervian (Campanian) of southern Limburg*, 174 pp. 10 plates, 5 text-figs., 4 tables. Drukkerij "Ernest van Aelst" te Maestricht.

Schijfsma, E. (1955). La position stratigraphique de *Globotruncana helvetica* Bolli en Tunisie. *Micropaleont.* **1**(4), 321–334. 13 text-figs.

Schönfelder, E. (1933). Die Kreideanhaufungen im Geschiebemergel des nordlichen Schleswig. ihre Fossilfuhrung und geologische Bedeutung. *Jber. niedersachs. geol. Ver.* **25**, (1932–1933), 85–128. Plates 13–15, 1 map.

Seibold, E. and Seibold, I. (1959). Foraminiferen der Bankund Schwamm-Fazies im unteren Malm Süddeutschlands. *Neues Jb. Geol. Palaont. Abh.* **109**(3), 309–438. Plates 7, 8, 20 text-figs., charts.

Seibold, E. and Seibold, I. (1960). Über Funde von Globigerinen an der Dogger/ Malm-Grenze Süddeutschlands. 21st Int. Geol. Congr. 6, 64–68. 1 text-fig.

Seiglie, G. A. (1958). Notas sobre algunos foraminíferos planctónicos del Cretácico Superior de la Cuenca de Jatibónico. Mems Soc. cub. Hist. nat. 'Felipe Poey' 24(1), 53–82. 7 plates, 2 text-figs.

Seiglie, G. A. (1959). Notas sobre algunas especies de Heterohelicidae del Cretácico superior de Cuba. Boln. Asoc. mex. Geol. petrol. 11(1–2), 51–62. Lamina 1–4, text-figs. 3, 4.

Seiglie, G. A. (1960). Una nueva especie de Heterohelicidae del Cretácico superior de Cuba. Mems. Soc. cub. Hist. nat. 'Felipe Poey' 24(2), 121–124. 3 text-figs.

Sellier de Civrieux, J. M. (1952). Estudio de la Microfauna de la Seccion-Tipo del Miembro Socuy de la Formacion Colon, Distrito Mara, Estado Zulia. Boln Geol. Minist. Minas Venez. 2(5), 231–310. 11 lams., 12 figs-texto.

Shaw, A. B. (1953). Preliminary survey of the foraminifera of the lower shale of the Niobrara formation in the Laramie Basin. Guidebook, Wyoming Geol. Assoc. 8th Ann. Field Conf., 47–55. 2 plates, 1 text-fig., 3 tabs.

Sidó, M. (1961). Mikropaläontologische Untersuchung der oberkreideschichten des Vékényer Tales. Annls Inst. geol. publ. hung. 49(3), 807–824. 4 plates.

Sidó, M. (1970). Globigerinelloides algerianus Cushman et Ten Dam a dunantuli apti kepzodmenyekben. Foldt. Kozl. 100(4), 388–391. 2 plates.

Sidó, M. (1971). A bakonyi es vertesi rotaliporas-turreliteszes margaosszlet Foraminifera-tarsulasai. Foldt. Kozl. 101, 44–52. 1 plate, 3 tabs.

Sigal, J. (1948). Notes sur les genres de foraminifères Rotalipora Brotzen 1942 et Thalmanninella, famille des Globorotaliidae. Revue Inst. fr. Petrole 3(4) 95–103. 2 plates.

Sigal, J. (1952). Aperçu stratigraphique sur la Micropaléontologie du Crétacé. Alger, 19th Int. Geol. Congr., Monographies Regionales, 1st ser.: Algerie no 26, 1–45. 46 figs.

Sigal, J. (1955). Notes micropaléontologiques nord-africaines. 1. Du Cénomanien au Santonien: zones et limites en facies pelagique. C. r. somm. Seanc. Soc. geol. Fr. 7/8, 157–160.

Sigal, J. (1956a). Notes micropaléontologiques nord-africans. 4. Biticinella breggiensis (Gandolfi) nouveau morphogenre. C. r. somm. Seanc. Soc. geol. Fr. 3/4, 35–57. 1 text-fig.

Sigal, J. (1956b). Notes micropaléontologiques malgaches. 2. Microfaunes albiennes et cénomaniennes. C. r. somm. Seanc. Soc. geol. Fr. 11, 210–214. 3 text-figs.

Sigal, J. (1959). Notes microapléontologiques Alpines. Les genres Schackoina et Leupoldina dans le Gargasien Vocontien. Etude de morphagénèse. Rev. Micropaleont. 2(2), 68–79. 53 text-figs.

Sigal, J. (1966a). Contribution a une monographie des Rosalines. 1. Le genre Ticinella Reichel, souche des Rotalipores. Eclog. geol. Helv. 59(1), 185–217. 6 plates, 1 tab.

Sigal, J. (1966b). Le concept taxinomique de spectre. Exemples d'application chez les Foraminifères. Propositions de règles de nomenclature. Mem. Hors-Serie no. 3 Soc. geol. Fr. 1–126. 10 plates, 7 text-figs., 1 tab.

Sigal, J. (1969). Contribution a une monographie des rosalines. 2. L'espèce Rotalipora appenninica (O. Renz, 1936), origine phyletique et taxinomie. (In Brönnimann, P. and Renz, H. H., eds), Proc. I Plankt. Conf. 2, 622–639. 2 plates. E. J. Brill, Leiden.

Sigal, J. and Dardenne, M. (1962). Correlations dans la craie du Bassin de Paris, Périmètre de Dammartin-en-Goële. *Annls Soc. geol. N.* **80**, 219–223. Plates 13,14, map.

Simon, W. and Bartenstein, H. (eds) (1962). *Leitfossilien der Mikropalaontologie.* 432 pp., 61 plates, 27 text-figs., 22 tabs. Gebruder Borntraeger, Berlin.

Skinner, H. C. (1962). Arkadelphia Foraminiferida. *Tulane Stud. Geol.* **1**(1), 1–72. 6 plates, 2 text-figs.

Sliter, W. V. (1968). Upper Cretaceous foraminifer from southern California and northwestern Baja California, Mexico. *Paleont. Contr. Univ. Kans.* **49**(7), 1–141. Plates 1–24. text-figs. 1–9, tabs 1–15.

Sliter, W. V. (1971). Predation on benthic foraminifers. *J. foramin. Res.* **1**(1), 20–29. 3 plates, 1 text-fig., 1 tab.

Sliter, W. V. (1972a). Upper Cretaceous planktonic foraminiferal zoogeography and ecology—eastern Pacific margin. *Palaeogeogr. Palaeoclimatol. Palaeoecol.* **12**(1–2), 15–31. 10 text-figs.

Sliter, W. V. (1972b). Cretaceous foraminifers—depth habitats and their origin. *Nature* **239**, 514, 515. 1 text-fig., 1 tab.

Sluis, J. P. van der (1950). *Geological, petrographical and palaeontological results of explorations, carried out from September* 1917 *till June* 1919 *in the Island of Ceram* by L. Rutten and W. Hotz, Ser. **3**: Geology, no. 3. Geology of East Seran. J. H. de Bussy, Amsterdam, 66 pp., 2 plates, 1 text-fig., 1 map, 2 tabs.

Smitter, Y. H. (1957). Upper Cretaceous foraminifera from Sandy Point, St. Lucia Bay, Zululand. *S. Afr. J. Sci.* **53**(7), 195–201. Figs. 1–26, 1 tab.

Solange, F. and Sigal, J. (1958). Les foraminifères du Crétacé inférieur vocontien (Note préliminaire).*C. r. somm. Seanc. Soc. geol. Fr.* 124–126. 4 text-figs.

Stelck, C. R. and Wall, J. H. (1954). Kaskapau foraminifera from Peace River area of western Canada. *Contr. Res. Coun. Alberta* **68**, 1–38. 2 plates, 5 text-figs.

Stenestad, E. (1968a). Three new species of *Heterohelix* Ehrenberg from the upper Senonian of Denmark. *Meddr dansk geol. Foren.* **18**(1), 64–70. Plates 1–3.

Stenestad, E. (1968b). Nogle Kridtlignende Kvartaeraflejringer i Nordjlland. *Meddr dansk geol. Foren.* **18**(3–4), 285–293. Plates 1, 2, text-figs. 1–7.

Stenestad, E. (1969). The genus *Heterohelix* Ehrenberg, 1843 (foraminifera) from the Senonian of Denmark. In (Brönnimann, P. and Renz, H. H., eds), *Proc. I Plankt. Conf.* **2**, 644–662. Plates 1–3, text-figs. 1–15.

Sturani, C. (1962). Il complesso sedimentario autoctono all'estremo nord-occidentale del Massiccio dell'Argentera (Alpi Marittime). *Memorie Ist. geol. Univ. Padova* **22** (1961–1962), 1–206. 14 plates, 31 text-figs., 1 tab.

Sturm, M. (1968). *Die Geologie der Flyschzone im Westen von NuBdorf/Attersee,* O.Oe, 302 pp., 11 pls., 9 text-figs., 8 tabs. Unveroff. Diss. Phil. Fak. Univ. Wien.

Sturm, M. (1969). Zonation of Upper Cretaceous by means of planktonic foraminifera, Attersee, Upper Austria. *Roczn. pol. Tow. geol.* **39**(1–3), 103–132. Plates 9–12. 3 text-figs., 2 tabs.

Sturm, M. (1970). Die planktonischen Foraminiferenfaunen des Steinbruches Sievering. In (Faupl, P., Grun, W., Lauer, G., Maurer, R., Papp, A., Schnabel, W. and Sturm, M.), Zur Typisierung der Sieveringer Schichten im Flysch des Wienerwaldes. *Jb. geol. Bundesanst., Wien* **113**(1), 73–158. 15 plates, 12 text-figs., 8 tabs.

Subbotina, N. N. (1949). Microfauna of the U.S.S.R. *Trudy, vses. nauchno-issled. geol.-razv. Inst.* n.s., **34**, 33–35. Plate II, (in Russian).

Subbotina, N. N. (1953). Foraminifères fossiles d'U.R.S.S., Globigerinidae, Globorotaliidae, Hantkeninidae. Traduction M. Sigal *In* Paris, *Bureau de Recherches Geologiques et Minieres* Trad. no **2239**. Pt. 1, 1–144, 15 plates, text-figs. 1–6; Pt. 2, 145–306, 25 plates, text-figs. 7–8.

Subbotina, N. N. (1964). Superfamily Globigerinidea. *In* Foraminifera from Cretaceous and Paleogene deposits of the West Siberian Lowland. *Trudy vses. neft. nauchno-issled. geol.-razv. Inst.* **234**, 249–254. Plates 54, 55 (in Russian).

Suleimanov, I. S. (1955). A new genus, *Gubkinella* and two new species of the family Heterohelicidae from the upper Senonian of southwestern Kyzyl-Kumy. *Dokl. Akad. Nauk. SSSR* **102**(3), 623–624. 2 text-figs. (in Russian).

Szulczewski, M. (1963). Stromatoloity z batonu wierchowego Tatr. *Acta geol. pol.* **13**(1), 125–148. 6 plates, 5 text-figs.

Takayanagi, Y. (1960). Cretaceous foraminifera from Hokkaido, Japan. *Sci. Rep. Tohoku Univ.*, *Ser.* 2 (*Geol.*) **32**(1), 1–154. 11 plates, 22 text-figs., 12 tabs.

Takayanagi, Y. (1965). Upper Cretaceous planktonic foraminifera from the Putah Creek subsurface section along the Yolo-Solano County line, California. *Sci. Rep. Tohoku Univ.*, *Ser.* 2 (*Geol.*) **36**(2), 161–237. Plates 20–29, 9 text-figs.

Takayanagi, Y. and Iwamoto, H. (1962). Cretaceous planktonic foraminifera from the Middle Yezo Group of the Ikushumbetsu, Miruto and Hatonosu areas, Hokkaido. *Trans. Proc. palaeont. Soc. Japan* n.s., **45**, 183–196. Plate 28, 1 tab.

Tamajo, E. (1960). Microfacies mesozoiche della Montagna della Busambra. *Riv. Miner. sicil.* **11**(63), 130–149. 14 text-figs., 1 tab.

Tappan, H. (1940). Foraminifera from the Grayson Formation of northern Texas. *J. Paleont.* **14**(2), 93–126. Plates 14–19.

Tappan, H. (1943). Foraminifera from the Duck Creek Formation of Oklahoma and Texas. *J. Paleont.* **17**(5), 476–517. Plates 77–83.

Tappan, H. (1951). Northern Alaska index foraminifera. *Contr. Cushman Fdn foramin. Res.* **2**(1), 1–8, Plate 1.

Tappan, H. (1960). Cretaceous biostratigraphy of northern Alaska. *Bull. Am. Ass. Petrol. Geol.* **44**(3), 273–297. Plates 1, 2, 7 text-figs.

Tappan, H. (1962). Foraminifera from the Arctic slope of Alaska. Part 3, Cretaceous Foraminifera. *Prof. Pap. U.S. geol. Surv.* **236**–C, 91–209. Plates 29–58, text-figs. 10–18.

Taylor, D. J. (1964). Foraminifera and the stratigraphy of the western Victorian Cretaceous sediments. *Proc. R. Soc. Vict.* **77**(2), 535–603. Plates 79–86. 7 text-figs.

Terquem, O. (1883). Cinquième mémoire sur les foraminifères du système oolithique de la zone à *Ammonites parkinsoni* de Fontoy (Moselle). *Bull. Soc. geol. Fr.* ser. 3, **11**, 339–406.

Terquem, O. (1886). Les foraminifères et les ostracodes du Fuller's Earth des environs de Varsovie. *Mem. Soc. geol. Fr.* ser. 3, **4**(2), 1–112. 12 plates.

Terquem, O. and Berthelin, G. (1875). Études microscopique des Marnes du Lias Moyen d'Essey-Les-Nancy, zone inférieure de l'assise a *Ammonites margaritatus*. *Mem. Soc. geol. Fr.* ser. **2**, 10(3), 1–126. 10 plates, 14 text-figs.

Thalmann, H. E. (1932). Die Foraminiferen-Gattung *Hantkenina* Cushman, 1924 und ihre regional-stratigraphische Verbreitung. *Eclog. geol. Helv.* **25**, 287–292.

Thalmann, H. E. (1933). On homonyms in foraminifera. *Contr. Cushman Lab. foramin. Res.* **9**(4), 96–98.

Thalmann, H. E. (1934). Die regional-stratigraphische Verbreitung der oberkreidischen Foraminiferen-Gattung *Globotruncana* Cushman, 1927. *Eclog. geol. Helv.* **27**, 413–428. 1 text-fig.

Thalmann, H. E. (1946). Additional homonyms in foraminifera erected since 1940. *Contr. Cushman Lab. foramin. Res.* **22**(4), 123–131.

Thalmann, H. E. (1959). New names for foraminiferal homonyms IV. *Contr. Cushman Fdn foramin. Res.* **10**(4), 130, 131.

Thalmann, H. E. and Ayala-Castañares, A. (1959). Evidencias micropaleontologicas sobre la edad Cretacico Superior de las "Pizzarras Necoxtla." *Paleont. mex.* **5**, 1–20. Plates 1–4, 2 text-figs.

Thomas, N. L. and Rice, E. M. (1927). Changing characters in some Texas species of Guembelina. *J. Paleont.* **1**(2), 141–144. Text-fig. 1.

Tilev, N. (1952). Étude des Rosalines maestrichtiennes (genre *Globotruncana*) due Sud-Est de la Turquie (Sondage de Ramandag). *Bull. Labs. Geol. Geogr. phys. Miner. Univ. Lausanne* **103**, 1–86. 24 text-figs.

Todd, R. (1970). Maestrichtian (Late Cretaceous) foraminifera from a deep-sea core of southwestern Africa. *Revta. esp. Micropaleont.* **2**(2), 131–154. 6 plates, 1 tab.

Todd, R. and Low, D. (1964). Cenomanian (Cretaceous) foraminifera from the Puerto Rico Trench. *Deep-Sea Res.* **11**(3), 395–414. 4 plates.

Tollman, A. (1960). Die Foraminiferenfauna des Oberconiae aus der Gosau des Ausseer Weissen bachtales in Steiermark. *Jb. geol. Bundesanst., Wien* **103**, 133–203. Tafeln 6–21, 2 abb.

Troelsen, J. C. (1955). *Globotruncana contusa* in the White Chalk of Denmark. *Micropaleont.* **1**(1), 76–82. Text-figs. 1, 2.

Trujillo, E. F. (1960). Upper Cretaceous foraminifera from near Redding, Shasta County, California. *J. Paleont.* **34**(2), 290–346. Plates 43–50, 3 text-figs.

Tutkovskii, P. A. (1887). Foraminifera from the Tertiary and Cretaceous deposits of Kiev; I—Foraminifera of the Cretaceous marl of Kiev. *Zap. Kiev. Obshch. Estest.* **8**(2), 345–360. Plates 3–7, 1 text-fig. (in Russian).

Tutkovskii, P. A. (1925). The fossil microfauna of the Ukraine, their geological significance and methods of investigation. *Trudy fiz.-mat. Vidd. vseukr. Akad. Nauk.* **1**(8), 1–24. Plates 1–42. (in Russian).

Tzankov, V., Kamenova, J., Simeonov, A., and Vaptzarova, Ya. (1964). The stratigraphy of the Upper Cretaceous between the valleys of the rivers Ossam and Danube. *God. sof. Univ. Fac. Geol. Geogr., Livre* 1, *Geol.* **57** (1962/1963), 217–240. Plates 1–8, 5 text-figs., 3 profiles (in Bulgarian).

Uguzzoni, M. P. M. and Radrizzani, C. P. (1967). I Foraminiferi delle Marne a Fucoidi. *Riv. ital. Paleont. Stratigr.* **73**(4), 1181–1256. Tav. 85–94.

Upton, C. (1898). Chalk under the microscope. *Proc. Cotteswold Nat. Fld. Club* **12**(3), 209–216. Plate A.

Vasilenko, V. P. (1961). Foraminifera from the Upper Cretaceous of the Mangyshlak Peninsula. *Trudy, vses, nachno-issled. geol.-razv. neft. Inst.* **171**, 1–487. 41 plates, 15 tabs., 40 text-figs. (in Russian).

Veljković-Zajec, K. (1954). A contribution to the knowledge of microfauna from the village of Zubetinac. *Zborn. Rad. geol. Inst.* **7**, 247–257. 3 plates. (in Serbian).

Veljković-Zajec, K. (1955). An account of the microfauna of Upper Cretaceous from the well drilling bechej 3 (Banat). *Zborn Rad. geol. Inst.* **8**, 321–331. 3 plates. (in Yugoslavian).

Vinogradov, C. (1960a). Limita Cretacic-Paleogen în bazinul văii Prahova. *Studii Cerc. Geol.-Geogr. Cluj.* **5**(2), 299–324. Plates 1–6, 2 tabs., 1 text-fig. (in Romanian).

Vinogradov, C. (1960b). Studiul Rotaliporelor si Praeglobotruncanelor din Ceno-manianul superior de la Bǎdeni. *Anal. Univ. C.I. Parhon Seria stiintelor naturii* 9(23), 31–42. 3 plates.

Visser, A. M. (1950). *Monograph on the foraminifera of the type-locality of the Maestrichtian (south-Limburg, Netherlands)*, 331 pp., 16 pls. Eduard Ijdo N.V., Leiden.

Visser, W. A. and Hermes, J. J. (1962). Geological results of the exploration for oil in Netherlands New Guinea. *Verh. geol.-mijnb. Genoot. Ned.* 20, 1–265. 85 text-figs., 18 enclosures.

Viterbo, I. (1965). Examen micropaléontologique du Crétacé du Maroc Meridional (Bassin cotier de Tarfaya). *Mem. Bur. Rech. Geol. Min.* 32, Colloque Int. Micro-paleont. (1963), 61–100. Plates 1–11, 2 text-figs., 3 tabs.

Vogler, J. (1941). Ober-Jura und Kreide von misol (Niederlandisch-Ostindien). *Palaeontographica* supplement-band IV, Abt. IV:4, 243–293. Taf. XIX–XXIV, 13 Textabb., 2 tab.

Voorthuysen, J. H. van (1951). Recent (and derived Upper Cretaceous) foraminifera of the Netherlands Wadden Sea (tidal flats). *Meded. geol. Sticht* n.s., 5, 23–32. Plates 1, 2, 1 map, 1 tab.

Voorwijk, G. H. (1937). Foraminifera from the Upper Cretaceous of Habana, Cuba. *Proc. Sect. Sci. K. ned. Akad. wet.* 40(2), 190–198. Plates 1–3.

Wall, J. H. (1967). Cretaceous foraminifer of the Rocky Mountain foothills, Alberta. *Bull. Res. Coun. Alberta* 20, 1–185. 19 plates, 4 tabs.

Weaver, D. W. *et al.* (1969). Geology of the Northern Channel Islands. *Spec. Publs. Soc. econ. Paleont. Miner., Tulsa* Pacific Sections, 1–200. 34 plates, 16 text-figs.

Weiss, L. (1955). Planktonic index Foraminifera of northwestern Peru. *Micro-paleont.* 1(4), 301–319. 3 plates, 1 text-fig., 2 tabs., 1 chart.

Weller, S. (1907). A report on the Cretaceous paleontology of New Jersey. *New Jers. geol. Surv., Paleont. Ser.* 4, 1–871. 61 plates.

Wernli, R. and Septfontaine, M. (1971). Micropaleontologie comparee du Dogger du Jura meridional (France) et des Prealpes Medianes Plastiques romandes (Suisse). *Eclog. geol. Helv.* 64(3), 437–458. 5 text-figs., 2 tabs.

Wessem, A. van (1943). *Geology and paleontology of central Camaguey, Cuba,* 91 pp., 3 pls. Drukkerijj J. van Bockhoven, Utrecht.

Wezel, F. C. (1965). Geologia della tavoletta Mirabella Imbàccuri (Prov. di Catania, Caltanisseta ed Enna, F. 272, I–NE). *Boll. Soc. geol. ital.* 84(7), 3–136. 2 plates, 55 text-figs.

White, M. P. (1928a). Some index foraminifera of the Tampico embayment area of Mexico. Part I. *J. Paleont.* 2(3), 177–215. Plates 27–29, text-fig. 1.

White, M. P. (1928b). Some index foraminifera of the Tampico embayment area of Mexico. Part II. *J. Paleont.* 2(4), 280–317. Plates 38–42.

White, M. P. (1929). Some index foraminifera of the Tampico embayment area of Mexico. Part III. *J. Paleont.* 3(1), 30 58. Plates 4, 5.

Wicher, C. A. (1949). O tumačenju starosti viših slojeva gornje krede kod Tampika (Meksiko) kao primer rasprostranjenja mikrofosila sirom sveta i o praktičnim posledicama koje proizlaze iz toga. *Glasnik Prirodn'achkog Muzeja Srpske Zeml'e, Ser. A, Min., Geol. Paleont.* Kn'iga 2, 49–105. Plates 2–8.

Wicher, C. A. (1956). Die Gosau-Schichten im Becken von Gams (Österreich) und die Foraminiferengliederung der höheren Oberkreide in der Tethys. *Palaont. Z.* **30**, 87–136. Abb. 1–7, Tafeln 12' and 13'.

Wicher, C. A. and Bettenstaedt, F. (1957). Zur Oberkreide-Gliederung der bayerischen Innviertel-Bohrungen. 6. *Ventrilabrella deflaensis*—ein Santon-Fossil. *Geologica bav.* **30**, 30–38. Abb. 3.

Wille-Janoschek, U. (1966). Stratigraphie und Tektonik der Schichten der Oberkreide und des Alttertiärs im Raume von Gosau und Abtenau (Salzburg). *Jb. geol. Bundesanst. Wien* **109**, 91–172. 3 abb., 11 tafeln.

Williams-Mitchell, E. (1948). The zonal value of foraminifera in the chalk of England. *Proc. Geol. Ass.* **59**(2), 91–112. Plates 8–10.

Williamson, W. C. (1848). On the some of the microscopical objects found in the mud of the Levant, and other deposits, with remarks on the mode of formation of calcareous and infusorial siliceous rocks. *Mem. Proc. Manchr. lit. phil. Soc.* 2nd ser., **8**, 1–128. 4 plates.

Witwicka, E. (1958). Stratigrafia mikropalaeontologiczna kredy górnej wiercenia w Chełmie. *Biul. Inst. geol.* **121** (Z Badan Mikropal., vol. 3). 177–267. Plates 8–19).

Woodward, A. (1894). The Cretaceous foraminifera of New Jersey. Part II. Original investigations and remarks. *Jl. N.Y. microsc. Soc.* **10**(4), 91–141.

Woodward, A. and Thomas, B. W. (1885). On the foraminifera of the Boulder-Clay, taken from a well-shaft 22 feet deep, Meeker County, central Minnesota. *Minn. geol. nat. Hist. Surv.* 13th Ann. Rept., 164–177. 4 plates, 38 figures.

Woodward, A. and Thomas, B. W. (1893). Microscopic fauna of the Cretaceous of Minnesota with additions from Nebraska and Illinois. *Minn. geol. nat. Hist. Surv.* 23–52. Plates C–E, 1 tab.

Woodward, A. and Thomas, B. W. (1895). The microscopial fauna of the Cretaceous in Minnesota, with additions from Nebraska and Illinois (foraminifera, radiolaria, coccoliths, rhabdoliths). *Minn. geol. nat. Hist. Surv.* Final Rept., 3(1), 23–54.

Wright, J. (1900). *In* Reade, T. M., A contribution to postglacial geology. *Geol. Mag.* n.s., dec. 4, **4**(3), 99–102. Plate 5.

Young, K. (1951). Foraminifera and stratigraphy of the Frontier Formation (Upper Cretaceous) Southern Montana. *J. Paleont.* **25**(1), 35–68. Plates 11–14, 6 text-figs.

Youssef, M. I. and Abdel-Aziz, W. (1971). Biostratigraphy of the Upper Cretaceous—Lower Tertiary in Farafra Oasis. Libyan Desert, Egypt. *Symposium on the Geology of Libya* (1969) 227–249. 5 plates, 4 text-figs., 1 tab.

Zakharova-Atabekyan, L. V. (1961). Revision of the systematics of Globotruncanidae and proposal of a new genus *Planogyrina* gen. nov. *Dokl. Akad. Nauk. armyan. SSR* **32**(1), 49–53. (in Russian).

Zamnatti-Scarpa, C. (1957). Studio di alcune "microfacies" del Bresciano. *Boll. Serv. geol. Ital.* **78**(4–5), 585–607. 3 plates, 1 text-fig., 1 tab.

Zanzucchi, G. (1955). Su una placca di argilla medio-cretacica nella "formazione ofiolitifera" del Monte Prinzera (Parma). *Atti Soc. ital. Sci. nat.* **94**(3–4), 369–377. Plate 28, map.

Ziegler, J. H. (1957). Die Fauna des Cardientones der Oberpfälz und die Bedeutung der Foraminiferen für seine Altersbestimmung (Coniac). *Geologica bav.* **30**, 55–86. 1 plate, 3 tabs.

Zittel, K. A. (1880). *Handbuch der Palaeontologie.* Band 1. *Palaeozoologie*, vol. 1, pt. 1. *Protozoa, Coelenterata, Echinodermata und Molluscoidea*, 765 pp., 558 text-figs., tabs. R. Oldenbourg, Munchen and Leipzig.

PLATES

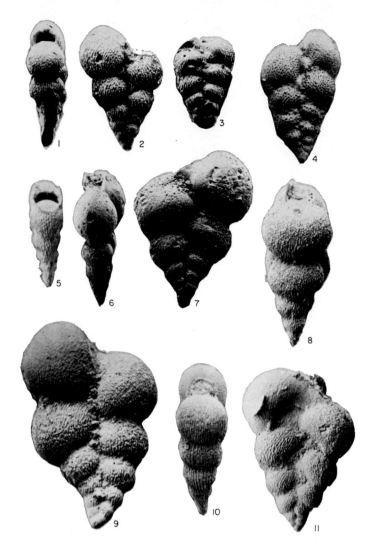

PLATE 1. 1. *Heterohelix americana* (Ehrenberg). Edge view, (×120). Ripley Formation. Wilcox Co., Alabama. 2. *Heterohelix americana* (Ehrenberg). Side view, (×120). Small initial coil. Ripley Formation. Wilcox Co., Alabama. 3. *Heterohelix americana* (Ehrenberg). Side view, (×120). Large initial coil. Ripley Formation. Wilcox Co., Alabama. 4. *Heterohelix carinata* (Cushman). Side view, (×120). Mooreville Chalk. Dallas Co., Alabama. 5. *Heterohelix carinata* (Cushman). Edge view, (×120). Hatcher Bluff Marl. Dallas Co., Alabama. 6. *Heterohelix glabrans* (Cushman). Edge view, (×120). Note angled edge. Prairie Bluff Chalk. Wilcox Co., Alabama. USNM 184675. 7. *Heterohelix glabrans* (Cushman). Side view, (×120). Prairie Bluff Chalk. Wilcox Co., Alabama. USNM 184674. 8. *Heterohelix pseudotessera* (Cushman). Edge view, (× 120). Hatcher Bluff Marl. Dallas Co., Alabama. USNM 184699. 9. *Heterohelix globulosa* (Ehrenberg). Side view, (×120). Mooreville Chalk. Dallas Co., Alabama. 10. *Heterohelix globulosa* (Ehrenberg). Edge view, (×120). Mooreville Chalk. Dallas Co., Alabama. USNM 184683. 11. *Heterohelix pseudotessera* (Cushman). Side view, (× 120). Note depressed panels along median line. Hatcher Bluff Marl. Dallas Co., Alabama. USNM 184698.

PLATE 2. 1. *Heterohelix moremani* (Cushman). Side view, (×258). Topotype of *Guembelina washitensis* Tappan. Grayson Formation. Denton Co., Texas. 2. *Heterohelix pulchra* (Brotzen). Side view, (×120). Note initial coil and relict apertures. Hatcher Bluff Marl. Dallas Co., Alabama. 3. *Heterohelix planata* (Cushman). Side view, (×120). Topotype. Taylor Marl. Red River Co., Texas. 4. *Heterohelix semicostata* (Cushman). Side view, (×192). Upper Taylor Marl. Lamar Co., Texas. 5. *Heterohelix semicostata* (Cushman). Oblique edge view, (×160). Upper Taylor Marl. Lamar Co., Texas. 6. *Heterohelix sphenoides* Masters. Side view, (×120). Holotype. Hatcher Bluff Marl. Dallas Co., Alabama.

675

PLATE 3. 1. *Heterohelix sphenoides* Masters. Oblique edge view, (×120). Paratype. Hatcher Bluff Marl. Dallas Co., Alabama. 2. *Heterohelix striata* (Ehrenberg). Side view, (×120). Hatcher Bluff Marl. Dallas Co., Alabama. 3. *Heterohelix striata* (Ehrenberg). Edge view, (×120). Hatcher Bluff Marl. Dallas Co., Alabama. 4. *Heterohelix stenopus* Masters. Side view, (×120). Hatcher Bluff Marl. Dallas Co., Alabama. 5. *Lunatriella spinifera* Eicher and Worstell. Side view, (×640). Metatype. Same specimen as fig. 6. Carlile Shale, Fairport Member. Hamilton Co., Kansas. 6. *Lunatriella spinifera* Eicher and Worstell. Side view, (×258). Metatype. Carlile Shale, Fairport Member. Hamilton Co., Kansas.

676

PLATE 4. 1. *Planoglobulina varians* (Rzehak). Edge view, (×96). Topotype. Note chamberlets. Gerhardsreiter Schichten. Siegsdorf, Bavaria. 2. *Planoglobulina varians* (Rzehak). Apertural view, (×96). Topotype. Note dumbbell-shaped chambers, chamberlets and supplementary chambers. Gerhardsreiter Schichten. Siegsdorf, Bavaria. 3. *Platystaphyla brazoensis* (Martin). Edge view, (×64). Note alignment of supplementary chambers in a plane. Prairie Bluff Chalk. Wilcox Co., Alabama. USNM 184749. 4. *Platystaphyla brazoensis* (Martin). Side view, (×64). Prairie Bluff Chalk. Wilcox Co., Alabama. USNM 184747 5. *Pseudoguembelina costellifera* Masters. Side view, (×120). Holotype. Prairie Bluff Chalk. Wilcox Co., Alabama. USNM 184726. 6. *Pseudoguembelina costata* (Carsey). Side view, (×120). Topotype. Corsicana Marl. Travis Co., Texas. USNM 184724. 7. *Pseudoguembelina excolata* (Cushman). Side view, (×120). Prairie Bluff Chalk. Wilcox Co., Alabama. USNM 184731. 8. *Pseudoguembelina excolata* (Cushman). Edge view, (×120). Prairie Bluff Chalk. Wilcox Co., Alabama. USNM 184730. 9. *Pseudoguembelina palpabra* Brönnimann and Brown. Side view, (×120). Prairie Bluff Chalk. Wilcox Co., Alabama. USNM 184733.

PLATE 5. 1. *Pseudoguembelina polypleura* Masters. Side view, (×120). Holotype. Prairie Bluff Chalk. Wilcox Co., Alabama. USNM 184734. 2. *Pseudoguembelina punctulata* (Cushman). Side view, (×120). Ripley Formation. Dallas Co., Alabama. USNM 184735. 3. *Pseudotextularia browni* Masters. Side view, (×120). Hatcher Bluff Marl. Dallas Co., Alabama. USNM 184742. 4. *Pseudotextularia browni* Masters. Edge view, (×120). Demopolis Chalk. Dallas Co., Alabama. USNM 184739.

678

PLATE 6. 1. *Pseudotextularia carseyae* (Plummer). Side view, (×120). Topotype. Kemp Clay. Milam Co., Texas. 2. *Pseudotextularia carseyae* (Plummer). Edge view, (×120). Note continuous costae. Hatcher Bluff Marl. Dallas Co., Texas. USNM 184738. 3. *Pseudotextularia elegans* (Rzehak). Edge view, (×120). Ripley Formation. Wilcox Co., Alabama. USNM 184744. 4. *Pseudotextularia elegans* (Rzehak). Side view, (×120). Ripley Formation. Wilcox Co., Alabama. USNM 184746. 5. *Ventilabrella austinana* Cushman. Side view, (×120). Hatcher Bluff Marl. Dallas Co., Texas. USNM 184710.

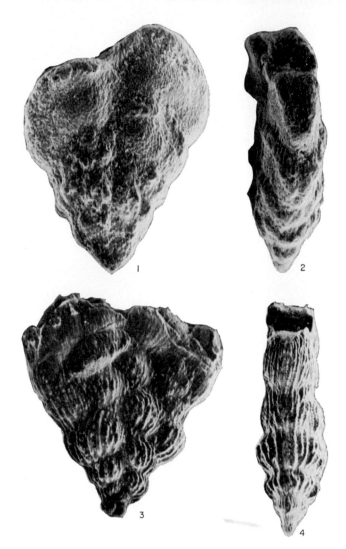

PLATE 7. 1. *Ventilabrella deflaensis* (Sigal). Side view, (×128). Ideotype. Biserial Stage. Mechta Barrache. Monts dela Haute Medjerda, Algeria. 2. *Ventilabrella deflaensis* (Sigal). Oblique edge view, (×128). Ideotype. Mechta Barrache. Monts dela Haute Medjerda, Algeria. 3. *Ventilabrella eggeri* Cushman. Side view, (×128). Note final irregularly shaped chambers. Lower Taylor Marl. Hill Co., Texas. 4. *Ventilabrella eggeri* Cushman. Edge view, (×128). Lower Taylor Marl. Hill Co., Texas.

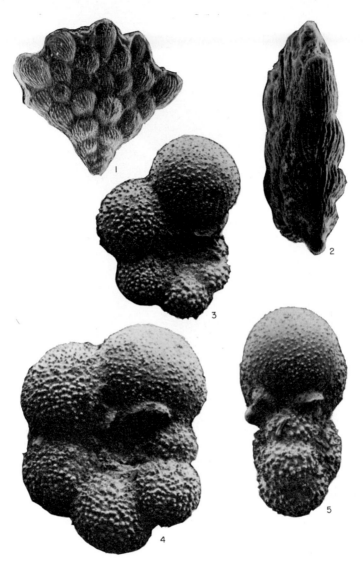

PLATE 8. 1. *Ventilabrella glabrata* Cushman. Side view, (×96). Lower Taylor Marl. Hill Co., Texas. 2. *Ventilabrella glabrata* Cushman. Edge view as seen from proloculus, (×128). Lower Taylor Marl. Hill Co., Texas. 3. *Globigerinelloides abberanta* (Netskaya). Umbilical view, (×176). Topotype of *Biglobigerinella multispina* Lalicker. Ephebic stage. Marlbrook Marl. Howard Co., Arkansas. USNM 184764. 4. *Globigerinelloides abberanta* (Netskaya). Umbilical view, (×176). Topotype of *Biglobigerinella multispina* Laiicker. Gerontic stage. Marlbrook Marl. Howard Co., Arkansas. USNM 184769. 5. *Globigerinelloides abberanta* (Netskaya). Peripheral view, (×176). Topotype of *Biglobigerinella multispina* Lalicker. Gerontic stage. Marlbrook Marl. Howard Co., Arkansas. USNM 184766.

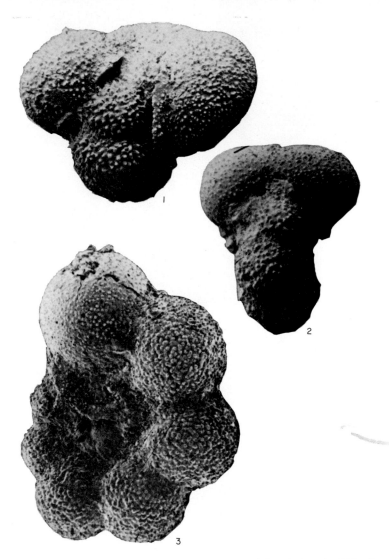

PLATE 9. 1. *Globigerinelloides abberanta* (Netskaya). Peripheral view, (×176). Gerontic stage having paired chambers, one with a double aperture. Demopolis Chalk. Dallas Co., Alabama. USNM 184772. 2. *Globigerinelloides abberanta* (Netskaya). Peripheral view, (×176). Gerontic stage with thick ultimate chamber. Hatcher Bluff Marl. Dallas Co., Alabama. 3. *Globigerinelloides alvarezi* (Eternod Olvera). Umbilical view, (×176). Hatcher Bluff Marl. Dallas Co., Alabama. USNM 184754.

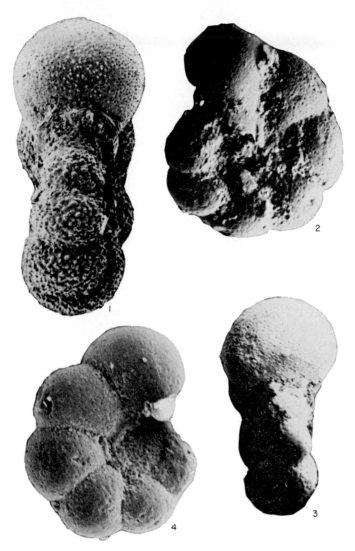

PLATE 10. 1. *Globigerinelloides alvarezi* (Eternod Olvera). Peripheral view, (×176). Hatcher Bluff Marl. Dallas Co., Alabama. USNM 184755. 2. *Globigerinelloides bentonensis* (Morrow). Umbilical view, (×128). Topotype. Greenhorn Limestone, Hartland Shale Member. Hodgeman Co., Kansas. 3. *Globigerinelloides bentonensis* (Morrow). Peripheral view, (×128). Topotype. Greenhorn Limestone, Hartland Shale Member. Hodgeman Co., Kansas. 4. *Globigerinelloides cushmani* (Tappan). Umbilical view, (×192). Topotype. Duck Creek Formation. Love Co., Oklahoma.

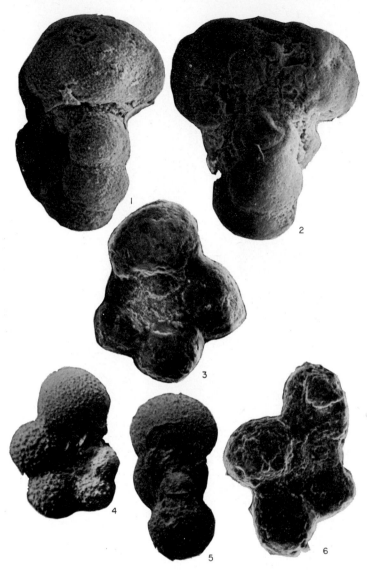

PLATE 11. 1. *Globigerinelloides cushmani* (Tappan). Oblique peripheral view, (×192). Topotype. Duck Creek Formation. Love Co., Oklahoma. 2. *Globigerinelloides cushmani* (Tappan). Peripheral view, (×192). Topotype. Gerontic stage with paired chambers. Duck Creek Formation. Love Co., Oklahoma. 3. *Globigerinelloides blowi* (Bolli). Umbilical view, (×320). Topotype of *Planomalina maridalensis* Bolli. Maridale Formation. Trinidad, W.I. 4. *Globigerinelloides escheri* (Kaufmann). Umbilical view, (×176). Ephebic stage. Hatcher Bluff Marl. Dallas Co., Alabama. USNM 184759. 5. *Globigerinelloides escheri* (Kaufmann). Peripheral view, (×176). Gerontic stage. Remnants of the ultimate chamber with paired apertures visible, penultimate chamber with a single aperture. Prairie Bluff Chalk. Wilcox Co., Alabama. 6. *Globigerinelloides saundersi* (Bolli). Umbilical view, (×192). Topotype. Maridale Formation. Trinidad, W.I.

PLATE 12. 1. *Globigerinelloides subcarinata* (Brönnimann). Peripheral view, (×176). Note angled periphery. Prairie Bluff Chalk. Wilcox Co., Alabama. USNM 184774. 2. *Globigerinelloides subcarinata* (Brönnimann). Umbilical view, (×176). Prairie Bluff Chalk. Wilcox Co., Alabama. USNM 184773. 3. *Globigerinelloides ultramicra* (Subbotina). Peripheral view. (×176). Ephebic stage. Ripley Formation. Dallas Co., Alabama. USNM 184779. 4. *Globigerinelloides ultramicra* (Subbotina). Umbilical view, (×176). Gerontic stage. Ripley Formation. Wilcox Co., Alabama. USNM 184780. 5. *Globigerinelloides ultramicra* (Subbotina). Peripheral view, (×176). Gerontic stage. Ripley Formation. Dallas Co., Alabama. USNM 184776. 6. *Hastigerinoides alexanderi* (Cushman). Umbilical view, (×176). Topotype. Austin Chalk. Grayson Co., Texas. 7. *Hastigerinoides alexanderi* (Cushman). Umbilical view, (×176). Topotype, Austin Chalk. Grayson Co., Texas.

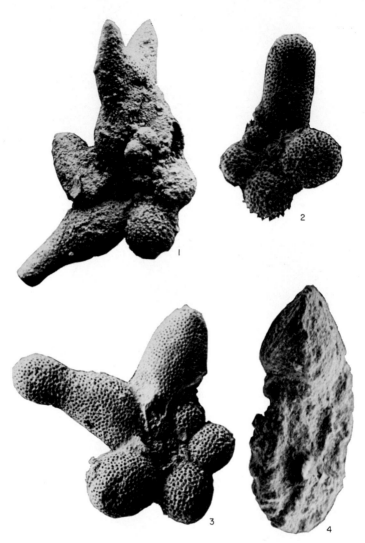

PLATE 13. 1. *Hastingerinoides alexanderi* (Cushman). Umbilical view, (×176). Topotype with bifurcated ultimate chamber. Small *Heterohelix* attached near umbilicus. Austin Chalk. Grayson Co., Texas. 2. *Hastigerinoides watersi* (Cushman). Umbilical view, (×176). Early ephebic stage. Hatcher Bluff Marl. Dallas Co., Alabama. 3. *Hastigerinoides watersi* (Cushman). Umbilical view, (×176). Topotype. Ephebic stage. Ultimate chamber broken at constriction. Austin Chalk. Grayson Co., Texas. USNM 184790. 4. *Planomalina buxtorfi* (Gandolfi) Oblique peripheral view, (×128). Topotype. Scaglia Bianca. Balerna, Switzerland.

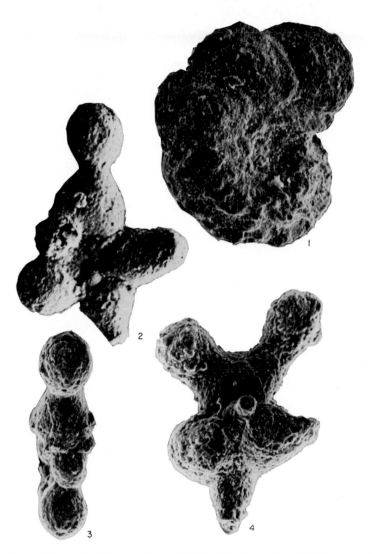

PLATE 14. 1. *Planomalina buxtorfi* (Gandolfi). Umbilical view, (×128). Topotype. Scaglia Bianca. Balerna, Switzerland. 2. *Leupoldina pentagonalis* (Reichel). Umbilical view, (×128). Metatype of *Schackoina pustulans* Bolli. Aptian. Trinidad, W.I. 3. *Leupoldina pentagonalis* (Reichel). Peripheral view, (×128). Same specimen as fig. 2. Aptian. Trinidad, W.I. 4. *Leupoldina cabri* (Sigal). Peripheral view. (×192). Metatype of *L. protuberans* Bolli. Aptian. Trinidad, W.I.

PLATE 15. 1. *Leupoldina cabri* (Sigal). Umbilical view, (×192). Metatype of *L. protuberans* Bolli. Aptian. Trinidad, W.I. 2. *Schackoina alberti* Massella. Umbilical view, (×258). Scaglia Rossa. Balerna, Switzerland. 3. *Schackoina bicornis* Reichel. Oblique peripheral view of last two chambers, (×320). Topotype. Scaglia Bianca. Balerna, Switzerland. 4. *Schackoina bicornis* Reichel. Umbilical view, (×258). Topotype. Scaglia Bianca. Balerna, Switzerland.

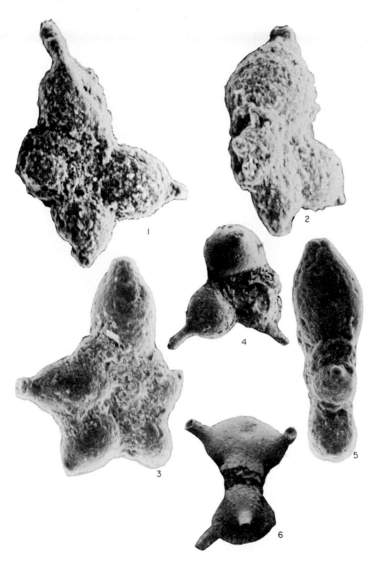

PLATE 16. 1. *Schackoina cenomana* (Schacko). Umbilical view, (×320). Topotype of *S. gandolfii* Reichel. Scaglia Rossa. Balerna, Switzerland. 2. *Schackoina cenomana* (Schacko). Oblique peripheral view, (×320). Topotype of *S. gandolfii* Reichel. Scaglia Rossa. Balerna, Switzerland. 3. *Schackoina masellae* Luterbacher and Premoli Silva. Umbilical view, (×320). Cenomanian. Patti, Sicily. 4. *Schackoina multispinata* (Cushman and Wickenden). Umbilical view, (×176). Topotype of *S. trituberculata* (Morrow). Greenhorn Limestone, Hartland Shale Member. Hodgeman Co., Kansas. 5. *Schackoina masellae* Luterbacher and Premoli Silva. Peripheral view, (×320). Cenomanian. Patti, Sicily. 6. *Schackoina multispinata* (Cushman and Wickenden). Peripheral view, (×176). Prairie Bluff Chalk. Wilcox Co., Alabama. USNM 184802.

PLATE 17. 1. *Schackoina moliniensis* Reichel. Oblique umbilical view, (×320). Topotype. Scaglia Rossa. Balerna, Switzerland. 2. *Schackoina multispinata* (Cushman and Wickenden). Umbilical view, (×176). Prairie Bluff Chalk. Wilcox Co., Alabama. USNM 184803. 3. *Schackoina sellaeforma* Masters. Umbilical view, (×176). Holotype. Hatcher Bluff Marl. Dallas Co., Alabama. USNM 184796. 4. *Schackoina sellaeforma* Masters. Oblique peripheral view, (×176). Paratype. Hatcher Bluff Marl. Dallas Co., Alabama. USNM 184797. 5. *Schackoina tappanae* Montanaro Gallitelli. Umbilical view, (×258). Greenhorn Limestone, Hartland Shale Member. Hodgeman Co., Kansas. 6. *Schackoina tappanae* Montanaro Gallitelli. Peripheral view, (×320). Greenhorn Limestone, Hartland Shale Member Hodgeman Co., Kansas.

690

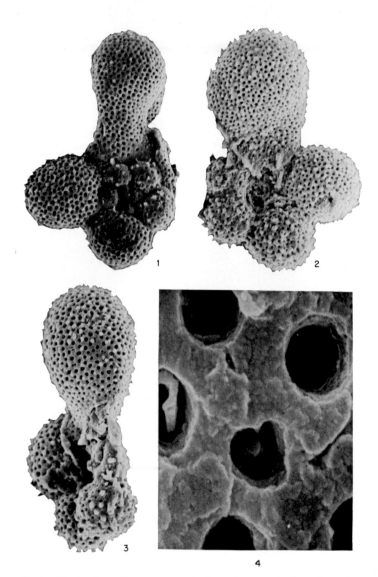

PLATE 18. 1. *Clavihedbergella moremani* (Cushman). Dorsal view, (×320). Topotype. Eagle Ford Shale. Hill Co., Texas. 2. *Clavihedbergella moremani* (Cushman). Ventral view, (×258). Topotype. Eagle Ford Shale. Hill Co., Texas. 3. *Clavihedbergella moremani* (Cushman). Oblique peripheral view, (×320). Topotype. Eagle Ford Shale. Hill Co., Texas. 4. *Clavihedbergella moremani* (Cushman). Chamber wall, (×6400). Topotype. Same specimen as fig. 3. Eagle Ford Shale, Hill Co., Texas.

691

PLATE 19. 1. *Clavihedbergella simplex* (Morrow). Ventral view, (×128). Topotpye. Greenhorn Limestone, Hartland Shale Member. Hodgeman Co., Kansas. 2. *Clavihedbergella simplex* (Morrow). Peripheral view, (×128). Topotype. Greenhorn Limestone, Hartland Shale Member. Hodgeman Co., Kansas. 3. *Clavihedbergella simplex* (Morrow). Dorsal view, (×128). Topotype. Greenhorn Limestone, Hartland Shale Member. Hodgeman Co., Kansas. 4. *Clavihedbergella subcretacea* (Tappan). Peripheral view, (×192). Lake Waco Formation, Cloice Member. Spring Valley, Texas. 5. *Clavihedbergella subcretacea* (Tappan). Dorsal view, (×192). Lake Waco Formation, Cloice Member. Spring Valley, Texas. 6. *Clavihedbergella subcretacea* (Tappan). Ventral view, (×128). Topotype. Duck Creek Formation. Love Co., Oklahoma.

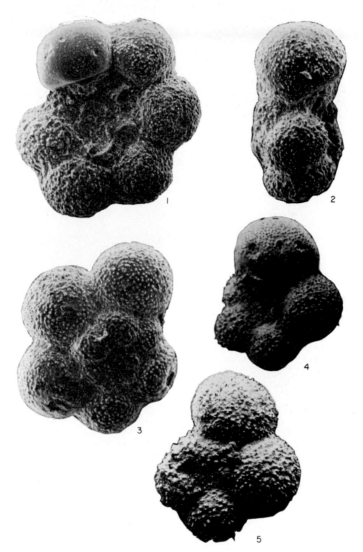

PLATE 20. 1. *Globigerina aprica* (Loeblich and Tappan). Ventral view, (×96). Remnants of portici suggesting accessory apertures. Austin Chalk, Gober Member, Fannin Co., Texas. 2. *Globigerina aprica* (Loeblich and Tappan). Peripheral view, (×96). Austin Chalk, Gober Member, Fannin Co., Texas. 3. *Globigerina aprica* (Loeblich and Tappan). Dorsal view, (×96). Austin Chalk, Gober Member. Fannin Co., Texas. 4. *Globigerina delrioensis* Carsey. Ventral view, (×176). Hatcher Bluff Marl. Dallas Co., Alabama. USNM 184814. 5. *Globigerina delrioensis* Carsey, Dorsal view, (×176). Neotype. Del Rio Shale. Travis Co., Texas. USNM 184813.

693

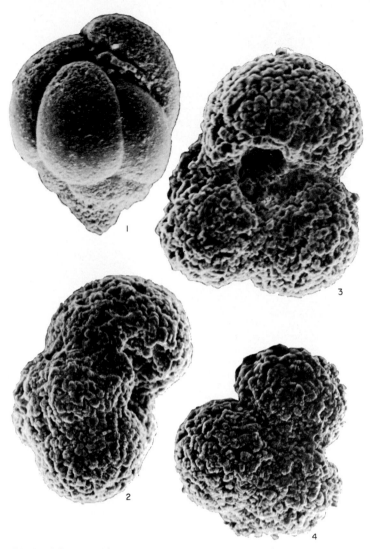

PLATE 21. 1. *Globigerina* (?) *graysonensis* Tappan. Side view, (×320). Topotype. Grayson Formation. Denton Co., Texas. 2. *Globigerina* (?) *helvetojurassica* Haeusler. Peripheral view, (×320). Topotype. Oxfordian. Birmenstorfer beds. Eisengraben, Switzerland. 3. *Globigerina* (?) *helvetojurassica* Haeusler. Ventral view, (×320). Topotype. Oxfordian. Birmenstorfer beds. Eisengraben, Switzerland. 4. *Globigerina* (?) *helvetojurassica* Haeusler. Dorsal view, (×320). Topotype. Oxfordian. Birmenstorfer beds. Eisengraben, Switzerland.

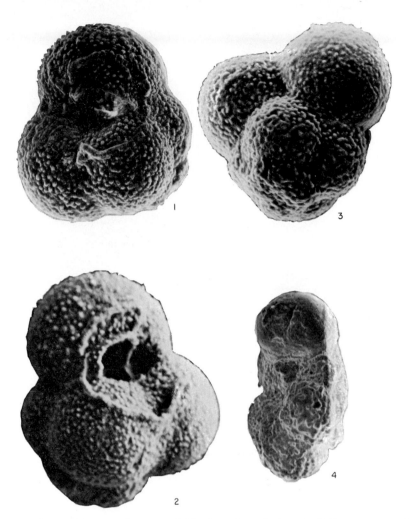

PLATE 22. 1. *Globigerina hoterivica* Subbotina. Ventral view, (×320). Topotype of *G. bathonica* Pazdrowa. Ultimate chamber missing. Bathonian. Ore-bearing clays. Ogrodzieniec, Poland. 2. *Globigerina hoterivica* Subbotina. Oblique ventral view, (×320). Topotype of *G. bathonica* Pazdrowa. Ultimate chamber missing. Bathonian. Ore-bearing clays. Ogrodzieniec, Poland. 3. *Globigerina hoterivica* Subbotina. Dorsal view, (×320). Topotype of *G. bathonica* Pazdrowa. Bathonian. Ore-bearing clays. Ogrodzieniec, Poland. 4. *Globigerina loetterlei* Nauss. Peripheral view, (×128). Greenhorn Limestone, Hartland Shale Member. Hodgeman Co., Kansas.

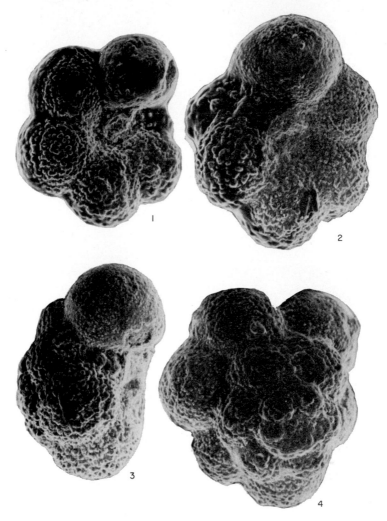

PLATE 23. 1. *Globigerina loetterlei*, Nauss. Ventral view (×128). Same specimen as Plate 22, fig. 4. Adhesive on periphery. Greenhorn Limestone, Hartland Shale Member. Hodgeman Co., Kansas. 2. *Globigerina paradubia* Sigal. Oblique ventral view, (×128). Greenhorn Limestone, Hartland Shale Member. Hodgeman Co., Kansas. 3. *Globigerina paradubia* Sigal. Peripheral view, (×128). Low-spired form. Greenhorn Limestone, Hartland Shale Member. Hodgeman Co., Kansas. 4. *Globigerina paradubia* Sigal. Dorsal view, (×128). Greenhorn Limestone, Hartland Shale Member. Hodgeman Co., Kansas.

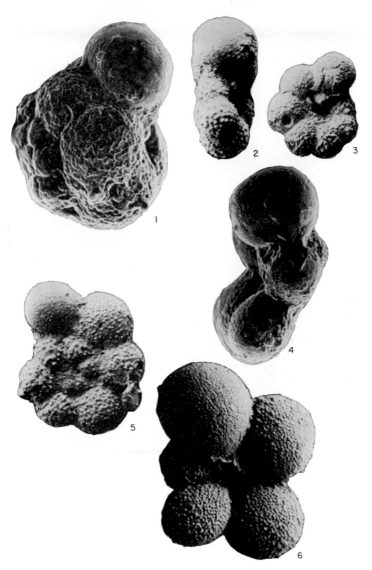

PLATE 24. 1. *Globigerina paradubia* Sigal. Peripheral view, (×128). High-spired form. Greenhorn Limestone, Hartland Shale Member. Hodgeman Co., Kansas. 2. *Globigerina planispira* Tappan. Peripheral view, (×176). Hatcher Bluff Marl. Dallas Co., Alabama. USNM 184821. 3. *Globigerina planispira* Tappan. Ventral view, (×176). Topotype. Note bore hole on antipenultimate chamber. Grayson Formation. Denton Co., Texas. USNM 184819. 4. *Globigerina simplicissima* (Magné and Sigal). Peripheral view, (×176). Greenhorn Limestone, Hartland Shale Member, Hodgeman Co., Kansas. 5. *Globigerina planispira* Tappan. Dorsal view (×176). Hatcher Bluff Marl. Dallas Co., Alabama. USNM 184822. 6. *Globigerina simplicissima* (Magné and Sigal). Ventral view, (×128). Greenhorn Limestone, Hartland Shale Member. Hodgeman Co., Kansas.

697

PLATE 25. 1. *Globigerina trocoidea* (Gandolfi). Ventral view, (×176). Topotype. Scaglia Variegata. Balerna, Switzerland. USNM 184823. 2. *Globigerina trocoidea* (Gandolfi). Oblique peripheral view, (×128). Topotype. Scaglia Variegata. Balerna, Switzerland. 3. *Globigerina trocoidea* (Gandolfi). Dorsal view, (×128). Topotype. Pseudoplanispiral form. Scaglia Variegata. Balerna, Switzerland. 4. *Globigerina washitensis* Carsey. Peripheral view, (×192). Low-spired form. Duck Creek Formation. Cooke Co., Texas.

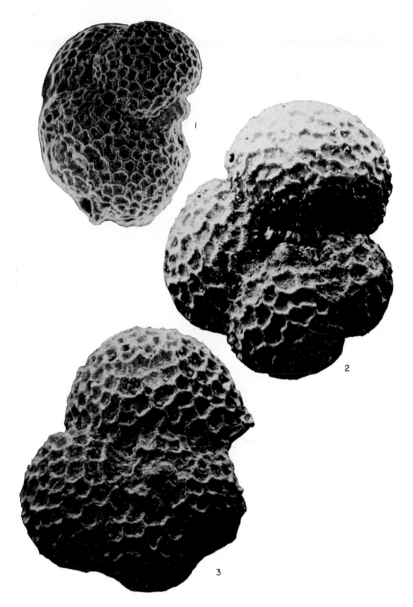

PLATE 26. 1. *Globigerina washitensis* Carsey. Peripheral view, (×128). High-spired form. Duck Creek Formation. Cooke Co., Texas. 2. *Globigerina washitensis* Carsey. Ventral view, (×192). Note irregular size and shape of polygons. Duck Creek Formation. Cooke Co., Texas. 3. *Globigerina washitensis* Carsey. Dorsal view, (×192). Note irregularity of polygons. Duck Creek Formation. Cooke Co., Texas.

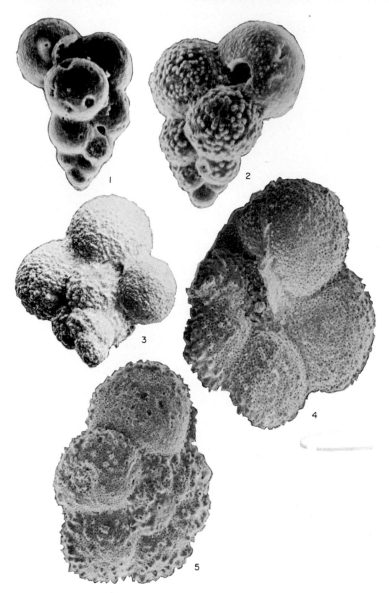

PLATE 27. 1. *Guembelitria cenomana* (Keller). Side view, (× 320). Gault Clay. Padderworth, Kent, England. 2. *Guembelitria cretacea* Cushman. Side view, (× 320). Topotype. Kemp Clay. Guadalupe Co., Texas. 3. *Guembelitria cenomana* (Keller). Side view, proloculus orientation, (× 258). Topotype of *Guembelitriella graysonensis* Tappan. Grayson Formation. Denton Co., Texas. 4. *Praeglobotruncana delrioensis* (Plummer). Oblique ventral view, (× 192). Ultimate chamber missing. Pepper Shale. McLennan Co., Texas. 5. *Praeglobotruncana delrioensis* (Plummer). Oblique dorsal view, (× 192). Pepper Shale. McLennan Co., Texas.

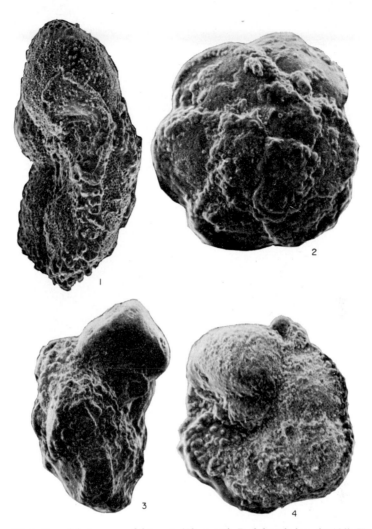

PLATE 28. 1. *Praeglobotruncana delrioensis* (Plummer). Peripheral view, (× 192). Kummer-form ultimate chamber missing. Pepper Shale. McLennan Co., Texas. 2. *Praeglobotruncana stephani* (Gandolfi). Dorsal view, (× 128). Greenhorn Limestone. Hartland Shale Member. Hodgeman Co., Kansas. 3. *Praeglobotruncana stephani* (Gandolfi). Peripheral view, (× 128). Note wide keel. Greenhorn Limestone. Hartland Shale Member. Hodgeman Co., Kansas. 4. *Praeglobotruncana stephani* (Gandolfi). Ventral view, (× 128). Greenhorn Limestone. Hartland Shale Member. Hodgeman Co., Kansas.

PLATE 29. 1. *Praeglobotruncana* sp. Dorsal view, (×128). Greenhorn Limestone. Hartland Shale Member. Hodgeman Co., Kansas. 2. *Praeglobotruncana* sp. Peripheral view, (×128). Greenhorn Limestone. Hartland Shale Member. Hodgeman Co., Kansas. 3. *Praeglobotruncana* sp. Ventral view, (×128). Greenhorn Limestone. Hartland Shale Member. Hodgeman Co., Kansas.

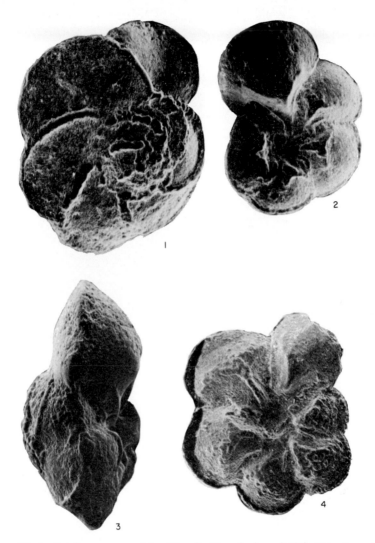

PLATE 30. 1. *Rotalipora appenninica* (Renz). Dorsal view, (×96). Topotype. Scaglia Bianca. Balerna, Switzerland. 2. *Rotalipora appenninica* (Renz). Ventral view, (×96). Topotype. *R. evoluta* Sigal form. Scaglia Bianca. Balerna, Switzerland. 3. *Rotalipora appenninica* (Renz). Peripheral view, (×96). Topotype. Scaglia Bianca. Balerna, Switzerland. 4. *Rotalipora cushmani* (Morrow). Ventral view, (×64). Topotype. Dark periphery is adhesive. Greenhorn Limestone. Hartland Shale Member. Hodgeman Co., Kansas.

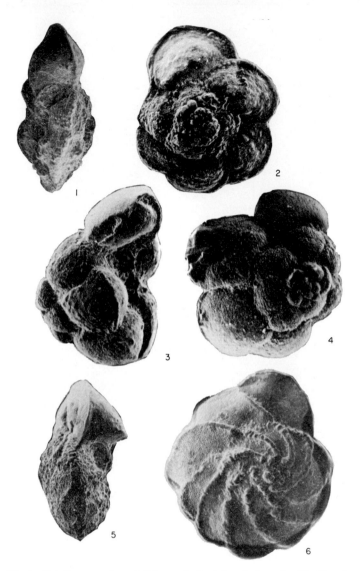

PLATE 31. 1. *Rotalipora cushmani* (Morrow). Peripheral view, (×64). Topotype. Low-spired. Greenhorn Limestone, Hartland Shale Member. Hodgeman Co., Kansas. 2. *Rotalipora cushmani* (Morrow). Dorsal view, (×64). Topotype. Low-spired. Greenhorn Limestone, Hartland Shale Member. Hodgeman Co., Kansas. 3. *Rotalipora cushmani* (Morrow). Peripheral view, (×64). Topotype. High-spired. Greenhorn Limestone, Hartland Shale Member. Hodgeman Co., Kansas. 4. *Rotalipora cushmani* (Morrow). Dorsal view, (×64). Topotype. High-spired. Greenhorn Limestone, Hartland Shale Member. Hodgeman Co., Kansas. 5. *Rotalipora greenhornensis* (Morrow). Peripheral view, (×64). Topotype. Greenhorn Limestone, Hartland Shale Member. Hodgeman Co., Kansas. 6. *Rotalipora greenhornensis* (Morrow). Dorsal view, (×64). Topotype. Greenhorn Limestone, Hartland Shale Member. Hodgeman Co., Kansas.

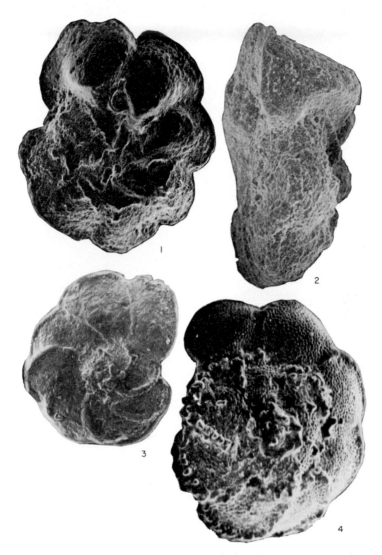

PLATE 32. 1. *Rotalipora deeckei* (Franke). Ventral view, (×96). Topotype of *R. gandolfii* Luterbacher and Premoli Silva. Scaglia Rossa. Balerna, Switzerland. 2. *Rotalipora deeckei* (Franke). Oblique peripheral view, (×128). Topotype of *R. gandolfii* Luterbacher and Premoli Silva. Scaglia Rossa. Balerna. Switzerland. 3. *Rotalipora deeckei* (Franke). Dorsal view, (×64). Topotype of *R. gandolfii* Luterbacher and Premoli Silva. Scaglia Rossa. Balerna, Switzerland. 4. *Rotalipora ticinensis* (Gandolfi). Dorsal view, (×128). Topotype of *R. klausi* Lehmann. Albian. Cotier Basin, Tarfaya, Morocco.

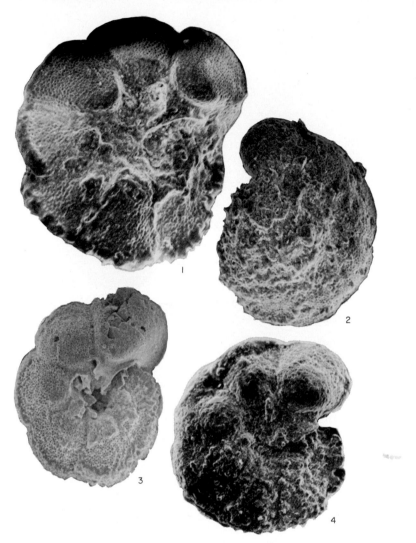

PLATE 33. 1. *Rotalipora ticinensis* (Gandolfi). Ventral view, (×128). Topotype of *R. klausi* Lehmann. Albian. Cotier Basin, Tarfaya, Morocco. 2. *Rotalipora ticinensis* (Gandolfi). Dorsal view, (×128). Topotype. Scaglia Bianca. Balerna, Switzerland. 3. *Rotalipora ticinensis* (Gandolfi). Ventral view, (×128). Main Street Formation. Johnson Co., Texas. 4. *Rotalipora ticinensis* (Gandolfi). Ventral view, (×128). Topotype. Dark area on the penultimate and antipenultimate chambers is adhesive. Scaglia Bianca. Balerna, Switzerland.

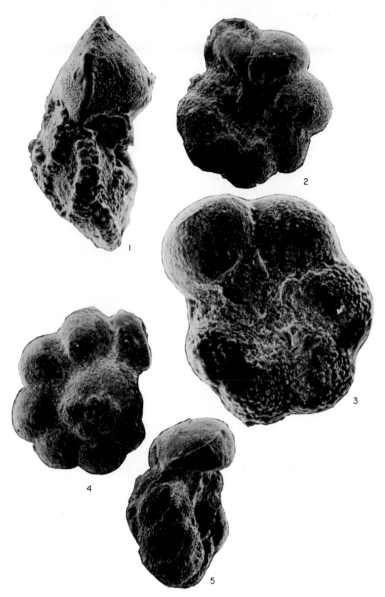

PLATE 34. 1. *Rotalipora ticinensis* (Gandolfi). Peripheral view, (×128). Topotype. Scaglia Bianca. Balerna, Switzerland. 2. *Rotalipora multiloculata* (Morrow). Ventral view, (×64). Topotype. Greenhorn Limestone, Hartland Shale Member. Hodgeman Co., Kansas. 3. *Rotalipora multiloculata* (Morrow). Ventral view, (×128). Topotype. Greenhorn Limestone, Hartland Shale Member. Hodgeman Co., Kansas. 4. *Rotalipora multiloculata* (Morrow). Dorsal view, (×64). Topotype. Greenhorn Limestone, Hartland Shale Member. Hodgeman Co., Kansas. 5. *Rotalipora multiloculata* (Morrow). Peripheral view, (×64). Topotype. Note the weakly developed keel. Greenhorn Limestone, Hartland Shale Member. Hodgeman Co., Kansas.

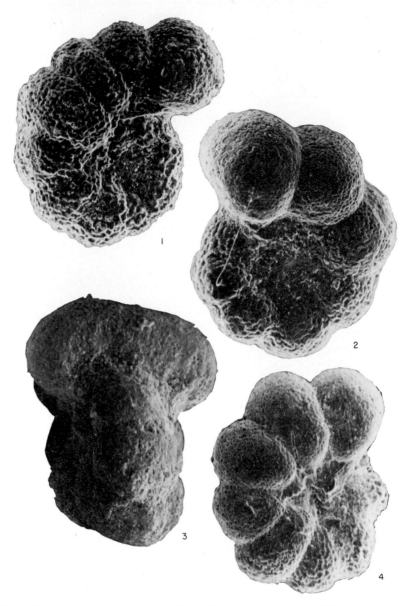

Plate 35. 1. *Biticinella breggiensis* (Gandolfi). Umbilical view, (×128). Topotype. Uncoiling. Scaglia Bianca. Balerna, Switzerland. 2. *Biticinella breggiensis* (Gandolfi). Umbilical view, (×128). Topotype. Scaglia Bianca. Balerna, Switzerland. 3. *Biticinella breggiensis* (Gandolfi). Oblique peripheral view, (×128). Note thickness of the ultimate chamber Scaglia Variegata. Balerna, Switzerland. 4. *Ticinella praeticinensis* Sigal. Ventral view, (×128). Scaglia Variegata. Balerna, Switzerland.

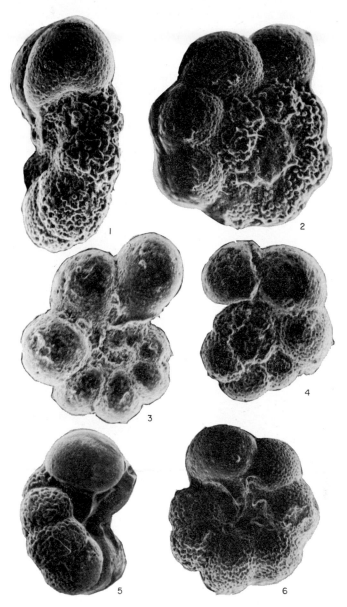

PLATE 36. 1. *Ticinella praeticinensis* Sigal. Peripheral view, (×128). Scaglia Variegata Balerna, Switzerland. 2. *Ticinella praeticinensis* Sigal. Dorsal view, (×128). Scaglia Varie, gata. Balerna, Switzerland. 3. *Ticinella primula* Luterbacher. Ventral view, (×128). Scaglia Bianca. Balerna, Switzerland. 4. *Ticinella primula* Luterbacher. Dorsal view, (×128). Scaglia Bianca. Balerna, Switzerland. 5. *Ticinella roberti* (Gandolfi). Oblique peripheral view, (×120). Scaglia Variegata. Balerna, Switzerland. 6. *Ticinella roberti* (Gandolfi). Ventral view, (×96). Scaglia Variegata. Balerna, Switzerland.

709

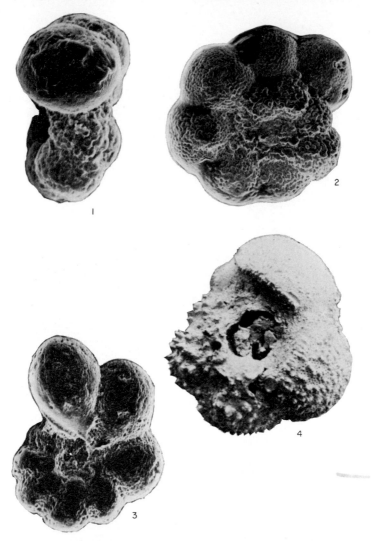

PLATE 37. 1. *Ticinella primula* Luterbacher. Peripheral view, (×128). Scaglia Bianca. Balerna, Switzerland. 2. *Ticinella roberti* (Gandolfi). Dorsal view, (×96). Scaglia Variegata. Balerna, Switzerland. 3. *Ticinella primula* Luterbacher. Ventral view, (×128). Scaglia Bianca. Balerna, Switzerland. 4. *Globotruncana aegyptiaca* Nakkady. Ventral view, (×120). Ripley Formation. Dallas Co., Alabama. USNM 184826.

PLATE 38. 1. *Globotruncana arca* (Cushman). Peripheral view, (× 120). Demopolis Chalk. Dallas Co., Alabama. USNM 184835. 2. *Globotruncana arca* (Cushman). Ventral view, (× 120). Demopolis Chalk. Dallas Co., Alabama. USNM 184836. 3. *Globotruncana calcarata* Cushman. Dorsal view, (× 120). Topotype. Pecan Gap Chalk. Collin Co., Texas. USNM 184840. 4. *Globotruncana arca* (Cushman). Dorsal view, (× 120). Demopolis Chalk. Dallas Co., Alabama. USNM 184834.

PLATE 39. 1. *Globotruncana calcarata* Cushman. Ventral view, (×120). Topotype. Tubulo-spines in both adapertural and abapertural positions on ultimate chamber. Pecan Gap Chalk. Collin Co., Texas. 2. *Globotruncana concavata* (Brotzen). Peripheral view, (×64). Hatcher Bluff Marl. Dallas Co., Texas. USNM 184847. 3. *Globotruncana concavata* (Brotzen). Dorsal view, (×120). Hatcher Bluff Marl. Dallas Co., Texas. USNM 184844.

PLATE 40. 1. *Globotruncana contusa* (Cushman). Peripheral view, (×128). Colon Shale. Km. 92, Hato Nuevo, Colombia. 2. *Globotruncana contusa* (Cushman). Ventral view, (×96). Colon Shale. Km. 92, Hato Nuevo, Colombia. 3. *Globotruncana contusa* (Cushman). Peripheral view, (×128). Colon Shale. Km. 92. Hato Nuevo, Colombia. 4. *Globotruncana contusa* (Cushman). Dorsal view, (×80). Prairie Bluff Chalk. Wilcox Co., Alabama. USNM 184849.

PLATE 41. 1. *Globotruncana coronata* Bolli. Dorsal view, (×64). Turonian. Edelbachgraben, Gosautal, Austria. 2. *Globotruncana coronata* Bolli. Ventral view, (×64). Turonian. Edelbachgraben, Gosautal, Austria. 3. *Globotruncana cretacea* (d'Orbigny). Ventral view, (×120). Secondary chamber built over breach in antipenultimate chamber. Hatcher Bluff Marl. Dallas Co., Alabama. USNM 184866. 4. *Globotruncana cretacea* (d'Orbigny). Dorsal view, (×120). Hatcher Bluff Marl. Dallas Co., Alabama. USNM 184862.

PLATE 42. 1. *Globotruncana cretacea* (d'Orbigny). Peripheral view, (×120). Hatcher Bluff Marl. Dallas Co., Alabama, 3. *Globotruncana elevata* (Brotzen). Dorsal view, (×120). Hatcher Bluff Marl. Dallas Co., Alabama. USNM 184950. 3. *Globotruncana elevata* (Brotzen). Peripheral view, (×120). Hatcher Bluff Marl. Dallas Co., Alabama. USNM 184951. 4. *Globotruncana elevata* (Brotzen). Ventral view, (×120). Hatcher Bluff Marl. Dallas Co., Alabama. USNM 184949.

PLATE 43. 1. *Globotruncana duwi* Nakkady. Dorsal view, (×120). Topotype of *G. skewesae* Brönnimann and Brown. Kemp Clay. Milam Co., Texas. 2. *Globotruncana duwi* Nakkady. Ventral view, (×120). Topotype of *G. skewesae* Brönnimann and Brown. Kemp Clay. Milam Co., Texas. 3. *Globotruncana duwi* Nakkady. Rear peripheral view, (×120). Topotype of *G. skewesae* Brönnimann and Brown. Kemp Clay. Milam Co., Texas. 4. *Globotruncana duwi* Nakkady. Oblique peripheral view, (×120). Topotype of *G. skewesae* Brönnimann and Brown. Kemp Clay. Milam Co., Texas. 5. *Globotruncana duwi* Nakkady. Ventral view, (×120). Topotype of *G. skewesae* Brönnimann and Brown. Kemp Clay. Milam Co., Texas.

PLATE 44. 1. *Globotruncana ellisi* Brönnimann and Brown. Dorsal view, (×120). Topotype. Kemp Clay. Williamson Co., Texas. 2. *Globotruncana ellisi* Brönnimann and Brown. Ventral view, (×120). Topotype. Kemp Clay. Williamson Co., Texas. 3. *Globotruncana ellisi* Brönnimann and Brown. Rear peripheral view, (×120). Topotype. Kemp Clay. Williamson Co., Texas. 4. *Globotruncana fornicata* Plummer. Dorsal view, (×120). Neanic individual. Hatcher Bluff Marl. Dallas Co., Alabama. USNM 184879. 5. *Globotruncana fornicata* Plummer. Peripheral view, (×120). Hatcher Bluff Marl. Dallas Co., Alabama. USNM 184878. 6. *Globotruncana fornicata* Plummer. Ventral view, (×120). Hatcher Bluff Marl. Dallas Co., Alabama. USNM 184880.

717

PLATE 45. 1. *Globotruncana gansseri* Bolli. Ventral view, (×120). Ripley Formation. Wilcox Co., Alabama. USNM 184887. 2. *Globotruncana gansseri* Bolli. Peripheral view, (×120). Ripley Formation. Wilcox Co., Alabama. USNM 184888. 3. *Globotruncana gansseri* Bolli, Dorsal view, (×120). Ripley Formation. Wilcox Co., Alabama. USNM 184886. 4. *Globotruncana havanensis* Voorwijk. Ventral view, (×120). Ripley Formation. Dallas Co., Alabama. USNM 184892. 5. *Globotruncana havanensis* Voorwijk. Peripheral view, (×120). Demopolis Chalk. Dallas Co., Alabama. USNM 184891. 6. *Globotruncana havanensis* Voorwijk. Oblique peripheral view, (×192). Topotype of *G. petaloidea* Gandolfi. Colon Shale. Km. 92, Hato Nuevo, Colombia.

PLATE 46. 1. *Globotruncana helvetica* Bolli. Ventral view, (×64). Ideotype of *G. carpathica* Scheibnerová. Turonian. Tunisia. 2. *Globotruncana helvetica* Bolli. Peripheral view, (×64). Ideotype of *G. carpathica* Scheibnerová. Turonian. Tunisia. 3. *Globotruncana linneiana* (d'Orbigny). Peripheral view, (×120). Hatcher Bluff Marl. Dallas Co., Alabama. USNM 184896. 4. *Globotruncana lapparenti* Brotzen. Ventral view, (×120). Arcola Limestone. Dallas Co., Alabama. USNM 184895. 5. *Globotruncana linneiana* (d'Orbigny). Dorsal view, (×120). Topotype of *G. canaliculata* (Reuss). Turonian. Edelbachgraben, Gosautal, Austria. USNM 184902. 6. *Globotruncana linneiana* (d'Orbigny). Ventral view, (×120). Hatcher Bluff Marl. Dallas Co., Alabama. USNM 184899.

719

PLATE 47. 1. *Globotruncana lunaris* Masters. Dorsal view, (×120). Holotype. Hatcher Bluff Marl. Dallas Co., Alabama. USNM 184905. 2. *Globotruncana lunaris* Masters. Peripheral view, (×120). Paratype. Hatcher Bluff Marl. Dallas Co., Alabama. USNM 184904. 3. *Globotruncana marginata* (Reuss). Peripheral view, (×120). Hatcher Bluff Marl. Dallas Co., Alabama. USNM 184909. 4. *Globotruncana marginata* (Reuss). Ventral view, (×120). Hatcher Bluff Marl. Dallas Co., Alabama. USNM 184920.

720

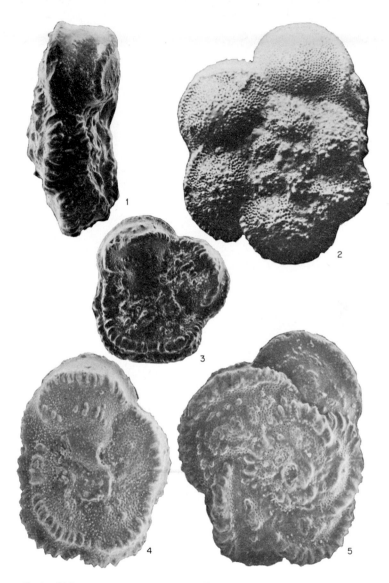

PLATE 48. 1. *Globotruncana mayaroensis* Bolli. Peripheral view, (×96). Maastrichtian. Habana, Cuba. 2. *Globotruncana marginata* (Reuss). Dorsal view, (×120). Hatcher Bluff Marl. Dallas Co., Alabama. 3. *Globotruncana mayaroensis* Bolli. Ventral view, (×64). Maastrichtian. Habana, Cuba. 4. *Globotruncana mayaroensis* Bolli. Ventral view, (×96). Maastrichtian. Habana, Cuba. 5. *Globotruncana mayaroensis* Bolli. Dorsal view, (×128). Maastrichtian. Habana, Cuba.

PLATE 49. 1. *Globotruncana nothi* Brönnimann and Brown. Ventral view, (×120). Demopolis Chalk. Dallas Co., Alabama. USNM 184924. 2. *Globotruncana nothi* Brönnimann and Brown. Dorsal view, (×120). Demopolis Chalk. Dallas Co., Alabama. USNM 184926. 3. *Globotruncana nothi* Brönnimann and Brown. Peripheral view, (×120). Demopolis Chalk. Dallas Co., Alabama. USNM 184925. 4. *Globotruncana obliqua* Herm. Peripheral view, (×64). Ripley Formation. Dallas Co., Alabama. USNM 184928. 5. *Globotruncana obliqua* Herm. Ventral view, (×64). Demopolis Chalk. Dallas Co., Alabama. USNM 184930. 6. *Globotruncana obliqua* Herm. Dorsal view, (×120). Ripley Formation. Dallas Co., Alabama. USNM 184927.

PLATE 50. 1. *Globotruncana pustulifera* Masters. Peripheral view, (×120). Paratype. Demopolis Chalk. Dallas Co., Alabama. USNM 184932. 2. *Globotruncana renzi* Gandolfi. Peripheral view, (×96). Topotype of *G. tarfayaensis* Lehmann. Coniacian. Wadi Zehar, Tarfaya, Morocco. 3. *Globotruncana pustulifera* Masters. Dorsal view, (×120). Paratype. Demopolis Chalk. Dallas Co., Alabama. USNM 184933.

PLATE 51. 1. *Globotruncana renzi* Gandolfi. Dorsal view, (×120). Demopolis Chalk. Dallas Co., Alabama. USNM 184829. 2. *Globotruncana rosetta* (Carsey). Peripheral view, (×120). Demopolis Chalk. Dallas Co., Alabama. USNM 184940. 3. *Globotruncana renzi* Gandolfi. Ventral view, (×120). Demopolis Chalk. Dallas Co., Alabama. USNM 184830.

PLATE 52. 1. *Globotruncana rosetta* (Carsey). Ventral view, (×120). Topotype. Upper Taylor Marl. Travis Co., Texas. USNM 184938. 2. *Globotruncana sigali* Reichel. Dorsal view, (×64). Topotype of *G. undulata* Lehmann. Santonian. Falaise, Tarfaya, Morocco. 3. *Globotruncana sigali* Reichel. Ventral view, (×64). Topotype of *G. undulata* Lehmann. Santonian. Falaise, Tarfaya, Morocco.

PLATE 53. 1. *Globotruncana sigali* Reichel. Oblique peripheral view, (×64). Topotype of *G. undulata* Lehmann. Santonian. Falaise, Tarfaya, Morocco. 2. *Globotruncana stuarti* de Lapparent). Peripheral view, (×120). Prairie Bluff Chalk. Wilcox Co., Alabama. USNM 184947. 3. *Globotruncana stuarti* (de Lapparent). Ventral view, (×120). Prairie Bluff Chalk. Wilcox Co., Alabama. USNM 184946.

PLATE 54. 1. *Globotruncana stuarti* (de Lapparent). Dorsal view, (×120). Prairie Bluff Chalk. Wilcox Co., Alabama. USNM 184945. 2. *Globotruncana trinidadensis* Gandolfi. Dorsal view, (×120). Ripley Formation. Wilcox Co., Alabama. USNM 184957. 3. *Globotruncanci trinidadensis* Gandolfi. Ventral view, (×120). Prairie Bluff Chalk. Wilcox Co., Alabama. USNM 184956.

PLATE 55. 1. *Globotruncana ventricosa* White. Peripheral view, (×120). Demopolis Chalk. Dallas Co., Alabama. USNM 184960. 2. *Globotruncana ventricosa* White. Rear peripheral view, (×120). Hatcher Bluff Marl. Dallas Co., Alabama. USNM 184961. 3. *Globotruncana ventricosa* White. Ventral view, (×120). Hatcher Bluff Marl. Dallas Co., Alabama. USNM 184962.

PLATE 56. 1. *Rugoglobigerina rugosa* (Plummer). Rear peripheral view, (×176). Topotype. Note costellae pattern, Kemp Clay. Milam Co., Texas. USNM 184983. 2. *Rugoglobigerina hexacamerata* Brönnimann. Dorsal view, (×176). Prairie Bluff Chalk. Wilcox Co., Alabama. USNM 184969. 3. *Rugoglobigerina hexacamerata* Brönnimann. Ventral view, (×128). Main Street Formation. Johnson Co., Texas, 4. *Rugoglobigerina rugosa* (Plummer). Natural section, (×128). Kemp Clay. Milam Co., Texas. 5. *Rugoglobigerina rugosa* (Plummer). Ventral view, (×176). Prairie Cluff Chalk. Wilcox Co., Alabama. USNM 184986.

729

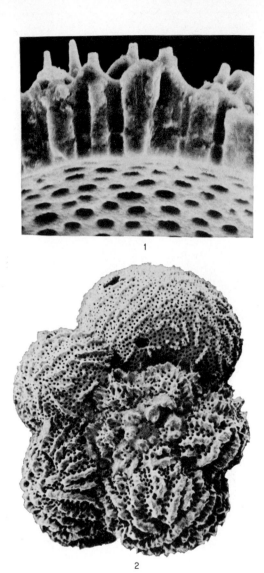

1

2

PLATE 57. 1. *Rugoglobigerina rugosa* (Plummer). Natural section, (×1280). Kemp Clay. Milan Co., Texas. 2. *Rugoglobigerina rugosa* (Plummer). Dorsal view, (×258). Note decreasing coarseness of costellae in direction of growth. Prairie Bluff Chalk. Wilcox Co., Texas. USNM 184985.

730

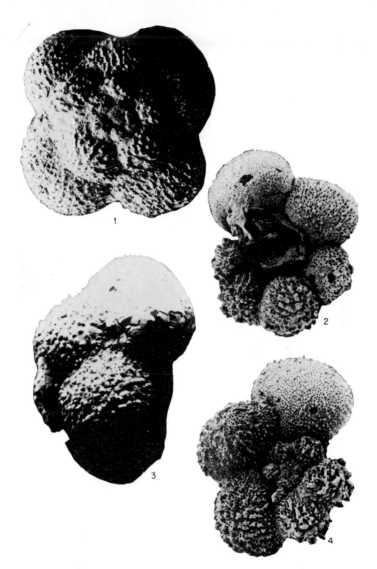

PLATE 58. 1. *Rugoglobigerina tradinghousensis* Pessagno. Dorsal view, (×176). Hatcher Bluff Marl, Dallas Co., Alabama. USNM 184989. 2. *Trinitella scotti* Brönnimann. Ventral view, (×120). Prairie Bluff Chalk. Wilcox Co., Alabama. USNM 184966. 3. *Rugoglobigerina tradinghousensis* Pessagno. Peripheral view, (×176). Hatcher Bluff Marl. Dallas Co., Alabama. USNM 184990. 4. *Trinitella scotti* Brönnimann. Dorsal view, (×120). Prairie Bluff Chalk. Wilcox Co., Alabama. USNM 184963.

6. A Zoogeographic and Taxonomic Review of Euthecosomatous Pteropoda

ALLAN W. H. BÉ[1] and RONALD W. GILMER[2]

[1]*Lamont-Doherty Geological Observatory* *of Columbia University, Palisades,* *New York 10964, U.S.A.*

[2]*Department of Zoology,* *University of California,* *Davis,* *California 95616, U.S.A.*

Euthecosomatous pteropods are integral members of the marine ecosystem, but their biology and ecology are not well understood. Some aspects of their feeding and swimming behaviour, natural diet, reproduction and shell microstructure are discussed in this review paper. A classification of twenty-eight common euthecosomatous species is presented and illustrated with scanning electron micrographs.

The majority of euthecosomatous species (21) inhabit the circumglobal warm-water belt. Only four species live in subantarctic–antarctic waters, of which three are also encountered in arctic–subarctic regions. Their distribution patterns are discussed and portrayed in zoogeographic maps.

Outline

Introduction

To most students of oceanology, the planktonic gastropods we shall discuss are known as "Pteropoda". To specialists in marine plankton they are more properly referred to as members of the Euthecosomata. To laymen they are described as "sea-butterflies", because the rapid movements of their wing-like fins aptly resemble their fluttering namesakes (Plate 1).

The first reference about these widely distributed molluscs appeared in 1676 when F. Martens described and illustrated an organism now known as *Clione limacina*, but it was not until 1804 that the eminent Georges Cuvier

733

established the order (later raised to class) of Pteropoda within the Mollusca. Between 1800 and 1840 many new pteropod species were described by Rang, d'Orbigny and Lesueur.

For most of the nineteenth century, zoologists could not agree on the taxonomic status of the Pteropoda. One school, led by Cuvier (1817) and Gegenbaur (1855), considered them as a distinct class of Mollusca equivalent in rank with the Pelecypoda, Gastropoda, Scaphopoda, and Cephalopoda. Another school believed that the Pteropoda should be placed within either the Cephalopoda or Gastropoda. Among the proponents that pteropods should be assigned to the cephalopods were Oken (1815), Escholtz (1829), Huxley (1853) and Lankester (1885).

De Blainville (1824) was the first worker to recognise that pteropods were gastropods related to the Opisthobranchia. He described in detail the anatomical differences and similarities between shelled and naked (non-shelled) pteropods, upon which he based the subdivisions of Thecosomata and Gymnosomata. However, he introduced many nomenclatural errors in his classification of molluscs, which caused much confusion among his contemporaries and subsequent zoologists. De Blainville's correct view that pteropods were indeed opisthobranch gastropods was not confirmed until some six decades later by Boas (1886) and Pelseneer (1888a, b).

The voyage of the HMS *Challenger* in 1872 ushered in the era of oceanographic explorations. Expeditions whose aim was scientific discovery were launched in rapid succession and among their many scientific fruits was the accumulation of large collections of planktonic animals and plants from all regions of the oceans. Subsequent expeditions which were especially notable for their plankton investigation were the extensive cruises of the following research vessels: *National* (1889), *Ingolf* (1895–1896), *Valdivia* (1898–1899), *Albatross* (1899–1900), *Siboga* (1899–1900), *Michael Sars* (1910), *Dana* (1921–1935), *Carnegie* (1928–1929), and *John Murray* (1933–1934).

The comparative anatomy, taxonomy and biogeography of marine plankton were intensively studied by leading biologists in Europe in the two decades before and after the turn of the 19th century. During this period the foundations of our present knowledge of the pteropods were laid by the works of Boas (1886), Pelseneer (1888a, b), Meisenheimer (1905, 1906a, b), Schiemenz (1906), Bonnevie (1913), Tesch (1904, 1913) and Vayssière (1915). More recent investigations are those by Tesch (1946, 1948), Massy (1932), Pruvot-Fol (1954), van der Spoel (1967, etc.), Lalli (1970, 1972), and Rampal (1973a). First-hand observations of their feeding and swimming behaviour, natural diet, buoyancy regulation, and oxygen consumption were made recently by Gilmer (1974), while scuba diving in the Florida Current as well as from laboratory experiments. The reader is referred to these above-mentioned studies for more detailed accounts of the biology of pteropods than we shall present here.

Some remarks on the biology and ecology of euthecosomatous pteropods

The wings of pteropods are uniquely adapted for enabling the animal to exist as a planktonic form. These wings are homologous to the gastropod foot and serve a dual function for swimming and feeding (Plate 1). By flapping its highly muscular wings in a dorsal-ventral direction, the animal can move about in the water column; during periods of feeding the ciliated tracts on the posterior portions of the wings provide currents for drawing particulate food into the mouth.

The Thecosomata are the only known opisthobranch gastropods possessing a ciliary mucus feeding habit. Euthecosomes feed by ingesting strings of mucus which contain food particles comprised of micro-organisms and organic matter. The mechanism of this unique ciliary feeding process was first described by Morton (1954b) for *Limacina retroversa.* In this organism water containing the potential food items is drawn through the mantle cavity by ciliary action onto mucus secreted inside the cavity by the pallial gland. The resulting strings of mucus and food pass through the exhalent mantle aperture near the buccal region of the foot and are carried by cilia to the mouth. Swimming motions are possible during feeding as the wings do not interfere with the transport of food from the mantle cavity, and as a result there is correspondingly little modification in the family Limacinidae towards the development of flotation mechanisms since members can feed simultaneously as they swim.

Gilmer (1974) has described a mucus-ciliary feeding mechanism for the family Cavoliniidae that was heretofore unknown. The process is similar to that described by Morton (1954b) for *Limacina,* though a more complex pathway for the movement of food to the mouth is involved. Yonge's notion (1926) that food collection in the mantle cavity is lost in higher thecosome families is incorrect (Gilmer, 1974). The Cavoliniidae have undergone secondary symmetry in which the mantle cavity has been rotated 180° from its position as it occurs in the family Limacinidae. Consequently, the mouth is now separated from the mantle cavity apertures by the greater expanse of the wings (see Plate 1). Food is collected on the pallial gland inside the mantle cavity, but must now move around or over the wings in order to enter the mouth. This action is accomplished by strong ciliary currents on the wings which draw the food strings towards the mouth from the region of the exhalent mantle aperture (Plate 1c). This mode of feeding requires the absence of swimming motions since they would obviously interfere with the transport of food strings around the wings. Consequently, an increase in gelatinous tissue characterises the family Cavoliniidae and provides these forms with an increased, and in some instances, neutral buoyancy that eliminates the need for continuous locomotory activity.

Through field observations by scuba diving, Gilmer (1974) has demonstrated that several of the genera in the Cavoliniidae have become adept at buoyancy regulation by development of various body appendages. All species of *Diacria* and *Cavolinia* possess large mantle appendages which emerge through the lateral clefts of their shells (Plate 1a, b, c). Ciliary action on these appendages produces currents which probably cleanse the mantle cavity from excess food particles. Gilmer also believes that the mantle appendages play a role in buoyancy control via ionic gradients within the mantle (Denton, 1964). In addition, *Cavolinia gibbosa*, *C. uncinata*, and *C. inflexa* possess a neutrally buoyant "pseudoconch"—a transparent, gelatinous structure secreted by the external mantle lobes of the dorsal shell surface (Plate 1c). It is conjectured that the pseudoconch also aids in suspending the thecosome body during feeding as well as at other times when the wings must remain stationary. Since *Creseis*, *Hyalocylis*, and *Styliola* lack such mantle appendages and pseudoconchs, their buoyancy regulation does not appear to be as efficient as that of *Cavolinia* or *Diacria* and thus the former sink slowly during feeding (Gilmer, 1974). It is not known whether the genera *Clio* and *Cuvierina* possess these gelatinous structures or not.

According to Boas (1886), Pelseneer (1888a, b), Morton (1954b), and Paranjape (1968), euthecosomatous pteropods feed on dinoflagellates, diatoms, coccolithophorids, radiolaria, silicoflagellates, foraminifera, tintinnids and infusoria. Gilmer (1974) has determined the size frequency distributions of food particles in mucus food strings of *Creseis acicula*, *Cavolinia uncinata*, and *C. tridentata* from the Florida Current. The food particles fall into three categories: 1, whole organisms (phytoplankton and protozoa cells greater than 5 μm and multicellular larvae); 2, non-living matter, including organic aggregates and detrital fragments from unicellular and multicellular organisms; and 3, unidentified particles smaller than about 5 μm. For all three species, particles less than 5 μm account for the largest percentage by number; 40 to 49% of the particles are less than 20 μm; 82 to 95% of all particles are less than 80 μm in size. With the exception of the maximum size of the particle ingested, Gilmer found no obvious adaptations for preferential size sorting of food prior to ingestion in these species. In both *Cavolinia* and *Creseis* the upper limit is approximately 200 μm and corresponds to the dimensions of the ciliated tract along which food must pass to enter the mouth.

In contrast to the predominantly herbivorous and detrital diet of the Euthecosomata, the shell-less pteropods belonging to the Gymnosomata are active carnivores and often prey upon thecosomes and organisms larger than protozoa. For example, the diets of the gymnosomes *Paedoclione doliiformis* and *Clione limacina* consist largely of the thecosomes *Limacina helicina* and *L. retroversa* (Lalli, 1970, 1972). Gymnosomes as well as some carnivorous bathypelagic thecosomes (e.g. *Limacina helicoides*) usually occur as solitary individuals, in contrast to the swarming behaviour of many herbivorous euthecosomatous species living near the surface (Bonnevie, 1913; Tesch, 1946, 1948).

The Thecosomata are prey for a variety of animals other than the gymnosomatous pteropods mentioned above. Lebour (1932) found that *L. retroversa* provides food for herring, *Sagitta* (chaetognatha) and probably mackerel. LeBrasseur (1966) recorded *L. helicina* as a constituent in the diet of salmon, particularly *Oncorhynchus keta* from the northeastern Pacific. Meisenheimer (1905) reported that *L. helicina* along with other thecosomes probably served as food for whales and sea birds. Dunbar (1942) even found several specimens of *L. helicina* in a ringed seal. Russell (1960) observed that four species of *Cavolinia* contributed to the food source of tuna species in the North Atlantic. Okutani (1960) recorded, in one instance, that the cephalopod *Argonauta boettgeri* preys on *Cavolinia tridentata*.

The shell of the Euthecosomata serves as a refuge from predators and unfavourable environments. The animal can withdraw into the shell quickly by contraction of its large retractor muscle. This action causes the animal to sink at a rate dependent upon the size and shape of the shell. But, for all its advantages the shell must also be a liability, because its relatively high density creates a negative buoyancy. Since perpetual swimming to counteract this negative force would require much energy, it is apparent that adaptations towards neutral buoyancy would minimise the energy necessary for maintaining the shell in the water column.

The Euthecosomata can avoid predation by either withdrawing into the shell and sinking or by out-swimming the attacker. Gilmer (1974) made *in situ* observations of swimming rates for five euthecosome species using a meter stick and measuring the distance travelled between two points over a timed interval. *Creseis acicula* averaged 12 cm/s; *Cavolinia longirostris*, 11 cm/s; *C. uncinata* and *C. tridentata* each averaged 14 cm/s; while the smaller form *Diacria quadridentata* averaged 7 cm/s. These values are significantly greater than those calculated from observing specimens in a tank (Kornicker, 1959). The differences indicate the difficulty of studying delicate oceanic organisms in the unnatural conditions of the laboratory.

The ability of pteropods to sink and swim rapidly may explain why they can perform great feats of vertical movements. Many species exhibit diurnal vertical migration, ascending at sundown and remaining at shallow depths during the night, and descending during the day-time over a depth range of several hundred meters.

Euthecosome species are protandric hermaphrodites (Bonnevie, 1916; Hsiao, 1939a, b; Morton, 1954a; van der Spoel, 1967). Male and female sex organs are carried in the same individual. The younger individuals tend to have only spermatocytes (sperm-producing tissue) and as growth continues oocytes (egg-producing tissue) develop producing a true hermaphroditic stage in which cells of both sex types are present. In most species, the older individuals no longer contain spermatocytes and function only as females (Morton, 1954a; van der Spoel, 1967).

It is not known to what extent either self-fertilisation or cross-fertilisation occurs among euthecosomes. In the reproduction process the sperm are stored in an accessory sac of the gonad referred to as the *bursa seminis*. If copulation occurs, sperm received from a mating partner are deposited in another gonad sac, the *receptaculum seminis*. When the oocytes become functional as eggs at a later stage, they are fertilised by the stored sperm in either or both of the accessory gonad sacs and the zygotes are released into the water as floating egg masses. After a period of development the egg masses break apart as the free-swimming veliger stage is reached. Hermaphroditism in these molluscs may be a solution to the general problem that sexually reproducing plankton encounter in the open ocean, where there is great risk of never encountering a mate.

The brood protection in *Limacina inflata* (Lalli and Wells, 1973) is an unusual development in pteropods. Adults of this species brood their embryos and early veligers in their mantle cavities and when the veligers attain a size of about 67 μm in shell diameter they are released. Some bathypelagic species such as *Limacina helicoides*, *Clio balantium* and *Clio chaptalii*, are viviparous and do not have spawn and free-swimming larval stages (Bonnevie, 1913; Tesch, 1946, 1948).

The protandric nature of pteropods and its relation to growth and seasons of spawning is not well-known for the majority of species. Redfield (1939) and Hsiao (1939a, b) investigated the reproductive history of a population of *Limacina retroversa* from the Gulf of Maine. They found that seasonality appears to affect growth and the rate of sexual succession between the male and female stages. Juveniles produced in April developed at faster rates than juveniles born in December, presumably due to the different environmental conditions encountered. The April juveniles developed both types of sex cells simultaneously while December breeding populations produced young which showed a gradual procession from a pure male stage to a true hermaphroditic stage in which both eggs and sperm were produced. April juveniles also initiated sexual development at an earlier stage (shell diameter 0.6 mm) as opposed to the December brood (shell diameter 0.8 mm) (Hsiao, 1939b). Redfield (1939) and Hsiao (1939a, b) also found that older individuals in which no further succession of the gonad was apparent, were produced from a juvenile population observed five months earlier and thus they concluded that optimum size and sexual maturity were reached in less than a year. Lebour (1933) came to similar conclusions for *L. retroversa* from the English Channel. Morton (1954a) found that *L. bulimoides* from the Benguela Current had a single peak in their breeding cycle with no evidence for a restitution of the gonad to the pure male phase once individuals had spent their oocytes.

Shell size and microstructure

The shells of species belonging to the family Limacinidae are generally smaller than those of the family Cavoliniidae. The former are normally less than 2.5 mm in length, except for the bathypelagic *Limacina helicoides* whose shell may reach 15 mm in diameter (Plate 4, figs 6a–c). A distinguishing feature of all *Limacina* species is the left-handed coiling direction of their shells. The shell aperture can be closed by an operculum, but this structure is commonly lost in adult specimens.

Most specimens of the family Cavoliniidae collected in plankton tows or bottom sediments have shells that are usually smaller than 10 mm in length. However, *Creseis acicula*, and many species of *Clio* can attain a maximum length close to 20 mm. The largest euthecosomatous species. *Clio balantium*, has a shell length of up to 30 mm (Plate 6, figs 18a–d).

Unlike the spirally coiled shell of the Limacinidae, the greatest majority of the cavoliniid shells are bilaterally symmetrical in the adult stages. An exception is *Styliola subula*, whose shell reflects a small degree of torsion. The protoconchs of the Cavoliniidae have an initially radial or biradial symmetry, and this symmetry is maintained in the adults of species of *Creseis*. However, a clear differentiation into ventral and dorsal sides is developed in the adult stages of *Clio*, *Cuvierina*, *Diacria* and *Cavolinia*. All species of the Cavoliniidae lack an operculum.

Surface ornamentation in the form of longitudinal and transverse striations and fine ridges can be used for intraspecific differentiation among species of the Limacinidae as well as the Cavoliniidae, but it is customary to use other gross morphological features for species identifications. Some euthecosomes have pigments in the periostracum and shells. *Limacina helicoides* has a characteristically chestnut brown shell and other species are also known to have shell pigmentation. This colouration is different from the pigmentation in the soft parts of some pteropod species collected at great depths. Hardy (1965, p. 147) observed that the brown soft parts of *Diacria trispinosa* and dark-violet tissues of *Clio polita* were visible through the transparent shells of specimens collected from deep waters off Scotland.

The shells of the Limacinidae and Cavoliniidae are constructed of aragonite, but their microstructures are strikingly different. The Limacinidae, with the possible exception of *Limacina inflata*, possess a crossed-lamellar shell microstructure (Plate 10, figs 1 and 2). In contrast, the Cavoliniidae have a helical microstructure that is unlike that of any other molluscan group (Plate 10, figs 3–7; Bé *et al.*, 1972). The crossed-lamellar and helical microstructures consist of first-order elongated rods, which in turn are made up of second-order blocks with approximate dimensions of 0.2 μm × 0.2 μm × 0.4 μm (Plate 10, fig. 7).

In *Cuvierina columnella* and other Cavoliniidae, the helical aragonite rods coil clockwise (as the spiral moves away from the observer) and their central axes are normal to the growth surface. Within *C. columnella's* average wall thickness of about 40 μm, the helix spiral makes four turns on the average; the helix radius at the outer shell surface is only 1.6 μm and increases to about

TABLE 1. Average density of euthecosomatous species in the Caribbean Sea collected by nets of 3 mesh sizes, expressed in number of individuals per thousand cubic metres of water. Relative abundances for the species are expressed in percentages in the three right-hand columns (data from Wells, 1973).

| Species | Specimens/1000 m³; | | | Rel. abundance (%) | | |
| | Net mesh size (No.) | | | Net mesh size (No.) | | |
	0	6	20	0	6	20
Cavolinia inflexa (Rang)	6.1	7.6	6.9	2.9	1.2	0.4
Cavolinia longirostris (Lesueur)	3.2	7.1	3.1	1.5	1.2	0.2
Clio pyramidata (Linné)	0.7	0.7	3.6	0.4	0.1	0.2
Creseis acicula Rang	4.4	6.6	16.6	2.1	1.1	0.9
Creseis virgula conica (Esch.)	48.1	188.2	384.5	23.0	31.0	21.7
Creseis virgula virgula Rang	6.1	16.9	34.8	2.9	2.8	2.0
Diacria quadridentata (Lesueur)	0.4	—	—	0.2	—	—
Diacria trispinosa (Lesueur)	5.1	8.3	12.5	2.4	1.4	0.7
Hyalocylis striata (Rang)	0.6	3.8	0.4	0.3	0.6	trace
Limacina bulimoides (d'Orbigny)	8.7	29.3	92.3	4.2	4.8	5.2
Limacina inflata (d'Orbigny)	103.6	257.3	1066.4	49.6	42.3	60.3
Limacina lesueuri (d'Orbigny)	2.4	2.4	12.0	1.2	0.4	0.7
Limacina trochiformis (d'Orbigny)	14.9	61.2	128.6	7.1	10.1	7.3
Styliola subula (Quoy and Gaimard)	4.3	16.9	6.0	2.1	2.8	0.3
Styliola n.sp.	0.2	1.5	0.8	0.1	0.2	0.1
	208.8	607.8	1768.5	100.0	100.0	100.0

14 μm towards the inner shell surface. The helical rods are nested in such a manner as to give omnidirectional continuity and flexibility, as well as maximum strength to the relatively thin, fragile shell. This is a decided advantage for an organism with a planktonic life-style.

The phylogenetic significance of these two contrasting microstructures has been considered by Bé *et al.* (1972) and Rampal (1973b). According to the former authors, the helical microstructure of the Cavoliniidae may indicate that they are evolutionary neomorphs (derived from ancestors with reduced or no shells) which have regained the ability to construct an exoskeleton on a new architectural plan. The Limacinidae, having a crossed-lamellar microstructure that is basically similar to other molluscan shell structures, is according to this viewpoint a more primitive group than the Cavoliniidae.

Inadequacies of field sampling

Apart from biological variability which is caused by natural vicissitudes, the density and species composition of a catch in a plankton net can be artificially influenced by several factors in the sampling method: i.e. mouth size of net, mesh size of net, towing speed and duration, and the reliability of the flowmeter in measuring the volume of water filtered (UNESCO, 1968).

McGowan and Fraundorf (1966) have shown that the species diversity of seven zooplankton groups (Thecosomata, Gymnosomata, Heteropoda, Nudibranchiata, Cephalopoda larvae, Euphausiacea and fish larvae) increases with an increase in the mouth size of the net. After a comparison of the effectiveness of nets ranging in mouth diameter from 20 to 140 cm, they concluded that the largest diameter net was the most efficient for sampling pteropods and other zooplankters. The lower diversity in the smaller nets was attributed to net avoidance by these motile organisms.

The mesh size of the net also has a strong influence upon the density and species composition of the plankton collected. Tesch (1946, 1948) noted the general under-representation of the small *Limacina* species in the Dana collections from the Atlantic and Indo-Pacific Oceans. Since many *Limacina* specimens are less than 1 mm in diameter, they would escape through the stramin S-nets (with 2 mm mesh aperture) used during the "Dana" cruises. Even hauls with P-nets (with 0.5 mm mesh aperture) failed to catch the small *Limacina* species in sufficient numbers. The fact that Tesch's (1946, 1948) distributional studies were primarily of the larger-sized family Cavoliniidae and family Peraclididae, indicates that coarse nets ($>$ 2 mm mesh-aperture) have a collecting bias in favour of the Cavoliniidae and against the Limacinidae. Finer nets ($<$ 300 μm mesh aperture) catch more of the Limacinidae, while their relatively slow towing speed allows the larger Cavoliniidae to avoid capture.

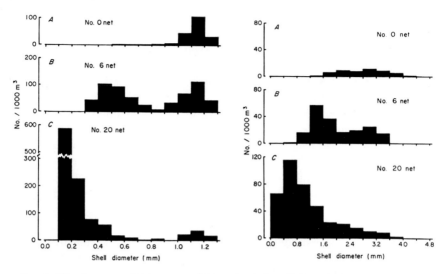

FIG. 1. Size frequency histograms of *Limacina inflata* (left) and *Creseis virgula conica* (right) collected in the Caribbean Sea by one-meter No. 0, 6, and 20 nets (Wells, 1973).

Wells (1973) has recently demonstrated the effects of different mesh sizes on estimates of the density of euthecosomatous pteropods in the Caribbean Sea off Barbados. Using three one-meter diameter nets with mesh sizes of no. 0 (569 μm aperture), No. 6 (239 μm) and No. 20 (76 μm), his data reveal some interesting trends from which some general conclusions may be drawn (Table 1). First, the density of total pteropods collected was inversely correlated with the mesh size. Second, of the fourteen species caught by all three nets, the seven largest species (*C. pyramidata, D. quadridentata, D. trispinosa, H. striata, S. subula, C. longirostris* and *C. inflexa*) were collected in low but uniform abundance. Third, the abundance of the seven smaller species increased as the mesh size decreased.

Size-frequency histograms for the two small and commonest species, *Limacina inflata* and *Creseis virgula conica* (Fig. 1), show that their densities were significantly higher in the fine-meshed nets and that the smallest individuals were not sampled by the coarse, No. 0 net. The inverse relationship between mesh size and the abundance of larger *L. inflata* individuals can possibly be caused by either net clogging or net avoidance in the No. 20 net collections. Wells' demonstration that the species composition and density of tropical pteropods can be strongly affected by the net's mesh size points to the difficulty of making quantitative comparisons of previous pteropod studies that have been obtained under a myriad of different sampling techniques.

TABLE 2. Historical development of the classification of Thecosomata (from Chen and Bé, 1964).

	Pelseneer (1888a)	Meisenheimer (1905)	Tesch (1913, 1946)	Morton (1958) McGowan (1960)
Order	Pteropoda	Pteropoda	Pteropoda	Thecosomata
Sub-order	Thecosomata	Thecosomata A. Euthecoso- mata	Thecosomata Tribe A. Euthecosomata	Euthecosomata
Family	Limacinidae	Limacinidae	Limacinidae	Limacinidae
Genera	*Limacina* *Peraclis*	*Limacina* *Peraclis* *Procymbulia*	*Limacina*	*Limacina*
Family	Cavoliniidae	Cavoliniidae	Cavoliniidae	Cavoliniidae
Genera	*Cavolinia* *Cuvierina* *Clio*	*Cavolinia* *Cuvierina* *Clio* *Creseis*	*Cavolinia* *Cuvierina* *Euclio* *Creseis*	*Cavolinia* *Cuvierina* *Clio* *Creseis*
	Subgenera *Creseis* *Styliola* *Hyalocylis* *Clio* s. str.	*Styliola* *Hyalocylis* *Diacria*	*Styliola* *Hyalocylis* *Diacria*	*Styliola* *Hyalocylis* *Diacria*
		B. Pseudo- thecosomata	Tribe B. Pseudo- thecosomata	Pseudothecoso- mata (Suborder)
Family	Cymbuliidae	Cymbuliidae	Cymbuliidae	
Genera	*Cymbulia* *Gleba* *Cymbuliopsis*	*Cymbulia* *Gleba* *Corolla*	*Cymbulia* *Gleba* *Corolla*	
Family Genus		Desmopteridae *Desmopterus*	Desmopteridae *Desmopterus*	
Family Genus			Peraclididae *Peraclis*	
Family Genus			Procymbuliidae *Procymbulia*	
Sub-order	Gymnosomata	C. Gymnosomata	Gymnosomata	Gymnosomata (Order)

Taxonomy

The classification of the Pteropoda has undergone several stages of development since Pelseneer's (1888a) systematic survey of the Thecosomata and Gymnosomata. Table 2 compares four classifications of the Thecosomata which have been in use during the past century. Meisenheimer's (1905, 1906a, b) division of the Thecosomata into Euthecosomata and Pseudothecosomata on the basis of differences of their feet and fins has been accepted by all later workers. Further refinements at the family and subfamily levels have been introduced by Tesch (1913, 1946, 1948), Bonnevie (1913), Morton (1958),van der Spoel (1967, 1972) and Rampal (1973a).

At present the taxonomic position of euthecosomatous pteropods within the Phylum Mollusca is as follows:

PHYLUM MOLLUSCA

CLASS Gastropoda
SUBCLASS Opistobranchia
ORDER Thecosomata de Blainville, 1824—(shelled pteropods)

SUBORDER Euthecosomata Meisenheimer, 1905
Family Limacinidae Gray, 1847
Family Cavoliniidae Fischer, 1883

SUBORDER Pseudothecosomata Meisenheimer, 1905
Family Cymbuliidae Cantraine, 1841
Family Desmopteridae Chun, 1889
Family Peraclididae Tesch, 1913
Family Procymbuliidae Tesch, 1913

ORDER Gymnosomata de Blainville, 1824—(naked pteropods) (with 6 families)

The taxonomic key presented in the following pages and illustrated in Plates 3–9 includes twenty-eight euthecosomatous species; seven in the family Limacinidae and twenty-one in the family Cavoliniidae, which we consider to be morphologically distinct and to occur commonly in the oceans.

A number of species (*Limacina cochlystyloides, Creseis chierchiae, Creseis caliciformis, Clio andreae, Clio scheelei, Clio orthotheca, Clio campylura*) are either so rarely found, or are considered morphological variants of commonly recognised species that we shall not include them in our taxonomic key.

Van der Spoel (1967, 1972) has recognised several formae or morphological variants within certain species of *Limacina, Creseis, Clio, Diacria, Cuvierina*, and *Cavolinia*. Some of these formae appear to have discrete distributions, as was shown by van der Spoel (1962) for the polytypic species *Clio pyramidata*. In this study, we shall treat the individual species broadly both in a morphological as well as a distributional sense in order that their identification as well as their occurrence in the oceans can be grasped with relative ease. Accordingly, our taxonomic key does not consider any subspecies or formae, other than those of *Creseis virgula* which can be distinctly differentiated from each other.

In the present study we shall use the generic name *Limacina* instead of *Spiratella*, and thereby concur with J. H. Fraser's argument (in: van der Spoel, 1972):

> The names *Limacina* Bosc, 1817 and *Spiratella* Blainville, 1817, are both in frequent use. Priority therefore depends on the date of publication. This has not been clearly established and both were issued in December 1817. Lamarck's description of *Limacina* dated 1824 has no priority. The name *Limacina* is used here pending any more definite information concerning the date as this name is based on a better description and is at present probably in more general use.

Table 3 presents the hierarchical subdivisions of the families, genera and species of euthecosomatous pteropods, and Plate 2 points out morphological features of the shells and the terminology that are used in the taxonomic key.

The daggers (†) following some of the species names in the taxonomic key refer to species whose intraspecific variation is known in some detail.

Taxonomic key to the euthecosomatous Pteropoda

	Plates and figures
1a. Shell spirally left-coiled . . . . . . Family Limacinidae (2–7)	Pls 3–4
1b. Shell not coiled; conical in early ontogeny, later becoming bilaterally symmetrical, straight or slightly curved dorsally . . . . Family Cavoliniidae (8–28)	Pl. 5–9

FAMILY LIMACINIDAE

2a. Shell apex depressed by subsequent, greatly expanding whorls; thickened rib develops on outer margin of adult shell (ca. 0.9 mm diameter) . . . . *Limacina inflata* (d'Orbigny, 1836)	Pl. 3, figs 1a–d
2b. Shell apex not depressed 3	
3a. Shell high-spired; height greater than or equal to maximum diameter 4	
3b. Shell low-spired; height less than maximum diameter 6	
4a. Shell height equals maximum diameter; callus deposit on inner side of aperture . . . . . *Limacina trochiformis* (d'Orbigny, 1836)	Pl. 3, figs 2a–d
4b. Shell height greater than maximum diameter 5	

TABLE 3. Hierarchical subdivisions of the families, genera and species of euthecosomatous pteropods.

				Plate	figs
FAMILY LIMACINIDAE	Limacina		L. inflata	3	1a–d
			L. trochiformis	3	2a–d
			L. retroversa	3	3a–d
			L. bulimoides	3	4a–d
			L. lesueuri	3	5a–d
			L. helicoides	4	6a–c
			L. helicina	4	7a–g
	Subfamily Cuvierininae	Cuvierina	C. columnella	5	8a–e
		Hyalocylis	H. striata	5	9a–b
		Styliola	S. subula	5	10a–d
	Subfamily Clioinae	Creseis	C. acicula	5	11a–b
			C. virgula constricta	5	12a–b
			C. virgula conica	5	13a–c
			C. virgula virgula	5	14a–c
FAMILY CAVOLINIIDAE		Clio	C. cuspidata	5	15a–d
			C. polita	6	16a–c
			C. chaptalii	6	17a–c
			C. balantium	6	18a–d
			C. antarctica	6	19a–c
			C. convexa	7	20a–e
			C. pyramidata	7	21a–c
			C. sulcata	7	22a–b
		Diacria	D. trispinosa	7	23a–d
			D. quadridentata	7	24a–e
	Subfamily Cavoliniinae	Cavolinia	C. longirostris	8	25a–d
			C. tridentata	8	26a–e
			C. inflexa	8	27a–g
			C. globulosa	9	28a–d
			C. gibbosa	9	29a–c
			C. uncinata	9	30a–c

5a. Umbilicus distinct; continuous longitudinal   Pl. 3, figs 3a–d
striations on shell; deep notches at sutures in
profile view . . . .
. . *Limacina retroversa* (Fleming, 1823)†

5b. Umbilicus closed; short discontinuous longi-   Pl. 3, figs 4a–d
tudinal striations on shell; shallow notches at
sutures in profile view . . . .
. . *Limacina bulimoides* (d'Orbigny, 1836)

6a. Shell chestnut brown; whorls expanding rapidly;   Pl. 4, figs 6a–c
large size (up to 15 mm diameter) . . .
. . . *Limacina helicoides* Jeffreys, 1877

6b. Shell translucent; whorls expanding regularly;
small size (less than 8 mm width) 7

7a. Umbilicus narrow; transversely lined callus   Pl. 3, figs 5a–d
deposit on inner side of aperture . . . .
. . . *Limacina lesueuri* (d'Orbigny, 1836)

7b. Umbilicus wide; callus deposit absent; trans-   Pl. 4, figs 7a–g
verse striations on shell of one subspecies .
. . . *Limacina helicina* (Phipps, 1774)†

FAMILY CAVOLINIIDAE

8a. Shell conical throughout ontogeny with circular,
oval or triangular cross-section . . . .
. . . . . Subfamily Clioinae (9–21)

8b. Shell conical in early ontogeny only; adult shell
is inflated with dorsal side larger than ventral
side and narrow, slit-like aperture . . .
. . . . Subfamily Cavoliniinae (22–28)

8c. Shell conical in early ontogeny and later dis-   Pl. 5, figs 8a–e
carded by adult; adult shell is bottle-shaped with
kidney-shaped aperture Subfamily Cuvierininae
. . . *Cuvierina columnella* (Rang, 1827)

Subfamily Clioinae

9a. Adult shell oval or circular in cross-section 10

9b. Adult shell triangular or quasi-triangular in
cross-section 15

10a. Adult shell with transversely raised growth in-   Pl. 5, figs 9a–b
crements appearing as ripples over entire shell
length; shell very thin and slightly curved dorsally . . . *Hyalocylis striata* (Rang, 1828)

10b. Adult shell smooth without transversely raised
growth increments 11

11a. Adult shell with longitudinal groove running   Pl. 5, figs 10a–d
obliquely along dorsal length; transverse and
longitudinal striations on shell surface   .   .
. *Styliola subula* (Quoy and Gaimard, 1827)*
11b. Adult shell without longitudinal groove; shell
surface without ornamentation   .   .   .   12
12a. Shell straight, long and pencil-like   .   .   .   Pl. 5, figs 11a–b
.   .   .   .   . *Creseis acicula* (Rang, 1828)
12b. Shell cross-section expanding more rapidly,
curved or straight   13
13a. Shell strongly flexed dorsally to an angle of   Pl. 5, figs 14a–c
about 50° dorsally   .   .   .   .
Creseis virgula (Rang, 1828) *virgula* (Rang, 1828)
13b. Shell straight or slightly curved   .   .   .   14
14a. Shell without constriction at protoconch   .   Pl. 5, figs 13a–c
.   .   .   . *Creseis*
virgula (Rang, 1828) *conica* Escholtz, 1829
14b. Shell with external constriction at protoconch   Pl. 5, figs 12a–b
.   .   .   . *Creseis virgula*
(Rang, 1828) *constricta* (Chen and Bé, 1964)
15a. Adult shell with long lateral spines (extensions   Pl. 5, figs 15a–d
of lateral ribs) and very wide aperture; mid-
dorsal ridge distinct; posterior end curved
dorsally; protoconch teardrop-shaped with
sharp point   .   . *Clio cuspidata* (Bosc, 1802)
15b. Shell without long lateral spines   .   .   .   16
16a. Posterior part of shell curved dorsally and shell
about equally compressed dorso-ventrally   17
16b. Posterior part of shell straight or curved ventrally; ventral side nearly flat; dorsal side greatly
arched   19
17a. Aperture width less than half the total shell   Pl. 6 figs 16a–c
length; oval in cross-section   .   .   .   .   .
.   .   .   .   . *Clio polita* (Pelseneer, 1888)
17b. Aperture width more than half the total shell
length   18
18a. Shell regularly curved; single-lined lateral ribs   Pl. 6, figs 17a–c
.   .   .   .   .   . *Clio chaptalii* Gray, 1850

---

* At time of submission of this paper, the description of a new species was brought to our attention: *Styliola sinecosta*, a new species of Pteropod (Opisthobranchia: Thecosomata) from Barbados, West Indies. Veliger, 16(3), 293–296 by F. E. Wells, Jr. (1974).

18b. Shell curved only at posterior end; double-lined   Pl. 6, figs 18a–d
gutter-shaped lateral ribs; large shell (up to 30
mm) . . . *Clio balantium* (Rang, 1834)

19a. Adult shell lateral margins nearly parallel .   Pl. 6, figs 19a–c
. . . . . . *Clio antarctica* Dall, 1908

19b. Adult shell lateral margins divergent . 20

20a. Shell posterior curved ventrally; double-lined   Pl. 7, figs 20a–e
lateral ribs . . . *Clio convexa* (Boas, 1886)

20b. Shell posterior straight; single-lined lateral
ribs 21

21a. Adult shell with three longitudinal ridges on   Pl. 7, figs 21a–c
dorsal side *Clio pyramidata* Linnaeus, 1767†

21b. Adult shell with nine longitudinal ridges on   Pl. 7, figs 22a–b
dorsal side . . *Clio sulcata* (Pfeffer, 1879)†

Subfamily Cavoliniinae

22a. Apertural margin thickened and spherical protoconch . . . . . . *Diacria* (23)

22b. Apertural margin not thickened and thimble-
shaped protoconch . . . *Cavolinia* (24)

23a. Shell flattened, with long, tapering posterior   Pl. 7, figs 23a–d
section (often broken off near septal process);
two prominent lateral spines . . . . .
. . *Diacria trispinosa* (de Blainville, 1827)†

23b. Shell inflated, biconvex, without lateral spines   Pl. 7, figs 24a–e
Diacria quadridentata (de Blainville, 1821)†

24a. Embryonic shell absent in adults (adult shell is   Pl. 8, figs 25a–d
posteriorly truncated), dorsal apertural lip has
channel-like fold . . . .
Cavolinia longirostris (de Blainville, 1821)†

24b. Embryonic shell present in adults and posterior
end curved or straight 25

25a. Shell's greatest width between lateral spines 26

25b. Shell's greatest width anterior to lateral spines
. . . . 28

26a. Shell somewhat flattened; greatest shell width   Pl. 8, figs 27a–g
between two lateral spines in the middle of the
shell; ventral, apertural margin projecting
straight forward . . . .
. . . *Cavolinia inflexa* (Lesueur, 1813)†

26b. Shell inflated; greatest shell width anterior to
lateral spines; ventral apertural margin curved
ventrally 27

27a. Shell posterior section straight, ventral apertural   Pl. 8, figs 26a–d
margin curved slightly . . . .
. . *Cavolinia tridentata* (Niebuhr, 1775)†

27b. Shell posterior section strongly curved dorsally;   Pl. 9, figs 30a–d
ventral margin strongly curved over aperture;
longitudinal and transverse ribs on ventral side
. . . *Cavolinia uncinata* (Rang, 1829)†

28a. Shell inflated, ventral side rounded; small pos-   Pl. 9, figs 28a–d
terior section *Cavolinia globulosa* (Gray, 1850)

28b. Shell's ventral side protruding and sharply   Pl. 9, figs 29a–d
angular; large posterior section . . . . .
. . *Cavolinia gibbosa* (d'Orbigny, 1836)†

Zoogeography and faunal provinces

The geographic distribution patterns of the thecosomatous pteropods in the
world's oceans have emerged from the studies of numerous investigators.
The first major work was that of Pelseneer (1888a), who reported the occurrences of 42 thecosomatous species from the world-wide plankton collection
of the *Challenger* expedition. He arranged the species in ten provinces:
Arctic, North Atlantic, South Atlantic, Indian Ocean, Australasian, West
Pacific, East Australian, North Pacific, Southeast Pacific and Antarctic.

Meisenheimer (1905, 1906a, b) published the first global maps of thecosome
distribution. These were based on the collections of the *Valdivia* expedition,
and on previously published data. His geographic divisions of pteropod
communities (see Fig. 2) agreed, essentially, with the three main pelagic regions
proposed by Giesbrecht (1892). Both Giesbrecht and Meisenheimer recognised
a central warm-water faunal region which separates the northern and the
southern cold-water faunal regions. According to Meisenheimer the boundaries between the warm- and cold-water regions are delineated by the 15°C
surface-water isotherm in the North Atlantic and by the 17°C isotherm in the
South Atlantic. Meisenheimer's zoogeographic divisions correspond closely
with Sverdrup *et al.*'s (1942) temperature-salinity-circulation patterns of the
upper water masses of the oceans (Fig. 3). A major difference between Meisenheimer's faunal regions and Sverdrup *et al.*'s physical divisions lies in the
delineation of their respective "Transition Zones", i.e. the northern parts of
Meisenheimer's North Atlantic and North Pacific Transition Zones fall within
Sverdrup *et al.*'s subarctic watermasses.

In reviewing the distribution patterns of pteropods, we have drawn upon

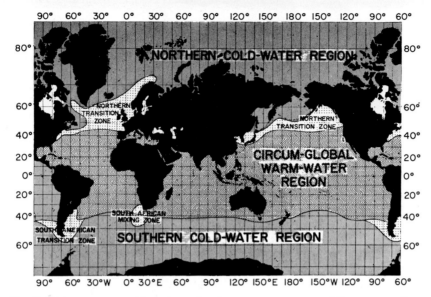

FIG. 2. Major zoogeographic regions of euthecosomatous pteropod faunas according to Meisenheimer (1905). The colder areas of the Northern Transition Zones include subarctic species. Compare with Fig. 3 and Table 4.

the extensive, but scattered, literature by a number of investigators. Only a few of these papers, e.g. McGowan's (1960) investigation of pteropods in the North and Equatorial Pacific, and Sakthivel's (1969, 1973a, b) studies in the Indian Ocean, contain data which are both quantitative and of ocean-wide significance. Most of the published descriptions of pteropod distribution only indicate absences or presences of the species at the sample localities. Such data and the great range of different sampling gear and methods used preclude any quantitative comparisons.

It should also be borne in mind that horizontal distributions are greatly influenced by variations in abundance due to patchiness, diurnal vertical migration and seasonal abundance. For example, cold-water species (*L. helicina, L. retroversa*) generally show sharper seasonal fluctuations in abundance than warm-water species (Fig. 4). According to Raymont (1963, p. 405) this indicates that the former have a shorter, more restricted reproductive season, whereas reproduction in warm-water species does not have such narrow time limits. Seasonal fluctuations in species composition should also be considered in non-synoptic distributional studies based on samples taken over a period of many years and during different seasons. Chen and Bé (1964) have observed that one group of pteropods (*Limacina inflata, L. bulimoides,*

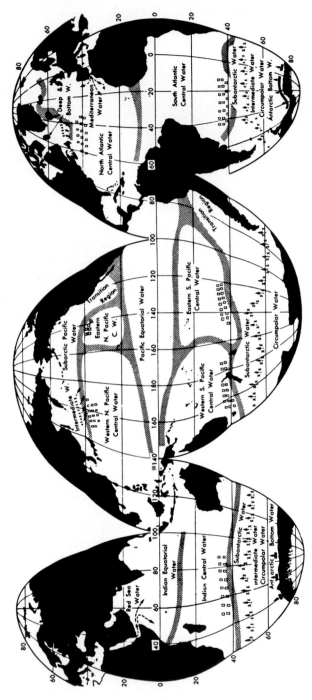

FIG. 3. The upper watermasses of the ocean and their approximate boundaries. Squares indicate the regions in which the central water masses are formed; crosses indicate the lines along which the antarctic and arctic intermediate waters sink (Sverdrup et al., 1942).

Styliola subula, Clio pyramidata) occurs predominantly between December and May in the Sargasso Sea and that it is replaced by another group (*Limacina trochiformis, Creseis acicula, Cavolinia inflexa*) between June and November.

The most up-to-date and comprehensive synthesis of pteropod distributions is that of van der Spoel (1967). In constructing our zoogeographic maps for individual species (Figs 5–31), we have probably perused the same published distributional data as was available to van der Spoel (see asterisked papers listed in the References). The main difference between our maps and those of van der Spoel lies in the interpolation between stations, the drawing of distributional limits, and our inclusion of more recently published data. Our maps were drawn to emphasise interconnections between species occurrences,

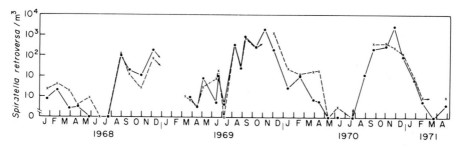

FIG. 4. Seasonal changes in the abundance of *Limacina* (= *Spiratella*) *retroversa*/m³ of water in St. Margaret's Bay, Nova Scotia. ·——·, inside the Bay; × – – – – ×, outside the Bay (Paranjape and Conover, 1973).

which often appear as isolated and widely separated patches or as present/ absent dots on the maps of individual investigators. In doing so, we may have been guilty of idealising the distribution patterns and overextending the actual distributional limits of many species. We have also attempted to delineate regions of high species abundance and/or high frequency of occurrence (heavier stippled areas on the maps) from areas of sparser or less frequent occurrence (in lighter stipples). These are partly based on published data and partly on our qualitative examination of plankton samples in the Lamont plankton collection. The shadings on the maps should not be interpreted as indicating the relative abundances between the pteropod species. In Rotramel's (1973) words, "such maps are merely a collection of spots or a portion of the map's surface enclosed by a line drawn through peripheral collecting stations in a free-wheeling game of connect-the-dots". Our maps are admittedly crude and semiquantitative and as such they are primarily intended as a general synthesis of the distributional data from the scattered literature as well as a first-order approximation of the zoogeographic patterns of euthecosomatous pteropods.

TABLE 4. Species composition of euthecosomatous pteropods in Meisenheimer's (1905) major world distributional subdivisions shown in Fig. 2.

I.   NORTHERN COLD-WATER REGION

   1. Arctic Province
     Limacina helicina

   2. Subarctic Province
     Limacina retroversa (present in North Atlantic, absent in North Pacific)
     Limacina helicina

II. CIRCUMGLOBAL WARM-WATER REGION

   1. Northern and Southern Subtropical Provinces (Central watermasses)
     Limacina lesueuri
     Limacina bulimoides
     Creseis virgula constricta (North Atlantic)
     Styliola subula
     Clio pyramidata
     Cavolinia gibbosa
     Cavolinia inflexa

   2. Tropical Province
     Limacina trochiformis
     Hyalocylis striata
     Cavolinia uncinata
     Clio convexa (Indo-Pacific only)
     Cavolinia globulosa (Indo-Pacific only)

   3. Warm-water Cosmopolitan Province (*Subtropical and †Tropical affinity)
     †*Limacina inflata*
     †*Creseis virgula virgula*
     Creseis virgula conica
     Creseis acicula
     †*Cuvierina columnella*
     Clio cuspidata
     Clio balantium
     †*Cavolinia longirostris*
     *Cavolinia tridentata
     Diacria trispinosa
     Diacria quadridentata

III. SOUTHERN COLD-WATER REGION

   1. Subantarctic Province
     Limacina retroversa
     Clio antarctica

   2. Antarctic Province
     Limacina helicina
     Clio sulcata

TABLE 4 (*continued*)

NORTHERN AND SOUTHERN TRANSITION ZONES
Limacina retroversa
Clio pyramidata
Clio antarctica (S. hemisphere only)

BATHYPELAGIC ZONE ($>$ 1000 m depth)
Limacina helicoides
Clio chaptalii (Mostly in tropical areas)
Clio polita (Wide distribution—subpolar to tropics)

In Table 4 we have listed the species composition of the euthecosomes for Meisenheimer's three major faunal regions which are shown in Fig. 2. The assignment of species to each region is based on Meisenheimer's species groupings as well as more recently published data (see asterisked papers in the References) and should also conform with the distribution patterns of individual species (Figs. 5–31).

Several broad generalisations can be made about the distribution of euthecosomatous pteropods:

1. Pteropods are widely distributed and abundant in all oceans, and the distribution patterns of individual species are closely related to the hydrographic conditions. Since most pteropods feed on phytoplankton and detritus, there is a close association between pteropod abundance, seasonal phytoplankton blooms, and nutrient levels. Pteropods tend to be abundant in active current systems in upwelling regions (e.g. Benguela Current, Somali Current, Peru Current, California Current) and they tend to be sparser in the oligotrophic central watermasses.

2. Within each of the major oceanic regions and provinces (see Fig. 2), the pteropod fauna is remarkably homogeneous. Twenty-one species and two subspecies listed in Table 4 are warm-water inhabitants which are found predominantly between 40°N and 40°S. The species composition in the so-called "Circumglobal Warm-water Region" is relatively similar in the Pacific, Indian and Atlantic Oceans, although faunal variations do exist. For instance, within the Warm-water Region, we can distinguish some species which are more abundant in the subtropical central waters than in the tropical areas. These are: *Limacina lesueuri, L. bulimoides, Creseis virgula constricta, Styliola subula, Clio pyramidata, Cavolinia gibbosa,* and *C. inflexa.*

A few species are more common in tropical than subtropical waters; these are: *Limacina trochiformis, Hyalocylis striata, Clio convexa, Cavolinia uncinata* and *C. globulosa.*

The majority of the warm-water species (11) however, live in both tropical and subtropical waters, and hence, are placed in a third subdivision, the "Warm-water Cosmopolitan Province" (see Table 4). Species in this group have the ability to adapt to a wider range of environmental conditions than species of the two previous groups. Active species interchange is likely between the northern and southern subtropical regions.

3. Within some local areas of the warm-water belt there may be differences in species diversity. For example, there is greater species diversity in the western than in the eastern basin of the Mediterranean (Meisenheimer, 1906). The Red Sea, being an enclosed basin, has a small number of pteropod species. The faunal diversity of the central watermasses of the oceans is generally lower than in the surrounding boundary currents. Finally, there are at least two species (*Cavolinia globulosa* and *Clio convexa*) which are known only from the Indo-Pacific and are absent in the Atlantic Ocean.

4. The low species diversity of the Cold-water Region contrasts sharply with the high diversity of the Warm-water Region. The Arctic Ocean is populated by a single species, *Limacina helicina*. In the Subarctic Province, *Limacina retroversa* is the dominant species, while *L. helicina* occurs in considerable numbers.

In the Antarctic Province *Clio sulcata* is present in addition to *L. helicina*, while in the Subantarctic Province *Clio antarctica* and *L. retroversa* form the bulk of the pteropod population. The continuous distributions of the cold-water species around Antarctica contrast with the interrupted distribution patterns of their northern hemisphere counterparts.

5. The occurrences of *Limacina helicina* and *L. retroversa* in both hemispheres is often cited as an example of bipolar distribution. One attempt of explaining the origin of their bipolarity suggests that the climatic warming during the recent geological past has caused the present-day separation of a once continuous distribution of the northern and southern cold-water populations along the eastern boundary currents of the Pacific and Atlantic.

6. The Transition Zones do not harbour any endemic species, but they are regions where mixing of subtropical and subpolar faunas takes place. Species which commonly co-occur in the transitional waters are *Clio pyramidata*, *Limacina retroversa*, *L. helicina* and various subtropical species which have been transported by western boundary currents.

Distribution patterns of individual species

POLAR SPECIES

Limacina helicina (Phipps)   (Plate 4, figs 7a–g; Fig. 5)
Limacina helicina is a well-known bipolar species which occurs in the Arctic and Antarctic regions and is commonly encountered in subpolar regions.

Details of intraspecific variation in this species are given in McGowan (1963).

It is the sole thecosomatous pteropod in the Arctic Ocean (Kramp, 1961; Harding, 1966; Hansen et al., 1971). In the North Atlantic, it is found in the subarctic waters of Hudson Bay, Baffin Bay, Davis Strait, Labrador Sea, Grand Banks, Greenland Sea and the Norwegian Sea (Kerswill, 1940; Bonnevie, 1913; Kramp, 1961; Chen and Bé, 1964).

In the North Pacific it is abundant in subarctic waters north of 45°N. It is less common in the Transition Zone, from where it is transported southward by the California Current as far as Point Conception (32°N) (McGowan, 1960, 1963).

In the southern hemisphere, Limacina helicina is most abundant in Antarctic waters between Antarctica and the Antarctic Polar Front (Meisenheimer, 1960b; Massy, 1932; Chen, 1968). It is less common in the subantarctic waters of the West Wind Drift and its northern limit is approximately coincident with the Subtropical Convergence (Hubendick, 1951; Chen, 1966, 1968). Van der Spoel (1967) reports its occurrence as far north as 10°S in the Brazil Current and Meisenheimer (1905) found it at 30°S near South Africa.

The occurrence of this species in the high latitudes of both hemispheres without any geographic continuity between the northern and southern populations, represents a classical example of bipolarity (Meisenheimer, 1905, 1906a, b).

This is a common epiplanktonic species and is rarely found below 300 m. Diurnal migration patterns for this species have been investigated by Hansen et al. (1971) in the Beaufort Sea. They presented strong evidence that Limacina helicina is the cause of the pycnocline scattering layer—an acoustic layer usually about 1 m thick which appears as a thin, continuous line at an average depth of 50 m on the echosounder.

Clio sulcata (Pfeffer)   (Plate 7, figs 22a–b; Fig. 6)

Clio sulcata is endemic to the Antarctic region. This species and *Limacina helicina* constitute the only two circum-Antarctic euthecosomatous pteropods which inhabit the waters between Antarctica and the Antarctic Polar Front (Pelseneer, 1888a; Meisenheimer, 1905, 1906b, Massy; 1920, 1932; van der Spoel, 1962; Okutani, 1963; Chen, 1966, 1968). Chen (1968) reported an unusually high density of *C. sulcata* in the upper 300 m of water near the South Sandwich Islands in May. Massy (1932) also noted high concentrations of this species near these islands. *C. sulcata* is generally much less frequently collected than *L. helicina* in the southern hemisphere.

Clio sulcata extends to about 45°S in the Indian Ocean and to 46°S in the Mid-Pacific (Meisenheimer, 1905). Massy (1920) found it as far north as 49°S near New Zealand.

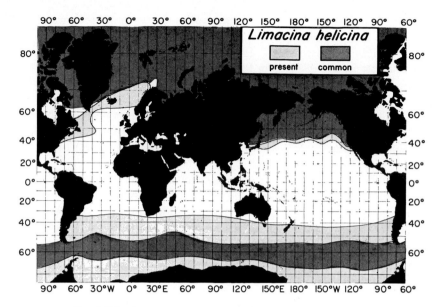

FIG. 5. Distribution of *Limacina helicina* in the world's oceans.

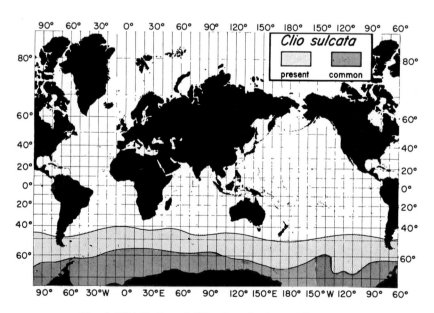

FIG. 6. Distribution of *Clio sulcata* in the world's oceans.

SUBPOLAR SPECIES

Limacina retroversa (Fleming)   (Plate 3, figs 3a–d; Fig. 7)

Limacina retroversa occurs in the subpolar and transitional waters of both hemispheres. Its total absence in the North Pacific, however, is an enigma. Details of intraspecific variation in this species are given in Tesch (1946, 1947).

In the North Atlantic, this species often appears in such dense swarms in the region between Newfoundland and Great Britain that it serves as a food source for whales and fish (Lebour, 1931; Hardy, 1924, 1926). The larger subspecies *L. retroversa balea*, whose finely sculptured shell has up to 10 whorls, is found in more northern latitudes (above 65°N). The smaller subspecies, *L. retroversa retroversa*, with a faintly sculptured shell consisting of 5 to 6 whorls, lives in more southern latitudes (40°–55°N). Tesch (1946) designates this latter subspecies as a useful indicator of the presence of Atlantic Water in the North Sea and in adjacent regions.

This species is transported by the Labrador Current and slope waters as far south as Cape Hatteras (34°N) in the North Atlantic (Myers, 1968; Chen and Hillman, 1970), and as far south as southern Portugal in the eastern North Atlantic (Bonnevie, 1913). Migrant populations from the Atlantic occur sporadically in the Mediterranean (van der Spoel, 1967), though Tesch (1946) and Bonnevie (1913) indicate that this species is rare below 50°N in the eastern North Atlantic.

In the southern hemisphere, *L. retroversa* is distributed in a continuous belt between about 38°S and Antarctica. It is abundant north of the Antarctic Polar Front between 42°S and 55°S (Baker, 1954; Chen, 1968).

L. retroversa is an epiplanktonic form and is most common in the upper 150 m of the water column. Bigelow (1926) reported an optimal summer depth zone of 20–30 m for this species in the Gulf of Maine, and noted that tows at sunrise or sunset during spring and summer yielded the highest number of specimens. This suggests a probable vertical diurnal migration pattern for this species.

Bigelow (1926) also reported that the maximum surface temperature range for this species lies between 2°C and 19°C, with an optimum range between 7°C and 12°C. *L. retroversa* can tolerate salinities ranging from 31‰ to 36‰.

This species exhibits distinct seasonal fluctuations in abundance with highest concentrations recorded between August and December off Nova Scotia (Paranjape and Conover, 1973; see Fig. 4), in August in the Labrador Sea and between May and October at Station Charlie at 52°45'N, 35°30'W (Chen and Bé, 1964). In the Gulf of Maine the geographic range of *L. retroversa* is restricted between March and April and is most extensive from early June to November (Bigelow, 1926). In coastal waters (100 m depth) the sporadic

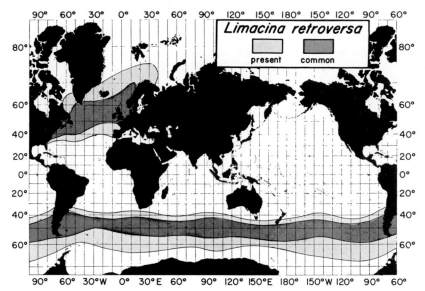

FIG. 7. Distribution of *Limacina retroversa* in the world's oceans.

occurrence of *L. retroversa* is attributed to the hydrographic conditions of this environment (Bigelow 1926, Redfield 1929).

Clio antarctica Dall   (Plate 6, figs 19a–c; Fig. 8)

Clio antarctica is an indicator species of subantarctic waters, occurring north of the Antarctic Polar Front as well as within the Southern Transition Zone (Meisenheimer, 1905, 1906b; Tesch, 1948; Chen, 1966). Meisenheimer (1905) notes the northern-most boundary of *C. antarctica* as occurring in the Brazil Current where it is carried as far north as 20°S.

Tesch (1948) reported the occurrence of this species in the upper water layers (20–40 m) below New Zealand, but it is also recorded at depths between 200 m and 1000 m in the Antarctic Intermediate Water north of the Antarctic Polar Front (Chen, 1966).

In comparison with the only other subantarctic species *Limacina retroversa*, *C. antarctica* has a patchier and lower density distribution.

Chen (1966) found this species common in water ranging from 2°C to 3°C and from 34.2‰ to 34.4‰ in salinity. A vertical diurnal migration pattern is not known for *C. antarctica*.

This species was referred to as *Clio australis* by Pelseneer (1888a) and Meisenheimer (1905, 1906b), and as *Clio pyramidata* forma *antarctica* by van der Spoel (1962, 1967).

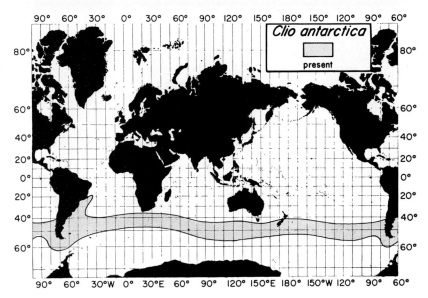

FIG. 8. Distribution of *Clio antarctica* in the world's oceans.

SUBTROPICAL SPECIES

Limacina lesueuri (d'Orbigny)   (Plate 3, figs 5a–d; Fig. 9)

Limacina lesueuri is a subtropical species and is predominantly found in the oligotrophic, central water-masses. It is one of the few warm-water pteropods which appears to avoid or sharply diminish in abundance in equatorial waters (Meisenheimer, 1905). It is generally less common and has a patchier distribution than the other warm-water Limacinidae, including *L. inflata*, *L. bulimoides* or *L. trochiformis* (Tesch, 1946; Moore, 1949; Wormelle, 1962; Chen and Bé, 1964; Rampal, 1968), but this may be partly an artifact, since *L. lesueuri's* larger size enables it to avoid capture by plankton nets more efficiently.

In the Atlantic, it is ubiquitous in subtropic regions but is more common in the western basin of the North Atlantic and along the west coast of Africa (Bonnevie, 1913; Tesch, 1946). It is generally not an abundant species. At Bermuda it comprised only 4.9% of *Limacina* species (Moore, 1949), and it is considered rare in the Gulf Stream by Wormelle (1962) and Myers (1968). In the Mediterranean, it is present in low numbers in the southwestern basin (Rampal, 1968).

In the Pacific it is scarce but is widely distributed over the whole warm-water region from 40°N to 40°S (Meisenheimer, 1905; Tesch, 1948). McGowan (1968) reported it as an oceanic form occurring on the outer fringes of the central portion of the California Current. *L. lesueuri* is found as far north as

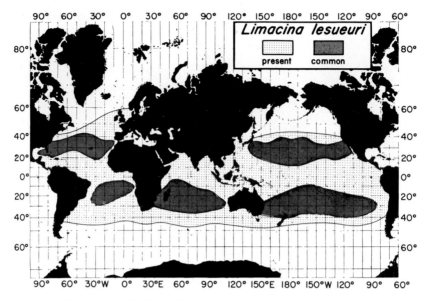

Fig. 9. Distribution of *Limacina lesueuri* in the world's oceans.

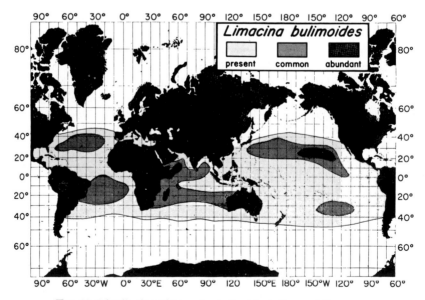

Fig. 10. Distribution of *Limacina bulimoides* in the world's oceans.

12°N in the Indian Ocean (Meisenheimer, 1905), but it is more common south of the equator with most frequent occurrences in the Mozambique Channel (35°S to 10°S) (Sakthivel, 1969).

This species is generally epiplanktonic though Wormelle (1962) collected some forms at depths of 600 m in the Florida Current. In the Florida Current *Limacina lesueuri* migrates to the surface at night and descends to mean day levels of 100 m or deeper (Wormelle, 1962). Moore (1949) also found a high correlation between moonlight night surface abundance and peak occurrences in April, June and December for this species near Bermuda.

This species has a temperature range from 13°C to 27°C and a salinity range from 35.7‰ to 36.42‰ (Williams, 1972; van der Spoel, 1967).

Limacina bulimoides (d'Orbigny)　(Plate 3, figs 4a–d; Fig. 10)

Limacina bulimoides is a very abundant subtropical species which occurs in highest concentrations in the central water-masses of all oceans. It is generally less abundant in the tropical waters and boundary currents, with the exception of high concentrations in the Somali Current (Sakthivel, 1968) and the Brazil Current (Boltovskoy, 1971b).

In the Atlantic, *L. bulimoides* is common in the upper layers of the Sargasso Sea (Tesch, 1946; Moore, 1949; Chen and Bé, 1964). It generally extends over the entire warm-water region from 45°N to 40°S (Meisenheimer, 1905; Massy, 1932; Morton, 1954a) and is widely distributed in the Gulf Stream (Wormelle, 1962; Myers, 1968). It is also present in both the eastern and western basins of the Mediterranean (Tesch, 1946; Rampal, 1968).

The geographical range of *L. bulimoides* in the Pacific extends from 40°N to 40°S (Pelseneer, 1888a; Meisenheimer, 1905). Fager and McGowan (1963) found it widely distributed in the Equatorial and Central Pacific water masses and in the Kuro Shio and its extension. In the Indian Ocean, Sakthivel (1969) found the highest concentrations of this species between 10°S and 12°N east of Somalia and noted its scarcity below 30°S.

L. bulimoides is reported by Moore (1949) and Wormelle (1962) to have a preferred depth range of 80 to 120 m in Bermuda waters and the Florida Current. Myers (1968) found this species absent from the upper 100 m of the Gulf Stream during daylight hours but noted its presence in these waters at night. Moore (1949) found a good correlation between night surface abundance and moonlight intensity for this species.

Van der Spoel (1967) recorded a temperature range of 13.8°C–27.8°C and a salinity range of 35.5‰ to 36.7‰ for the Florida Current populations.

Styliola subula (Quoy and Gaimard)　(Plate 5, figs 10a–d; Fig. 11)

Styliola subula is a subtropical species which is most abundant in the central water masses between 50°N and 45°S, and in the western boundary currents of all oceans (Meisenheimer 1905; Tesch 1946; McGowan, 1960; Sakthivel, 1969). It is not commonly encountered in the eastern tropical areas of the Atlantic, Pacific and Indian Oceans (Meisenheimer, 1905; Massy, 1920; McGowan, 1960).

The predominance of populations of *S. subula* in the cooler, subtropical waters of both hemispheres and the limited communication between the eastern tropical regions is a good example of a "bi-subtropical" or "anti-subtropical" distribution pattern (McGowan, 1960).

In the North Atlantic Chen and Bé (1964) described this species as sub-tropical cold-tolerant and found that it preferred deeper waters during the summer months. In the Florida Current Wormelle (1962) records evidence for diurnal vertical migration of 300 m for juveniles and 400 m for adults, with a mean day level at 234 m and a mean night level at 81 m. The juveniles remained at 50–200 m below the adults at night. Off Cape Hatteras Myers (1968) records the absence of adults from the upper 75 m during the day, though juveniles are always present in the surface waters.

This species is common to both the western and eastern basins of the Mediterranean (Tesch, 1946; Menzies, 1958; Rampal 1968). Menzies found a distinct avoidance of surface waters by this species during the day.

Tesch (1948) recorded this species as abundant in the Pacific and Indian Oceans between 40°N and 40°S and could not support Meisenheimer's contention (1905) that this species avoided the equatorial waters in this region. Sakthivel (1969, 1973b) collected *Styliola subula* as far north as 7°N in the Indian Ocean, but found it to be most abundant in the western equatorial region and between 20°S to 40°S and between Madagascar and Australia. Stubbings (1938) also found this species to be abundant in the Indian Ocean between 25°S and 34°S. In the Pacific it is present along the outer fringes of the entire California Current (McGowan, 1968).

Chen and Bé (1964) recorded an optimum temperature range of between 18°C and 22°C for this species, a maximum temperature range between 14.2°C and 27.7°C, and a salinity range from 35.5‰ to 36.7‰.

Clio pyramidata Linnaeus   (Plate 7, figs 21a–c, Fig. 12)

Clio pyramidata has a wide range of morphological variation and the different forms have been considered as different species, subspecies or formae by various investigators (Meisenheimer, 1905; Tesch, 1913, 1946, 1947, 1948; van der Spoel, 1962, 1967). The morphological intergradations and overlapping distributions of these variants indicate that they are genetically linked as a cline and that they belong to a polytypic species. Data on intraspecific variation in this species are given by van der Spoel (1962, 1963, 1967, 1969c).

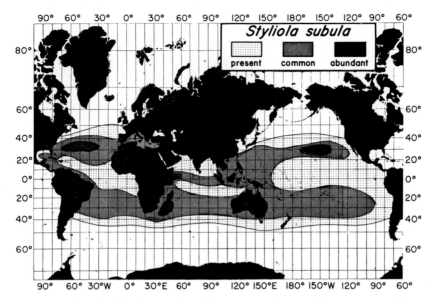

FIG. 11. Distribution of *Styliola subula* in the world's oceans.

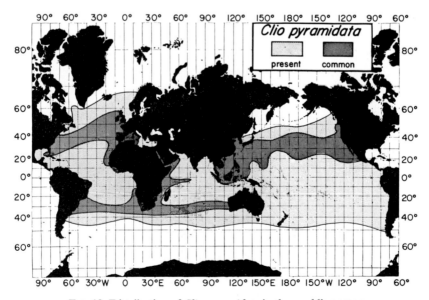

FIG. 12. Distribution of *Clio pyramidata* in the world's oceans.

In the present study, we consider *Clio pyramidata* to include four sub-species: *Clio pyramidata pyramidata*, *Clio pyramidata lanceolata*, *Clio pyramidata lata*, and *Clio pyramidata angusta*. Van der Spoel (1962, 1963, 1967) included, in addition, *convexa*, *martensii*, *excisa*, *sulcata*, and *antarctica* as formae of *Clio pyramidata*. We prefer to follow Tesch's (1946, 1948) classification, in which *sulcata* and *antarctica* are distinct species of *Clio*. Our arguments for assigning the rank of species to *Clio convexa* have already been included under the discussion of this species.

Clio pyramidata is the most common and widespread species of the genus *Clio*. Its distribution ranges from about 63°N to 40°S in the Atlantic Ocean (Tesch, 1946), illustrating its eurythermic nature. It occurs abundantly in the Transition Zone of the North Atlantic and extends as far north as Iceland (Tesch, 1946; Kramp, 1961). It is also very common in the northern Sargasso Sea during the winter months (Chen and Bé, 1964). Populations in the equatorial and South Atlantic waters belong to *Clio pyramidata lanceolata* (van der Spoel, 1967).

Clio pyramidata ranges from 50°N to 45°S in the Indo-Pacific Ocean (Tesch, 1948; McGowan, 1960; Sakthivel, 1969). It is most abundant in the central waters of the North Pacific and is rare in the eastern tropical Pacific (McGowan, 1960). Several varieties extend throughout the California Current (McGowan, 1968). It is commonly found off the Somali coast in the western equatorial Indian Ocean, the Bay of Bengal and the northern Arabian Sea to 20°N (Sakthivel, 1969).

Moore (1949) listed a winter maximum for this species off Bermuda with a corresponding low in mid-summer. Chen and Bé (1964) concluded that *C. pyramidata* was a fairly cold-tolerant species which inhabited the Sargasso Sea year-round, descending to greater depths during the warmest months.

Stubbings (1938) found a mean day level of 400–500 m for this species in the Indian Ocean and a somewhat uniform distribution from 0 to 1500 m during the night. Wormelle (1962) reported that this species is 6 to 100 times more abundant in night-time tows compared with day tows—indicative of extensive diurnal migration. *C. pyramidata* has an affinity for avoiding surface waters and is considered both an epiplanktonic and a mesoplanktonic form.

Smith and Teal (1973) demonstrated that the respiration of *C. pyramidata* is influenced strictly by temperature over its normal depth range (upper 500 m) below which the combined effects of pressure and temperature cause a significant increase in metabolic activity.

The temperature and salinity ranges for this eurythermic-euryhaline species lie between 7°C and 27.7°C and 35.5‰ and 36.7‰, respectively (Chen and Bé, 1964; van der Spoel, 1967).

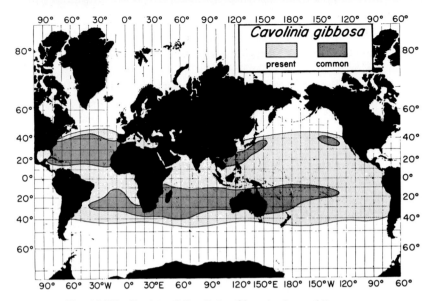

FIG. 13. Distribution of *Cavolinia gibbosa* in the world's oceans.

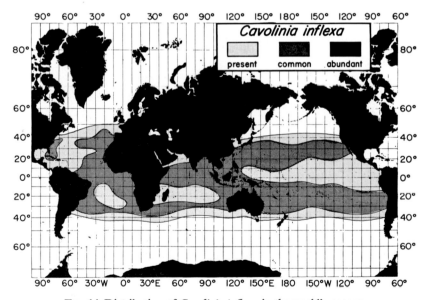

FIG. 14. Distribution of *Cavolinia inflexa* in the world's oceans.

Cavolinia gibbosa (d'Orbigny)   (Plate 9, figs 29a–d; Fig. 13)

Cavolinia gibbosa is a warm-water species with a patchy, but widespread distribution in the subtropical regions. Van der Spoel (1967) provides information concerning its intraspecific variation.

In the North Pacific, it appears more frequently in the higher latitudes (30°N to 40°N) than in the lower latitudes (McGowan, 1960). In the South Pacific, Indian Ocean, and Atlantic Ocean, *C. gibbosa*'s occurrences are predominantly in the subtropical belts, while it is conspicuously rare in the equatorial regions (Meisenheimer, 1905; Tesch, 1946, 1948; Sakthivel, 1969).

Tesch (1946) recorded *C. gibbosa* as occurring in both the eastern and western basins of the Mediterranean, though Menzies (1958) and Rampal (1968) found it common only in the middle and eastern regions.

This species is not commonly encountered, which may account for the lack of knowledge concerning its vertical distribution. It is not known whether this species exhibits a distinct diurnal vertical migration pattern. Chen and Bé (1964), Menzies (1958), and Wormelle (1962) found no evidence that it was more abundant either at night-time or day-time, although Moore (1949) found a strong correlation between night surface abundance and moon-light intensity.

According to Chen and Bé (1964), *C. gibbosa* occurs in the northern Sargasso Sea in water temperatures between 16.8°C and 27.9°C and salinities between 35.5‰ and 36.7‰.

Cavolinia inflexa (Lesueur)   (Plate 8, figs 27a–g; Fig. 14)

Cavolinia inflexa is a subtropical species which occurs commonly in boundary currents and near land masses. It is probably the most frequently encountered species of the genus. Its intraspecific variation is described by Tesch (1946) and van der Spoel (1967, 1968a).

In the Pacific Ocean, it is most abundant in the subtropical latitudes and appears to avoid the equatorial current system (McGowan, 1960). It is present in the offshore regions of the southern California Current (McGowan, 1968). In the Indian Ocean, *C. inflexa* is common in the Indo-Malayan waters (Tesch, 1948), but greater densities probably occur in the subtropical latitudes between 20°S and 40°S (Meisenheimer, 1905) and in the Somali upwelling area (Sakthivel, 1969).

In the Atlantic Ocean, *C. inflexa* is common between 40°N and 30°S (Meisenheimer, 1905; Tesch, 1946). High concentrations occur in the eastern North Atlantic near the Canary Islands (Massy, 1909; Tesch, 1946) and in the western Mediterranean (Rampal, 1968).

C. inflexa is common only during summer months in the Transition Zone of the western North Atlantic (Chen and Bé, 1964). Bonnevie (1913) obtained the highest numbers of this species between 0 and 250 m in the North Atlantic. It is not common in the Florida Current; Wormelle (1962) reported *C. inflexa* at a mean day level of 88 m and a night-time mean at 98 m. Further offshore she found a mean day level at 174 m and 164 m for night-time. Menzies (1958) described a somewhat different vertical distribution for this species from the Mediterranean, where he found *C. inflexa* between 400 and 800 m. Van der Spoel (1967) suggests that this discrepancy may be due to the different localities but Menzies' data are somewhat questionable since oblique tows were used which could introduce sampling error for vertical distribution data.

The known temperature range for this species is 16°C–28°C and the salinity range varies from 35.5‰ to 36.6‰ (Chen and Bé, 1964).

TROPICAL SPECIES

Limacina trochiformis (d'Orbigny)   (Plate 3, figs 2a–d; Fig. 15)

Limacina trochiformis is a ubiquitous warm-water species, whose peak abundance occurs in tropical regions especially in the Florida Current, Gulf Stream, Somali Current, the southeastern Arabian Sea, Bay of Bengal, and the eastern tropical Pacific (Pelseneer, 1888a; Tesch, 1904; Meisenheimer, 1905; Wormelle, 1962; Chen and Bé, 1964; Myers, 1968; McGowan, 1968; Sakthivel, 1969). Although it prefers upwelling regions and the close proximity of land masses in the lower latitudes, *L. trochiformis* is transported to the middle latitudes (45°N and 40°S) via boundary currents, including the Gulf Stream, Mozambique Current, and Kuro Shio (Fager and McGowan, 1963; Chen and Bé, 1964; Chen and Hillman, 1970). *L. trochiformis* is often associated with *Creseis virgula conica* or *Creseis acicula* in the Gulf Stream and is considered to be an indicator species of this water mass by Chen and Hillman (1970). It was the most common species in the upper 25 m of the Sargasso Sea (Myers, 1968), although Moore (1949) found only a few specimens off Bermuda.

In the Indian Ocean, *L. trochiformis* is widespread and is the third most abundant species of *Limacina*. In this ocean it is most abundant off the Somali coast between 0° and 10°N and 45°E to 55°E (Sakthivel, 1969).

L. trochiformis is abundant in both the eastern and western sectors of the Mediterranean (Rampal, 1968).

Wormelle (1962) found a diurnal migration pattern for this species in the Florida Current with a mean day level of 165 m versus 99 m as a night-time mean. Off Cape Hatteras, Myers (1968) found no clear migratory pattern, though he recorded maximum concentrations at 50 m between 1200 and 1700 h. He also found that no adults occurred below 100 m at any time.

The temperature range for this species is 13.8°C–27.9°C and the salinity range is 35.5‰–36.8‰ (Chen and Bé, 1964; van der Spoel, 1967).

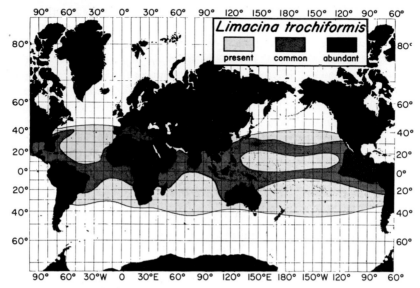

FIG. 15. Distribution of *Limacina trochiformis* in the world's oceans.

Hyalocylis striata (Rang)   (Plate 5, figs 9a–b; Fig. 16)

Hyalocylis striata is a tropical species which is also frequently encountered in the subtropical regions between 35°N and 30°S (Meisenheimer, 1905; Tesch, 1946, 1948). There is agreement between authors concerning McGowan's statement (1960) that this species is adapted to warmer water conditions but achieves abundance only in those areas of lateral or vertical water movement where mixing of cooler and presumably richer water takes place.

In the North Pacific, *H. striata* is carried by the Kuro Shio and the North Pacific Current to the California region (Tesch, 1948; Fager and McGowan, 1963; McGowan, 1968). It is quite common in the eastern tropical Pacific (Tesch, 1948).

In the Indian Ocean, this species occurs predominantly in tropical waters of the Arabian Sea, the Bay of Bengal, and the Indonesian archipelago (Tesch, 1904; Stubbings, 1938; Sakthivel, 1969). It is transported along the African coast by the Mozambique Current but is rare south of 20°S (Sakthivel, 1969).

In the Atlantic, *H. striata* is most common between 30°N and 10°S (Tesch, 1946). This species is relatively rare in the Sargasso Sea (Moore, 1949; Chen and Bé, 1964). Moore (1949) and Williams (1972) reported winter maxima for this species from Bermuda and the west coast of Florida respectively.

H. striata is irregularly distributed in the Gulf of Mexico, and extremely dense concentrations are recorded in the northeast of the Gulf by Hughes

(1968). In the Mediterranean this species is restricted to, and is densely concentrated in the eastern Mediterranean (Meisenheimer, 1905; Menzies, 1958;
Rampal, 1968).

Wormelle (1962) recorded a mean day level of 283 m and a mean night level
of 94 m for this species. Bonnevie (1913) found a day maximum between 100 m
and 250 m in the North Atlantic and Stubbings (1938) found this species from
200 m to 500 m in the Indian Ocean. In the Gulf Stream *H. striata* has a
maximum concentration at 50 m during the night, while neither juveniles nor
adults occur below 100 m at any time (Myers, 1968). Menzies (1958) found a
bimodal distribution with one peak near the surface and one between 200 and
400 m at both day and night stations in the Mediterranean.

Van der Spoel (1967) gives a temperature range between 17.5°C and 27.8°C,
and a mean salinity value of 36.2‰ for this species.

Cavolinia uncinata (Rang)   (Plate 9, figs 30 a–d; Fig. 17)

Cavolinia uncinata is a tropical species. Its intraspecific variation is discussed
by van der Spoel (1969b, 1971).

In the Pacific Ocean it is most frequent in the eastern and western tropical
regions between 30°N and 20°S, and in the Kuro Shio region (Tesch, 1948;
McGowan, 1960). It also occurs along the outer fringes of the southern portion of the California Current (McGowan, 1968).

In the Indian Ocean, *C. uncinata* is more commonly observed in the tropical
belt than in the high latitude areas (Tesch, 1948). Sakthivel (1969) found this
species in the Bay of Bengal but considered it to be rare in the Arabian Sea and
in waters south of the equator. This is one of the few pteropod species to
inhabit the Red Sea (van der Spoel, 1971).,

This species is more common in the Atlantic than in either the Pacific or
Indian Oceans. It frequently occurs between 30°N and 30°S but extends to
40° lat in either hemisphere (Meisenheimer, 1905; Tesch, 1946). It is not common
near Bermuda (Moore, 1949) and it was rare in Wormelle's samples
(1962) from the Florida Current, Williams (1972) found it off West Florida
only during March and May.

C. uncinata is generally epiplanktonic though it has been recorded from
depths below 700 m (Wormelle, 1962). This author also reported a greater
abundance of this species in night tows over day tows, indicating a pattern of
diurnal vertical migration.

Williams (1972) observed a temperature range of 17.4°C–23.1°C and a
salinity range from 36.01‰ to 36.13‰ for *C. uncinata*. Gilmer (1974) found this
species in the surface waters of the Florida Current only during months when
the surface temperature exceeded 25°C.

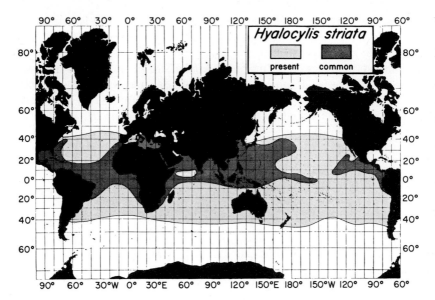

FIG. 16. Distribution of *Hyalocylis striata* in the world's oceans.

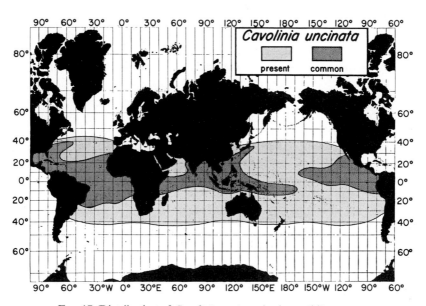

FIG. 17. Distribution of *Cavolinia uncinata* in the world's oceans.

Cavolinia globulosa (Gray)   (Plate 9, figs 28a–d; Fig. 18)

Cavolinia globulosa is a predominantly tropical species of the Indian and Pacific Oceans and occurs between 30°N and 30°S. It is absent in the Atlantic Ocean (Tesch, 1948). Its highest concentrations occur in the Indonesian archipelago (Tesch, 1948) and in the North and South Equatorial Currents of the Indian Ocean (Sakthivel, 1969). This author found that this species was uncommon in the Bay of Bengal, the Arabian Sea and the central region of the southern Indian Ocean. Stubbings (1938) recorded it as common in the Bay of Bengal and in the eastern part of the Indian Ocean. It is rare in the equatorial Pacific waters (McGowan, 1960).

Tesch (1948) proposed that the rigid stenothermic habit of this species inhibits it from rounding the Cape of Good Hope. There are sporadic reports of its occurrence in the tropical Atlantic (van der Spoel, 1967; Deevey, 1971) but these are unsubstantiated.

Stubbings (1938) found a daytime distribution between 500 and 600 m for this species. Tesch (1948) commonly found it above 100 m in the Indonesian Archipelago and Indian Ocean.

Clio convexa (Boas)   (Plate 7, figs 20a–e; Fig. 19)

This species was considered as a variety of *Clio pyramidata* by Boas (1886) and a forma by van der Spoel (1967, 1973). McGowan (1960) proposed a new species name, *Clio* "teschi", but it is still unpublished. The present authors agree with McGowan (1960, p. 105) that its shell morphology is sufficiently distinct from *Clio pyramidata* that it warrants recognition as a separate species (see Plate 7, figs 20a–e, 21a–c). However, since *Clio* "teschi" has not been published, we propose to elevate the variety or forma *convexa* to the rank of species.

Clio convexa differs from *Clio pyramidata* in having double-lined lateral ribs (Plate 7, fig. 20e) in contrast to the sharp, single-lined lateral ribs of *Clio pyramidata*. The protoconch of *Clio convexa* is shorter with a blunt cusp, whereas that of *Clio pyramidata* is longer and more elongate and has a sharp cusp. The posterior part of *Clio convexa* is curved ventrally, whereas it is straight and flat in *Clio pyramidata*.

Clio convexa is predominantly a tropical species restricted to the southern regions of the Indo-Pacific Oceans. Van der Spoel (1967) considers it to occupy the southern regions to 30°S in the East and West Pacific, where it commonly occurs with *Clio pyramidata lanceolata*.

WARM-WATER COSMOPOLITAN SPECIES

Limacina inflata (d'Orbigny)   (Plate 3, figs 1a–d; Fig. 20)

Limacina inflata is one of the most common warm-water cosmopolitan

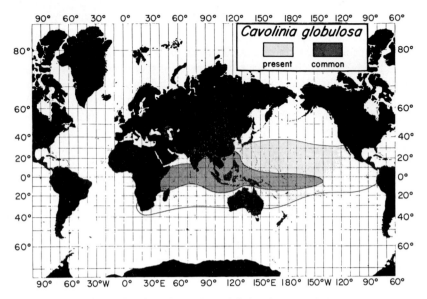

FIG. 18. Distribution of *Cavolinia globulosa* in the world's oceans.

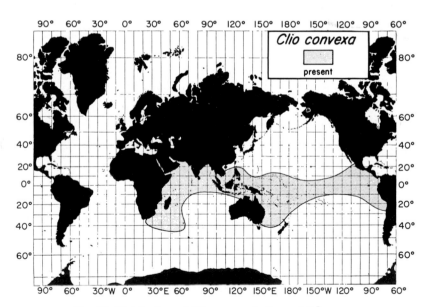

FIG. 19. Distribution of *Clio convexa* in the world's oceans.

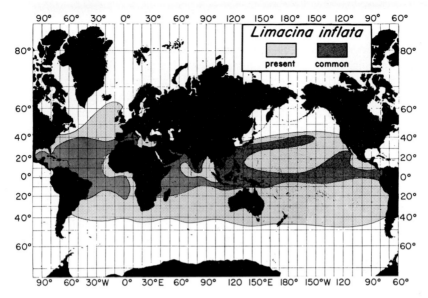

FIG. 20. Distribution of *Limacina inflata* in the world's oceans.

pteropods and is widely distributed in the tropical and subtropical regions of all oceans. Its poleward boundaries lie within the Transition Zones (Meisenheimer, 1905; Bonnevie, 1913; Tesch, 1946, 1948; McGowan, 1960, 1968; Sakthivel, 1969).

In the Atlantic Ocean, *L. inflata* is very abundant in the Sargasso Sea, Gulf Stream, Caribbean Sea, the equatorial region, the Brazil Current, and the central South Atlantic (Moore, 1949; Wormelle, 1962; Chen and Bé, 1964; Myers, 1968; Boltovskoy, 1971b). It is less common along the eastern Atlantic margin (Bonnevie, 1913; Tesch, 1946). In the Mediterranean, *L. inflata* is predominantly found in the western basin (Tesch, 1946; Rampal, 1968).

Limacina inflata is abundant in the western Indian Ocean, the Arabian Sea, off the coasts of India, and in the South Equatorial Current between Madagascar and the Timor Sea (Sakthivel, 1973a). It is not frequently encountered in the Bay of Bengal and the North Equatorial Current, possibly because of the relatively low salinities in these regions (Sakthivel, 1973a).

Limacina inflata exhibits distinct seasonal fluctuations in abundance (Wormelle, 1962; Moore, 1949). In the Indian Ocean, it is more abundant during the southwest monsoon (mid-April to mid-October) than during the northeast monsoon (mid-October to mid-April) (Sakthivel, 1973a). This author also found peak concentrations in July and August off the Somali Coast.

In the Sargasso Sea, *L. inflata*'s maximum density occurred in January (Moore, 1949). In the Florida Current the seasonal variation was somewhat bi-modal, having peak concentrations in January and July (Wormelle, 1962).

This species is an epiplanktonic species living primarily in the upper 300 m of water. The surface waters exhibit sharp differences in the abundance of this species between day and night due to its diurnal vertical migration. Wormelle (1962) recorded a mean day level of 236 m and night level of 232 m for this species in the Florida Current. Sakthivel (1973a) found it rare during the day in surface waters of the central Indian Ocean, but quite abundant in this region at night.

Limacina inflata has a surface temperature range between 14°C and 28°C and a surface salinity range between 35.5‰ and 36.7‰ in the Sargasso Sea (Chen and Bé, 1964).

The relative "success" of this species may be attributed to its unique possession of a protective brood pouch, in which the embryos and early veligers develop (Lalli and Wells, 1973).

Creseis virgula (Rang)   (Plate 5, figs 12a–b, 13a–c; 14a–c; Fig. 21)

The distribution map of *Creseis virgula* (Fig. 21) includes the two subspecies *Creseis virgula virgula* (Plate 5, figs 14a–c) and *C. virgula conica* (Plate 5, figs 13a–c). Another subspecies, *C. virgula constricta* (Plates 5, figs 12a, b), is so far only recorded from the Sargasso Sea (Chen and Bé, 1964). Intraspecific variation within this species is discussed by Tesch (1913, 1948) and van der Spoel (1967).

Creseis virgula's distribution pattern is very similar to that of *Creseis acicula*, except that the former appears to be more restricted to the lower latitudes and, therefore, has more tropical affinities than *C. acicula*.

In the Pacific, Fager and McGowan (1963) characterised *C. virgula* as inhabiting the Equatorial and Central Pacific water masses and suggested that it is transported by the Kuro Shio and its extension to the middle latitudes. Both varieties are found south of 34°N in the California Current (McGowan, 1968). Meisenheimer (1905) and Tesch (1948) found it over the entire tropical Indo-Pacific from 35°N to 35°S. In the Indian Ocean, Sakthivel (1969) found peak occurrences of this species north of 10°S in the equatorial zone, off the Somali Coast, the Bay of Bengal, and the Gulf of Oman. It is also present in the northern Arabian Sea (Stubbings, 1938). Russell and Colman (1935) encountered great numbers of this species in the upper 25 m both inside and out of the Great Barrier Reef near Australia.

In the Atlantic, *C. virgula* is transported to middle latitudes by the western boundary currents including the Gulf Stream (Wormelle, 1962; Myers, 1968) and the Brazil Current (Boltovskoy, 1971b). Tesch (1946) attributes its range along the American coast to 41°N to the Gulf Stream, but it is more generally distributed between 35°N and 35°S (Pelseneer, 1888a; Tesch, 1946; Meisenheimer, 1905; van der Spoel, 1967).

Chen and Hillman (1970) found *C. v. conica* to be an indicator of both Gulf Stream and Sargasso Sea water in the Cape Hatteras region throughout the year. There is also some evidence that *C. v. conica* dominates the warm-water layer which overlies the eastern North American slope water in July. *C. v. conica* is also indicative of the warm, low-salinity water of the West Florida Estuarine Gyre (Williams, 1972).

Wormelle (1962) recorded a mean day level of 206 m and a mean night level of 98 m in the Florida Current at the "10-mile station", while she found a mean day level of 319 m versus 167 m mean night level at a station 40 miles offshore. Stubbings (1938) found a mean day level of 50 m for *Creseis virgula* in the Indian Ocean.

Chen and Bé (1964) found that this species was abundant throughout the year in Bermuda waters, and extended into the colder transition regions only during warmer months (June–November). Menzies (1958) found this to be the most abundant cavoliniid species in the Mediterranean, with higher concentrations in the eastern region.

The temperature and salinity ranges vary somewhat between subspecies. Williams (1972) listed a temperature range from 7°C to 31°C and a salinity range from 34.95‰ to 36.68‰ for *C. virgula conica*. *Creseis virgula virgula* is somewhat more stenothermic with a temperature range from 15°C to 27.9°C (Chen and Bé, 1964; van der Spoel, 1967).

Creseis acicula (Rang)   (Plate 5, figs 11a–b; Fig. 22)

Creseis acicula is a warm-water cosmopolite which is, with the exception of the central water masses, almost equally abundant in the tropical and subtropical regions between 45°N and 40°S of all oceans (Meisenheimer, 1905; Tesch, 1946; Hida, 1957; van der Spoel, 1967). In low and middle latitudes, *C. acicula* is commonly encountered in the active current systems and areas bordering land masses, while it appears distinctly sparser in the gyres of the Sargasso Sea, the central South Atlantic and the central Indian Ocean (Tesch, 1946; McGowan, 1960; Chen and Bé, 1964; Sakthivel, 1969). In the California Current it only occurs south of 27°N (McGowan, 1968). It is the most abundant species in the Great Barrier Reef area (Russell and Colman, 1935).

Creseis acicula and its close relative *Creseis virgula* inhabit depths which are probably closer to the surface than most of the other euthecosomatous species. Tesch (1946) considered *C. acicula* a surface dweller. And according to Myers (1968) the majority of the *Creseis acicula* population in the western

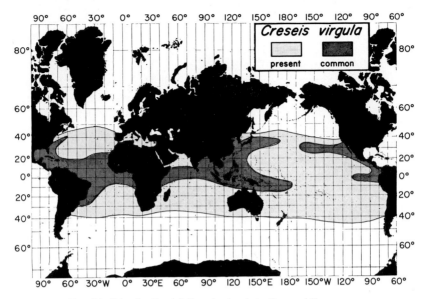

FIG. 21. Distribution of *Creseis virgula* in the world's oceans.

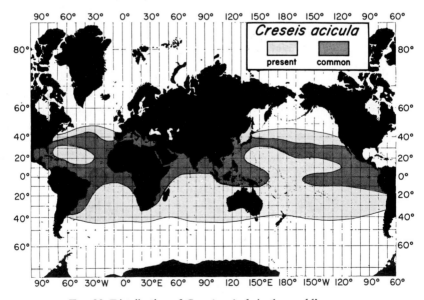

FIG. 22. Distribution of *Creseis acicula* in the world's oceans.

North Atlantic off Cape Hatteras lives in the upper 50 m throughout the day. Moore (1949) did not find a correlation between night surface abundance and moonlight intensity for this species. Bonnevie (1913) found its greatest abundance in the upper 50 m for the North Atlantic, but recorded individuals living as deep as 750 m. Stubbings (1938) found a daytime concentration at 250 m with a spread of 500 m in the Indian Ocean. In the Florida Current, *C. acicula* inhabits the shallowest depths of any euthecosome with a mean day level of 157 m and a nighttime optimum at 52 m (Wormelle, 1962).

This species is the most abundant euthecosome in the Sargasso Sea during the summer months (Chen and Bé, 1964). Moore (1949) reported peak concentrations near Bermuda in late autumn and again in spring but in rather low numbers.

Van der Spoel (1967) lists temperature and salinity ranges of 10°C–27.9°C and 35.5‰–36.7‰ respectively. Chen and Bé (1964) estimated an optimum temperature range for this species between 24°C and 27°C. *C. acicula* occurs in Mandapan Lagoon in the Indian Ocean at temperatures between 26°C and 33°C and salinities between 25‰ and 45‰ (Tampi, 1959).

Cuvierina columnella (Rang)   (Plate 5, figs 8a–e; Fig. 23)

Cuvierina columnella is a warm-water cosmopolite with a rather patchy, but widespread distribution in tropical and subtropical waters. It occurs commonly in the western equatorial Pacific and in a belt between about 20°N and 30°N in the North Pacific (McGowan, 1960). It is absent in the central and eastern equatorial Pacific, but occurs along the outer fringes of the entire California Current (McGowan, 1968). Data concerning the intraspecific variation in this species is given by van der Spoel (1970a).

In the Indian Ocean, *C. columnella* is common in the equatorial waters, especially between 5°N and 5°S, and in the Somali Current, but is rare in the Bay of Bengal and the Arabian Sea (Sakthivel, 1969).

This species is widespread in the tropical Atlantic between 40°N and 35°S. It appears to be more abundant in the Sargasso Sea and Gulf Stream than in the eastern sector of the North Atlantic (Tesch, 1946). Bonnevie (1913) and Meisenheimer (1905) found that it was common along the west coast of Africa. In the Mediterranean Sea, it is present only near the Straits of Gibraltar (Rampal ,1968; Tesch, 1946).

Cuvierina columnella's distribution pattern points to its greater affinity to the tropical regions and its transport to the middle latitudes via the western boundary currents.

This species is usually found in subsurface waters. Bonnevie (1913) reported highest numbers occurring between 100 and 250 m. In the Florida Current, the peak concentrations lie between 150 and 250 m, with some evidence of

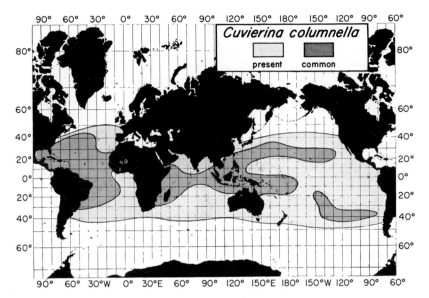

FIG. 23. Distribution of *Cuvierina columnella* in the world's oceans.

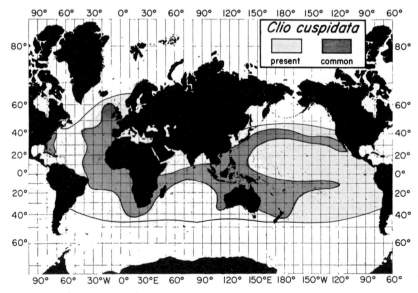

FIG. 24. Distribution of *Clio cuspidata* in the world's oceans. The denser shading shows an idealised interconnection between scattered sample localities at which this species were recorded.

diurnal migration (Wormelle, 1962). Moore (1949) found a positive correlation between night surface abundance and moonlight intensity for this species.

Van der Spoel (1967) lists a temperature range from 17.9°C to 26.2°C.

Clio cuspidata (Bosc)   (Plate 5, figs 15a–d; Fig. 24)

Clio cuspidata is a rare, but widespread warm-water cosmopolite. In the North Pacific it has been recorded from only 15 localities between 44°N and the equator (McGowan, 1968). It occurs along the outer fringes of the southern regions of the California Current. Tesch (1948) collected it along the east coast of New Zealand as far south as 45°S.

This species has been collected more frequently on the eastern than on the western part of the Atlantic Ocean (van der Spoel, 1967). It is however, consistently present in the Florida Current and Gulf Stream (Meisenheimer, 1905; Wormelle, 1962; Myers, 1968). *C. cuspidata* has been collected as far north as 60°N, south of Iceland, and from localities south of the Cape of Good Hope (Meisenheimer, 1905, 1906; Bonnevie, 1913; Tesch, 1946). Tesch (1946) found it to be common throughout the Indian Ocean, but it was considered rare by Stubbings (1938) and Sakthivel (1969).

Menzies (1958) and Rampal (1968) considered this an abundant species in the Mediterranean with high concentrations in both the eastern and western basins.

The denser shading in the distribution map in Fig. 24 shows the idealised interconnections between sample localities where *C. cuspidata* has been observed.

According to McGowan (1960), *C. cuspidata* has been collected in surface waters and therefore is not a bathypelagic species. Its sparsity may be in part due to its ability to avoid towed nets. Bonnevie (1913) found high concentrations of this species between 50 m and 250 m but reported it from depths of up to 1500 m. Wormelle's specimens (1962) occurred between 0 m and 362 m in the Gulf Stream. Stubbings (1938) reported a daytime range between 400 m and 900 m in the Indian Ocean.

Clio balantium (Rang)   (Plate 6, figs 18a–d; Fig. 25)

Clio balantium is the largest euthecosomatous pteropod species. The specimen shown in Plate 6, figs 18a–d, has a length of 30 mm which is close to the maximum recorded size.

This warm-water cosmopolite is encountered rarely in all oceans. Its large size may indicate greater motility and hence greater ability to avoid capture.

Tesch (1946, 1948) reported that its main occurrence is between 30°N and 20°S in the Atlantic, and between 40°N and 40°S in the Indo-Pacific. However, McGowan (1960) is of the opinion that it is more common, if not restricted,

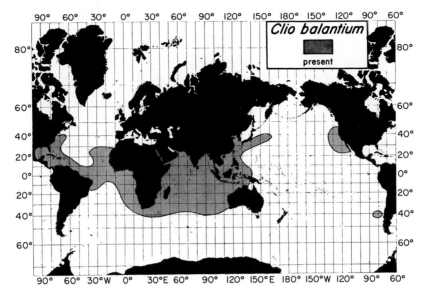

FIG. 25. Distribution of *Clio balantium* in the world's oceans.

between 35°N and 46°N in the Transition Zone of the North Pacific. It also occurs sporadically in the California Current (McGowan, 1968).

Judging from the records of Meisenheimer (1905, 1906), Massy (1932), Tesch (1946) and Wormelle (1962), this species is more common in the eastern North Atlantic and is carried by the Equatorial Current and Gulf Stream into the western regions. It is absent in the Sargasso Sea (Moore, 1949; Chen and Bé, 1964; van der Spoel, 1967), and it has not been reported from the Mediterranean.

Wormelle (1962) collected *C. balantium* between 365 m and 730 m at night and between 137 m and 247 m at dusk in the Florida Current.

Clio balantium was referred to as *Clio recurva* by Tesch (1913), Wormelle (1962) and van der Spoel (1967).

Cavolinia longirostris (de Blainville) (Plate 8, figs 25a–d; fig. 26)

Cavolinia longirostris is a warm-water cosmopolitan species, whose patchy occurrences are most frequently recorded in tropical regions. In the Pacific Ocean, it is most abundant in the eastern and western tropical areas including the Indonesian archipelago (Tesch, 1946), and the Kuro Shio and its extensions (McGowan, 1960). Intraspecific variation in this species is discussed by van der Spoel (1971).

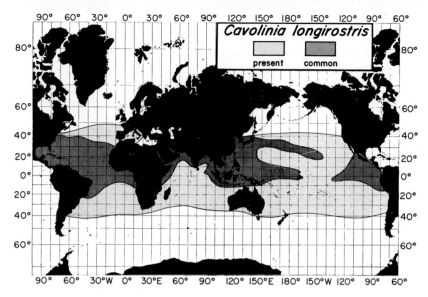

FIG. 26. Distribution of *Cavolinia longirostris* in the world's oceans.

In the Indian Ocean, *C. longirostris* is common in tropical regions but rare south of 20°S (Sakthivel, 1969). Tesch (1948) considered it abundant along the west coast of Sumatra, and Stubbings (1938) found high numbers of this species in the Arabian Sea.

In the Atlantic Ocean, it is abundant in the tropical waters, where it often occurs in large patchy swarms (Tesch, 1946). It is rare in the Sargasso Sea and in the Bermuda waters (Moore, 1949; Chen and Bé, 1964; Deevey, 1971). Tesch (1946) and Rampal (1968) recorded this species in the eastern and western basins of the Mediterranean Sea.

There is only a suggestion that this species exhibits diurnal vertical migration (Wormelle, 1962; Myers, 1968). Wormelle (1962) reported a mean day level of 215 m as opposed to a mean night level of 76 m. Stubbings (1938) reported an unusual vertical distribution optimum at 2000 m for this species during the day. However, since oblique tows were used, this observation is somewhat tenuous.

Russell and Colman (1935) found optimum concentrations of this species in autumn near the Great Barrier Reef. In the Florida Current, this species was present in the surface waters during the warmer months (July to November) (Gilmer, 1974).

Van der Spoel (1967) gives a temperature range from 17.4°C to 27.8°C and a salinity range from 36.2‰ to 36.8‰ for this species.

Cavolinia tridentata (Niebuhr)   (Plate 8, figs 26a–d; Fig. 27)

Cavolinia tridentata is a warm-water cosmopolite but is extremely patchy in its occurrences. The species may possibly have a greater preference for subtropical rather than tropical regions. In the Pacific, there are a total of 69 records of this species and most of them occur in the subtropical latitudes (Tesch, 1948; McGowan, 1960). It occurs only on the outer fringes of the central and southern portions of the California Current and is never common (McGowan, 1968).

In the Indian Ocean, this species was mainly observed in the higher latitudes (20°S to 40°S) of the eastern sector, as well as in the Indonesian archipelago (Meisenheimer, 1905; Tesch, 1904; 1948; Stubbings, 1938).

In the North Atlantic, Tesch (1946) noted that *C. tridentata* avoids the equatorial waters and is more common in the subtropical region. It occurs in both the eastern and western basins of the Mediterranean but is slightly more abundant in the latter (Tesch, 1946; Rampal, 1968).

Cavolinia tridentata's large size (up to 20 mm) probably enables it to avoid towed nets more successfully than the other smaller *Cavolinia* species.

Van der Spoel (1974) discussed intraspecific variation in this species.

Wormelle (1962) found a greater abundance of this species at night, indicating a probable diurnal migration. It was observed only during the winter months in the upper 30 m of the Florida Current by Gilmer (1974).

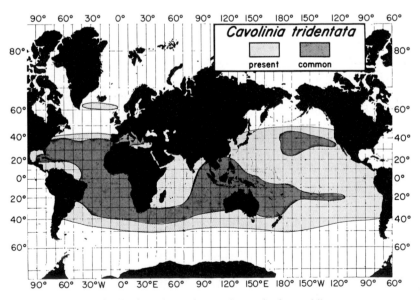

Fig. 27. Distribution of *Cavolinia tridentata* in the world's oceans.

Diacria trispinosa (de Blainville)   (Plate 7, figs 23a–d; Fig. 28)

Diacria trispinosa is a warm-water cosmopolitan species which occurs abundantly in the tropical regions, warm boundary currents, and areas bordering land masses and islands—such as the Caribbean Sea, W. Mediterranean Sea, Arabian Sea, Bay of Bengal, and the Indonesian and Pacific archipelagoes (Meisenheimer, 1905; Tesch, 1904, 1946, 1948; Sakthivel, 1969). Its peak abundance in the Kuro Shio and western Pacific coincides with the relatively high primary productivity that results from the mixing of subarctic and subtropical waters (McGowan, 1960). In the California Current, it occurs along the outer fringes of the northern and central regions (McGowan, 1968).

In the Indian Ocean, *D. trispinosa*'s highest density occurs along the east coast of Africa and Arabia (Sakthivel, 1969).

In the Atlantic Ocean, it occurs commonly from 45°N to 35°S (Meisenheimer, 1905; Bonnevie, 1913; Tesch, 1946), but it has been reported as far north as the west coast of Ireland (55°N) by Vane (1961) and Vane and Colebrook (1962). In the Sargasso Sea and Florida Current it is usually present in small numbers (Moore, 1949; Wormelle, 1962; Chen and Bé, 1964; Myers, 1968). Moore (1949) found a winter maximum for this species near Bermuda and a good correlation between night surface abundance and moonlight intensity. Williams (1972) reported it off the shelf of West Florida between November and February at depths from 30 m to 190 m. Wormelle's Florida

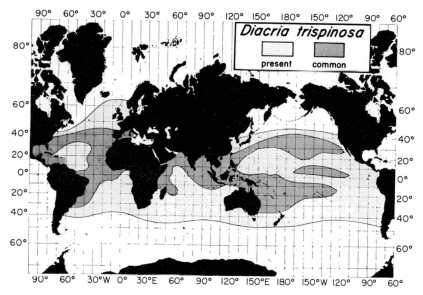

FIG. 28. Distribution of *Diacria trispinosa* in the world's oceans.

Current data (1962) give a mean day level at 219 m and a nighttime mean at 187 m.

Smith and Teal (1973) found a pressure effect on respiration in this species above 20 atm (200 m) and a maximum range to 500 m.

Williams (1972) lists a temperature range from 9.1°C to 28°C and a salinity range from 34.98‰ to 36.68‰ for *D. trispinosa*. Intraspecific variation in this species is discussed by Tesch (1946) and Panhorst and van der Spoel (1974).

Diacria quadridentata (de Blainville)   (Plate 7, figs 24a–e; Fig. 29)

Diacria quadridentata is a warm-water cosmopolitan species which has a rather patchy, but widespread distribution. It is more stenothermic and has a narrower latitudinal distributional range than *Diacria trispinosa*. Its intraspecific variation is discussed by van der Spoel (1968b, 1969a, 1971).

In the Pacific and Indian Oceans its highest densities occur in equatorial currents and boundary currents but is rare beyond 30°S (Tesch, 1948; Sakthivel, 1969). It is less common in the eastern than in the western tropical Pacific (McGowan, 1960). It occurs only along the southern fringe of the California Current (McGowan, 1968).

In the Atlantic Ocean, the highest numbers of this species occur immediately south of the equator (Tesch, 1946), and as far north as 35°N (Bonnevie, 1913).

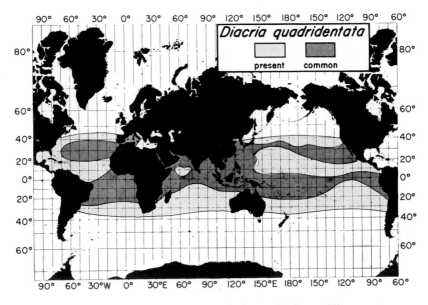

FIG. 29. Distribution of *Diacria quadridentata* in the world's oceans.

Meisenheimer (1905) reported it as abundant in the Brazil Current as far south as 25°S.

Its preference for boundary currents is evidenced by its rarity in the Sargasso Sea (Moore, 1949; Chen and Bé, 1964) and its high relative abundance of 8.5% of the total pteropod population in the Florida Current (Wormelle, 1962). Off Cape Hatteras it occurred in peak concentrations during January and was regarded as an indicator of Gulf Stream water (Chen and Hillman, 1970). Gilmer (1974) found it to be most common in the surface waters of the Florida Current in March and April.

Stubbings (1938) reported a double pattern of diurnal migration for this species in the Arabian Sea with a mean night level at 700 m and a mean near the surface during sunrise and sunset. Wormelle (1962) found a mean day level of 169 m and 134 m at night with some evidence of diurnal vertical migration.

Williams (1972) listed temperature and salinity ranges of 19.0°C to 25.5°C and 35.7‰ to 36.25‰, respectively, for this species.

BATHYPELAGIC SPECIES

Limacina helicoides Jeffreys (Plate 4, figs 6a–c; Fig. 30)

Limacina helicoides is an ovoviviporous bathypelagic species which is rather rarely encountered, but apparently has a very wide distribution. Its occurrence has been reported from isolated stations south of Iceland (van der Spoel, 1964), in the western North Atlantic (Bonnevie, 1913), off the western coasts of Africa (Meisenheimer, 1905, and van der Spoel, 1970), east of New Zealand and Tasman Sea (Tesch, 1948), the Mediterranean Sea, the Pacific Sub-antarctic waters and the Drake Passage (Chen, 1968; Bé, unpublished data).

Tesch (1948) noted that *L. helicoides* has been collected most frequently in the Atlantic Ocean, very rarely in the Pacific Ocean and that it has not yet been reported from the Indian Ocean.

L. helicoides occurs between 500 m and 1500 m depth in the North Atlantic according to Murray and Hjort (1912, p. 589).

Smith and Teal (1973) found that pressure had no effect on respiration rates to depths of 2000 m (200 atm), but that it increased significantly at temperatures above 10°C.

Clio polita (Pelseneer) (Plate 6, figs 16a–c; Fig. 31)

Clio polita is a bathypelagic dweller and is very rarely encountered in the upper few hundred meters of water.

Clio polita has been collected from a total of 47 localities in the Atlantic and only one locality in the Pacific (Gulf of Panama). With only a few exceptions, most specimens were obtained from depths greater than 1000 m. Most

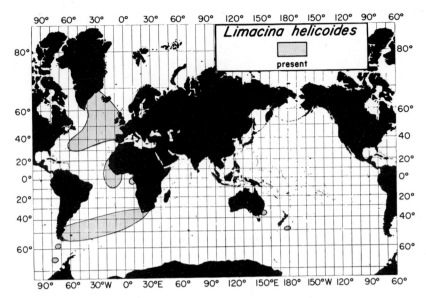

FIG. 30. Distribution of *Limacina helicoides* in the world's oceans.

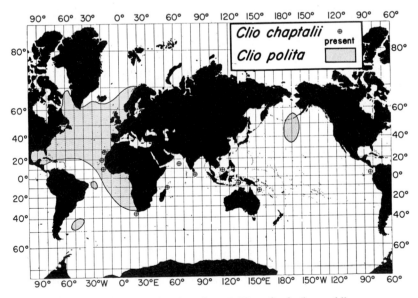

FIG. 31. Distribution of *Clio chaptalii* and *Clio polita* in the world's oceans.

of the occurrences are from areas north of 30°N or south of 30°S (Tesch, 1946; McGowan, 1960).

Clio chaptalii Gray (Plate 6, figs 17a–c; Fig. 31)

Clio chaptalii is an even rarer species than *C. polita*. A total of twelve specimens have been collected in the Indo-Pacific from six scattered tropical localities (Tesch, 1946, 1948; McGowan, 1960). In the Atlantic, Tesch (1946) recorded its occurrence at eleven localities between 25°N and the equator. The greatest majority of *C. chaptalii* specimens have been collected from depths greater than 1000 m.

Acknowledgements

We wish to gratefully acknowledge the assistance of Saijai Tuntivate in taking the scanning electron micrographs, and Fritz Goro and Leroy Lott in taking the light photomicrographs in Plates 3 to 9; Drs J. C. and E. Braconnot (Station Zoologique, Villefranche-sur-mer) for the photograph of *Styliola subula* in Plate 1.

We have benefited from the comments and suggestions of Dr. Carol M. Lalli and Dr. Fred E. Wells, Jr. (McGill University), Dr. S. van der Spoel (University of Amsterdam), Dr. William M. Hamner (University of California, Davis), Dr. Jeannine Rampal (Laboratoire de Biologie animale (plancton), Université Aix Marseille), Dr. Mark Shih (National Museum of Natural Sciences, Ottawa), Dr. Roger Batten (American Museum of Natural History, New York), and Dr. O. Roger Anderson (Lamont-Doherty Geological Observatory).

Support from the U.S. National Science Foundation (Grants GA 31,388X; GB 22,851 and GB 33,417) and from the National Geographic Society is also gratefully acknowledged. This contribution was submitted in May 1974, and is Lamont-Docherty Geological Observatory contribution no. 2388.

References

The references marked with an asterisk (*) contain data on horizontal, vertical and seasonal distributions upon which the maps in Figs 5–31 are based.

*Austin, H. (1971). The characteristics and relationships between the calculated geostrophic current component and selected indicator organisms in the Gulf of Mexico Loop Current System, 369 pp. Ph.D. Dissertation. Florida State University.

*Baker, A. de C. (1954). The circumpolar continuity of antarctic plankton species. *Discovery Rep.* **27**, 201–218.

Bé, A. W. H., MacClintock, C. and Chew-Currie, D. (1972). Helical shell structure and growth of the pteropod *Cuvierina columnella* (Rang) (Mollusca, Gastropoda), *Biomineralization Res. Rept.* **4**, 47–79.

*Bigelow, H. B. (1926). Plankton of the offshore waters of the Gulf of Maine. *U.S. Bur. Fish. Bull.* **40**(2), 509 pp.

Blainville, M. H. de (1824). *Dictionnaire des Sciences Naturelles*, Vol. 32, 567 p.

*Boas, J. E. V. (1886). Spolia Atlantica. Bidrag til Pteropodernes. Morfologi og systematik samt til kundskaben om deres geografiske udbredelse. *Vidensk. Selsk. Skr., 6 Raekke, naturvidensk. mathemat. Afd.* IV, 1, 1–231.

*Boltovskoy, D. (1971a). Contribucion al conocimiento de los Pteropodos Thecosomados sobre la plataforma continental Bonaerense. *Rev. Museo de la Plata* (Nueva Serie) *Sec. Zool.* **11**, 121–136.

*Boltovskoy, D. (1971b). Pteropodos Thecosomados del Atlantico Sudoccidental. *Malacologia* **11**(1), 121–140.

*Bonnevie, K. (1913). Pteropoda from the Michael Sars North Atlantic Deep-Sea Expedition. *Rep. Sci. Res. "Michael Sars" North Atl. Deep-sea Exp.* 1910, **III**(1), 1–69.

Bonnevie, K. (1916). Mitteilungen über Pteropoden. I. Beobachtungen über den Geschlechtsapparat von *Cuvierina columnella* (Rang). *Jenaische Zeitschr. Naturwiss.* **54**(2), 245–276.

*Chen, C. (1966). Calcareous zooplankton and the Scotia Sea and Drake Passage. *Nature* **212**(5063), 678–681.

*Chen, C. (1968). Zoogeography of thecosomatous pteropods in the West Antarctic Ocean. *Nautilus* **81**, 94–101.

*Chen, C. and Bé, A. W. H. (1964). Seasonal distributions of euthecosomatous pteropods in the surface waters of five stations in the western North Atlantic. *Bull. Mar. Sci. Gulf. Carrib.*, **14**(2), 185–220.

*Chen, C. and Hillman, N. S. (1970). Shell-bearing pteropods as indicators of water masses off Cape Hatteras, North Carolina. *Mar. Sci. Bull.* **20**(2), 350–367.

Cuvier, G. (1817). *Le règne animal distribué d'après son organisation, pour servir de base à l'histoire naturelle des animaux*. Vol. 1–4, Paris.

*Deevey, G. B. (1971). The annual cycle in quantity and composition of the zooplankton of the Sargasso Sea off Bermuda. I. The upper 500 m. *Limno. and Oceanogr.* **16**(2), 219–240.

*Della Croce, N. and Frontier, S. (1966). Thecosomatous pteropods from the Mozambique Channel. *Boll. Dei. Mus. e Degli Ist. Biol. dell'Univ. di Genova* **34**(207), 107–113.

Denton, I. J. (1964). The buoyancy of marine molluscs. *In* (Wilbur, K. and Yonge, C. M., eds), *Physiology of Mollusca*, Vol. 1, 425–434. Academic Press, New York.

Dunbar, M. J. (1942). Marine macroplankton from the Canadian Eastern Arctic. II. Medusae, Siphonophora, Ctenophora, Pteropoda, and Chaetognatha. *Can. J. Res.* **20D**(3), 71–77.

Eschscholtz, F. (1829). *Zoologischer Atlas*, Vols. 1–5. Reimer, Berlin.

*Fager, E. W. and McGowan, J. A. (1963). Zooplankton species groups in the North Pacific. *Science* **140**(3566), 453–460.

*Frontier, S. (1963a). Plancton récolté en mer d'Arabie, Golfe Persique et Golfe d'Aden. I. Données generales—repartition quantitative, *Cah. Orstom—Oceanographie* **3**, 17–29.

*Frontier, S. (1963b). Zooplancton récolté en mer d'Arabie, Golfe Persique et

Golfe d'Aden. II. Ptéropodes, Systematique et repartition. *Cah. Orstom—Océanographie* **6**, 233–254.

*Frontier, S. (1963c). Hétéropodes et ptéropodes récoltés dans le plancton de Nosy-Bé. *Cah. Orstom—Océanographie* **6**, 213–227.

*Frontier, S. (1963d). Présence de *Creseis chierchiae* (Boas) dans l'Ocean Indien. *Cah. Orstom—Océanographie* **6**, 229–232.

*Frontier, S. (1966). Liste complementaire des ptéropodes du plancton de Nosy-Bé (Madagascar). *Cah. Orstom—Océanographie* **4**(2), 141–146.

*Frontier, S. (1968). Données sur la faune pelagique vivant au large des côtes du Gabon du Congo et de l'Angola (0° à 18°S; 6°E à la côte). Hétéropodes et ptéropodes. *Cah. Orstom—Océanographie* **417**, 1–11.

Giesbrecht, W. (1892). Systematik und Faunistik der pelagischen Copepoden des Golfes von Neapel und der angrenzenden Meeresabschnitte. *Fauna und Golf. Neapel* **19**, 831 pp.

Gegenbaur, C. (1855). *Untersuchungen über Pteropoden und Heteropoden. Ein Beitrag zur Anatomie und Entwicklungsgeschichte dieser Thiere*, 228 pp. Engelmann, Leipzig.

Gilmer, R. W. (1974). Some aspects of feeding in thecosomatous pteropod molluscs. *J. Exp. Mar. Biol. Ecol.* **15**, 127–144.

*Hansen, W., Bulleid, E. and Dunbar, M. J. (1971). Scattering layers, oxgyen distribution and the copepod plankton in the upper 300 metres of the Beaufort Sea. *Mar. Sci. Centre Ms. Rep.* **20**, 84 pp. McGill University.

*Harding, G. C. (1966). Zooplankton distributions in the Arctic Ocean with notes on life cycles. M.Sc. Thesis, Marine Science Centre, McGill University.

Hardy, A. C. (1924). The herring in relation to its animate environment. Part I. The food and feeding habits of the herring with special reference to the east coast of England. *Min. Agric. Fish., Fish. Invest. Ser. II*(3), 1–53.

Hardy, A. C. (1926). The herring in relation to its animate environment. Part II. Report on the trials with the plankton indicator. *Min. Agric. Fish., Fish. Invest. Ser. II* **8**(7), 1–13.

Hardy, A. C. (1956). *The Ocean Sea*. (Part 1, *The World of Plankton*, 335 pp.) Houghton Mifflin Co., Boston.

Herman, Y. (1965). Étude des sediments quaternaires de la mer rouge. *Ann. Inst. Oceanogr. Monaco* **42**, 339–415.

*Herman, Y. (1971). Vertical and horizontal distribution of pteropods in Quaternary sequences. *In* (Funnell, B. M. and Riedel, W. R., eds), *The Micropaleontology of Oceans*, 463–486. Cambridge University Press, London.

Herman, Y. and Rosenberg, P. E. (1969). Pteropods as bathymetric indicators, *Mar. Geol.* **7**, 169–173.

*Hida, T. S. (1957). Chaetognaths and pteropods as biological indicators in the North Pacific. *U.S. Fish and Wildlife Serv., Spec. Sci. Rep. Fisheries* **215**, 1–13.

Hoefs, J. and Sarnthein, M. (1971). O^{18}/O^{16} ratios and related temperatures of Recent pteropods shells (*Cavolinia longirostris* Lesueur) from the Persian Gulf. *Mar. Geol.* **10**, 11–16.

Hsiao, S. C. T. (1939a). The reproductive system and spermatogenesis of *Limacina retroversa*. *Biol. Bull. Mar. Biol. Lab., Woods Hole* **76**, 7–25.

Hsiao, S. C. T. (1939b). The reproduction of *Limacina retroversa* (Fleming). *Biol. Bull. Mar. Biol. Lab., Woods Hole* **76**, 280–303.

Hubendick, B. (1951). Pteropoda with a new genus. *Furth. Zool. Res. Swed. Antarct. Exp.* 1901–1903, **4**(6), 1–10.

*Hughes, W. A. (1968). The thecosomatous pteropods of the Gulf of Mexico, 59 pp. M.S. Thesis, Texas A&M University.

Huxley, T. H. (1853). On the morphology of the cephalous mollusca, as illustrated by the anatomy of certain Heteropoda and Pteropoda collected during the voyage of H.M.S. "Rattlesnake" in 1846–50. *Phil. Trans. Roy. Soc., London* **143**, 29–65.

*Kerswill, C. J. (1940). The distribution of pteropods in the waters of eastern Canada and Newfoundland. *J. Fish. Res. Bd. Can.* **5**(1), 23–31.

Kornicker, L. S. (1959). Observations on the behavior of the pteropod *Creseis acicula* Rang. *Bull. Mar. Sci. Gulf Carib.* **9**(3), 331–336.

*Kramp, P. L. (1961). Pteropoda. *Medd. Grønland*, **81**(4), 1–12.

Lalli, C. M. (1970). Structure and function of the buccal apparatus of *Clione limacina* (Phipps) with a review of feeding in gymnosomatous pteropods. *J. Exp. Mar. Biol. Ecol.* **4**, 101–118.

Lalli, C. M. (1972). Food and feeding of *Paedoclione doliiformis* Danforth, a neotenous gymnosomatous pteropod. *Biol. Bull.* **143**(2), 392–402.

Lalli, C. M. and Wells, F. E., Jr. (1973), Brood protection in an epipelagic thecosomatous pteropod, *Spiratella* ("*Limacina*") *inflata* (d'Orbigny). *Bull. Mar. Sci.* **23**, 933–941.

Lankester, E. R. (1885). Mollusca. *Encyclopaedia Britannica*, 9th ed., 690–691.

Lebour, M. V. (1931). *Clione limacina* in Plymouth waters. *J. Mar. Biol. Ass. U.K.* **17**, 785–791.

Lebour, M. V. (1932). *Limacina retroversa* in Plymouth waters, *J. Mar. Biol. Ass. U.K.* **18**, 123–129.

LeBrasseur, R. J. (1966). Stomach contents of salmon and steelhead trout in the northeastern Pacific Ocean. *J. Fish. Res. Bd. Can.* **23**(1), 85–100.

Martens, F. (1676). *Spitzbergische oder Grönlandische Reisebeschreibung, gethan im Jahre* 1671, p. 169, pl. P, fig. f. Hamburg, 1675.

*Massy, A. L. (1909). The Pteropoda and Heteropoda of the coast of Ireland. Fisheries, Ireland, Scientific Investigations 1907. II. *Dept. Agric. Techn. Instruct. for Ireland* 1–52.

*Massy, A. L. (1920). Eupteropoda (Pteropoda: Thecosomata) and Pterota (Pteropoda: Gymnosomata). *Brit. Ant.* (*Terra Nova*) *Exp.* 1910, *Nat. Hist. Rept., Zoology* **II**(9), (Mollusca 3), 203–232.

*Massy, A. L. (1932). Mollusca: Gastropoda. Thecosomata and Gymnosomata. *Discovery Rep*, **3**, 267–296.

*McGowan, J. A. (1960). The systematics, distribution and abundance of the Euthecosomata of the North Pacific, 197 pp. Ph. D. dissertation. University of California, San Diego.

*McGowan, J. A. (1963). Geographical variation in *Limacina helicina* in the North Pacific. *Syst. Ass. Publ. No.* 5, *Speciation in the Sea*, 109–128.

*McGowan, J. A. (1967). Distributional atlas of pelagic molluscs in the California Current region. *CalCOFI Atlas No.* 6, *State of California Marine Research Committee*. 218 pp.

*McGowan, J. A. (1968). The Thecosomata and Gymnosomata of California. *Veliger* **3**, 103–129.

*McGowan, J. A. (1971). Oceanic biogeography of the Pacific, *In* (Funnell, B. M.

and Riedel, W. R., eds), *The Micropaleontology of Oceans*, 3–74. Cambridge University Press, London.

McGowan, J. A. and Fraundorf, V. J. (1966). The relationship between size of net used and estimates of zooplankton diversity. *Limnol. Oceanogr.* **11**(4), 456–469.

*Meisenheimer, J. (1905). Pteropoda. *Wiss. Ergebn. Tiefsee Exped. "Valdivia"* **9**(1), 1–314.

*Meisenheimer, J. (1906a). Die tiergeographischen Regionen des Pelagials, auf Grund der Verbreitung der Pteropoden. *Zool. Anz.* **28**, 155–163.

*Meisenheimer, J. (1960b). Die Pteropoden der deutschen Süd-polar Expedition 1901–1903. *Deutsch. Südpol. Exp.* 1901–1903, *IX (Zool.)* **1**(2), 92–152.

*Menzies, R. J. (1958). Shell-bearing pteropod gastropods from Mediterranean plankton (Cavoliniidae). *Pubbl. Staz. Zool. Napoli* **30**, 381–401.

*Moore, H. B. (1949). The zooplankton of the upper waters of the Bermuda area of the North Atlantic. *Bull. Bingham Oceanogr. Collect.* (Peabody Mus. Nat. Hist. Yale Univ.) **12**(2), 1–97.

Moore, R. C., Lalicker, C. G. and Fischer, A. G. (1952). *Invertebrate Fossils*, 766 pp. McGraw-Hill, New York.

*Morton, J. E. (1954a). The pelagic Mollusca of the Benguela current. I. First survey R.R.S. "William Scoresby" March 1950 with an account of the reproductive system and sexual succession of *Limacina bulimoides*. *Discovery Rep.* **27**, 163–199.

Morton, J. E. (1954b). The biology of *Limacina retroversa*. *J. Mar. Biol. Ass. U.K.* **33**, 297–312.

Morton, J. E. (1958). *Molluscs*, 219 pp. Hutchinson & Co., London.

Murray, J. and Hjort, J. (1912). *The Depths of the Ocean*, 821 pp. Macmillan Co., London.

*Myers, T. D. (1968). Horizontal and vertical distribution of thecosomatous pteropods off Cape Hatteras, 224 pp. Ph.D. dissertation. Duke University.

Oken, L. (1815). *Lehrbuch der Naturgeschichte* **3**. Zoologie, Abt. 1, 2. Leipzig und Jena.

Okutani, T. (1960). *Argonauta boettgeri* preys on *Cavolinia tridentata*. *Venus* **21**, 39–41.

Okutani, T. (1964). Thecosomatous pteropods collected during the second cruise of the Japanese expedition of deep sea. *Jap. J. Malac.* **22**, 336–341.

Panhorst, W. L. and van der Spoel, S. (1974). Notes on the adult and young stages in *Diacria* (Gastropoda, Pteropoda). *Basteria* **38**, 19–26.

Paranjape, M. A. (1968). The egg mass and veligers of *Limacina helicina* Phipps. *Veliger* **10**(4), 322–325.

*Paranjape, M. A. and Conover, R. J. (1973). Zooplankton of St. Margaret's Bay 1968–1971. *Fish. Res. Board of Canada. Tech. Rept.* no. 401, 82 pp.

*Pelseneer, P. (1888a). Report on the Pteropoda collected by H.M.S. "Challenger" during the years 1873–1876. II. The Thecosomata. *Rep. Sci. Res. Voy. H.M.S. "Challenger" during the years* 1873–1876. *Zoology* **23**(1), 1–132.

*Pelseneer, P. (1888b). Report on the Pteropoda collected by H.M.S. "Challenger" during the years 1873–1876. III. Anatomy. *Rep. Sci. Res. Voy. H.M.S. "Challenger" during the years* 1873–1876. *Zoology* **23**(2), 1–97.

Pruvot-Fol, A. (1954). Mollusques opisthobranches. *Faune de France* **58**, 1–457. Lechevalier, Paris.

*Rampal, J. (1963). Ptéropodes thécosomes de pêches par paliers entre les Baléares, la Sardaigne et la Côte Nord-Africaine. *Comm. Int. Explor. Mer Médit.* **17**(2), 637–639.

Rampal, J. (1965a). Variations morphologiques au cours de la croissance d'*Euclio cuspidata* (Bosc). (Ptéropode thécosome). *Bull. Inst. Océanogr. Monaco* **65**(1360), 1–12.

Rampal, J. (1965b). Ptéropodes thécosomes indicateurs hydrologiques. *Rev. Trav. Inst. Pêches Marit.* **29**(4), 393–400.

*Rampal, J. (1967). Repartition quantitative et bathymétrique des ptéropodes thécosomes récoltés en Méditerranée Occidentale au nord du 40ᵉ parallèle remarques morphologiques sur certaines espèces. *Rev. Trav. Inst. Pêches Marit.* **31**(4), 403–416.

*Rampal, J. (1968). Les ptéropodes thécosomes en Méditerranée. *Comm. Int. Explor. Sci. Mer Médit. Monaco* 1–142.

Rampal, J. (1973a). Clès de détermination des ptéropodes thécosomes de Méditer-anée et de l'Atlantique Euroafricain. *Rev. Trav. Inst. Pêches Marit.* **37**(3), 369–381.

Rampal, J. (1973b). Phylogenie des ptéropodes thécosomes d'après la structure de la coquille et la morphologie du manteau. *C.R. Acad. Sci. Paris* **277**, 1345–1348.

Raymont, J. E. G. (1963). *Plankton and Productivity in the Oceans*, 660 pp. Pergamon Press, Oxford,

Redfield, A. C. (1939). The history of a population of *Limacina retroversa* during its drift across the Gulf of Maine. *Biol. Bull.* **76**(1), 26–47.

*Rodriguez, D. G. (1965). Distribucion de pteropods en Veracruz. *Ver. Univ. Nac. Autonoma. Mex. Inst. Biol. An.* **36**, 249–251.

Rotramel, G. L. (1973). The development and application of the area concept in biogeography. *Syst. Zool.* **22**(3), 227–232.

*Russell, F. S. and Colman, J. S. (1935). The Zooplankton. IV. The occurrence and seasonal distribution of the Tunicata, Mollusca, and Coelenterata (Siphonophora). *Sci. Rep. of the Great Barrier Reef Exp.* 1928–1929 **2**(7), 203–276.

Russell, H. D. (1960). Heteropods and pteropods as food of fish genera, *Thunnus* and *Alepisaurus*. *Nautilus* **74**, 46–56.

*Sakthivel, M. (1969). A preliminary report on the distribution and relative abun-dance of Euthecosomata with a note on the seasonal variation of *Limacina* species in the Indian Ocean. *Bull. Natn. Inst. Sci., India* **38**, 700–717.

*Sakthivel, M. (1973a). Studies on *Limacina inflata* d'Orbigny (Thecosomata, Gastropoda) in the Indian Ocean. *In* (Zeitzschel, B., ed.), *The Biology of the Indian Ocean*, 383–397. Springer-Verlag, New York.

*Sakthivel, M. (1973b). Biogeographical change in the latitudinal boundary of a bisubtropical pteropod *Styliola subula* (Quoy et Gaimard) in the Indian Ocean. *In* (Zeitzschel, B., ed.), *The Biology of the Indian Ocean*, 401–404. Springer-Verlag, New York.

*Sarnthein, M. (1969). The plankton-benthos ratio of mollusks in the Recent sediments of the Persian Gulf. *In* (Brönniman, P. and Renz, H. H., eds), *Proc. I Plankt. Conf.*, **2**, 594–598. E. J. Brill, Leiden.

*Sarnthein, M. (1971). Oberflächensedimente im Persischen Golf und Golf von Oman. II. Quantitative Komponentenanalyse der Grobfraktion. *"Meteor" Forsch. Ergebn.* **C**(5) 1–113.

*Schiemenz, P. (1906). Die Pteropoden der Plankton Expedition. *Ergebn. Plankton-Exp. Humboldtstiftung* II(F) (b), 1–30.

Smith, K. L. and Teal, J. M. (1973). Temperature and pressure effects on respiration of thecosomatous pteropods. *Deep-Sea Res.* **20**, 853–858.

Spoel, S. van der (1962). Aberrant forms of the Genus *Clio* Linnaeus, 1767, with a review of the Genus *Proclio* Hubendick, 1951 (Gastropoda, Pteropoda). *Beaufortia* **9**(107), 173–199.

Spoel, S. van der (1963). A new forma of the species *Clio pyramidata* Linnaeus, 1767 and a new resting-stage of *Clio pyramidata* Linnaeus, 1767 forma *sulcata* (Pfeffer, 1879) (Gastropoda, Pteropoda). *Beaufortia* **10**(114), 19–28.

*Spoel, S. van der (1964). Notes on some pteropods from the North Atlantic. *Beaufortia* **19**(121), 167–176.

*Spoel, S. van der (1967). *Euthecosomata, a group with remarkable developmental stages* (Gastropoda, Pteropoda), 375 pp. J. Noorduyn en Zoon, N.V., Gorinchem.

Spoel, S. van der (1968a). The shell and its shape in Cavoliniidae (Pteropoda, Gastropoda). *Beaufortia* **15**(206), 185–189.

Spoel, S. van der (1968b). A new form of *Diacria quadridentata* (Blainville, 1821), and shell growth in this species (Gastropoda, Pteropoda). *Vedensk. Medd. Dansk Naturh. Foren.* **131**, 217–224.

Spoel, S. van der (1969a). *Diacria quadridentata* forma *danae* from the Atlantic. *Basteria* **33**(5–6), 105–107.

Spoel, S. van der (1969b). Two new forms of *Cavolinia uncinata* (Rang, 1829) (Pteropoda, Gastropoda). *Beaufortia* **16**(220), 185–198.

Spoel, S. van der (1969c). The shell of *Clio pyramidata* L, 1767 forma *lanceolata* (Lesueur, 1813) and forma *convexa* (Boas, 1886) (Gastropoda, Pteropoda). *Vidensk. Medd. Dansk naturh. Foren.* **132**, 95–114.

*Spoel, S. van der (1970a) Morphometric data on Cavoliniidae, with notes on a new form of *Cuvierina columnella* (Rang, 1827) (Gastropoda, Pteropoda). *Basteria* **34**(5–6), 103–151.

*Spoel, S. van der (1970b). The pelagic Mollusca from the "Atlantide" and "Galathea" expeditions collected in the East Atlantic. *Atlantide Report* No. 11, 99–139.

*Spoel, S. van der (1971). New forms of *Diacria quadridentata* (de Blainville, 1821), *Cavolinia longirostris* (de Blainville, 1821) and *Cavolinia uncinata* (Rang, 1829) from the Red Sea and the East Pacific Ocean (Mollusca, Pteropoda). *Beaufortia* **19**(243), 1–20.

*Spoel, S. van der (1972). Pteropoda Thecosomata. *Cons. Internat. Explor. Mer.* Zooplankton Sheet 140–142, 12 pp.

Spoel, S. van der (1973). *Clio pyramidata* Linnaeus, 1767 forma *convexa* (Boas, 1886) (Mollusca, Pteropoda), *Bull. Zool. Mus. Univ. Amsterdam* **3**(3), 15–20.

Spoel, S. van der (1974). Geographical variation in *Cavolinia tridentata* (Mollusca, Pteropoda). *Bijdragen Dierkunde* **44**(1), 100–112.

*Stubbings, H. G. (1938). Pteropoda. The John Murray Exp. 1933–1934. *Sci. Rep.* **5**(2), 3–33.

Sverdrup, H. U., Johnson, M. W. and Fleming, R. H. (1942). *The Oceans. Their Physics, Chemistry and General Biology*, 1087 pp. Prentice-Hall, Englewood Cliffs, N.J.

*Taki, I. and Okutani, I. (1962). Planktonic Gastropoda collected by the training vessel Umitaka-Maru, from the Pacific and Indian Oceans in the course of her antarctic expedition, 1956. *J. Fac. Fish. Anim. Husb. Hiroshima Univ.* **4**, 81–97.

Tampi, P. R. S. (1959). The ecological and fisheries characteristics of a salt water lagoon near Mandapam. *J. Mar. Biol. Ass. India* **1**(2), 113–130.

*Tesch, J. J. (1904). The Thecosomata and Gymnosomata of the Siboga Expedition. *Siboga Rep.* **52**, 1–92.

*Tesch, J. J. (1907). The Pteropoda of the Leyden Museum. *Notes from the Leyden Museum* **29**, 181–203.

*Tesch, J. J. (1913). Pteropoda. *In: Das Tierreich*, **36**, I–XVI, 1–154, 108 figs. Friedländer, Berlin.

*Tesch, J. J. (1946). The thecosomatous pteropods. I. The Atlantic. *Dana Rep.* **5**(28), 1–82.

*Tesch, J. J. (1947). Pteropoda Thecosomata. *Cons. Intern. Expl. Mer.* Zooplankton Sheet **8**, 1–6.

*Tesch, J. J. (1948). The thecosomatous pteropods. II. The Indo-Pacific. *Dana Rep.* **5**(30), 1–45.

*Thiriot-Quiévreux, C. (1968). Variations saisonnières des mollusques dans le plancton de la région de Banyuls-sur-mer (zone sud du Golfe du Lion), Novembre 1965—Décembre 1967). *Vie et Milieu* Serie B: Oceanographie, **19**(1B), 35–83.

*Tokioka, T. (1955). On some plankton animals collected by the Syunkotu-Maru in May-June 1954. IV. Thecosomatous Pteropods. *Publ. Seto Marine Biol. Lab.* **5**(1), 59–74.

Troost, D. G. and Spoel, S. van der (1972). Juveniles of *Cavolinia inflexa* (Lesueur, 1813) and *Cavolinia longirostris* (de Blainville, 1821), their discrimination and development (Gastropoda, Pteropoda). *Bull. Zool. Mus., Univ. Amsterdam* **2**(2), 221–235.

UNESCO (1968). *Zooplankton Sampling. Monographs on Oceanographic Methodology*, **2**, 1–176. Imprimeries Populaires, Geneva.

*Vane, F. R. (1961). Contributions towards a plankton atlas of the northeastern Atlantic and the North Sea. Part III. Gastropoda. *Mar. Ecol. Bull.* **5**, 98–101.

*Vane, F. R. and Colebrook, J. M. (1962). Continuous plankton records: contributions towards a plankton atlas of the northeastern Atlantic and the North Sea. Part VI: The seasonal and annual distribution of the Gastropoda. *Mar. Ecol. Bull.* **5**(50), 247–253.

*Vayssière, A. (1915). Mollusque eupteropodes (ptéropodes thécosomes) provenant des campagnes des yachts "Hirondelle" et "Princesse Alice" (1885–1913). *Rés. Camp. Sci. accomplies sur son yacht par Albert I Prince souverain de Monaco* **47**, 3–226.

*Wells, F. E. Jr. (1973). Effects of mesh size on estimation of population densities of tropical euthecosomatous pteropods, *Mar. Biol.* **20**, 347–350.

Wells, F. E. Jr. (1974). *Styliola sinecosta*, a new species of Pteropod (Opisthobranchia: Thecosomata) from Barbados, West Indies. *Veliger* **16**(3), 293–296.

*Williams, S. W. (1972). The temporal and spatial variation of selected thecosomatous pteropods from the Florida Middle Ground, 203 pp. M.S. Thesis. Florida State University.

*Wormelle, R. L. (1962). A survey of the standing crop of plankton of the Florida current. VI. A study of the distribution of the pteropods of the Florida current. *Bull. Mar. Sci. Gulf Carib.* **12**(1), 93–136.

Yonge, C. M. (1926). Ciliary feeding mechanisms in the thecosomatous pteropods. *J. Linn. Soc.* **36**, 417–429.

PLATES

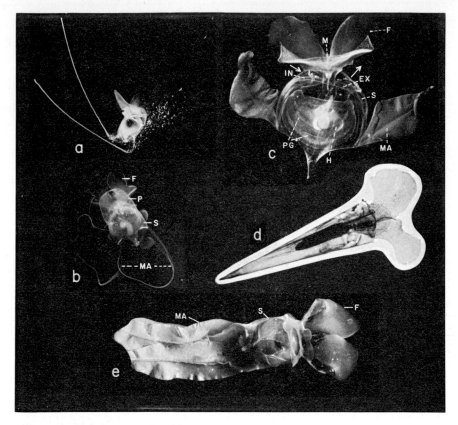

PLATE 1. Living pteropods with extended wing-like fins and mantle appendages. (F, wings; MA, mantle appendages; M, mouth; S, shell).

    fig. a. *Cavolinia uncinata* (Rang) feeding *in situ* at 15 min the Florida Current. The motionless feeding posture is common to all *Cavolinia* and *Diacria* species. Note the conspicuous field of mucus strings emerging from the exhalent mantle aperture; they represent the initial process of food collection in the Cavoliniidae. fig. b. Same specimens as in fig. a, showing the position of the gelatinous "pseudoconch" (P), present also in *Cavolinia gibbosa* (d'Orbigny) and *C. inflexa* (Lesueur). P, pseudoconch. fig. c. Living specimen of *Cavolinia tridentata* (Niebuhr) *platea* (Tesch), shell length of 11 mm. Arrows indicate entry (IN) and exit (EX) of water currents moving through the mantle cavity across the pallial gland (PG) where food particles from the water are entangled in mucus for transport to the mouth (M). H, heart; PG, pallial gland. fig. d. Living *Styliola subula* (Quoy & Gaimard) with prominent digestive gland (elongated dark organ). fig. e. Adult *Cavolinia tridentata* (Niebuhr), shell length of 14 mm, with long mantle appendages indicative of old age.

    (fig. a, by L. P. Madin; figs b, c, and e by R. W. Gilmer; fig. d by J. C. and E. Braconnot.)

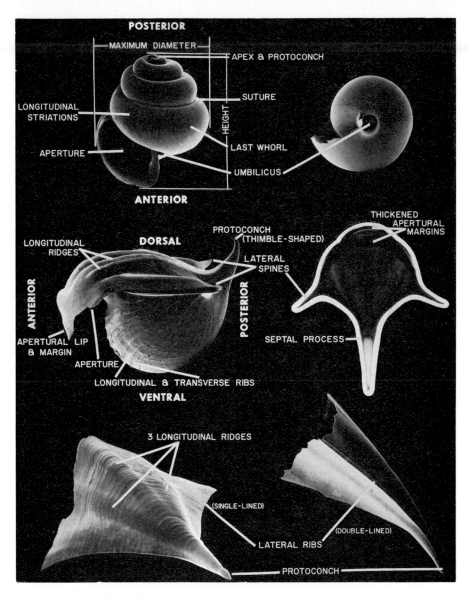

PLATE 2. Morphological terms used in taxonomic key.
Upper left, *Limacina retroversa;* upper right, *Limacina helicina;* middle left, *Cavolinia uncinata;* middle right, *Diacria trispinosa;* lower left, *Clio pyramidata;* lower right, *Clio convexa.*

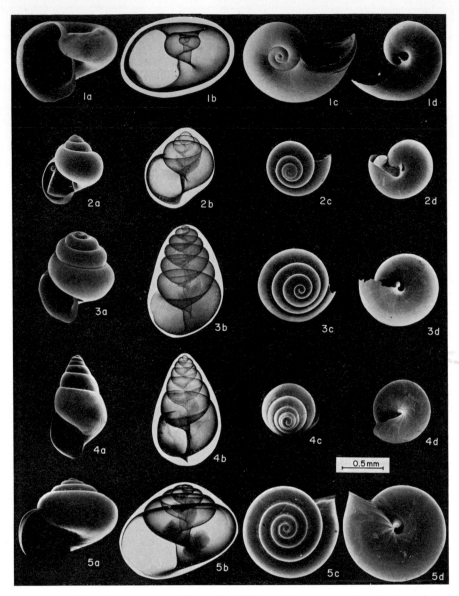

PLATE 3. All specimens are magnified ×23. All illustrations are scanning electron micrographs, except figs 1b–5b which are projection X-ray micrographs. Vertical rows a and b show apertural views; row c shows apical views, row d shows umbilical views.

figs 1a–d. *Limacina inflata* (d'Orbigny). a, b. 35°27′N, 72°04′W. c. 4°43′N, 52°05′E. d. 35°00′N, 48°00′W; figs 2a–d. *Limacina trochiformis* (d'Orbigny). 34°20′N, 75°45′W; figs 3a–d. *Limacina retroversa* (Fleming). a, c, d. 39°29′N, 72°23′W. b. 35°00′N, 75°00′W; figs 4a–d. *Limacina bulimoides* (d'Orbigny). 35°00′N, 48°00′W; figs 5a–d. *Limacina lesueuri* (d'Orbigny). 35°12′N, 41°46′W.

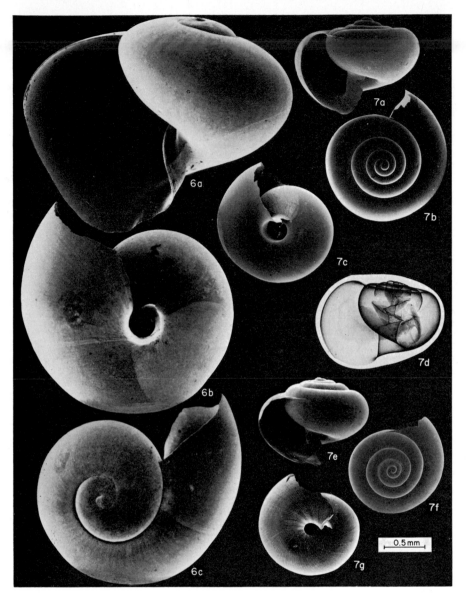

PLATE 4. All specimens are magnified ×23.

figs 6a–c. *Limancina helicoides* Jeffreys. North Atlantic; figs 7a–g. *Limacina helicina* (Phipps). a–c. North Pacific: 39°38′N, 133°41′W. e–f. South Pacific: 49°56′S, 150°01′W.

PLATE 5. All specimens are magnified ×8.5, except as otherwise indicated.
figs 8a–e. *Cuvierina columnella* (Rang). a, b, d, e. 22°22′N, 64°13′W. b. (×5). c. Bay of
Bengal; figs 9a–b. *Hyalocylis striata* (Rang). a. 22°22′N, 64°13′W (×6). b. 35°26′N,
72°04′W; figs 10a–d. *Styliola subula* (Quoy and Gaimard). a, c. 35°00′N, 48°00′W. b. 34°19′N,
75°45′W; protoconch (×19). d. 07°04′S, 80°40′E; figs 11a–b. *Creseis acicula* (Rang). a.
08°18′S, 96°14′W, b. 34°00′N, 75°50′W; figs 12a–b. *Creseis virgula* (Rang) *constricta* (Chen
and Bé). a, b. 33°01′N, 56°20′W. b. protoconch (×33); figs 13a–c. *Creseis virgula* (Rang)
conica Escholtz. a, b, c. 35°30′N, 74°30′W. c. protoconch (×31); figs 14a–c. *Creseis virgula*
(Rang) *virgula* (Rang). a. 02°47′N, 35°42′W. b, c. 11°59′N, 44°21′E; figs 15a–d. *Clio cus-
idatap* (Bosc). a–d. North Atlantic; (×4). d. protoconch (×24).

PLATE 6.

figs 16a–c. *Clio polita* (Pelseneer). a. North Atlantic (×3.5). b. North Atlantic (×5.5). c. North Atlantic (×17); figs 17a–c. *Clio chaptalii* Gray. Equatorial Atlantic (×3.5); figs 18a–d. *Clio balantium* (Rang). Indian Ocean off South Madagascar (×2); figs 19a–c. *Clio antarctica* Dall. a, b, c. 58°54′S, 75°52′W. a, c. (×8.5). b. protoconch (×14).

803

PLATE 7. All specimens are magnified (×8.5) except where otherwise indicated.
figs 20a–e. *Clio convexa* (Boas). a, b, c, d, e. Tablas Strait, Philippines. d. protoconch
(×34); figs 21a–c. *Clio pyramidata* Linnaeus. a, c. 22°22′N, 64°13′W. c. protoconch (×38).
b. 18°47′N, 140°45′E; fig. 22a–b. *Clio sulcata* (Pfeffer). a. 59°40′S, 24°33′W (×2). b. 64°47′S,
82°22′W; figs 23a–d. *Diacria trispinosa* (de Blainville). a, c, d. 34°27′N, 69°31′W. b. 18°47′N,
140°45′E. d. protoconch (×18); fig. 24a–e. *Diacria quadridentata* (de Blainville). a, c, d, e.
27°17′N, 77°08′W (×17). b. 27°46′N, 177°59′W.

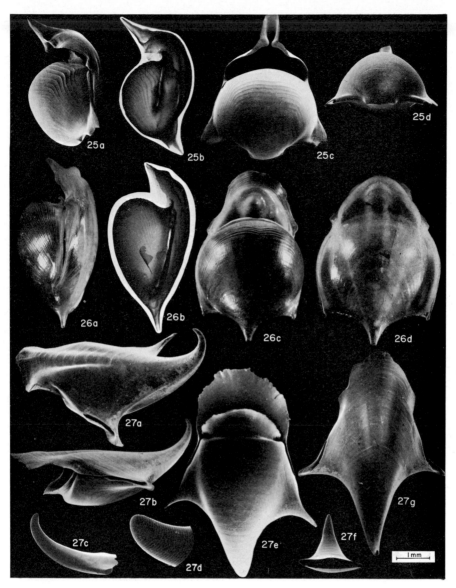

PLATE 8. All specimens are magnified ×8.5, except as otherwise indicated.
figs 25a–d. *Cavolinia longirostris* (de Blainville). a, c, d. 31°40′N, 78°40′W. b. 27°17′N, 77°08′W; figs 26a–d. *Cavolinia tridentata* (Niebuhr). a, c, d. Gulf Stream off Florida (×2.4), b. 39°05′N, 12°11′E (×2.4); figs 27a–g. *Cavolinia inflexa* (Lesueur). a, b, e, g. 39°05′N, 12°11′E. c, d, f. 35°30′N, 74°30′W. c, f. juvenile (×17). d. protoconch (×82).

PLATE 9. All specimens are magnified ×8.5.
figs 28a–d. *Cavolinia globulosa* (Gray). a, b, c, d. Bay of Bengal; figs 29a–d. *Cavolinia gibbosa* (d'Orbigny). a, c, d. 39°05′N, 12°11′E; figs 30a–d. *Cavolinia uncinata* (Rang). a, c, d. Bay of Bengal.

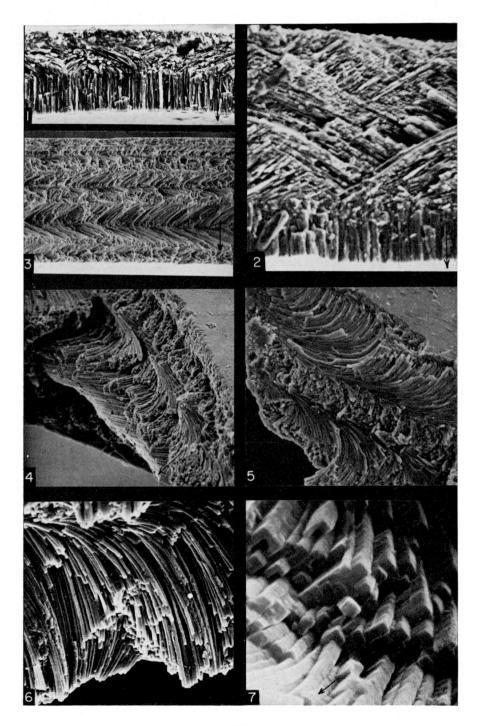

PLATE 10. Types of shell microstructures of euthecosomatous pteropods. (Arrows in figures point to growth surface).

fig. 1. Crossed-lamellar and prismatic structures in *Limacina bulimoides* (×1450) (N4052/P236-4b); fig. 2. Crossed-lamellar and prismatic structures in *Limacina lesueuri*. The prismatic structure in the lower parts of figs 1 and 2 is columellar myostracum which is produced by the columellar muscle in the last whorl of adult shells (×3360) (N 2784/P64-3); fig. 3. Helical microstructure in *Cuvierina columnella*. In this vertical section, the helix spiral makes at least four turns, as deduced from the number of "ridges" and "valleys". Note that the helices are smaller towards the outer shell surface (top) and larger towards the inner shell surface (×550) (N 1797/P119); fig. 4. Oblique view of fragmented shell of *Cuvierina columnella*, showing helical microstructure (×1280) (N1586/P104); fig. 5. Surface of *Cavolinia longirostris* shell, fractured parallel to outer shell surface and showing crescentic segments of helices (×1220) (N 1307/P93); fig. 6. Fracture parallel to *Cuvierina columnella's* shell surface, showing morphology of individual rods (×3080) (N 1544/P107); fig. 7. Oblique view of *Cuvierina columnella's* curved aragonite rods, showing that they are made up of second-order blocks. Note the L-shaped cross-section of some rods; the long leg of each "L" points to growth surface (×15400) (N 2003/P121).

General Index

Absolute abundance, total planktonic foraminifera, 40
Abundance
  death assemblages of planktonic foraminifera, 41
  living planktonic foraminifera, 16
  planktonic foraminifera, 51
Albatross cores, 128
  expedition, 734
Accumulation rates, palaeontology and, 119
Allele, 322
Allopatic populations, 322
Alternation of generations, 306, 322
Anatomy, marine plankton, 734
Antarctic Ocean
  Bottom Water, 104, 170
  palaeoclimatic reconstruction, Imbrie-Kipp method, 124
  polycystine radiolaria, 812
  seasonal distribution, 816
  Quaternary palaeoclimatology, 126, 127
Antarctic province
  euthecosomatous pteropods, 754, 756
  planktonic foraminifera, 36
Antipenultimate, 322
Antitropicality, planktonic foraminifera, 40
Apertural flap, 322
  lip, 322
Aperture
  accessory, 322
  primary, 322
  relict, 322
  supplementary, 322
  types, 332
  umbilical, 322
  umbilical-equatorial, 322
  umbilical-extraumbilical, 322
Aragonite, euthecosomatous pteropod shells, 740
Aragonitic pteropod shells, 103
Arctic Ocean
  Canadian coast

planktonic foraminifera, 29
euthecosomatous pteropods, 754, 756
planktonic foraminifera, 30, 36
Quaternary palaeoclimatology, 126, 127
Areal distribution
  planktonic foraminifera, 32
  surface sediments, 51
Argentine Basin, bottom currents, sedimentation and, 120
Assemblages
  faunal, northeast Atlantic palaeoclimatic variations and, 136
  in palaeoclimatic reconstruction, 123
Atlantic Ocean
  central waters, palaeosalinity, 131
  Deep Water, 104
  equatorial, differential solution in, 105
  faunal curves, oxygen isotope curves and, 133
  foraminiferal fauna, 156
  lysocline, 104
  pre-Holocene Quaternary sediments, 108
  Quaternary palaeoclimatology, 131, 132
  sediments, *Globorotalia truncatulinoides d'Orbigny* coiling direction in, 117, 118
  solution effects, glaciation and, 110
  synchronous variations of oxygen isotope ratios, 156
  undercurrent, 26
  equatorial cores, *Globorotalia menardii/ G. tumida* in, 108
  Globigerinoides ruber in, 109
  equatorial core tops, *Globigerinoides ruber* in, 103
  Globorotalia tumida in, 102
  lysocline depth, 106
  north, basin, for palaeoclimatic reconstruction, 121
  deglacial polar water retreat from, 150–152
  palaeogene planktonic foraminifera, 209

i 1

Taxonomic Index